Pions to quarks
Particle physics in the 1950s

Pions to quarks

Particle physics in the 1950s

Based on a Fermilab symposium

EDITORS

LAURIE M. BROWN
Northwestern University

MAX DRESDEN
SUNY at Stony Brook

LILLIAN HODDESON
*University of Illinois at Urbana–Champaign
and Fermilab*

EDITORIAL ASSISTANT

MAY WEST
Fermilab

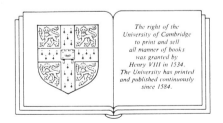

The right of the
University of Cambridge
to print and sell
all manner of books
was granted by
Henry VIII in 1534.
The University has printed
and published continuously
since 1584.

CAMBRIDGE UNIVERSITY PRESS
Cambridge
New York   New Rochelle   Melbourne   Sydney

Published by the Press Syndicate of the University of Cambridge
The Pitt Building, Trumpington Street, Cambridge CB2 1RP
32 East 57th Street, New York, NY 10022, USA
10 Stamford Road, Oakleigh, Melbourne 3166, Australia

First published 1989

Printed in the United States of America

Library of Congress Cataloging-in-Publication Data

Pions to quarks: particle physics in the 1950s : based on a Fermilab symposium /
editors, Laurie M. Brown, Max Dresden, Lillian Hoddeson;
editorial assistant, May West.

    p. cm.

Based on the lectures and discussions of historians and physicists at the Second
International Symposium on the History of Particle Physics, held at Fermilab on
May 1–4, 1985.

Includes indexes.

ISBN 0-521-30984-0

1. Particles (Nuclear physics) – History – Congresses.
I. Brown, Laurie M. II. Dresden, Max, 1918–   III. Hoddeson, Lillian.
IV. International Symposium on the History of Particle Physics
(2nd : 1985 : Fermilab)
QC793.16.P56   1989
539.7'21'09045 – dc19   88–25644

British Library Cataloguing-in-Publication Data

Pions to quarks.
1. Elementary particles. Theories, history
I. Brown, Laurie M. (Laurie Mark), 1923–
II. Dresden, Max   III. Hoddeson, Lillian
539.7'21'01

ISBN 0-521-30984-0 hard covers

Contents

Contributors

Luis W. Alvarez
Bldg. 50B, Room 5239
Lawrence Berkeley Laboratory
1 Cyclotron Road
Berkeley, CA 94720

Edoardo Amaldi
Experimental Physics
CERN
CH-1211
Geneva 23, Switzerland

Ugo Amaldi
Experimental Physics
CERN
CH-1211
Geneva 23, Switzerland

John P. Blewett
310 West 106th Street
New York, NY 10025

Laurie M. Brown
Department of Physics and Astronomy
Northwestern University
Evanston, Illinois 60201

Owen Chamberlain
Department of Physics
University of California
Berkeley, CA 94720

Geoffrey F. Chew
Department of Physics
University of California
Berkeley, CA 94720

William Chinowsky
Bldg. 50A, Room 2129
Lawrence Berkeley Laboratory
1 Cyclotron Road
Berkeley, CA 94720

Ernest D. Courant
Bldg. 902-A
Brookhaven National Laboratory
Upton, NY 11973

Richard H. Dalitz
Department of Theoretical Physics
University of Oxford
1 Keble Road
Oxford OX1 3NP, England

Max Dresden
Department of Physics
State University of New York
Stony Brook, NY 11790

Val L. Fitch
Physics Department – Jadwin Hall
Princeton University
P.O. Box 708
Princeton, NJ 08544

William B. Fowler
Fermilab
P.O. Box 500
Batavia, IL 60510

Allan D. Franklin
Department of Physics and Astrophysics
University of Colorado
Boulder, CO 80309

Shuji Fukui
School of Human Sciences
Sugiyama Jogaku–en University
37–234, Iwasaki–Takenoyama
Nissin, Aichi, 470-01 Japan

Peter Galison
Program in History of Science
Bldg. 200, Room 31
Stanford University
Stanford, CA 94305

Murray Gell-Mann
Department of Physics and Astronomy
California Institute of Technology
Pasadena, CA 91125

Gerson Goldhaber
Bldg. 50A, Room 2160
Lawrence Berkeley Laboratory
1 Cyclotron Road
Berkeley, CA 94720

John L. Heilbron
Office for History of Science and
   Technology
University of California, Berkeley
470 Stephens Hall
Berkeley, CA 94720

Armin Hermann
Universität Stuttgart
Historisches Institut
Abteilung für Geschichte der
   Naturwissenschaften und Technik
Seidenstraße 36
7000 Stuttgart 1, West Germany

Lillian Hoddeson
Department of Physics
University of Illinois
Urbana, IL 61801

Robert Hofstadter
Department of Physics
Stanford University
Stanford, CA 94305

Lawrence W. Jones
Department of Physics
University of Michigan
Ann Arbor, MI 48109

Donald W. Kerst
Department of Physics
University of Wisconsin
Madison, WI 53706

Michiji Konuma
Department of Physics
Keio University, Yogami Campus
Yokohama 223, Japan

Leon Lederman
Fermilab
P.O. Box 500
Batavia, IL 60510

Robert E. Marshak
202 Fincastle Drive
Blacksburg, VA 24060

Louis Michel
IHES, Bures
Inst. Hautes Etudes Scientifiques
35, Route de Chartres
F-91440 Bures-sur-Yvette, France

Yoichiro Nambu
Enrico Fermi Institute
University of Chicago
5640 S. Ellis Avenue
Chicago, IL 60637

Yuval Ne'eman
University of Tel Aviv
Ramat Aviv
Tel Aviv, Israel

Abraham Pais
Department of Physics
Rockefeller University
York Avenue and 66th Street
New York, NY 10021

Donald H. Perkins
Nuclear Physics Laboratory
University of Oxford
Keble Road
Oxford OX1 3RH, England

Oreste Piccioni
Physics Department
University of California, San Diego
La Jolla, CA 92093

Andy Pickering
Department of Sociology and Program
    in STS
326 Lincoln Hall
University of Illinois
Urbana, IL 61801

Bruno Pontecorvo
Joint Institute for Nuclear Research
DUBNA
Head Post Office – P.O. Box 79
101 000 Moscow, USSR

Helmut Rechenberg
Max-Planck-Institut für Physik und
    Astrophysik
Fohringer Ring 6
Postfach 40 12 12
D-8000 Munich 40, Germany

Frederick Reines
Department of Physics
University of California
Irvine, CA 92917

Arthur Roberts
University of Hawaii
Dumand Project
2505 Correa Road
Honolulu, HI 96822

George Rochester
Department of Physics
University of Durham
Science Laboratories, South Road
Durham DH1 3LE, England

Robert G. Sachs
Enrico Fermi Institute
University of Chicago
5640 S. Ellis Avenue
Chicago, IL 60637

Abdus Salam
International Center for Theoretical
    Physics
Strada Costeria 11
P.O. Box 586
34100 Trieste, Italy

Matthew Sands
160 Michael Lane
Santa Cruz, CA 95060

Silvan S. Schweber
Martin Fisher School of Physics
Brandeis University
Waltham, MA 02154

Robert W. Seidel
Bradbury Science Museum
Los Alamos National Laboratory
Los Alamos, NM 87544

Jack Steinberger
CERN
EP Division
CH-1211
Geneva 23, Switzerland

E. C. G. Sudarshan
Department of Physics
University of Texas at Austin
Austin, TX 78712

Daniel Sullivan
Allegheny College
Meadville, PA 16335

Valentine L. Telegdi
Laboratorium für Kernphysik
Eidg. Tech. Hochschule
Honggerberg
CH-8049
Zurich, Switzerland

Sam B. Treiman
Department of Physics
Princeton University
Princeton, NJ 08544

Kameshwar C. Wali
Physics Department
Syracuse University
Syracuse, NY 13210

Robert L. Walker
Box 56
Tesuque, NM 87574

D. Hywel White
Los Alamos National Laboratory
MS H846
Group MP4
Los Alamos, NM 87544

Arthur S. Wightman
Physics Department
Princeton University
Princeton, NJ 08544

Robert R. Wilson
916 Stewart Avenue
Ithaca, NY 14850

Chen Ning Yang
Institute for Theoretical Physics
State University of New York
Stony Brook, NY 11794

Foreword

LEON M. LEDERMAN

Born 1922, New York City; Ph.D., Columbia University, 1951; high-energy
particle physics; Nobel Prize, 1988, for the discovery of the neutrino beam
method and the demonstration of the doublet structure of the leptons
through the discovery of the muon neutrino; Fermi National Accelerator
Laboratory

This volume is the second in a series based on the lectures and discussions of historians and physicists at a Fermilab international symposium
on the history of particle physics. The first volume, *The Birth of Particle
Physics* (New York: Cambridge University Press, 1983), was based on a
symposium held in May 1980 that traced the emergence of the field out of
cosmic-ray and nuclear physics in the 1930s and 1940s and also examined the
extent to which relativistic quantum field theory could serve as a theoretical
structure for the new area. All the lectures were given by physicists; historians
of science participated in panels and in audience discussions, but did not serve
as speakers.

The present book, more complex than the first, grew out of the International Symposium on Particle Physics in the 1950s: Pions to Quarks, held at
Fermilab on 1–4 May 1985. Whereas the first symposium and volume dealt
with the birth and infancy of particle physics in the 1930s and 1940s, the
focus of this book is the period in which the field found its identity, became
established, and turned into a big science – the adolescence of particle
physics, with all the problems and anxieties that term evokes. At the second
Fermilab history symposium, and in the present volume, historians of science
played a major role, delivering talks and preparing chapters based on their
presentations.

The book is divided into ten parts, each containing chapters based on
symposium talks, a few also contain articles on topics that could not be
addressed in the three short days of the symposium, but which the editors felt
deserve a place in the volume. In addition, in their extensive Introduction,
the editors have made reference to other material that was overlooked or
dealt with too briefly in the other chapters. Even so, many omissions still

exist – there is room for many detailed historical studies that probe more deeply into the subject than was possible here. For those embarking on such historical enterprises, this volume is intended as a source book. For those simply eager to learn about particle physics in the fifties, we hope this volume offers interesting reading.

Most of the readers of this book will fall into one of three categories. The one into which I fit consists of the older scientists, the participants – I shudder to hear us referred to as "living sources." The second category includes the historians, the professionals who are to bring illumination and broader perspectives. The reader will note a tension in the way certain of the physicists and historians refer to each other, a reflection of the different professional objectives of these two groups of scholars in looking back on past developments. It was an explicit aim of the organizers of the symposium, Laurie Brown, Max Dresden, and Lillian Hoddeson, to bring this difference to the attention of those attending the meeting and to the readers of this volume. The members of the two camps need to make peace and become sensitized to each other's different concerns and approaches in the interest of reconstructing the past. The third group consists of the people who are now doing particle physics. I think, in some sense, the immediate benefits of this volume are to those in this third class of audience – to help them raise their consciousness about the fact that the field they work in has a culture and a history, to which they contribute in their everyday work. I noticed that many of those who work at Fermilab were quite moved by our symposium on the 1950s – by this reflective pause in our pursuit of high-energy particle physics. We are somewhat uneasily aware that we may be in the midst of some kind of a revolutionary change in our concept of how the world works. If it is true that particle physics is central to what might be the third scientific revolution in the twentieth century, then this historical endeavor to unravel our intellectual heritage looms as even more important and more relevant than we believed.

I would like to close these remarks with a small moral story about those in the first component of the audience, the aging scientists, about their interactions with historians, and about their possible contributions to a volume such as this one. The story has to do with a new member of the golf club who complained to the manager that golf balls were very expensive and that the caddies were not diligent enough and kept losing the golf balls. The manager of the club said to this new member: "The only solution is for you to use the best caddy we have in this business, old Pete. He has the finest eyesight of any caddy ever, anywhere." The new member said, "Well that's fine." The manager cautioned, "The only difficulty is he's getting on in years. He's over 80 and he can't carry the clubs. You'll have to carry the clubs yourself." Well, that seemed curious, but on the other hand the golf balls were very expensive. And so this new member of the club said, "Okay, I'll carry the clubs. Let's go, old Pete!" Pete replied, "Yes Sir!" The new member said, "Are you ready?" Pete said, "Yes Sir, I'm ready! Sorry I can't carry the clubs." The new member said, "That's alright, I'll carry the clubs, you just watch the ball." As he addressed the ball, he asked old Pete, "Ready?" Pete answered, "Yes Sir! I'm ready!" He hit the ball. "Do you see it?" the new member asked. Pete replied, "Yes, I see it. . .I see it. . .okay, I see it." And so the new member said, "Good! Where is it?" To which Pete confessed, "I forget." As this story underlines, there is no time to lose in holding history symposia and compiling volumes such as this one. The reader will observe in these pages that many recollections have already been rewritten by the passage of time, and a few have already been totally wiped out.

Editors' acknowledgments

It is with great pleasure that we take this opportunity to express our thanks to everyone who helped to realize the 1985 Fermilab International Symposium on Particle Physics in the 1950s and the present volume. Space permits us to thank individually only a few of those who contributed to these efforts.

We are most deeply indebted to three without whose support the symposium and this volume could not have been undertaken: Leon Lederman, Judy Zielinski Schramm, and May West. Lederman, Fermilab's director, felt strongly enough about the importance of recording the history of particle physics in the 1950s to urge us to organize a Fermilab symposium on this theme; he offered encouragement and advice at every turn. Schramm, then the director's assistant at Fermilab, arranged the physical and financial aspects of bringing symposium participants from all over the world to the symposium, helping to accommodate them and make their stay productive, and working with patience and good humor to solve their problems. Our greatest thanks go to May West, of Fermilab's library staff, whose steady devotion to bringing to culmination both the symposium and the proceedings served as the backbone of those undertakings. In the midst of other pressing responsibilities, she coordinated many of the tasks of seeing that the symposium ran smoothly. For the volume, West organized a large number of the detailed editing jobs, including copyediting, correspondence with authors, checking references, and liaison with Cambridge University Press.

The long list of others to whom we are particularly grateful includes the following:

> Malcolm and Eleanor MacFarlane, for fruitful fund-raising suggestions

Sue Grommes, for generating or correcting computer text files and for her help at the symposium registration desk

Adrienne Kolb, for compiling the symposium program booklet, help at the symposium Registration Desk, and editorial support

Kate Pientak, for assisting Schramm and West in arranging the symposium

Gordon Baym, for all varieties of general support and for technical assistance

Alice McLerran, for coordinating the transfer of computer text files from authors to the Fermilab Cyber

Edith Brown, for help with computer file transfers

Judith Nicholls, for expert consultation on computer problems

Angela Gonzales, for designing the symposium poster and this volume's jacket

Nicholas P. Samios, for allowing us to use the photograph on which both the poster and the book jacket are based

Saundra Cox, for mounting the symposium exhibit of material kindly supplied by George Rochester on cosmic-ray physics during our period

Bj Bjorken, for generously contributing the use of his home for the symposium May Wine Buffet and 1950s party

John Barry, Cynthia Sazama, Barbara Burwell, and the capable Fermilab cafeteria staff, headed by Peggy McAuliff, for exceptional service and delicious meals at the symposium

Margaret Pearson and Rick Fenner, for public information services

Avril Quarrie and Gretta Anema, for organizing the Family and Friends Activities at the symposium

Tony Frelo, for photography at the symposium

Arthur Roberts, for providing the participants in the symposium some lighthearted yet reflective moments through singing several of his songs about particle physics, and for allowing us to reprint excerpts in this volume*

Drasko Jovanovic and Rolland Johnson, for giving tours of Fermilab to visitors at the symposium

Chuck Anderson and his Site Services staff, particularly Robert Armstrong, for security services, and Richard Skokan, for audio help

Raeburn Wheeler, for computer services

Pam Naber and the Fermilab Housing Office, for invaluable housing service

Linda Braddy and her staff at the Fermilab Children's Center, for

---

* All song excerpts © Arthur Roberts.

providing certain members of the symposium staff with necessary child care

Brigette Brown, for general assistance and for handling of microphones during the symposium

Morris Low and Catherine Westfall, for help with microphones at the symposium

Peter Galison, for help in planning the historians' contribution to the symposium

Ellen Carr Lederman, for miscellaneous general advice and support

M. Lynn Stevenson, for help with arranging the program.

For help in selecting speakers and topics, we would like to thank the Organizing Committee of the symposium: Jeremy Bernstein, James D. Bjorken, James W. Cronin, Gerald Holton, Leon M. Lederman, Victor F. Weisskopf, Robert R. Wilson, and ourselves. For meeting with the three of us at Fermilab on 1 April 1983 in a preliminary symposium planning session, we thank Richard Carrigan, Yoichiro Nambu, Philip Livdahl, Albert Wattenberg, Gordon Baym, Drasko Jovanovic, Leon Lederman, Lawrence Jones, Mark Oreglia, Jonathan Schonfeld, Francis Cole, and Robert Schlutter.

For hosting the symposium and for many varieties of support, ranging from secretarial assistance to office and auditorium space, we wish to thank Fermilab, an institution devoted to high-energy physics research, which is supported by the U.S. Department of Energy and operated by the Universities Research Associates (URA). For necessary funding for the symposium or for the proceedings preparation we are pleased to thank the Alfred P. Sloan Foundation, the AUA Trust Fund, and the History and Philosophy of Science Program of the National Science Foundation.

Photographs of the symposium

The photographs on the following pages are courtesy of the Fermilab Photography Department.

Symposium participants.

Chen Ning Yang.

Lillian Hoddeson conversing with Emilio Segrè.

Victor Weisskopf, John Heilbron, and Peter Galison.

Donald Perkins and George Rochester.

Kameshwar C. Wali, Louis Michel, and Max Dresden at Symposium party.

Laurie Brown at the podium.

Helmut Rechenberg and Yoichiro Nambu.

Andy Pickering, Alan Morton, James Cushing, and Catherine Westfall.

Sam Treiman and Abraham Pais.

Judy Zielinski Schramm, Ellen and Leon Lederman, May West, Rocky and Adrienne Kolb.

Robert Walker, Charles Peyrou, and Edoardo Amaldi.

Ernest Courant and Robert Marshak at the May Wine buffet.

Abbreviations

AdA	(Annello d'Accelerazione) – Italian collider
ADONE	Electron–positron (e^+e^-) 1.5-GeV/beam storage ring
AEC	Atomic Energy Commission
AG	alternating gradient
AGS	alternating-gradient synchrotron
ANL	Argonne National Laboratory
AUI	Associated Universities Incorporated
BCS	Bardeen–Cooper–Schrieffer (theory of superconductivity)
BNL	Brookhaven National Laboratory
Caltech	California Institute of Technology
CAR	Canonical Anticommutation Relations
CCR	Canonical Commutation Relations
CEA	Cambridge electron accelerator
CERN	European Center for Nuclear Research
CVC	conserved vector current
DOD	Department of Defense
EAS	extensive air shower
Fermilab	Fermi National Accelerator Laboratory
FFAG	fixed-field alternating gradient
G.E.	General Electric (corporation)
GM	Geiger–Müller (counter)
GNS	Gelfand–Naimark–Segal
LBL	Lawrence Berkeley Laboratory (formerly UCRL)
LEP	large electron–positron collider
LHC	lepton–hadron collider (Amaldi, in Chapter 12)
linac	linear accelerator
LSZ	Lehmann–Symanzik–Zimmerman
MED	Manhattan Engineer District
MIT	Massachusetts Institute of Technology

MTA	materials-testing accelerator
MURA	Midwestern Universities Research Association
NAL	National Accelerator Laboratory (now Fermilab)
NSF	National Science Foundation
ONR	Office of Naval Research
PCAC	partially conserved axial vector current
PS	proton-synchrotron
QCD	quantum chromodynamics
QED	quantum electrodynamics
Q.E.D.	*quod erat demonstrandum* (end of proof)
QFT	quantum field theory
rf	radio frequency
rms	root mean square
SLAC	Stanford Linear Accelerator
SSC	superconducting supercollider
S matrix	scattering matrix
UCLA	University of California at Los Angeles
UCRL	University of California Radiation Laboratory (at Berkeley)
UCSD	University of California at San Diego
USC	University of Southern California
WAG	Western Accelerator Group
ZGS	zero-gradient synchrotron

Notation

General

A	axial vector interaction (as in $V - A$); also amplitude
$\mathscr{A}$	amplitude
B	baryon number
BeV	billion electron volt (10^9 eV)
C	charge conjugation
d	deuteron
E	energy
e, e^+	electron, positron
eV	electron volt
$F(q), F_1, F_2$, etc.	form factor (q = momentum transfer)
GeV	giga electron volts (10^9 eV)
g	grain density (in nuclear emulsion)
$g_A, g_{\pi nn}$, etc.	coupling constants
$\hbar$	Dirac $h\,(h/2\pi)$
$\mathbf{I}$	isospin (or isotopic spin) operator
I_1, I_2, I_3	isospin components
I, I_0	ionization
i	$\sqrt{-1}$, imaginary unit
J	angular momentum
J/ψ	a new particle
K^*	kaon-pion resonance
K_L, K_S	kay-long, kay-short
$K_{\pi 3}$, etc.	decay modes of K mesons
$K, K^0, \bar{K}^0, K^\pm, K_1, K_2$	K meson, kaon
kV	kilovolt
$l, l^\pm$	lepton
l, l_{12}	orbital angular momentum
M or m	general particle mass

MeV	million (mega) electron volts (10^6 eV)
MW	megawatt
m_e	electron mass
$m\pi$	pion mass
m_μ	muon mass
N	nucleon
$n, \bar{n}$	neutron, antineutron
$p, \bar{p}$	proton, antiproton
p or P	momentum
p_t or p_T	transverse momentum
Q	electric charge operator, also decay energy (in Chapter 4)
q	momentum transfer
R	range (of moving charged particle)
R_0	nuclear force range (in Chapter 5)
S	strangeness
SO(7)	rotation group in seven dimensions
SU(2), SU(2) $\times$ SU(2)	special unitary groups
SU(3)	special unitary group in three dimensions
T	time-reversal operator
t	time
V	vector interaction
$V - A$	vector minus axial vector interaction
$V^\pm, V, V_1^0$	V particle
$W, W^\pm$	charged vector meson
Y	hyperon; also hypercharge
Z	atomic number
Z, Z^0	neutral vector meson
$\alpha, \bar{\alpha}$	anisotropy (in Chapter 4)
β	electron in β decay
Γ_μ^A	axial vector operator (in Chapter 27)
γ	photon; also γ ray
γ_μ, γ_5	Dirac matrices
$\Delta, \Delta^{++}, \Delta^\pm, \Delta^0$	delta particle
Δ	change or difference, as in ΔS
$\partial_\mu A_\mu$	4-divergence of A_μ
η	eta meson
$\theta, \bar{\theta}, \theta^0, \bar{\theta}^0, \theta^\pm$	theta meson
θ, ϕ	angles
$\varkappa$	kappa meson
Λ	lambda particle
$\mu, \mu^\pm$	muon
μm or μ	micrometer or micron
μsec	microsecond
$\nu, \bar{\nu}$	neutrino, antineutrino (also ν_μ, ν_e, ν_τ)
ν	number of betatron oscillations per revolution (in Chapter 13)
Ξ	xi particle (or "cascade particle")
$\pi, \pi^0, \pi^\pm$	pi meson, pion
$\varrho(r)$	density function
ϱ^0	meson (in Chapter 7)

$\Sigma^{\pm},\Sigma^{0},\Sigma$	sigma hyperon
σ	sigma meson; also cross section, "σ-star"
$\tau,\bar{\tau},\tau^{0},\bar{\tau}^{0},\tau^{\pm}$	tau meson
τ,τ_{π}, etc.	lifetime
χ	meson (in Chapter 4)
χ^{+}	a mode of K decay
Ω^{-}	omega-minus hyperon
ω	omega meson
3-3 resonance	three-three resonance
$0^{-},2^{+}$, etc.	spin-parity symbols
$\vec{v}_{e}$	"forward" arrow, denotes scalar product (in Chapter 32)
$\overleftarrow{P},\bar{\sigma},\overleftarrow{J}$	"backward" arrow, denotes axial vector (in Chapter 32)

Chapter 42

C^{*}	as in C^{*} algebra
f	test function
$\mathscr{F},\mathscr{F}_{s},\mathscr{F}_{a}$	Fock space (s = symmetric, a = antisymmetric)
H	Hamiltonian
$\mathscr{H}$	Hilbert space
I	inversion
$L^{2}(R^{n})$	square integrable in a space of n dimensions
L_{+}	restricted Lorentz group
$L_{+}(\mathbb{C})$	complex linear transformation of Lorentz group
$\lim_{t\to\pm\infty}$	limit as t approaches $+\infty$ or $-\infty$
P^{0}	energy operator
P^{μ}	energy-momentum operator
$P_{+}^{\uparrow}$	restricted Poincaré space
$SL(2,C)$	special linear group in two dimensions, complex
s	spin
$\mathscr{S}$	real vector space; test function space
U,V	unitary transformation
$\mathscr{U}^{(n)},\hat{\mathscr{U}}^{(n)}$	sequence of tempered distributions
Z	renormalization constant
Δ^{+}	propagator
Θ,Θ^{-1}	PCT operators
Λ	scale factor
$\phi,\phi(f)$	Hilbert space operator
ψ_{0}	physical vacuum
$0(4)$	rotation group in four dimensions
$\sum_{j=1}^{n}$	sum over j from $j=1$ to $j=n$
$\bigoplus_{n=0}^{\infty}$	direct sum of spaces
$\prod_{j=1}^{s}$	product symbol
$\otimes$	direct product

$\in$	belonging to
$\in \| R^4$	belonging to four-dimensional space
$\subseteq$	is a member of the set
$\Box, \Box_{y_i}$	d'Alembertian operator
$\{a, A\}$	group element
$p, x + iy$	4-vectors

PART I

INTRODUCTION

1 Pions to quarks: particle physics in the 1950s

LAURIE M. BROWN

Born 1923, New York City; Ph.D., Cornell University, 1951;
theoretical physics; Northwestern University

MAX DRESDEN

Born 1918, Amsterdam; Ph.D., University of Michigan, 1946;
theoretical physics; State University of New York at Stony Brook

LILLIAN HODDESON

Born 1940, New York City; Ph.D., Columbia University, 1966; history of
modern physics; University of Illinois at Urbana–Champaign and Fermilab

Between 1947 and 1963, a period freely referred to here as "the fifties," elementary particle physics was growing up and undergoing the triumphs and pains of its adolescence. Beginning with the discovery of charged pions in cosmic rays, the period ended with the proposition that the basic constituents of the hadrons are quarks. The period witnessed the vindication of quantum field theory in renormalized quantum electrodynamics (QED), only to see it rejected as a theory of the strong and weak interactions. It saw the concept of symmetry emerge as a fundamental characteristic of basic physics, followed by its downfall in the parity revolution. Researchers discovered particles that were so different from any known previously that they labeled them "strange," and they found unexpected properties in familiar interactions. The fifties was a time of intellectual turmoil, of shifting attitudes and changing theoretical fashions.

For particle physics, as for science in general, the period was a time of transition, as energy was redirected from the problems of war to the concerns of peace. Profiting from their participation in the radar and atomic bomb projects (we are speaking now mainly of the United States, but that is where the shift began), physicists planned and carried out cooperative scientific research on a scale previously unknown in peacetime and drew on the prestige that derived from their wartime successes to obtain the necessary funding. Elementary particle physics, then called, significantly, "high-energy nuclear physics," served as a prototype for other large-scale scientific projects, such as the space programs in the United States and the Soviet Union.

At first, the new particles (Robert Oppenheimer called them "the subnuclear zoo") had their "photographs" taken, mainly by cosmic-ray zoologists, using cloud chambers and nuclear emulsion. Soon it became clear that a more detailed study of the animals would require the controlled, intense high-energy particle beams of new accelerators. With the efforts of a new breed of physicist, the accelerator specialist, "high-energy nuclear" physics became "elementary particle" physics. Cosmic-ray workers returned to their traditional role of supplying challenging puzzles to the geophysicist, the astronomer, and the cosmologist.

Until the midfifties, fast particles were obtained either from the cosmic rays or from accelerators of several hundred million electron volts (1 MeV being 10^6 eV) operated at individual universities. Detectors were ionization counters, as well as the simple visual devices, cloud chambers and photographic emulsions. Nevertheless, those relatively simple devices yielded most important discoveries! Beginning in the midfifties, particle accelerators in the giga-electron-volt (1 GeV = 10^9 eV) range were designed, built, and operated by a large staff of physicists, engineers, and technicians working at national laboratories. At the same time, sophisticated large-scale equipment for detection and analysis of particles was developed and utilized at the national laboratories by university groups.

The exciting and tumultuous period of the fifties invites comparison with a period that began about fifty years earlier with the discoveries of x rays, the electron, and radioactivity, setting the stage for Ernest Rutherford's discovery of the nuclear atom in 1911. Together with the revolutionary theories of Max Planck and Albert Einstein, Rutherford's atom, as interpreted by Niels Bohr in 1913, led to a quantum description of atomic structure and to relativistic quantum field theory. Half a century later, the pion, the muon, and the *V* particles revealed an unsuspected richness of fundamental objects and hinted at the existence of universal types of interaction and new internal quantum numbers that were suggestive of a new "spectroscopy." Ironically, this very richness eventually led to questioning whether or not *any* particles were fundamental. Each of the periods that we have been considering ended with the emergence of imperfect but highly influential models – Bohr's atomic model in 1913 and the "naïve" quark model in 1963 – both of which had arbitrary features and were inconsistent with the dynamical theories current at the time. Neither model had a fundamental justification, but each had significant predictive power and became the basis for further conceptual advances.

In recent years, when theory called for new particles (such as the *W* and *Z*), experiment obligingly provided them, but in the fifties experiment outran theory and produced surprise after surprise. Neither the muon nor the strange particles were expected, nor were they welcomed, for the most part, for they destroyed what might have been a consensus for a new unification. Without the muon, physicists had anticipated a closed system in which the electron, proton, and neutron were the constituent particles of matter, while the photon and Yukawa meson were field quanta that carried the electromagnetic and strong interactions. Add the neutrino for weak interactions, and complete the picture by including the antiparticles of the fermions. The muon changed all that. To paraphrase a famous query of Isidor Isaac Rabi: Who ordered the new generations of particles? That question still haunts us.

The extraordinary successes scored by renormalized QED ill prepared the physics community for the failure of field theory to account for other fundamental processes, although, in spite of its failures, field theory was thought to be *qualitatively* correct. Thus, new programs were undertaken, notably *S*-matrix and dispersion theory, to isolate and abstract the successful features of field theory while circumventing its problems. The mathematical foundations of quantum field theory were studied, but that work did not appreciably affect the mainstream of physics. A new type of field theory emerged, based on non-commuting gauge fields. Proposed by Chen Ning Yang and Robert L. Mills (and also by Ronald Shaw in a doctoral thesis at Cambridge University), the non-Abelian gauge theory was to have great impact in later decades, but during the fifties it was regarded as only a formal curiosity.

In contrast, the dramatic demonstration of the nonconservation of parity in the midfifties was immediately accepted by physicists, even though some were dismayed. "I do not believe that the Lord is a weak left-hander," Wolfgang Pauli wrote to Victor Weisskopf on 17 January 1957.[1] By the end of the decade, particle physicists had become used to broken symmetry and had developed an elegant way to see how it could come about.

Swords into plowshares: entering the 1950s

World War II interrupted careers, absorbed creative energies, and uprooted the lives of many physicists, although most of them were spared the war's worst horrors. In the combatant countries, science became the handmaiden of war. Normal scientific communication and research in pure physics almost ceased, with a few notable exceptions. Among the latter was Werner Heisenberg's relativistic *S*-matrix theory, proposed in 1942 and discussed in this symposium by Helmut Rechenberg in Chapter 39. Heisenberg's theory was one of the earliest attempts to replace quantum fields by a basic description in terms of only directly observable quantities. The *S* matrix was fashionable for a few years after the war, then was dropped for a time, and was revived again in the late 1950s. Other wartime achievements, such as the Japanese two-meson theory, the measurement of the lifetime of the cosmic-ray meson, and the Rome experiments on the capture rates of slow mesons, were treated in our first volume.[2] Also mentioned there is the strong-coupling-meson theory, first proposed by Gregor Wentzel in Zurich.[3]

Some physicists not engaged in weapons development, but living under wartime bombardment, managed to continue research, even though time and resources were limited.[4] George D. Rochester and Cecil F. Powell were among the British physicists assigned to train students during the war. Rochester spent his nights at Manchester running a fire brigade, his days teaching and doing cosmic-ray research with the Hungarian expatriate physicist Lajos

Jánossy. Continuing this work after the war, Rochester and Clifford C. Butler discovered the *V* particles in 1947, as Rochester recalls in Chapter 4. At Bristol, Powell was carrying on nuclear research with photographic emulsions. In 1945 he was joined by Giuseppe P. S. Occhialini, an Italian expatriate physicist who had been working in Brazil. Soon Occhialini brought over his co-workers: Cesare M. G. Lattes and Ugo Camerini, another expatriate Italian. In 1947, Lattes, Occhialini, and Powell discovered the $\pi-\mu-e$ decay chain, establishing the existence of the pion.

Although it took the particle physics community several years to appreciate the importance of the *V* particles (some reasons for this are discussed by Rochester), the pion discovery had immediate far-reaching consequences. Lattes has given his personal account of it in our earlier volume,[5] and Donald H. Perkins, who published a report of what was probably the first example of a stopping negative pion producing a nuclear interaction, gives another perspective in Chapter 5. The discovery of the pion was treated in our first symposium, but because pion physics became a major theme of the fifties, we review here the events leading up to that discovery.

By 1947 it had become evident that the cosmic-ray meson (now called the muon) had a longer mean life and a smaller mass than the Yukawa meson, the postulated carrier of nuclear forces, was expected to have. Even more puzzling, the cosmic-ray meson seemed to interact only weakly with nuclear matter; it penetrated the atmosphere, and many meters of earth as well. The Rome experiment on meson capture was analyzed early in 1947 to show that the discrepancy from the expected interaction strength was a factor of 10^{12}![6] The solution to this puzzle was revealed within months by the improved nuclear-emulsion technique. There were actually two "mesons," decaying in cascade: The strongly interacting one, the pion, was produced primarily at high altitude and decayed rapidly into the weakly interacting muon.

To see the details of cosmic-ray interactions, the photographic emulsion was made sensitive to lightly ionizing particles, and a method was found for the uniform photographic development of thick emulsions. Microscopic scanning of plates and track measurement were systematized.[7] Beyond those technical aspects, an important element of success was the adventurous spirit and exuberant esprit de corps of the emulsion workers.[8] In 1946, Occhialini left emulsions of the new type, created for this work by Ilford, Ltd., under the direction of C. Waller, for exposure at 3,000 m altitude at the French cosmic-ray observatory on the Pic-du-Midi in the Pyrenees. "When they were recovered and developed in Bristol," according to Powell, "it was immediately apparent that a whole new world had been revealed."[9]

Among the new events were stopping mesons (that is, particles of intermediate mass, as determined by their ionization and scattering at the end of their range) producing nuclear "stars." The first such event was published by Perkins, of Imperial College, London, who found it in an Ilford emulsion that had been flown in an airplane at 30,000 feet for several hours.[10] In addition to the star-producing mesons, the Bristol group found two events showing $\pi-\mu$ decay.[11] The electron from the muon decay could not be seen in those early events, because the new emulsions were not yet sensitive to particles giving minimum ionization. The following year, Lattes, returning to Brazil, stopped in Berkeley and participated in the observation in nuclear emulsion of the first artificially produced pions.[12]

Butler and Rochester observed the first *V* particles in a cloud chamber crossed by a thick lead bar, operated between the poles of an eleven-ton electromagnet, which had been brought to Manchester by Patrick M. S. Blackett when he moved there from Birkbeck College, London, in 1938.[13] In October 1946, they observed the first *V*, a neutral particle decaying just below the lead into two charged particles, both of them lighter than a proton. The second *V*, observed in May 1947, was a charged-particle track that showed a sharp kink and change of ionization occurring above the lead. Both *V* particles appeared to have masses about half that of the proton. Curiously, no more *V* particles were observed after that for more than two years, although at the end of 1948 the Bristol group found a particle of unusual mass in a new Kodak electron-sensitive emulsion.[14] Rochester has described this two-year hiatus as "tantalizing and embarrassing for the Manchester group."[15] To try to obtain more *V* events, they decided to build a new chamber and large magnet that could be taken apart and transported to the Pic-du-Midi, to be operated in the much more intense cosmic rays at high altitude.[16] Before that, however, a letter from Carl D. Anderson at the California Institute of Technology (Caltech) to Blackett, dated 28 November 1949, ended the suspense:

> Rochester and Butler may be glad to hear that we have about 30 cases of forked tracks similar to those they described in their article in *Nature* about two years ago, and so far as we can see now their interpretation of these events as caused by new unstable particles seems to be borne out by our experiments.[15]

After this somewhat uncertain start, Powell's "new world" began to be intensively explored by cosmic-ray physicists on mountain peaks, in airplanes, and with high-altitude balloon flights. The track patterns observed in cloud chamber and emulsion, in addition to the *V*'s, testified to a variety of decay schemes, of possibly many new particles. The taxonomy and the number of species of the subnuclear zoo were proving difficult to establish. At the same time, the scientific importance of the 1947 discoveries was emphasized. (Their value was already recognized: the 1948, 1949, and 1950 Nobel Prizes in physics went, respectively, to Blackett, Hideki Yukawa, and Powell. On the other hand, Occhialini, who had been closely associated with Blackett and Powell in their most important discoveries, was suitably honored only in 1982, when he shared the Wolf Prize with George Uhlenbeck.) As regards the nonstrange particles, by mid-1949 it was tentatively established that the cosmic-ray meson, now known as the muon, has spin $\frac{1}{2}$ and decays into an electron and two neutrinos.[17] Its mass was determined to be 217 ± 4 electron masses, within a few percent of its presently accepted value.

In December 1951, an entire conference in Bristol was devoted to "*V*-Particles and Heavy Mesons." By that time, good evidence existed for neutral as well as charged *V* particles, both heavier and lighter than the proton. Charged particles with mass about half that of the proton had at least two types of decay: $\varkappa$ mesons, decaying into pion + muon + neutrino, and τ mesons, decaying into three pions. The Bristol conference standardized the nomenclature for the new particles and also led to collaborative efforts for making balloon exposures of emulsion stacks. The unpublished conference proceedings contain some doggerel by Perkins that concludes:

> So counter-control your cloud chamber
> And up with emulsions sky high,
> We'll find mesons in increasing number,
> And understand all, by and by.*

The third, fourth, and fifth Rochester conferences on high-energy nuclear physics, held at the University of Rochester early in 1953, 1954, and 1955, had sessions on cosmic rays and on the "new particles," which began to be analyzed in terms of *associated production* and *strangeness*, which we discuss later.

Two European conferences are particularly memorable to cosmic-ray physicists, one held in France and the other in Italy. Bagnères de Bigorre is a French resort near the Pic-du-Midi, where a conference was held in 1953 that concentrated on the strange particles. In Chapter 4, Rochester says that it was "attended by all the leading cosmic-ray particle physicists in the world [and] was unique in timing and scope."[18] He cites as the "main impact" the presentation by Robert W. Thompson of the findings of his cloud-chamber group, which used accurate measurements and ingenious analysis to establish the decay of a light V^0 particle into a π^+ and a π^- (the θ mode).[19] This presentation, together with the analysis of the τ-meson decay by Richard Dalitz, was the beginning of what became known as the "tau–theta puzzle."

The last conference at which cosmic-ray research made a major contribution to particle physics was held in Pisa in the summer of 1955. The results presented there, mainly from the G-stack (93 kg of emulsion, flown from a balloon, and measured and analyzed by a large collaboration of laboratories, twenty-one European and one Australian), showed that the particles having similar masses, close to half a protonic mass, and with different decay modes, the *K* mesons, had masses that were actually equal within a few tenths of 1 percent. The same conclusions could be drawn even more strongly from

---

*   Perkins claims this "originated as a joint effort of several people in Bristol."

recent results of the Berkeley Bevatron accelerator, presented at the same conference. "Thereafter," as Perkins says in Chapter 5, "accelerators took over," or, to quote Charles Peyrou's less cryptic commentary, "We had survived the arrival of the [Brookhaven] Cosmotron because of the impetus we had, but now the story was finished."[20]

Pion physics and nucleon structure

Modern particle physicists tend to forget that their subject grew out of both cosmic-ray *and* nuclear physics and was once called *high-energy nuclear physics*. In July 1958, the Eighth Annual International Conference on High Energy Physics was held at the European Center for Nuclear Research (CERN), in Geneva, Switzerland, continuing the series of Rochester Conferences that began in 1950, described by Robert Marshak in Chapter 45. Aside from the shift of venue from central New York to central Europe, another distinction worth noting is that the word "nuclear" was dropped from the title of the conference. From that time on, the field was called *high-energy* or *particle physics*, and what was then known as *high-energy nuclear physics* became present-day *intermediate-energy physics*.

It was to study nuclear forces that the postwar high-energy accelerators, made possible by the newly discovered principle of phase stability, were built. The discovery of the charged pions in the cosmic rays in 1947 led immediately to their production at the accelerators that could produce them: Edwin M. McMillan's electron-synchrotron and Ernest O. Lawrence's 184-inch synchrocyclotron, both at the Radiation Laboratory at Berkeley.[21] Yukawa's meson was the accepted quantum of nuclear forces, and meson interactions were considered fundamental to understanding nuclei.

Not until 1950 were neutral pions detected, first at the Berkeley synchrocyclotron, later at the electron-synchrotron and in high-altitude cosmic rays.[22] Roughly equal numbers of neutral and positive pions were produced in a hydrogen target by 330-MeV x rays obtained from the electron machine. The equal production was puzzling, for one imagined that the x-ray photons interacted directly with the charged mesons of the "cloud" around the proton, while a neutral meson could be "liberated" only by the less likely action of the photon on the charge of the massive proton core. A possible resolution of the paradox was suggested by postulating excited nucleon states, the so-called isobars, which are a characteristic feature of strong-coupling meson theory.[23]

Herbert Anderson has given an insightful historical account of the experiments on pion scattering that were begun in 1951 at the University of Chicago synchrocyclotron.[24] (Anderson, with John Marshall, was largely responsible for the construction of the machine, which at 450 MeV had a 100-MeV advantage over the Berkeley synchrocyclotron.) Scattering of mesons in hydrogen qualitatively confirmed some of the main predictions of meson theory: Rising at the start with energy, the scattering cross section rapidly attained its "geometrical" value; it had a strong *p*-wave threshold behavior, consistent with the pion being pseudoscalar.[25] Again the charge ratio was unexpected, the π cross section being larger than the π^-, even though π^- had more channels for interaction than π^+ (namely, charge exchange and radiative capture, in addition to elastic scattering).

As in the meson photoproduction case, the explanation of the scattering behavior was an excited nucleon state, or, in the language of scattering, a pion–nucleon resonance.[26] Specifically, assuming that the scattering was dominated by an intermediate pion–nucleon state of spin $\frac{3}{2}$ and isospin $\frac{3}{2}$ (which became known as the 3-3 resonance), it was predicted that for pion energy of about 120 MeV, at the peak of the resonance the ratio of cross sections of π^+, π^0, and π^- would be $9:2:1$. The measured cross sections were in good agreement with these ratios and with the predicted energy dependences. An account of the early work on the 3-3 resonance as seen in photoproduction, and of two higher-lying pion–nucleon resonances found in the 1950s, is given by Robert L. Walker in Chapter 6.

The idea that the nucleon is a structure of finite size, as opposed to the point-like electron, originated with the measurement of the neutron's magnetic moment in 1933. The neutron is not a "Dirac particle," for a neutral Dirac particle has zero magnetic moment. The isotopic partner of the neutron, the proton, also has a large anomalous moment, in addition to its Dirac moment. The theoretical argument is, however, a far cry from actually measuring the size and shape of the nucleon's charge and magnetic-moment distributions. In 1955, Robert Hofstadter began that ambitious program, using for his tool the precise measurement of electron scattering. Hofstadter tells the story of his successful efforts, which won him a Nobel Prize in 1961, in Chapter 7.[27] A commentary on the interpretation of the experiments comprises Chapter 8, by Robert G. Sachs and Kameshwar C. Wali, which concerns the relationship between the measured form factors and the distribution of charge and magnetic moment in the nucleon.

Revolution in the laboratory

Accelerators

The fifties saw a remarkable change in the particle physicist's laboratory, which received new tools, greatly increased funding, and new institutional settings. By 1960, costly accelerators had replaced cosmic rays as the principal sources of high-energy particles. Large bubble chambers, spark chambers, and scintillation counters replaced cloud chambers and nuclear emulsion as the principal detectors. National funding agencies, established in the wake of World War II, were generously supporting the new technology, bringing the physicist into a new complex relationship with the government. As a consequence of the increased size and cost of the accelerators and detectors, the principal setting for leading experiments was shifting from university laboratories to new large facilities based at national or international laboratories serving "users" from numerous smaller institutions. These dramatic changes marked the beginning of the transition to "big science" particle physics.

This transition was in part based upon the remarkable achievements of physicists and engineers during World War II. Technical strides in microwave technique, electronics, vacuum technology, cryogenics, and computing, made in programs such as the MIT Radiation Laboratory and the Manhattan Project, were widely exploited by particle physicists in the design and construction of accelerators and detectors.[28] They even thought of incorporating surplus radar equipment into new accelerators, such as the Berkeley linear accelerator.[29] The wartime practice of having industry provide materials for weapons research and development was adapted to postwar research needs.

Because the United States led in wartime research, American particle physicists in the postwar period had a strong advantage over their European and Japanese colleagues in developing advanced technology.[30] The Manhattan Engineering District (MED) established a precedent for well-funded large research projects, and the network of cooperation between leading American physicists and military leaders continued after the war and supported particle physics. For example, as Robert Seidel recounts in Chapter 34, General Leslie Groves, the military head of the MED, sponsored the completion of Lawrence's 184-inch cyclotron.[31] Directly out of the MED grew the three federal funding agencies that supplied the bulk of support for American particle physics in the post–World War II decades: the Atomic Energy Commission (AEC), the Office of Naval Research (ONR), and the National Science Foundation (NSF), as discussed by Silvan S. Schweber in Chapter 46.[32]

The U.S. government's willingness to fund postwar science derived from an increased appreciation for the "practical," especially military, value of science, and a bargain was struck, with the government providing research funds to advance the physicists' scientific programs. The inseparability of nuclear physics from its military implications was underlined in a leading textbook by Milton S. Livingston and John P. Blewett, two of the main accelerator architects:

> At the end of World War II, when physicists returned to their laboratories, the enhanced status of nuclear physics was immediately evident. The exciting and dangerous development of atomic energy, with its tremendous implications for national security, stimulated strong popular support for spending government funds on building still larger and higher-energy accelerators. With such impetus the new synchrocyclotrons were rapidly developed.[33]

The bargain was perhaps most evident at the Berkeley laboratories, as emphasized by John Heilbron in Chapter 3 and Seidel in Chapter 34. The Bevatron was built not only to produce antiprotons and controlled high-energy particle beams for research but also to study nuclear forces that might lead to new weapons, to keep the Berkeley engineering staff together, and to train manpower for defense purposes. Berkeley accelerator physicists explicitly contributed to defense work in the early 1950s in the Materials Testing Accelerator (MTA), a huge accelerator, discussed by Seidel in Chapter 34 and by Blewett in Chapter 10, that was intended to produce fissionable material. Berkeley's involvement in that project slowed construction on the Bevatron, enabling the Brookhaven Cosmotron to be the first machine to accelerate particles above 1 GeV.

Schweber suggests in Chapter 46 that the work done by theoretical physicists for the military during the 1950s may have influenced the content of their scientific research, pointing out that during that period American theoretical physics had a decided empirical orientation, in contrast to the more fundamental orientation of European particle theory. However, it should be recalled that American (as well as British) scientists have a long history of empiricism in their research, in contrast to a greater concern with abstraction on the Continent.[34]

After World War II, leading members of the particle physics community readily applied the administrative expertise they had gained in wartime.[35] Thus, Luis Alvarez, a group leader at both the MIT Radiation Laboratory and Los Alamos, brought large specialized staffs to bear on building and instrumenting a large hydrogen bubble chamber.[36] Many working relationships established during the tense World War II years continued into the 1950s particle physics community. For example, Robert Bacher and Hans Bethe, who had been division leaders at Los Alamos, set up the postwar Cornell nuclear physics program, bringing in a number of their wartime Los Alamos colleagues. When Bacher moved to Caltech a few years later, he again recruited former colleagues from his Los Alamos days.

Technical developments in accelerators from 1945 to 1960 are treated in this volume by Matthew Sands, Blewett, Ernest D. Courant, Donald W. Kerst, and Seidel and by a panel of experts in Chapters 9, 10, 11, 13, 34, and 12, respectively. Events in the forties were dominated by the discovery of the phase stability principle – in 1944 by Vladimir Veksler in the Soviet Union and independently in 1945 by McMillan in the United States. That discovery enabled particles in a circular accelerator to enter the relativistic regime.[37] With the help of AEC and ONR funding, this principle made possible the postwar synchrocyclotrons, electron-synchrotrons, and linear accelerators. Examples at Berkeley were the synchrocylotron, based on Lawrence's 184-inch cyclotron, which operated in 1946 at 380 MeV, McMillan's electron-synchrotron, which operated in 1949 at 320 MeV, and the 32-MeV proton linear accelerator built by Alvarez and Wolfgang Panofsky.[38]

Even before 1950, as Robert R. Wilson says in Chapter 12, the need for higher-energy machines was apparent, since accelerators in the 300-MeV range could only just barely do meson physics. Economic constraints, however, dictated a new type of construction. Synchrocyclotrons, like ordinary cyclotrons, require a great deal of iron inside the orbit, ruling out their extension to the giga-electron-volt range. Synchrotrons allow more economical construction, since their magnetic fields are confined to a toroidal region, rather than a cylindrical region, but to have operated an electron-synchrotron at 1 GeV would have consumed too much power because of synchrotron radiation. Thus, thoughts turned in the late 1940s to the proton-synchrotron in the 1–10-GeV range. In the United States, Berkeley and Brookhaven submitted proposals to the AEC. In England, Marcus Oliphant proposed a 1.3-GeV synchrotron for Birmingham. The AEC decided in 1948 to fund a 6-GeV synchrotron in Berkeley, the Bevatron, as well as a 3-GeV synchrotron at Brookhaven, the Cosmotron. Those two accelerators, the first large machines to be funded by the AEC, began a trend in which larger and larger grants for accelerators would be funneled into fewer and fewer geographic locations.[39] The Cosmotron achieved its first 1.3-GeV beam in May 1952; Oliphant's 1-GeV synchrotron began work in 1953; and the Bevatron came on line in 1954.

As these synchrotrons were turning on, Courant, Livingston, and Hartland Snyder, working at Brookhaven in the summer of 1952, made the major accelerator innovation of the 1950s. They invented a new method of focusing particle beams by a sequence of alternating converging and diverging magnetic lenses, called "alternating-gradient" (AG) or "strong" focusing. This scheme permitted great reduction in the size of magnets, thus strongly cutting the cost of large accelerators.[40] (Unbeknown to the Brookhaven team, this invention had also been made by Nicholas Christofilos in Greece two years earlier.) Brookhaven then proposed a 30-GeV-range AG accelerator, the alternating-gradient synchrotron (AGS), and CERN modified the design of its proton-synchrotron, the PS, also in the 30-GeV range, to make use of strong focusing. The PS came on line at 26 GeV in 1959, and the AGS at 33 GeV in 1960. Meanwhile, Wilson had built the first strong-focusing machine at Cornell, a 1.3-GeV electron-synchrotron, which came on line in 1954.

Another principle that contributed to a reduction in the cost of circular accelerators was the "separated-function" magnet, conceived in 1952 by Toshio Kitagaki at Tohoku University, and independently by Milton White at Princeton. Realizing that intermittent focusing is sufficient in a circular AG machine, they suggested cutting costs by using different magnets for focusing (quadrupole lenses) and bending the beam (dipole lenses).[41]

Detailed study of AG focusing resulted in many innovations, not only in the techniques of building larger circular accelerators but also in the design of linear accelerators and in the design of experiments. The Courant–Livingston–Snyder paper on strong focusing introduced quadrupole-focusing magnets, which came to be used in forming intense, highly focused secondary beams for experiments.[42] This approach was later applied in the discovery of the antiproton, and Blewett (see Chapter 10) realized that the AG concept could be used in the linear accelerator.[43]

Kerst tells in Chapter 13 how the possibilities of higher-energy machines operating with AG focusing stimulated interest by midwestern American physicists in building their own large accelerator, resulting in the formation of the Midwestern Universities Research Associates (MURA).[44] MURA pioneered many accelerator innovations of the 1950s, including the fixed-field alternating-gradient (FFAG) magnet accelerator, beam stacking, colliding beams, computer control of beams, radio-frequency (rf) modulation of cavities, the application of Liouville's theorem to accelerator problems, spiral-ridge alternating-gradient focusing, and rf beam injection and extraction.[45] The FFAG accelerator was conceived at MURA in 1954 by Keith Symon (and independently by Snyder, A. A. Kolomensky, L. H. Thomas, and Tihiro Ogawa). In this machine, the particles spiral outward, focused by a fixed nonlinear magnetic field having alternating gradients, and follow stable orbits for a wide range of particle momenta, making possible an intense beam. In 1956, MURA proposed an accelerating system composed of two accelerators in the 15–20-GeV range, with oppositely directed and interacting beams.[46]

Colliding beams to achieve higher center-of-mass energy had been suggested in the 1940s by both Wilson and Rolf Wideröe, but the idea did not then appear viable because it was not clear how to accumulate a sufficiently intense beam for such collisions. In 1956, the MURA group (Kerst, Symon, Andrew Sessler, and others) showed that sufficient beam concentration could be achieved by "stacking" a number of high-intensity pulses in a ring. (See Kerst in Chapter 13 and Sands in Chapter 9.)[47] In the same year, Gerard O'Neill and, independently, Donald Lichtenberg, Roger Newton, and Marc Ross, as well as W. M. Brobeck, proposed the idea of colliding beams in a separate storage ring having constant guide field.[48] Practical colliding-beam development has used storage rings, and the role of the FFAG in colliding-beam development was to remain only historical.

In the midst of this rush of accelerator advance based on strong focusing, the American particle physics community was startled to learn that the government had authorized the construction at Argonne National Laboratory of a conventional weak-focusing machine, the Zero-Gradient Synchrotron (ZGS). As Lawrence Jones explains in Chapter 12, this decision was a cold-war response to Veksler's announcement in 1955 that a 10-GeV proton accelerator was under construction at Dubna, near Moscow. Washington's resolve quickly to build an American accelerator of higher energy than the Dubna machine set up severe tensions in the American accelerator physics community, whose members were appalled by the government's neglect of the pathbreaking strong-focusing concepts then being developed.[49] Among the other cold-war actions that alienated some physicists from the government were the security clearance requirements at national laboratories, insufficient declassification of research reports, and the Oppenheimer hearings.

In Europe, accelerator building was vigorously pursued in the late fifties, with six accelerators of energy between 1 and 30 GeV becoming operant during 1956–9, and eleven more giga-electron-volt-range accelerators under construction. The first in the series of international conferences on high-energy accelerators was held in 1956 at Geneva. Accelerator development was now an independent profession practiced by a new specialist, the "accelerator physicist."

Detectors

Throughout the boom in accelerator building and designing in the 1950s, a revolution was also taking place in the means of detecting high-energy particles, as Peter Galison discusses in Chapter 14.[50] The visual detectors in 1950 were cloud chambers and nuclear emulsions; the electronic counting detectors included ionization chambers as well as proportional counters, Geiger counters, and early forms of scintillation counters.

Attempts to extend the usefulness of the cloud chamber led to the high-pressure cloud chamber [50 atmospheres (atm) pressure] and the diffusion cloud chamber, whose advantage is its continuous sensitivity. The cloud chamber is limited by being an intrinsically low-density device, in which one sees only a small part of the track of a fast charged particle, and by its considerable dead time, during which it cannot detect particles. Ralph Shutt's group at the Cosmotron, as William B. Fowler discusses in Chapter 22, developed the diffusion chamber, invented by Alexander Langsdorf in the late 1930s, into a tool capable of demonstrating the associated production of "strange" particles.[51] Fowler points out that the high-pressure diffusion cloud chamber helped to motivate the invention of the hydrogen bubble chamber and that the cryogenic expertise involved in the latter development was applied to the first superconducting accelerator.[52]

The bubble chamber, invented by Donald Glaser in 1952, and discussed here in Chapter 14 by Galison and in Chapter 19 by Alvarez, effectively replaced the modified cloud chambers during the second half of the decade. Enabling with its higher density a far better measurement of the range of particles than the cloud chamber, and being capable of operating in a magnetic field to measure particle momentum,[53] the bubble chamber was a versatile tool. When filled with liquid hydrogen, one could directly see proton interactions. By the latter part of the decade, the bubble chamber had been developed into a large-scale experimental tool through efforts by Alvarez, Jack Steinberger, and others.

Analyzing the statistically more significant data samples produced by the bubble chamber required computers, which turned the data analysis into an enterprise consisting of many workers.[54] Alvarez's seventy-two-inch hydrogen bubble chamber, costing $2 million, required a $1-million computer and a sizable staff to build and use the device. Thus, tabletop experiments gave way to experiments conducted by dozens of researchers and costing millions of dollars.

Efforts to obtain good spatial resolution with high triggering rate led to the scintillation-counter array (hodoscope) and later to the spark chamber, discussed by Galison in Chapter 14 and by Shuji Fukui in Chapter 15. The scintillation detector grew out of a series of observations, including Hartmut Kallmann's γ-ray counting in naphthalene (mothball material) in Germany in the late 1940s and Hofstadter's use of sodium iodide in the United States in 1948, as discussed by him in Chapter 12. A descendant of the particle counters, the spark chamber was based on wartime advances in electronic timing circuits.

The 1950s also saw refinement of the nuclear-emulsion technique, which gave precise information on mass, energy, and the modes of interaction and decay of particles. Emulsion provided good angular and spatial resolution because of the small size of its grains, but it had the disadvantage of requiring measurement with high-powered microscopes.[55] Advances in the 1950s included finer grains and thicker emulsions, often stacked, which gave good range data. Emulsion experiments, with cosmic-ray exposures on mountaintops and balloon flights, were often international collaborations in this period.

Important advances were also made in calorimetry and Cerenkov counters, as mentioned by the panel in Chapter 12. The transistor, invented in 1947, began to enter detector circuits in the late 1950s; however, these circuits became commercially available to particle physics experiments only during the 1960s. The counter and spark chamber were much improved by better electronics. Wilson's application of the Monte Carlo technique to solving the differential equations of cascade showers allowed an understanding of shower detectors for high-energy photons.

National and international facilities

With the development of large accelerators and detectors in the 1950s arose new institutions having far-reaching implications: national and international high-energy research laboratories. In the United States, national laboratories emerged out of the MED network. The Argonne Laboratory, which was built in March 1943 for the research of scientists working with Enrico Fermi at the Chicago Metallurgical Laboratory, entered the high-energy field in 1958, when plans were developed for the 12.5-GeV ZGS. The MED established Brookhaven National Laboratory on Long Island in 1946 for peacetime research, as discussed by Courant in Chapter 11, but as historian Allan Needell explains, also to satisfy "the U.S. government's desire to keep careful control over the technology of nuclear reactors."[56] Initially, Brookhaven was to make available a nuclear reactor for research with neutrons, a 700-MeV synchrocyclotron and a particle accelerator of at least 1 GeV. The last project grew into the Cosmotron.[57] The formation, in July 1946, of the Associated Universities Incorporated (AUI), a group of American East Coast universities, to run the Brookhaven Laboratory marked the beginning of consortium management of large national laboratories in the United States.

The national laboratory at Stanford was established in the early 1960s, on roots planted in the 1950s.[58]

The growth of interuniversity laboratories outside the United States was slow relative to that in the United States, because it was necessary to make up for the devastating setbacks of World War II. However, the principal European accelerator laboratory CERN was planned in the early 1950s, as Edoardo Amaldi and Armin Hermann relate in Chapters 35 and 36. These plans grew out of both scientific and political interests.[59] By 1957, roughly twenty laboratories in Europe were engaged in high-energy physics. Although for two decades the United States maintained its leadership in particle physics, by 1980 the European efforts were yielding outstanding results. International theoretical institutions were also created, such as the International Center for Theoretical Physics in Trieste, discussed by Abdus Salam in Chapter 37, which was planned in the 1950s. This establishment of international laboratories reinforced a new trend toward collaboration as well as competition.

Interesting interactions began to take place between the different large laboratories. For example, as discussed by Blewett and Courant in Chapters 10 and 11, the research at Brookhaven that led to strong focusing grew out of a request by CERN for assistance in designing its new high-energy proton accelerator. Large laboratories also brought a new life-style to the university-based high-energy experimentalist, who, after teaching classes and attending committee meetings, would travel to one of the large facilities and work there with students as a "user." The effects of user life-style on families, universities, and the psyches of physicists are yet to be examined in detail.[60] At Argonne, and later at other national laboratories, the user community gained power through new user organizations. User groups also provided a lobbying mechanism to help increase the funding for high-energy physics research and thus had a political role as well as a scientific role.

The user groups also lobbied within the high-energy physics community for the establishment of a "truly" national laboratory. At the close of the war, the Radiation Laboratory at Berkeley, the principal prewar accelerator laboratory, planned to have much of its research conducted by visiting scientists. But in practice, most of the machine time was reserved for Berkeley physicists, and by the late 1950s the outside researchers were openly disturbed by this policy. Many experimentalists also became annoyed with Brookhaven's policy of limiting usage primarily to universities in the immediate area. A movement began to promote the idea of a "truly national laboratory" (TNL, a pun on BNL, or Brookhaven National Laboratory). This movement culminated in the creation in the late 1960s of the first National Accelerator Laboratory (NAL), which was renamed Fermilab in 1974.[61,62]

Attempts to classify the new particles

After a set of phenomena has been selected for scientific study (e.g., the solar system, atoms, or molecules) and its general characteristics observed, the next step toward understanding and mathematical modeling, as opposed to mere description, usually is the identification of the *constituents* of the system (sun and planets, nucleus and electrons, and atoms, respectively, for the examples given). Finally, a theory is made, which is provisional insofar as it deals with a certain level of experience and may be superseded by a theory based upon a deeper level of structure. Observation of this deeper level then initiates a new cycle. Such is the "three-stage methodology" of the Japanese theorist Mituo Taketani, who calls the stages "phenomenological," "substantialistic," and "essentialistic." Whether or not his methodology, proposed in the mid-1930s, has universal validity in the history of science, the development of particle physics over the past fifty years has followed this pattern in an exemplary manner. Perhaps the major achievement of our historical period was identifying and classifying the members of this particular substantialistic stage, to use Taketani's language.

The main principles used to classify particles (except for strangeness) originated in earlier decades, but their deeper meaning and overriding importance were manifested only in the 1950s. The main characteristics employed in the classification of the fundamental particles are listed in Table 1.1, together with the decade in which each first made an impact. The Fermi field theory, involving the exchange of an electron–neutrino pair, a universal theory of weak and strong interactions, was the main fundamental theory of nuclear forces in the mid-1930s, but by 1940 Yukawa's meson essentially replaced the Fermi field as the carrier of the strong interaction. Electron–muon universality had to await the demonstration of the leptonic character of the sea-level cosmic-ray meson in the Rome muon-capture experiments of 1947. And while the idea of a universal Fermi interaction surfaced in the late 1940s, the meaning of "universality" was doubtful before the establishment of parity nonconservation (1957), the *V–A* interaction (1958), and the Cabibbo angle (1963).*

Finally, the notion that "elementary" particles might really be composites is traceable to the Fermi–Yang model of the pions as nucleon–antinucleon bound states, a model accounting for the pion's triplet isospin and its otherwise rather surprising negative intrinsic parity.[63] The Fermi–Yang model of 1949 directly inspired the Sakata model of the midfifties and influenced the quark models of the 1960s.[64] In Murray Gell-Mann's first publication on *The Eightfold Way* in 1961, he introduced a fundamental triplet as a pedagogical device for "mathematical purposes," namely, for the construction of group representations, but lest anyone take those constituents to be strongly interacting, he called them "leptons."[65] [Gell-Mann probably arrived at SU(3) via

---

*   *Ed. note:* A truly universal weak interaction requires the same form of interaction for muon, electron, and strange-particle weak interactions; hence the importance of the universal *V–A*. It would also need to explain the suppression, by an order of magnitude, of strange-particle decay rates relative to β decay. That was supplied by the Gell-Mann–Levy–Cabibbo angle theory.

Table 1.1. *Fundamental particle characteristics and decade of first impact*

Characteristic	Decade
Electric charge, mass, spin, magnetic moment	1920s
Anomalous magnetic moment	1930s
Particle–antiparticle distinction	1930s
Lepton–hadron distinction	1930s
Meson–baryon distinction	1930s
Isospin	1930s
Intrinsic parity	1940s
Electron–muon universality	1940s
Baryon constituents	1940s
Associated production and strangeness	1950s
Universal Fermi interaction	1960s
SU(3), etc., and quarks	1960s

an earlier proposed "global symmetry" scheme,[66] while Yuval Ne'eman's SU(3) was motivated by a generalized gauge invariance principle.[67]]

Associated production

The peculiar property of the V particles was the contrast between their long lifetimes, which showed that they were decaying by the weak interaction, and their relatively copious production, which showed that they were strongly interacting particles. This peculiarity was well established by 1951, although the particles were not yet named "strange." The observation of so few V particles after their discovery in 1947 proved to be a misleading clue to their production cross section, which is actually large. By 1951, it was concluded that in a high-energy collision, the probability of producing a V particle was more than 1 percent of that of producing a pion. And as Abraham Pais argued, "if one would consider the same mechanism which produces them to be instrumental for their decay, one would estimate lifetimes – of the order of 10^{-21} sec."[68] That estimate did not accord with the observed lifetimes of the V particles, which were longer than about 10^{-10} sec. (An object moving with nearly light velocity traverses 3 cm in 10^{-10} sec, which is why the first sightings of V particles were in a cloud chamber, and not in nuclear emulsion.)

Thus, when standard quantum theoretical thinking was applied to the V particles, there arose the serious problem of reconciling their copious production and their long mean lives. To inhibit their decay seemed more promising than to enhance their production (theoretically speaking), and the large suppression factor needed, of order 10^{11}, meant that a relatively strict selection rule was called for, one that ruled out electromagnetic as well as strong decay.

The resolution of this puzzle was not long in appearing. At a symposium held in Tokyo on 7 July 1951 to consider possible explanations for the peculiar behavior of the V particles, several groups proposed theoretical models and published them shortly afterward. As Kazuhiko Nishijima later described them (at the Wingspread International Conference in 1984), "These models were all different, but there was one thing in common. They all assumed that the V particles were produced in pairs."[69] Soon afterward, and independently, Pais also produced a model of the same type that incorporated what he called the "even–odd rule."[70]

The idea that V particles must be produced in pairs, which became known as *associated production*, was surprisingly difficult to verify. As Fowler points out in Chapter 22, during 1953–4 only nine examples were observed in Shutt's continuously sensitive high-pressure hydrogen diffusion cloud chamber at the Brookhaven Cosmotron. On the basis of these few events, however, Bruno Rossi concluded in January 1955 that "there is good evidence for associated production in hydrogen."[71] More associated events were observed in cloud chambers with metal plates, but the presence of unassociated multiple events made them harder to interpret. The liquid-hydrogen bubble chamber soon replaced the high-pressure diffusion cloud chamber, and as Steinberger relates in Chapter 20, a two-day exposure to the Cosmotron negative-pion beam yielded fifty-five associated-production events. By that time, a more detailed description of the production and decay characteristics had been formulated in terms of the concept of "strangeness."

Strangeness

The V-parity selection rule (or, equivalently, Pais's even–odd rule) forbids interactions connecting an even number to an odd number of V particles, such as the decay of a V particle into non-V particles, other than weakly. However, it does not forbid a production process such as two neutrons going to two lambda (Λ) particles. (We shall now begin to use the modern name Λ for the particle V_1^0 that decays $\Lambda \to p + \pi^-$ or $\Lambda \to n + \pi^0$, with a Q value of 37 MeV.) Two-Λ production was not seen, even though its energy threshold is low compared with other observed production processes.[72] Furthermore, in 1952 a particle labeled Ξ (Greek xi) was found to decay as $\Xi \to \Lambda + \pi^-$. Its lifetime was long, and because Λ decay is also weak, the decay of Ξ into ordinary particles was by two successive stages of weak interaction (hence the name *cascade particle*). The even–odd rule required, at most, one weak decay to non-V particles.[73]

These suggestive facts led Gell-Mann and Nishijima in 1955 to propose a new quantum number, called *strangeness*, S, to replace the even–odd rule. S is conserved additively in strong and electromagnetic interactions. A process that violates this law of conservation is either entirely forbidden or at most weak (it may be doubly weak, etc.) For example, the Λ is assigned $S = -1$, so that its decay to $p + \pi^-$, having $S = 0$, is weak. The Ξ is assigned $S = -2$, so that its decay occurs in *two* stages of $\Delta S = -1$. Finally, the process $n + n \to \Lambda + \Lambda$ would also have $\Delta S = -2$; hence, it would be doubly weak.[74]

There is more to the strangeness concept. It predicts the electric charges of the members of a given particle multiplet. The quantity S (more exactly, $S/2$) is introduced by both Nishijima and Gell-Mann as a displacement in the relationship between charge Q, baryon number B, and the third component of the isospin vector I_3. The relation that holds for nucleons and pions, $Q = I_3 + B/2$ is modified to read[75,76] $Q = I_3 + (B + S)/2$. The same relation, with $S = 0$, known since the 1930s, was used in the analysis of the pion scattering experiments in the early 1950s, for example, in connection with the 3-3 resonance.

Introducing S in this way preserves the most important advantages of the isospin formalism. First, it enumerates the charge states in an isotopic multiplet (so that the negative Ξ particle mentioned earlier, for example, with $S = -2$, has $I_3 = -\frac{1}{2}$ by the foregoing formula; thus, it must have an isotopic partner with $I_3 = +\frac{1}{2}$ and $Q = 0$). Second, the strangeness-allowed ($\Delta S = 0$) strong-interaction processes are charge-independent (i.e., isospin-invariant), implying the existence of a number of relationships among isospin amplitudes (like the $9 : 2 : 1$ ratios of the 3-3 resonance mentioned earlier). The strangeness idea is elegant, economical, and powerful, and it proved to be a natural and simple way to understand the occurrence of a number of isospin multiplets and to categorize their interactions.

Beginning in 1955, many experiments performed in Berkeley at the 4.5-GeV Bevatron used stacks of nuclear emulsion, electronic counters, and, later, hydrogen bubble chambers. These experiments did much to confirm and to fill out the strangeness scheme, finding new strange resonances analogous to the nonstrange pion–nucleon resonances, and also a kaon–pion meson resonance, called K^*.[40] This story is told by Gerson Goldhaber in Chapter 16 and by Alvarez in Chapter 19.[77] They followed upon another major discovery at the Bevatron, namely, the first observations of the antiproton and the antineutron, proving the existence of antimatter. These experiments are commented on by Owen Chamberlain and by Oreste Piccioni in Chapters 17 and 18.

K mesons

The heavy mesons, having mass about 500 MeV/c^2, are called K mesons or *kaons*. It was not known in 1953 how many different K mesons there were, because the masses and mean lives inferred from different decay modes, while similar, did not agree precisely. K mesons have provided challenging experimental and theoretical problems from the time of their discovery to the present. In the 1950s, the two outstanding problems posed by K mesons were *particle mixing* and the *tau–theta* ($\tau-\theta$) *puzzle.**

By 1954, several charged-K-meson decay modes were known to exist:

---

* *Ed. note:* Considering the K meson's eccentric disregard for the "laws" of physics, the K could well stand for Franz Kafka, the novelist!

decay into two pions, into three pions, or into leptons (with or without a pion). For the neutral K meson, only one decay mode was unambiguously identified: $\theta^0 \rightarrow \pi^+ + \pi^-$. (The first V particle was a θ^0!) Other possible neutral-K-meson decay modes, similar to those of the charged K, were also indicated. The two- and three-pion decay modes (of both charged and neutral K mesons), labeled θ and τ, respectively, became the ingredients of the τ–θ puzzle. The notion of particle mixing concerns all neutral K mesons, independent of their decay modes, but we shall, for simplicity, refer here only to θ and τ mesons.

The symmetry operation of charge conjugation (C) transforms a particle to its antiparticle. [This statement assumes the invariance of all natural laws under C. As discussed in the next section, C invariance is actually violated, but CP, where P is the parity operator, is a good symmetry in strong, electromagnetic, and ordinary weak (i.e., not superweak) interactions. When the latter are included, the operation that exchanges particle and antiparticle is CPT, T being the time-reversal operator. An operation closely related to C is the G parity, C accompanied by a rotation about the second axis in isospin space. Invariance under G of the strong interaction has as one of its consequences that states consisting of an even number of pions are not strongly connected to states consisting of an odd number of pions.[78]]

Gell-Mann and Pais in 1955 questioned the behavior of the neutral kaon under the transformation C, the symmetry operation of charge conjugation. They considered the two possibilities: Was the kaon its own antiparticle (like the neutral pion), or did it have an antiparticle distinct from itself (as does the neutron)?

According to the strangeness scheme, K^+ and K^- have $S = +1$ and -1, and therefore $I_3 = +\frac{1}{2}$ and $-\frac{1}{2}$, respectively. That strongly suggests that there are two distinct *neutral* kaons: K^0 (with $S = +1$, $I_3 = -\frac{1}{2}$) and $\bar{K}^0$ (with $S = -1$, $I_3 = +\frac{1}{2}$). However, certain decay modes of K^0 are eigenstates of C – for example, the θ mode, consisting of two pions, which has $C = (-1)^s$, where s is the spin of the K. But the neutral kaons are *not* eigenstates of C; instead, the eigenstates are the linear combinations

$$K_1 = (K^0 + \bar{K}^0)/\sqrt{2} \quad \text{and} \quad K_2 = (K^0 - \bar{K}^0)/\sqrt{2}$$

If one of those states has the C permitting its decay into two pions, the other has not, and should therefore have a different (and longer) mean life.

These results, distinguishing the *produced* particles K^0 and $\bar{K}^0$, which have definite strangeness, from the *decaying* particles K_1 and K_2, which have definite C, were derived by Gell-Mann and Pais in 1955.[79] Further beautiful consequences, derived by Pais and Piccioni, involve interference effects arising from the strong interaction with matter of these particle mixtures.[80]

As described by William Chinowsky in Chapter 21, Leon Lederman and his group at the Cosmotron set out to look for the long-lived neutral K meson, the K_2 defined earlier, using a large (36-inch) cloud chamber expanded in a strong magnetic field and exposed to a 1.9-GeV negative-pion beam. They found a particle decaying into three pions with a mean life near 5×10^{-8} sec (the current value), compared with the K_1 mean life of just under 10^{-10} sec.[81] The theoretical predictions of the long-lived K^0 and of the interference properties of the particle mixture, which were subsequently verified in detail by experiment, were remarkably uncharacteristic of particle physics during the 1950s. In most other cases, experiment provided surprises, forcing theory to accommodate to it. (Another exception to the rule was Yoichiro Nambu's prediction of the ω meson to explain the nucleon's isoscalar form factor.[82])

The question of the number of different K mesons came into focus as the $\tau-\theta$ puzzle at the Sixth Annual Rochester Conference in 1956, after years of effort by cosmic-ray and accelerator physicists (see Val Fitch's experimentalist's perspective in Chapter 31) and through the masterful and painstaking analysis of Dalitz. So sharply was the question posed at Rochester that within months the puzzle was solved, with rich consequences for physics, as will be discussed in the next section. Oppenheimer opened the session on "Theoretical Interpretations of New Particles" at Rochester, commenting:

> There are the five objects $K_{\pi3}$, $K_{\pi2}$, $K_{\mu2}$, $K_{\mu3}$, K_{e3}. They have equal, or nearly equal, masses, and identical, or apparently identical lifetimes. One tries to discover whether in fact one is dealing with five, four, three, two, or one particle. Difficult problems arise no matter what assumption is made.

The simplest solution would have been a single K meson [meaning, of course, a single isospin multiplet (K^+, K^0) and its antiparticle multiplet $(\bar{K}^0, K^-)$], so that Oppenheimer's "five objects" would have been merely five different decay modes of one meson. The problem with this solution was that the spin and parity assignments deduced from different decay modes did not agree with each other. The difficulty with making them different particles was that no known principle required them to have equal mean lives, as well as equal masses. Assuming that the usual space–time symmetries were obeyed in the decay, including the reflection symmetries C, P, and T, the decay mode $\theta \rightarrow 2\pi$ would allow only the spin-parity assignments 0^+, 1^-, 2^+, and so forth. However, the observation of $\theta \rightarrow 2\pi^0$ excluded the odd spin values, whereas any spin greater than zero would allow the unobserved decay $\theta^+ \rightarrow \pi^+ + \gamma$. This narrowed down the θ-mode spin parity to 0^+.

The τ mode, the three-pion decay mode, was a special study of Dalitz, who analyzed the data concerning it with a method that he invented and began to apply in 1953.[83] By the beginning of 1954, he had found that either 0^- or 1^+ was compatible with the data (and neither of these was compatible with the spin parity of θ). His report at the Fifth Rochester Conference in 1955 concluded:[84]

1. If the spin of the meson is less than 5, it cannot decay into two π mesons.
2. If the spin is small, the parity is certainly odd, and the spin value could be 0, 2, 4, or 6.

At the Sixth Annual Rochester Conference, discussing the phase space distribution of the τ decay (Dalitz plot), he stated:

> The simple-minded interpretation is that the distribution is uniform. This would point to a τ meson of spin-parity 0^-, though other possibilities, such as 2^-, are not excluded. The establishment or exclusion of a 2^- distribution requires much more information than is presently available.[85]

To this, Oppenheimer made the cryptic response, "The τ meson will have either domestic or foreign complications. It will not be simple on both fronts."[88] It would appear to the unbiased observer (if such a near-mythical creature exists) that Dalitz had been telling the world, or whoever would listen, for two years either that θ and τ are not the same particle or that some symmetry property (parity?) is being violated – telling them, however, without actually *saying* it in so many words.

Toward a universal weak interaction: parity violation, V–A, and two neutrinos

By April 1956, at the time of the Sixth Annual Rochester Conference, the experimental data on K-meson decays, and especially Dalitz's masterly analysis of the three-pion (τ-meson) decay mode, confronted physicists with a puzzling picture: What appeared to be a single particle nonetheless decayed into states of opposite parity. During the previous year, it had been suggested that there might be two or more K mesons of closely similar, though not identical, properties. However, as discussed by Marshak in Chapter 45, the experimental results reported at Rochester left little room for such possibilities.[87] In the discussions, Richard P. Feynman asked, as the question was posed to him by Martin Block, if it was possible that "parity is not conserved." The conference summary continued: "Yang stated that he and Lee looked into the matter without arriving at any definite conclusions."[88]

Soon afterward, Tsung Dao Lee and Yang examined the experimental basis for believing in parity conservation. They concluded that there was substantial experimental support for parity conservation in strong and electromagnetic interactions, but little in the case of weak interactions. In May 1956 they came to the startling recognition that all the β decay experiments carried out up to that time had never tested parity inversion invariance! Their analysis showed convincingly that there was no experimental evidence either to confirm or to refute parity invariance in the weak interactions (other than the $\tau-\theta$ puzzle), and they suggested several experiments that could settle the issue.[89]

It is worth noting that in spite of the incisive analysis of Lee and Yang, few physicists were confident that this would provide a way out of the $\tau-\theta$ paradox. In 1982, Yang recalled his own doubts, as well as those of Chien-Shiung Wu, who collaborated on one of the first parity experiments:

> At that time (September 1956) I was not betting on parity nonconservation; Lee was not betting on parity nonconservation, I don't think anybody was

> really betting on parity nonconservation...Miss Wu was thinking that even
> if the (experimental) result did not give parity nonconservation, it was a good
> experiment. It should be done because beta decay did not previously yield
> any information about left–right symmetry.[90]

That was hardly the comment of a person waiting anxiously to spread a new scientific gospel.

Valentine Telegdi reported that as early as August 1956 he had become intrigued by the ideas of Lee and Yang, and he began to study the $\pi-\mu-e$ decay chain in nuclear emulsion, where he expected to find large effects if indeed the $\tau-\theta$ puzzle were to be explained by the nonconservation of parity. However, he was aware that he was gambling, and he received little encouragement from his colleagues.[91] Pauli, in a letter to Weisskopf, called the Lee–Yang analysis *verdienstlich* (i.e., commendable)[92], but he did not think that parity-nonconserving interactions had anything to do with reality.[93] Lev Landau considered parity nonconservation to be complete nonsense as late as October 1956, but a few months later he changed his mind and submitted a paper containing the two-component neutrino theory.[94]

The proposal that parity might not be conserved evoked strong negative reactions among many physicists, who believed that invariance principles were among the most reliable ones for interpreting data and constructing theories. It was not particularly upsetting for an invented symmetry like isospin not to be exact, but it was hard to accept the violation of what was thought to be an obvious space–time symmetry. Pauli was astounded when he learned that experiments in β decay and μ decay unequivocally showed the violation of parity invariance.

What particularly bothered Pauli was that there appeared to be no physical reason why parity was conserved in strong and electromagnetic interactions, but not in the weak.[95] He considered that to be the central issue (and it has not been satisfactorily resolved to this day). Eugene Wigner recalled in 1982: "It was a great shock to me when the lack of validity of these [parity inversions and time reversal] was proved."[96] To the members of the generation that created quantum mechanics, parity nonconservation was a rude shock, and they never fully came to terms with it.

One of the barriers to identifying parity nonconservation as a general feature of the weak interactions was that their "universal" character was in question. The β decay theory of Fermi had introduced the first new fundamental interaction after gravitation and electromagnetism. It employed a Hamiltonian density that was proportional to the product of the field operators of the four participating particles: neutron, proton, electron, and neutrino.[97] Fermi's β decay interaction was constructed in analogy to electromagnetism, and it is technically called a "vector" interaction (it is the scalar product of two relativistic four-vectors). However, Fermi was aware that relativity also allowed four other forms of interaction (scalar, pseudoscalar, axial vector, and tensor), or any linear combination of the five interaction forms. Experiment was to show what combination nature preferred, and that might even vary from nucleus to nucleus.

Experiments that were done by 1956 did not point to a single, simple, universal choice for the β decay Hamiltonian, nor was it clear to what extent other processes (e.g., muon decay) belonged to the general category of weak interactions describable by Fermi's theory. In 1947, Bruno Pontecorvo had suggested that weak interactions might always involve neutrinos; more specifically, he argued that the capture of a muon by a nucleus (which had recently been observed) should result in the release of a neutrino.[98] The observation that the capture rates of muons and electrons (appropriately scaled) were nearly the same hinted at a possible $\mu-e$ universality, as was noted by a number of authors at about that time.[99] Decays of the V particles were apparently the first weak processes observed in which no neutrino was involved.[100] Given these indications of the possible existence of a larger class of weak interactions, it was surprising that the key to the $\tau-\theta$ puzzle came out of an analysis of the already well-explored field of β decay.

The classical β decay experiments analyzed by Yang and Lee measured energy and angular distributions and electron–neutrino correlations, obtaining information about rates, selection rules, and type of interaction, but not parity. In 1956, a difficult electron–neutrino correlation experiment on ^{6}He led to the belief (later disproved) that the β decay interaction was a combination of scalar and tensor forms.[101] Lee and Yang constructed a class of parity-non-conserving theories and investigated their consequences.[89] In such theories, the predicted angular distributions contain pseudoscalar quantities that change sign upon the substitution of a left-handed for a right-handed co-ordinate system. The absence of such pseudoscalar terms, in fact, defines a parity-conserving theory.

The early experiments to test the ideas of Yang and Lee are discussed by Telegdi in Chapter 32. The first of these, by a Columbia University–U.S. Bureau of Standards collaboration, measured the angular distribution of the electrons emitted in the decay of polarized ^{60}Co nuclei and showed that the electrons are preferentially emitted in a direction opposite to the nuclear spin.[102] The projection of the electron velocity on the nuclear spin *is* a pseudoscalar quantity; hence, the experiment dramatically demonstrated the nonconservation of parity, and furthermore the observed effect was maximal (i.e., as large as possible). Further theoretical analysis, prompted by a letter from Reinhard Oehme to Lee and Yang, showed that not only parity invariance P but also charge conjugation invariance C had to be violated in the decay.[103]

Another consequence of parity-nonconserving interaction is that the electrons emitted in β decay are longitudinally polarized, to a degree measured by the ratio of their velocities to the velocity of light. Thus, relativistic β particles are almost completely polarized.[104] This helps to explain why some of the early electron scattering experiments, which had assumed the β rays to be un-polarized, failed to agree with the theory, as discussed by Allan Franklin in

Chapter 29. The resolution of the puzzle of the θ and τ decays was found to be merely the recognition that K meson decay was a weak interaction, where obviously (by now!) parity was not expected to be conserved.

Lee and Yang considered other processes that exhibited the violation of P and C invariance, such as the successive weak interactions in the $\pi-\mu-e$ decay chain. They showed that the muon produced in the decay of a pion would be longitudinally polarized and that the muon's subsequent decay would give an asymmetric angular distribution to its decay electron. Richard L. Garwin, Lederman, and Marcel Weinrich set out to test this idea, using a stopping beam of positive muons, obtained from the decay of pions produced by the Columbia University cyclotron at Nevis, New York. (Positive pions were used because slow negative mesons usually are captured by nuclei, rather than decaying.) They found a large electron angular asymmetry, showing that the muons were strongly polarized, and establishing the nonconservation of parity beyond a doubt.[105]

As early as 1956, Telegdi, at Chicago, stimulated by the Lee–Yang work in preprint form, decided to study the same $\pi-\mu-e$ decay chain in nuclear emulsion. As in the experiment of Garwin and associates, the main objective was to detect the polarization of the muon. However, in an emulsion, a slow positive muon can easily pick up an electron and form muonium, an exotic atom analogous to hydrogen. Because of the large magnetic moment of the electron, this atom can precess in a magnetic field (e.g., the fringing field of the cyclotron in which the exposure was made); thus, magnetic shielding was done with care. The time that was required and the relatively slow pace of data analysis using nuclear emulsions caused the publication of the Chicago results to be delayed until the time of the Columbia cyclotron experiment.[106] The results confirmed those of Wu and associates and Garwin and associates, showing P and C to be violated, while consistent with CP conservation.

This last property, CP invariance or "combined inversion" invariance, holds to good accuracy, but James Cronin and Fitch in 1964 found that it was violated in certain rare (perhaps superweak) processes.[107] Before that, however, Lee and Yang, Salam, and Landau (independently) observed that a two-component version of neutrino theory would allow a natural formulation of a CP-conserving, but P- and C-violating, weak interaction.[108] Because of its masslessness, such a neutrino has inevitably a "handedness" to its spin. It was found that in nature all neutrinos spin in a left-handed sense relative to their direction of flight, whereas antineutrinos are right-handed. To restore the symmetry in passing from particle to antiparticle (C), one must also change the handedness (also called chirality) by a parity transformation (P). An ingenious experiment, performed by Maurice Goldhaber, demonstrated the left-handedness of the β decay neutrino.[109] An early experiment by Steinberger hinted at an asymmetry in the decay of the Λ hyperon (later predicted by Lee and Yang) and was later conclusively confirmed by him, once again giving evidence of parity nonconservation.[110]

The correct form of the four-fermion weak interaction was eventually established by experiment to be a mixture of the "vector" (V) and "axial vector" (A) forms of interaction.[111] Before that, however, several theoretical groups had speculated, on grounds of symmetry, that the interaction should be $V–A$.[112] The $V–A$ theory yields a two-component theory of left-handed neutrinos, and it is CP-invariant. The version of Feynman and Gell-Mann went considerably further. It led to new interesting physical insights by postulating that the weak interaction has a current–current form (analogous to the electromagnetic interaction of two fundamental charges), the V current being conserved (as is the electromagnetic current vector) and the A current being nearly conserved.

Each of the postulated weak currents has contributions from the hadrons (both the strange and the nonstrange hadrons) and from the leptons. This, too, is analogous to QED; the main difference is the lack of a field to carry the interaction, analogous to the photon. (The currently "standard" electroweak theory restores this analogy as well.) It is easy to show that the A-current cannot be conserved, as that would have unacceptable physical consequences (e.g., the pion decay would be forbidden). The partial conservation of the axial vector current is related to the relatively small value of the pion mass, as discussed by Sam Treiman in Chapter 27. (More precisely, the four-divergence of the current is proportional to the mass.)

Important exact results follow from the exact conservation of the vector current (CVC), and important approximate results follow from the approximate conservation of the axial vector current (PCAC). Applied to hadrons, one sees that the vector coupling constant (analogous to the electric charge in QED) is renormalized exactly as it is for leptons (again analogous to QED). The axial vector coupling is only "slightly" renormalized. These conservation principles, used in conjunction with the symmetry properties of the hadrons [isospin, SU(3)], permit the calculation of many weak processes in terms of a few "matrix elements," which also determine strong and electromagnetic processes. (This powerful technique, based on the Wigner–Eckart theorem, is well known in atomic and nuclear physics.)

Applied to the decay of strange particles in unmodified form, the "universal" weak interaction turns out to give rate discrepancies as large as a factor of twenty or forty. There is also a small (2 percent) discrepancy in the rate of muon decay, if the β decay coupling is used. Both of these illnesses are cured, however, if one adopts the modification first proposed by Gell-Mann and M. Levy and later applied by N. Cabibbo.[113] The SU(3) currents of nonstrange and strange particles are added, but unlike the case of the electromagnetic currents, the nonstrange current is given a coefficient a, the strange current b, where $a^2 + b^2 = 1$. (Cabibbo puts $a = \cos \theta$, and θ is called the "Cabibbo angle.")

The fifties saw the realization of the direct detection of the neutrino as a particle. After an experimental tour de force taking years of effort, Frederick Reines and C. L. Cowan, Jr. cabled Pauli in 1956 that the particle he had suggested in 1930 had at last been detected.[114] Reines tells this story in Chapter 24. An equally trailblazing experiment at Brookhaven by a Columbia

University group found that the neutrino associated with the muon (as in $\pi-\mu$ decay) is different from that of β decay, thus establishing the validity of "lepton number" and its conservation, needed to explain the forbiddenness of reactions like the decay of a muon into three electrons.[115] The two-neutrino experiment involved the first large-scale use of spark chambers, whose development is discussed by Fukui in Chapter 15, and it demonstrated the feasibility of experiments with neutrino beams.

The monumental discoveries in the weak interaction in the 1950s confirmed their importance and demonstrated again their surpassing ability to surprise. In themselves, they constitute a revolution in the science of elementary particles. Revolutions in science used to be paced by the centuries. Now they seem to occur every decade, but we should not take them for granted.

Summary

The period 1947–63 was well marked near both ends by pioneering experimental discoveries and theoretical constructions. The year 1947 revealed the existence of the Yukawa meson, the pion, and the first of the strange particles. It witnessed the first triumphs of renormalized QED. The muon appeared as a second-generation lepton. The beginning of the 1960s saw two neutrinos, new pseudoscalar and vector mesons, and *CP* violation. On the theoretical side appeared spontaneous symmetry breakdown, Regge poles, current algebra, and fractionally charged quarks. In February 1964, the experimental observation of the Ω^- hyperon established "flavor" SU(3) as the internal group of the hadrons. Flavor is now regarded as a mere mnemonic for quark structure, but the success of flavor SU(3) in the sixties strongly emphasized the value of internal symmetry considerations. The theory of symmetries, group theory, became a dominant theoretical theme of the sixties, together with the generalized *S* matrix. In the form of renormalized gauge theory, either exact or spontaneously broken, symmetry is currently even more dominant. (The first renormalized gauge theory, QED, was established in 1947.)

In the midfifties came the temporarily unsettling revelation that parity invariance (P) and charge conjugation invariance (C) were violated in weak interactions. It was pointed out that the principle of *CPT* invariance, involving also time-reversal invariance (T), is theoretically valid under the minimal assumption of microscopic causality. For a time it was thought that *CP* could replace *C* in defining the particle–antiparticle transformation, since invariance under *T* seemed to be secure. In 1964, even that bastion fell, and a new type of weak force was revealed that still lies outside of today's Standard Model. Not least important, the fifties was the decade that established the neutrino's properties. In spite of its great elusiveness, the neutrino was finally detected directly. It was found to have only two components (not four, as do other fermions), and two kinds of neutrinos were identified, corresponding to the electron and muon generations.

In terms of theoretical particle dynamics, the period began promisingly with the triumph of renormalized QED, which still affords the most precise agreement of theory and experiment of any scientific field. But quantum field theory applied to the weak and the strong nuclear interactions led only to failure. The weak interaction was unrenormalizable; although the first approximation in perturbation theory gave generally good results, higher approximations inevitably diverged. In strong interactions, the results were far worse: Even in renormalizable versions (which did exist), the perturbation series failed *ab initio* because its expansion parameter, the strong-coupling constant, is too large.

Thus, the only known candidates for fundamental theories of the strong and weak interactions were largely abandoned. Only zealots worked on quantum field theory, while other theorists tried to make end runs around it or were content to do pure phenomenology. For the strong interactions, the *S*-matrix program of Heisenberg, developed in wartime and in the immediate postwar period, was revived in the midfifties. That approach tried to retain general features of field theory (such as relativity, unitarity, crossing symmetry, causality), but eschewed a detailed dynamic description of the time development of the system. *S*-matrix theorists generalized Heisenberg's approach and incorporated analyticity ideas that were borrowed from the 1926 optical researches of Hendrik Kramers and Ralph Kronig, further extended in 1959 by Tullio Regge. By 1961, the *S*-matrix was a major industry in theoretical physics. Terms such as "bootstrap" (Chew), "double-dispersion relations" (Mandelstam), and "Regge poles and cuts" were added to the physics vocabulary.

Theories of the weak interactions followed a different path to avoid the pitfalls of field theory. The discovery of *P* and *C* violation made possible and stimulated beautiful new experiments that established the properties of the weak interaction and of the neutrino. To analyze the accumulating experimental data, theorists provisionally accepted the use of first-order perturbation results (tree diagrams) and ignored the infinite higher-order corrections (loop diagrams). They concentrated instead on the form of the interaction term that would appear in the future mathematically consistent theory. Interaction currents were found to have mixed vector and axial vector character, conventionally denoted as V–A; the mixing of those two currents of opposite parity is the origin of parity nonconservation. Furthermore, the V current is conserved, and the A current partially conserved. Thus, it was possible to relate probability amplitudes in the weak interactions of hadrons to corresponding amplitudes of conserved currents that entered in the strong and electromagnetic interactions of hadrons, including the nuclei. After the introduction of SU(3) symmetry in 1961, current algebra in 1962, the Cabibbo angle in 1963, and quarks in 1964, the idea of a *universal Fermi interaction*, a mere speculation in the 1940s, became a meaningful reality.

Quantum field theory, in the fundamental sense, was not completely abandoned. In addition to its continued application in the course of normal science (e.g., to positron interactions and other electromagnetic phenomena), its deeper mathematical properties were investigated by the "axiomatists," who were trying to place field theory on consistent mathematical foundations. As part of this program, new forms of field theory were developed, especially by German theorists, that were closer to the *S*-matrix theory. Finally, the virtues of the Yang–Mills massive-vector non-Abelian gauge theory of 1954, supplemented by the Nambu–Jona–Lasinio spontaneous symmetry-breaking method of 1961, began to be appreciated. It was extended in the 1970s to become the electroweak theory and the quantum chromodynamics of today's Standard Model.

The fifties was a well-marked historical period in particle physics, not only in the intellectual sphere but also in terms of technical developments, social organization, and funding. Technically, we note the change from cosmic rays to accelerators as the major source of high-energy particles. Accelerator energies went from several hundred million electron volts to several billion electron volts, with experiments increasingly using selected secondary beams (typically mesons). Detectors were revolutionized, moving from relatively small-scale counter arrays, cloud chambers, and nuclear emulsion to large bubble chambers, scintillation counters, and spark chambers. Electronic analysis of data increasingly displaced pencil, paper, slide rule, and nomograms.

Small experimental groups, based almost entirely in university physics departments, sometimes using near-amateurs as assistants, were replaced to a large extent by larger groups operating at national laboratories and drawing upon engineers and other professional technical assistants. Accompanying the shift to large national facilities, new user organizations represented researchers working on experiments away from their home institutions. The high-energy nuclear physicists of the immediate postwar period became elementary particle physicists, and new specialties formed within this new discipline: accelerator physics, bubble-chamber physics, and computerized data analysis.

There was a remarkable increase in the scale of funding for these scientific enterprises, and with the entry of the acronymic funding agencies, ONR, AEC, and NSF, came science advisory committees, bureaucrats to man the new funding agencies, and, of course, politics. By the end of the period, the bargain of accepting research funds offered out of motives related to military posture and national pride was taken for granted.

There is a feeling now, after the discoveries of *W* and *Z* particles in the eighties, that there is an experimental lull. Theory, which is unable at present to explain in detail a vast accumulation of "low-energy" data (such as the classical pion experiments), is becoming more speculative and mathematical. In the fifties, on the other hand, experiment led the theory, producing major surprises to which theory could only react. This raises the historical question: When experiment dominates, does the available technique, especially equipment, hence funding, play a larger role than usual in the scientific process?

Does this influence extend so far that it affects the content of scientific theories?

Many physicists would forcefully reject the notion that scientific objectivity could be threatened by that kind of external influence. But the charge has been made, at the Fermilab symposium, and also elsewhere, that such influence was effective in the fifties. As an example, it was asked whether or not theorists had turned to phase-shift analysis, to *S*-matrix methods, and then in the sixties to Regge pole phenomenology, neglecting fundamental field theory, because the profession had accepted large sums to make experiments that were turning out huge quantities of data, calling for immediate analysis and scientific justification. If this phenomenon existed, was it correlated with national situations, so that in countries with direct involvement in accelerator physics there was one kind of theory, while in countries without accelerators there was another? Or was there rather an internal dynamic which dictated that fundamental physics shift its conceptual basis? Questions such as these demand detailed examination by historians of particle physics.

A number of strong-willed scientists take the view that there is no philosophically valid way to speak of the development of scientific ideas per se, but only of the development of the ideas of scientists as individuals. We do not agree with this position, for even the most individualistic of scientists is engaged in leading the thoughts of and reacting to *other* individuals, and that is just another way of describing a *social* process.

Notes

1 Letter, 17 January 1957, Pauli Letter Collection, CERN, Geneva.
2 *The Birth of Particle Physics*, edited by Laurie M. Brown and Lillian Hoddeson (Cambridge University Press, 1983), especially Chapters 4, 11, 13 and 23.
3 Note 2 (Chapters 22 and 23). For an account of strong-coupling theory, see Wolfgang Pauli, *Meson Theory of Nuclear Forces* (New York: Interscience, 1946).
4 The University of Rome meson-capture experiments were carried out in part while bombs were falling on the central railroad station a few blocks away.
5 Note 2 (Chapter 19).
6 M. Conversi, E. Pancini, and O. Piccioni, "On the Disintegration of Negative Mesons," *Phys. Rev. 71* (1947), 209–10; E. Fermi, E. Teller, and V. Weisskopf, "The Decay of Negative Mesotrons in Matter," *Phys. Rev. 71* (1947), 314–15.
7 C. F. Powell, P. H. Fowler, and D. H. Perkins, *The Study of Elementary Particles by the Photographic Method* (New York: Pergamon Press, 1959). This is a comprehensive treatise on nuclear emulsions, including their history.
8 This spirit persisted through the 1950s. See C. O'Ceallaigh, "A Contribution to the History of C. F. Powell's Group in the University of Bristol 1949–65," in *Colloque International sur l'Histoire de la Physique des Particules, J. Phys. (Paris) (Suppl.)* 43 : 12 (1982), 185–90. There is considerable overlap between the contents of this excellent conference, held at Paris in July 1982, and the two Fermilab symposia on the history of particle physics.
9 "Fragments of an Autobiography," in *Selected Papers of Cecil F. Powell*, edited by E. H. S. Burhop, W. O. Lock, and M. G. K. Menon (Amsterdam: North Holland, 1972), pp. 7–28.
10 D. H. Perkins, "Nuclear Disintegration by Meson Capture," *Nature (London) 159* (1947), 126–7; G. P. S. Occhialini and C. F. Powell, "Nuclear Disintegrations Produced by Slow Charged Particles of Small Mass," *Nature (London) 159* (1947), 186–90.

11 C. M. G. Lattes, H. Muirhead, G. P. S. Occhialini, and C. F. Powell, "Processes Involving Charged Mesons," *Nature (London) 159* (1947), 694–7.

12 E. Gardner and C. M. G. Lattes, "Production of Mesons by the 184-Inch Berkeley Cyclotron," *Science 107* (1948), 270–1.

13 G. D. Rochester and C. C. Butler, "Evidence for the Existence of New Unstable Particles," *Nature (London) 160* (1947), 855–7. A possible forerunner was a track found earlier: Louis Leprince-Ringuet and Michael Lheritier, "Existence probable d'une particule de masse 990*m* dans le rayonnement cosmique," *C. R. Acad. Sci. 219* (1944), 618–20.

14 R. Brown, U. Camerini, P. H. Fowler, H. Muirhead, and C. F. Powell, "Observations with Electron-Sensitive Plates Exposed to the Cosmic Radiation," *Nature (London) 163* (1949), 47–51, 618–20.

15 Quoted in George D. Rochester, "The Early History of the Strange Particles," in *Early History of Cosmic Ray Studies*, edited by Yataro Sekido and Harry Elliot (Dordrecht: D. Reidel, 1985), pp. 299–322.

16 See C. C. Butler, "Early Cloud Chamber Experiments at the Pic-du-Midi," in note 8 (pp. 177–84). In the same volume, see C. Peyrou, "The Role of Cosmic Rays in the Development of Particle Physics," pp. 7–68, and G. D. Rochester, "Observations on the Discovery of the Strange Particles," pp. 169–76.

17 See note 14 and Jack Steinberger, Chapter 20 of this volume. Also see J. Steinberger, "On the Range of the Electrons in Meson Decay," *Phys. Rev. 74* (1948), 500–1, E. P. Hincks and B. Pontecorvo, "The Penetration of μ-Meson Decay Electrons and Their Bremsstrahlung Radiation," *Phys. Rev. 75* (1949), 698–9; Robert B. Leighton, Carl D. Anderson, and Aaron J. Seriff, "The Energy Spectrum of the Decay Particles and the Mass and Spin of the Mesotron," *Phys. Rev. 75* (1949), 1432–7.

18 Peyrou (note 16) said that this conference was responsible for "serious" physicists beginning to take seriously the new (strange) particles. Dalitz called it a "major event in the lives of all the physicists who took part in it." See his article, "Strange Particle Theory in the Cosmic Ray Period," in note 8 (pp. 195–206), and Chapter 30 of this volume.

19 Also presented at the Third Rochester Conference, December 1952, in *High Energy Nuclear Physics. Proceedings of the Third Annual Rochester Conference, December 18–20, 1952*, edited by H. P. Noyes, M. Camac, W. D. Walker (New York: Interscience, 1953), pp. 39–40. See also Robert W. Thompson, "On the Discovery of the Neutral Kaons," in note 2 (pp. 251–60).

20 See Peyrou, note 16.

21 See note 12 for some details of the first observations.

22 R. Bjorklund, W. E. Crandall, B. J. Moyer, and H. F. York, "High Energy Photons from Proton–Nucleon Collisions," *Phys. Rev. 77* (1950), 213–18; J. Steinberger, W. K. H. Panofsky, and J. Steller, "Evidence for the Production of Neutral Mesons by Photons," *Phys. Rev. 78* (1950), 802–5; A. G. Carlson, J. E. Hooper, and D. T. King, "Nuclear Transmutations Produced by Cosmic-Ray Particles of Great Energy. V. The Neutral Mesons," *Philos. Mag. 41* (1950), 701–24.

23 Y. Fujimoto and H. Miyazawa, "Photo-Meson Production and Nucleon Isobars," *Prog. Theor. Phys. 5* (1950), 1052–4; Keith A. Bruckner and K. M. Case, "Neutral Photomeson Production and Nucleon Isobars," *Phys. Rev. 83* (1951), 1141–7. For strong-coupling theory, see note 3.

24 H. L. Anderson, "Early History of Physics with Accelerators," in note 8 (pp. 101–62); see also Herbert L. Anderson, "Meson Experiments with Enrico Fermi," *Rev. Mod. Phys. 27* (1955), 269–72.

25 The value zero for the spin of the charged pion was obtained by arguments based on detailed balance, relating the production of π^+ mesons in proton–proton collisions to the absorption of π^+ mesons in deuterium. Its odd parity is obtained by comparing the captures of π^- in hydrogen and deuterium: W. K. H. Panofsky, R. L. Aamodt, and J. Hadley, "The Gamma-Ray Spectrum Resulting from the Capture of Negative μ-Mesons in Hydrogen and Deuterium," *Phys. Rev. 81* (1951), 565–74; W. Chinowsky and J. Steinberger, "Absorption of Negative Pions in Deuterium: Parity of the Pion," *Phys. Rev. 95* (1954), 1561–4. The

charge-independence hypothesis implied that all three pions have the same spin and parity.

26 Keith A. Bruckner, "Meson–Nucleon Scattering and Nucleon Isobars," *Phys. Rev. 86* (1952), 106–9. An energy-dependent cross section showing a resonance peak was first obtained by J. Ashkin, J. P. Blaser, F. Feiner, J. Gorman, and M. D. Stern, "Total Cross Sections of 135 MeV to 250 MeV Negative Pions in Hydrogen," *Phys. Rev. 93* (1954), 1129–30.

27 Robert Hofstadter, "The Electron-Scattering Method and Its Application to the Structure of Nuclei and Nucleons," in *Nobel Lectures, Physics, 1942–62* (Amsterdam: Elsevier, 1964), pp. 560–81. As the title of the lecture indicates, the award was made for Hofstadter's study of nuclei as well as nucleons. The earliest of such experiments was E. M. Lyman, A. O. Hanson, and M. B. Scott, "Scattering of 15.7 MeV Electrons by Nuclei," *Phys. Rev. 84* (1951), 626–37.

28 These advances are presently being explored by L. Hoddeson, C. Westfall, Roger Meade, and Paul Henriksen in the Los Alamos History Project, to be published in 1990. For advances made at the MIT Radiation Laboratory, see Henry E. Guerlac, *Radar in World War II* (New York: American Institute of Physics, 1987), and Paul Henriksen, "The MIT Radiation Laboratory and Solid-State Physics" (unpublished). Further references on the effects of World War II on the physics community are included in note 10 of Chapter 46.

29 Alvarez discusses the advantages of using obsolete radar equipment in a 1945 proposal for an electron linear accelerator: L. Alvarez, "A Proposal for Accelerating Electrons to Very High Energies," 14 April 1945, AM-2290, Los Alamos Report Library. However, little of this surplus equipment was actually used. See Luis W. Alvarez, *Adventures of a Physicist* (New York: Basic Books, 1987), p. 156.

30 For a good discussion, see Robert Seidel, "A Home for Big Science: The Atomic Energy Commission's Laboratory System," *Historical Studies in the Physical Sciences 16 : 1* (1986), 135–75. For both financial and military reasons, other countries were not able to support accelerator and detector development on a comparable level in the postwar period. Japanese nuclear physicists were for some time forbidden to engage in nuclear research, and many turned to cosmic-ray studies. The fact that numerous results of nuclear research remained classified turned Italian physicists away from the area, and they, too, turned typically to cosmic rays. These countries also did not have available the large store of surplus radar parts, as Fukui reminds us in Chapter 15.

31 See also Seidel, note 30, and R. Seidel, "Accelerating Science: The Postwar Transformation of the Lawrence Radiation Laboratory," *Historical Studies in the Physical Sciences 13 : 2* (1983), 375–400.

32 See also Daniel J. Kevles, *The Physicists* (New York: Knopf, 1978), Chapters 21–23.

33 M. Stanley Livingston and John P. Blewett, *Particle Accelerators* (New York: McGraw-Hill, 1962), p. 3.

34 Note 2 (Introduction, pp. 3–36). See also Gerald Holton, "The Formation of the American Physics Community in the 1920s and the Coming of Albert Einstein," *Minerva 19* (1981), 569–81; "Success Sanctifies the Means," in *Transformation and Traditions in the Sciences*, edited by E. Mendelsohn (Cambridge University Press, 1984), pp. 155–73; "The Migration of Physicists to the United States," in *The Muses Flee Hitler*, edited by J. C. Jackman and C. M. Borden (Washington: Smithsonian Institution, 1983), pp. 169–88.

35 See, e.g., Kevles, note 32 and Holton, note 34.

36 P. Galison, "Bubble Chambers and the Experimental Workplace," in *Observation, Experiment, and Hypothesis in Modern Physical Science*, edited by P. Achinstein and O. Hannaway (Cambridge; Mass.: MIT-Bradford, 1985), pp. 309–73.

37 E. M. McMillan, "The Synchrotron – A Proposed High Energy Particle Accelerator," *Phys. Rev. 68* (1945), 143–4; V. Veksler, "A New Method of Acceleration of Relativistic Particles," *Dok. Akad. Nauk SSSR 43* (1944), 329–31; English translation in *C. R. Acad. Sci. USSR 44* (1944), 393–6.

38 A bibliography of historical attempts to discuss accelerator history in this period is included in Chapter 46 of this volume. Further references can be found in note 4 of L. Hoddeson, "Establishing KEK in Japan and Fermilab in the U.S.: Internationalism, Nationalism and

High Energy Accelerators," *Social Studies of Science 13* (1983), 1–48; and C. Westfall, "The First Truly National Laboratory: The Birth of Fermilab" (dissertation, Michigan State University, 1988).

39 See Seidel, "Accelerating Science," note 31.

40 Ernest D. Courant, M. Stanley Livingston, and Hartland Snyder, "The Strong-Focusing Synchrotron – A New High Energy Accelerator," *Phys. Rev. 88* (1952), 1190–6.

41 T. Kitagaki, "A Focusing Method for Large Accelerators," *Phys. Rev. 89* (1953), 1161–2; M. G. White, "Preliminary Design Parameters for a Separated-Function Machine," Princeton, N. J., 3 March 1953, unpublished internal report.

42 See note 40. Also, Owen Chamberlain, "Optics of High-Energy Beams," in E. Segrè, G. Friedlander, and W. Meyerhof (eds.), *Annu. Rev. Nucl. Sci. 10* (1960), 161–92.

43 J. P. Blewett, "Radial Focusing in the Linear Accelerator," *Phys. Rev. 88* (1952), 1197–9.

44 For an account of the major achievements of the MURA group, see the recorded group interview with Kerst, Kent Terwilliger, Frederick Mills, Francis Cole, Keith Symon, Donald Young, Stanley Snowden, Cyril Curtis, and Curtis Owen, conducted by Hoddeson with Donald Moyer at Fermilab, 29 May 1979; also, interviews with Lawrence Jones and Frederick Mills at Fermilab, by Hoddeson, in Fermilab History Collection (FHC).

45 See F. T. Cole, "Historical Overview of High-Brightness Accelerators."

46 *Program and Proposal for Midwestern Universities Research Association High-Energy Research Facility*, presented to the Atomic Energy Commission, 5 April 1956, FHC.

47 D. Kerst, "Properties of an Intersecting-Beam Accelerating System," in *CERN Symposium on High Energy Accelerators and Pion Physics, Geneva, June 11–23, 1956*, edited by E. Regenstreif (Geneva: CERN, 1956), pp. 36–9. In beam stacking, a series of accelerations are made, and the particles are brought up to some given energy. By putting the different accelerated beams next to one another in energy, they overlap somewhat in configuration space, and one can build up a higher density. These beams of slightly different energy can be stored in the same physical region, in which one thus has a very large circulating beam, which can be stored in a ring and later collided.

48 Gerard K. O'Neill, "Storage Ring Synchrotron: Device for High-Energy Physics Research," *Phys. Rev. 102* (1956), 1418–19.

49 The bitter political controversy between Argonne and the MURA group has been amply discussed in the literature, e.g., by Daniel Greenberg, *The Politics of Pure Science* (New York: New American Library, 1967), and by Leonard Greenbaum, *A Special Interest* (Ann Arbor: University of Michigan Press, 1971).

50 An excellent text surveying all the detectors in use during the 1950s, giving some of the history and most of the referencing to original papers, is David M. Ritson (ed.), *Techniques of High Energy Physics* (New York: Interscience, 1961).

51 Alexander Langsdorf, Jr., "A Continuously Sensitive Diffusion Cloud Chamber," *Rev. Sci. Instrum. 10* (1939), 91–103; D. H. Miller, E. C. Fowler, and R. P. Shutt, "Operation of a Diffusion Cloud Chamber with Hydrogen at Pressures up to 15 Atmospheres," *Rev. Sci. Instrum. 22* (1951), 280.

52 In that development, Fowler was a principal participant. See L. Hoddeson, "The First Large-Scale Application of Superconductivity: Building the Fermilab Energy Doubler, 1972–83," *Historical Studies in the Physical and Biological Sciences 18 : 1* (1987), 25–54.

53 P. Galison, "Bubble Chambers and the Experimental Workplace," in *Observation, Experiment and Hypothesis in Modern Physical Science*, edited by P. Achinstein and O. Hannaway (Cambridge, Mass.: MIT-Bradford, 1985), pp. 309–73; D. A. Glaser, in *Nobel Lectures, Physics, 1942–62* (Amsterdam: Elsevier, 1964), pp. 529–51. See also L. Alvarez, in *Nobel Lectures, Physics, 1963–70* (Amsterdam: Elsevier, 1972), pp. 241–90.

54 Luis W. Alvarez, "Recent Developments in Particle Physics," *Science 165* (1969), 1071–91; Galison, note 36; also J. Heilbron, R. W. Seidel, and B. W. Wheaton, *Lawrence and His Laboratory: Nuclear Science at Berkeley, 1931–1961* (Berkeley: Office for History of Science and Technology, 1982). For another perspective, see J. Krige and D. Pestre, "The Choice of CERN's First Large Bubble Chamber for the Proton Synchrotron (1957–1958)," *Historical*

Studies in the Physical Sciences 16 : 2 (1986), 255–79; also see A. Hermann, J. Krige, U. Mersits, and D. Pestre, *History of CERN, Vol. I* (Amsterdam: North Holland, 1987); and the series of working papers of the CERN History Project.

55 A well-documented history of nuclear emulsions is that by Mildred Widgoff, "Nuclear Emulsions," in Ritson, note 50 (pp. 115–205).

56 For an excellent discussion of the reactor roots of Brookhaven, see A. Needell, "Nuclear Reactors and the Founding of Brookhaven National Laboratory," *Historical Studies in the Physical Sciences 14 : 1* (1983), 93–122. See also Needell, "Sources for the History of Physics at Brookhaven National Laboratory" (New York: American Institute of Physics, 1977), and N. F. Ramsey, "Early History of Associated Universities and Brookhaven National Laboratory," *Brookhaven Lecture Series 55* (30 March 1966), FHC. Ramsey says that the idea for AUI came in part out of a "mood of discouragement, jealousy and frustration." He and Rabi felt that "Columbia had made great contributions to the war effort but with little scientific benefit in return."

57 The synchrotron project was eventually dropped.

58 See W. Panofsky, "The Evolution of SLAC and Its Program," *Phys. Today 36 : 10* (1983), 34–41.

59 For more detailed information, see especially *Studies in CERN History* which includes approximately 18 working papers by the research team of the CERN History Project, Armin Hermann, John Krige, Dominique Pestre, Ulrike Mersits, and Lanfranco Belloni.

60 Interesting beginnings have been made by S. Traweek, "Downtime, Uptime, Spacetime and Power: An Ethnography of the High-Energy Physics Community in Japan and the United States," dissertation, University of California at Santa Cruz, 1982.

61 Leon Lederman, "The Truly National Laboratory (TNL)," Part II of an internal report of the Accelerator Department (AGS) Study, no. 6, Brookhaven, 25 June 1963, FHC.

62 Hoddeson, note 38.

63 E. Fermi and C. N. Yang, "Are Mesons Elementary Particles?" *Phys. Rev. 76* (1949), 1739–43.

64 Shoichi Sakata, "On a Composite Model for the New Particles," *Prog. Theor. Phys. 16* (1956), 686–8. Composite models for the V particles were proposed even earlier by M. A. Markov in *Rep. Acad. Sci. USSR* (1955) and by M. Goldhaber in *Phys. Rev. 101* (1956). The U(3) symmetry of the Sakata model is explicitly discussed in Mineo Ikeda, Shuzo Ogawa, and Yoshio Ohnuki, "A Possible Symmetry in Sakata's Model for Boson-Baryons System," *Prog. Theor. Phys. 22* (1959), 715–24 (Part I), and ibid. *23* (1960), 1073–99 (Part II).

65 Murray Gell-Mann, "The Eightfold Way: A Theory of Strong Interaction Symmetry," unpublished California Institute of Technology Laboratory Report CTSL-20, 1961; reprinted in Murray Gell-Mann and Yuval Ne'eman, *The Eightfold Way* (New York: W. A. Benjamin, 1964), pp. 11–57. Leptons are fermions that do not interact strongly.

66 M. Gell-Mann, "Models of the Strong Couplings," *Phys. Rev. 106* (1957), 1296–300; Julian Schwinger, "A Theory of Strong Interactions," *Ann. Phys. (N.Y.) 11* (1957), 407–34.

67 C. N. Yang and R. L. Mills, "Conservation of Isotopic Spin and Isotopic Gauge Invariance," *Phys. Rev. 96* (1954), 191–5; J. J. Sakurai, "Theory of Strong Interactions," *Nuovo Cimento 19* (1961), 165–71.

68 A. Pais, "Some Remarks on the V Particles," *Phys. Rev. 86* (1952), 663–72.

69 Kazuhiko Nishijima, "From Isospin to Strangeness," talk presented at Wingspread International Conference, 29 May–1 June 1984, in *Fifty Years of Weak Interactions*, edited by David Cline and Gail Riedasch (Madison: University of Wisconsin, 1986), pp. 325–31. The Japanese papers were published in *Prog. Theor. Phys. 6* (1951), as follows: Y. Nambu, K. Nishijima, and Y. Yamaguchi, "On the Nature of V particles," 615–19, 619–22; K. Aizu and T. Kinoshita, "On a Possible Model of the V Particles," 630–1; H. Miyazawa, "A Model for V Particles," 631–3; S. Oneda, "Note on the Theory of V Particles and τ Mesons," 633–5.

70 Pais, note 68. See also his Chapter 23 in this volume and his book *Inward Bound* (Oxford: Clarendon Press, 1986), pp. 517–23. A more general formulation, applicable to production, scattering, decay, etc., was to give the V particles a new quantum number (called V parity by

Nishijima), a multiplicative quantum number -1, and to give the pions and the nucleons a V parity of $+1$. Then the V parity of the interaction term in the Hamiltonian, obtained by multiplying these numbers, would be $+1$ for strong and electromagnetic interactions, but -1 for V-parity-violating weak interactions.

71 Bruno Rossi, in *High Energy Nuclear Physics, Proceedings of the Fifth Annual Rochester Conference*, edited by H. P. Noyes, E. M. Hafner, G. Yekutieli, and B. J. Raz (New York: Interscience, 1955), p. 130.

72 Murray Gell-Mann has emphasized this point in his contribution "Strangeness," note 8 (pp. 395–402), a lively account of this youthful achievement.

73 R. Armenteros, K. H. Barker, C. C. Butler, A. Cachon, and C. M. York, "The Properties of Charged V-Particles," *Philos. Mag. 43* (1952), 597–612.

74 Nishijima, note 69, has emphasized this point, and also the explanation by strangeness for the predominance of K^+ mesons over K^- mesons at relatively low energies of production. (Namely, $K^+ + K^-$ has a higher threshold energy than K^+ plus a negative-strangeness hyperon.)

75 Tadao Nakano and Kazuhiko Nishijima, "Charge Independence for V Particles," *Prog. Theor. Phys. 10* (1953), 581–2; K. Nishijima, "Some Remarks on the Even–Odd Rule," *Prog. Theor. Phys. 12* (1954), 107–8; K. Nishijima, "Charge Independence Theory of V Particles," *Prog. Theor. Phys. 13* (1955), 285–304.

76 M. Gell-Mann, "Isotopic Spin and New Unstable Particles," *Phys. Rev. 92* (1953), 833–4.

77 Luis Alvarez's work with the hydrogen bubble chamber was very important in discovering the "strange resonances," i.e., excited states of hyperons and heavy mesons that decay strongly. This extended the notion of pion–nucleon resonances into the strange-particle field. Because Alvarez does not provide references in his Chapter 19, we give some here: M. Alston, L. W. Alvarez, P. Eberhard, M. L. Good, W. Graziano, H. K. Ticho, and S. G. Wojcicki, "Some Considerations on the Recently Found Evidence for a $\pi \varLambda$ Resonance," *Phys. Rev. Lett. 5* (1960), 520–4; Alston et al., "Resonance in the $K-\pi$ System," *Phys. Rev. Lett. 6* (1961), 300–2; Alston et al., "Study of Resonances of the $\Sigma-\pi$ System," *Phys. Rev. Lett. 6* (1961), 698–702; G. Alexander, G. R. Kalbfleisch, D. H. Miller, and G. A. Smith, "Production of Strange-Particle Resonant States by 2.1 BeV/c π-Mesons," *Phys. Rev. Lett. 8* (1962), 447–50; P. Bastien, M. Ferro-Luzzi, and A. H. Rosenfeld, "Sigma Decay Modes of Pion–Hyperon Resonances," *Phys. Rev. Lett. 6* (1961), 702–5; G. M. Pjerrou, D. J. Prowse, P. Schlein, W. E. Slater, D. H. Stork, and H. K. Ticho, "Resonance in the $\varXi \pi$ System at 1.53 GeV," *Phys. Rev. Lett. 9* (1962), 114–17; L. Bertanza, V. Brisson, P. L. Connolly, E. L. Hart, I. S. Mittra, G. C. Moneti, R. R. Rau, N. P. Samios, I. O. Skillicorn, S. S. Yamamoto, M. Goldberg, L. Gray, J. Leitner, S. Lichtman, and J. Westgard, "Possible Resonances in the $\varXi \pi$ and $K\bar{K}$ Systems," *Phys. Rev. Lett. 9* (1962), 180–3.

78 L. Michel, "Selection Rules Imposed by Charge Conjugation," *Nuovo Cimento 10* (1953), 319–39; T. D. Lee and C. N. Yang, "Charge Conjugation, a New Quantum Number G, and Selection Rules Concerning a Nucleon–Antinucleon System," *Nuovo Cimento 3* (1956), 749–53.

79 M. Gell-Mann and A. Pais, "Behavior of Neutral Particles under Charge Conjugation," *Phys. Rev. 97* (1955), 1387–9.

80 A. Pais and O. Piccioni, "Note on the Decay and Absorption of the θ^0," *Phys. Rev. 100* (1955), 1487–9. These effects were observed at Berkeley in 1961: R. H. Good, R. P. Matsen, F. Muller, O. Piccioni, W. M. Powell, H. S. White, W. B. Fowler, and R. W. Birge, "Regeneration of Neutral K Mesons and Their Mass Difference," *Phys. Rev. 124* (1961), 1223–39.

81 K. Lande, E. T. Booth, J. Impeduglia, L. M. Lederman, and W. Chinowsky, "Observation of Long-Lived Neutral V Particles," *Phys. Rev. 103* (1956), 1901–4; K. M. Lande, L. M. Lederman, and W. Chinowsky, "Report on Long-Lived K^0 Mesons," *Phys. Rev. 105* (1957), 1925–7. Evidence for the long-lived K^0 was also obtained at the Bevatron in 1956, using nuclear emulsion: W. F. Fry, J. Schnepps, and M. S. Swami, "Evidence for a Long-Lived Neutral Unstable Particle," *Phys. Rev. 103* (1956), 1904–5.

82 Yoichiro Nambu, "Possible Existence of a Heavy Neutral Meson," *Phys. Rev. 106* (1957), 1366–7.

83 R. H. Dalitz, "On the Analysis of τ Meson Data and the Nature of the τ Meson," *Philos. Mag. 44* (1953), 1068–80; R. H. Dalitz, "Decay of τ Mesons of Known Charge," *Phys. Rev. 94* (1954), 1046–51; E. Fabri, "A Study of τ Meson Decays," *Nuovo Cimento 11* (1954), 479–91.

84 R. H. Dalitz, in Noyes et al., note 71 (p. 140).

85 R. H. Dalitz, in *High Energy Nuclear Physics, Proceedings of the Sixth Annual Rochester Conference*, edited by J. Ballam, V. L. Fitch, T. Fulton, K. Huang, R. R. Rau, and S. B. Treiman (New York: Interscience, 1956), p. VIII-20.

86 J. R. Oppenheimer, in Ballam et al., note 85 (p. VIII–22).

87 L. Alvarez, in Ballam et al., note 85 (p. VIII–8). For suggestions of two K mesons: T. D. Lee and C. N. Yang, "Mass Degeneracy of the Heavy Mesons," *Phys. Rev. 102* (1956), 290–1; T. D. Lee and J. Orear, "Speculations on Heavy Mesons," *Phys. Rev. 100* (1955), 932–3.

88 Note 85 (p. VIII–27). Comments on the weak interactions during this period, including the attitudes of various physicists, can be found in recent publications, such as the following: Abraham Pais, *Inward Bound* (Oxford University Press, 1986); Robert P. Crease and Charles C. Mann, *The Second Creation* (New York: Macmillan, 1986); R. P. Feynman, *"Surely You're Joking Mr. Feynman!"* (New York: Norton, 1985). See also the papers and discussions of the Wingspread International Conference, note 69, and see note 8 International Colloquium on the History of Particle Physics). Valuable accounts are found also in *Adventures in Experimental Physics*, Gamma Volume, edited by Bogdan Maglic (Princeton: World Science Education, 1973), and in the Nobel Prize addresses of Tsung Dao Lee and Chen Ning Yang. Recent additions to this literature are in *C. N. Yang, Selected Papers 1945–1980, With Commentary* (San Francisco: Freeman, 1983), pp. 26–37, and in T. D. Lee, *Selected Papers*, Vol. 3, edited by G. Feinberg (Boston: Birkhauser, 1986), pp. 475–509.

89 T. D. Lee and C. N. Yang, "Question of Parity Conservation in Weak Interactions," *Phys. Rev. 104* (1956), 254–8.

90 C. N. Yang, note 8 (p. 450).

91 V. L. Telegdi, in *Adventures in Experimental Physics*, note 88 (pp. 131–6), and in Chapter 32 of this volume.

92 W. Pauli to V. F. Weisskopf, 27 January 1957, in CERN Letter Collection.

93 W. Pauli to C. S. Wu, 19 January 1957, quoted in *Adventures in Experimental Physics*, note 88 (p. 122).

94 L. Landau, "On the Conservation Laws for Weak Interactions," *Nucl. Phys. 3* (1957), 127–31.

95 W. Pauli, "Zur alter und neue Geschichte des Neutrinos," in *Collected Scientific Papers*, Vol. 1, edited by R. Kronig and V. F. Weisskopf (New York: Wiley-Interscience, 1964), p. 13. In this lecture delivered in 1957, Pauli stated that neutrino couplings had been subsumed under weak interactions.

96 Eugene Wigner, note 8 (p. 448).

97 E. Fermi, "Versuch einer Theorie der β-Strahlen," *Z. Phys. 88* (1934), 161–71. Fermi's was a contact interaction of four spin-$\frac{1}{2}$ particles (fermions). The first weak-interaction theory with an "intermediate boson" (as in the current "Standard Model") was a part of Yukawa's meson theory: H. Yukawa, "On the Interaction of Elementary Particles, I," *Proc. Phys.-Math. Soc. Japan 17* (1935), 48–57.

98 B. Pontecorvo, "Nuclear Capture of Mesons and the Meson Decay," *Phys. Rev. 72* (1947), 246–7.

99 G. Puppi, "On Cosmic Ray Mesons," *Nuovo Cimento 5* (1948), 587–8; T. D. Lee, M. Rosenbluth, and C. N. Yang, "Interaction of Mesons with Nucleons and Light Particles," *Phys. Rev. 75* (1949), 905; J. Tiomno and J. A. Wheeler, "Charge-Exchange Reaction of the μ-Meson with the Nucleus," *Rev. Mod. Phys. 21* (1949), 153–65.

100 The first two *V* particles observed were probably a K^0 and a K^+, each of them decaying to two pions. The next observations included $\Lambda \rightarrow p + \pi^-$. See notes 15 and 16.

101 B. M. Rustad and S. L. Ruby, "Gamow–Teller Interaction in the Decay of He^6," *Phys. Rev.* 97 (1955), 991–1002.

102 C. S. Wu, E. Ambler, R. W. Hayward, D. D. Hoppes, and R. P. Hudson, "Experimental Test of Parity Conservation in Beta Decay," *Phys. Rev.* 105 (1957), 1413–15.

103 T. D. Lee, Reinhard Oehme, and C. N. Yang, "Remarks on Possible Noninvariance under Time Reversal and Charge Conjugation," *Phys. Rev.* 106 (1957), 340–5.

104 H. Frauenfelder, R. Bobone, E. van Goeler, N. Levine, H. R. Lewis, R. N. Peacock, A. Rossi, and G. DePasquali, "Parity and the Polarization of Electrons from Co^{60}," *Phys. Rev.* 106 (1957), 386–7.

105 Richard L. Garwin, Leon M. Lederman, and Marcel Weinrich, "Observations of the Failure of Conservation of Parity and Charge Conjugation in Meson Decays: The Magnetic Moment of the Free Muon," *Phys. Rev.* 105 (1957), 1415–17. See also R. Garwin in *Adventures in Experimental Physics*, note 88 (pp. 124–30).

106 Jerome I. Friedman and V. L. Telegdi, "Nuclear Emulsion Evidence for Parity Nonconservation in the Decay Chain $\pi^+ \rightarrow \mu^+ + e^+$," *Phys. Rev.* 105 (1957), 1681–2.

107 J. H. Christenson, J. W. Cronin, V. L. Fitch, and R. Turlay, "Evidence for the 2π Decay of the K_2^0 Meson," *Phys. Rev. Lett.* 13 (1964), 138–40.

108 T. D. Lee and C. N. Yang, "Parity Nonconservation and a Two-Component Theory of the Neutrino," *Phys. Rev.* 105 (1957), 671–5; A. Salam, "On Parity Conservation and Neutrino Mass," *Nuovo Cimento* 5 (1957), 299–301; L. D. Landau, "On the Conservation Laws for Weak Interactions," *Nucl. Phys.* 3 (1957), 27–31; R. Gatto, "K^0 Decay Modes and the Question of Time Reversal of Weak Interactions," *Phys. Rev.* 106 (1957), 168–9.

109 M. Goldhaber, L. Grodzins, and A. W. Sunyar, "Helicity of Neutrinos," *Phys. Rev.* 109 (1958), 1015–17.

110 F. Eisler, R. Plano, A. Prodell, N. Samios, M. Schwartz, J. Steinberger, P. Bassi, V. Borelli, G. Puppi, G. Tanaka, P. Woloschek, V. Zoboli, M. Conversi, P. Franzini, I. Mannelli, R. Santangelo, V. Silvestrini, D. A. Glaser, C. Graves, and M. L. Perl, "Demonstration of Parity Nonconservation in Hyperon Decay," *Phys. Rev.* 108 (1957), 1353–5.

111 Lincoln Wolfenstein, "Muon Capture in Complex Nuclei" in *Proceedings of the International Conference on Fundamental Aspects of the Weak Interactions*, edited by G. C. Wick and W. J. Willis (Upton, N, Y.: Brookhaven National Laboratory, 1964), pp. 292–5.

112 E. C. G. Sudarshan and R. E. Marshak, in *Proceedings of the Padua–Venice Conference on Mesons and Newly-Discovered Particles*, edited by N. Zanichelli (Bologna: 1958); reprinted in *The Development of Weak Interaction Theory*, edited by P. K. Kabir (New York: Gordon & Breach, 1963), pp. 118–28. See also E. C. G. Sudarshan and R. E. Marshak, "Chirality Invariance and the Universal Fermi Interaction," *Phys. Rev.* 109 (1958), 1860–2; R. P. Feynman and M. Gell-Mann, "Theory of the Fermi Interaction," *Phys. Rev.* 109 (1958), 193–8; J. J. Sakurai, "Mass Reversal and Weak Interactions," *Nuovo Cimento* 7 (1958), 649–60. See citations in note 88 for various versions of priority, etc.

113 M. Gell-Mann and M. Levy, "The Axial Vector Current in Beta Decay," *Nuovo Cimento* 16 (1960), 705–26; N. Cabibbo, "Unitary Symmetry and Leptonic Decays," *Phys. Rev. Lett.* 10 (1963), 531–3.

114 F. Reines and C. L. Cowan, Jr., "Detection of the Free Neutrino," *Phys. Rev.* 92 (1953), 830–1; C. L. Cowan, Jr., F. Reines, F. B. Harrison, H. W. Kruse, and A. D. McGuire, "Detection of the Free Neutrino: A Confirmation," *Science* 124 (1956), 103–4. See also F. Reines, note 8 (pp. 237–60).

115 G. Danby, J.-M. Gaillard, K. Goulianos, L. M. Lederman, N. Mistry, M. Schwartz, and J. Steinberger, "Observation of High-Energy Neutrino Reactions and the Existence of Two Kinds of Neutrinos," *Phys. Rev. Lett.* 9 (1962), 36–44.

2 Particle physics in the early 1950s

CHEN NING YANG

Born 1922, Hofei, Anhwei, China; Ph.D., University of Chicago, 1948; theoretical physics; Nobel Prize, 1957, for the theory of nonconservation of parity in weak interactions; State University of New York at Stony Brook

What I shall discuss in this chapter is a collection of unconnected remarks about particle physics in the early 1950s.

One of the earliest international conferences on particle physics took place 17–22 September 1951 at the University of Chicago. I believe it was organized by Samuel Allison and his colleagues to celebrate the fiftieth birthday of Enrico Fermi. But Fermi did not want such a celebration, so the conference was named the International Conference on Nuclear Physics and the Physics of Fundamental Particles. It was supported by the Office of Naval Research (ONR) and the Atomic Energy Commission (AEC), and there were approximately 200 participants. In the proceedings of that conference, edited by Jay Orear, Arthur Rosenfeld, and Robert A. Schluter, one finds that the first session was begun with a talk by Fermi. Table 2.1 shows the list of elementary particles as presented by Fermi in that talk.

It is interesting that Fermi had no doubt about the existence of the anti-proton, the antineutron, and the neutrino. The absence of the antineutrino in the table should not be taken as reflecting on Fermi's commitment to the Majorana neutrino. (I recall he regarded that as an unsettled question.) However, the V, τ, and K particles were in total confusion. Also notice that the names pion and muon had already been adopted in Fermi's talk.

Five more people spoke at that session, on various experiments: Luis W. Alvarez, "Recent Work at Berkeley on Elementary Particles"; Edoardo Amaldi, "On the Scattering of μ-Mesons of 200–1000 MeV Kinetic Energy by Nuclei"; Herbert Anderson, "Scattering of Pions by Liquid Hydrogen"; Gilberto Bernardini, "The Interaction of Pions with Nuclei"; and T. H. Johnson, "Scattering of π^- in Hydrogen." It is noteworthy that these experiments used detectors that included counters, nuclear emulsions, and cloud chambers.

Table 2.1. *Fermi's list of twenty-one elementary particles*

e	electron	μ^-	negative muon
e^+	positron	μ^+	positive muon
P	proton	G	graviton
$\bar{P}$	antiproton	V^+	positive V particle
N	neutron	V^-	negative V particle
$\bar{N}$	antineutron	V^0	neutral V particle
γ	photon	τ^+	positive τ meson
π^+	positive pion	τ^-	negative τ meson
π^-	negative pion	$\varkappa^+$	positive $\varkappa$ meson
π^0	neutral pion	$\varkappa^-$	negative $\varkappa$ meson
		ν	neutrino

Note: Fermi expressed a belief in the existence of anti-nucleons.
Source: Proceedings of the International Conference on Nuclear Physics and the Physics of Fundamental Particles, edited by J. Orear, A. H. Rosenfeld, and R. A. Schluter (University of Chicago Press, 1951).

A few months later the second Rochester conference was held, 11–12 January 1952, with notes taken by A. M. L. Messiah and Pierre Noyes. In between these conferences, Fermi, Herbert Anderson, and collaborators had made measurements on pion–nucleon scattering, and Keith A. Bruckner had distributed his preprint on the 3-3 resonance.[1] I remember that at the conference Fermi was quite excited about the new experimental results and about his letter to Richard P. Feynman on phase shifts, which was reproduced as Appendix 3 in the conference proceedings. Another highlight of the meeting was Robert W. Thompson's report on the θ particle, which everybody recognized as a careful and beautiful piece of experimental work. There was also Abraham Pais's report on the "megalomorphian zoology," which was the beginning of the associated-production idea. The title of Robert Marshak's concluding talk, "To Be or Not to Be (Astonished), That Is the Question," aptly summarizes the general confusion in those days of cosmic-ray particle physics, when systematic studies were very difficult and when one was never sure what was the question that one should explore.

In looking over records of old conferences, I noticed that physicists born before 1905 seemed to have general reservations in the period 1945–55 whether or not quantum mechanics were applicable inside of "the electron radius." For example, at the Shelter Island conference, 2–4 June 1947, Robert Oppenheimer began by distributing to each participant a memorandum entitled "The Foundations of Quantum Mechanics: Outline of Topics for

Discussion." In this memorandum he talked about the puzzlement arising from the Marcello Conversi, Ettore Pancini, and Oreste Piccioni experiment and wrote that

> To date, no completely satisfactory understanding of this discrepancy exists, nor is it clear to what extent it indicates a breakdown in the customary formalism of quantum mechanics.[2]

At the Chicago conference of 1951, Fermi touched upon the same reservations, as can be seen from the record of his talk:

> Theoretical research may proceed on two tracks: 1. Collect experimental data, study it, hypothesize, make predictions, and then check. 2. Guess; if nature is kind and the guesser clever he may have success. The program I recommend lies nearer to the first track. It is desirable to arrange experimental data so as to exhibit most clearly the features which come from fundamental particle interactions taking place at "contact," namely within about 10^{-13} cm. This may be done by assuming quantum mechanics holds in regions outside "contact" (there is little doubt in my mind that it does), and using it to remove from consideration phenomena which do not depend on what happens in the "contact" volume. The result is a compressed expression of experimental results, in which the nature of fundamental interactions between particles may be more easily discernible.

We see here that Fermi was cautious about what happens inside of 10^{-13} cm. On the other hand, in that same period, younger physicists, those of my generation, seemed to have very little inclination to question the validity of quantum mechanics. It is interesting to inquire into the origin of this difference in a priori worries between the different generations.[3]

Another general point that may be profitable to discuss is Fermi's recommendation that one follow the empirical track to do theoretical research.[4] Paul A. M. Dirac, on the other hand, definitely preferred a different track, as seen from the following quotations:

> The modern physical developments have required a mathematics that continually shifts its foundation and gets more abstract...[5]

> Quite likely these changes will be so great that it will be beyond the power of human intelligence to get the necessary new ideas by direct attempts to formulate the experimental data in mathematical terms.[5]

> The method is to begin by choosing the branch of mathematics which one thinks will form the basis of the new theory. One should be influenced very much in this choice by considerations of mathematical beauty.[6]

In the summer of 1982, Dirac and I were both at an Erice conference,[7] and I asked him what branch of mathematics – analysis, algebra, geometry, or topology – is likely to supply the basic ingredient for the next breakthrough in theoretical physics. He answered, unambiguously and without reservation: analysis.

To return to the particles: There was great confusion at the 1951 Chicago conference whether or not some of them really existed. At the end of Fermi's

Table 2.2. *Excerpt from list of particles by Anderson*

Name Daughters	Q	Lifetime	Remarks
Hyperons			
$\Lambda^0 \to p + \pi^-$	$+ 35 \pm 2$ MeV	$\sim 3 \times 10^{-10}$ sec	Some evidence also for $Q \sim 20$ and $Q \sim 75$
$\Lambda^+ \to n + \pi^+$	$+ (?)135$ MeV		6–7 cases in emulsion
$[\Lambda^+ \to p + \pi^0$	Consistent with 135 MeV	$<3 \times 10^{-10}$ sec	Not universally accepted]
$Y^- \to \Lambda^0 + \pi^-$	$+(65 \pm 12)$ MeV		Explains some cascades
$[Y^- \to (?)^- + \pi^+$	$+12$ MeV		$(?)^-$ is approximately of nucleonic mass, but no evidence it is p^-]

Note: See Table 21.1 for full listing.
Source: Proceedings of the Fourth Annual Rochester Conference on High Energy Nuclear Physics, edited by H. P. Noyes, E. M. Hafner, J. Klarmann, and A. E. Woodruff (University of Rochester, 1954).

talk, there was a discussion:

> E. McMillan asked if the τ and the $\varkappa$ mesons are in good standing now. Fermi replied yes, somewhat. Then there was a general demand for Fermi's definition of V, τ, and $\varkappa$ particles with the implication that there were differences of view among the delegates.

I would like to comment here that the published proceedings of those days captured more of the flavor and the tone of the discussion than the dry conference reports of today.

By the fourth Rochester conference, 25–27 January 1954, the identity of the new particles had become better established. Table 2.2 is a reproduction of part of the list of particles given by Carl D. Anderson at that conference.

Looking at the exuberant growth of our knowledge of particle physics in the early 1950s, I recognize the following characteristics:

a. Rapid development was taking place, with experiments usually done in a few months and rarely lasting more than one year. Figure 2.1 gives a plot that illustrates the growth of the field in the 1950s.

b. Many new ideas were constantly being explored, a sizable fraction of which turned out to be fruitful.

c. Generally, the whole field of particle physics was driven in the early 1950s by experimental discoveries. This presents a very important characteristic difference from today's particle physics. Figure 2.2 illustrates the strands of development of particle physics in 1945–57. Not included in this diagram are developments in accelerator and detector designs, without which, of course, particle physics could not have made such rapid progress.

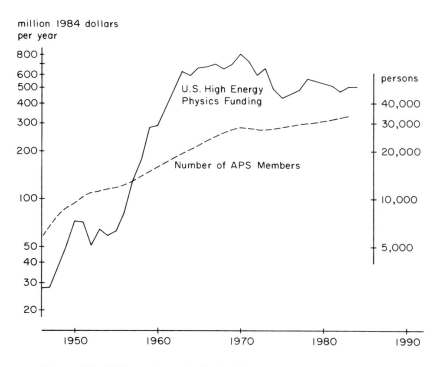

Figure 2.1. U.S. particle physics funding per year and number of members of the American Physical Society. The funding curve was calculated from data in the following sources: (a) for the period 1946–63: 1959 report of the Beams Panel of the president's Science Advisory Committee, *U.S. Bureau of Labor Statistics Handbook*, and data of DOE; (b) for the period 1963–84: October 1983 hearing of the Committee on Science and Technology of the House of Representatives, and 1984 report of the Elementary Particle Physics Panel (Perl panel) of the Physics Survey Committee. The number of members of the American Physical Society was taken from the *Bulletin of the American Physical Society 28* (1983), 1454.

Let me conclude by mentioning one episode in the early conceptual development of gauge fields that remains isolated in Figure 2.2. Historically, it was Hermann Weyl who first attempted to introduce a space–time-dependent scale variation to incorporate electromagnetism into general relativity.[8] He wrote three papers on the subject in 1918–19. Albert Einstein[9] criticized Weyl's ideas. The objection was as follows. Consider two clocks brought around loops L_1 and L_2 as illustrated in Figure 2.3. Einstein said that if Weyl's idea of space–time-dependent scale variations was right, these clocks, originally identical, would go at different speeds after they were brought around their respective loops: "the length of a common ruler (or the speed of a common clock) would depend on its history." Weyl tried to

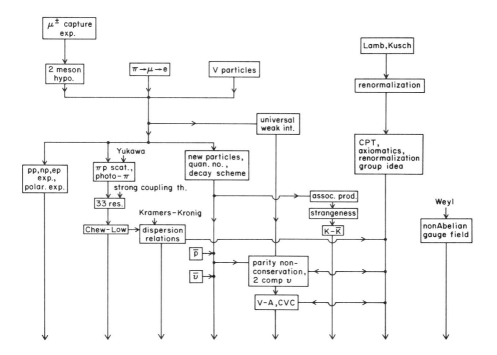

Figure 2.2. Strands of development in particle physics, 1945–54.

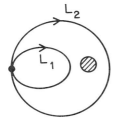

Figure 2.3. Paths of clocks in Einstein's discussion of Weyl's gauge theory.

explain away this difficulty by differentiating "between the determination of a magnitude in nature by 'persistence' (*Beharrung*) and by 'adjustment' (*Einstellung*)."[10] But Einstein's objection cannot be explained away by such arguments. It was only after quantum mechanics, in which the number $i = \sqrt{-1}$ entered in an essential way, and Weyl's *scale* variation evolved into a *phase* change, that the gauge field idea took root.[11]

With the replacement *scale* → *phase*, Einstein's objection is no longer relevant: When two originally identical clocks are brought around two different loops, they acquire different phases. But that does not affect their respective speeds. In fact, the phase difference produces no observable effects unless one can produce an interference between the two clocks, which, to say the least, is very difficult. Notice that if we substitute electrons for Einstein's clocks, the phase difference of the electrons after paths L_1 and L_2 is measurable, and the measurement is precisely the Aharonov–Bohm experiment.[12] In other words, if in Figure 2.3 there is a magnetic flux in the shaded area, then Einstein's *gedanken* experiment in fact is equivalent to the Aharonov–Bohm experiment.

Notes

1 H. L. Anderson, "Early History of Physics with Accelerators," in Colloque International sur l'Histoire de la Physique des Particules, *J. Phys.* (Paris) (*Suppl.*) *43:12* (1982), 101–59.

2 Robert E. Marshak, "Particle Physics in Rapid Transition: 1947–1952," in *The Birth of Particle Physics*, edited by Laurie M. Brown and Lillian Hoddeson (Cambridge University Press, 1983), pp. 376–401.

3 See, in this connection, the discussion among Schweber, Dirac, and Weisskopf in *The Birth of Particle Physics*, note 2 (pp. 265–7).

4 Fermi's style of choosing problems in physics generally emphasizes the empirical roots of physics. See H. L. Anderson's remarks in *The Birth of Particle Physics*, note 2 (p. 268).

5 P. A. M. Dirac, "Quantised Singularities in the Electromagnetic Field," *Proc. R. Soc. London, Ser. A 133* (1931), 60–72.

6 P. A. M. Dirac, "Relation between Mathematics and Physics," *Proc. R. Soc. Edinb. 59:2* (1938–9), 122–9.

7 A. Zichichi (ed.), *Gauge Interactions* (New York: Plenum Press, 1984), p. 38.

8 C. N. Yang, *Selected Papers 1945–1980, With Commentary* (San Francisco: Freeman, 1983), p. 525.

9 A. Einstein, remarks at the end of H. Weyl, *Sitzungsberichte d. k. Preuss. Akad. d. Wiss.* (1918), p. 465.

10 H. Weyl, "Electricity and Gravitation," *Nature* (*London*) *106* (1921), 800–2.

11 For a description of this history, see Yang, note 8 and C. N. Yang, *Proceedings of The International Symposium on Foundations of Quantum Mechanics* (Tokyo: Physical Society of Japan, 1984), p. 5.

12 Y. Aharonov and D. Bohm, "Significance of Electromagnetic Potentials in the Quantum Theory," *Phys. Rev. 115* (1959), 485–91.

3   An historian's interest in particle physics

J. L. HEILBRON

Born 1934, San Francisco; Ph.D., University of California, Berkeley, 1964;
history of the physical sciences since the Renaissance; University of
California, Berkeley

The symposium on which this volume is based was the fourth large
meeting sponsored by physicists and devoted to the history of nuclear or
particle physics. Many of the historians who attended earlier meetings came
away with the impression that their hosts considered them to be passive
receptacles for the true stuff of history (namely, reminiscences and recollections) or active clerks, able to look up bibliography, spell, and get dates
right.[1] This was a violation of parity between physicists and historians in the
study of the past, and it resulted, of course, in weak interactions. I shall
take the opportunity provided by the democratic action of this symposium's
organizers to say a few words about the methods and objectives of historians.
I shall then mention a few aspects of the history of particle physics particularly
interesting to historians.

Although physics and history aim at different things, there are important
parallels between them. Classical physics may have its analogue in dynastic
history, the concern with rulers, courts, diplomacy, and wars characteristic of
the historians of the nineteenth century, a concern that eventually made some
people as impatient, and showed itself as limited, as the physics of mass
points. While physicists extended their work to the domains of relativity and
the quantum, historians brought social, economic, and institutional forces to
center stage. Historians of physics underwent this revolution – which took
them from exclusive concern with great men and battles of ideas to consideration of the milieu as well as the content of science – about twenty-five
years ago. Experience with nuclear and particle physics in the 1950s assisted
the realization that modern science is the creation of society as well as of
scientists.

Like physics, history has its applied side, and, again like physics, it can be very dangerous when misapplied. We base our actions on our understanding of our past, whether directly experienced or mediated by books or colleagues. I need not give examples of the manipulation and mobilization of social forces by myths represented as history.

All this is by way of saying that history has its own demands, standards, changes, and challenges, even dangers – in short, its own intellectual base – with significant formal analogies to physics. Max Planck used to emphasize the analogies. In his opinion, pure physics was closer to history than to engineering.[2] Both aim at understanding the causes of things.

Insofar as historians may be said to have a particular goal, it is to understand the connection of events from a wider perspective than any of the historical actors, however well placed they were, could have attained. This aspiration does not imply a feeling of superiority to the actors, nor any special wisdom. It does imply the obligation and the patience to study a large quantity and broad range of sources – which should not be only literary – from and about the past.

The richer the sources, the wider, in principle, the perspective they open and support. The industrious historian who has looked at private correspondence, government papers, foundation and university reports, newspapers, patent applications, court records, architectural monuments, scientific apparatus, motion pictures, painted neckties, and literary T-shirts – as well as the scientific literature – necessarily sees connections that the people being studied could not have known.

Some of these connections may appear farfetched, and sometimes the lust for originality – from which the historian also suffers – creates grotesque associations. More often, the overly ambitious historian may offer a third-order correction before finding the first approximation. But even in these cases it is not licit to reject the proposed connections merely because they did not leave a trace in the memories of the historical actors or because they do not now appear to make good scientific sense.

From the point of view just sketched, one can understand that most historians do not consider the unsupported recollections of former participants very good evidence about events in the distant past. The problem of partial observation is in this case compounded by failing and selective memory.

A striking instance of the fallibility of even short-term memory was recorded some years ago in a survey of parents asked to chronicle major events, such as teething, talking, and toilet training, in the development of their toddlers. The answers deviated systematically from the clinical histories. In the case of toilet training, the mothers misremembered by an average of fourteen weeks, and the fathers by an average of twenty-two weeks. All the deviations were in the direction of expectations instilled by the authority of Dr. Benjamin Spock. By a similar and perhaps stronger mechanism, scientists may find it difficult to give an account of important events or discoveries in which they have participated other than what appears in textbooks and other semiofficial representations.[3]

Let us take an example from physics. About twenty years ago, James Franck submitted to an interview about the experiments he had done with Gustav Hertz – experiments that brought them the Nobel Prize. Franck described in detail how Niels Bohr's work had inspired theirs, just as described in the textbooks and in the Nobel proceedings.[4] The interviewer pointed out that Franck and Hertz had begun their experiments, which were directed toward measuring ionization potentials, in 1911, before Bohr had taken up atomic theory, and that the results they had got by 1914, when they became acquainted with Bohr's theory, seemed to them a definitive refutation of it. Franck at first declined to believe this account. When shown his own papers and correspondence, he broke off the interview in great distress. By the next scheduled meeting, he had regained his customary good spirits. Not so his interviewer.

This is not to say that reminiscence and spontaneous recollection have no value to the historian. They may provide leads to important sources and connections, and they always present information about the socialization of the discipline. Franck's recollections of his work with Hertz were worthless as testimony about what happened, but significant as an indication of his understanding, as an elder statesman, of the development of science.

The historian of recent science is dependent on scientists not for recollections necessary to historical reconstruction but for instruction in science. The student of Galileo's mechanics can prosecute his inquiries quite competently without having laid eyes on any living scientist. The student of Dirac's electron theory is dead on departure if he or she has not had the same sort of training that professional physicists receive. The consequent socialization may leave historians of recent physics with a conflict of identity that does not afflict those who stick to Galileo. This uncertainty may be reinforced by acquaintance with successful older physicists, whose experience so far exceeds the experience of their historians as to make reference to wider perspectives appear ludicrous. The one side therefore tends to assume an ascendancy that the other too frequently concedes.

This psychological fact may inhibit historians of recent science in carrying out one of their main duties, the establishment of an appropriate historical periodization. Parceling out history into episodes and eras is not merely a convenience for dividing a book into chapters. The punctuation of history carries with it the worldview of its grammarians. A decision about when the Middle Ages ended or the Renaissance began reveals what the decider considers to be medieval, and what representative of modernity, what definitive of culture and what unimportant, what the causes and what the consequences of change.

Physicists tend to periodize by the dates of publication of great papers. A new epoch accordingly begins in 1900, with Planck's presentation of his radiation formula; it runs to 1925, when the papers of Werner Heisenberg, Max Born, and Pascual Jordan on matrix mechanics appeared, and reaches a high point in 1913, when Bohr's theory of the atom became public. But this is to mark an epoch by the same criterion used to determine priority: date of publication. Historians should consider other criteria, suggested by a longer and broader view. For example, they might prefer a time when the community of physicists identified the quantum as a menace to classical physics, or they might prefer a punctuation that does not give pride of place to breakthroughs in theory. Institutional or social or engineering considerations might recommend partitions around 1870, 1915 or 1920, 1940, 1950, and so on.

We must have reasons for our epochs. They are our own retrospective creations. Does the subject "particle physics in the 1950s" make good historical sense? For full credit, an answer should not merely list the discoveries and inventions of the time, but state why they differ qualitatively (if they do) from those made in earlier or later periods, and why they – and not other factors – should be taken as a basis of periodization. By basing our periodization on several factors, and by taking a long view, we may be able to integrate the history of physics into general history. That is the goal, as challenging as a moon shot and as elusive as the truth, toward which many of us are working.

I take it for granted that historians of recent physics have an interest in recent physics: in its theories, experiments, and instruments, in its contributions to knowledge. I shall therefore confine myself to aspects of particle physics of particular concern to historians of the modern type, for whom social and institutional context is at least as important as disciplinary content.

That cornucopia of physics in the 1950s, the Bevatron, is a convenient object for illustrating the methods of these historians. Why was the machine built? The usual answer from physicists is that the machine came into existence to make antiprotons, or to make particle physicists independent of cosmic rays. This was the sort of answer given by Ernest Lawrence and the Atomic Energy Commission (AEC) in announcing the discovery of the antiproton: The Bevatron was built to do pure physics, to make antimatter.[5] With the detection of the negative proton in 1955, it realized its purpose and justified its expense.

To this account, historians would add some or all of the following:

1. The AEC built the Bevatron in order to investigate nuclear forces in the hope that they might be exploited in new sorts of weaponry.
2. The AEC did not build the Bevatron so much to knock the nucleus to pieces as to provide an opportunity to keep the experienced engineering staff at Berkeley together for mobilization in a national emergency.
3. The AEC cared little about particle physics, but much about maintaining good cheer at Berkeley, which was the only one of the Manhattan Engineer District's installations untouched by the severe decline in morale and staff suffered by the district immediately after the war.
4. The Bevatron, despite its uniqueness in energy, is best understood

as only the biggest of the many redundant accelerators commissioned at universities in the immediate postwar years by the Manhattan Engineer District, the Office of Naval Research, and the AEC.[6]

The general purpose of these machines, according to an internal AEC report of 1947, was to train personnel for employment in the laboratories of federal agencies or their contractors – it being supposed that university accelerator laboratories would attract the best talent and train it in ways useful to government.[7] The connection between accelerator building and personnel supply comes out forcefully in the eloquent recommendation regarding the Superconducting Supercollider by Sheldon Glashow and Leon Lederman.[8] They mention as spin-off from the project "physicists accustomed to solving the unsolvable." Such physicists would no doubt be useful in Star Wars laboratories.

There is good documentary evidence for the several answers I have given to the question, Why the Bevatron? Curiously, the least frequently mentioned before 1955 was the making of antiprotons or the enlargement of the human spirit by the hunt for fundamental particles. The design energy of the Bevatron rose and fell with political processes and financial circumstances, not with calculations of production thresholds. Robert Seidel tells a part of this story in Chapter 34 of this book.

To the historian, all of the causes I have suggested for the building of the Bevatron are right. Each characterizes a different interest group: the physicists, the agencies, the engineers, various strata in the AEC and the Congress, and graduate students made upwardly mobile by fellowships and job opportunities. The answer to the question, Why the Bevatron? requires a tale in the style of *Rashomon*, or the story of the blind men and the elephant, in which each character describes what he experienced from a limited perspective. It would not be possible for the participants in this meeting to tell such a tale for the Bevatron, since the points of view of government, engineering, and the military do not seem to be represented here.

An historian might want to integrate the several causes of the building of the Bevatron – and thereby the principal causes of the support of particle physics in the 1950s – by identifying as their connecting link a trade-off, more or less explicit, between physicists and government. According to this account, university physicists received generous support for research that was of little, if any, direct concern to the sponsoring agency, in return for supplying trained manpower and technical advice. Most of the Ph.D.s trained with AEC funds at university accelerator laboratories did go to work for the AEC in one way or another – some 75 percent of the graduates of 1953 for example.[9] The trade-off applied to the community, not to each physicist separately. A symbol of the arrangement was the fissioning of the Radiation Laboratory of the University of California into its purer and more applied branches: into its Berkeley and Livermore laboratories.

In parallel with their investigations of the social role of big machines, historians would wish to study the socialization of the men and women who worked with and around them. By "socialization" I mean the effects of training, working conditions, the award system, and so forth on the attitudes of physicists toward their discipline and its goals. These attitudes include expectations about level of support and also political tone, which differed markedly between, say, Brookhaven and Berkeley.

The award system offers a useful probe into this socialization. In their interest in prizes, and dissatisfaction when they do not get them, physicists may be unusual. General Leslie Groves thought so. In his experience, physicists were more conscious of rank than the military; he hesitated to appoint Robert Oppenheimer head of Los Alamos, he said, not because he thought him a security risk but because Oppenheimer did not have a Nobel Prize, and consequently, in the general's judgment, might lack authority.[10] The award of the Nobel Prize for the antiproton almost opened the practices of the community of particle physicists to public scrutiny. Historians might regret that it did not. We often find the study of priority disputes, and the system of credit and reward, particularly instructive about the general circumstances of the science of any period.

The Nobel Prize does not seem well adapted to experimental particle physics. The one is set up for the individual, and serves the cult of personality; the other requires teamwork and group allegiance. Here is the way the head of physics research at Brookhaven in 1956, Samuel Goudsmit, who was trained as a physicist of the old school, described the new men he wanted for his laboratory:

> In this new type of work experimental skill must be supplemented by personality traits which enhance and encourage the much needed cooperative loyalty. Since it is a great privilege to work with the Cosmotron, I feel that we now must deny its use to anyone whose emotional build-up might be detrimental to the cooperative spirit, no matter how good a physicist he is. . . . I shall reserve the right to refuse experimental work in high energy to any member of my staff whom I deem unfit for group collaboration. I must remind you that it is, after all, not you but the machine that creates the particles and events which you are now investigating with such great zeal. The designers and builders of the Cosmotron get little credit for this. That we are favored with the opportunity to use this accelerator is for most of us a mere matter of luck and not of selection or intent, a circumstance which should fill us with humility.[11]

Preliminary surveys suggest that the number of distinct prizewinners is considerably less than the number of prizes awarded, and that the disparity increases the higher the honor. The top prizes tend to be won by people who have already won top prizes. What is the purpose of these redundant awards? They can scarcely be said to stimulate the growth of science or to instill a productive competitiveness. Is the number of prizes desired a confirmation of Groves's diagnosis of the peculiar rank-consciousness of physicists? Is there such a consciousness? If so, what consequences did and does it have for the direction and pace of research? How did and does the hierarchy of rank affect the peer-review system? How do prizewinners correlate with laboratory directors and high government advisers?

A last example of matters interesting to historians about particle physics since the war is its difference in tone from the European physics of the interwar years. An indication of the difference may be found in terminology. In the 1930s, the tradition of growing neologisms on Greek roots was still vigorous, and in the naming of the deuteron, for example, the several competing suggestions came forward with the certification of philologists and elaborate justification from scientists.[12] The appeal to Greek roots had its rationale in the ease with which neologisms so grown can be assimilated into all Western languages and the neutrality they might convey about the long-term meaning of the phenomena.

What do we have in the 1950s? "Strangeness," a word barely utterable in Romance languages and expressive of a surprise only briefly felt; and latterly we have had quarks, with their colors, flavors, tops, and bottoms. Does the new terminology express cynicism or disdain by particle theorists toward their own creations? Or does it express an enhanced consciousness of the transience of theoretical constructs?

Indifference to Greek may be taken as a symbol of a lack of classical culture, a pragmatic approach to life and science, and depreciation of philosophy. These, and big machines, are supposed to be the American way. Historians are interested in assessing the consequences for theory and theorizing of the fact that the American style is "unphilosophical" – to use a word Wolfgang Pauli applied critically to Heisenberg, and Heisenberg applied critically to Americans.[13] More about these matters will be found in Silvan Schweber's Chapter 46.

The Americanization of physics may not be rooted in the nature of things. How and why it occurred are questions of much current interest to historians. Among the significant episodes under study is the rehabilitation of European physics under American sponsorship after World War II. The most striking international response to American initiatives, CERN, is the subject of a large historical project directed by Armin Hermann (see Chapter 36).

Several of the aspects of the history of particle physics in the 1950s that I have mentioned – the machine–personnel trade-off, the connection with the military and industry, the politics of big science – seem to have structural analogies to the contemporary situation, to the Strategic Defense Initiative and the Supercollider. It might be useful and instructive to discuss parallels and antiparallels between the history we are gathered to review and the present we hope to survive.

Notes

1 Roger H. Stuewer (ed.), *Nuclear Physics in Retrospect. Proceedings of a Symposium on the 1930s* (Minneapolis: University of Minnesota Press, 1979); Colloque International sur

l'Historie de Physique des Particules, *J. Phys. (Paris)* (*Suppl.*) 43:12 (1982); Laurie M. Brown and Lillian Hoddeson (eds.), *The Birth of Particle Physics* (Cambridge University Press, 1983).

2 J. L. Heilbron, *The Dilemmas of an Upright Man: Max Planck as Spokesman for German Science* (Berkeley: University of California Press, 1986), p. 113.

3 David F. Musto, "Studies on the Accuracy of Oral Interviews," *National Colloquium on Oral History Proceedings 3* (1969), 171–2.

4 Interview of James Franck by T. S. Kuhn, 10 July 1962, pp. 7–11, Archive for History of Quantum Physics, Office for History of Science and Technology, University of California, Berkeley.

5 Press release, Office of Public Information, University of California, Berkeley, 19 October 1955.

6 These alternatives are discussed and documented in a forthcoming history of the Lawrence Berkeley Laboratory.

7 Robert W. Seidel, "Accelerating Science: The Postwar Transformation of the Lawrence Radiation Laboratory," *Historical Studies in the Physical Sciences 13* (1983), 375–400.

8 Sheldon L. Glashow and Leon M. Lederman, "The SSC: A Machine for the Nineties," *Phys. Today 38:3* (1985), 29–37.

9 Thomas H. Johnson, "Physics Research Programs of the U.S. Atomic Energy Commission," AEC news release, 30 April 1953, pp. 12–13.

10 Leslie R. Groves, *Now It Can Be Told. The Story of the Manhattan Project* (New York: Harper, 1962), p. 62.

11 S. A. Goudsmit, Memorandum, 1956, Brookhaven National Laboratory Archieves.

12 Roger H. Stuewer, "The Naming of the Deuteron," *Am J. Phys. 54* (1986), 206–18.

13 Pauli to Bohr, 11 February 1924, in Wolfgang Pauli, *Wissenschaftlicher Briefwechsel 1*, edited by A. Hermann, K. von Meyenn, and V. F. Weisskopf (Berlin: Springer-Verlag, 1979), p. 143; J. L. Heilbron, "The Earliest Missionaries of the Copenhagen Spirit," *Revue d'histoire des sciences 38* (1985), 195–230 (especially 206).

PART II

PARTICLE DISCOVERIES IN COSMIC RAYS

The beginning is easy to recite for us,
The ending is nowhere in sight for us.
And though the answers may some day be nearer,
Things will get worse, before they get clearer.
– © Arthur Roberts, "Some People Don't Know
Where to Stop" (1952)

4 Cosmic-ray cloud-chamber contributions to the discovery of the strange particles in the decade 1947–1957

GEORGE D. ROCHESTER

Born 1908, Wallsend, England; Ph.D., University of Durham, 1937;
experimental studies of cosmic rays; University of Durham

Introduction

This chapter is a personal account of the most significant contributions by cosmic-ray physicists to the discovery and study of the strange particles in the ten years that followed the discovery in 1947 of the *V* particles in conventional cloud chambers. [*V* particle is a generic term invented by Patrick M. S. Blackett and Carl D. Anderson, and generally adopted in the early days, to describe the strange particles whose decay resulted in characteristic V-shaped tracks (see Figures 4.4, 4.9, and 4.10).] For fuller accounts of the subject, one should see the many excellent reviews in *Progress in Cosmic Ray Physics*, *Progress in Physics*, supplements to *Il Nuovo Cimento*, and *Reviews of Modern Physics*, and in numerous papers in the proceedings of cosmic-ray and particle physics conferences. Two long historical reviews by Charles Peyrou[1] and myself[2] give the main references.

I am most grateful to the many cosmic-ray physicists who have allowed me to use their figures and photographs in this chapter. Though references to the sources of some of the figures have been made in the text, I would like to thank specifically the following: Professor C. D. Anderson and the CIT group for Figure 4.7; Professor H. S. Bridge for Figures 4.13 and 4.21; Sir Clifford Butler for Figures 4.2 and 4.6; Sir Clifford Butler and the Manchester Pic-du-Midi group for Figures 4.8, 4.9, 4.12, and 4.18; Professor W. Fretter and the Berkeley group for Figure 4.14; Professor Leprince-Ringuet, Dr. C. Peyrou, and the Ecole Polytechnique group for Figure 4.20; Professor Reynolds, Dr. A. L. Hodson, and the Princeton group for Figure 4.22; Professors B. Rossi and H. S. Bridge and the MIT group for Figures 4.19 and 4.21; Professor R. W. Thompson for Figure 4.16; Professor Thompson and the Indiana group for Figures 4.15, 4.17, and 4.23.

Finally, I wish to thank the head of the Durham Physics Department for the use of many of the departmental facilities.

Significant factors in the discovery of the strange particles in cloud chambers

Two special factors contributed to the success achieved with cloud chambers in the discovery and investigation of strange particles and, in certain areas, made their contribution unique, namely, penetrating-shower selection and counter control.

Penetrating-shower selection

Historically, the *V* particle work grew out of the detailed study of the so-called penetrating showers (ps) by Lajos Jánossy and his associates before and during the war.[3] To Jánossy, such a shower meant a shower of penetrating particles (i.e., muons), but we now know that it was simply a high-energy interaction consisting mainly of nucleons, pions, muons, electrons, and photons. Jánossy's work was important in the present context because he devised a simple arrangement of Geiger–Müller counters in a mass of lead, the counters being arranged in a tight coincidence (fivefold to sevenfold in different experiments), so that they selected the rare nucleonic interactions and excluded the much more frequent electron showers, knock-on events, and so on. With hindsight this seems simple and obvious, but it became of great importance when it was discovered that *V* particles were also produced in such interactions. Indeed, in a fascinating paper presented at the 1980 Fermilab historical symposium, Anderson ruefully reflected that he missed the *V* particles in 1946 because he had failed to study Jánossy's papers.[4] (A portrait of Jánossy is shown in Figure 4.1.)

The total energy of a typical high-energy interaction selected by the Jánossy arrangement is not known precisely, but for *V*-particle work it has been estimated to be in the range 5–15 GeV.[5]

The counter-controlled cloud chamber

Of equal importance was the counter-controlled cloud chamber, a well-known device of great historical significance in cosmic-ray research, by which an event could be made to trigger a cloud chamber, and, so to speak, take a photograph of itself, thus enabling rare events to be picked out in the presence of a large unwanted background. The counter-controlled chamber in a magnetic field proved to be ideal for the discovery and study of neutral unstable strange particles, and by a strange twist of fortune the ps-counter cloud-chamber setups turned out to be of just the right dimensions to observe decaying particles with lifetimes in the range 10^{-10} to 10^{-8} sec. Moreover, apart from the obvious fact that with cloud chambers the signs and momenta of the charged particles in the shower could be determined, counter control implied time association and therefore the possibility of observing the decays of neutral particles created at considerable distances above the chamber. This was not possible with the principal rival technique, the nuclear emulsion, for

Figure 4.1. Lajos Jánossy (1912–78).

although thousands of two-pronged stars looking like V^0 decays were present in exposed emulsions, it was seldom possible to find out which were decays and which small nuclear interactions. Only when exact data for one form of V^0 decay (the V_1^0 or Λ^0 particle) became available was it possible to find examples in emulsions;[6] indeed, the first decay, later termed the V_2^0, could not have been found in the nuclear emulsions available in 1947, for none were electron-sensitive, and the charged decay products of the V_2^0 were relativistic, and ionized close to minimum. The difficulty of identifying V_2^0 decays in emulsion is shown by the fact that in a well-known review by Robert W. Thompson in 1956, only four examples were reported.

The first *V* particle

The cloud-chamber photographs of penetrating showers taken by our group from 1941 onward in Manchester convinced Blackett at the end of the war that the best equipment available in the laboratory, namely, his magnet cloud chamber, should be turned over to penetrating-shower work. Blackett himself decided not to continue his cosmic-ray researches, but began a difficult project working on the design and construction of a high-speed wooden rotor to test his theory of the connection between rotation and magnetism. Nonetheless, he continued to take a keen interest in cosmic rays, for the subject remained a major component of the departmental research program.

The Blackett–Wilson cloud chamber was old-fashioned and not suitable for penetrating-shower work, and so with the help of a new young recruit to cosmic rays, Clifford C. Butler, and later (for a short period) Keith Runcorn, a new team was set up that modernized the cloud chamber, making it deeper

Figure 4.2. Clifford Butler adjusting the *V*-particle cloud chamber in Manchester [*Source*: C. C. Butler].

and introducing many new features that stemmed from wartime develop-ments (e.g., flash-tube lighting instead of tungsten lamps). (A picture of Butler working on the chamber is shown in Figure 4.2.) The main purpose of this experiment was to study the particles in penetrating showers, to try to identify the slow particles that looked muonlike, and to examine the pene-trating particles, which we assumed, following Jánossy and Walter Heitler, to be typical sea-level penetrating particles. The purpose of the 3-cm lead plate across the chamber was to help identify penetrating particles. Later, it was to play a significant role in the reasoning that led to the conclusion that the *V* events were decays and not nuclear interactions. Wherever possible, mass measurements by the ionization-and-momentum method were made on all slow particles, the ionization being estimated visually by comparison with electron tracks on the same photographs, and the momentum by curvature in the magnetic field.

In the years 1946–8, many thousands of photographs were taken, and many slow mesons were seen. The data, now only of historical interest, were published in a series of papers by Butler and myself.[7–10] For the identification

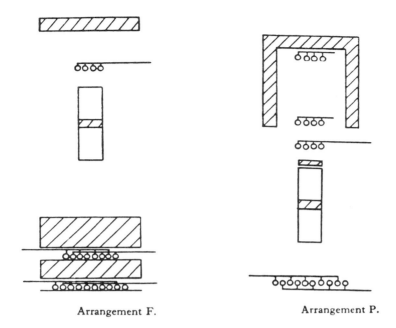

Arrangement F.   Arrangement P.

Figure 4.3. Diagram showing the counter arrangements and cloud chamber used in the original *V*-particle work [*Source*: Rochester and Butler (1948)[7]].

of muons and pions, ionization–momentum curves were drawn using the mass ratio given by Bristol at the time, namely, $m_\pi : m_\mu = 1.65$. This value, together with the arbitrariness of visual estimates, although carefully carried out by independent observers, put most of the mass measurements very close to the muon curve and well away from the pion curve. It was therefore thought, erroneously, that the creation of muons was being observed. The matter was cleared up in later papers by Butler and associates and K. H. Barker and Butler, where the slow and fast particles were shown to be similar to those found at Bristol in high-energy interactions, namely, pions and protons.[11,12]

However, the most important feature of the early work was undoubtedly the discovery of the first *V* particles on 15 October 1946 and 23 May 1947. The counter arrangements are shown in Figure 4.3 and the first V^0 in Figure 4.4. Blackett's contribution, apart from training us in the technique of counter control and giving us the use of his large magnet, was of vital importance in that he took a keen personal interest in this work, insisting on seeing every event of interest. Indeed, I recall that when shown the first V^0 event he said immediately that he had never seen anything like it before and was convinced it was entirely new. Moreover, he took the data and worked in parallel with us, trying out various decay schemes. Tribute should also be paid to the high level of criticism in the laboratory that ensured that vital points,

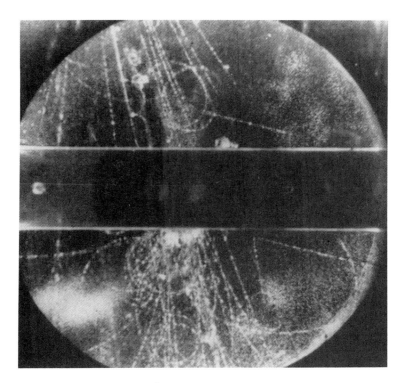

Figure 4.4. The original V^0 decay observed on 15 October 1946. The V is below the plate and to the right. Note that this photograph is one of a stereoscopic pair [*Source*: Reprinted by permission from Rochester and Butler, *Nature (London) 160* (1947),[13] 855; copyright © 1947 Macmillan Magazines Ltd.].

such as whether the events represented decays or nuclear interactions, were thoroughly discussed in numerous colloquia.

Copies of the V particle publication, together with high-quality photographic prints, were sent to many of the leading particle and theoretical physicists in the world, but there was little comment except from Jánossy, Enrico Fermi, Heitler, Cecil F. Powell, Bruno Rossi, and John Wheeler. Jánossy, in a letter to me dated 6 November 1947, pointed out what we had already noticed, that the high value of the transverse momentum of the V^+ decay was convincing evidence for a mass much greater than the pion. The V particle work was published in December 1947.[13] The original analysis has been discussed so fully elsewhere that it will not be repeated here.[2]

The years of frustration

After the early discoveries that promised so much, there followed several frustrating years, a period of strain for Butler and myself, when no further examples of the *V* particles were found. Indeed, the next *V* particles to be found by any Manchester group were found in the summer of 1950. How the deadlock was broken by Manchester will be described in due course, but in retrospect, even in the year 1948, which at the time seemed bleak indeed, two events occurred that were to inaugurate the enormous developments of 1949 and later. The first was the symposium on cosmic rays held in honor of Robert A. Millikan's eightieth birthday at the California Institute of Technology in June 1948, and the second was the production of the first electron-sensitive emulsion in November by the research staff of English Kodak Ltd.

Blackett sent me to the Millikan conference, where among the participants were almost all the leading experimental and theoretical physicists in cosmic-ray physics. Many references were made to the possible existence of particles of mass between the pion and the proton, but the principal new discovery that attracted most attention was a particle with a mass of possibly 1,000 m_e forming a star, found by Louis Leprince-Ringuet in nuclear emulsion.[14] I talked in detail about penetrating showers and showed many of the best examples of mesonlike particles and penetrating particles.[9] I referred to the *V* particle discoveries, but, as pointed out to me by Abraham Pais recently, I did not make much of the *V* events, and indeed there is probably little doubt that this reflected my disappointment at the absence of new examples. Robert Oppenheimer was very much the conference leader. In his summary at the end, he pointed out that more and more intermediate-mass particles were being found, and he made it clear that he was impressed by the French work.

Nonetheless, later events showed that the *V*-particle work did attract the attention of some scientists, notably Anderson, Rossi, and Robert B. Brode. Indeed, Anderson immediately switched the magnet cloud chamber he and his group had been using to measure the momentum spectrum of the electrons from muon decay and began using it for *V* particle work in Pasadena and on White Mountain. The sequel was an exciting letter from Anderson to Blackett dated 28 November 1949, which included the following paragraph:

> Rochester and Butler may be glad to hear that we have about 30 cases of forked tracks similar to those they described in their article in *Nature* about two years ago, and so far as we can see now their interpretation of these events as caused by new unstable particles seems to be borne out by our experiments.

The significance of this work will be assessed shortly, but first a brief comment on the second event, the advent of the first electron-sensitive emulsion, the Kodak NT4. Some of this emulsion was made available to Powell in Bristol, and immediately he sent a block to the Jungfraujoch, where it was exposed in the Aletschgletscher under a block of lead. On development a few weeks later, there appeared the first example of the τ decay (i.e., the decay of a charged particle into three charged pions). This outstandingly important event was published by R. Brown and associates on 15 January

1949[15] and is described by Donald Perkins in Chapter 5. The τ meson proved to be a rare particle, but its significance at that time was twofold: In the first place, it confirmed the existence of particles of mass about 1,000 m_e; second, it showed that these particles had different modes of decay.

The move to the Pic-du-Midi

Numerous discussions about the absence of V particles took place in Manchester in the years 1947–9, well before the news of the Anderson discoveries reached us. Three possible causes were identified:

1. The producing layers of lead in the Rochester–Butler setup were too far from the chamber.
2. The chamber was too small.
3. The rate of high-energy nucleons at sea level was too low.

Blackett had foreseen that a larger chamber at high altitude might be needed, and in 1947 he had formed a new group under the direction of J. A. Newth to design and build a large magnet cloud chamber suitable for operation at the Jungfraujoch, where the nucleon rate was expected to be some fifteen times greater than at sea level. A new magnet was necessary because the largest piece of the Blackett magnet was too heavy for transport up the Jungfrau. The Newth project was an amibitious one and unfortunately progressed slowly. It did not become fully operational until 1951. The slowness was worrisome, as was the possibility that the new emulsions might capture the field. Indeed, I recall that Giuseppe P. S. Occhialini, on one of his frequent visits to Manchester, facetiously remarked that the days of the conventional cloud chamber were numbered!

The net result was that Blackett asked me to form an emulsion team and asked Butler to modify our old equipment as far as possible and look into the possibility of moving the whole setup to a mountain site in France. Part of my assignment was to examine an idea that originated with Blackett and Occhialini after dinner one Sunday evening (and, as often happened, was expounded with great enthusiasm by Blackett in the laboratory on the following Monday morning!). The idea was to expose emulsions in a powerful magnetic field to see if the magnetic deflection of fast charged cosmic-ray particles could be detected in the presence of scattering.

The history of the choice of the Pic-du-Midi as the mountain site has been given already in a fine paper by Butler.[16] Again Occhialini was behind the move, partly because as an expert speleologist he knew the mountain well, as well as the director, J. Rösch. Also, Occhialini did not believe that the potential usefulness of exposing emulsions in magnetic fields could be evaluated at sea level alone, as Nora Page and I were then doing in Manchester on a Blackett-type magnet being built at the works of Metropolitan Vickers. Blackett was not easily convinced, but after much correspondence, he sent J. J. Braddick and Butler to France to look at the Pic and other sites, such as the Leprince-Ringuet station on the Aiguille-du-Midi near Mont Blanc.

Figure 4.5. The Pic-du-Midi in summer [*Source*: C. C. Butler].

In the end, in spite of inadequate power supplies and the difficulties of access, particularly in winter, the Pic was chosen, a major factor undoubtedly being the enthusiasm of the director and the confidence he inspired. A picture of the observatory, which is 2,867 m above sea level, is shown in Figure 4.5. Eventually, after much weighing of the pros and cons, the Manchester chamber and magnet were dismantled in August 1949, and by 4 October the magnet was on the mountain. Blackett visited the Pic in September 1949 (Figure 4.6), by which time Rösch had fulfilled his promise to get an electric cable up the mountain. The magnet was energized late in November and used all winter for emulsion experiments, most of the exposures being made by one of my colleagues at Manchester, Stephen Goldsack. Moreover, by this time, it had been realized that the whole emulsion project was one of some complexity and difficulty, and in consequence expert help was enlisted on both the experimental and theoretical sides, on the emulsion side by bringing in the Brussels group, which then included Occhialini and Constance C. Dilworth, and on the theoretical side, J. E. Moyal, a statistics specialist in Manchester, and Y. Goldschmidt-Clermont. The results were published in 1950 by Dilworth and associates and found useful, if limited, application later at CERN when magnetic fields of 200 kG became available.[17]

Figure 4.6. A group on the Pic-du-Midi in September 1949. Occhialini is at the extreme left; on the right are Blackett (with the pipe), Cosyns, and Rösch. The other three members of the group are Labordens, Madame Labordens, and Ladormen [*Source*: C. C. Butler].

The early Anderson work

Early in 1950, Anderson sent the manuscript of a paper on the *V* particles (which appeared later in the same year under the names of A. J. Seriff, Robert B. Leighton, C. Hsiao, Eugene W. Cowan, and Anderson), in which details were given of the equipment and the main results up to that time.[18] The penetrating-shower trigger was a sixfold coincidence consisting of two trays of counters above the chamber and one below, a trigger requiring three particles in the top tray, two in the second, and one in the third. Above the top tray was a 20-cm-thick transition layer of lead, and between the first and second layers a further 6 cm of lead.

The paper listed the observation of six V^0 particles in Pasadena, and twenty-four V^0 and four V^+ particles at White Mountain, 3,200 m above sea level. Apart from the confirmation of the original discoveries, the statistical weight of the new observations showed the *V* particles were not exceptionally rare and, as had been surmised, increased markedly in frequency with altitude. Again, many of the secondaries were clearly heavier than electrons, some being in the mass range 200–400 m_e, and some around the proton mass. Finally, it seemed that many of the decays were two-body decays and were

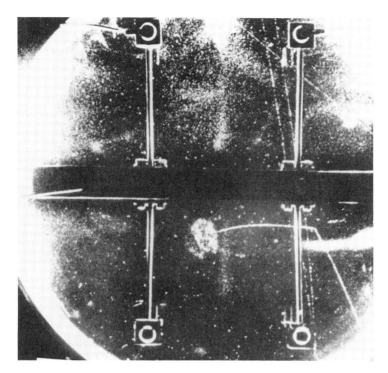

Figure 4.7. An early example of a V^0 decay obtained by the Anderson group [*Source*: C. D. Anderson]. Because of distortion, identification is difficult, but the decay of a V_1^0 particle to a pion and a proton is the most probable interpretation.

the products of a neutral particle of lifetime about 3×10^{-10} sec. Anderson never published any of his photographs because of gas distortion, but he allowed John G. Wilson and me to include some in our atlas.[19] An example is shown in Figure 4.7.

The Manchester V particle work on the Pic-du-Midi

The cloud chamber and its associated equipment went up the Pic in the spring of 1950 and started operating in the summer. The chamber was the same as used in Manchester, but the lead and counters were arranged to maximize the observation of V particles. Moreover, greater care was taken to isolate the chamber thermally from the magnet, the chamber temperature being kept constant to $0.25°C$ by a water jacket fed from a thermostated supply. This markedly reduced convection currents in the chamber. The chamber, counters, and lead are shown diagrammatically in Figure 4.13(1), the trigger being at first a sixfold set (i.e., $A_1 B_3 C_2$), but later, and more

Figure 4.8. The Pic-du-Midi cloud chamber and magnet [*Source*: C. C. Butler].

frequently, fourfold ($A_2 B_2$) and fivefold ($A_1 B_2 C_2$) sets. The thickness of lead above the chamber was typically 15 cm, the chamber being of illuminated diameter 28′cm and depth 7 cm. At first a 2-cm lead plate was placed across the chamber, but later this was dispensed with, or a thinner plate was used. The magnetic field was 7,500 G for a power of 12 kW, and the maximum detectable momentum (MDM) for a track 6 cm in length was 8×10^9 eV/c. The resetting time was 6 min, but later it was reduced to 2.5 min. A general view of the magnet and chamber is shown in Figure 4.8.

In the first six months, July 1950 to January 1951, forty-three V decays (of which thirty-six were V^0 and seven were $V^{\pm}$ decays) were observed. This rate increased markedly when the resetting time was reduced, and, indeed, at a much later date (1953), Blackett carried out an analysis of many V particle experiments and showed that with a resetting time of 2 min, the Pic chamber could detect ten V events per week.[5] This analysis was based on an assumed penetrating-shower counting rate of eleven per hour (i.e., 200 showers per

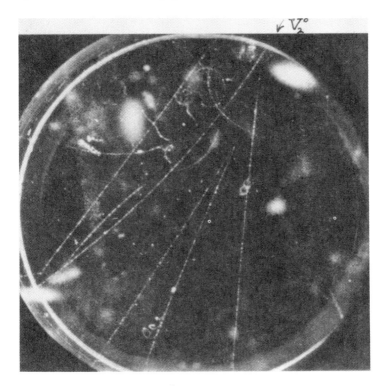

Figure 4.9. A Pic-du-Midi V_2^0 decay [*Source*: Armenteros et al. (1951)[21]].

day), of which one-third were penetrating showers of mean energy 10^{10} eV, and the experimental fact that 3 percent of such showers had V events. The Pic setup proved, therefore, to be a very efficient collector of V particles.

By the summer of 1951, many of the Pic decays had been examined in detail, and it was confirmed that many of the secondaries of the decays seemed to be protons and light mesons and were the products of two-body decays. It was therefore assumed that there were two types of V^0 particles, termed V_1^0 and V_2^0, that decayed according to the schemes

$$V_1^0 \to p \ + \pi^- \tag{4.1}$$

$$V_2^0 \to \pi^+ + \pi^- \tag{4.2}$$

An example of V_2^0 decay is shown in Figure 4.9.

The Pic-du-Midi results were published in 1951 in two important papers by R. Armenteros and associates in which the evidence for the schemes (4.1) and (4.2) was set out.[20,21] In the second paper, a new method of analysis was introduced that proved of far-reaching importance when developed later by Thompson of Indiana. The Manchester analysis made use of the general properties of two-body decays, following the original work of J. Blaton and

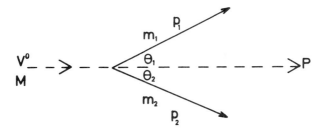

Figure 4.10. A two-particle decay in the laboratory system.

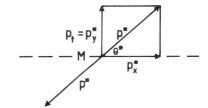

Figure 4.11. A two-particle decay in
the center-of-mass system.

J. Podolanski. Referring to Figures 4.10 and 4.11, in which are defined the
quantities involved in a two-particle decay in the laboratory and center-of-
mass systems, a parameter α can be defined, as in (4.3), that allows examination of the anisotropy in the center-of-mass system

$$\alpha = \frac{p_1\cos\theta_1 - p_2\cos\theta_2}{p_1\cos\theta_1 + p_2\cos\theta_3} \tag{4.3}$$

where the quantities on the right-hand side of this expression are the
longitudinal components of the momenta of the secondary particles in the
laboratory system. It is readily shown that α can also be expressed as

$$\alpha = \frac{p_1^2 - p_2^2}{P^2} = \frac{m_1^2 - m_2^2}{M^2} + 2p^*\cos^*\left(\frac{1}{M^2} + \frac{1}{P^2}\right)^{1/2} \tag{4.4}$$

The average value of α taken over many decays of the same type of particle is

$$\bar{\alpha} = \frac{m_1^2 - m_2^2}{M^2} \tag{4.5}$$

A typical plot of α versus P for the Pic-du-Midi results is shown in Figure
4.12, from which it is clear, in spite of the fairly wide spread of the experi-

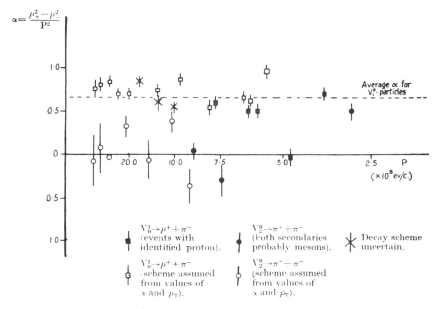

$$\alpha = \frac{p_+^2 - p_-^2}{P^2}$$

Figure 4.12. The plot of α versus P for early Pic-du-Midi data [*Source*: Armenteros et al. $(1951)^{21}$].

mental values, that the observed V^0 decays cluster about two of the lines corresponding to the decay schemes (4.1) and (4.2) for which the values of $\bar{\alpha}$ are 0.69 and zero. This work was the first to give a quantitative description of the main V^0 particles, and it has proved to be essentially correct, apart from an error in the derived mass of the V_2^0 particle. Quite independently, the Thompson group in Indiana found nine examples of V^0 decays at ground level in a twelve-inch-diameter magnet cloud chamber.[22] Three of the positive decay fragments had masses close to the proton mass, and several of the light negative decay products were roughly of mesonic mass. This group introduced for the first time the Q value (i.e., the kinetic energy of the decay products with the decay products at rest). For two of the three cases of V_1^0 decay, the Q values were 31 ± 5 MeV and 34 ± 10 MeV, close to the presently accepted Q values.

Later work provided tests of the hypothesis of two-body decay, the most important being the following:

1. an examination of the coplanarity of the plane of the decay products with the point of decay,
2. the transverse momentum balance of the decay products, and
3. the uniqueness of the Q values.

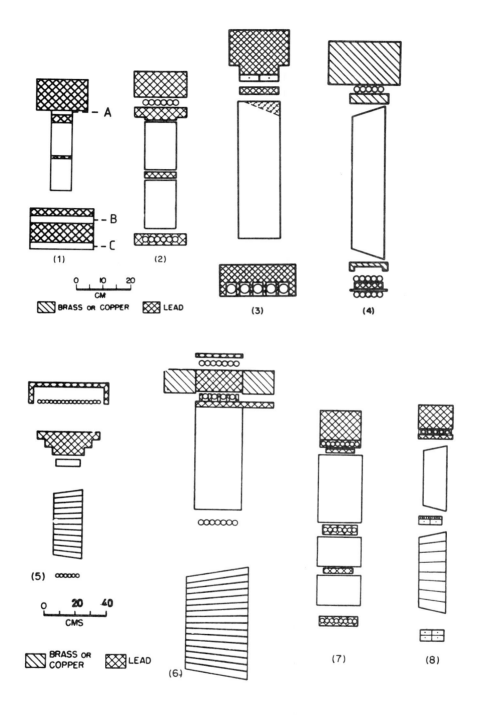

- A

- B

- C

(1)   (2)

0   10   20
      CM

⬛ BRASS OR COPPER   ⬛ LEAD   (3)   (4)

(5) ⬤⬤⬤⬤⬤

0   20   40
      CMS

⬛ BRASS OR
   COPPER   ⬛ LEAD   (6)   (7)   (8)

Figure 4.13. Schematic representation of the cloud chambers used by various groups in the study of unstable particles [*Source*: Bridge (1956)[23]]. All diagrams show side views of the experimental arrangements; illuminated

The *V* particle explosion

The realization that the particle world was infinitely richer than had ever been envisaged led immediately to an immense expansion in *V* particle work, especially in the United States. The older groups were expanded rapidly, new groups were established, and the machine builders began to design for operation in the GeV region. A glance at any of the U.S. physics conference literature of the period will show that for at least seven years in the 1950s, every particle meeting devoted a major session to the "strange" particles, as they had come to be called. The impact of the discovery of the new particles on one famous series of conferences, the Rochester conferences, is brought out admirably in Chapter 45 by Robert Marshak.

Lack of space makes it impossible to explore this aspect of particle history, and therefore only the highlights will be presented here. The magnitude of the cosmic-ray effort alone may be gauged by the fact that at the peak of output, the yield achieved was of the order of several thousand *V* particles per year.

A survey of the principal cloud-chamber equipment in the period 1950–7

Some of the cloud chambers and counter sets used by the major groups in the years following the Manchester and Pasadena work are displayed in a series of diagrams shown in Figure 4.13, reproduced with minor modifications from a fine summary by H. S. Bridge.[23]

Of the few cloud-chamber groups not included in Bridge's summary, mention should be made of the fine multiplate-chamber work of W. B. Fretter and his co-workers, at Berkeley and at altitude, and of the German group at

volumes are shown, and the disposition of the counters and absorbers forming the detection system is indicated. (1) Cloud chamber with magnetic field originally used by Rochester and Butler, and subsequently used at the Pic-du-Midi by other members of the Manchester group.[24] The chamber is cylindrical in shape. The counters are not shown. (2) Double-cloud-chamber arrangement with magnetic field used at Pasadena (220 m) by York and associates.[25] The chambers are rectangular and are 30 cm in width. (3) Chamber with magnetic field employed by Astbury, Newth, and others of the Manchester group at the Jungfraujoch (3,460 m).[26] The chamber is 50 cm in width. (4) Chamber with magnetic field employed by Thompson and associates at Indiana (230 m).[27] The chamber is 27 cm in width. (5) The multiplate chamber and counter set used by the MIT group at Echo Lake. (6) The double-chamber arrangement used by the Ecole Polytechnique group at the Pic-du-Midi.[29] (7) Multiple chamber used by the CIT group at Pasadena (220 m). All of the chambers operate in a magnetic field, and the experimental arrangement can be varied to include a multiplate chamber if desired (Leighton 1954).[30] The chambers are 55 cm in width. (8) The complex chamber and counter arrangement used by the Princeton group at Echo Lake. Note the change in scale between parts (1)–(4) and (5)–(8).

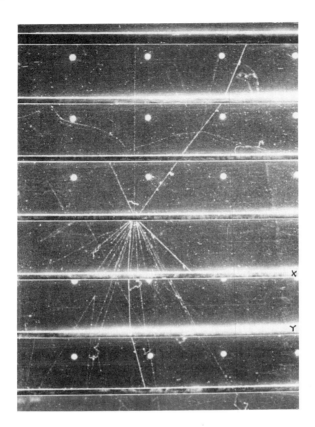

Figure 4.14. A Fretter multiplate cloud-chamber picture of a high-energy interaction in which a V^0 particle is created. The V^0 is between plates X and Y [*Source*: photo by W. B. Fretter, in Rochester and Wilson, *Cloud Chamber Photographs* ... (1952),[19] reprinted by permission; copyright © 1952 Pergamon Press plc.].

Göttingen led by Martin Deutschmann. An example of Fretter's work is shown in Figure 4.14. Other contributions will be noted later in the text.

Special attention will be paid to the work of three groups, the Thompson group at Bloomington, Indiana, the MIT group at Echo Lake, and the Ecole Polytechnique Pic-du-Midi group, because of their great historical significance. The Thompson group made an outstanding contribution to the V^0 problem, and the other groups to the nature of charged V particles.

The Indiana cloud chamber

The Indiana work stemmed from the construction and operation of the large magnet cloud chamber shown in section in Figure 4.13(4) and from the top in Figure 4.15, in which an unusually high order of measurement ac-

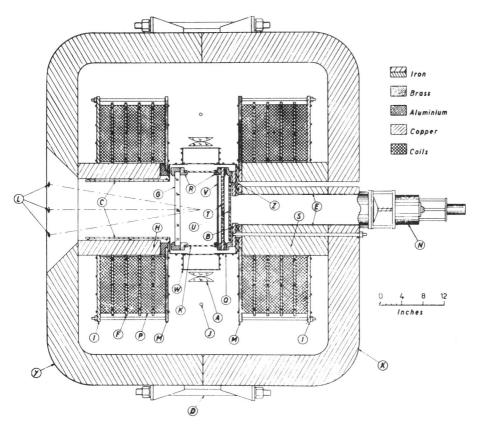

Figure 4.15. Top view of the large Indiana cloud chamber [*Source*: Thompson et al. (1956)[32]].

curacy was achieved. The details of the apparatus were given at the Bagnères conference in 1953 and in considerable detail in a later publication.[32] The chamber was inside a large iron yoke of annealed low-carbon steel (X and Y), with two pole pieces (H) around which were copper pancakes, water-cooled and arranged so that zero-field tests at full power could be carried out to measure chamber distortion. The field was very uniform and was accurately measured at many representative points. Expansion was by a flexible rubber diaphragm, as in the Manchester chambers, but a novel and an important feature was a space $\frac{11}{16}$ inch thick between the holey plate (V) and the plate stop (T) of the diaphragm. This undoubtedly contributed to the remarkably low level of distortion, as did the very effective cooling of the coils. No-field runs showed that the MDM was 5×10^{10} eV/c, the highest ever to be achieved in any conventional Wilson cloud chamber.[33] The impact of the Thompson work will be considered in the next paragraph.

Bagnères de Bigorre: a landmark conference

Many excellent cosmic-ray particle conferences were held in the period 1947–53, but by common consent the Bagnères de Bigorre conference, devoted entirely to the strange particles and attended by all the leading cosmic-ray particle physicists in the world, was unique in timing and scope.

By 1953, an immense amount of data had been obtained on the strange particles by the leading emulsion and cloud-chamber groups, and the situation was one of enormous fascination, but also of considerable confusion. On the emulsion side, as Perkins has described, there was clear evidence for a particle of mass about 970 m_e, the principal data coming from the τ-meson work. There were, however, other candidates of roughly the same mass decaying by other modes than the τ mode, and even an outsider of mass about 1,500 m_e. On the cloud-chamber side, there was also clear and accurate evidence from the Thompson group for a V^0 particle of mass about 970 m_e, and also for particles of mass roughly the same as the proton. The questions to be answered were what particles existed and whether or not there was one particle of mass 970 m_e decaying by different modes. It was here that the Indiana contributions, both on the observational and analytical sides, were decisive. The main development, an extension of the Manchester analysis, was outlined at the Rochester conference late in 1952 and then in 1953 in the *Physical Review*.[34]

The main impact at Bagnères was the presentation of the Indiana results for the V_1^0 and the V_2^0 decays on a p_t/α plot, a procedure justified by the experimental fact that nearly all V^0 decays seen in cloud chambers were relativistic (i.e., $\beta \sim 1$).[36] In Figure 4.16 this corresponds to the shaded area. The Indiana Bagnères curves for the V^0 decays are shown in modified form in Figure 4.17, and it is at once apparent that because of the high accuracy of the experimental data, the two curves for the main types of V^0 particle are clearly delineated. The data cluster closely about the theoretical ellipses corresponding to $\bar{\alpha} = 0$ for the V_2^0 particle and $\bar{\alpha} = 0.69$ for the V_1^0 particle for the decays given by (4.2) and (4.1). Two-body decays for mixtures of pions and muons for the V_2^0 particle are not excluded, but $V_2^0 \rightarrow \mu^+ + \mu^-$ is, as is decay to particles heavier than the light mesons. For decay to pions, and the experimental value of Q, the mass of the V_2^0 was found to be $971 \pm 10\ m_e$. Later work gave a value of $966 \pm 10\ m_e$. Note that three-body decays are excluded by this form of analysis. Two-body decay was also confirmed by the three tests noted earlier. Good evidence for the coplanarity test was presented by Bridge and Rossi.[37] The Q value of the V_1^0 decay was found to be 37 MeV, a value confirmed by the MIT group. As stated earlier, a more precise value of 36.92 MeV was found by G. Friedlander and associates in 1954 from emulsion observations. It will be observed that one point (no. 328) was completely off both the V_1^0 and V_2^0 curves, and it was assumed that this possibly represented a three- (or more) body decay of a V^0 particle. Many such cases were found by the Leighton group in Pasadena. A full discussion of so-called anomalous V^0 decays was given by Thompson in 1956.[35]

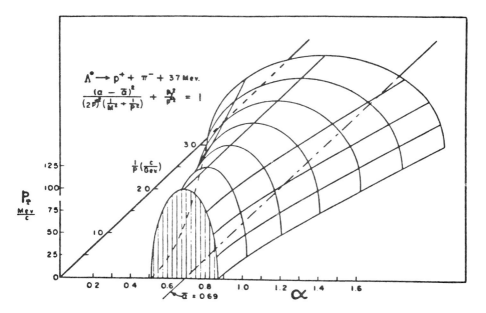

Figure 4.16. Isometric drawing of the Q surface for the V_1^0 (Λ^0) decay in $(\alpha, p_t, 1/P)$ space [*Source*: after Thompson (1956)[35]].

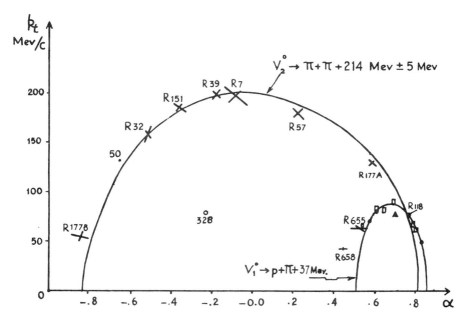

Figure 4.17. The Indiana plot of p_t versus α for the V_1^0 and V_2^0 particles as presented at the Bagnères conference.

The Bagnères conference was also noteworthy for the decision to rationalize the symbols used to describe the new particles. The new classification, later published by Edoardo Amaldi and associates,[38] was the following:

A. *Groups*

Light mesons (*L* mesons): π, μ, and so forth

Heavy mesons (*K* mesons): all particles heavier than pions and lighter than protons

Hyperons (*Y* particles): all particles with mass between the neutron and the deuteron

B. *"Christian Names"*

Capital letters for hyperons, and small Greek letters for mesons

1. *Hyperons*

Λ^0: particle previously known as V_1^0 and characterized by decay scheme $\Lambda^0 \to p + \pi^-$

(It was at this conference that Leprince-Ringuet suggested the name "hyperon." Now the town of Bagnères de Bigorre has *l'Avenue de l'Hyperon!*)

2. *Heavy Mesons*

$\tau \to 3\pi$

$\varkappa \to \mu + 2$ neutral particles

$\chi \to \pi + 1$ neutral particle

θ^0: particle known previously as V_2^0 characterized by $\theta^0 \to \pi^+ + \pi^-$; later it became $K_{\pi 2}^0$ and then K_s^0

On the theoretical side, the Bagnères conference was notable for initiating the famous θ–τ puzzle as a direct result of the identity of the masses of the θ^0 and τ^+ mesons, a problem that was not solved until the breakdown of parity in weak interactions was proved. On the debit side, the conference did nothing to resolve the problem of associated production, so confidently predicted for years by American and Japanese theoretical physicists.

A number of groups initiated measurements of the lifetimes of the V^0 particles, given later by G. D. James as 1.26 ($\pm_{0.28}^{0.25}$) $\times 10^{-10}$ sec for the θ^0 meson and by D. I. Page as 3.7×10^{-10} sec for the Λ^0 hyperon.[39,40] Most groups used the maximum-likelihood procedure of M. S. Bartlett of Manchester.[44]

Charged strange particles

Two types of charged strange particles were identified in the early days of *V*-particle work, namely, charged *K* mesons, of which the second *V* particle of the original two was probably the first example, and charged hyperons, the first example of which was found by R. Armenteros and associates at the Pic-du-Midi and published in 1952.[42] Both emulsion and cloud-chamber workers found that the charged *K* mesons were a complex group, and their successful identification required both techniques. By contrast, because of the characteristic signatures of their decays, charged hyperons

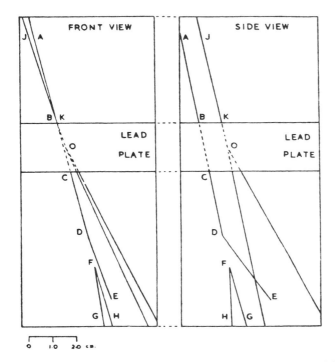

Figure 4.18. The two orthogonal projections reconstructed from the original stereoscopic photographs of the Manchester Ξ^- hyperon [*Source*: Armenteros et al. (1952)[42]].

were identified and classified on a few examples. For this reason, charged hyperons will be considered first.

Charged hyperons

Figure 4.18 shows the two orthogonal projections reconstructed from the original photographs of the Manchester charged hyperon. The simplest interpretation of the event was that it represented a cascade decay of the form

$$V^- \rightarrow V^0_{1 \text{ or } 2} + \text{negative meson}$$

The possibility of a chance coincidence was not fully eliminated, but the striking confirmation by the Pasadena groups, Leighton and associates[43] and Anderson and associates[44] in 1953, and the very fine photographs of Cowan[45] and of Fretter and W. B. Friesen[46] in 1954 made certain a decay of the form

$$\Xi^- \rightarrow \Lambda^0 + \pi^- + 65 \text{ MeV}$$

As Peyrou has remarked, everyone at Bagnères was convinced even by the small number then discovered, and no one seemed to be much bothered that they were all of negative sign! The most exact data came from Cowan and

Fretter and Friesen, for their events were obtained in conditions that allowed drop counts.

The first examples of another form of superproton, as they were often termed, the Σ hyperon, were found in a cloud-chamber photograph by the Pasadena group, Carl M. York and associates, in 1953 and by A. Bonetti and associates of the Milano and Geneva emulsion group.[47,48] The decay mode was

$$\Sigma^+ \rightarrow p^+ + \pi^0 + 110 \text{ MeV}$$

The alternative mode $\Sigma^+ \rightarrow n + \pi^+$ was also found in emulsions.

The superprotons, Σ^- and Σ^0, were found at the Cosmotron. Lifetimes were of the order of 10^{-10} sec. The hyperons, Λ^+ and Λ^-, were never found, in agreement with theory.

Charged K mesons

The Bristol τ meson, or $K_{\pi3}$, proved to be singularly elusive, for other examples were slow to appear. Indeed, two years elapsed before the next τ mesons were discovered. The next significant charged K-meson discoveries also came from the emulsion technique and were made at Bristol in 1951 by C. O'Ceallaigh, who found examples of two- and three-body decays of K mesons to muons and neutral particles, classified as $K_{\mu2}$ and $K_{\mu3}$. In 1952 and at the Bagnères conference, M. G. K. Menon and O'Ceallaigh suggested a third form, a two-body decay symbolized as $K_{\pi2}$. Because only a few examples had been found, many particle physicists were somewhat skeptical, and, indeed, the unraveling of the many modes of decay and the determination of their relative frequencies of occurrence proved to be very arduous. The Menon–O'Ceallaigh analysis proved to be correct, but, somewhat unexpectedly, the solution of the problem of the K-meson group came not from emulsions but from the cloud-chamber work of the MIT and Ecole Polytechique groups. In retrospect, this is now seen to be due to the fact that the secondaries of the commoner forms of charged K mesons were of such long range that they could not easily be identified and measured accurately in the small blocks of emulsion used at the time, a point made by Bonetti in his discussion of the results of the G-stack collaboration.[49]

The MIT work grew out of a long study by Rossi and his collaborators of cosmic-ray events in multiplate chambers. Reference has already been made to the work of this group on the Λ^0 hyperon, and in the course of their work, many examples were found of stopping particles, termed by them S particles. Many of the S particles decayed, emitting charged secondary particles that were clearly π or μ mesons. An example is reproduced in Figure 4.19. Small showers were also occasionally seen associated with these decays, indicating the emission of either γ rays or a π^0 meson. The absence of nuclear interactions indicated that at least some of the charged particles were muons.[50,51]

The Ecole Polytechnique two-chamber setup (Figure 4.13) was first ad-

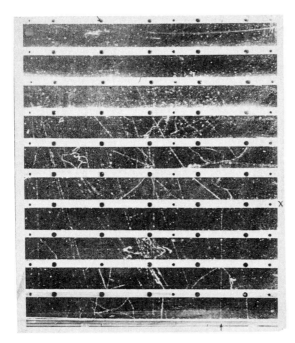

Figure 4.19. An MIT example of an S particle that stops in plate X and emits a penetrating secondary particle [*Source:* Annis et al. (1952)[50]].

vocated by Peyrou in 1949 and then later in 1950 by Peyrou and B. Gregory, when Leprince-Ringuet agreed to provide the necessary resources for building and installation on the Pic-du-Midi. During the most fruitful period of operation, the multiplate chamber (the lower chamber) had fifteen 1-cm copper plates, equivalent to a total stopping power of 133 g cm^{-2}, sufficient to stop the most energetic secondary expected. The setup was triggered by a ps set, and many K mesons were seen to stop and decay. The first few results were presented at the Bagnères conference, and all the measured masses were consistent with the mass of the τ meson.

As results accumulated and the Ecole and the MIT results were examined more carefully, it became possible to distinguish certain well-marked features of the charged secondaries of the decaying K mesons. The Ecole results are illustrated in Figure 4.20. In these diagrams, the ranges of the particles are plotted against the event number, whether the particle stops or not, the stopping region for particles that stopped being indicated by heavy black lines. The emission of γ rays or showers is also indicated, and, where known (as in many of the Ecole cases), the measured mass of the primary. For the longest-range secondaries, the results of both the MIT and the Ecole experiments are combined in Figure 4.21.

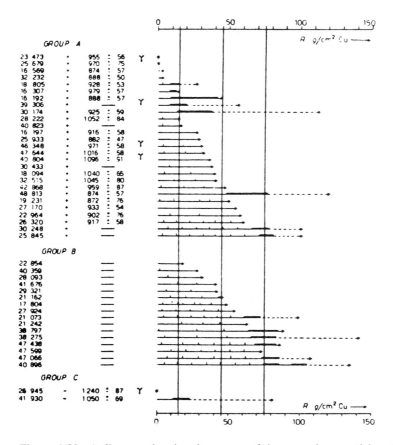

Figure 4.20. A diagram showing the ranges of the secondary particles of the *K* mesons observed by the Ecole Polytechnique team at the Pic-du-Midi [*Source*: Armenteros et al. (1955)[52]]. The solid arrows indicate the amounts of matter crossed before the secondaries left the bottom chamber. The thick lines show the range limits for secondaries that stopped. A γ indicates a γ ray correlated with the *K* decay. There are eight $K_{\mu2}$, one $K_{\pi2}$ and five $K_{\pi3}$.

The first and most striking feature of both experiments is that the stopping secondaries fall into three well-separated and distinct groups, the first around a range of 15 g cm^{-2} of copper, the second at 45 g cm^{-2}, and the third at about 75 g cm^{-2}. These were identified as the charged secondaries of τ ($K_{\pi3}$), χ ($K_{\pi2}$), and a new one, $K_{\mu2}$. From the experimental behavior (a secondary of unique range!) and calculation, the long-range particles were identified as muons, and the decay as two-body. The mass of the primary was known to be less than 1,000 m_e, and no pions of the observed range could be emitted by a primary particle of such a mass. Again, observation and calculation suggested that the recoil particle was very light, presumably a neutrino. All the evidence, then, was completely consistent with a decay of the following type:

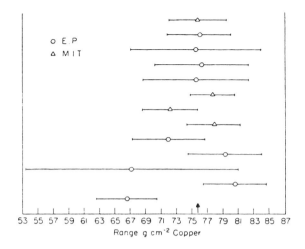

Figure 4.21 The distribution of range limits for stopped secondaries of the longest range from K mesons observed by the Ecole Polytechnique and MIT groups [*Source*: Bridge (1956)[23]].

$$K^+_{\mu 2} \rightarrow \mu^+ + \nu$$

It is noteworthy that all the observed $K_{\mu 2}$ particles were positive. Fuller accounts of this important work have been given by Gregory and associates,[29] Armenteros and associates,[52] Bridge and associates,[53] Bridge,[23] and Peyrou.[1]

That the χ (or $K_{\pi 2}$) decay was to $\pi^+ + \pi^0$ rather than $\pi^+ + \gamma$ was proved by the fact that the observed showers were not, in general, in a direct line with the π^+ particle. The form of the decay was established by a most remarkable photograph obtained by the Princeton group at Echo Lake, A. L. Hodson and associates,[31] and shown in Figure 4.22, in which the π^0 decays into two Dalitz pairs, namely,

$$K^+_{\pi 2} \rightarrow \pi^+, \qquad \pi^0 \rightarrow (e^+ + e^- + e^+ + e^-)$$

Associated production

As is well known, experimental evidence for the associated production of strange particles was first found at the Brookhaven Cosmotron in 1953 using artificially accelerated negative pions and hydrogen as target.[54] Examples were found in cosmic rays from 1954 onward, and all observations proved to be in good accord with the strangeness theory of Murray Gell-Mann and Kazuhiko Nishijima. This theory also accounted for many puzzling facts found by cosmic-ray workers, such as the strong positive excess of K mesons and the absence of such hyperons as Λ^+, Λ^-, and Ξ^+.

It is pertinent to inquire why associated production was not seen earlier. The reason is most likely the extensive use of materials of high Z, mostly lead, and the selection of events of high multiplicity, leading to a considerable confusion of particles. This point was strongly made by G. D. James and

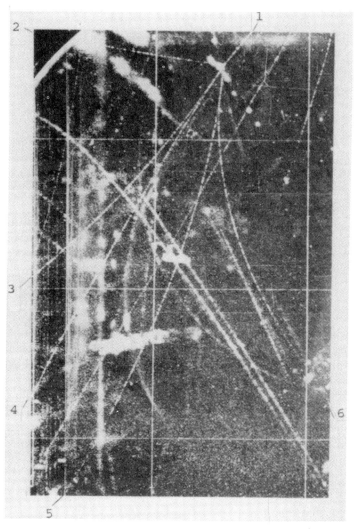

Figure 4.22. The complete $K_{\pi 2}$ decay: $K_{\pi 2} \rightarrow \pi^+$, $\pi^0 \rightarrow (e^+ + e^- + e^+ + e^-)$. In the figure, the $K_{\mu 2}$ is marked 1, the π^+ is marked 2, and the electrons are 3, 4, 5, and 6 [*Source*: Hodson et al. (1954)[31]].

R. A. Salmeron in 1955, coupled with the suggestion that production in light materials should be looked for.[55] The idea was tested in the Manchester Jungfraujoch chamber by looking for *V*-particle production in a plate across the chamber made of either copper or graphite. In some 55,000 photographs of penetrating showers, ten examples of associated production of *K* mesons were found, including the first example of the production of two neutral kaons of opposite strangeness.[56] Of the many other examples found by other

Figure 4.23. An example of associated production [*Source*: Thompson et al. (1954)[58]]. Tracks 1 and 2 are the secondaries from the decay of the $K_{\pi2}^0$ meson, and 3 and 4 are from the decay of the Λ^0 hyperon. Track 5 is spatially unrelated. The event is fully consistent with the reaction $\pi^- + p \rightarrow K^0 + \Lambda^0$.

cosmic-ray workers, mention might be made of the observation in 1955 of the production of $\Xi^- + K^0 + K^0$ by J. D. Sorrels, Leighton, and Anderson.[57]

In many of the cosmic-ray examples, however, convincing proof was lacking that the particles were those assumed. But there were some well-proven cases, such as the event reproduced in Figure 4.23 and published by Thompson and associates in 1954.[58] In this case, the product particles were unambiguously identified as Λ^0 and K^0 particles, first, by the clear identification of tracks 1 and 2 as arising from the K^0 meson and 3 and 4 from the Λ^0 hyperon by the determination of P_t and α and the positions of the values for the two particles on the well-established Q-curve plots (see Figures 4.16 and 4.17), and, second, by the Q values, which for the two particles were $Q(\pi, \pi) = 223 \pm 10$ MeV and $Q(p, \pi) = 37 \pm 4$ MeV, very close to the accepted values. Every measurement was then fully consistent with the production reaction

$$\pi^- + p \rightarrow K^0 + \Lambda^0$$

Notes

1 C. H. Peyrou, "The Role of Cosmic Rays in the Development of Particle Physics," in Colloque International sur l'Histoire de la Physique des Particules, *Phys. (Paris) (Suppl.)* *43:12* (1982), 7–67; G. D. Rochester, "Observations on the Discovery of the Strange Particles," ibid., 169–76.

2 George D. Rochester, *Early History of Cosmic Ray Studies: Personal Reminiscences with Old Photographs*, edited by Y. Sekido and H. Elliot (Dordrecht: Reidel, 1985), pp. 299–321.

3 L. Jánossy and P. Ingleby, "Penetrating Cosmic Ray Showers," *Nature (London) 145* (1940), 511; L. Jánossy, "Penetrating Cosmic-Ray Showers," *Proc. R. Soc. London, Ser. A 179* (1942), 361–76.

4 Carl D. Anderson, with Herbert L. Anderson, "Unraveling the Particle Content of Cosmic Rays," in *The Birth of Particle Physics*, edited by Laurie M. Brown and Lillian Hoddeson (Cambridge University Press, 1983), pp. 131–54.

5 P. M. S. Blackett, "*V*-Particles and the Cloud Chamber," *Nuovo Cimento (Suppl.) 11* (1954), 264–89.

6 M. W. Friedlander, D. Keefe, M. G. K. Menon, and M. Merlin, "On the Mass of the Λ^0-Particle," *Philos. Mag. 45* (1954), 433–42.

7 G. D. Rochester and C. C. Butler, "The Penetrating Particles in Cosmic-Ray Showers: I. Heavily-Ionizing Particles," *Proc. Phys. Soc. London 61* (1948), 307–12.

8 G. D. Rochester and C. C. Butler, "The Penetrating Particles in Cosmic-Ray Showers: II. The Lightly-Ionizing Penetrating Particles in Penetrating Showers," *Proc. Phys. Soc. London 61* (1948), 535–41.

9 G. D. Rochester, "The Penetrating Particles in Cosmic-Ray Showers," *Rev. Mod. Phys. 21* (1949), 20–6.

10 G. D. Rochester, "Observations on the Ionizing Particles in Cosmic Ray Showers," in *Cosmic Radiation, The Colston Papers* (London: Butterworth, 1949), pp. 111–15.

11 C. C. Butler, W. G. V. Rosser, and K. H. Barker, "Some Properties of Penetrating Cosmic-Ray Showers and Star Phenomena Seen in the Cloud Chamber," *Proc. Phys. Soc. London, Sect. A 63* (1950), 145–64.

12 K. H. Barker and C. C. Butler, "The Nuclear Interaction Length of the Particles in Penetrating Cosmic-Ray Showers," *Proc. Phys. Soc. London, Sect. A 64* (1951), 4–9.

13 G. D. Rochester and C. C. Butler, "Evidence for the Existence of New Unstable Elementary Particles," *Nature (London) 160* (1947), 855–7.

14 L. Leprince-Ringuet, "Photographic Evidence for the Existence of a Very Heavy Meson," *Rev. Mod. Phys. 21* (1949), 42–3.

15 R. Brown, U. Camerini, P. H. Fowler, H. Muirhead, C. F. Powell, and D. M. Ritson, "Observations with Electron-Sensitive Plates Exposed to Cosmic Radiation. II. Further Evidence for the Existence of Unstable Charged Particles of Mass $\sim$1000 m_e and Observations on Their Mode of Decay," *Nature (London) 163* (1949), 82–7.

16 C. C. Butler, "Early Cloud Chamber Experiments at the Pic-du-Midi," in Colloque International sur l'Histoire de la Physique des Particules, *J. Phys. (Paris) (Suppl.) 43:12* (1982), 177–84.

17 C. C. Dilworth, S. J. Goldsack, Y. Goldschmidt-Clermont, and F. Levy, "The Magnetic Deflection of Fast Charged Particles in the Photographic Emulsion," *Philos. Mag. 41* (1950), 1032–57.

18 A. J. Seriff, R. B. Leighton, C. Hsiao, E. W. Cowan, and C. D. Anderson, "Cloud-Chamber Observations of the New Unstable Cosmic-Ray Particles," *Phys. Rev. 78* (1950), 290–1.

19 G. D. Rochester and J. G. Wilson, *Cloud Chamber Photographs of the Cosmic Radiation* (London: Pergamon Press, 1952), pp. 104–7.

20 R. Armenteros, K. H. Barker, C. C. Butler, A. Cachon, and A. H. Chapman, "Decay of V-Particles," *Nature (London) 167* (1951), 501–3.

21 R. Armenteros, K. H. Barker, C. C. Butler, and A. Cachon, "The Properties of Neutral V-Particles," *Philos. Mag. 42* (1951), 1113–35.

22 R. W. Thompson, H. O. Cohn, and R. S. Flum, "Cloud Chamber Observations of the Neutral V-Particle Disintegration," *Phys. Rev. 83* (1951), 175.

23 H. S. Bridge, "Experimental Results on Charged K-Mesons and Hyperons," in *Progress in Cosmic Ray Physics* (Amsterdam: North Holland, 1956), pp. 3, 143–252.

24 G. D. Rochester and C. C. Butler, "The New Unstable Cosmic-Ray Particles," *Rep. Prog. Phys. 16* (1953), 364–407.

25 C. M. York, R. B. Leighton, and E. K. Bjornerud, "Cloud-Chamber Study of Charged V Particles," *Phys. Rev. 95* (1954), 159–70.

26 J. S. Buchanan, W. A. Cooper, D. D. Millar, and J. A. Newth, "Measurements of Forty-Four Charged V-Events," *Philos. Mag. 45* (1954), 1025–42.

27 Y. B. Kim, J. R. Burwell, R. W. Huggett, and R. W. Thompson, "Magnetic Cloud Chamber Study of $V^\pm$ Events," *Phys. Rev. 96* (1954), 229–31.

28 H. S. Bridge, H. Courant, B. Dayton, H. C. De Staebler, B. Rossi, R. Stafford, and D. Willard, "Recent Results on S-Particles," *Nuovo Cimento 12* (1954), 81–9.

29 B. Gregory, A. Lagarrigue, L. Leprince-Ringuet, F. Müller, and C. Peyrou, "Étude des mesons K chargés, au moyen de deux chambres de Wilson super-posées," *Nuovo Cimento 11* (1954), 292–309.

30 R. B. Leighton, "Loeb Lectures" (1954).

31 A. L. Hodson, J. Ballam, W. H. Arnold, D. R. Harris, R. R. Rau, G. T. Reynolds, and S. B. Treiman, "Cloud-chamber Evidence for a Charged Counterpart of the θ^0 Particle," *Phys. Rev. 96* (1954), 1089–95.

32 R. W. Thompson, J. R. Burwell, and R. W. Huggett, "The θ^0-Meson," *Nuovo Cimento (Suppl.) (Ser. 10) 4* (1956), 286–318.

33 R. W. Thompson, A. V. Buskirk, L. R. Etter, C. J. Karzmark, and R. H. Rediker, "An Unusual Example of V^0 Decay," *Phys. Rev. 90* (1953), 1122.

34 R. W. Thompson, A. V. Buskirk, L. R. Etter, C. J. Karzmark, and R. H. Rediker, "The Disintegration of V^0 Particles," *Phys. Rev. 90* (1953), 329. See Robert W. Thompson, "On the Discovery of the Neutral Kaons," in *The Birth of Particle Physics*, edited by Laurie M. Brown and Lillian Hoddeson (Cambridge University Press, 1983), pp. 251–60.

35 R. W. Thompson, "Decay Processes of Heavy Unstable Neutral Particles," in *Progress in Cosmic Ray Physics* (Amsterdam: North Holland, 1956), pp. 255–337.

36 R. W. Thompson, A. V. Buskirk, H. O. Cohn, C. J. Karzmark, and R. H. Rediker, "The Disintegration Schemes of Neutral V Particles," in *Proc. Conf. Int. Ray Cosmique* (1953), Bagnères de Bigorre (University of Toulouse), pp. 30–5.

37 H. S. Bridge and B. Rossi, "Observations on the New Unstable Particles with a Multiplate

Cloud Chamber," in *Proc. Conf. Int. Ray Cosmique* (1953), Bagnères de Bigorre (University of Toulouse), pp. 21–6.

38 E. Amaldi, C. D. Anderson, P. M. S. Blackett, W. B. Fretter, L. Leprince-Ringuet, B. Peters, C. F. Powell, G. D. Rochester, B. Rossi, and R. W. Thompson, "Symbols for Fundamental Particles," *Nature (London) 173* (1954), 123.

39 G. D. James, "Some Notes on the Production of *V*-Particles," *Nuovo Cimento (Suppl.) (Ser. 10) 4* (1956), 325–32.

40 D. I. Page, "A New Estimate of the Lifetime of Λ^0-Particles," *Philos. Mag. 45* (1954), 863–8.

41 M. S. Bartlett, "On the Statistical Estimation of Mean Lifetimes," *Philos. Mag. 44* (1953), 249–62; "Estimation of Mean Lifetimes from Multiplate Cloud Chamber Tracks," *Philos. Mag. 44* (1953), 1407–8.

42 R. Armenteros, K. H. Barker, C. C. Butler, A. Cachon, and C. M. York, "The Properties of Charged *V*-Particles," *Philos. Mag. 43* (1952), 597–612.

43 R. B. Leighton, E. W. Cowan, and V. A. J. van Lint, in "Recent Measurements on Charged *V*-Particles and Heavy Mesons in Pasadena," in *Proc. Conf. Int. Ray Cosmique* (1953), Bagnères de Bigorre (University of Toulouse), pp. 97–101.

44 C. D. Anderson, E. W. Cowan, R. B. Leighton, and V. A. J. van Lint, "Cascade Decay of *V* Particles," *Phys. Rev. 92* (1953), 1089.

45 E. W. Cowan, "A *V*-Decay Event with a Heavy Negative Secondary, and Identification of the Secondary *V*-Decay Event in a Cascade," *Phys. Rev. 94* (1954), 161–6.

46 W. B. Fretter and E. W. Friesen, "Cascade Decay of a Negative Hyperon," *Bull. Am. Phys. Soc. 29:6* (1954), 18.

47 C. M. York, R. B. Leighton, and E. K. Bjornerud, "Direct Experimental Evidence for the Existence of a Heavy Positive *V*-Particle," *Phys. Rev. 90* (1953), 167.

48 A. Bonetti, R. Levi Setti, M. Panetti, and G. Tomasini, "Observation of the Decay at Rest of a Heavy Particle," *Nuovo Cimento 10* (1953), 345–7; "On the Existence of Unstable Charged Particles of Hyperprotonic Mass," ibid., 1736–43.

49 A. Bonetti, "On the Composition of the *K*-Particle Decay Spectrum," *Nuovo Cimento (Suppl.) (Ser. 10) 4* (1956), 419–24.

50 M. Annis, H. S. Bridge, H. Courant, S. Albert, and B. Rossi, "*S*-Particles," *Nuovo Cimento 9* (1952), 624–7.

51 H. S. Bridge, C. Peyrou, B. Rossi, and R. Stafford, "Cloud-Chamber Observations of the Heavy Charged Unstable Particles in Cosmic Rays," *Phys. Rev. 90* (1953), 921–23.

52 R. Armenteros, B. Gregory, A. Hendel, A. Lagarrique, L. Leprince-Ringuet, F. Muller, and C. Peyrou, "Further Discussion of the K_μ Decay Mode," *Nuovo Cimento (Suppl.) (Ser. 10) 1* (1955), 915–41.

53 H. S. Bridge, H. DeStaebler, Jr., B. Rossi, and B. V. Sreekantan, "Evidences for Heavy Mesons with the Decay Processes $K_{\pi2} \rightarrow \pi + \pi^0$ and $K_{\mu2} \rightarrow \text{U} + \gamma$ from Observations with a Multiple Cloud Chamber," *Nuovo Cimento (Suppl.) (Ser. 10) 1* (1955), 874–87.

54 W. B. Fowler, R. P. Shutt, A. M. Thorndike, and W. L. Whittemore, "Production of V_1^0 Particles by Negative Pions in Hydrogen," *Phys. Rev. 91* (1953), 1287.

55 G. D. James and R. A. Salmeron, "The Production of Hyperons and Heavy Mesons," *Philos. Mag. 46* (1955), 571–86.

56 W. A. Cooper, H. Filthuth, J. A. Newth, G. Petrucci, R. A. Salmeron, and A. Zichichi, "Examples of the Production of (K^0, K^0) and (K^+, K^0) Pairs of Heavy Mesons," *Nuovo Cimento (Suppl.) (Ser. 10) 5* (1957), 1388–97.

57 J. D. Sorrels, R. B. Leighton, and C. D. Anderson, "Associated Production of Ξ^- with Two θ^0 Particles," *Phys. Rev. 100* (1955), 1457–9.

58 R. W. Thompson, J. R. Burwell, R. W. Huggett, and C. J. Karzmark, "Evidence for Double Production of V^0 Particles," *Phys. Rev. 95* (1954), 1576–9.

5   Cosmic-ray work with emulsions in the 1940s and 1950s

DONALD H. PERKINS

Born 1925, Hull, England; Ph.D., University of London, 1948; experimental high-energy physics; University of Oxford

The photographic method

The use of photographic plates to record ionizing radiations dates back to Henri Becquerel, who in 1896 discovered radioactivity from the blackening of plates by uranium salts.[1] In 1910–11, M. Kinoshita showed that it was possible to record individual tracks due to α particles at both verticle and tangential angles to the emulsion plane.[2] During the 1920s and 1930s, experiments by M. Blau and H. Wambacher in Germany, G. Zhdanov in the USSR, H. J. Taylor in England, and R. Wilkins and H. Rumbaugh and A. Locher in the United States recorded particles both from cosmic rays and from accelerators, on occasion achieving sensitivity to protons by use of organic sensitizing dyes (e.g., pinakryptol yellow).[3–7] The general nonreproducibility of the results, however, placed the technique under something of a cloud. Indeed, in 1935, Taylor, at Cambridge, using Ilford R1 and R2 emulsions, concluded that it was "impossible to deduce with any accuracy the energy of individual particles from range in emulsion." Similar remarks were voiced by M. Stanley Livingston and Hans Bethe in 1937.[8]

This situation was transformed for two reasons. First, Cecil F. Powell and F. Fertel in 1943 showed that the energy resolution in the study of (d, p) reactions in boron, using the proton range in emulsion, was at least as good as that obtained with counters and absorbers (Figure 5.1); the method *could* be made quantitative.[9] Second, in 1945 the Ministry of Supply in London (in connection with the nuclear program) set up a panel under Joseph Rotblat, including chemists from the photographic firms of Ilford Ltd. and Kodak Ltd., with the specific task of stimulating the production of thick "nuclear research emulsions" of high sensitivity. This was achieved by increasing the halide/gelatin ratio from 1 : 6 to 1 : 1 and the use of sensitizers.

By mid-1946, the Ilford B and C series had been produced in thicknesses up to 200 μm, sensitive to charged particles of ionization $I \geq 6I_0$, where I_0 is the

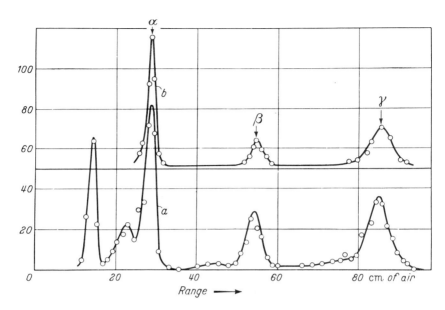

Figure 5.1. Results in the range spectrum of protons emitted at 90° from the (*d, p*) reaction on boron. The upper curve is that deduced from measurements by Cockroft and Lewis with counters and absorbers; the lower is the range distribution of protons in Ilford "half-tone" emulsions, by Powell, Champion, and Fertel in 1943 [*Source*: Powell et al., *Study of Elementary Particles* ... (1959),[41] reprinted by permission; copyright © 1959 Pergamon Press plc.].

minimum ionization. (The letters A, B, C, . . . , K referred to decreasing grain size, and the numbers 0–5 referred to increasing sensitivity.) In 1948, Kodak produced the NT4 emulsion, and Ilford the G5 emulsion, both sensitive to minimum ionization, I_0. Subsequently, emulsions with similar properties were produced commercially by Nikfi in the USSR and Fuji in Japan, but the sensitivity and other characteristics of nuclear research emulsions have remained essentially unchanged since 1948.

Discovery of the pion
The main triumph of the emulsion method was the discovery of the pion in 1947. We recall that in 1935, Hideki Yukawa had postulated a quantum of the nuclear field with a Compton wavelength equal to the nuclear-force range, R_0, and hence a mass $mc^2 = \hbar c/R_0 \sim 150$ MeV.[10] Sin-itiro Tomonaga and Gentaro Araki had pointed out that if such a quantum had negative charge, it would be brought to rest by ionization loss and then be captured in an atomic Bohr orbit, from which it would suffer rapid nuclear absorption, leading to annihilation and disintegration of the nucleus.[11] The famous experiment of Marcello Conversi, Ettore Pancini, and Oreste Piccioni was

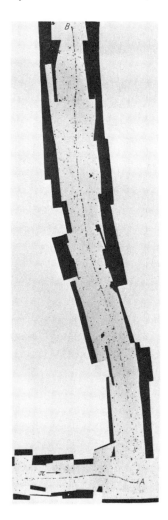

Figure 5.2. The first complete $\pi \to \mu$ decay event, recorded in C2 emulsion at Pic-du-Midi [*Source*: Reprinted by permission from Lattes et al., *Nature (London) 159* (1947),[16] 1127–8; copyright © 1947 Macmillan Magazines Ltd.].

carried out in Rome toward the end of World War II. With the aid of an electromagnet, they observed separately the fates of positive and negative mesons, that is, particles of intermediate mass (between electron and proton), stopping in carbon and iron blocks.[12] Surprisingly, in carbon, virtually all the negative mesons decayed. In 1947, Robert Marshak and Bethe postulated the hypothesis that there must be *two* mesons: The parent, strongly interacting meson was to be identified with the Yukawa force quantum, and it decayed to a weakly interacting daughter meson – what had been observed by Conversi and associates and their predecessors in counter and cloud-chamber experiments.[13] The final, vital piece of the experimental jigsaw was the discovery in 1947 of $\pi \to \mu$ decay, at Bristol, by C. M. G. Lattes, H. Muirhead, G. P. S. Occhialini, and Powell (Figures 5.2 and 5.3), and the so-called σ mesons,

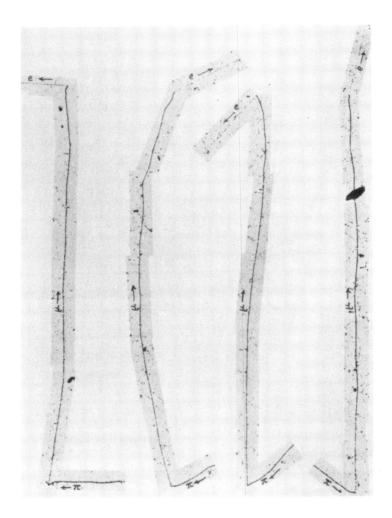

Figure 5.3. Four $\pi \to \mu \to e$ decays recorded at Bristol in Kodak NT4 emulsion in 1948 [*Source*: Powell et al., *Study of Elementary Particles* ... (1959),[41] reprinted by permission; copyright © 1959 Pergamon Press plc.]. The constancy of range of the muon is well illustrated.

giving rise to nuclear interactions at the end of the range, by Perkins in London and Occhialini and Powell in Bristol (Figure 5.4)[14,15] The pions (the positives giving the $\pi^+ \to \mu^+$ decay and the negatives, π^-, the σ stars) were indeed the long-sought Yukawa quanta, and the muons the weakly interacting daughter mesons, forming the hard or penetrating component of cosmic rays at sea level.

Although these developments all occurred in 1947, the logical sequence (if there was one) was not really as I have indicated. Table 5.1 shows the chrono-

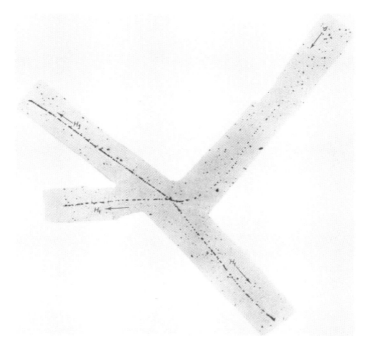

Figure 5.4. First σ star (nucleon disintegration following capture of a π^-) observed in Ilford B1 emulsion exposed in an aircraft [*Source*: Reprinted by permission from Perkins, *Nature (London) 159* (1947),[14] 126–7; copyright © 1947 Macmillan Magazines Ltd.].

Table 5.1. *1947*

Date	Authors	Publication	Subject
Jan. 25	Perkins[36]	*Nature 159*, 126	First "σ star" (π^-); $m = 100-300\, m_e$
Feb. 1	Conversi et al.[12]	*Phys. Rev. 71*, 209	Negative mesons; decay in carbon (π^-)
Feb. 8	Occhialini and Powell[15]	*Nature 159*, 93	Six σ stars; $m = 350 \pm 100\, m_e$
May 24	Lattes et al.[16]	*Nature 159*, 694	Two $\pi-\mu$ decays
Sept. 15	Marshak and Bethe[13]	*Phys. Rev. 72*, 506	Two-meson hypothesis
Sept. 15	Weisskopf[17]	*Phys. Rev. 72*, 510	High rate of production of mesons, to be reconciled with W.I.
Oct. 4	Lattes et al.[18]	*Nature 160*, 453	644 mesons; 105 σ stars ($40\ \pi-\mu$); 11 complete $\pi \rightarrow \mu$; 499 ϱ mesons ($\mu^\pm$)
Dec. 20	Rochester and Butler[28]	*Nature 160*, 855	V particles

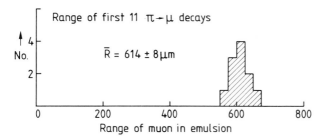

Figure 5.5. Range distribution of muons for the first eleven complete $\pi \to \mu$ decays observed by Bristol group in C2 emulsion in 1947 [*Source*: based on Lattes et al. (1947)[18]].

logical order of published papers in 1947. First was the publication of a single σ star (Figure 5.4) in B1 emulsion, showing that some mesons did induce nuclear disintegrations, to be confirmed two weeks later with six events, by Occhialini and Powell. The paper of Conversi and associates appeared in the *Physical Review* between the dates of these two *Nature* publications. I had heard, third-hand, of the Conversi result before it was published, and I realized that there was a discrepancy between the result of that experiment and the emulsion σ event (never mind the Yukawa theory). The observation of the first two $\pi-\mu$ decays, in May, was a great and exciting surprise, but not (to me at least) an obvious solution to the difficulties.[16] The secondary μ particle was presumably positive and not obviously related to the negative mesons of Conversi stopping in carbon. The Marshak and Bethe paper appeared in September, next to one by Victor Weisskopf, pointing out the difficulties and a solution in terms of two generically related mesons.[17]

Properties of pions and muons

The experimental situation on the emulsion data did not really become clear until the Bristol paper of October 1947, describing the observation of 644 mesons, of which 105 were σ events (giving nuclear stars), 40 were $\pi-\mu$ decays (11 of which had a complete muon track in emulsion), and the remaining 499 were ϱ mesons [which came to rest and did nothing, i.e., gave no secondary of $I \geq 6I_0$].[18]

First, the range distribution of the secondary muon in the 11 complete $\pi-\mu$ decays is given in Figure 5.5 and is clear proof of two-body decay, $\pi \to \mu^+$ neutral. It is important to state that meson *decay* was not the only possible interpretation, and F. C. Frank at Bristol considered that it could be due to a process of nuclear capture, with reemission of the same meson, and an energy release of 4 MeV. Powell decided that it must be a decay process, and he was right (helped in the knowledge that mass estimates from grain counting in the tracks gave $m_\pi > 1.3m_\mu$). Yet, not much later, in 1957, Luis W. Alvarez was to discover exactly such a type of reaction in the form of muon-catalyzed HD fusion in a hydrogen bubble chamber, where a muon is absorbed and reemitted (with energy of a few MeV) many times (up to 170 times, in the latest experiments in DT mixtures).[19]

Second, from the observed numbers of $\pi-\mu$ decays and σ events, and the geometric efficiencies, the true π/σ ratio was found to be ~1, within 20 percent, as might be expected if π and σ really correspond to π^+ and π^-. The more numerous ϱ mesons were thought to be due to incoming μ^+ and μ^- mesons from π decay in the atmosphere. The existence of σ stars and the observation (Figure 5.6) of pions produced in nuclear disintegrations were clear proofs that the pions had strong interactions, candidates for the long-sought Yukawa quantum.

Positive identification of the charges of the various mesons in cosmic rays had to await the beautiful and conclusive experiments of I. Barbour, in 1949, at Chicago.[20] C2 emulsions were flown in a balloon between the poles of a permanent magnet of 13 kG, and the sign of charge and momenta were determined from the magnetic deflections in the thin air gap between two facing emulsions (Figure 5.7). The particles in $\pi-\mu$ decay were shown to be positive (π^+, μ^+), the σ particles negative (π^-), and the ϱ's equally positive and negative $(\mu^\pm)$. Similar results were obtained in mountain exposures by C. Franzinetti (in 1950).[21] Of course, the charges of the μ, σ, and ϱ mesons were found also in the accelerator experiments at Berkeley.[22]

The mass values of the mesons observed in the cosmic-ray work with emulsions up to 1950 are summarized in Table 5.2, together with the results using accelerators. The three main methods relied on variations of grain density with range (g, R), Coulomb scattering with range (α, R), and momentum (from magnetic deflection) with range (p, R). The earliest measurements by the (g, R) method gave a wrong result for m_π/m_μ, but by and large the early cosmic-ray and accelerator values are in good accord with modern data.

The pion lifetime, τ_π, was first measured reliably by J. R. Richardson at the Berkeley cyclotron in 1948.[23] Attempts in cosmic rays were at least heroic.[24] They exposed emulsions in cocoa tins tied to a vertical pole stuck in the Aletsch glacier, Jungfraujoch. By comparing the rate of locally produced pions with that of pions entering the emulsions from outside, the lifetime could in principle be determined. The result (60 $\pm$ 30 nsec) was too low by a factor of 4. The identity of the neutral particle in the decay $\pi \rightarrow \mu +$ neutral was investigated by C. O'Ceallaigh.[25] He searched back along a cone opposite to the muon direction in 250 decays, but found no e^+e^- pairs in a total length that, had the decay been of the form $\pi \rightarrow \mu + \gamma$, should have given five pairs.

Finally, I want to mention cosmic-ray emulsion experiments on the neutral pion. The discovery of the π^0 was actually made in 1950 by R. Bjorklund and associates and Jack Steinberger at the Berkeley machine,[26] but a Bristol experiment by G. Carlson, J. Hooper, and D. T. King was contemporaneous.[27] Neutral pions were observed via the conversion of γ rays (from $\pi^0 \rightarrow 2\gamma$)

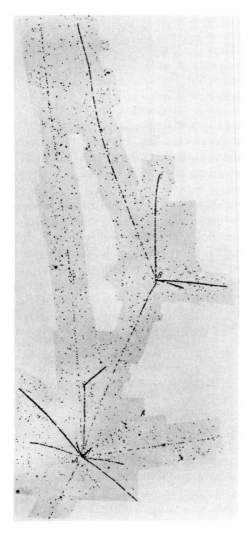

Figure 5.6. Production of a σ meson, and its subsequent nuclear capture, in a cosmic-ray disintegration (C2 emulsion, Bristol, 1947) [*Source*: Powell et al., *Study of Elementary Particles* ... (1959),[41] reprinted by permission; copyright © 1959 Pergamon Press plc.].

to e^+e^- pairs, pointing toward the origin of nuclear interactions (Figure 5.8). The energy spectrum of these pairs is shown in Figure 5.9. Suppose that E_1 and E_2 are two energies that correspond to equal intensities on either side of the maximum. Then it is a remarkable property (independent of the pion spectrum) that $(E_1 E_2)^{1/2} = m_{\pi^0}/2$. The value of m_{π^0} obtained was unfortunately too high (about 290 m_e) because the wrong value was used for the

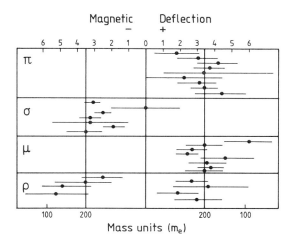

Figure 5.7. Results of Barbour on magnetic deflection of cosmic-ray mesons in C2 emulsion, showing that π's and μ's from π–μ decay are positive, σ mesons (π^-) are negative, and ϱ mesons ($\mu^\pm$) are equally positive and negative [*Source*: Barbour (1949)[20]].

Table 5.2. *Masses in MeV/c^2*

Sources	Year		m_π	m_μ	m_π/m_μ
Cosmic rays					
Lattes et al.[16]	1947	(g, R)			1.65 ± 0.15
Goldschmidt et al.[37]	1948	(a, R)	139 ± 6	103 ± 4	1.35 ± 0.10
Lattimore	1948	(a, R)	148 ± 40	—	—
Brown et al.[29]	1949	(g, R)	148 ± 10	—	—
Barbour[20] (balloon)	1949	(p, R)	138 ± 12	112 ± 13	1.23 ± 0.20
Franzinetti[21] (Jungfraujoch)	1950	(p, R)	144 ± 4	111 ± 2	1.30 ± 0.05
Camerini et al.[24]	1948	(a, g)	145 ± 4	—	—
Accelerator					
Gardner and Lattes[22]	1948	(p, R)	160 ± 8	—	1.32 ± 0.01
Bowker[38]	1950	(g, R)	135 ± 12	—	—
		(g, R)	143 ± 7	—	—
Van Rossum	1950	(g, R)	159	103	—
Bradner et al.[39]	1950	(p, R)	143 ± 3	—	—
		(p, R)	142 ± 4	108 ± 3	—
Bishop et al.[40]	1949	(p, R)	141 ± 3	—	—
Present values			139.567	105.6594	1.321

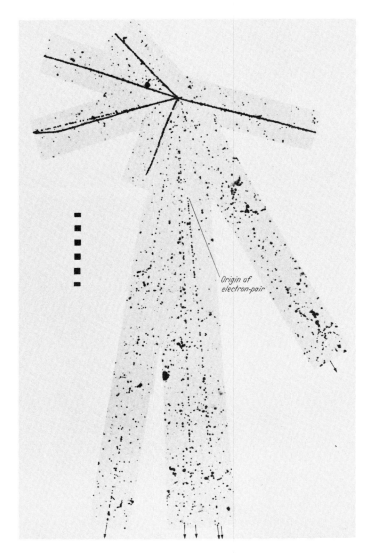

Figure 5.8. Production of an e^+e^- pair from materialization of a γ-ray from the decay $\pi^0 \to 2\gamma$. Because of the short π^0 lifetime, the pion points at the vertex of the parent interaction (G5 emulsion) [*Source*: Powell et al., *Study of Elementary Particles* ... (1959),[41] reprinted by permission; copyright © 1959 Pergamon Press plc.].

"scattering constant" employed in deducing the e^+e^- pair energies from the Coulomb scattering of the electrons. Nevertheless, this was a brilliant experiment, and it gave valuable confirmation that neutral as well as charged pions were produced in the cosmic rays.

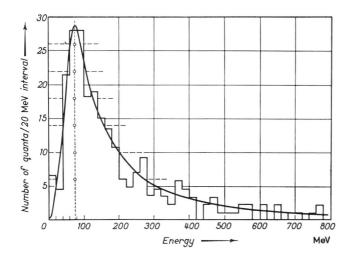

Figure 5.9. An e^+e^- pair spectrum observed in G5 emulsion exposed in a balloon, as measured by Carlson, Hooper, and King[27] in 1950. It has the property that for equal intensities on either side of the peak, the energies E_1 and E_2 are related by $(\sqrt{E_1 E_2})^{1/2} = m_{\pi^0}/2$ [*Source*: Carlson et al. (1950)[27]].

K particles and hyperons

The cloud-chamber discovery of V particles with masses between those of the pion and the proton by George Rochester and Clifford Butler in 1947 was mirrored by observations, from 1949 to 1953, of new heavy particles in cosmic-ray experiments with emulsions.[28]

Figure 5.10 shows the first τ-decay event, observed in 1949 by R. Brown and associates.[29] The three secondaries are coplanar, consistent with a three-body decay, and one is clearly a slow π^-, giving a σ star. If all the secondaries are pions, energy-momentum conservation and the angles between the secondaries give a Q value of 75 MeV or a mass $m_\tau = 980 \pm 20\ m_e$. The τ track itself is very long (3,000 μm), and measurements on it provided a mass estimate $m_\tau = 1,080 \pm 160\ m_e$. Within a year or so, several other τ decays had been found in various laboratories.

Other examples of heavy-meson decay events were soon found. Figure 5.11 shows an event, one of two published by O'Ceallaigh (1951) at Bristol, detailing the decay of a so-called $\varkappa$ particle into a muon of range 1,100 μm, and thus not due to a pion decay.[30] Measurements on the particle $\varkappa$ gave a mass of $1,125 \pm 140\ m_e$ (this particular decay mode of the kaon would now be called $K_{\mu 3}$). Other decay modes, such as $\chi\ (= K_{\pi 2}, K \to \pi^+ \pi^0)$ and $K_\beta\ (= K_{e3})$, were observed.

In addition, charged hyperprotonic particles, heavier than a proton, and later called hyperons, were observed in emulsion experiments. The first two

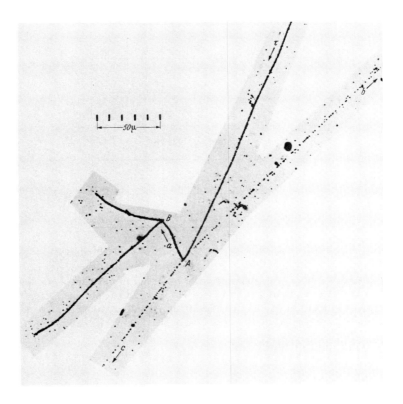

Figure 5.10. First τ-meson decay [*Source*: Reprinted by permission from R. Brown et al., *Nature (London) 163* (1949),[29] 82–7; copyright © 1949 Macmillan Magazines Ltd.]. The short-range secondary is a π^-, giving a σ star. The coplanarity of the three secondaries indicates three-body decay, and if all are pions, the mass $m_\tau \approx 980\ m_e$ in this case.

"*J* events," observed by A. Bonetti, are shown in Figure 5.12.[31] The first shows the decay to a lightly ionizing particle (assumed to be $\Sigma^+ \rightarrow n\pi^+$), and the second, to a proton ($\Sigma^+ \rightarrow p\pi^0$). The year 1953 was also notable for the first observation of a hypernucleus (with a bound Λ hyperon replacing a neutron) by M. Danysz and Jerzy Pniewski.[32]

The worldwide data on kaons and hyperons from cosmic-ray experiments using cloud-chamber and emulsion techniques were brought together at several conferences, of which the most notable, for me, was that at Bagnères de Bigorre in the Pyrenees in 1953. The prevailing atmosphere was one of excited confusion. In particular, there was a confrontation involving the cloud-chamber results of B. Gregory and Charles Peyrou at the Ecole Poly-technique (who found the $K_{\mu 2}$ decay mode and assigned a mass of 916 m_e) and the emulsion results from Bristol, yielding for the $\varkappa$ ($= K_{\mu 3}$) and χ

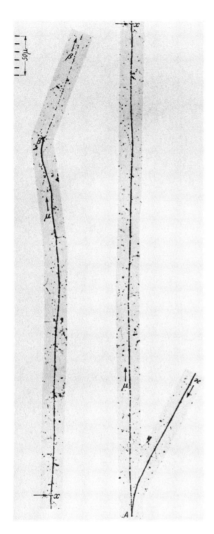

Figure 5.11. First example of $\varkappa$ decay [*Source*: O'Ceallaigh (1951)[30]]. The range of the muon secondary is 1,100 μm, and the mass of the parent particle was estimated to be 1,125 ± 140 m_e.

($= K_{\pi 2}$) decays masses in the region of 1,050 m_e. Had these results been averaged, the similarity to the τ mass (by then 966 m_e) and the θ_0 mass ($K_0 \rightarrow \pi^+ \pi^-$) would have been apparent – an easy enough remark to make with the benefit of hindsight. Another two years had to pass before the precision and reliability of the measurements were to finally force the conclusion that these were all different modes of one and the same particle.

Technical and sociological developments

So far, I have been describing the advances in physics, but none of these would have been possible without several important technical innova-

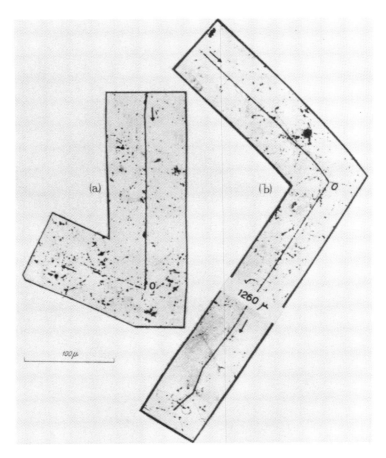

Figure 5.12. Two examples of Σ-hyperon decays, observed by Bonetti and associates in Milano. The one at upper left shows decay to a lightly ionizing particle (assumed to be $\Sigma^+ \to n\pi^+$), and the other, to a proton ($\Sigma^+ \to p\pi^0$) (*Source*: Bonetti et al. (1953)[31]].

tions. The first of the new emulsions were produced in thicknesses of 50 μm, 100 μm, and 200 μm. The obvious drawback of these was that the probability that a long-range secondary could be contained and brought to rest inside the emulsion layer was small. The use of much thicker emulsions (up to 1,000 μm) became possible with the invention of the temperature-cycle method of C. Dilworth and Occhialini in 1948.[33] The emulsions were soaked for a long period in developer at low temperature (so that it could diffuse uniformly through the layer) and then heated so that the chemical development reaction proceeded uniformly with depth. The second innovation was that of stripped emulsions in 1952 – a method previously used, it turned out, by Kinoshita in

Figure 5.13. Moments before the dawn launch of a polyethylene balloon near Bristol in the early 1950s. The hydrogen filling tube is visible at the bottom, together with the emulsion stack in a bamboo cage. All eyes are on rubber balloons sent aloft to gauge wind speed [*Source*: Bristol University, Bristol, England].

1915.[34] Emulsions were poured and dried on glass backing as usual, then stripped off and assembled like a stack of cards, so as to form a continuously sensitive medium of arbitrarily large volume. After exposure, the individual emulsion layers were mounted on glass for processing. The advantages of this approach, using 400-μm-thick layers, when the range of the muon in $K_{\mu 2}$ decay is 20 cm, are obvious.

A necessary innovation was the instigation of the manufacture, in Bristol and elsewhere (Figure 5.13), of very large polyethylene balloons, to take advantage of the much higher intensities of cosmic rays in the stratosphere than were available at mountain altitudes. These hydrogen-filled balloons were produced in large quantities and could easily float loads of 100 kg or more at altitudes of 30 km. Flying these balloons over England ran into considerable difficulties with the authorities because of the air-traffic hazard when a balloon was in the air, as well as the danger when the parachute and load descended in a highly populated country. (On one occasion, a failed balloon came down on the steam locomotive of the Bristol–London express. It and the equipment were confiscated by British Rail because we refused to pay the fine. Presumably they still have it!) There were also problems with weather and the high probability of losing such balloons overseas in normal wind conditions. Eventually, all operations had to be transferred to the Mediterranean area, particularly Sardinia, and recovery made at sea with the help of the Italian navy.

The balloon-flying expeditions in the Mediterranean had unforeseen and highly providential effects for high-energy physics in Europe. Many flights had to be made, with highly organized tracking and recovery procedures, and these were beyond the resources of a single university group. Large collaborations were formed, with balloon fabrication centered at Bristol and Padua, but preparation of stacks, radar, radiosonde and tracking equipment, and so forth, the responsibilities of many laboratories. The Sardinia expedition of 1953 involved twenty-two laboratories from twelve countries: Bristol, Brussels, Bern, Caen, Catania, Copenhagen, Dublin, Ecole Normale, Ecole Polytechnique, Geneva, Göttingen, London, Lund, Milan, Oslo, Padua, Rome, Sydney, Turin, Trondheim, Uppsala, Warsaw. Thirty flights were made, with the minimum "share" for any laboratory to join fixed at £1,000. The processed emulsion stacks were divided between laboratories. This implied that tracks in events found in one group had to be traced through the emulsions to those of a second, third, or even fourth group. Could this possibly work? It did, and it was a triumph of international collaboration and organization.

The largest and final stack of emulsions in this program was the famous G-stack (G for gigantic), with fifteen liters of emulsion (63 kg) flown for six hours at 27 km over northern Italy in 1954.[35] The flight ended with parachute failure, and the aluminum container came down in free-fall and hit a rock in the Apennines, tearing apart the container and shattering the bottom 10 percent of the emulsion stack. Yet the remainder was usable and provided definitive information on the kaon decay modes. Figure 5.14 shows the energy spectrum of secondaries deduced from particle ranges and other measurements. In particular, it provided the following results for the masses:

$$M(K_{\pi 2}) = 494.7 \pm 1.5 \text{ MeV}$$
$$M(K_{\mu 2}) = 495.7 \pm 1.5 \text{ MeV}$$
$$M(K_{\pi 3}) = 494 \text{ MeV}$$

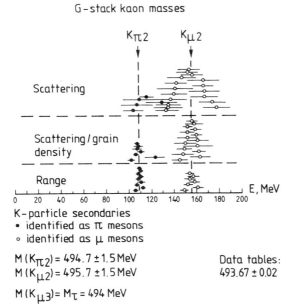

M ($K_{\pi 2}$) = 494.7 ± 1.5 MeV   Data tables:
M ($K_{\mu 2}$) = 495.7 ± 1.5 MeV   493.67 ± 0.02

M ($K_{\mu 3}$) = M_τ = 494 MeV

Figure 5.14. Energies of muons and pions observed in $K_{\mu 2}$ and $K_{\pi 2}$ decays in the G-stack collaboration of 1955. The various measurements are from scattering (top), scattering and grain density (middle) of incomplete secondaries, and range (bottom) of complete secondaries. The average kaon mass values for the decay schemes $K_{\pi 2} \rightarrow \pi + \pi^0$ and $K_{\mu 2} \rightarrow \mu + \nu$ are given [*Source*: Davies et al. (1955)[35]].

in satisfactory agreement with the present-day value of the charged-kaon mass, 493.67 ± 0.02 MeV.

These results were contemporaneous with those from the Berkeley Bevatron, using the same technique with artificially produced kaons. Thereafter, accelerators took over.

In retrospect

This short factual account of the discovery of pions and kaons in cosmic-ray emulsion experiments will mean little to historians, who perhaps are more concerned with motives and methods and the sociology of the scientific method. Perhaps I can give my personal impression of this scientific era in retrospect.

The main discovery of the emulsion method was that of the pion. In the 1950s, this was seen as crucial, since the pion was identified as *the* fundamental, Yukawa quantum of nuclear force. Nowadays, the pion does not enjoy this privileged position; it is simply a combination of u and d quarks (and antiquarks), just as are the common neutron and proton. In this sense, the discovery of the V particles and associated production was more important, since it led to a new quantum number, strangeness, and eventually to a new quark constituent (the s-quark). The discovery of the muon, in the 1930s and 1940s, is also seen today as more fundamental than that of the pion. It heralded a new lepton flavor, although the existence of this "heavy electron" is today just as much a puzzle as it was then. As I. I. Rabi once said, "Who ordered that?"

What can we say in retrospect about the emulsion technique? It came, by a fortuitous combination of circumstances, just at the right time in human affairs. It was a simple technique, requiring modest photographic processing equipment, a few microscopes, and a dedicated team of scanning personnel. It was tailor-made for international collaboration. In the 1940s and 1950s, small groups, with the meager resources available in universities in a Europe emerging from the catastrophe of World War II, were provided with an instrument allowing them to easily contribute at the forefront of physics research. The impact for European collaboration on a major scale, in the formation of CERN in 1953, hardly needs to be emphasized. These early collaborative efforts, motivated as much by altruism as by self-interest, were to my mind *the* great achievement of the emulsion technique. Nowadays, the emulsion technique is still contributing to physics in a small but important way. The best limit on $\nu_\mu - \nu_\tau$ mixing comes from a Fermilab emulsion neutrino experiment. Contributions on charm decays and on B-meson decays are other examples.

Finally, I have to comment about cosmic-ray physics, then and now. There is no question that the discoveries in cosmic rays from 1947 to 1953 led the way and stimulated the building of large accelerators. Without that impetus, high-energy physics could not possibly have developed as it has done, since it was in the early 1950s and later that the funding authorities gladly provided large resources in the belief that high-energy accelerators were vital for the nuclear (energy) program. Nor was this entirely illusory: The example cited already, of muon-induced fusion, was a by-product of high-energy physics, and had the Fermi constant been only ten times smaller than it is, one could speculate that all our world energy problems would be over for evermore (in more than one sense, however, for the evolution of main-sequence stars like the sun would have been correspondingly slowed, and we would not be here now).

Today, cosmic rays still produce intense excitement. Just two weeks ago, the Soudan I and Nusex proton decay experiments announced a correlation of underground muons in direction and phase with the pulsar Cygnus X3. If this is genuine, it appears very possible that these experiments are indicating the first real signs of new physics beyond the standard model.

Notes

1 Henri Becquerel, "Sur les radiations emises par phosphorescence," *C. R. Acad. Sci. 122* (1896), 420–1; "Sur les radiations invisibles emises par les corps phosphorescents," *C. R. Acad. Sci. 122* (1896), 501–3; "Emission de radiations nouvelles par l'uranium metallique," *C. R. Acad. Sci. 122* (1896), 1086–8.

2 S. Kinoshita, "Photographic Action of the α-Particles," *Proc. R. Soc. London*, Ser. A 83 (1910), 432–53.

3 M. Blau, "Mitteilungen aus dem Institut für Radiumforschung, Nr. 179. Ueber die photographische Wirkung natürlicher H-Strahlen," *Akad. Wiss. Wien 134* (1925), 427–36; M. Blau and H. Wambacher, "Disintegration Processes by Cosmic Rays with the Simultaneous Emission of Several Heavy Particles," *Nature (London) 140* (1937), 585.

4 A. Jdanoff, "Les traces des particules H et α dans les emulsions sensibles a la lumière," *J. Phys. Radium 6* (1935), 233–41.

5 H. J. Taylor, "Tracks of α-Particles and Protons in Photographic Emulsions," *R. Soc. London*, Ser. A 150 (1935), 382–94.

6 T. R. Wilkins, "Further Observations of Cosmic-Ray Tracks in Photographic Emulsions," *Phys. Rev. 50* (1936), 1099.

7 L. H. Rumbaugh and G. L. Locher, "Neutrons and Other Theory Particles in Cosmic Radiation of the Stratosphere," *Phys. Rev. 49* (1936), 855; "Search for Nuclear Cosmic-Ray Particles in the Stratosphere, Using Photographic Emulsions," ibid., 889.

8 M. S. Livingston and H. A. Bethe, "Nuclear Physics. C. Nuclear Dynamics, Experimental," *Rev. Mod. Phys. 9* (1937), 245–390.

9 C. F. Powell and F. Fertel, "Energy of High Velocity Neutrons by the Photographic Method," *Nature* (London) *144* (1939), 115.

10 H. Yukawa, "Interaction of Elementary Particles. Part I," *Proc. Phys.-Math. Soc. Japan 17* (1935), 48–57.

11 S. Tomonaga and G. Araki, "Effect of the Nuclear Coulomb Field on the Capture of Slow Mesons," *Phys. Rev. 58* (1940), 90–1.

12 M. Conversi, E. Pancini, and O. Piccioni, "Disintegration of Negative Mesons," *Phys. Rev. 71* (1947), 209–10.

13 R. E. Marshak and H. A. Bethe, "On the Two-Meson Hypothesis," *Phys. Rev. 72* (1947), 506–9.

14 D. H. Perkins, "Nuclear Disintegration by Meson Capture," *Nature (London)* 159 (1947), 126–7.

15 G. P. S. Occhialini and C. F. Powell, "Multiple Disintegration Processes Produced by Cosmic Rays," *Nature (London) 159* (1947), 93–4.

16 C. M. G. Lattes, H. Muirhead, G. P. S. Occhialini, and C. F. Powell, "Processes Involving Charged Mesons," *Nature (London) 159* (1947), 694–7.

17 V. F. Weisskopf, "On the Production Process of Mesons," *Phys. Rev. 72* (1947), 510.

18 C. M. G. Lattes, G. P. S. Occhialini, and C. F. Powell, "Obstructions on the Tracks of Slow Mesons in Photographic Emulsions. 1. Existence of Mesons of Different Mass," *Nature (London) 160* (1947), 453–6.

19 L. W. Alvarez, H. Bradner, F. S. Crawford, Jr., J. A. Crawford, P. Falk-Variant, M. L. Good, J. D. Gow, A. H. Rosenfeld, F. Solmitz, M. L. Stevenson, H. K. Ticho, and R. D. Tripp, "Catalysis of Nuclear Reactions by μ Mesons," *Phys. Rev. 105* (1957), 1127–8.

20 I. Barbour, "Magnetic Deflection of Cosmic-Ray Mesons Using Nuclear Plates," *Phys. Rev. 76* (1949), 320–1.

21 C. Franzinetti, "On the Mass of Charged Particles of the Cosmic Radiation," *Philos. Mag. 41* (1950), 86–106.

22 E. Gardner and C. M. G. Lattes, "Production of Mesons by the 184-Inch Berkeley Cyclotron," *Science 107* (1948), 270–1.

23 J. R. Richardson, "The Lifetime of the Heavy Meson," *Phys. Rev. 74* (1948), 1720–1.

24 U. Camerini, H. Muirhead, C. F. Powell, and D. M. Ritson, "Observations on Slow Mesons of the Cosmic Radiation," *Nature (London) 162* (1948), 433–8.

25 C. O'Ceallaigh, "The Nature of the Neutral Particle Emitted in the Decay of the π-Meson," *Philos. Mag. 41* (1950), 838–48.

26 R. Bjorklund, W. E. Crandall, B. J. Meyer, and H. F. York, "High Energy Photons from Proton–Nucleon Collisions," *Phys. Rev. 77* (1950), 213–18; J. Steinberger, W. K. H. Panofsky, and J. Steller, "Evidence for the Production of Neutral Mesons by Photons," *Phys. Rev. 78* (1950), 802–5.

27 A. G. Carlson, J. E. Hooper, and D. T. King, "Nuclear Transmutations Produced by Cosmic-Ray Particles of Great Energy. V. The Neutral Mesons," *Philos. Mag. 41* (1950), 701–24.

28 G. D. Rochester and C. C. Butler, "Evidence for the Existence of New Unstable Elementary Particles," *Nature (London) 160* (1947), 855–7.

29 R. Brown, U. Camerini, P. H. Fowler, H. Muirhead, C. F. Powell, and D. M. Ritson, "Observations with Electron Sensitive Plates Exposed to Cosmic Radiation. II. Further Evidence for the Existence of Unstable Charged Particles of Mass ~1000 m_e and Observations on Their Mode of Decay," *Nature (London) 163* (1949), 82–7.

30 C. O'Ceallaigh, "Masses and Modes of Decay of Heavy Mesons. I. K-Particles," *Philos. Mag. 42* (1951), 1032–9.

31 A. Bonetti, R. L. Setti, M. Panetti, and G. Tomasini, "Observation of the Decay at Rest of a Heavy Particle," *Nuovo Cimento 10* (1953), 345–7; "On the Existence of Unstable Charged Particles of Hyperprotonic Mass," ibid., 1736–43.

32 M. Danysz and J. Pniewski, "Delayed Disintegration of a Heavy Nuclear Fragment. I," *Philos. Mag. 44* (1953), 348–50.

33 C. C. Dilworth, G. P. S. Occhialini, and R. M. Payne, "Processing Thick Emulsions for Nuclear Research," *Nature (London) 162* (1948), 102–3.

34 S. Kinoshita and H. Ikeuti, "Tracks of α-Particles in Photographic Films," *Philos. Mag. 29* (1915), 420–5.

35 J. H. Davies, D. Evans, P. E. François, M. W. Friedlander, R. Hillier, P. Iredale, D. Keefe, M. G. K. Menon, D. H. Perkins, C. F. Powell, J. Bøggild, N. Brene, P. H. Fowler, J. Hooper, W. C. G. Ortel, M. Scharff, L. Crane, R. H. W. Johnston, C. O'Ceallaigh, F. Anderson, G. Lawlor, T. E. Nevin, G. Alvial, A. Bonetti, M. di Corato, C. Dilworth, R. L. Setti, A. Milone, G. Occhialini, L. Scarsi, G. Tomasini, M. Ceccarelli, M. Grilli, M. Merlin, G. Salandin, and B. Sechi, "On the Massess and Modes of Decay of Heavy Mesons Produced by Cosmic Radiation (G-Stack Collaboration)," *Nuovo Cimento (Ser. 10) 2 : 5* (1955), 1063–103.

36 D. H. Perkins, "Nuclear Disintegration by Meson Capture," *Nature (London) 159* (1947), 126–7.

37 Y. Goldschmidt-Clermont, D. T. King, H. Muirhead, and D. M. Ritson, "Determination of the Masses of Charged Particles Observed in the Photographic Plate." *Proc. Phys. Soc. London 61* (1948), 183–94.

38 J. K. Bowker, "Observations on Grain Counting and the Photographic Emulsion," *Phys. Rev. 78* (1950), 87.

39 H. Bradner, F. M. Smith, W. H. Barkas, and A. S. Bishop, "Range–Energy Relation for Protons in Nuclear Emulsions," *Phys. Rev. 77* (1950), 462–7.

40 A. S. Bishop, H. Bradner, and F. M. Smith, "Improved Mass Values for π^-, π^+, and μ^+ Mesons," *Phys. Rev. 76* (1949), 588.

41 C. F. Powell, P. H. Fowler, and D. H. Perkins, *Study of Elementary Particles by the Photographic Method* (Oxford: Pergamon, 1959).

PART III

HIGH-ENERGY NUCLEAR PHYSICS

We had pi mesons, mu mesons;
Some folks thought it too few mesons,
Went out and discovered some new mesons –
   Some people don't know where to stop.
Now a few mesons might do mesons
A good turn by turning to glue mesons,
But how can you use twenty-two mesons?
   Some people don't down where to stop.
    – © Arthur Roberts, "Some People Don't
               Know Where to Stop" (1952)

6 Learning about nucleon resonances with pion photoproduction

ROBERT L. WALKER

Born 1919, St. Louis, Missouri; Ph.D., Cornell University, 1948; experimental high-energy physics; California Institute of Technology (emeritus)

One of the most active areas of research in particle physics during the 1950s and 1960s involved the discovery of new particles and the determination of their quantum numbers. That activity produced a great proliferation in the number of particle states, which again raised the question of what an elementary particle is and which stimulated the development of quark models and other classification schemes.

One of the methods used to identify new particle states was to observe and study resonances in scattering reactions. This chapter concerns the first three resonances found by this method, resonances in the πN interaction. For orientation, the total cross section for π^+ photoproduction, $\gamma p \to \pi^+ n$, is shown as a function of energy in Figure 6.1. The three peaks result mainly from the three resonances whose history I shall discuss. We now know that several other resonances exist in the energy region of the second and third peaks, but these were found much later by sophisticated partial wave analyses of very accurate πN scattering data.

Before reviewing the discovery of these resonances, I want to say a little about the experimental features that enable us to recognize a resonance. I consider only simple, fairly obvious resonances, not those that lie so hidden in the data that it takes a detailed computer analysis to find them. First, we expect to see a peak, or some evidence for a peak, in the cross section as a function of energy. Figure 6.1 illustrates this behavior. Second, and more important, we need some evidence that the scattering amplitude, which is a complex number, moves around in the complex plane in a manner characteristic of a resonance as the energy varies. This behavior is illustrated in Figure 6.2 for a purely elastic "textbook" example of a resonance, the first resonance of Figure 6.1. For this simple example, the amplitude $|A|$ is a

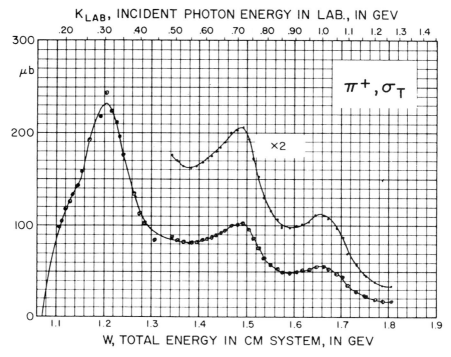

Figure 6.1. Total cross section for π^+ photoproduction from protons [*Source*: Walker (1969)[36]].

maximum at the resonance energy, and the phase shift δ passes through 90°.

In spite of the fact that the first resonance obeys these simple criteria spectacularly well, it took a long time for this resonance to gain general acceptance. The problem was partly due to the fact that the early relevant data were obtained with accelerators whose energies were too low to permit measurements at energies significantly above the peak.

The first resonance, P_{33}

Contributions to the discovery of the famous "3-3 resonance"[1] came from πN scattering experiments, from pion photoproduction experiments, and from theory. The theoretical contributions were remarkable because they seemed to require initially only the slightest amount of help from the experimental data. The photoproduction experiments depended, of course, on the availability of accelerators capable of producing "high-energy" photons. Those were electron-synchrotrons of approximately 300 MeV energy constructed soon after the war at Berkeley, Cornell, and MIT, followed a little later by a 500-MeV machine at Caltech.

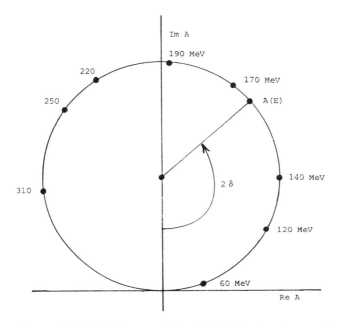

Figure 6.2. Scattering amplitude A(E) plotted on the complex plane. As the energy E increases, A(E) moves on a circle for a purely elastic resonance, becoming imaginary at the resonance energy, where the scattering phase shift $\delta = 90°$. The points indicated by 120 MeV, 140 MeV, and so forth, show schematically the experimental values of the πN scattering amplitude at the energies indicated (laboratory energy of the pion).

The first theoretical papers invoking a nucleon isobar, or resonance, to account for some experimental facts were published by Yoichi Fujimoto and Hiromari Miyazawa[2] in 1950 and by Keith A. Brueckner and Kenneth M. Case[3] in 1951. The experimental facts[4,5] were these: The total cross section for neutral-pion photoproduction was approximately equal to that for charged pions, with a value near 10^{-28} cm^2 for a bremsstrahlung spectrum having maximum energy 320 MeV, and the neutral cross section rises much more steeply with energy than does the charged cross section. The theorists pointed out that these facts were in conflict with weak-coupling perturbation theory, but that they could be explained qualitatively by the existence of nucleon isobars, which was one of the characteristic features of strong-coupling theory. A calculation by Brueckner and Case of the charged and neutral cross sections, using a nucleon isobar at 250 MeV, is reproduced in Figure 6.3.

Some months later, Brueckner[6] made a more specific isobar proposal in order to provide a theoretical interpretation of certain "apparently anomalous" features of the earliest data on pion–nucleon scattering.[7] Brueckner's interpretation was based on a nucleon isobar having $J = I = \frac{3}{2}$, the lowest of

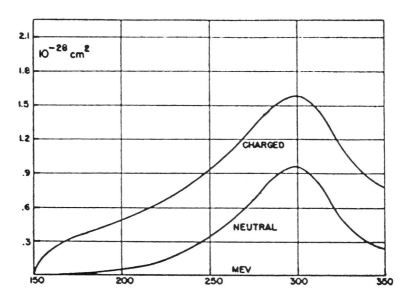

Figure 6.3. Calculation by Brueckner and Case of the charged and neutral photoproduction cross sections assuming a nucleon isobar at 250 MeV [*Source*: Brueckner and Case (1951)[3]].

those predicted by strong-coupling theory. The experimental observations considered by Brueckner were the following:

1. The $\pi^- p$ total cross section increased rapidly between 60 and 200 MeV (the laboratory energy of the incident pion).
2. At 60 MeV, the $\pi^+ p$ cross section was larger by a factor of 1.6 than the $\pi^- p$ cross section.
3. The π^- and π^+ cross sections for deuterium were equal, in support of the hypothesis of charge symmetry of the interaction.

More experimental data kept coming from the Chicago group[8] at that time, and while Brueckner's paper was still in proof, he was able to add a note and a figure showing total cross sections as a function of energy for both π^+ and π^-, in good agreement with his predictions (Figure 6.4). In his note he pointed out that the ratios of observed cross sections $\sigma(\pi^+ p \rightarrow \pi^+ p) : \sigma(\pi^- p \rightarrow \pi^- p) : \sigma(\pi^- p \rightarrow \pi^0 n)$ were approximately $9 : 1 : 2$, as would be the case if the scattering occurred predominantly in the $I = \frac{3}{2}$ state.

Only two months later, Brueckner and Kenneth M. Watson published a paper[9] extending the theory and phenomenological analysis to photoproduction. Cross sections at 90° in the laboratory for $\gamma p \rightarrow \pi^+ n$ obtained by J. Steinberger, A. S. Bishop, and Leslie J. Cook[10,11] and for $\gamma p \rightarrow \pi^0 p$ obtained by Al Silverman and Martin Stearns[12] and W. K. H. Panofsky,

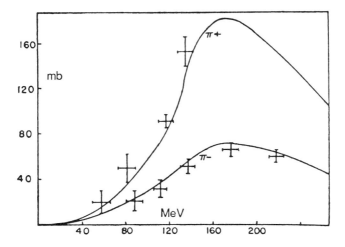

Figure 6.4. Total cross sections for scattering of π^+ and π^- mesons in hydrogen (including charge exchange) [*Source*: Brueckner (1952)[6]].

Steinberger, and J. Steller[13] were in good agreement with their theory incorporating the $J = I = \frac{3}{2}$ resonance. In particular, the rapid rise of the π^0 photoproduction cross section (higher-order contact at threshold) found a natural explanation in terms of this P_{33} resonance (Figure 6.5).

Although the arguments of Brueckner and Watson were persuasive, the experimentalists (notably Enrico Fermi) remained skeptical; the question of the existence of the resonance was one to be decided by further experiment. Work on both pion scattering and photoproduction continued, of course, as it would have even without this stimulation. Unfortunately, the energies of the accelerators with which the early data were obtained were not sufficient to obtain data significantly above the proposed resonance. One would have liked to see a peak in the energy dependence of the cross section, and it was also important to see how the angular distribution changed when the energy varied from below to above the resonance.

The energy limitation was removed when the Caltech electron-synchrotron began initial operation at 500 MeV in 1952. The first experiment performed with it showed a clear peak in the 90° cross section for π^0 photoproduction, at a photon energy near 300 MeV (Figure 6.6).[14] This behavior gave support to the resonance hypothesis.

At the same time, work on pion scattering continued at the Chicago cyclotron, and angular distributions were obtained at energies up to 135 MeV.[15,16] Separate distributions were measured for $\pi^+ p \rightarrow \pi^+ p$, $\pi^- p \rightarrow \pi^- p$, and $\pi^- p \rightarrow \pi^0 n$. As an example, the data at 135 MeV are shown in Figure 6.7. The Chicago experimenters analyzed their data in terms of s- and p-wave scattering phase shifts and thus began a period of many years when the newest sets

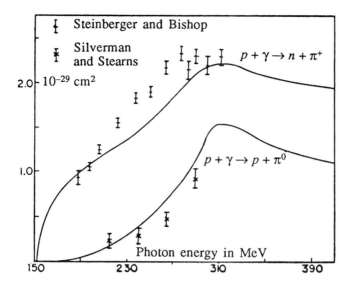

Figure 6.5. Calculations by Brueckner and Watson of the differential cross sections at 90° in the laboratory system for π^+ and π^0 photoproduction [*Source*: Brueckner and Watson (1952)[9]].

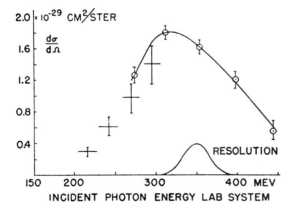

Figure 6.6. Differential cross section at 90° in the laboratory system for photoproduction of neutral pions in hydrogen [*Source*: Walker et al. (1953)[14]]. The crosses at lower energies are data of Silverman and Stearns.[12]

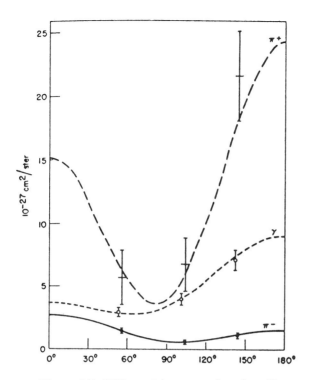

Figure 6.7. Differential cross sections for πN scattering at 135 MeV [*Source*: Anderson et al. (1953)[16]].

of phase shifts were reported at nearly every conference on particle physics. The problem of multiple solutions had already appeared in this 1953 paper of H. L. Anderson and associates in the form of a "Yang solution," which was just as satisfactory in fitting the data as the "first" or "Fermi solution."[16] The "first solution" showed a large phase shift of 38° at 135 MeV for the p state, with $J = I = \frac{3}{2}$ (P_{33}), implying a strong interaction in this state, consistent with the resonance idea. However, as mentioned earlier, the energy range investigated was not high enough to see this phase shift pass through 90°, which would have provided the classical proof of resonance, and there also remained the question about the other solution. The authors remained open-minded, and among their conclusions they stated "that our data do not extend far enough in energy to support the resonance hypothesis."

Meanwhile, photoproduction data above, as well as below, the proposed resonance were being obtained at the Caltech synchrotron. At the American Physical Society meeting in Albuquerque in September 1953, data were presented from two experiments on π^+ photoproduction.[17,18] They showed a striking peak in the energy dependence, and the angular distributions changed character from backward to forward peaking as the energy varied from below to above the peak. This behavior is illustrated in Figure 6.8. Both features supported the resonance hypothesis. For example, the change in forward–backward asymmetry could be caused by an s–p interference if the phase of the dominant p-wave term passed through 90° while the phase of the interfering s wave remained small and did not change sign.[19]

By this time, many people believed in the existence of the resonance, but a real proof was still lacking. Fermi, for one, still argued that the experimental facts might be explained in some other way and that such possibilities should be considered. I remember clearly a group conversation with Fermi, while sitting on the lawn during the Albuquerque meeting, in which he argued that point of view.

Later, pion–nucleon scattering data at higher energies came from Chicago,[20] from the Carnegie Institute of Technology,[21] and from Dubna.[22] With them came the inevitable phase-shift analyses, and much computer time was expended on this enterprise. More accurate data were accumulated, and theoretical aid came in the form of dispersion relations, to help eliminate some of the phase-shift solutions. Eventually the solution in which the P_{33} state displays a beautiful and unmistakable resonance behavior triumphed over all others. Thus ended a long story that began with a suggestion from theorists based on a slender hint from experimental data.

The second and third resonances, D_{13} and F_{15}

Now I shall review briefly the early history of the "second" and "third" resonances, which are clearly visible as peaks in the total cross section for π^+ photoproduction shown in Figure 6.1. The first indication of a possible resonance in this energy region came from data on πN total cross sections. In 1956, Rodney L. Cool, Oreste Piccioni, and D. Clark published data obtained at the Brookhaven Cosmotron showing a broad peak in the π^-p cross section, but none for π^+p (Figure 6.9).[23] Therefore, this peak was clearly a result of the isospin-$\frac{1}{2}$ part of the interaction. From the height of the peak, they argued that if it represented a resonance, the angular momentum must be $\frac{5}{2}$ or larger.

The data of Cool, Piccioni, and Clark showed that there was something interesting in the $I = \frac{1}{2}$ interaction, but did little to clarify what it was. They did not even resolve the two resonances, for some reason I do not understand. For comparison with this feature, Figure 6.10 shows data obtained nearly three years later at the Bevatron by H. C. Burrowes and associates,[24] in which the two peaks are clearly resolved.

After the work of Cool, Piccioni, and Clark, most of the data bearing on the discovery and elucidation of the second and third resonances came from photoproduction experiments. In 1952 the Cornell group built a new 1.2-GeV synchrotron, and during 1955–6 the Caltech machine was modified so as to be able to reach energies of 1.2–1.5 GeV. When photoproduction studies were

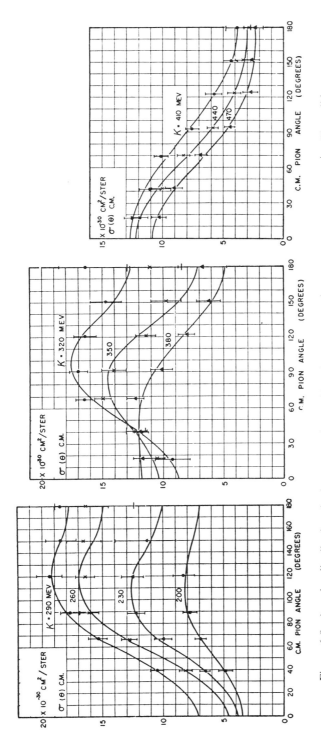

Figure 6.8. Angular distributions in the center-of-mass (C.M.) system for π^+ photoproduction, $\gamma p \to \pi^+ n$. The solid curves are fits to the data of the form $\sigma(\theta) = A + B\cos\theta + C\cos^2\theta$ [*Source*: Walker et al. (1955)[17]].

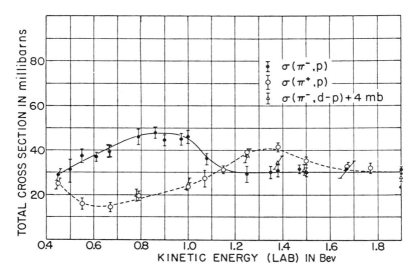

Figure 6.9. Total cross sections for $\pi^+ p$ and $\pi^- p$ interactions [*Source*: Cool et al. (1956)[23]].

extended into this energy region, two new "resonances" were seen, and knowledge about their properties gradually emerged.

The Cornell data were presented in a series of letters to the editor, submitted in April 1958. Photoproduction of π^0 at center-of-mass angles 52°, 90°, and 125°, and at laboratory photon energies from 450 to 950 MeV, was measured by John W. DeWire, H. E. Jackson, and Raphael Littauer,[25] and similar data of somewhat less accuracy were obtained by Peter C. Stein and K. C. Rogers.[26] Data on π^+ photoproduction at similar angles and energies were obtained by M. Heinberg and associates.[27] Both reactions provided evidence for a possible resonance near 700 MeV photon energy, and Robert R. Wilson[28] interpreted the data in this way, assigning quantum numbers $I = \frac{1}{2}$ and $J = \frac{3}{2}$ to the resonance. He pointed out that its energy was too low to correspond to the peak observed by Cool, Piccioni, and Clark. His interpretation was extended by Ronald Peierls,[29] who argued from the angular distributions that the parity of this state must be odd, so that it would be a D state with $I = \frac{1}{2}$ and $J = \frac{3}{2}$ (D_{13}). That assignment was indeed correct.

At about the same time, a parallel program of experiments was carried out at Caltech. Detailed measurements of π^0 photoproduction were published by James I. Vette,[30] and initial π^+ data were presented by F. P. Dixon and Walker[31] showing a nice peak in the 90° cross section at 700 MeV laboratory photon energy.

All of these results mentioned so far concerned the so-called second resonance at a total center-of-mass energy of 1,520 MeV, although Vette noted

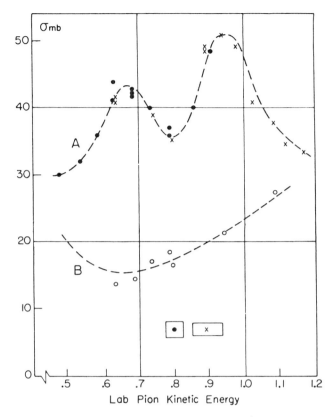

Figure 6.10. Total cross sections for $\pi^+ p$ and $\pi^- p$ interactions [*Source*: Burrowes et al. (1959)[24]].

that the angular distribution for π^0 production changed rather suddenly near 900 MeV. This behavior was later ascribed to the third resonance. A few months after their first letter, Dixon and Walker[32] presented rather complete angular distributions for π^+ photoproduction at several energies from 600 to 1,000 MeV. At 1,000 MeV, this distribution contained a major component of the form $1 + 6\cos^2\theta - 5\cos^4\theta$, which would result from a $D_{5/2}$ state excited by magnetic quadrupole photons or from an $F_{5/2}$ state excited by electric quadrupole. They pointed out that this behavior could be explained by a resonance in one of these states and that the resonance energy would be consistent with the peak observed by Cool, Piccioni, and Clark. Three of the angular distributions given by Dixon and Walker are shown in Figure 6.11.

In a paper based on his doctoral thesis, Peierls presented an analysis in which he argued that the second and third peaks represented resonances whose quantum numbers could be determined by the existing data on πN

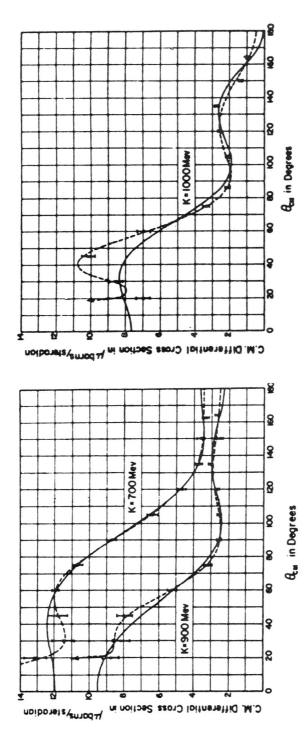

Figure 6.11. Three of the angular distributions for π^+ photoproduction [*Source*: Dixon and Walker (1958)[32]].

scattering and photoproduction.[33] His arguments were based mainly on the photoproduction angular distributions and on new measurements (the first of this type) of the polarization of recoil proton in π^0 photoproduction.[34,35] The polarization measurement at 700 MeV provided additional evidence for the D_{13} assignment already preferred for the second resonance. The angular distributions were used to decide the parity of the third resonance, thereby making the choice F_{15} rather than D_{15}.

Since 1960, our knowledge of πN scattering has improved enormously because of many new experiments with increasing accuracy, some involving polarization measurements that provide information not obtainable from differential cross-section measurements alone. These data have been fed into continuing programs of partial wave analyses, and surprising numbers of resonances have been identified in the πN system. The role of pion photoproduction then shifted to a determination of the electromagnetic couplings of the resonances. Information about these couplings has played an important role in the development of quark models, but an account of that would take us well beyond the subject of this chapter.[36]

Notes

1 In the notation P_{33}, D_{13}, F_{15}, etc., the first index is $2I$, and the second is $2J$, where I is the isospin and J the total angular momentum. The letters P, D, F, etc., indicate the orbital angular momentum of the πN system and thereby its parity.

2 Y. Fujimoto and H. Miyazawa, "Photomeson Production and Nucleon Isobar," *Prog. Theor. Phys.* 5 (1950), 1052–4.

3 Keith A. Brueckner K. M. Case, "Neutral Photomeson Production and Nucleon Isobars," *Phys. Rev.* 83 (1951), 1141–7.

4 J. Steinberger and A. S. Bishop, "Preliminary Results on the Production of Mesons by Photons on Carbon and Hydrogen," *Phys. Rev.* 78 (1950), 494–5.

5 J. Steinberger, W. K. H. Panofsky, and J. Steller, "Evidence for the Production of Neutral Mesons by Photons," *Phys. Rev.* 78 (1950), 802–5.

6 Keith A. Brueckner, "Meson–Nucleon Scattering and Nucleon Isobars," *Phys. Rev.* 86 (1952), 106–9.

7 The data considered by Brueckner were obtained at the Columbia and Chicago cyclotrons and were presented by J. Steinberger and H. L. Anderson in talks at American Physical Society meetings and through private communication. The only published report cited by Brueckner is C. Chedester, P. Isaacs, A. Sachs, and J. Steinberger, "Total Cross Sections of π-Mesons on Protons and Several Other Nuclei," *Phys. Rev.* 82 (1951), 958–9.

8 These data were published in the following letters to the editor: H. L. Anderson, E. Fermi, E. A. Long, R. Martin, and D. E. Nagle, "Total Cross Sections of Negative Pions in Hydrogen," *Phys. Rev.* 85 (1952), 934–5; E. Fermi, H. L. Anderson, A. Lundby, D. E. Nagle, and G. B. Yodh, "Ordinary and Exchange Scattering of Negative Pions by Hydrogen," *Phys. Rev.* 85 (1952), 935–6; H. L. Anderson, E. Fermi, E. A. Long, and D. E. Nagle, "Total Cross Sections of Positive Pions in Hydrogen," *Phys. Rev.* 85 (1952), 936.

9 K. A. Brueckner and K. M. Watson, "Phenomenological Relationships between Photomeson Production and Meson–Nucleon Scattering," *Phys. Rev.* 86 (1952), 923–8.

10 J. Steinberger and A. S. Bishop, "Preliminary Results on the Production of Mesons by Photons on Carbon and Hydrogen," *Phys. Rev.* 78 (1950), 494–5. Further results were published later: J. Steinberger and A. S. Bishop, "The Production of Positive Mesons by Photons," *Phys. Rev.* 86 (1952), 171–9.

11 A. S. Bishop, J. Steinberger, and Leslie J. Cook, "Production of Positive Mesons by Photons on Hydrogen," *Phys. Rev. 80* (1950), 291.

12 A. Silverman and M. Stearns, "Production of Neutral Mesons by γ-Rays Incident on Hydrogen," *Phys. Rev. 83* (1951), 206 (abstract). Later publication: A. Silverman and M. Stearns, "Production of π^0 Mesons by γ-Rays on Hydrogen," *Phys. Rev. 88* (1952), 1225–30.

13 W. K. H. Panofsky, J. N. Steinberger, and J. Steller, "Further Results on the Production of Neutral Mesons by Photons," *Phys. Rev. 86* (1952), 180–9.

14 R. L. Walker, D. C. Oakley, and A. V. Tollestrup, "Photoproduction of Neutral Mesons in Hydrogen at High Energies," *Phys. Rev. 89* (1953), 1301–2.

15 H. L. Anderson, E. Fermi, D. E. Nagle, and G. B. Yodh, "Angular Distribution of Pions Scattered by Hydrogen," *Phys. Rev. 86* (1952), 793–4.

16 H. L. Anderson, E. Fermi, R. Martin, and D. E. Nagle, "Angular Distribution of Pions Scattered by Hydrogen," *Phys. Rev. 91* (1953), 155–68.

17 R. L. Walker, J. G. Teasdale, and V. Z. Peterson, "Photoproduction of Positive Mesons from Hydrogen: Magnetic Spectrometer Method," *Phys. Rev. 92* (1953), 1090 (abstract); R. L. Walker, J. G. Teasdale, V. Z. Peterson, and J. I. Vette, "Photoproduction of Positive Pions in Hydrogen-Magnetic Spectrometer Method," *Phys. Rev. 99* (1955), 210–19.

18 A. V. Tollestrup, J. C. Keck, and R. M. Worlock, "Angular Distribution of Positive Photomesons from Hydrogen: Counter Telescope Method," *Phys. Rev. 92* (1953), 1090 (abstract); A. V. Tollestrup, J. C. Keck, and R. M. Worlock, "Photoproduction of Positive Pions in Hydrogen-Counter Telescope Method," *Phys. Rev. 99* (1955), 220–8.

19 See, for example, Murray Gell-Mann and Kenneth M. Watson, "The Interactions Between π-Mesons and Nucleons," *Annu. Rev. Nucl. Sci. 4* (1954), 219–70; K. M. Watson, J. C. Keck, A. V. Tollestrup, and R. L. Walker, "Partial Wave Analysis of the Experimental Photomeson Cross Section," *Phys. Rev. 101* (1956), 1159–72.

20 M. Glicksman, "Scattering of 217 MeV Negative Pions by Hydrogen," *Phys. Rev. 94* (1954), 1335–44; Horace D. Taft, "Scattering of 217 MeV Positive Pions on Hydrogen," *Phys. Rev. 101* (1956), 1116–27.

21 For example, J. Ashkin, J. P. Blaser, F. Feiner, and M. O. Stern, "Pion–Proton Scattering at 150, 170, and 220 MeV," in *Proceedings of the CERN Symposium on High Energy Accelerators and Pion Physics, 1956*, edited by E. Regenstreif (Geneva: CERN, 1956), pp. 225–9.

22 For example, A. I. Mukhin, E. B. Ozerov, B. M. Pontecorvo, E. L. Grigoriev, and N. A. Mitin, "Positive Pion–Proton Scattering at the Energies 176, 200, 240, 270, 307, and 310 MeV," in *Proceedings of the CERN Symposium on High Energy Accelerators and Pion Physics, 1956*, edited by E. Regenstreif (Geneva: CERN, 1956), pp. 204–24.

23 R. Cool, O. Piccioni, and D. Clark, "Pion–Proton Total Cross Sections from 0.45 to 1.9 BeV," *Phys. Rev. 103* (1956), 1082–97.

24 H. C. Burrowes, D. O. Caldwell, D. H. Frisch, D. A. Hill, D. M. Ritson, R. A. Schluter, and M. A. Wahlig, "Pion–Proton Resonance Scattering near 900 MeV," *Phys. Rev. Lett. 2* (1959), 119–21.

25 J. W. DeWire, H. E. Jackson, and Raphael Littauer, "High Energy Photoproduction of π^0 Mesons from Hydrogen," *Phys. Rev. 110* (1958), 1208–9.

26 P. C. Stein and K. C. Rogers, "Photoproduction of π^0 Mesons from Hydrogen at 500–900 MeV," *Phys. Rev. 110* (1958), 1209–11.

27 M. Heinberg, W. A. McClelland, F. Turkot, W. M. Woodward, R. R. Wilson, and D. M. Zipoy, "Photoproduction of π^+ Mesons from Hydrogen in the Region 350–900 MeV," *Phys. Rev. 110* (1958), 1211–12.

28 Robert R. Wilson, "Possible New Isobaric State of the Proton," *Phys. Rev. 110* (1958), 1212–13.

29 Ronald F. Peierls, "Photopion Cross Sections and a Second Resonance," *Phys. Rev. Lett. 1* (1958), 174–5.

30 J. I. Vette, "Photoproduction of Neutral Pions at Energies 500 to 940 MeV," *Phys. Rev. 111* (1958), 622–31.

31 F. P. Dixon and R. L. Walker, "Photoproduction of Single Positive Pions from Hydrogen in the 500–1,000 MeV Region," *Phys. Rev. Lett. 1* (1958), 142–3.

32 F. P. Dixon and R. L. Walker, "Photoproduction of Single Positive Pions from Hydrogen in the 600 to 1,000 MeV Region," *Phys. Rev. Lett. 1* (1958), 458–60.

33 Ronald F. Peierls, "Higher Resonances in Pion–Nucleon Interactions," *Phys. Rev. 118* (1960), 325–35.

34 P. C. Stein, "Polarization of Recoil Protons from the Photoproduction of π^0 Mesons from Hydrogen," *Phys. Rev. Lett. 2* (1959), 473–5.

35 P. L. Connolly and R. Weill, "Polarization in the Photoproduction of Pions," *Bull. Am. Phys. Soc. 4* (1959), 23.

36 For an early summary, see R. L. Walker, "Single Pion Photoproduction in the Resonance Region," in *Proceedings of the Fourth International Symposium on Electron and Photon Interactions at High Energies*, edited by D. W. Braben (Liverpool: Daresbury Nuclear Physics Laboratory, 1969), pp. 23–41.

7 A personal view of nucleon structure as revealed by electron scattering

ROBERT HOFSTADTER

Born 1915, New York City; Ph.D., Princeton University, 1938; experimental physics; Nobel Prize, 1961, for studies of electron scattering in nuclei and discoveries concerning the structure of the nucleons; Stanford University

I was graduated from City College of New York in 1935 after majoring in physics and mathematics. In my early years in college I was stimulated by my teachers Irving Lowen and Mark Zemansky. From both I learned to appreciate clarity and precision in thinking. I believe I owe to Zemansky the opportunity of entering the Princeton Graduate School in 1935. I suppose he wrote good letters of recommendation to the Princeton Admission Committee and to the General Electric Company's board of advisors for their Charles A. Coffin Foundation Fellowship. I had the Coffin Fellowship for the year 1935–6, and a stipulation of that fellowship was that I had to be involved in research, even in my first year of graduate study. This condition weighed pretty heavily on me, and as a result of advice from Harry Smyth, then chairman of the Physics Department, I started working with Louis Ridenour on a Wilson cloud-chamber project. Earlier, I had given some theoretical assistance to Walker Bleakney in his design of a new mass spectrograph operating on a cycloidal trajectory principle. Because of the time I spent on research, I really had much less time than I wanted for course work, in particular in the quantum mechanics course I took from Eugene Wigner.

In my second year, now with an assistantship to E. U. Condon, I really tried hard to become a theorist. I was strongly influenced by Condon, and now that I can think back about it, I should have been. I knew Condon very well at that time, and I believe he was, and remains in my memory, one of the brightest and most brilliant physicists I have ever met. I say this after long experience as a physicist. His elegance of thought and expression, his ideas, his spirit, and the beauty of his blackboard work, which he carried out with his left-handed writing, made deep and long-lasting impressions on me. Moreover, I loved his sense of humor. He was writing his book on *The Theory of Atomic Spectra* with George Shortley at the time, and he asked me to help in making some energy-level calculations. After a while, I guess he thought that I was not cut out to be a theorist, or else, and more probably, he needed experimental help in checking predictions he had made in his work on molecular physics.

He asked me to work experimentally with R. Bowling Barnes in the infrared laboratory Barnes had just built at Princeton. Condon was very interested in Barnes's work, and so I couldn't say no – or at least I didn't at that time, mostly, I suppose, because I was in awe of the great physicists at Princeton, and I felt they, and he, couldn't be wrong. As it developed, Barnes left Princeton almost immediately to found a company that subsequently did very well. That left me and Bob Brattain, an older graduate student, and Lyman Bonner, a research associate, as the senior people in the infrared project. We were joined a bit later by Bob Herman, who entered Princeton as a new graduate student.

In 1937, unexpectedly, Condon himself left Princeton to direct research at Westinghouse, and so I was left without an advisor or sponsor. I continued to work in infrared spectroscopy, more or less independently on my thesis work, with occasional encouragement from Rudolf Ladenburg. After three years at Princeton, that is, in 1938, I received my master's and Ph.D. degrees on a study of the hydrogen bond in light and heavy deuterated formic acid. I was one of the first, if not the first, to measure the distance between oxygen and hydrogen ions in the famous "hydrogen bond." This earned me a footnote in Linus Pauling's celebrated book on the *Nature of the Chemical Bond*, whose second edition appeared in 1940. Herman and I worked together on this problem, and Bob continued with it in his subsequent research.

I knew Frederick Seitz at the time, both because he had worked with Barnes and Condon and because he was a former graduate student of Wigner. Seitz came frequently to Princeton to consult with Wigner. Since I had been a student in a number of Wigner's classes, Eugene knew me pretty well, and in that close community I saw Fred on a number of occasions. This acquaintance led to Fred's invitation to me to join his project at the General Electric Research Laboratory, where he was working on the theory of luminescence. In the summer of 1938, I joined Seitz at Schenectady, where I worked on the photoconductivity of willemite (zinc orthosilicate). I became interested in this new field, and when I returned to Princeton for an additional year as a postdoctoral Proctor Fellow in 1938–9, I took up further research on this project with Herman, who was then finishing his Ph.D. work in infrared spectroscopy. Together we discovered in willemite the crystal warm-up dark currents that helped establish the theory of electron release from deep traps in crystals. The work on willemite was my introduction to the solid-state field and led to my familiarity with the alkali halide and F-center literature, which were to become so very important for my later work on nuclear detectors. However, I was really fascinated by nuclear physics and particle physics, particularly after reading Hans Bethe and Robert Bacher's wonderful 1936 article in the *Reviews of Modern Physics* and, later, Bruno Rossi and Ken Greisen's 1941 article on cosmic-ray theory.

When Seitz took a professorial position at the University of Pennsylvania, he and Gaylord Harnwell invited me to go along in 1939 as a Harrison Fellow. I suppose Seitz thought I would work in solid state, but I chose to do nuclear physics, this time with Ridenour, who had also joined Harnwell at Pennsylvania as a professor of physics. Ridenour was in charge of building a large pressurized Van de Graaff generator for nuclear physics studies. Leonard Schiff joined the Pennsylvania Physics Department in 1940, and he and I became immediate friends. Very shortly thereafter, preparations for war work became a vital necessity, and Schiff and I did a project together on a method of measuring the purity of helium gas. We published a paper on this subject in the *Review of Scientific Instruments.*

I was rather independent, and I believe I must have irritated some people in the Physics Department. I could see that I would not do well at Pennsylvania. I therefore moved to City College of New York for one semester in 1941. Shortly afterward, the United States entered World War II. I moved into war work at this time by making an unsolicited application to the National Bureau of Standards in Washington, D. C., where I was accepted for work on the optical proximity fuse, with Joseph Henderson and Seth Neddermeyer, who both became lifelong friends.

The optical fuse suffered by comparison with the radio proximity fuse of Merle Tuve and Harry Diamond, and the future for it looked bleak. I therefore resigned, in the middle of the war, and went to work at the Norden Laboratory, a branch of the wartime Norden Company that made the famous bombsight. It was through Don Bayley, a physicist friend from Pennsylvania, that I was able to obtain this job without immediately being assigned to military duty. At Norden, I worked on a radar altitude controller to serve as part of an automatic pilot for aircraft control.

When the war was over, I was offered, and accepted, a position in 1946 as an assistant professor at Princeton. I had a heavy teaching schedule, but started to do research on detecting γ-rays. I intended that this work should be useful for the cyclotron research that was directed at Princeton by Milton White. I started by developing a thallium halide crystal conduction counter.

In the middle of that effort, word came through Martin Deutsch in 1947 that Hartmut Kallmann in Germany had made a scintillation counter for γ-rays with naphthalene as the crystal detector. My previous exposure to the solid-state field enabled me to discover quickly in 1948 that NaI(Tl) and other activated alkali halides made far better detectors than naphthalene, or even anthracene, which had just been discovered by P. R. Bell. I became rather well known in nuclear physics circles immediately after I published several papers on the properties of the NaI(Tl) scintillator detector. My second graduate student, Jack McIntyre, and I then made the startling discovery in

1950 that allowed us to see sharp γ-ray lines with NaI(Tl), and so we established that γ-ray spectroscopy could be done with our detector. I described the history of the NaI(Tl) discovery and what it has meant to science and medicine in an IEEE publication in 1975.[1]

In spite of this achievement, Princeton did not offer me a promotion to an associate professorship. While I had just begun in 1950 to realize this would be the case, I was offered simultaneously two positions in California. One was at Stanford, where Schiff was the new chairman of the Physics Department, and the other was at the Radiation Laboratory at Berkeley. I believe that Simon Sonkin made the suggestion to Schiff that I might be available. Edwin McMillan offered me the second position at Berkeley with a promise to appointment as associate professor in the following year. McMillan knew me because I had spent the previous summer (1949) at his electron-synchrotron laboratory, where McIntyre and I had carried out an experiment on the Compton effect at several hundred MeV. McIntyre and I brought the scintillation-counter technique to Berkeley that summer, and there were several appreciative physicists who realized the potential value of this new technique using our inorganic and organic scintillators. Among them were Luis Alvarez, Ernest Lawrence, Jack Steinberger, and Pief Panofsky, and, at Stanford, Feiix Bloch and Schiff. It wasn't easy to choose between these two good offers, but I wanted a pure academic position immediately, and therefore I took the Stanford offer of an associate professorship. I left Princeton in late August of 1950, and McIntyre joined me at Stanford in September of 1950.

On the car trip to Stanford, my wife and I stopped at St. Louis to see Hilda and Eugene Feenberg, who were at Washington University. When Eugene asked me what I was going to do at Stanford, I replied that I was going to work on the new electron linear accelerator, which Bill Hansen started and Ed Ginzton was completing, and I was going to do high-energy physics of the kind I had already done at Berkeley. I knew I wanted to detect high-energy electrons and γ-rays, and I had the means to detect electromagnetic showers with our new crystals of NaI(Tl). Sometime during our conversation, Eugene remarked, "Why not do electron diffraction like the earlier work on atoms?" This suggestion appealed to me, and I thought about it on the rest of the auto trip to California. I didn't know that Schiff had already proposed electron scattering studies of nuclei and hydrogen in 1949.[2]

When I arrived at Stanford, I knew I wanted to use large NaI(Tl) crystals and other crystals as total-absorption energy detectors. In addition, I now wanted to do electron scattering on nuclei similar to G. P. Thomson's work on electron diffraction in crystals. I could start working immediately with large NaI(Tl) crystals by using cosmic rays. Thus, graduate student Alvin Hudson and I studied the energy loss of cosmic-ray muons in NaI(Tl).[3] As soon as the linear accelerator worked even a little, I asked graduate student Asher Kantz to begin a study of electromagnetic showers. This he did, and together we published some of the nicest data I have ever taken.[4,5]

Figure 7.1. The 180° double-focusing magnetic spectrometer, supported on a gun mount, that was used in the early Stanford work on electron scattering. Target chamber and monitor are also shown [*Source*: Hofstadter, Nobel Lecture, 1961[39]; © Les Prix Nobel 1961].

At the same time, I tried to develop the electron scattering tools we would need to study nuclear structure. The large NaI(Tl) crystals could not be used because of pile-up due to the extremely low duty cycle of the linear accelerator, and so I followed the early work of E. M. Lyman, A. O. Hanson, and M. B. Scott at 15.7 MeV at Illinois by using a magnetic spectrometer.[6] I needed a very good double-focusing magnetic spectrometer, and so I used the 180° design of the Caltech nuclear physics group. This kind of detector could perform well even with the terrible duty cycle of the linear accelerator. Through Bloch and Ginzton's ONR support and a $5,000 grant from the Research Corporation, I managed to construct not one, but two of them, and they were mounted on an obsolete naval gun mount I obtained from the U.S. Navy. A gun mount was used because it was precise, could carry the five-ton weight, and was cost-free. The one shown in Figure 7.1 was assigned to graduate student Harry Fechter and research associate McIntyre, who had come from Princeton with me. This spectrometer was reserved for electron scattering

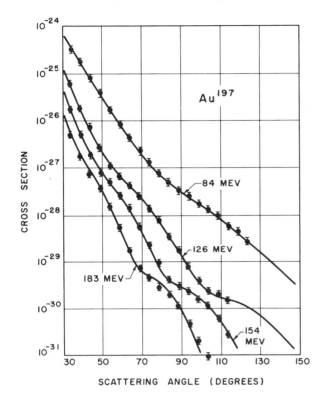

Figure 7.2. Some of the electron scattering experimental results in ^{197}Au and (solid lines) theoretical fittings by D. G. Ravenhall and D. R. Yennie [*Source*: Hofstadter et al. (1953)[7]]. Point-size gold nuclei would give theoretical curves several factors of ten above the fits for the various energies.

studies. The other spectrometer was assigned to graduate student Dick Helm, with the problem of observing the γ-rays following capture of muons in so-called μ-mesic atoms. The linear accelerator's duty cycle proved to give too much neutron background for this study, and besides, James Rainwater and Val Fitch at Columbia were able to obtain very nice data, while we were still struggling with the background problem. So Helm and I abandoned this effort, and he joined me in the subsequent electron scattering investigations.

The electron scattering work was successful immediately, and strong deviations from point-size scattering were seen in the case of gold (Figure 7.2).[7] In this same experiment, scattering from individual protons in hydrogen was observed for the first time. The protons in this case were hydrogen nuclei in polyethylene (CH_2), and the proton elastic peak could easily be separated from the carbon peak because of nuclear recoil. At about this time the first

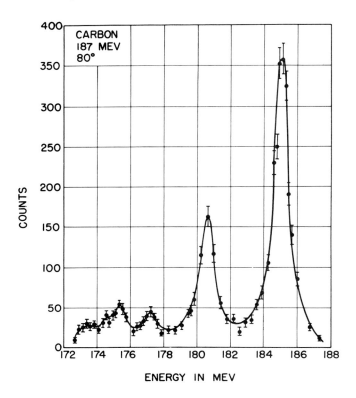

Figure 7.3. Elastic and inelastic electron scattering from carbon at 80° and 187 MeV [*Source*: Fregeau and Hofstadter (1955)[10]].

use of a synchrotron in electron scattering was reported by R. W. Pidd and associates[8] at electron energies of 30–45 MeV.

Proton and deuteron elastic peaks were clearly separated in CH_2 and CD_2, as described in a paper[9] following up the earlier results.[7] Although the main subject of this paper was concerned with nuclei, such as Be, C, Ta, Au, and Pb, the point was made that the proton and deuteron elastic scattering cross sections could themselves be measured relative to carbon in CH_2 or CD_2 as a standard. In a subsequent paper it was shown for the first time that inelastic electron scattering could be observed from known energy levels in the beryllium nucleus, and later in carbon.[10] Figure 7.3 shows a typical result in carbon from the work of graduate student J. H. Fregeau and Hofstadter.

An experiment specifically designed to study finite size effects in the proton was carried out in early 1955 by graduate student R. W. McAllister and me.[11] In this work, scattering from individual protons could be observed because a gaseous hydrogen target was used. Deviations from the Rosenbluth point-

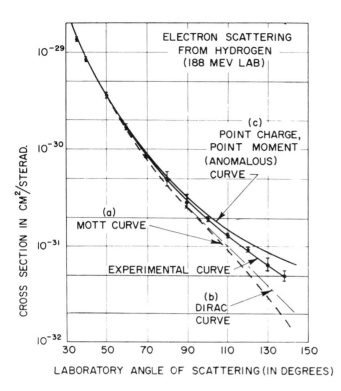

Figure 7.4. Some of the first experimental results on electron scattering from the proton [*Source*: Hofstadter and McAllister (1955)[11]].

charge behavior were observed immediately. Figure 7.4 shows some of these first experimental results. I quote a remark from this paper:

> Nevertheless, if we make the naive assumption that the proton charge cloud and its magnetic moment are both spread out in the same proportions, we can calculate simple form factors for various values of the proton "size." When these calculations are carried out we find that the experimental curves can be represented very well by the following choices of size. At 188 MeV the data are fitted accurately by an rms radius of $(7.0 \pm 2.4) \times 10^{-14}$ cm. At 236 MeV, the data are well fitted by an rms radius of $(7.8 \pm 2.4) \times 10^{-14}$ cm.

It was also noted that if this interpretation was correct, the Coulomb law of force was not violated at distances as small as 7×10^{-14} cm. These conclusions were subsequently tested at higher energies by graduate student E. E. Chambers and me and were verified in much greater detail. Deviations by a factor of 10 from point-charge scattering were easily seen with a new larger magnetic spectrometer capable of reaching 500 MeV scattered-electron

Figure 7.5. A view of the 1,000-MeV magnetic spectrometer, to the right, and a smaller 500-MeV spectrometer at the left [*Source*: Hofstadter, Nobel Lecture, 1961[39]; © Les Prix Nobel 1961].

energy. A still larger spectrometer, to the right in Figure 7.5, was used in most of the subsequent studies.

The Rosenbluth formulation is given in equations (7.1)–(7.3):

$$\frac{d\sigma}{d\Omega} = \sigma_{NS}\left\{ F_1^2 + \frac{\hbar^2 q^2}{4M^2 c^2}\left[2(F_1 + \varkappa F_2)^2\tan^2\frac{\theta}{2} + \varkappa^2 F_2^2 \right]\right\} \tag{7.1}$$

with

$$\sigma_{NS} = \left(\frac{e^2}{2E}\right)^2 \frac{\cos^2(\theta/2)}{\sin^4(\theta/2)} \frac{1}{1 + (2E/Mc^2)\sin^2(\theta/2)} \tag{7.2}$$

and

$$q = \frac{(2E/\hbar c)\sin(\theta/2)}{[1 + (2E/Mc^2)\sin^2(\theta/2)]^{1/2}} \tag{7.3}$$

in which F_1 and F_2 are the Dirac and Pauli (anomalous magnetic moment) form factors respectively, q is the momentum transfer, M is the mass of the nucleon, and other quantities have their conventional meanings. F_1 and F_2 are functions of q.

In the work of Chambers and Hofstadter, the best-fitting phenomenological model of the charge and magnetic-moment distributions corresponded to an exponential model.[12] This model could be obtained by inverting the non-relativistic form factor

$$F(q) = \frac{4\pi}{q} \int_0^\infty \varrho(r) (\sin qr) r \, dr \qquad (7.4)$$

to obtain the density function, $\varrho(r)$, from the known values of phenomenological form factors F_1 and F_2. Although we realized that the nonrelativistic result could not hold at the short distances within the proton, nevertheless the form factor in equation (7.5) was used by us in a phenomenological sense and corresponds to a spherical exponential model, with rms radius a, as follows:

$$F(q) = \frac{1}{[1 + (q^2 a^2/12)]^2} \qquad (7.5)$$

This model subsequently has been called a "dipole" model, and its applications have now survived the last thirty years. More about this later. Incidentally, McAllister and I were the first to determine the size and charge distribution of the α particle at the same time that we were doing the proton studies.

In the interpretation of the early work on nucleon form factors, we were greatly aided by the theoretical work and insights of Don Yennie and Geoff Ravenhall, who not only developed the precise phase-shift method of calculating from electron scattering the charge distributions of heavy nuclei but also clarified our notions of the form factors of the nucleons themselves.[13] Schiff's graduate student, V. Z. Jankus, calculated electron–deuteron cross sections in anticipation of elastic and inelastic studies of the deuteron.[14] The experimental work on the deuteron was carried out by Mason Yearian and me in a study designed to find the magnetic form factor of the neutron.[15] Jankus's inelastic-continuum calculations were used to determine the approximate form factors of the neutron. A method of determining neutron form factors was first suggested in a review article I wrote in 1956, in which the neutron and proton were assumed to act incoherently in the inelastic breakup of the deuteron by the scattered electron.[16] An example of the deuteron's inelastic continuum is shown in Figure 7.6.

Subsequently, both the incoherent method ("area" method) and the amplitude method ("peak" method) were used to measure the two neutron form factors. Again, in this early work, the dipole model (exponential distribution) could fit the magnetic data for the neutron with an rms radius lying between 8.0 and 9.0×10^{-14} cm. In the inelastic continuum we could observe the constituent behavior of neutron and proton in the deuteron, and the inelastic spectrum brought out clearly the momentum distribution of the nucleons in the bound state.

I want to emphasize that the electron scattering method brought out the physics of the nucleons in the deuteron in a very clear fashion, and this made our understanding and interpretation quite clear. Similar studies were made

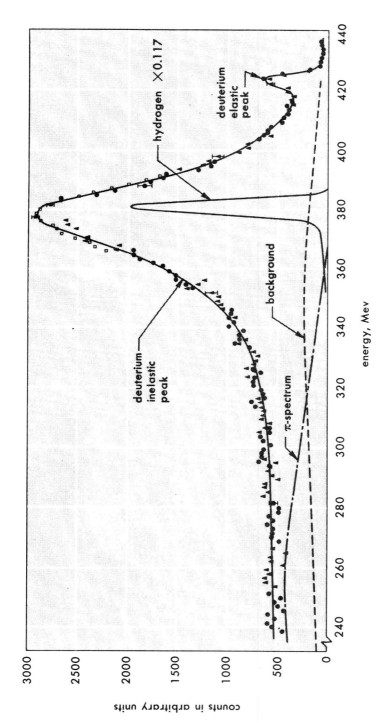

Figure 7.6. The inelastic continuum of the deuteron, the elastic peak, and the region of meson production. A comparison proton peak is also shown [*Source:* Hofstadter (1956)[16]].

of the nucleons within the α particle and in beryllium. At this time, many of the phenomena of nuclear physics and elementary particle physics were being revealed simultaneously by the electron scattering method.

Following these experimental discoveries, a number of theoretical developments took place. One of the most important was that of Yoichiro Nambu, who predicted that the proton and neutron form factors could be explained by the existence of a new heavy neutral meson.[17] Although Nambu called this meson "ϱ_0," it is now called the ω meson and was subsequently discovered by a Berkeley group in 1961 as a three-pion resonance decay produced by $\bar{p}p$ events at 1.61 GeV/c.[18] Nambu also developed a mass-spectral representation of the nucleon's form factor. This representation has been in use ever since. Two later theoretical developments by Geoffrey Chew and associates[19] and by P. Federbush and associates[20] represent the basic works on the use of dispersion-theory methods in developing a modern theory of electromagnetic structure of the nucleon. The mass-spectral representations of the form factors were justified in these two papers. In place of the often-tried and rather unsuccessful methods of perturbation theory, these newer methods showed that the nucleon form factors could be written simply as a *sum*, in terms of the masses of the intermediate states, particularly of those states with the lightest masses. The meson of Nambu was one of these lightest states, with a presently known mass of 783 MeV. A two-pion resonance was suspected from a number of separate strong-interaction experiments at this time (1959–61) and was shown to be a possible contributor to the electromagnetic form factors of the nucleon. Typical theoretical papers pointing to such a possibility were written by W. R. Frazer and J. R. Fulco,[21] Gregory Breit,[22] J. Bowcock and associates,[23] J. J. Sakurai,[24] and S. Bergia and associates.[25] Clear experimental evidence for the two-pion resonance, now called the ϱ meson at 770 MeV, was not long in appearing.[26–28]

The proton and neutron "sizes" were determined very early in our work in the fifties, and the detailed shape of the form factors was worked on by our group subsequently for several years. Some of our last data[29–33] were taken at a time when I could see clearly that the next great advances would require a much larger electron linear accelerator that could provide much higher energies. Even earlier than this I had already suggested building a one-mile machine at Stanford and was strongly supported in this suggestion by Ginzton. Our latest data, obtained at the end of the fifties, reinforced this early idea. More could be said about the origin of the SLAC accelerator, which was built subsequently in the sixties, but this is not the occasion for doing so.

At the beginning of the sixties, other groups started to work on the proton and neutron form factors, notably the Cornell group.[34] In any case, some of our very last data are shown in Figures 7.7 and 7.8 for the proton and the neutron.[35] Figure 7.7 shows F_1 and F_2, and Figure 7.8 shows the modified form factors F_{ch} and F_{mag}, sometimes also indicated by G symbols, which are preferred by some authors. The modified F's (or G's) can be interpreted more

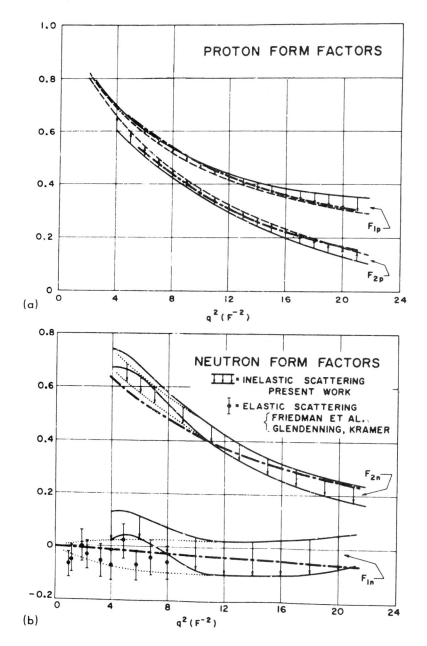

Figure 7.7. Form factors F_1 and F_2 for both proton and neutron [*Source*: de Vries et al. (1962)[35]].

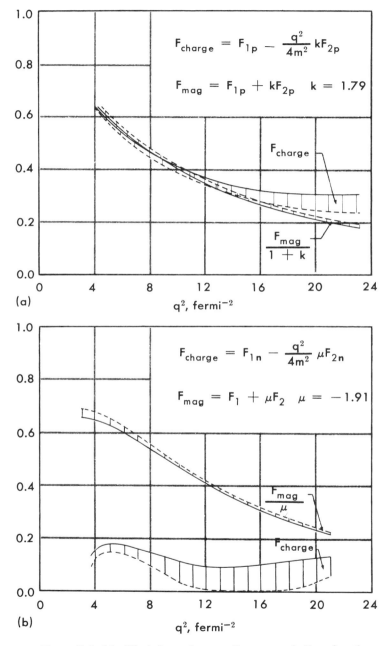

Figure 7.8. Modified form factors $F_{\text{ch(arge)}}$ and F_{mag} for the proton and neutron [*Source*: de Vries et al. (1962)[35]].

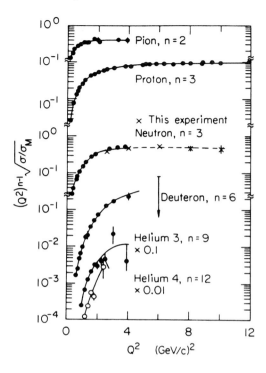

Figure 7.9. Recent magnetic form factors for proton and neutron [*Source:* Rock et al. (1982)[37]].

easily in terms of charge and magnetic-moment concepts.[36] Lately, the F_1 and F_2 have also returned to favor. Appropriate definitions of the modified form factors are shown in Figure 7.8. We have pointed out[35] that at least one other heavy meson was needed to produce the very best fits of our data. The ϕ meson was discovered a bit later, with a mass close to 1,020 MeV.

It is appropriate to make a comment on very recent elastic nucleon form factors, which cover q^2 regions twelve to twenty-five times those of our data of the fifties. Figure 7.9 shows recent magnetic proton and neutron elastic form factors.[37] Note the scale, because our old data lie below or near 1 GeV^2. The form factors in the figure are expressed in terms of a product of $(q^2)^{n-1}$ and the square root of the ratio of the cross sections to the Mott (σ_m) cross section. Here one studies deviations from q^4 behavior ($n = 3$). The power $n = 3$ corresponds to the asymptotic region of our old dipole form factor, and also to the number of elementary quark constituents of the nucleon. The deviations from the dipole model are small, as seen in Figure 7.10, but are now being investigated extensively.[38] Such deviations provide information on the details of the quark–gluon vertex and the pointlike character of quarks and their

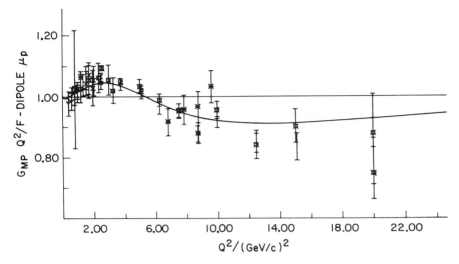

Figure 7.10. Study of the proton's magnetic form factor compared with a dipole model, which is represented by the horizontal line at ordinate 1.00 [*Source*: Gari and Krumpelmann, "Semiphenomenological Synthesis ...," *Z. Phys. A322*(4):689; © 1985 Springer–Verlag].

interactions. It is astonishing to me that our old exponential (i.e., dipole) model has held up so well over the years.

Our experiments in the fifties ended a fruitful era in my career as a physicist, and I am tempted to make some nostalgic remarks about those times. The research went easily and straightforwardly. The Office of Naval Research was very kind in supporting us during those years. I was lucky, and I experienced a complete decade in which I, my graduate students, and my colleagues could study so many of the fundamental properties of nuclei and nucleons in very small and intimate groups, rarely exceeding three physicists. With such small groups and with experimental apparatus that just a few of us could design, build, and handle comfortably in our experiments, we were able to enjoy what might be thought of as a monopoly, albeit a temporary one, in the fundamental field of nucleon and nuclear structure. Those were certainly good times for us, and because I never liked working in a highly competitive atmosphere, they were right for me. It is difficult to know if something of this sort can happen again, but I think it will be a pity if fundamental particle research cannot be done once more by small groups of physicists, such as we were in the fifties.

Notes

1 R. Hofstadter, "Twenty-five Years of Scintillation Counting," *IEEE Trans. Nucl. Sci. 22* (1975), 13–25; G. J. Hine, "The Inception of Photoelectric Scintillation Detection Commemorated after Three Decades," *J. Nucl. Med. 18* (1977), 867.

2 L. I. Schiff, Microwave Laboratory technical report no. 102, Stanford University, November 1949.

3 A. Hudson and R. Hofstadter, "Energy Loss of Cosmic-Ray Mu-Mesons in Sodium Iodide Crystals," *Phys. Rev. 88* (1952), 589–96.

4 A. Kantz and R. Hofstadter, "Electron-Induced Showers in Copper," *Phys. Rev. 89* (1953), 607–17; D. G. Ravenhall and D. R. Yennie, "Results of a Phase Shift Calculation of High-Energy Electron Scattering," *Phys. Rev. 96* (1954), 239–40.

5 A. Kantz and R. Hofstadter, "Large Scintillators, Cerenkov Counters for High Energies," *Nucleonics 12* (1954), 36–43.

6 E. M. Lyman, A. O. Hanson, and M. B. Scott, "Scattering of 15 · 7-MeV Electrons by Nuclei," *Phys. Rev. 84* (1951), 626–34.

7 R. Hofstadter, H. R. Fechter, and J. A. McIntyre, "Scattering of High-Energy Electrons and the Method of Nuclear Recoil," *Phys. Rev. 91* (1953), 422.

8 R. W. Pidd, C. L. Hammer, and E. C. Raka, "High-Energy Electron Scattering by Nuclei," *Phys. Rev. 92* (1953), 436–7.

9 R. Hofstadter, H. R. Fechter, and J. A. McIntyre, "High-Energy Electron Scattering and Nuclear Structure Determinations," *Phys. Rev. 92* (1953), 978–87.

10 J. A. McIntyre, B. Hahn, and R. Hofstadter, "Inelastic Scattering of 190-MeV Electrons in Beryllium," *Phys. Rev. 94* (1954), 1084–5; J. H. Fregeau and R. Hofstadter, "High-Energy Electron Scattering and Nuclear Structure Determinations. III. Carbon-12 Nucleus," *Phys. Rev. 99* (1955), 1503–9.

11 R. Hofstadter and R. W. McAllister, "Electron Scattering from the Proton," *Phys. Rev. 98* (1955), 217–18.

12 E. E. Chambers and R. Hofstadter, "Structures of the Proton," *Phys. Rev. 103* (1956), 1454–63.

13 D. R. Yennie, M. M. Levy, and D. G. Ravenhall, "Electromagnetic Structure of Nucleons," *Rev. Mod. Phys. 29* (1957), 144–57.

14 V. Z. Jankus, "Calculation of Electron–Deuteron Scattering Cross Sections," *Phys. Rev. 102* (1956), 1586–91.

15 M. R. Yearian and R. Hofstadter, "Magnetic Form Factor of the Neutron," *Phys. Rev. 102* (1958), 552–64.

16 R. Hofstadter, "Electron Scattering and Nuclear Structure," *Rev. Mod. Phys. 28* (1956), 214–54.

17 Y. Nambu, "Possible Existence of a Heavy Neutral Meson," *Phys. Rev. 106* (1957), 1366–7.

18 B. C. Maglic, L. W. Alvarez, A. H. Rosenfeld, and M. L. Stevenson, "Evidence for a $T = 0$ Three-Pion Resonance," *Phys. Rev. Lett. 7* (1961), 178–82.

19 G. F. Chew, R. Karplus, S. Gasiorowicz, and F. Zachariasen, "Electromagnetic Structure of the Nucleon in Local-Field Theory," *Phys. Rev. 110* (1958), 265–76.

20 P. Federbush, M. L. Goldberger, and S. B. Treiman, "Electromagnetic Structure of the Nucleon," *Phys. Rev. 112* (1958), 642–65.

21 W. R. Frazer and J. R. Fulco, "Effect of a Pion–Pion Scattering Resonance on Nucleon Structure," *Phys. Rev. Lett. 2* (1959), 365–8.

22 G. Breit, "The Nucleon–Nucleon Spin-Orbit Potential," *Proc. Nat. Acad. Sci. U.S.A. 46* (1960), 746–53.

23 J. Bowcock, W. N. Cottingham, and D. Lurie, "Effect of a Pion–Pion Scattering Resonance on Low-Energy Pion–Nucleon Scattering," *Nuovo Cimento (Ser. 10) 16* (1960), 918–38.

24 J. J. Sakurai, "Theory of Strong Interactions," *Ann. Phys. (N.Y.) 11* (1960), 1–48.

25 S. Bergia, A. Stanghellini, S. Fubini, and C. Villi, "Electromagnetic Form Factors of the Nucleon and Pion–Pion Interaction," *Phys. Rev. Lett. 6* (1961), 367–71.

26 I. Derado, "Experimental Evidence for the Pion–Pion Interaction at 1 GeV," *Nuovo Cimento (Ser. 10) 15* (1960), 853–5.

27 D. Stonehill, C. Baltay, H. Courant, W. Fickinger, E. C. Fowler, H. Kraybill, J. Sandweiss, J. Sanford, and H. Taft, "Pion–Pion Interaction in Pion Production by $\pi^+ - p$ Collisions," *Phys. Rev. Lett. 6* (1961), 624–5.

28 A. R. Erwin, R. March, W. D. Walker, and E. West, "Evidence for $\pi-\pi$ Resonance in the $I = 1, J = 1$ State," *Phys. Rev. Lett.* 6 (1961), 628–30.

29 R. Hofstadter, F. Bumiller, and M. R. Yearian, "Electromagnetic Structure of the Proton and Neutron," *Rev. Mod. Phys. 30* (1958), 482–97.

30 F. Bumiller, M. Croissiaux, and R. Hofstadter, "Electron Scattering from the Proton," *Phys. Rev. Lett. 5* (1960), 261–3.

31 R. Hofstadter, F. Bumiller, and M. Croissiaux, "Splitting of the Proton Form Factors and Diffraction in the Proton," *Phys. Rev. Lett. 5* (1960), 263–5.

32 R. Hofstadter, C. de Vries, and R. Herman, "Dirac and Pauli Form Factors of the Neutron," *Phys. Rev. Lett. 6* (1961), 290–3.

33 R. Hofstadter and R. Herman, "Electric and Magnetic Structure of the Proton and Neutron," *Phys. Rev. Lett. 6* (1961), 293–6.

34 R. R. Wilson, K. Berkelman, J. M. Cassels, and D. N. Olson, "Scattering of High-Energy Electrons by Protons," *Nature (London) 188* (1960), 94–7; D. N. Olson, H. F. Schopper, and R. R. Wilson, "Electromagnetic Properties of the Proton and Neutron," *Phys. Rev. Lett. 6* (1961), 286–90.

35 C. de Vries, R. Hofstadter, and R. Herman, "Neutron Form Factors and Nucleon Structure," *Phys. Rev. Lett. 8* (1962), 381–5.

36 J. D. Walecka, "Spectral Functions in the Static Theory," *Nuovo Cimento 11* (1959), 821–36; F. J. Ernst, R. G. Sachs, and K. C. Wali, "Electromagnetic Form Factors of the Nucleon," *Phys. Rev. 119* (1960), 1105–14; R. G. Sachs, "High Energy Behavior of Nucleon Electromagnetic Form Factors," *Phys. Rev. 126* (1962), 2256–60; K. J. Barnes, "Choice of Electromagnetic Form Factors," *Phys. Lett. 1* (1962), 166–8; L. L. Foldy, "The Electromagnetic Properties of Dirac Particles," *Phys. Rev. 87* (1952), 688–93.

37 S. Rock, R. G. Arnold, B. Bosted, B. T. Chertok, B. A. Mecking, I. Schmidt, Z. M. Szalata, R. C. York, and R. Zdarko, "Measurement of Elastic Electron–Neutron Cross Sections up to $Q^2 = 10(\text{GeV}/c)^2$," *Phys. Rev. Lett. 49* (1982), 1139–42.

38 M. Gari and W. Krumpelmann, Report T. P. II/247, Ruhr University, Bochum, Germany; published as "Semiphenomenological Synthesis of Meson and Quark Dynamics and the EM Structure of the Nucleon," *Z. Phys. A322*(4) (1985), 689–93.

39 R. Hofstadter, "The Electron Scattering Method and Its Application to the Structure of Nuclei and Nucleons," Nobel Lecture, December 1961, Les Prix Nobel en 1961 (Stockholm: Nobel Foundation, 1962), pp. 114–35.

8 Comments on electromagnetic form factors of the nucleon

ROBERT G. SACHS

Born 1916, Hagerstown, Maryland; Ph.D., Johns Hopkins University, 1939; theoretical physics; University of Chicago

KAMESHWAR C. WALI

Born 1927, Bijapur, India; Ph.D., University of Wisconsin, 1959; theoretical physics; Syracuse University

These comments are intended to provide some historical background to a point that Robert Hofstadter touched on in Chapter 7. The point concerns the remark following his equation (7.4), to the effect that the nonrelativistic physical interpretation of the form factors as a Fourier transform of the charge density "could not hold at the short distances within the proton." One of us (R.G.S.) took an interest in this specific problem after hearing Hofstadter raise the same issue in a presentation of his work many years ago. At that time he called for some valid physical interpretation of the form factors in the relativistic region, corresponding to this nonrelativistic interpretation. As Hofstadter has often suggested, it is useful to have a physical insight into the meaning of the form factors in order to further their interpretation in terms of the structure of the nucleon.

After the static magnetic moments, the electron–nucleon ($e-N$) interaction in the limit of $q^2 = 0$ (q is the momentum transfer) was the first measurement that gave information about the extended structure of the nucleon. Leslie L. Foldy made the interesting observation that this $e-N$ interaction could be attributed to the anomalous magnetic moment of the neutron.[1] His calculation was based on a relativistic Hamiltonian, which he then extended by assuming that a complete effective relativistic Hamiltonian could be written as a power series in the d'Alembertian operator.[2] George Salzman was the first to point out that this was equivalent to the expansion of the matrix element of the nucleon current density in a power series in q^2, which could in turn be written as a linear combination of two functions, later defined as the form factors $F_1(q^2)$ and $F_2(q^2)$. These ultimately became known as the Dirac and the Pauli form factors, respectively. In the language of F_1 and F_2, the $e-N$ interaction could be expressed as the sum of two terms: One term was the charge radius $\langle r^2 \rangle_1$ associated with $F_1(q^2)$, and the other, the charge radius $\langle r^2 \rangle_2$ associated with $F_2(q^2)$.[3] The experimental value of the $e-N$ interaction was almost completely accounted for by $\langle r^2 \rangle_2$, implying that $\langle r^2 \rangle_1$, associated with the Dirac form factor, was extremely small, or if due to an extended source, the extended structure contained mostly neutral constituents. This result, which was consistent with Foldy's original interpretation, seemed paradoxical in the light of meson theories, in which $n \to p + \pi^-$ in addition to $n \to n + \pi^0$; also, these theories led to a sizable $\langle r^2 \rangle_1$ for the proton; it seemed to require fortuitous cancellations due to relativistic corrections or some other cause.[4]

Subsequently, Donald R. Yennie and associates showed that the combinations ($\varkappa$ is the anomalous magnetic moment of the nucleon in nuclear magnetons, M is the nucleon mass)

$$F_1(q^2) - \frac{q^2}{4M^2} \varkappa F_2(q^2) \qquad \text{and} \qquad F_1(q^2) + \varkappa F_2(q^2)$$

appear naturally as the form factors of the electromagnetic vertex of a spin-$\frac{1}{2}$ particle when one considers the problem from a group-theoretic point of view.[5] The two combinations were the two nonvanishing (spin-nonflip and spin-flip) amplitudes. They defined the first combination as the "charge" form factor and remarked on the ambiguities inherent in the identification of the electric-charge form factor.

But the relation between the momentum-space form factors and the spatial distributions of charge and magnetization remained unspecified in Yennie's analysis. Hofstadter's question (In what sense are the relativistically invariant momentum-space form factors related to the three-dimensional distributions of charge and magnetization?) remained unanswered. The difficulty was that these are static distributions defined in the rest frame, while the matrix element of the current density is taken between nucleon states of two different momenta, **p** and **p**′, corresponding to two different rest frames.

What was needed was a generalization of the notion of a static distribution in the rest frame to include the relativistic case, and the answer was found to lie in the choice of the Breit (or "brick wall") frame **p**′ = −**p** to replace the rest frame in defining "static" distributions. This frame had already been shown by Yennie and associates to be the crucial one for determining the symmetry properties of the matrix element. It was shown[6,7] that the combinations

$$F_{ch} = F_1 - \frac{q^2}{4M^2} \varkappa F_2 \qquad \text{and} \qquad F_{mag} = F_1 + \varkappa F_2$$

are the physically meaningful electromagnetic form factors, in the sense that they are the Fourier transforms of the spatial charge and magnetization distributions inside the nucleon in the Breit frame.

The electric-charge radius $\langle r^2 \rangle_{ch}$ was then related to $F'_{ch}(0)$. The measured $e-N$ interaction had to be compared directly with $F'_{ch}(0)$, not with either

$\langle r^2 \rangle_1$ or $\langle r^2 \rangle_2$. When theoretical calculations of $\langle er^2 \rangle$ in fixed-source meson theories were compared to $\langle r^2 \rangle_{ch}$ rather than $\langle r^2 \rangle_1$, there was no longer any contradiction.[6] This immediately removed the Foldy paradox mentioned earlier. Further, the high-momentum-transfer behavior of $F_{ch}(q^2)$ and of $F_{mag}(q^2)$ based on this physical interpretation[7] placed constraints on the high-momentum-transfer behavior of $F_1(q^2)$ and $F_2(q^2)$. The latter constraints played an important role in phenomenological models of form factors based on vector dominance.

Notes

1 L. L. Foldy, "The Electron–Neutron Interaction," *Phys. Rev. 83* (1951), 688.
2 L. L. Foldy, "The Electromagnetic Properties of Dirac Particles," *Phys. Rev. 87* (1952), 688–93.
3 G. Salzman, "Neutron–Electron Interaction," *Phys. Rev. 99* (1955), 973–9.
4 B. D. Fried, "The Electron–Neutron Interaction as Deduced from Pseudoscalar Meson Theory," *Phys. Rev. 88* (1952), 1142–9; S. B. Treiman and R. G. Sachs, "Neutron–Electron Interaction in Cutoff Theory," *Phys. Rev. 103* (1956), 435–9; see Salzman, note 3.
5 D. R. Yennie, M. M. Levy, and D. G. Ravenhall, "Electromagnetic Structure of Nucleons," *Rev. Mod. Phys. 29* (1957), 144–57 (Appendix).
6 F. J. Ernst, R. G. Sachs, and K. C. Wali, "Electromagnetic Form Factors of the Nucleon," *Phys. Rev. 119* (1960), 1105–14. It should be noted that Walecka also contributed to the physical interpretation; he showed that the F_{ch} and F_{mag} measure the interaction with static electric and magnetic fields, respectively. J. D. Walecka, "Spectral Functions in the Static Theory," *Nuovo Cimento 11* (1959), 821–36.
7 R. G. Sachs, "High-Energy Behavior of Nucleon Electromagnetic Form Factors," *Phys. Rev. 126* (1962), 2256–60.

PART IV

THE NEW LABORATORY

It will cost a billion dollars, ten billion volts 'twill give,
It will take five thousand scholars seven years to make it live . . .
   Take away your billion dollars, take away your tainted gold,
   You can keep your damn ten billion volts, my soul will not be sold.
         – © Arthur Roberts, "Take Away Your Billion Dollars"
         (also known as the Brookhaven National Anthem) (1946)

By toil and sweat the Soviets have reached ten billion volts,
Shall we downtrodden physicists submit? No, no – revolt! . . .
Oh, if we outbuild the Russians, it will be because we spend
Give, oh give those billion dollars, let them flow without an end.
         – © Arthur Roberts, "Ten Years Later" (1956)

Don Glaser watched and wondered as the bubbles rose within,
And said, how do those bubbles know just where they should begin?
And once the crucial question framed, the answer too was slated
Beneath our hero's onslaught fierce to be elucidated . . .
The bubble chamber now is made of sterner stuff than beer,
It's run by corporations huge on megabucks per year,
The beam dumps pulse, the cameras click, the great computers chatter,
Vast cohorts grind the answers out; some do, some do not matter . . .
         – © Arthur Roberts, "Birth of the Bubble Chamber," (1956)

9 The making of an accelerator physicist

MATTHEW SANDS

Born 1919, Oxford, Massachusetts; Ph.D., Massachusetts Institute of
Technology, 1948; experimental physics; University of California, Santa
Cruz (emeritus)

"Go west...."

As the decade of the 1950s began, there was still no accelerator with
an energy as high as 1 GeV.[1] Still, the stage was set for an almost explosive
development. The new principle of phase-stable acceleration, proposed by
Vladimir Veksler and Edwin McMillan in 1945, had already been put to use
in a large proton-synchrocyclotron at Berkeley (600 MeV) and in a rash
of electron-synchrotrons with energies near 300 MeV.[2] Indeed, by 1950, the
first two large proton-synchrotrons were under construction: the 3-GeV Cosmotron at the Brookhaven National Laboratory, and the 6-GeV Bevatron at
the Berkeley Radiation Laboratory.

These developments were, however, of little personal concern to me. I was,
at the time, a young faculty member at the Massachusetts Institute of
Technology, working on cosmic-ray research in the laboratory of Bruno
Rossi.[3] However, early in 1950, I began to give a little thought to the
possibility of switching at least part of my research to particle physics, using
the recently completed MIT electron-synchrotron.

Such thoughts were brusquely terminated in April of that year, when I was
forced, on one day's notice, to abandon my position at MIT and flee the state
of Massachusetts to avoid legal prosecution by an irate former wife. A day
later I found myself in New York City without money or job, and with quite
an uncertain future.

Fortunately, during World War II, I had become something of an expert in
the design of electronic instruments (under the tutelage of Joseph Kiethley at
the Naval Ordnance Laboratory and of William Higinbotham at the Los
Alamos Laboratory), and that knowledge now came to my rescue.[4] I was
immediately able to obtain work in the Electronics Department of the

recently founded Brookhaven National Laboratory, while I made inquiries about a more long-term job. I had a quick response from Robert Walker, a friend and colleague from the Los Alamos days. Walker had recently completed his graduate degree at Cornell University and had joined Robert Bacher at the California Institute of Technology. Bacher, whom I had also known at Los Alamos (as the head of the division in which I worked), was then organizing a high-energy physics group and preparing to build a high-energy electron-synchrotron). After an interview with Bacher at the spring meeting of the American Physical Society in Washington, D.C., I was invited to join the new Synchrotron Laboratory at Caltech, where I was to have a primary responsibility for building electronic instrumentation.

So it happened that – like many "young men" before me – I headed west to start a new life, arriving in Pasadena on 1 July 1950, where I was to begin a thirteen-year association that turned out to be the most productive and rewarding period of my scientific life.

The Caltech synchrotron (1950–2)

The first two years at Caltech were busy. In addition to building up a modern electronics laboratory and designing counting equipment for particle experiments, I found myself involved in building the Caltech synchrotron. We were, in a first phase, to build a 0.5-GeV electron accelerator using as a base the cast-off magnet, vacuum chamber, and a few other components that had served as a one-quarter-scale test model of the Berkeley Bevatron.

The founding of the Caltech Synchrotron Laboratory was, I suppose, greatly facilitated by the fact that we got "free" the hand-me-down equipment from Berkeley, as well as a cavernous empty laboratory building – ideal for our purposes – that had served for over a decade as the Optics Shop for building the great 5-m Hale Telescope of the Palomar Observatory. It was necessary only to get some outside support for building some of the remaining components: coils and power supplies for the magnet, an injector, an rf system, and the controls. This initial support – as well as ongoing support for the future – was provided by the three-year-old Atomic Energy Commission (for which Bacher was one of the initial commissioners).

At first, my contribution to the accelerator construction was to provide the electronic controls for one or another major component of the accelerator. But in time my involvement grew. I felt that I needed to understand more of the details about the accelerator to make appropriate designs for the controls. Also, when all of the pieces were in place and it came time to make the accelerator work, all of the six physicists in the laboratory were drawn into the process. It became customary then, and for several years to follow, that the "chief operator" of the synchrotron on each shift should be one of the research physicists.

I still remember some of the struggles we had to get the machine going. Beyond the usual problems with balky components, our main source of difficulty was the very low strength of the magnetic guide field (6 G) at

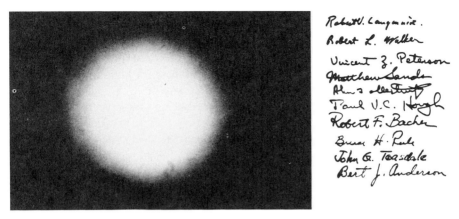

Figure 9.1. Signed photograph of spot produced by the Caltech synchrotron bremsstrahlung beam.

injection. The disturbances due to remanent magnetism, and due to eddy-current effects from the metal vacuum chamber and from the thick magnet laminations, were very large and had to be corrected carefully. Later we were to have serious problems of another and more welcome sort, namely, from high-current effects in the accelerated beam. These we never did understand. It was not until many years later that such collective effects began to be a major area of study in accelerator physics, an area in which I was to participate.

Two years to the day after my arrival in California – on 1 July 1952 – we got our first accelerated beam. I can recall the sense of excitement on seeing everything finally work and knowing that we had achieved the highest-ever energy of an electron beam. Immediately, everyone gathered around the control desk; then a piece of x-ray film was exposed to the high-energy bremsstrahlung beam, and everyone signed the photo (Figure 9.1). Besides Bacher, Walker, and me, there were our physicist colleagues: Robert Langmuir (who had worked on a small synchrotron at General Electric), Vincent Peterson and John Teasdale (graduates of the Berkeley Radiation Laboratory), and Alvin Tollestrup (from the Caltech nuclear physics laboratory), and our two engineers, Bruce Rule and Bert Anderson. Paul Hough was present as a summer visitor.

Italian sojourn, 1952–3

Unfortunately, I missed out on some of the excitement of the early photoproduction research with the synchrotron. Miscalculating the likelihood that I would be kept on at Caltech beyond the end of my initial two-year research appointment, I had applied for and received one of the newly created Fulbright Scholarships, which would take me to Italy for the 1952–3 academic year. So a few months after the successful turn-on of the syn-chrotron, I was installed at the University of Rome, where I intended to do some serious thinking about the state of particle physics. It was not to be.

On my arrival in Rome I heard from Edoardo Amaldi about the evolving plans for the creation of a joint European laboratory for high-energy physics – which was to become CERN. Consideration was being given to building a large proton-synchrotron using the new alternating-gradient (AG) principle that had been put forth some months earlier by Ernest Courant, M. Stanley Livingston, and Hartland Snyder.[5] I had read the original paper, and I re-membered having had some intuitive misgivings about the extreme gradients that had been suggested. I began to look into the matter a little more deeply. It then occurred to me that magnet imperfections would lead to a resonant growth of the betatron oscillations if their frequency was in integer multiple of the revolution frequency. At the large frequencies (of many hundreds) that were being contemplated, it would, I felt, be difficult to maintain the "tune" safely between two integer values. Weaker focusing would have to be used. Fortunately, Bruno Touschek had also just arrived in Rome. Perhaps because of his earlier work with Rolf Wideröe on a betatron, he became interested in my problem, and we collaborated on a paper on the integral resonances, which was published in *Nuovo Cimento* early in 1953.[6]

While the paper was in press, we heard that a group in England had also discovered the resonance problem and had also realized that there would be half-integer resonances.[7] (It is my impression that no one outside of the USSR has ever read our paper.) Because of the resonance problem, the two large proton accelerators being designed at Brookhaven and at CERN were at the end constructed with more moderate gradients. Touschek and I had many informal collaborations after that – most important on the problem of electron storage rings generally, and, more particularly, on the organization of a summer school on that subject in Varenna in 1969.[8]

Shortly before my arrival in Rome, some of the senior particle physicists in Italy (Amaldi and Gilberto Bernardini in Rome, and others) had begun to work toward obtaining a high-energy accelerator for Italy. I had once sug-gested in earlier correspondence with Amaldi that they could probably get the highest energy on a limited budget by building an electron-synchrotron. Early in 1953 a small group of physicists from several Italian universities was assembled under the leadership of Giorgio Salvini (then at Pisa) to study a possible project for a synchrotron with an energy near 1 GeV. In February of 1953 the group met at Pisa for a two-week workshop, and I was prevailed on to join them and to give a series of lectures – in my very newly minted Italian – on the theory and technique of electron-synchrotrons. The preparation of these lectures forced me for the first time to think through in a somewhat coherent way all that I had been learning about accelerators.

The Italian group went on to design and build a 1.2-GeV electron-synchrotron (completed in 1958), and in the process to create the Laboratori

Nazionali di Frascati outside of Rome – the laboratory that in 1960 began the pioneering work on electron–positron colliding beams, first with the storage-ring AdA and later with ADONE. At that first workshop in Pisa, I developed a number of lifelong friendships that led later to many pleasant and fruitful collaborations. The names that come to mind are Ruggiero Querzoli, Fernando Amman, Carlo Bernardini, Giorgio Corazza, Enrico Persico, and Mario Puglisi, among many others.

I finished off my year in Europe by accepting a request from Francis Perrin of the French Commissariat à l'Energie Atomique to spend the month of July 1953 at the Saclay laboratory. There I worked with a small group that had been assembled by Stan Winter (and included Henri Bruck, Robert Levy-Mandel, and José Taillet, among others) to lay out a preliminary design for an accelerator that was to become, five years later, their 3-GeV proton-synchrotron SATURNE.

Caltech – phase II

By the time of my return to the United States in the summer of 1953, I had been appointed to the Caltech faculty and took up my responsibilities: teaching courses in electronics, electromagnetic theory, optics, and others; designing electronic instruments; studying the photo production of pions;[9] and thinking about phase II of the synchrotron. Let me recount one aspect of this last work.

Our program had, from its inception, called for using the synchrotron at its phase I energy of 0.5 GeV for a few years, and then engaging in a major reconstruction that would provide a beam energy three times higher: 1.5 GeV. In 1953 I was asked to take an overall look at the contemplated design and to consider, in particular, whether or not the guide field should be based on the new AG principle. After a bit of study, I concluded that AG focusing would offer very little advantage for our particular machine – probably not enough to warrant taking a chance on possible unforeseen problems. We decided to proceed with a "weak-focusing" design. (That not very adventurous decision turned out to be a fortunate one.)

In going over the design, however, I discovered a disturbing fact: There would be an extremely strong radiation damping of the internal oscillations of the bunch.[10] The damping (whose rate is proportional to the cube of the electron energy in a given ring) would set in rather suddenly at some intermediate energy and reduce the initial bunch dimensions of some tens of centimeters to a submicrometer size by the end of the acceleration time. Such a small bunch would give greatly enhanced synchrotron radiation – so much so, in fact, that it would be impossible to reach the design energy, if indeed the beam would survive at all. When I pointed out this problem to my colleagues, no one was particularly worried. Their response was in the tradition of Ernest O. Lawrence: "We'll figure out how to make it work somehow."

I continued to stew about it, however, until I was hit by an inspiration.

Clearly, the radiation-damping theory that had been worked out by Julian Schwinger and others was incomplete.[11] The classical theory gives a smooth, continuous loss of energy, whereas we know that the radiation is emitted in photons of discrete energy. It follows that an accelerating electron loses energy discontinuously. The *average* rate of loss (as predicated correctly by the classical theory) will give *damping*, but the *quantum fluctuations* in loss will excite the same oscillations. The bunch dimensions will approach a stationary size at which there is a balance between the classical damping and the quantum excitation. I was able to show that this minimum bunch length would be of the order of centimeters.[12] We were saved. (Similar results were obtained independently and at nearly the same time in the USSR by A. A. Sokolov and I. M. Ternov, who were studying solutions to the Dirac equation in cylindrical magnetic fields.[13])

In 1956 the Caltech synchrotron was put into operation in its phase II. We immediately reached an energy of 1.2 GeV and recaptured the world championship for electron energy that we had earlier lost to Cornell. It was at once evident that my ideas about quantum excitation were correct. Indeed, it took us a couple of years to reach the design goal of 1.5 GeV, because the initial radio-frequency accelerating system did not provide a voltage quite high enough to contain the quantum-induced oscillations.[14]

Later the theory of radiation damping and excitation was extended to AG guide fields and to transverse oscillations by K. Robinson of the Cambridge Electron Accelerator and by A. A. Kolomensky of the Lebedev Institute.[15,16] These radiation effects turn out to be dominant in determining the behavior of stored beams in modern electron storage rings. It is still surprising to me that in an electron storage ring – which may have a diameter measured in tens of meters to tens of kilometers (clearly "classical" dimensions) – the dimensions of the circulating bunches are proportional to the quantum constant $\hbar$.

Accelerator heyday

The second half of the decade (1956–9) was clearly the heyday of the accelerator builder. Six accelerators with energies greater than 1 GeV, and ranging up to 30 GeV, were put into operation: proton accelerators at Dubna, Saclay, and CERN, and electron accelerators at Frascati and Orsay, besides ours at Caltech. Eleven more giga-electron-volt-range accelerators were under construction: proton machines at Moscow, Brookhaven, Princeton, Daresbury, and Argonne, and electron machines at Cambridge (Massachusetts), Lund, Tokyo, Kharkov, Tomsk, and Hamburg.[17]

This intense activity was highlighted by the initiation of a series of international conferences on high-energy accelerators, the first of which took place in 1956 at Geneva, hosted by CERN. One of the exciting aspects of that first conference was that I was able to meet for the first time my colleagues in accelerator science from the USSR and to hear firsthand reports on their work, which had been declassified only a short time earlier (for the "Atoms for Peace" conference in 1955). I remember, particularly, warm contact with Veksler, V. P. Dzhelepov, V. V. Vladimirski, A. M. Baldin, Kolomensky, S. M. Rubchinski, and F. A. Vodopianov.

Also in this brief period, two new ideas became public that were to be of great importance. The first was the concept of the fixed-field alternating-gradient (FFAG) guide field, in which a highly nonlinear, but *static*, magnetic field would provide stable orbits for a wide range of particle momenta (20 to 1 or more).[18] In such a device, particles could be accelerated rapidly – say in milliseconds (the time required to modulate a radio-frequency cavity) rather than in the more usual seconds (the time to ramp a large magnet). An accelerator with such a magnet promised a large increase – perhaps 10^3 or more – in the intensity of the output beam, and also might provide the possibility of "stacking" beams of very high current at the top energy. The FFAG concept was introduced and studied in detail at the Midwestern Universities Research Association (MURA), whose work led in the late 1950s to a proposal for a high-intensity 10- or 12-GeV proton accelerator. Among those working in the MURA group, I recall especially Donald Kerst, Keith Symon, Francis Cole, Jackson Laslett, Fred Mills, Lawrence Jones, and Kent Terwilliger.

The possibility of stationary, stacked beams of high current led to the second important new idea. Kerst and his collaborators proposed that sufficient concentrations of protons could be accumulated so that two such beams could be put into collision and yield a workable interaction rate – and the notion of a practical colliding-beam machine was born.[19] (Many had noted earlier the great advantage in center-of-mass energy that colliding beams of relativistic particles could provide over fixed-target machines, but the practical possibility had been elusive.)

The colliding-beam idea was quickly adapted to electrons by Gerard O'Neill of Princeton University,[20] and the first colliding-beam project was initiated by him with collaborators at Stanford University: C. Barber, B. Gittelman, and B. Richter. Soon thereafter, Touschek and his co-workers at Frascati (Bernardini, R. Bizzari, Corazza, G. Ghigo, and Querzoli) built the first operating electron–positron storage ring – the 250-MeV machine AdA[21] (a ring that unfortunately never reached the hoped-for collision rates).

I followed these developments with some interest, but with no direct involvement – although I did consult with O'Neill on the quantum effects to be expected in an electron storage ring. All that changed in 1959.

The MURA summer study, 1959

By 1959, with the completion of the CERN proton-synchrotron (PS) – soon to be followed by the Brookhaven AGS – the maximum available proton energy had reached approximately 30 GeV. For reasons never very clear to me, the interest of the high-energy physics community seemed to be turning toward machines of higher *intensity* rather than of higher energy. (In part, I suppose, the new 30-GeV machines seemed to be very large; in part, there were groups pushing certain high-intensity accelerator projects; and there may have been other, physics-related, considerations as well.)

An example of the public attitude is to be found in the report "A Proposed Federal Program on High-Energy Accelerator Physics," released by the White House in May 1959, which had been written by a distinguished panel chaired by E. R. Piore, with members J. W. Beams, Hans Bethe, Leland Haworth, and McMillan.[22] The conclusions were that

> The most urgent...needs at present are to extend *electron energies* [and] to improve *both electron and proton intensities*.... (emphasis added)

The report went on to recommend construction of the Stanford two-mile linear accelerator (SLAC) and continued support for the MURA work toward a high-intensity FFAG machine.

I hasten to add that the panel had not consulted me. Indeed, I did not generally share their opinion. It appeared to me that 30 GeV was still a "small" energy for particle physics. With collisions of 30-GeV protons on protons at rest, the available "collision energy" is only about 6 GeV, or about 3 GeV per nucleon – not a lot larger than the collision energy of 1 GeV we had available at our little Caltech synchrotron. And it seemed to me – on rather naïve, intuitive thinking – that if we were really to explore the inner workings of the proton, we would need to have collision energies much larger than the proton's rest energy of 1 GeV.

A colliding-beam machine with, say, 10 GeV in each beam would give a 20-GeV collision energy – a significant step above the 30-GeV machines – but seemed at that time to be somewhat speculative and, in addition, would not give high-energy secondary beams. On the other hand, if a proton beam of, say, 300 GeV were available, it would give a collision energy greater than 20 GeV, and high-energy secondary beams as well.

Only quite a bit later did I find out that an earlier (1958) advisory panel (of the National Science Foundation) not only had endorsed the Stanford and MURA projects but also had made the observation that "a 200 BeV accelerator, though expensive, may eventually be worthwhile."[23] This consideration seems to have been discarded in 1959.

At the time, I also had some concerns about large FFAG accelerators. They appeared to be technically very complex, demanded high precision, and would not, apparently, be very flexible in exploitation. Also, their realization might encounter unforeseen basic difficulties. Their cost would be large, probably something like $100 million – similar to that for SLAC, a number that seemed large – one and one-half times the annual federal support for high-energy physics at the time.

It was in this setting that I was invited to participate in a two-week study at MURA in the summer of 1959. Some forty or fifty physicists involved with accelerators (both their construction and use) were invited to study the evolving MURA proposal for a large FFAG facility and to offer criticism and suggestions for its improvement.

After arriving at Madison and hearing the initial "briefings" on the status of the MURA program, it appeared to me interesting to challenge the "conventional wisdom" that the logical next step in proton accelerators should be toward high intensity. I wanted to understand whether it was indeed true that a "really high-energy" proton-synchrotron (say 100 GeV or even higher) would cost that much more or, indeed, whether such a machine might not be built at a cost comparable to what was being contemplated for a 12-GeV FFAG machine. Drawing on the unusual resources available at the study – Courant, Brück, Hildred Blewett, Robinson, among others, were particularly helpful – I tackled the problem.

It became apparent right away that a straightforward extension of the existing 30-GeV designs – based on a linac injector and an AG guide field – to an energy of 100 GeV would probably be possible, but rather expensive. Magnets similar in size to the 30-GeV ones would be necessary to contain the lateral spread of the injected beam. And if low injection fields were to be avoided (our bugaboo at Caltech), an expensive, high-energy linac injector would be needed.

It then occurred to me that I could get around many of the difficulties by introducing a new concept that I dubbed the "cascade synchrotron." A low-energy linac would inject into an intermediate synchrotron, the "booster," which would provide a high-energy beam for injecting into the large, "main-ring" synchrotron. The high injection energy obtained this way would mean high initial magnetic fields in the main ring, and the rather large decrease in the lateral dimensions of the bunch that would occur during acceleration in the booster ring (adiabatic damping of phase space) would permit the use of a much smaller aperture (and therefore much cheaper magnets) in the main ring. Suddenly, a really large step in energy – say to 300 Gev – seemed possible. Indeed, the cascade principle probably would make sense only for an energy at least that high.

By the end of the two weeks, the general outline of a design had been filled out and written down.[24] A conventional linac would provide a 50-MeV beam to the booster, which would deliver a 10-GeV beam to the main ring, which would accelerate it to 300 GeV. The booster magnets would have the same cross section as that of the 30-GeV AGS, but would have only one-third the circumference. The main ring would have a much smaller cross section, but a circumference ten times that of the AGS. The complete main-ring magnet might have about the same mass as the AGS, and therefore, perhaps, a similar cost. The rather drastic decrease in the proposed magnet dimensions is shown in Figure 9.2 (which is taken from a talk I gave a little later[25]). It compares the cross section of a proposed main-ring magnet with those of the AGS (30 GeV) and Cosmotron (3 GeV).

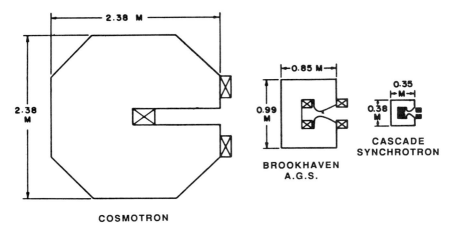

CImage labels:
2.38 M
2.38 M
COSMOTRON

0.85 M
0.99 M
BROOKHAVEN
A.G.S.

0.35 M
0.38 M
CASCADE
SYNCHROTRON

Figure 9.2. Comparison of the dimensions of the proposed cascade synchrotron magnet with those of the Brookhaven Cosmotron and the AG5 [*Source*: Sands (1961)[25]].

At the same time, an attempt was made, with the help of my friends, to make a rough cost estimate. Our first cut gave about $80 million for the accelerator proper (excluding any experimental facilities). To my dismay, the largest chunk was for the tunnel to house the accelerator!

When these ideas were exposed to the other participants of the MURA summer study, a certain amount of skepticism was expressed:

> It won't work. . . . The aperture is too small; you'll never get the beam around the first time. . . . You'll never be able to align such a large ring. . . . The magnets won't stay put. . . . How will you transfer such a high-energy beam between the two rings? . . . How can you modulate such a high-voltage rf system? . . . and so on.

I had worried about these matters and had answers (many of which were based on new ideas that had been invented at the summer study):

> I believe it will. . . . We'll use position monitors and snake the beam around. . . . You won't *need* to have long-range alignment in an AG ring. . . . We'll build an automatic, self-alignment system for the magnets. . . . It's already planned for the AGS fast beam extraction. . . . We'll use mechanically-tuned, high-Q cavities. . . and so on.

After weathering the criticisms, I became more convinced that the cascade synchrotron was on the right track. It is now amusing, in retrospect, to compare my shocking proposal for a small-aperture magnet to the magnet cross sections that were adopted years later for the Fermilab 400-GeV main ring and for the 1,000-GeV Tevatron (Figure 9.3).

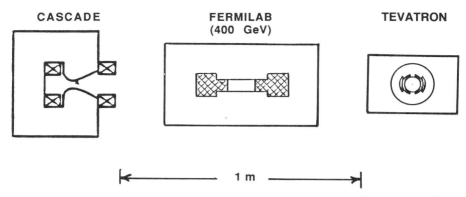

Figure 9.3. Comparison of the dimensions of the proposed cascade synchrotron magnet with those of the Fermilab 400-GeV ring and Tevatron [*Source*: Sands (1961)[25]].

Later work

Having satisfied my curiosity, I returned to work with our 1.5-GeV accelerator at Caltech. But the ideas germinated, and eight or nine months later Walker, Tollestrup, and I, with encouragement from Murray Gell-Mann, decided to take seriously the notion of a super-high-energy accelerator and to begin a serious study of a possible design. We were joined by others at Caltech: J. Matthews, Peterson, Rule, and E. Taylor. We invited consultants from elsewhere: Blewett, Courant, R. Hulsizer, Robinson, and Snyder. We later expanded our efforts by involving physicists from other universities in our area (UCLA, USC, and UCSD) in an informal consortium called the Western Accelerator Group (WAG). I remember particularly the participation of Harold Ticho, Carl York, Robert Richardson, Oreste Piccioni, and Robert Swanson. Invitations to Stanford and Berkeley to join forces were politely declined.

As our work progressed, I became more and more convinced of the desirability, the technical feasibility, and the acceptable cost of a high-energy cascade accelerator – and that it should become part of our national program.[25] By mid-1961, interest in pushing toward higher energies was growing, and programs aimed toward a new generation of high-energy machines had also been initiated at Berkeley and at Brookhaven – and there were even the now perennial discussions of a possible "world machine."

A detailed account of these events, of the ensuing political maneuvers, of the death of WAG, and of the relation, if any, of these matters to the eventual creation of Fermilab would take us too far into the next decade. That story will have to wait for some possible future "Conference on Particle Physics in the 1960s."

Let me just end my story with a little personal anecdote. A short while ago, as I was looking through some old files (at the request of Lillian Hoddeson), I came across a letter of April 1961 from Robert Wilson, then at Cornell, in regard to a contemplated visit to Caltech. As an aside, he wrote:

> I have been watching your efforts with the 300 GeV machine with open-mouthed admiration. It seems to me that you are working on the right problem and at the right time, and I am sure that something will come of it all.

He was right. Ten years later, almost to the day, the new Fermilab, under Wilson's leadership, produced the world's first protons with an energy of 300 GeV – and even higher.

Notes

1 In 1950, the abbreviation "BeV" (billion electron volts) was in general use in the United States. By 1960, the internationally accepted alternative "GeV" (giga-electron-volt = 10^9 eV) had taken over.
2 Electron-synchrotrons with energies near 300 MeV were in operation at Berkeley, at Purdue University, at Cornell University, and at the Massachusetts Institute of Technology.
3 For example: M. Sands, "Low Energy Mesons in the Atmosphere," *Phys. Rev. 77* (1950), 180–93.
4 William C. Elmore and M. Sands, *Electronics – Experimental Techniques* (New York: McGraw-Hill, 1949).
5 E. Courant, M. S. Livingston, and H. Snyder, "The Strong-Focusing Synchrotron – A New High Energy Accelerator," *Phys. Rev. 88* (1952), 1190–6. The AG principle had been invented independently (1950) by N. Christofilos in Greece, but had not been published.
6 M. Sands and B. Touschek, "Alignment Errors in the Strong-Focusing Synchrotron," *Nuovo Cimento 10* (1953), 604–13.
7 J. B. Adams, M. G. N. Hine, and J. D. Lawson, "Effect of Magnet Inhomogeneities in the Strong-Focusing Synchrotron," *Nature (London) 171* (1953), 926–7.
8 B. Touschek (ed.), *Physics with Intersecting Storage Rings*, proceedings of the International School of Physics "Enrico Fermi" Course XLVI (New York: Academic Press, 1971).
9 For example: M. Sands, J. G. Teasdale, and R. L. Walker, "Negative-to-Positive Ratio of Photomesons from Deuterium," *Phys. Rev. 95* (1954), 592–3; M. Bloch and M. Sands, "Photoproduction of Pion Pairs in Hydrogen," *Phys. Rev. 108* (1957), 1101–2, and *Phys. Rev. 113* (1959), 305–15.
10 The mechanism is described in my lectures in *Physics with Intersecting Storage Rings*, note 8.
11 D. Iwanenko and I. Pomeranchuk, "On the Maximal Energy Attainable in a Betatron," *Phys. Rev. 65* (1944), 343; J. Schwinger, "Electron Radiation in High Energy Accelerators," *Phys. Rev. 70* (1946), 798–9; N. H. Frank, "The Stability of Electron Orbits in the Synchrotron," *Phys. Rev. 70* (1946), 177–83; L. I. Schiff, "Production of Particle Energies Beyond 200 MeV," *Rev. Sci. Instrum. 17* (1946), 6–14; D. Bohm and L. Foldy, "The Theory of the Synchrotron," *Phys. Rev. 70* (1946), 249–58.
12 M. Sands, "Synchrotron Oscillations Induced by Radiation Fluctuations," *Phys. Rev. 97* (1955), 470–3.
13 A. A. Sokolov and I. M. Ternov, *Zh. Eksp. Teor. Fiz. 30* (1956), 207, 1161.
14 M. Sands, "Observation of Quantum Effects in an Electron Synchrotron," *Nuovo Cimento 15* (1960), 599–605.
15 K. W. Robinson, "Radiation Effects in Circular Electron Accelerators," *Phys. Rev. 111* (1958), 373–80.

16 A. Kolomenskij and A. N. Lebedev, "The Theory of Electron Motion in Cyclic Accelerators in Presence of Radiation," *Nuovo Cimento (Suppl.)* 7 (1958), 43–60.

17 These machines were reported in one or another of the following proceedings of international conferences: *CERN Symposium on High Energy Accelerators and Pion Physics*, edited by E. Regenstreif (Geneva: CERN, 1956); *International Conference on High Energy Accelerators and Instrumentation*, edited by L. Kowarski (Geneva: CERN, 1959); *International Conference on High Energy Accelerators*, edited by M. Hildred Blewett (Upton, N.Y.: Brookhaven, 1961); *International Conference on High Energy Accelerators*, edited by A. A. Kolomensky, A. B. Kusnetsov, and A. N. Lebedev (Dubna: JINR, 1963).

18 Reports on the FFAG developments will also be found in the conference proceedings mentioned in note 17.

19 MURA staff, "The MURA Two-Way Electron Accelerator," in *Proceedings, International Conference on High Energy Accelerators and Instrumentation*, edited by L. Kowarski (Geneva: CERN, 1959), pp. 71–4.

20 G. O'Neill, "Experimental Methods for Colliding Beams," in *Proceedings, International Conference on High Energy Accelerators and Instrumentation*, edited by L. Kowarski (Geneva: CERN, 1959), pp. 23–8.

21 C. Bernardini, G. F. Corazza, G. Ghigo, and B. Touschek, "The Frascati Storage Ring" and "A 250 MeV Electron-Positron Storage Ring: The AdA," in *Proceedings, International Conference on High Energy Accelerators*, edited by M. Hildred Blewett (Upton, N.Y.: Brookhaven, 1961), pp. 256–61.

22 White House press release, 17 May 1959, "An Explanatory Statement on Elementary Particle Physics and a Proposed Federal Program in Support of High Energy Accelerator Physics," report of a special panel by the president's Science and Advisory Committee and the General Advisory Committee to the Atomic Energy Commission.

23 "Supplement to the Report of the Advisory Panel on High-Energy Accelerators to the National Science Foundation," 7–8 Aug. 1958 (NSF 58–29), L. J. Haworth, chairman, H. L. Anderson, H. R. Crane, B. T. Feld, E. J. Lofgren, L. I. Schiff, F. Seitz, R. Serber, M. G. White, R. R. Wilson.

24 M. Sands, "Ultra-High Energy Synchrotrons," MURA internal report no. 465, June 1959; M. Sands, "A Proton Synchrotron for 300 GeV," CTSL-10, September 1960 (report of the Caltech Synchrotron Laboratory).

25 M. Sands, "Design Concepts for Ultra-High Energy Synchrotrons," in *Proceedings, International Conference on High Energy Accelerators*, edited by M. Hildred Blewett (Upton, N.Y.: Brookhaven National Laboratory, 1961), pp. 145–54.

10 Accelerator design and construction in the 1950s

JOHN P. BLEWETT

Born 1910, Toronto, Canada; Ph.D., Princeton University, 1936; accelerator physics; Brookhaven National Laboratory (retired)

The spectacular progress in high-energy physics during the 1950s and the 1960s was due in large part to the availability of four large particle accelerators: the Cosmotron, the Bevatron, the AGS, and the CERN PS.* Although the history of the design, construction, and first operation of each of these machines is an adventurous saga in itself, relatively little attention has been devoted to it.

Many of the machine builders who could detail these histories for posterity are no longer with us. Ernest Lawrence, Leland Haworth, Stanley Livingston, George Kenneth Green, Nicholas Christofilos, Hartland Snyder, John Adams, Frank Goward, and many others are gone, leaving very little beyond journal articles in the way of records of their eventful careers.[1] For the most part, all were too busy with the future to spend much time recording the past. Livingston and I provided a sketchy history of the development of the accelerator art in our book *Particle Accelerators*.[2] But very many of the colorful events in accelerator history remain unrecorded.

In this chapter I attempt to provide a little background for accelerator history during the 1950s. At Brookhaven, we were in touch with all important accelerator projects and observed at first hand many of the unsung tribulations that beset accelerator builders, as well as the uplifting moments when their machines performed as they should.

---

*   *Ed. note:* All were proton accelerators: Cosmotron (1952), 3 GeV, Brookhaven, New York; Bevatron (1954), 6.2 GeV, Berkeley, California; AGS (alternating-gradient synchrotron, 1960), 33 GeV, Brookhaven; CERN PS (proton-synchrotron, 1959), 28 GeV, Geneva, Switzerland.

Autobiographical note

My career covers the first thirty-two years of Brookhaven's existence. I joined the staff in the summer of 1947, Brookhaven's first year. Previously I had been on the staff of the Research Laboratory of the General Electric Company, where I worked during the war on radar and radar counter-measures. After the war, I was involved in studies of synchrotron radiation (including the first verification of the existence of that radiation), in accelerator design, and in the analysis of electron orbits using G.E.'s first big computer. That was an entirely mechanical device occupying a room eighty-five feet long. My attempt to analyze orbits in synchrotrons was defeated by backlash in the mechanical integrators.

During my years at Brookhaven, I spent a good deal of time visiting CERN, including half a year during 1953–4, the time of CERN's establishment in Geneva.

The Cosmotron – design and construction

In 1950, the Cosmotron was still two years away from 3-GeV operation. Mesons were newly discovered, and their properties were still unknown. The antiproton was yet to be discovered. High-energy physics was still relatively simple and comprehensible.

At Brookhaven, work on the Cosmotron was centered in the "accelerator project." Livingston, the project's first chairman, had left to return to his professorship at MIT. Milton White had come from Princeton to take over for a short time, then had been replaced during the later days of construction by George Collins. Leadership from the director's office was provided by Haworth, who acted as chairman during the gaps between designated chairmen. Continuity in direction of the actual design and construction of the Cosmotron was provided throughout by Green and myself. The initial design of the machine was described in the paper of Livingston, John Blewett, Green, and Haworth.[3]

One of our best moves was the encouragement and support of a young theorist, Ernest Courant, later to become famous as one of America's most accomplished accelerator theorists. He is best known as one of the inventors of alternating-gradient focusing, or "strong focusing" (see Chapter 11 by Courant), but his achievements in the solution of many other accelerator problems are of almost equal importance. During the Cosmotron days he worked with another young theorist, Nelson Blachman; Blachman drifted away from accelerator science by 1952 and became noted elsewhere for his work on information theory. He is now to be found on the West Coast, at GTE Sylvania.

By 1950, Courant and Blachman had solved three major problems associated with the design of the Cosmotron. Could we include ten-foot straight sections to ease the problems of beam injection, extraction, and targeting? They said yes. How good a vacuum did we need to avoid gas scattering of the beam during acceleration? Their answer indicated an achievable vacuum.[4] Finally, what would be the effects of random errors in magnetic field? The magnet was to be made up of 288 laminated steel blocks – what were the mechanical and magnetic tolerances on their construction? Granted these tolerances, what had to be the dimensions of the vacuum chamber so that the remaining random errors in field would not cause excursions in the accelerating beam that would drive it into the vacuum-chamber walls? Courant and Blachman assumed severe but reasonable tolerances and called for a magnet gap nine inches high and three feet wide. This meant magnet blocks eight feet high and eight feet across.

The Cosmotron, designed by a fresh, young, and imaginative group, embodied a number of new approaches. I mention two for which I was responsible. First was the design of the magnet.[5] In previous years at the G.E. Research Laboratory I had been intrigued by the possibilities of air-core magnets. Playing with air-core designs, and then with the advantages of enclosure in iron, I came up with a C-shaped design for the magnet cross section, with the main exciting coil buried in the slot that was the magnet aperture, and return windings out of the way above and below the opening of the gap. Previously, the accepted magnet design (the so-called H magnet) had coils wrapped around magnet poles and provided magnetic-flux returns in steel structures that completely enclosed the magnet gap. It was evident that my C magnet would begin to be affected by magnetic-saturation effects at a slightly lower magnetic field than would the H magnet. But the C magnet's advantages of accessibility were evidently so valuable that I had no difficulty in persuading the group to accept it.

Now came a period for study of the detailed behavior of the field pattern as guidance for a modeling and mechanical design program. My former wife, Hildred Blewett, a talented mathematical physicist, set up a "relaxation program" – a procedure for solving the field equations that involved approaching the solution by a long and tedious sequence of approximations. Our best approach to a computer was a "computing bureau" in New York City that consisted of twenty-three operators with mechanical desk computers, supervised by a mathematician. Hildred finally decided, as I remember, that supervision of this operation would take longer than would doing it herself.

By the beginning of 1950, the magnet blocks, built to our specifications by Bethlehem Steel, had all been delivered. We decided to set up a test stand (Figure 10.1 shows the stand in the process of assembly) on which each magnet block could be run to full field and measured precisely.[6] Then the blocks would be arranged around the accelerator's circumference in such a sequence that the random variations in their fields would have the least effect on the accelerated beam. As can be seen from the picture, that was no small job. It was done carefully and precisely, and it had a profound effect on the success of the Cosmotron.

My second contribution was the radio-frequency (rf) accelerating system.[7]

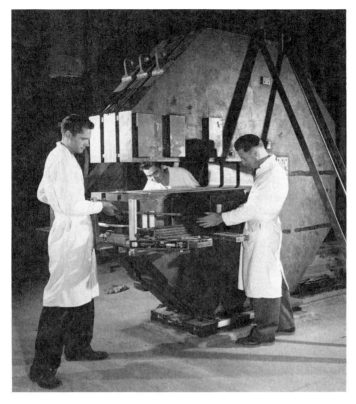

Figure 10.1. Cosmotron magnet test stand [*Source*: Blewett et al. (1953)[6]].

Here the main problem arose from the fact that the proton velocity, and hence its frequency of revolution around the Cosmotron, had to go through a large change. At the injection energy of 3 or 4 MeV, the proton velocity is less than one-tenth of the velocity of light; at 3 GeV, the final energy, it is 97 percent of the velocity of light. The frequency of the accelerating electric field must thus change by more than a factor of ten during the 1-sec accelerating cycle. How to make an accelerating system present a tolerable impedance to an rf amplifier over such a frequency range was not clear. What was required was a material having ferromagnetic properties, but one that would not support disastrous electric eddy currents. Iron was out, even in thin laminations; iron powders and very thin permalloy tape structures were studied; all gave little promise of tolerable properties. Then, in discussions at the Bell Laboratories, I learned of a new development at the Philips Gloeilampenfabrieken in Eindhoven, Holland. There the physical chemistry group had just evolved the new substance "Ferroxcube" (a ferromagnetic insulator), which was exactly what we needed. I made a hasty trip to Eindhoven and was delighted with what I saw; I established contact with North American Philips in Irvington, New York, and placed orders for Ferroxcube samples for testing at Brookhaven. By now, the reader no doubt has recognized Ferroxcube as the first of the ferrites now omnipresent in a multitude of forms in TV sets, computers, and all sorts of electronic equipment.

Philips appreciated the fact that they were the sole supplier, and for the large quantities of ferrite that we needed, they asked prices that were difficult for us to meet. So we encouraged the General Ceramics and Steatite Corporation of New Jersey to pursue a ferrite development of their own. The result was "Ferramic," not as good as the product of sophisticated Philips, but competitive. The next time a Philips representative visited me, I had a rod of Ferramic lying on my desk. Philips's price came down, and the final rf cavity for the Cosmotron was loaded about half with Ferroxcube and half with Ferramic. In this connection, it is proper to mention the active cooperation of Martin Plotkin throughout the development of the ferrite-loaded cavity. He made many of the first measurements and studies of ferrites in the United States, and he was responsible for the detailed design and construction of the final cavity.

The rf accelerating signal, programmed to match the frequency of revolution of the accelerating protons, was generated at low level in a circuit involving an ingenious use of saturating ferrites, by A. I. Pressman, with later contributions by Ed Rogers. Between the low-level signal generator and the accelerating cavity was a versatile power amplifier designed by Luke Yuan. Yuan was one of the few accelerator builders whose goal was to join the group of high-energy physicists who would use the Cosmotron when it came into operation. With Sam Lindenbaum of Columbia and Brookhaven, he planned to study the scattering of mesons. Did an increase in cross section at lower energies indicate an approach to a resonance to be observed at energies available with the Cosmotron? (Indeed, it did.) This was the first of many contributions made to high-energy physics by the team of Yuan and Lindenbaum.[8] During the years since construction of the Cosmotron, Yuan has joined me in many explorations of the desirability and achievability of higher and higher energies. On this subject we have written much and have organized many studies and conferences.[9]

The injector for the Cosmotron was a machine of some historic interest. It was a pressurized 4-MeV Van de Graaff electrostatic accelerator built for us by the High Voltage Engineering Corporation. This corporation, founded by R. J. Van de Graaff and several associates, had just been formed, and ours was their first order. From our end, the machine construction and operation were supervised by Clarence Turner, the man who had built the Van de Graaff injector for Luis Alvarez's first linear accelerator.

The ion optical system that carried the proton beam from the Van de Graaff injector into the aperture of the Cosmotron magnet was designed by Garry Cottingham, with some help from me. The beam, as delivered by the injector, needed an optical transformation to make it acceptable in the Cosmotron. After some thought, we decided that this could be done by letting the return windings of one of the deflecting magnets generate a field at right angle to its main field. We had invented a quadrupole – we called it "the four-wire lens." That was in 1950.

Birmingham and Berkeley

In the competition to complete the Cosmotron and be the first in the human race to reach and pass billion-volt energies, we had two competitors. At the University of Birmingham, England, Mark Oliphant had made an early proposal (1943) for construction of a proton-synchrotron. He had not appreciated the principle of phase stability, however, and there was no assurance that the machine would work. When phase stability was discovered in 1945, the proposal was resurrected, and a machine was started at Birmingham, aiming for a peak energy of 1 GeV. Given adequate support, that could well have been the first operating proton-synchrotron. Funds were short at Birmingham; the staff, though few in number, strove valiantly to be first to a billion volts, but they could not bring their machine on line until 1953.

Our other competitor was much more formidable. At the University of California Radiation Laboratory (UCRL, now the Lawrence Berkeley Laboratory), construction was in progress on a proton-synchrotron – the Bevatron – aimed at 5 or 6 GeV, the energy chosen to be sufficient to produce the antiproton if it existed[10] (see Chapter 34 by Robert Seidel). The laboratory had a splendid history of invention and construction of successful cyclotrons, a powerful staff of physicists and engineers, and a backing of experienced administrators, building space, machine shops, and all of the other amenities that we were just assembling at Brookhaven. They more than overwhelmed our slight advantage in building a smaller machine, and we regarded UCRL with awe and something of a feeling of doom.

At UCRL, machine design was guided mainly by experience and by model studies; not much detailed theoretical analysis had previously been found profitable. Brookhaven's choice of aperture was based on a very careful study of the effects of the choice of mechanical tolerances, but the Berkeley approach was to build a magnet with a very large aperture, observe the excursions of the protons, and then reduce the aperture as far as possible by insertion of pole pieces. The aperture chosen for the initial tests was four feet, to be compared with Brookhaven's choice of nine inches. As a check, a quarter-scale model of the Bevatron was built, including shutters with which the vertical aperture could be closed down until the beam disappeared.

With typical UCRL expedition, the quarter-scale model was built and tested. A beam was injected, the shutters were gradually closed, and, at an aperture considerably larger than that scaled to the aperture we had chosen for the Cosmotron, the beam disappeared. Lawrence was quoted as suggesting that Brookhaven's funds should be stopped, and we proceeded with construction even more nervously than before. Now, much later, we know that the point at which the Bevatron model's beam disappeared was quite consistent with the mechanical tolerances that had been met in the construction of the model's magnet, but then that was far from clear. UCRL's quarter-scale model was then dismantled and shipped to Caltech, where it was converted into an electron-synchrotron and enjoyed a long and honorable career.

At that point, an event of profound significance occurred. There was deep concern in the nuclear community about a possible shortage in the nation's reserve supply of fissionable material. At Berkeley, Alvarez, who had recently designed and built the first proton linear accelerator, and Lawrence conceived the idea that the needed materials could be produced artificially by a huge linear accelerator. The final machine was to be located beside the Mississippi River, whose water would be needed for cooling. The essential front end of the machine, however, was to be constructed at the Livermore Weapons Laboratory of the University of California. It was a huge structure – a copper-lined steel vacuum tank sixty feet in diameter and sixty feet long – so big that a spur of the railroad ran into its midst for delivery of large parts. It was called the Materials Testing Accelerator – the MTA. We knew nothing about this project; it was completely secret.[11]

The significance of the MTA for us was that its construction, with the haste deemed appropriate, required the services of all of the skilled accelerator hands engaged in construction of the Bevatron. So far as we could see, for no visible reason, construction of the Bevatron stopped, and its staff disappeared. We asked no questions, but were thankful for the decreased pressure.

1952 – Operation of the Cosmotron

Early in 1952 the Cosmotron was ready for its first test. It is shown in Figure 10.2. Beam was injected, acceleration was turned on, and the beam circulated around the ring as it should. Everything appeared to be working – the beam was accelerating to an energy slightly higher than the injection energy – but then, in a rather fuzzy fashion, it disappeared. Could it be that we should have believed the conclusions from the Berkeley quarter-scale model? We checked all components of the machine and finally found a small burned-out component in the rf accelerating system. Then, one afternoon in May of 1952, Hildred and Yuan and I returned from lunch and turned the machine on for yet another try, and, to our delight, the beam survived – its energy went straight up well past a billion volts. I am credited with the profound remark: "Well I'll be damned, a billion volts!" We made the national headlines, including a nice spread in *Life* magazine. Within the week we had visitors from Berkeley full of generous congratulations. When they went home, the aperture of the Bevatron went from four feet to one foot.

During the latter half of 1952, we carried out a highly worthwhile project for which credit must be given to Collins. He insisted that while the memories

Figure 10.2. The Cosmotron [*Source*: BNL].

of our labors were fresh, we should write papers about the machine and its components and provide posterity with a detailed and accurate history and description of the Cosmotron. Other projects were put aside, and the Cosmotron's builders were confined to their offices until the writing was done. The result was the "Cosmotron issue" of the *Review of Scientific Instruments*.[12] Edited by Hildred Blewett, and with a foreword by Haworth, the Cosmotron issue appeared in September 1953 and included a total of twenty-seven papers. We take some pride in this contribution to the archives of high-energy physics – particularly since most of the original records of the Cosmotron's construction seem to have disappeared.

The completion of the Cosmotron was a major technical achievement. It was the first proton-synchrotron, the first accelerator to accelerate particles to energies above a billion electron volts (GeV), and the first synchrotron to provide external beams of primary particles. The beam intensity, originally 10^{10} protons per pulse, was increased by more than a factor of 100 during its life. The demand for machine time increased from ten shifts per week in 1953 to twenty-one shifts by 1961. The machine concluded an honorable career of sixteen years when, superseded by the AGS, it was shut down in 1968.

Strong focusing

The story of the invention of strong focusing (now called alternating-gradient focusing) has been told many times. Courant, one of its inventors, describes the historic summer of 1952 in Chapter 11. My part in that invention lay in the realization that the original concept of a strong-focusing synchrotron would be equally useful in the linear accelerator, which appeared as a companion paper to the paper of Courant, Livingston, and Snyder in the December 1952 *Physical Review*.[13]

That work led to what I think was the first application of strong focusing in an accelerator. At Berkeley, Alvarez had built the first proton linear accelerator. He faced the problem that the accelerating fields in the linear accelerator are defocusing. Alvarez coped with this problem by including "grids" in the accelerating electrodes. But mechanical and electrical problems forced the grid design into a very inefficient form, and as a result, the beam intensity was limited to very low levels. When news of our discovery reached Berkeley, Alvarez and his team quickly removed the grids and replaced them with electrostatic quadrupoles. Without major reconstruction, they were not able to reach optimum parameters, but it was possible to show the major improvement now possible in linear accelerator performance. This work was described in a paper presented by Craig Nunan and Robert D. Watt at a Brookhaven conference in December of 1952. A more formal report appeared in 1955 in the "linear accelerator issue" of the *Review of Scientific Instruments*.[14]

Nick Christofilos

Early in 1953 we were startled to learn that strong focusing had been invented a couple of years earlier in Athens by Christofilos, a Greek engineer. At Brookhaven, our first awareness of Christofilos occurred two or three years before the invention of strong focusing at Brookhaven. He was then engaged in the installation of elevators in Athens, but, fascinated by physics problems, was spending much of his spare time in the American Library in Athens reading the *Physical Review* and related journals. He was particularly intrigued by the problems in design of particle accelerators.

After some thought, he dreamed up a scheme that he thought would solve the focusing problems that seemed about to limit the energy achievable in synchrotrons to 10 GeV or so. He thought that the scheme might have military applications, and so he did not publish it. But he did take out a patent, which he sent to the University of California Radiation Laboratory, where he hoped that it might come to the attention of such people as Lawrence. It was read by a member of the Berkeley staff, who concluded that it was unsound. UCRL sent it to us for confirmation of this opinion; I read it and agreed that Christofilos was trying to violate Maxwell's equations. These conclusions were sent to Christofilos, who had become sufficiently erudite that he immediately appreciated the unsoundness of his scheme. He retired into himself and soon reemerged with a new idea: strong focusing. Once again he took out a patent and sent it to Berkeley. This time the Berkeley reaction was, not unreasonably, "another unsound scheme from Greece." The document went, unread, into someone's files.

A couple of years later, Christofilos visited America. His first stop was at the New York Public Library for a look at the latest *Physical Review*. He was confronted by our papers on strong focusing and concluded that his idea had been stolen; so his next step was a trip out to Brookhaven. I was delegated to show him around. Knowing only of his first, unsound invention, I looked forward to a tiresome and pointless day. He showed me his second invention. That was a real shocker. Evidently he had discovered the principle of strong focusing two years before the Brookhaven invention!

At Berkeley, when the Brookhaven publications appeared, someone recalled Christofilos's second communication. It was resurrected and appreciated as essentially the same invention as ours. A letter was written to us pointing this out. Unfortunately, I am not sure whether this reached Brookhaven before or after Christofilos's visit. In any case, I was not aware of it at that time.

Our relations with Christofilos speedily became more friendly. We published an acknowledgment of his priority.[15] And he looked at our various projects with great interest – so that when we offered him a job, he accepted with alacrity. He joined our staff in June 1953. At that point I went off to spend six months in Europe helping in the establishment of the CERN laboratory. In the meantime, Christofilos became active in Brookhaven's Accelerator Department. The staff rapidly came to appreciate the fact that he was extremely bright, very enthusiastic, and full of ideas, but not the easiest person in the world to organize into the system. On my return from CERN, I was greeted by Green, the department chairman, with the news that "We have decided to put Christofilos into your Linear Accelerator group."

So I told him about the problems of linac design. He was much interested in our attempts to incorporate permanent-magnet quadrupoles for focusing of the linac beam and made several ingenious suggestions. But the problem that really intrigued him was the shaping of the "drift tubes," the accelerating electrodes whose shape must be chosen so that the whole linac cavity structure is resonant at the design radio frequency. Christofilos believed that those shapes could be calculated, rather than designed by construction of a tedious series of models. Those readers skilled in the art of calculating electromagnetic-field patterns in systems with cylindrical symmetry know well that an intimate acquaintance with Bessel functions is an essential prerequisite. He had never heard of Bessel functions, but he acquired an advanced mathematics text and rapidly became expert. Then he worked out a design procedure that was indeed the one that we used in the final linac design.[16]

Before long, Christofilos became restless. His mind was aboil with new ideas. He was intensely interested in the possibilities of controlled nuclear fusion. While still in Greece he had evolved a new method for containment of the hot plasma needed to sustain a fusion reaction. That was later to become the "Astron" project. At that time, all such programs were highly classified. There was some debate whether or not Brookhaven should sponsor new secret projects just when we were attempting to declassify our research reactor and make Brookhaven a completely open laboratory. Finally it was

Figure 10.3. Nicholas Christofilos [*Source*: BNL].

decided that Christofilos could pursue his ideas at the Livermore Weapons Laboratory in California. He left us in November of 1956, never to return, and he died at Livermore in 1972. Those who came to know Christofilos regarded him with affection and admiration (Figure 10.3).

CERN

We were well aware of the impending organization of CERN. A year earlier, while the Cosmotron was still in the final construction stages, we had had a visit from Ed Regenstreif, who was sent by Pierre Auger of UNESCO in Paris to study American work as a guide in the formation of the international laboratory that was to become CERN. We expected another visit in August of 1952; so we were indeed delighted that we had a new invention to offer.

As described by Courant in Chapter 11, the original impetus that led to the invention of strong focusing was our desire to help the budding CERN proton-synchrotron (PS) group. Because we knew that they planned a scaled-up Cosmotron, and because we expected them to visit us and ask for advice, we made a study of ways in which the Cosmotron magnet design could be improved. Livingston's suggestion that reversal of the back legs of some magnets could defer overall saturation effects to higher fields at the expense of local alternating gradients led to the analysis out of which fell the strong-focusing discovery. An idea that Livingston hoped might give a 20 percent improvement in performance thus led to a major discovery.

Our official visitors were Odd Dahl of Norway and Goward from England, head and deputy head of the CERN PS group. With these two came a consultant to CERN, Rolf Wideröe of Brown-Boveri, Switzerland. The reader will remember that Wideröe is really the father of the accelerator art. His 1927 paper described the first operating linear accelerator; it was this paper that inspired Lawrence to invent the cyclotron.[17] Wideröe was still full of volatile enthusiasm, and indeed, so far as I know, he still is. He sent me his latest reprint in 1984.

Dahl, Goward, and Wideröe were delighted with the news about strong focusing, as well as with the success of the Cosmotron. They returned to Europe and initiated work on analysis of the new idea. Very quickly the British team of Adams (later to be CERN's director general), Mervyn Hine, and John Lawson completed a study of alternating-gradient focusing. The study confirmed the Brookhaven conclusions, but pointed out the existence of dangerous resonances that could occur and possibly destroy the accelerated beam if careful attention were not paid to a suitable choice of machine parameters.[18]

Collaboration between CERN and Brookhaven was close from that time on. Courant, Hildred, and I were invited to an organizing meeting in Paris later in 1952. We then were invited to come to CERN for an extended period to help with the initial problems of the new laboratory. Hildred and I agreed to spend six months at CERN beginning in July of 1953. Our first station was in Bergen, Norway, where we were to assist Dahl in organizing the CERN PS project. Also in Bergen were Kjell Johnsen, later to be the director of CERN's pioneering Intersecting Storage Ring colliding-beam project, and Jan Andersen, a graduate student at the University of Bergen. The other proton-synchrotron project group members were scattered over Europe – orbit theorists in England and Germany, radio-frequency acceleration studies in Heidelberg, and a magnet-modeling program in Paris. An inner executive group was set up with meetings in Bergen. It was to guide the planning for the proton-synchrotron; it consisted of Dahl, Goward, W. Gentner (of Freiburg), and the two Blewetts.

This group spent a good deal of time discussing project organization and costs; we pressed continually for a move to Geneva, where the project could be pulled together and begin to advance. My main activity in Bergen, however, was CERN PS magnet model no. 2. Model no. 1 was being built in Paris with great care, but was proceeding rather slowly, and it seemed that basic magnet-design information could be obtained rather more quickly. It happened that Andersen had been involved at the university in the construction of a small betatron built in the university's machine shops. A stack of iron magnet laminations had been left over; Andersen and I sheared them to shape to be built into a roughly one-fifth-scale model of the final CERN PS magnet. It was finished in less than two weeks. By the end of the summer, Hildred, Andersen, and I had measurements that yielded the basic parameters of the CERN PS magnet.

In the fall of 1953, the move to Geneva of the PS group took place. Professor Extermann, head of the Institut de Physique of the University of Geneva, had promised office and laboratory space for the group. But on the great moving day, when we all assembled on the steps of the Institut de Physique, it turned out that Extermann had no memory of his promise. We could, he said, occupy chairs in his library, but that was to be the extent of his support. Consternation reigned for some time. Then Goward undertook to attempt to reason with our reluctant host. Behind closed doors the discussion proceeded for an interminable period. Finally, Goward's dogged British determination won the day, and, grudgingly, we were given offices.

We did not need any further such problems, for we knew that the CERN Council, mistrusting the qualifications of the newly assembled, rather young, and nationally heterogeneous PS group, had decided to give it a public examination. For that purpose, an international audience of accelerator experts would be assembled, and the CERN group would present in detail its design for the CERN PS, inviting criticism or outright disapproval if such were appropriate. The PS group had about three weeks in which to integrate itself and prepare for this grueling ordeal.

It was a period of intense concentration and preparation. We American visitors felt a little like mother hens watching over a brood of chicks continually in danger of annihilation. But splendid basic contributions were provided by all. Shining in my memory were, for example, the teamwork of Adams and Hine on proton-orbit analysis – Hine providing brilliant intuitions and Adams checking their correctness. Johnsen provided a mathematical treatment of the dreaded phase transition and showed that it could be passed without danger. Pierre Denis, the one Swiss member of the PS team, and I visited the Secheron plant, a Geneva manufacturer of railroad equipment, and talked them out of a dc generator to provide power for final measurements on our magnet model.

The dreaded day finally arrived, 26 October 1953, and with it the array of visiting experts. We expected a fine performance by our PS group, and we were not disappointed.[19] All performed beautifully, and the visitors were easily convinced that the CERN group's design was sound and as complete as could be expected at this early date. Work proceeded expeditiously, but at a less hectic pace. The group expanded and prepared for a move to CERN's final site on the French border. Hildred and I soon returned to Brookhaven.

The Electron Analog

Both at CERN and at Brookhaven, lengthy debates were held during 1953–4 about the "phase transition." The question was whether or not the synchrotron could be made to pass safely through a discontinuous radio-frequency phase shift that had to happen at an energy of a few GeV in most strong-focusing proton machines. Should a model be built to test our ability to pass through the phase transition?

I was strongly opposed to this project. As mentioned in the previous section, Johnsen's theory said that there would be no problem in passing the phase transition, and I believed him. At CERN, the majority thought as I did, with a vocal minority in the person of Regenstreif. Then, while I was still at CERN, I learned that Brookhaven had decided to proceed with an electron model to be called the Electron Analog. I wrote a strong letter home protesting, claiming that a phase-transition problem did not exist and complaining that the project would set the AGS completion date back so far that the CERN PS would be finished before the AGS. But at Brookhaven the model was believed to be necessary, probably more for political than scientific reasons. It was built, using electrostatic deflecting and focusing fields. It was completed during 1955, and as I had predicted, it worked. Also as I had predicted, the CERN PS came into operation a year before the AGS.

MURA and Argonne

At Brookhaven, we followed with great interest the emergence of the Midwestern Universities Research Association (MURA) (see Chapter 13 by Donald Kerst). When the group was first formed, a dozen of its members were welcomed at Brookhaven for an intensive three-week course on accelerator design in the summer of 1953.[20] I was away at CERN during that period, but I have often heard from former MURA members about the concentrated and valuable orientation they received from Green and others of the Brookhaven staff. Our interactions with MURA were frequent throughout its life. We visited Madison often, keeping abreast of MURA inventions and the many contributions made by Kerst, K. R. Symon, and their talented associates to the basic design and understanding of accelerators.

During the same period, the design was proceeding at Argonne for the zero-gradient synchrotron (ZGS). We were appalled by this project and by Argonne's failure to take advantage of the new strong-focusing principle. Haworth, our director, was continually in touch with Norman Hilberry, Argonne's new director. Presumably as a result of their discussions, Hilberry invited a group of us to come to Argonne and present our views. We accepted, and late in 1956 or early in 1957 four of us, Haworth, Green, Hildred, and I, made the trip to Chicago. I don't have a record of the attendance, but, as I remember, we met with Hilberry, Wally Zinn, Jack Livingood, Roger Hildebrand, Courtenay Wright, Albert Crewe, and others. We described our work, now well advanced, on the AGS. We explained to the group that for the money they planned to spend, they could achieve three times the energy, and we offered to expand our production order for magnets to whatever extent they might desire, to supply an Argonne alternating-gradient machine. The response to our presentation, which came primarily from the representatives of the University of Chicago, was thoroughly hostile. We were represented as interfering officiously with a splendid project and were told in so many words to mind our own business. In some confusion, we replied, "We only came because you invited us," and we left. We went on to Madison, where we enjoyed a much more hospitable reception from the MURA group.

The Alternating-Gradient Synchrotron (AGS)

Most of us were confident that the Electron Analog would work, and work proceeded on the 30-GeV AGS. A famous architect-engineering firm, Stone and Webster, was hired to help with foundation and building design. With this team, a group headed by Jack Lancaster [later to be chief engineer of the Very Large Array (VLA) at the Associated Universities Incorporated radioastronomy laboratory in New Mexico] made extensive tests on the properties of Long Island sand. We learned that it had not been compacted by the most recent glacier and wondered how it would behave under load. There was reason for concern, because orbit theory predicted that rather close mechanical tolerances had to be met to ensure satisfactory operation of the machine. We dug pits to learn what was under the surface sand and found, mostly, just more sand, with occasional "clay lenses." We did a soil-loading test in which we piled up several hundred tons of Cosmotron concrete-shield blocks and measured how much the earth sank, then unloaded the ground, measured again, then piled the blocks on again and looked for long-term sinking of the surface. Finally, we decided to be extra safe and support the machine and the linac injector on steel I-beam piles driven fifty feet into the ground. This worked very well, but in the light of later experience, the approach may have been unnecessarily conservative.

Hildred presided over the design of the AGS magnets, and Cal Lasky, also of Brookhaven, was in charge of their manufacture at the Baldwin-Lima-Hamilton Corporation. With a team of inspectors, he spent most of his time at the factory measuring, revising welding procedures, and instructing the factory staff. Similar efforts went into the magnet coils built at the National Electric Coil Company. Finally, magnets and coils began arriving. All were stacked on the floor of the new AGS experimental area ("the target building") and were carefully measured, then distributed around the ring in a sequence planned to minimize the effects of their small deviations from mean values of such parameters as remanent field.

The 50-MeV linac injector was my responsibility. My linac team tested several novel types of traveling-wave linac structures. Some were quite ingenious, but finally we fell back on a sophisticated version of the Berkeley standing-wave linac built by Alvarez. Christofilos, besides his drift-tube calculations, made many other contributions to the linac design. As a test of the details of our design, we built a three-drift-tube model simulating the section of the linac at about the 30-MeV point. Cottingham presided over this operation. He estimated that about 125 kilowatts (kW) of pulsed power would be adequate to power the model, and we set it up with its power supply on the second floor of one of the seven barracks buildings that housed the AGS project. Various measuring devices indicated that the design rf accelerating field had been reached. This corresponded to about 700,000 volts (V) between drift tubes. Unfortunately, we did not then appreciate the need for extreme cleanliness in high-powered rf systems, and we left some films of machining oil inside the drift tubes. This oil spread itself on the surfaces of the drift tubes in thin films that, under high fields, emitted electrons in copious quantities and in turn generated lots of 700,000-V x rays. Finally, it took about 500 kW to bring the model up to full field. Of this, 375 kW were going into x rays, and it was not safe to linger anywhere in the building. Needless to say, the final linac was kept scrupulously clean.

We made a rather detailed exploration of the possibility of using permanent-magnet quadrupoles for focusing in the linac. Ferrite rings were magnetized by discharging a large condenser bank into a step-down very-high-secondary-current transformer that powered a four-turn magnetizing coil inside the rings. After the magnetizer exploded several times, blasting pieces of ferrite through the walls of the barracks, we tamed it by wrapping it with heavy piano wire. The permanent-magnet idea was a good one and has since been resurrected at Los Alamos and elsewhere, but we finally lost our nerve over our inability to vary permanent-magnet fields, and we installed pulsed electro-magnet quadrupoles capable of having their fields varied from outside.

The high-powered rf system needed to generate in the linac the fields to ac-celerate the protons presented some problems. Pulsed power was necessary at a level of 5 million watts (MW) at a frequency of 200 million cycles per second (MHz). No power amplifier existed that would deliver so much power in a single package at this frequency. Smaller sources were available, but we hoped to avoid the headaches of combining the outputs of several amplifiers at the single input to the linac tank. A large klystron seemed a reasonable development, and we easily persuaded the Eimac Corporation on the West Coast to undertake the project. In 1955 they were ready to predict delivery by early 1957. But the project was plagued by continuing problems and delays, and early in 1958 we canceled it and ordered a pair of conventional triodes from the French Thomson–Houston Company. We planned to combine the outputs of these tubes in what was formally called a "waveguide hybrid junction" – more popularly known as a "magic tee." We then would make a waveguide-to-coaxial transformation to introduce the power into the tank.

The magic tee was a horrendous gadget; three four-foot-wide rectangular brass pipes and massive auxiliary gear had to be tuned by cranking nine large tuners with no real assurance about which should be tuned first or how to know when the setting was correct. We put in a heroic month, two teams of us covering the day and night in twelve-hour shifts. Sal Giordano, one of our brightest electrical engineers, with Vinnie Racaniello, one of our cleverest technicians, turned the cranks for twelve hours; then I took over with Frank Toth, our chief linac technician. Finally we established the correct procedure and fed full power into the tank; in no time, we had an accelerated 50-MeV beam ready for injection into the AGS.

The AGS was ready for its first test. The synchrotron magnets were in place and aligned precisely, and the magnet power supply had been tested on magnet pulses. Enough controls were installed for a first test. The rf system was not yet quite under control, so the first tests would involve injection into the ring in the hope that as the magnetic field increased, the beam would make a number of revolutions, spiraling inward and finally being intercepted by the inner wall of the vacuum chamber.

Happily, that is exactly what happened after some adjustments. A couple of months later, the rf was in operation, and in July of 1960 the beam was taken without difficulty through the dreaded phase transition and accelerated to 30 GeV.

Green and Haworth – a tribute

Green and Haworth were two great men whose contributions to high-energy physics have been, and are, insufficiently appreciated. Green was the technical guiding spirit throughout the design and construction of both the Cosmotron and the AGS. His command of the disciplines of physics, electrical engineering, mechanical engineering, and civil engineering was without match. He knew and understood every detail of our accelerators and very often caught possible errors in construction long before they were noted by anyone else. Withal, he was a very kind and patient man. I look back with deep appreciation on my thirty years of association with him.

Haworth was Brookhaven's director throughout the laboratory's formative years. It was his wisdom, guidance, and administrative skill that made possible the construction of the Cosmotron and the AGS. His was a deep understanding of the science of the construction and use of particle accelerators. His continual interest in our progress was much more than academic. Many technical suggestions came from him – for example, the idea of using electric fields in the Electron Analog. Also, it is not generally known that he was the first to suggest the FFAG technique. Modestly, he refrained from publishing and, thereafter, gave all of the credit to the MURA team. In summary, he was a man with whom it was a pleasure and a privilege to be associated.

Notes

1 Two of these articles are as follows: E. O. Lawrence and M. S. Livingston, "The Production of High Speed Light Ions without the Use of High Voltages," *Phys. Rev. 40* (1932), 19–35; L. J. Haworth, L. D. P. King, C. T. Zahn, and N. P. Heydenburg, "An Apparatus for Low Voltage Nuclear Research," *Rev. Sci. Instrum. 8* (1937), 486–93.

2 M. Stanley Livingston and John P. Blewett, *Particle Accelerators* (New York: McGraw-Hill, 1962).

3 M. S. Livingston, J. P. Blewett, G. K. Green, and L. J. Haworth, "Design Study for a Three-Bev Proton Accelerator," *Rev. Sci. Instrum. 21* (1950), 7–22.

4 Nelson M. Blachman and Ernest D. Courant, "Scattering of Particles by the Gas in a Synchrotron," *Phys. Rev. 74* (1948), 140–4.

5 J. P. Blewett, M. H. Blewett, G. K. Green, W. H. Moore, and L. W. Smith, "Magnet. Part I – Design," *Rev. Sci. Instrum. 24* (no. 9, the "Cosmotron issue") (1953), 737–42.

6 M. H. Blewett, J. M. Kelly, and W. H. Moore, "Magnet. Part IV – Testing Program for the Individual Blocks of the Magnet," *Rev. Sci. Instrum. 24* (no. 9, the "Cosmotron issue") (1953), 760–4.

7 J. P. Blewett, "Radio-Frequency System. Part I – Design Principles," *Rev. Sci. Instrum. 24* (no. 9, the "Cosmotron issue") (1953), 779–81.

8 S. J. Lindenbaum and Luke C. L. Yuan, "Total Cross Section of Hydrogen for 150- to 750-MeV Positive and Negative Pions," *Phys. Rev. 100* (1955), 306–23.

9 For example, J. W. Bittner (ed.), *Summer Study on Storage Rings, Accelerators and Experimentation at Super-High Energies*, Brookhaven National Laboratory, June 10–July 19, 1963 (Chairman, J. P. Blewett, and L. C. L. Yuan) (BNL 7534).

10 A detailed history of the early beginnings of the Lawrence Berkeley Laboratory is presently being prepared by John Heilbron, Robert Seidel, and Bruce Wheaton.

11 See *Particle Accelerators*, note 2 (p. 344). The MTA is referred to as "the Mark I Accelerator."

12 M. H. Blewett (ed.), "The Cosmotron," *Rev. Sci. Instrum. 24* (1953), 723–870.

13 J. P. Blewett, "Radial Focusing in the Linear Accelerator," *Phys. Rev. 88* (1952), 1197–9.

14 Luis W. Alvarez, Hugh Bradner, Jack V. Franck, Hayden Gordon, J. Donald Gow, Lauriston C. Marshall, Frank Oppenheimer, Wolfgang K. H. Panofsky, Chaim Richman, and John R. Woodyard, "Berkeley Proton Linear Accelerator," *Rev. Sci. Instrum. 26* (1955), 111–33.

15 E. D. Courant, M. S. Livingston, H. S. Snyder, and J. P. Blewett, "Origin of the 'Strong-Focusing' Principle," *Phys. Rev. 91* (1953), 202.

16 N. C. Christofilos, "Method of Computation of Drift Tube Shapes," in *CERN Symposium on High Energy Accelerators and Pion Physics*, edited by E. Regenstreif (Geneva: CERN, 1956), pp. 176–8.

17 R. Wideröe, "Ueber ein neues Prinzip zur Herstellung hoher Spannungen," *Archiv. Elektrotech. 21* (1928), 387–406.

18 J. B. Adams, M. G. N. Hine, and J. D. Lawson, "Effect of Magnet Inhomogeneities in the Strong-Focusing Synchrotron," *Nature (London) 171* (1953), 926–7.

19 "Lectures on the Theory and Design of an Alternating-Gradient Proton Synchrotron" (Geneva: CERN-PS Group, 1953).

20 See BNL Annual Report, July 1, 1954.

11   Early history of the Cosmotron and AGS at Brookhaven

ERNEST D. COURANT

Born 1920, Göttingen, Germany; Ph.D., University of Rochester, 1943; theory of particle accelerators; Brookhaven National Laboratory and State University of New York at Stony Brook

In 1946, before the Atomic Energy Commission (AEC) could be formed, the Manhattan Engineering District established Brookhaven National Laboratory, a laboratory devoted exclusively to peacetime research. The mission was to make available in the Northeast certain facilities that were too large to be supported by any single university. Specifically, these facilities were

>   a nuclear reactor for research with neutrons,
>   a 700-MeV synchrocyclotron, and
>   a particle accelerator to reach or surpass the then-unheard-of energy
>   of a billion electron volts (proton-synchrotron).

The reactor was, I think, the first built purely for scientific research purposes; it was decommissioned some years ago and replaced by a more modern reactor, which is still operating. The building of the cyclotron was given up after a year or so of design work; instead, the Nevis synchrocyclotron (somewhat smaller) was built by Columbia University.

As for the proton-synchrotron: Edwin McMillan and Vladimir I. Veksler had independently discovered the principle of phase stability, which was first applied to the synchrocyclotron and the electron-ring synchrotron. But for really high energy, the synchrocyclotron would become too massive to be attractive, since it requires the entire interior of the top energy orbit to be filled with magnetic field. The electron-synchrotron requires magnetic field only at the other radius and thus is more economical, but the phenomenon now known as synchrotron radiation made for very large power require-

Work performed under the auspices of the U.S. Department of Energy.

ments at high energy. These requirements seemed to limit the energy to some 300 MeV (a rather elastic limit, which has since increased considerably, but really only by applying massive radio-frequency power).

The best way of getting to the highest conceivable particle energies was then, and is still, the use of protons. In a proton-synchrotron, unlike the electron-synchrotron, the radio frequency has to be modulated with high precision in order to track the magnetic field and at the same time keep the orbit at constant radius. This complication is outweighed by not having to worry about synchrotron radiation.

Several proposals for proton-synchrotrons appeared, notably one for a 1-GeV machine at Birmingham, England (Mark Oliphant, J. S. Gooden, and G. S. Hide), and one for a 10-GeV machine by William Brobeck at Berkeley.[1,2] Discussions among Leland Haworth at Brookhaven, Ernest Lawrence at Berkeley, and the AEC authorities led to the decision that Brookhaven and Berkeley, instead of competing for the 10-GeV prize, would each build a smaller proton-synchrotron, one around 3 GeV and one at 6 GeV. Haworth chose the smaller size, with the hope of finishing faster; in later years he often said that that was the best decision he had ever made. The reader will learn more about the 6-GeV machine at Berkeley, the Bevatron, in other chapters in this volume.

An accelerator project was set up at Brookhaven under M. Stanley Livingston. In the spring of 1947, he, Brookhaven's director, Philip M. Morse, and the personnel director, R. A. Patterson, visited Cornell, where I was doing postdoctoral work under the guidance of Hans Bethe. They invited me to Brookhaven for the summer, and the next year I joined for good. Among others in the accelerator group were George Kenneth Green, John P. and Hildred Blewett, and a young theorist named Nelson M. Blachman, with whom I worked on several of the theoretical problems of the proposed machine, named the Cosmotron.

One important problem we tackled was the beam blowup and loss due to scattering by the gas in the vacuum chamber – we worked out criteria for this and concluded that a vacuum of 10^{-6} mm Hg was desirable, if not strictly necessary.[3] A second was the dynamics of particle oscillations, both transverse (betatron oscillations) and longitudinal (synchrotron oscillations), as modified by the fact that this machine, unlike the earlier electron-synchrotrons and cyclotrons, had straight sections between the circular arcs (i.e., noncircular orbits).[4] David M. Dennison and Theodore H. Berlin[5] had tackled a similar problem at Michigan; I think Robert Serber was also involved. We derived a matrix formalism for handling the spatially periodic force fields seen by the particles, and we found that the frequencies of the oscillations are more complicated to calculate than in the circular case, that the amplitudes of oscillations are modulated, and that there might be, especially if the straight sections were long, a "transition energy" at which the stable and metastable phase equilibrium points that give phase stability exchange roles. But we saw that in the

Cosmotron, with its rather short straight sections, the last problem would be avoided.

The more practical people worked hard on the magnets, vacuum systems, radio frequency, and so forth, and by the spring of 1952 the machine was finished. On 20 May 1952, a beam was injected into the Cosmotron and accelerated to 1.3 GeV, by far the highest energy ever attained by artificial acceleration. Soon we surpassed that record and achieved 2.3 GeV in June, and 3 GeV (the design energy) the next year, when pole-face-correcting windings were installed. Soon important physics was done with this new facility. I shall only mention the verification of the Pais–Gell-Mann hypothesis of associated production of hyperons, which was accomplished by William B. Fowler, Ralph P. Shutt, Alan Thorndike, and William L. Whittemore in 1953,[6] as described by Fowler and William Chinowsky in this volume.

In the meantime, CERN was being formed in Europe. A delegation of Europeans (Odd Dahl, Frank Goward, and Rolf Wideröe) was expected to visit us to see if they could pick up some good pointers from us. They were planning, as the centerpiece of their new international laboratory, to build a proton-synchrotron even bigger than the Bevatron, around 10 GeV. As I recall, Livingston set up a study group especially to enable us to tell them not only what we had done but also what one might do better. One problem bothering him was that the magnets of the Cosmotron all faced outward; therefore, negative secondary beams were easily obtained, but positive secondaries tended to hit the inside wall of the machine. In addition, magnet saturation effects tended to reduce the usable "good-field" region at the fields corresponding to top energy. Therefore, it would be better to alternate the magnet sectors, with some having the back legs on the inside and others on the outside.

I pointed out that this might have a drawback: The focusing gradients might easily be different in the inward and outward sectors, especially in the fringing fields. From my earlier work with Blachman on straight sections, I knew how to attack this problem mathematically: Set up matrices for the focusing action of each sector, and multiply them together.

Almost at once I saw that the alternating gradients could enhance stability rather than weaken it! With the right parameters, the stability could be made stronger than in the conventional case. Livingston saw at once that this was something fundamentally new and that the focusing could be pushed to make it much stronger, so that it would be possible to make the magnet aperture really small. That, in turn, makes the magnets – and other components – much cheaper, and so one can go to higher energies than without "strong" focusing. We published a design with 1-inch aperture for 30 GeV.[7] Hartland Snyder explained the new results in terms of optical principles, and we and Blewett[8] saw that the same principle could also be used without bending magnets (e.g., with just quadrupoles) to focus beam lines and to replace the grids that were then thought necessary in proton linear accelerators. So we had something to tell our European visitors when they came.

Two difficulties soon became apparent: Imperfections of the magnets, differences between supposedly equal units, could lead to resonant beam blow-up. That was pointed out especially by John B. Adams, Mervyn Hine, and John D. Lawson in England, and for a while led to great pessimism.[9] But we soon saw that resonances could be avoided by staying between them – albeit at the cost of tightened tolerances and somewhat enlarged aperture allowances. The second difficulty was that the transition energy (i.e., the change in the position of phase stability), first found (as an academic curiosity) in the studies on straight sections, now came right in the middle of the interesting energy range. Fortunately, we saw right away, thanks to the earlier work with Blachman, that at transition energy the beam tends to be sharply bunched, making it reasonably easy to jump from the old to the new stable phase point. But that seemed awkward, and many people were skeptical.

While all this excitement was going on, some red-faced people at Berkeley dug up and sent us what they had thought was a crank letter from Greece, a letter they had received a couple of years earlier. It turned out that an engineer named Nicholas Christofilos in Athens had thought up essentially the same scheme after reading about plans for the Bevatron. We soon saw that he deserved full credit,[10] and we hired him at Brookhaven. He worked on the Electron Analog and on linac focusing; later he moved to Livermore to work on fusion and on weapons ideas.

Because of the worries about resonances and about transition energy, we proposed an electron analog accelerator that would also go through transition. (Robert R. Wilson built a 1-GeV strong-focusing electronsynchrotron at Cornell, but that one did not have the transition problem and therefore could not model it.) The energy of the analog was just a few MeV, rather than GeV; it used electrostatic rather than magnetic guide and focusing fields. It was built in about eighteen months and met our highest expectations. We saw that the transition problem could be managed just as per theory. The analog also gave a very beautiful demonstration that resonances existed and that low-order nonlinear resonances were more important than had been thought, but that nonlinearity could also stabilize beam blowup.[11]

On 9 September 1953, Haworth formally proposed to the AEC that the alternating-gradient synchrotron (AGS) be built, including the analog. The project was approved early in 1954. The formal proposal was a five-page letter (plus a few graphs), rather than the 500-page books customary nowadays. Approval took four months, rather than four years. In the meantime, CERN went ahead with a similar machine. They did not take the detour via the analog; that may (or may not) be the reason why the CERN PS had a beam in 1959, a year or so before the AGS in 1960. But it was always a friendly and collaborative rivalry between us and CERN, and it remains so to this day.

Notes

1 M. L. Oliphant, J. S. Gooden, and G. S. Hide, "The Acceleration of Charged Particles to Very High Energies," *Proc. Phys. Soc. London 59* (1947), 666–77.
2 W. M. Brobeck, "Design Study for a 10-Bev Magnetic Accelerator," *Rev. Sci. Instrum. 19* (1948), 545–51.
3 N. M. Blachman and E. D. Courant, "Scattering of Particles by the Gas in a Synchrotron," *Phys. Rev. 74* (1948), 140–4.
4 N. M. Blachman and E. D. Courant, "Dynamics of a Synchrotron with Straight Sections," *Rev. Sci. Instrum. 20* (1949), 596–607.
5 D. M. Dennison and T. H. Berlin, "Stability of Orbits in the Racetrack," *Phys. Rev. 69* (1946), 542–3.
6 W. B. Fowler, R. P. Shutt, A. M. Thorndike, and W. L. Whittemore, "Production of Heavy Particles by Negative Pions," *Phys. Rev. 93* (1954), 861–7.
7 E. D. Courant, M. S. Livingston, and H. S. Snyder, "The Strong-Focusing Synchrotron – A New High-Energy Accelerator," *Phys. Rev. 88* (1952), 1190–6.
8 J. P. Blewett, "Radial Focusing in the Linear Accelerator," *Phys. Rev. 88* (1952), 1197–9.
9 J. B. Adams, M. G. N. Hine, and J. D. Lawson, "Effect of Magnetic Inhomogeneities in the Strong-focusing Synchrotron," *Nature (London) 171* (1953), 926–7.
10 E. D. Courant, M. S. Livingston, H. S. Snyder, and J. P. Blewett, "Origin of the Strong-Focusing Principle," *Phys. Rev. 91* (1953), 202.
11 E. D. Courant, "Non-Linearities in the AG Synchrotron," in *Proceedings of the CERN Symposium on High Energy Accelerators and Pion Physics*, vol. 1, edited by E. Regenstreif (Geneva: CERN, 1956), pp. 254–61.

12   Panel on accelerators and detectors in the 1950s

LAWRENCE W. JONES (CHAIRMAN),
LUIS W. ALVAREZ, UGO AMALDI,
ROBERT HOFSTADTER, DONALD W. KERST,
AND ROBERT R. WILSON

Jones: The discussions are organized as follows: first some remarks
on the subject of particle accelerators in the 1950s, then some remarks on
particle detectors. Although people like Luis Alvarez bridge both fields,
generally the two fields are more or less discrete. Robert R. Wilson deals with
aspects of particle accelerators from the Cornell perspective; in a separate
chapter in this volume, Donald Kerst summarizes the MURA developments
in the 1950s;* Robert Hofstadter discusses detectors, in particular the development of solid inorganic scintillators. Ugo Amaldi and Alvarez then take
up other aspects of particle detectors. The panel concludes with a general discussion, including responses to questions from the floor.

It may be useful at this point to summarize the accelerator and detector
technologies that were available at the beginning of the 1950s, and then
briefly catalog the conceptual advances and inventions during that decade.
Electron acceleration techniques evolved rapidly during the 1940s, beginning
with the invention of the betatron by Kerst and, following the war, the development of the electron-synchrotron by Edwin McMillan. Meanwhile,
William Hansen and others invented and developed the electron linear accelerator, substantially in the form it has retained to the present. The proton accelerators before 1950 were at first high-voltage machines, developed by John
Cockcroft and Ernest T. S. Walton, then electrostatic generators developed
by Robert J. Van de Graaff and Raymond G. Herb, followed by Ernest O.
Lawrence's cyclotron. Postwar developments included the synchrocyclotron
of McMillan and Vladimir Veksler and the Alvarez proton linear accelerator.

---

*   Most of Kerst's contribution to this panel has been incorporated in Chapter 13 and
     does not appear in the present chapter.

As the 1950s began, proton-synchrotrons with energies of over a billion volts (1 GeV) at Brookhaven, Berkeley, and abroad were under construction.

Particle detectors in the year 1950 included gas counters (ionization chambers, proportional counters, and Geiger counters) and early scintillation counters using RCA 931 photomultipliers and organic crystal scintillators of anthracene and stilbene. The Wilson cloud chamber and nuclear emulsions were the only visual techniques available at that time.

Accelerator developments in the 1950s were paced by two milestones: the invention of alternating-gradient (AG) focusing in 1952 by Ernest Courant, M. Stanley Livingston, and Hartland Snyder and the proposal for colliding beams by Kerst. The AG, or strong-focusing, principle spawned application not only to larger proton- and electron-synchrotrons but also to charged-particle beam transport at all energies. It led to the development of fixed-field alternating-gradient (FFAG) ideas by Keith Symon, Tihiro Ohkawa, and A. A. Kolomenskii. Although no high-energy machine was ever built using this concept, it has been incorporated in all cyclotrons built since 1960. As a footnote, it has been recognized that elements of the AG focusing principle were anticipated in part by Llewelyn H. Thomas a decade earlier.

The colliding-beam concept built on the development of an understanding of the ensemble of particles in the phase space defined by the parameters of an accelerator or storage ring. Whereas earlier orbit theories focused on single-particle stability, the work of Symon, Andrew Sessler, and others formulated the particle motion in all six phase-space dimensions in canonical variables to which Hamiltonian mechanics and Liouville's theorem could be applied. It was in this context that Kerst realized that interesting values of what came to be known as luminosity could be achieved through beam stacking in a dc ring. Long straight sections, low-β insertions, and synchrotron radiation damping for electrons were subsequently considered, further enhancing the attractiveness of the colliding-beam concept.

This decade also saw the beginning of serious application of the digital computer to accelerator design. At the beginning of the 1950s, we designed machine parameters by multiplying matrices on our slide rules and using commercial mechanical (Marchand or Frieden) calculators; by the end of the decade, those marvels of high-speed computing, the IBM 650s and 704s, were cranking out reams of accelerator computations.

Detector advances during the 1950s were dominated by the invention of the bubble chamber by Donald Glaser and its adaptation to liquid hydrogen by Alvarez and his group. However, there were also major advances in scintillation-counter technology, including the development of plastic and liquid organic scintillators and the application by Hofstadter of sodium iodide and other inorganic crystal scintillators. Improved photomultipliers and electronics aided the development and exploitation of the scintillation counter as well as the Cerenkov counter.

The success of the bubble chamber inspired counter physicists to seek devices which could approach the bubble chamber in spatial resolution while at the same time achieving the time resolution and triggering capability of counter techniques. This led to the development of the luminescent, or scintillation, chamber (building on the early work of E. K. Zavoiskii), and later of the spark chamber, which was just making its appearance at the end of the decade.

A recitation of detector techniques would not be complete without mention of calorimetry. Although the first true "calorimeter" may have been Kerst's device for calibrating the 300-MeV betatron beam from the temperature rise in a lead absorber, during the 1950s there were developed total-absorption calorimeters for electrons and γ particles, both in large single crystals of sodium iodide and lead glass and in sampling devices using alternating lead plates and plastic scintillator. The hadron calorimeter was introduced during this period by V. S. Murzin and N. L. Grigorov for cosmic-ray experiments. It was in the study of electromagnetic cascades in detectors that Wilson introduced the Monte Carlo technique, which has grown to such widespread current use in experimental particle physics.

Finally, the development of particle beams, including not only strong-focusing lenses (as noted earlier) but also electrostatic separators, was a significant evolution of this decade. It is interesting to note that there were almost no advances in the use of gas counting techniques in the 1950s. Multi-wire proportional chambers and drift chambers could have been, but were not, built at that time.

Wilson: My remarks are very informal. At Cornell, we had a 300-MeV electron-synchrotron built starting in 1947, and although we did some good research, I had been smarting under two difficulties. One was the π^0 production by γ rays (see Chapter 20), an experiment done by Jack Steinberger. With very similar equipment we had been desperately trying to measure the same thing at Cornell, but our energy then was only 200 MeV. You may notice on Steinberger's cross-section curve that the cross section was essentially zero there, and we observed no events.

The next development was when the Caltech group went over the peak of the resonance and down the other side (Robert Walker, Chapter 6). Again we were lingering just on the bottom side of or close to the top of the 300-MeV resonance. I felt that our energy was too small and that we should be doing something about it. What I had in mind was an energy of 1,000 MeV.

At that time, I had just gone to the organizational meeting of CERN at Copenhagen in June of 1952. One thing that impressed me there was a report by E. Regenstreif about the work that had been going on with the Cosmotron. He reported that some slits – some blanks – were put in to cut down the aperture and that one could cut the aperture all the way from the large original design down to two inches vertically and something like ten inches horizontally, and the beam would remain just the same.

So, as I was flying back and looking at the midnight sun returning from

Copenhagen that June, I began, instead of scaling up, as the Brookhaven people wanted to do, to scale down. Scaling the Cosmotron down to 1 GeV meant that you would come out with a ridiculously small aperture of about a third of an inch by perhaps three inches. That seemed very small. The magnet would also be very small in cross section. Well, I was impressed by that, and, always conservative, I increased the aperture in my mind to something like one-and-a-half inches by five inches, thinking that it's hard to make it much smaller.

In those days, one could simply decide: "Well, we'll build the magnet." Still, it was a small magnet, just about that size. We decided to make the biggest radius we could in the synchrotron room, which was something like six meters, maybe a little less than that, and get up to 1 GeV with a field of about 10,000 gauss, standard for iron. The idea, then, would be to use all of the power supplies, all of the controls, and all of the ancillary things, except for injection. I called up high-voltage engineering, and they said, sure, they had a Van de Graaff of 2 MeV. "How much is it?" I asked. Fifty thousand dollars is what I recall they said. "Fine, I'll buy it!" And I hung up. To me, that was really living, at the time. We did things very informally, as far as the Office of Naval Research was concerned. I think we were talking to William Wright and Jackson Laslett in that office; they were very nice people to get along with, and they encouraged me to go ahead in this particular way. So it was going to cost very little.

On the other hand, I was in for a penny, but before being in for a pound, I thought I should visit Brookhaven and check up on Regenstreif's story. As I went to Ken [George Kenneth] Green and to other people, no one would confess that they had made any such measurement there, and that this whole scheme of mine was just based on thin air. Well, nevertheless, it seemed like a good idea, and I was very, very pleased when the strong-focusing idea came along. I thought, 'Well, that'll be a lot of fun."

I had other ideas to make the aperture larger with what I called "Ubangi lips" on the aperture that would saturate, bringing the aperture down at high fields, but this sounded like a lot more fun. It was, I think, before the resonance phenomena were recognized. When we first had a die made (because of those other pole tips, which I thought we could put in later on and fix things up), I had it so that you could insert any kind of pole tips into a much larger gap, so that you didn't have to commit yourself.

Well, that meant that when the strong-focusing business came along, we could put those kinds of poles in. We were so inexpert that the first time we put in the pole tips, we weren't even in the necktie pattern, for those of you who remember. I think we were outside of it. It was easy then to have some new stampings made that would dilute the other stampings, and bring us into the necktie pattern very nicely.

Actually, with strong focusing it wasn't so courageous to go in that direction, because our 2-MeV injector was well above the transition energy.

Therefore, we didn't have to go through the unknown parts that Courant talked about in building the electron model at Brookhaven. It was a much easier thing that we were attempting to do at Cornell.

The other thing I remember is that we had a tradition of keeping physics going no matter what, and so until the energy of the new machine got above the energy of the 300-MeV machine, we kept our physics going full blast, and then turned it down after we had exceeded that energy.

I recall, too, that we did suffer from this haphazard manner of proceeding. I remember we built it practically by ourselves. The total cost was well within a million dollars, and probably closer to $100,000, although who was counting at the time? We were pretty cavalier about that, too, just because it seemed such a simple thing to do. What was done was done by my colleagues at Cornell. We did all of the stacking of the laminations, all of the making of the coil, and in fact, with our small shop there, the whole thing was produced within our laboratory. The hard part, of course, came when we had to make it work: to find out that we weren't in the lattice, to have the news of the various resonances come in that we had to avoid, which meant new pole tips. We would go around happily looking, tracing the beam around, and eventually it came around more and more. But it was hard work, first of all to get the machine to work at all. As soon as it got to 500 MeV, we started doing experiments with it, because it was our tradition to do whatever was required to do an experiment, and when we got to 500 we were beginning to break new ground, so it was consistent that we would have started our experimental program.

I remember when the energy got up to 700 MeV, though, we were immediately rewarded, because putting in a detector – we had a simple meson detector – we immediately saw the next two resonances above the 3-3 resonance: the next two bumps that were shown in the curve that Walker has shown earlier.

Jones: I must remark here that there was a meeting in April 1953 in Chicago which drew together people who later formed the MURA organization. We were given many theoretical lectures by Courant and his friends from Brookhaven about the strong-focusing principle. Wilson got up, and he pulled out of his briefcase a lamination with a strong-focusing profile on it, and people nearly jumped out of their chairs. He announced proudly that later that year he expected to have a strong-focusing synchrotron operating. I don't remember the time scale, but I think your boldness was widely appreciated and applauded.

Wilson: We did have an awful lot of trouble, but I think it was working at the end of 1953, certainly early in 1954; I can't exactly remember.

Jones: There was a question this morning about the interaction between astronomy and the accelerator physics theories. There were indeed a few. In the first place, one of the primary betatron resonances, the two-thirds resonance, was identified as the reason for a gap in the asteroid belt, in fact, not by us, but by astronomers long before. Let me recall one amusing inter-action in which I was personally involved. I had come back from a meeting in Wisconsin and was telling people at the lunch table about this negative-mass instability, whereby the mutual repulsion of electrons or protons by Coulomb forces in a circulating, homogeneous ring of particles causes them to clump up into little clumps and then drives an instability. George Uhlenbeck, who was very wise, stroked his long chin and puffed on his cigar, and said, "You should read the prize paper of Maxwell on the rings of Saturn." This is, of course, identically the opposite-sign force, mutual gravitational attraction, which is responsible for dispersing the particles in the rings of Saturn to be a homogeneous ring. And indeed, the central fields, the gravitational attraction to the planet or the magnetic guide field, in both cases were analogous.

Hofstadter: I originally thought I would talk primarily about sodium iodide, but at least at the beginning I'll talk about scintillation counters in general. Let me say the following: One shouldn't forget a certain number of names that have to do with the subject of scintillations. I'm not going to take time to go into details, but we should remember that there was such a thing as Crook's spintheroscope, and that Julius Elster and Hans Friedrich Geitel a long time ago – about 1900 – used scintillations; Ernest Rutherford used scintillations to do the scattering of α particles. Gamma rays were always difficult to detect. Geiger counters were used for a long time, with very low efficiencies. Then, during the last war, a Dutch physicist by the name of P. J. van Heerden invented a conduction counter which used silver chloride. That amounted to a solid ionization chamber.

When I was at Princeton in 1947, I thought it was a good idea, and I started to do some experimentation on silver chloride myself. As you probably know, these ideas eventually led to the use of germanium and lithium drift-type conduction detectors, which are not scintillation detectors. But they are very important.

Now, the scintillation detector per se, as we know it, came along in 1947, when Martin Deutsch brought back news to the United States that Hartmut Kallmann in Germany had used mothball material in front of photomultipliers and observed γ-ray counts. This was the first time that one really had a solid to use for detecting γ-rays. And many of you will remember that every nuclear physics laboratory smelled of mothballs in those days.

Just a year later, I discovered the sodium iodide detector, along with a lot of other ones – cesium iodide, and rubidium iodide, and whatever. I have made a list of ten important experiments that were done starting in the 1950s, and ending in 1959, which have used scintillation counters in a principal way in their findings.

The first experiment concerns the properties of the neutral pion, which Steinberger talked about (see Chapter 20). Then positronium was discovered, using sodium iodide detectors. Then pion scattering was done with scintilla-tion counters at Chicago by Herbert Anderson and Enrico Fermi and others.

Then μ-mesic atoms were done by Val Fitch and James Rainwater using sodium iodide detectors. In our experiments at Stanford we used Cherenkov and scintillation counters at the top of our magnetic spectrometers, to look at nuclear sizes and charge distributions. The antiproton experiment used those two scintillation counters mentioned by Owen Chamberlain (see Chapter 17), and indeed there was also a Cerenkov counter in that same experiment for timing velocity measurements. The nucleon structure experiments also used a whole series of scintillation counters. Then the violation of parity in weak interactions was demonstrated; that experiment was done with anthracene as a β counter and sodium iodide as a γ-ray counter. The Mössbauer effect came about through the use of the sodium iodide scintillation counter. The proof of the existence of the antineutrino was done by Frederick Reines with liquid scintillation counters. So you see all of these things came as a result of the invention of the scintillation counter. I regret very much that Kallmann never received any recognition for this very, very important development. I have estimated, for example, that in the case of sodium iodide, there are somewhere between ten thousand and fifty thousand papers in physics that report experiments which have used sodium iodide. And I have an independent judgment on that by Stanley S. Hanna, a nuclear physicist, who said it's certainly in the 10^4 range. It's just fantastic to see what the scintillation counter has done when one considers the number of applications of scintillation detectors and sodium iodide detectors in medicine, geology, chemistry, nutrition studies, and so forth. I think Ugo Amaldi is going to mention something about liquid scintillation counters; I might have something to say on that a little bit later.

In the 1950s, a number of us attempted to go from small detectors to very large detectors, so that we could detect electromagnetic showers, and later even strongly interacting particles.

Alvarez: Bob, I'd just like to pick up on what you said about Kallmann and his great contribution to particle physics and the fact that he never received any recognition at all. I have in my life made four nominations for Nobel Prizes, and one of those was Kallmann, a man I've never met. I have an enormous respect for his impact on our field. He's dead now.

Hofstadter: Yes, Luis, I did that too, but I didn't know that you had done it.

Ugo Amaldi: I have three comments which concern the development of scintillation detectors during this decade: one at the beginning, one toward the middle, and one toward the end of the decade.

The first point is related to Hofstadter's remarks and has to do with liquid scintillators. Although it is not well known, liquid scintillators based on naphthalene were used for the first time in 1949 by Mario Ageno and associates in the Instituto Superiore di Sanita in Rome (which I later joined in 1958).[1] Then there was the work done here in the States by George Reynolds and Giorgio Salvini, using terphenyl in benzene, and xylene, and at this stage the efficiency was already very high.[2] Kallmann and Milton Furst did a very complete study over two years, in which they understood fully how a three-component mixture could work.[3] There is a main solvent, a secondary solvent, and a solute; energy is transferred from one to the other component. In these experiments, the light yield could be measured accurately, as better photo-tubes were available. What I find extraordinary is that, probably because of all the work that had been done before in solids (both inorganic and organic), by 1952, the field was understood, and people could use the results. As is well known, many experiments have been done since then using this type of scintillator.

My second example is from around 1955. I have been studying the literature on the origin of sampling calorimeters because recently I have been working with total-absorption calorimeters and written a review article about them. The first paper I found was in the *Physical Review* in 1954.[4] Later on, in 1956 or 1957, the Russians developed the sampling hadron calorimeter for cosmic rays. Unfortunately, at that time nobody would care, because the energy of the running accelerators was not high enough to have a useful measurement of energy using a hadron calorimeter. The name of Murzin is very well known, in particular since he wrote a very beautiful article in 1957 in *Progress of Cosmic Ray Physics*. I want to remark that in a certain sense, this use of sampling calorimeters goes back to a paper by W. Blocker, R. W. Kenney, and W. K. H. Panofsky, who in 1950 made measurements with the first quanto-meter for γ rays.[5] The measurement was in fact based on an ionization chamber utilizing the principle of sampling for measuring total energies.

The last example occurred toward the end of the decade and had to do with our chairman. Jones gave a talk on instrumentation at the 1959 CERN conference showing a kind of detector which could be used for colliding beams and was based on the principle of what was called at that time "luminescent chamber" or "scintillation chamber." The first person who observed light from plastic fibers with an image intensifier was a Russian physicist, Zavoiskii. Later it was seen in the States. At that time, as you may remember, there was no fast detector which could work in the background of a possible colliding-beam machine. So one was looking for a fast detector, and the idea Jones proposed was to use a double layer. The first layer would be fibers of plastic scintillators sixty centimeters in diameter, and then around it cesium iodide looked at by image intensifiers. (Nowadays, a cesium iodide array is being built, in fact, at Cornell for covering the full solid angle in the CLEO detector.) In that way, one could do what was not possible to do with existing techniques, and what would become possible when the proportional wire chamber would be invented fifteen years later. In the same conference, Gerard O'Neill, who was mentioned earlier, presented the scheme of a colliding-beam accelerator.

I would like to remark in conclusion that those ideas current at the end of the 1950s may be the basis of the detectors of the future. I was in Aspen this summer discussing future detectors for the Superconducting Supercollider (SSC), and we are thinking about upgraded detectors for the hadron collider which could be installed in the Large Electron–Positron storage ring (LEP) tunnel (LHC). Both in the case of SSC and LHC, we have a very short time between bunch crossings, not more than tens of nanoseconds. So one cannot easily use wire chambers, because the drift time is too large, and people are thinking again of using scintillating fibers. Of course, nowadays the intensifier tubes are much better, and the idea is that one can get a better resolution now with glass fibers of a new hundred microns in diameter than by using a proportional chamber. So my last remark is that those ideas at the end of the fifties will be the ideas which will be working for the SSC ten years from now. This would be another step in the Darwinian substitution of detectors – as an example, consider the bubble chamber. That was a big field, an essential development, but now they are not in use at the new colliding-beam accelerators. Now we use wire chambers, but maybe for the new machines they will not be fast enough, and we have to go back to ideas which were really developed in the 1950s.

Jones: Let me make a further remark concerning the scintillation chamber. In the late 1950s, many of us were almost jealous of the great success of the bubble chamber. We were also seeking ways to combine the good spatial resolution of the bubble chamber and other visual techniques with the fast, triggerable properties of electronic counters. The scintillation (or luminescent) chamber appeared to offer hope of such a combination of properties, and several groups worked on such devices. Martin Perl and I, at Michigan, focused on imaging solid sodium iodide crystals with lenses onto image-intensifier cathodes, but most other groups pursued methods using scintillating plastic fibers. Those groups included Reynolds at Princeton, Dave Caldwell, then at MIT, Arthur Roberts, then at Rochester, and Ken Lande and Al Mann at Pennsylvania; Roy Kerth at Berkeley was also interested. Perl and I were invited to bring our chambers to the Bevatron to try to do physics. I recall that period with pleasure; we had a lot of fun with those chambers. But we actually did two respectable experiments which led to *Physical Review* publications about 1960 and to two Ph.D. theses. Amaldi was correct that Zavoiskii first published a track photograph; it was an α particle in sodium iodide under a microscope. But to our knowledge, the Russians never carried it further. The development of the spark chamber made possible the same objectives we had been seeking with the scintillation chamber, and we all happily embraced that newer device.

I would like to comment on Amaldi's discussion of the sampling calorimeter. I learned of the Russian calorimeter work from Giuseppe Cocconi and others at CERN in the early 1960s and later built and used hadron calorimeters for a series of cosmic-ray studies (1965–70). Although Amaldi is correct that this device was not interesting to most physicists doing fixed-target physics at accelerators in those days, I saw the calorimeter as a way of opening up the high-energy physics of neutrons. We built and used the first calorimeters at accelerators in neutron experiments at Brookhaven, Berkeley, and, later, Fermilab starting in 1966.

Alvarez: I think that I was put on this panel just for balance because there wasn't anyone else who'd had bubble-chamber experience. I talked to you this morning as a strong bubble-chamber advocate. As Jones just said, the bubble chamber was the instrument of choice for a period of less than ten years. It has essentially disappeared now, although there still are people doing bubble-chamber work.

I'd like to stand back just for a moment and look at the field of detectors, track-forming detectors particularly, the way my friends in paleontology do. As some of you know, I spend a lot of time these days talking with paleon-tologists, and the favorite thing they talk about is the range of a species or a genus or a family – that is, where it first appears in the stratographic record and where it goes out. That's called the range. And if you look at track-forming detectors, you can see the range of the cloud chamber. The first one, which came in around 1910, went out sometime in the early 1950s, or maybe a little earlier. Then the emulsions came in, had a very wonderful, short, very productive life, and then they kind of disappeared, too. And then the bubble chamber came in suddenly, did great things, and went extinct. I like to feel that I was a member of the class of dinosaurs – I worked with bubble cham-bers, and then, POW! We got hit by an asteroid, and now the track detectors of choice are, obviously, the new electronic things. In our laboratory, the one I know best is, of course, the time-projection chamber. I saw the most beauti-ful pictures of dE/dx versus momentum that someone showed me, with fan-tastic resolution. You can resolve muons and pions and kaons and protons – even a few deuterons – just beautifully, except where the tracks overlap. That is, by far, the most fantastic detector I've ever seen, and I'm delighted that it exists. I congratulate the people who developed it, and I don't feel a bit bad about the fact that my old favorite thing is no longer of much use.

Jones: I wonder if I could ask Wilson to remark on the Monte Carlo technique, which is neither fish nor fowl. It really doesn't fit under accelerators or detectors or, indeed, high-energy physics, and yet it's important to all of them.

Wilson: Well, I certainly did not invent the Monte Carlo method, but the use that I put it to may have been, as in the case of the synchrotron I just told you about, a case of benign ignorance. The Monte Carlo method was developed at Los Alamos by Stanley Ulam. But there I believe it was primarily used to simulate the solutions of differential equations. Misunder-standing how it was used, I just assumed that what one would do is follow some stochastic process, flipping a coin each time, and finding out how the event would develop in time. So it seemed to me that it would be natural in the case of photons or electrons incident on a detector (that was the problem I was concerned with) to use the Monte Carlo method and just follow each event through its development stochastically. Then follow a hundred or a thousand events, take an average, and that would be a good and very simple calculation. My son had an erector set, and I believe there was an oatmeal can, a cylinder, so I made a graph which I could put on the oatmeal cylinder with some Scotch tape, and with a cursor across it I would start the electric motor. The oatmeal can with the graphs on it would whirl, and I plotted the curves so that you could read off energy on one coordinate and what had happened to an electron or a photon as it fell into various little channels along the side. It was a very simple device, and did give fairly accurate information.

Kerst: I would like to make a comment having to do with detectors and their connection with colliding beams. When we understood how much phase space there was available, we felt that we could load particles to intensities where it was sensible to talk about a significant particle yield from colliding beams. I'm not sure the rest of the crowd felt this, but I was a little cool on it, because the detection problem looked bad. You see, we weren't thinking about big, long, convenient straight sections. We were trying to avoid anything as radical as that. In fact, that conservatism had to go to make colliding beams useful; you needed a long straight section. But colliding beams put the burden on the detectors. The detectors had proliferated in complexity and with adequateness. I had thought that the good thing about colliding beams was that all your junk would not go straight forward, but the yield would be spread out, so in that sense, detection might be easy. But when the PS began to operate at CERN, people were so extremely skillful at separating the beams and handling the beams that the colliding beams, in my estimation, went down. I know that fellows like Jones and others were thinking about a lot of experiments you could do with colliding beams. To me it looked like protons on protons, and then you go home. But it certainly isn't that way, and it isn't that way because detectors are so superb these days.

Hofstadter: I would like to make one remark following on Ugo Amaldi's talk about the liquid scintillators, because I remember very well when Reynolds started to work on that, and when Salvini came to Princeton; I was there at that time. I had just sent in a paper on terphenol and dibenzol scintillation counters – these were inorganic solid crystals. Salvini and Reynolds were using benzene, as I remember, and I suggested that they should use xylene. There are three kinds of xylene: orthoxylene, metaxylene, and paraxylene. I knew from literature I had read that orthoxylene was by far the best, so I suggested this to Reynolds and Salvini. Reynolds came into my laboratory where I was doing the sodium iodide work on spectroscopy, and we got some orthoxylene. We dropped a little bit of the terphenol in, and there were the biggest pulses that anybody had ever seen with a liquid scintillator. That's a little bit to be added to the history of liquid scintillation counters.

Amaldi: It seems to me that our roundtable will be incomplete if we forget about Cerenkov counters. I want to concentrate not on lead glass, but on other types of Cerenkov counters. In looking through the literature I found one interesting thing: the most difficult one was built first. In fact, R. H. Dicke, in 1947, made a focusing Cerenkov counter; it had a long lucite pipe, and then a kind of funnel for the photomultiplier. It was a complicated instrument and was built in 1947 – ten years after the explanation by James Franck and I. E. Tamm, and fourteen years after the discovery of Cerenkov. Only three years later, at the beginning of the decade we are discussing, J. V. Jelley constructed a nonfocusing water Cerenkov counter which has been used by many people later in experiments. Only a few years later, in 1953, W. Galbraith and Jelley observed Cerenkov light from cosmic rays in the atmosphere. So, from the time that the first counter was made in 1947 to the time when people looked at the sky and saw this light was only six years.

Hofstadter: Somebody had done it before Dicke – Ivan Getting.

Amaldi: But it was not done with photomultipliers.

Jones: I'd like to come back to a remark after Matthew Sands's talk about what I might call the Russian connection. Perhaps I could ask Alvarez to reflect on this. There was an Atoms for Peace meeting in 1955 where Veksler announced to the world the existence of the 10-GeV Dubna machine, and then after the meeting in 1956 at CERN, Alvarez and a few others went on to Russia.

Alvarez: No, just the opposite, we had come from Russia to Geneva in 1956.

Jones: But you wrote a beautiful *Physics Today* article on that.

Alvarez: That is because my diary was published in two installments in *Physics Today* in 1956. I have said to people that when I got to Moscow in 1956, I was much more surprised than I would have been if I'd found myself on the back side of the moon. This was in the coldest part of the cold war. I only knew one person who'd ever been to Moscow – I guess I knew that Victor Weisskopf had been there in 1936. But I only knew one other person who'd been there. It was just extraordinary to find yourself walking through Red Square. I remember poor Wilson almost got killed. He stuck his hand in a door on the subway, because the door was closing – he wanted to stop it – and there wasn't any automatic thing to keep... [laughter].

Jones: That brings to mind that, following that revelation of the Dubna machine and the consternation that the Bevatron was no longer the highest-energy machine in the world, at least from the Midwest perspective, there was a certain panic in Washington, and a decision was made to dash ahead, in spite of the AGS being under construction, with a quick, rapidly built machine that would be at least over 10 GeV. That, I believe, precipitated the decision to build the Argonne ZGS at 12.5 GeV. Following Joseph Stalin's death in 1953, Richard Crane at Michigan dubbed this the Stalin Memorial Accelerator. I think it's a tribute to the differences in style that some people in England (with Nimrod), and at Argonne, chose to go the safe, conservative way and eventually succeeded, only rather later and more expensively than one might have hoped, while others like Wilson said, "Damn the torpedoes, full speed ahead," and succeeded with new concepts in a very impressive way.

Alvarez: I don't think anybody should have panicked after seeing the Dubna machine. It was a real dog.

Amaldi: I have a quotation from John Adams, which is somewhat related to what has been said. We mentioned him already today in connection with the discovery of the possible difficulties of strong focusing. I found in the proceedings of the 1956 CERN conference the following sentence by him: "The evolution of accelerators bears some resemblance to the evolution of the large prehistoric monsters, in that their size increases until a better way is found of achieving the same result."

Jones: I think I liked better Wilson's comparing accelerators to cathedrals. I might also make another remark on Sands's talk earlier. I'm sorry O'Neill isn't here. He was indeed a vigorous contributor to the concept of colliding beams. As Kerst mentioned, we had also talked about storage rings, and in fact, some time earlier than that, we were aware that O'Neill was also thinking along that line. The problem that hung up the MURA group, and indeed I think was a real problem with proton machines until the first strong-focusing machines existed and worked, was the problem of being able to transfer beams out of machines with preservation of phase space. With the weak-focusing machines, in spite of Wilson's optimistic understanding, the beams were rather large. There seemed no way that one could preserve the phase-space density and flip beams back and forth from accelerator to storage ring and so forth, without unacceptably diluting the phase space. O'Neill, in fact, coupled his invention of storage rings to a Liouville-theorem-violating concept of foils, which were somehow to dampen the phase space. Now, of course, Sands realized that with electron machines, that wasn't necessary, that synchrotron radiation did it for you. But the essential problem in the middle and late 1950s with storage rings was this beam handling. Of course, now we see at this laboratory beams passed from ring to ring to ring to ring, with complete equanimity.

Albert Wattenberg: I was very pleased to see that Ugo Amaldi had found the reference to the Pugh–Frisch–Gomez paper on the first use of scintillation counters, actually, which were plastic. It isn't just that it was the first write-up of a, so to speak, calorimeter, but Otto Frisch had gone to Edwin H. Land of Polaroid – and Land had referred him to an absolutely fabulous chemist, Mark Hyman, who really developed the possibility of very-high-purity, long-optical-path-length plastic scintillation materials. These long-optical-path-length scintillation plastics made possible a lot of the very, very large experiments that were done both at the AGS and at CERN, as well as other experiments.

There is a question I would like to address to Hofstadter: We make most of our own apparatus in almost everything we've done, and I think it's very rare that we haven't done something ourselves in the laboratory. The one big exception, where we're all very dependent, or had been dependent, on industry, was the photomultiplier. I was wondering if he had comments on the history of the photomultiplier that have never been brought out.

Hofstadter: Well, I think that the story of the photomultiplier has been presented. George Morton of RCA wrote a long article on it. Vladimir Zworykin is the inventor of the photomultiplier. I think in my recollection it may be that Dicke was one of the very first to use the photomultiplier, when he was doing the Cerenkov detector. Also, Marietta Blau and somebody else used zinc sulfide and a photomultiplier. They originally used those little 931A photomultipliers which you could pick up in the junkyard, because they were used as noise generators in World War II.

Alvarez: RCA wasn't told that they were used as a noise generator, they just got big orders for these 931s! They were terribly upset when they found out what they were used for.

Wilson: We should remember James Allen, who searched for the mass of the neutrino with his own homemade multipliers – not photomultipliers, but multipliers, with beryllium electrodes.

Hofstadter: Yes, there was a Swiss physicist, too, who made his own photomultipliers, in fact, some very big ones. I forget his name now.

V. Telegdi: Well, I was very happy to have Ugo Amaldi say something that involved Kallmann. You know, we are all very indebted to the inventor of sodium iodide, who is sitting here, because we all know the physics that he has done, with it and otherwise. Now, Kallmann has not done any particle physics, but he invented the organic scintillator. And I think that his reputation and his acknowledgment among particle physicists is remarkably low for this great accomplishment. And in fact, the remark that Hofstadter made about the zinc sulfide is a very pertinent one. The great recognition of Kallmann and others was to have a scintillator that doesn't only scintillate but transmits its own light. Zinc sulfide does not. That is the interesting aspect of it.

Another thing that might interest the historians is how Kallmann managed to live through the war. He's Jewish, and he nevertheless did research in Germany in the laboratory all during the war. And in fact, toward the end of the war, I think he noticed some lab coat fluoresce, and that's what got him started on naphthalene. Now, this plastic scintillator is extremely interesting, because when terphenol and other crystals came out, it was extremely difficult – I remember Fermi, Anderson, and company doing the experiments at the cyclotron, and they used those 931s, which at that time were not used as noise generators, but they were available from industry because they were using them as headlight dimmers in the Cadillac! That was why they made them. It was very, very difficult to get crystals larger than 2 cm by 2 cm. Very difficult. I think the crystals that Anderson and company used were about an inch by an inch, glued on the side of a 1-P21. And so the plastic and the liquid really made things possible. And as far as the plastic is concerned, I remember very, very distinctly that Marcel Schein had a Chinese research associate, and this man made the first plastic scintillator that we had ever heard of. Of course, it did not have Hyman's quality, but it sure was easier to use.

Now, as far as storage rings, I don't know whether you remember anything that Rolf Wideröe did along these lines. When I was a graduate student in Zurich, I measured the energy of a betatron that Wideröe had built, and while I was working there, Wideröe told me, "You know, all these accelerators are wrong. They are all wasting energy. What you need is a synchroclash – a collision of two beams of opposite sign, if possible within the same ring." I think he has patents on it.

Kerst: We didn't mention Wideröe, that's right. I didn't know anything about Wideröe, except in a CERN meeting, all of a sudden, he popped up and said, "I've got a patent on it!" It was an electrostatic device. In an electrostatic device, you can run the same particle either way. I don't know what the patent looks like, but Amaldi says he's seen the literature, and it's bona fide. But let me correct one thing. Synchroclash won't work with betatrons. ("Synchroclash" is a word invented because there used to be something called "synchromesh" for automobile gearshifts, and "synchroclash" is one of these jargon things that built up.) If you have a synchrotron, you have bunches. And if you are synchronized to clash somewhere else where your detector isn't, it's no good.

Telegdi: I can argue that I, being Hungarian, and he, being Norwegian, we didn't use the word "synchroclash." Anyhow, I have one more remark, and then I have finished. The first use in particle physics of a photomultiplier was not of a photomultiplier, but of a particle multiplier. The rest of the structure was done by a well-known Hungarian experimental physicist by the name of Zoltan Lajos Bay in 1938, I think.

Jones: At the CERN 1956 conference on accelerators, Wideröe made an extended comment during the discussion in which he argued the electrostatic containment of counterrotating beams. The written rebuttal to Wideröe by Lawrence appears in the proceedings as only the brief sentence: "Even with fields as low as 10 kV/cm it is difficult to prevent sparks over long periods of time." I recall very distinctly, however, the withering oral putdown by Lawrence, where he noted that even a one-gauss magnetic field was equivalent to 300 volts per centimeter, and that an electrostatic guide-field collider of any interesting energy would lead to a ring of totally unreasonable size.

Amaldi: I want to complete what was said before on the use of scintillating materials. The first physicists who looked at a scintillator with a photomultiplier were S. C. Curran and W. Baker; that was classified work from 1944 until 1947. I think that the Swiss group was Blau and B. Dreyfus, who in 1945 looked at a screen of microcrystalline zinc sulfide with a photomultiplier.

Hofstadter: And John W. Coltman and Fitz-Hugh Marshall, also, at Westinghouse, in about 1947.

Amaldi: Another point I want to remark concerns Wideröe. Some of you may know that Bruno Touschek had been working with him in Hamburg in the years 1941–5. I'm pretty convinced that the idea that Touschek got later, of *e*-plus and *e*-minus colliding beams, came from his knowledge many years earlier that Wideröe had suggested this electrostatic colliding-beam facility.

Wilson: I might add that at Los Alamos we used a bank of 931s to detect the γ rays from the first atomic explosion. It was covered, and we didn't know what we were measuring, but assumed that it was the glass – that the glass was either making Cerenkov light or scintillating. We didn't know which one it was.

Wolfgang Paul: The first multiplier I saw was in 1937 when I was a graduate student at Berlin. It was used by an astrophysicist, K. O. Kiepenheuer, for measuring the light intensity coming from stars. The multiplier was built by a physicist named [Georg] Weiss; it was made by the Telefunken company in Germany. There were some available at that time.

Kiepenheuer also used another device which was very interesting. He used a photocathode in vacuum, accelerated the photoelectron to about 10 kV, and sent it through a thin window into a Geiger counter, counting the photoelectrons in that way. That was in 1935.

The second remark concerns the patent of Wideröe. The patent was in 1943, and in it he not only described electrostatic rings for electrons but also described storage rings using magnetic fields and wrote about electron–proton and proton–proton collisions. He had about three or four different magnetic configurations to accomplish this. He even utilized electrostatic focusing.

E. Segrè: I want to say something concerning the Monte Carlo method, and Wilson's remarks. In 1935–6, Fermi was studying the motion of neutrons in hydrogenous media. If you look at that paper, you will be quite surprised at how he used different mathematical methods and how he always would make approximations. At that time we were rather surprised, and we thought it was his great intuition. Many years later, he told me in Los Alamos that his great intuition was that early in the morning, between four o'clock and eight o'clock, with a little hand-driven adding machine, he was calculating the motion of neutrons in hydrogenous media, just taking cross section, mean free paths, density, and so on, whatever you had to take, such as absorption and scattering cross sections. And then he saw from that what happened, and that suggested to him the analytical method. But he really invented the Monte Carlo method around that time. He never published it. That is what I heard from him.

Notes

1 M. Ageno, M. Chiozzotto, and R. Querzoli, "On the New Technique of Scintillation Counters. II," *Accad. Naz. Lincei* 6 (1949), 626–31; M. Ageno, M. Chiozzotto, and R. Querzoli, "Scintillations in Liquids and Solutions," *Phys. Rev.* 79 (1950), 720.

2 George T. Reynolds, F. B. Harrison, and G. Salvini, "Liquid Scintillation Counters," *Phys. Rev. 78* (1950), 488.

3 Hartmut Kallmann and Milton Furst, "Fluorescence of Solutions Bombarded with High Energy Radiation (Energy Transport in Liquids)," *Phys. Rev. 79* (1950), 857–70; Hartmut Kallmann and Milton Furst, "Fluorescence of Solutions Bombarded with High Energy Radiation (Energy Transport in Liquids). Part II," *Phys. Rev. 81* (1951), 853–64.

4 G. E. Pugh, D. H. Frisch, and R. Gomez, "Efficient, Fast, Energy Sensitive γ-Ray Counters for Use above 50 MeV," *Rev. Sci. Instrum. 25* (1954), 1124–6.

5 Wade Blocker, Robert W. Kenney, and Wolfgang K. H. Panofsky, "Transition Curves of 330-MeV Bremsstrahlung," *Phys. Rev. 79* (1950), 419–28.

13 Accelerators and the Midwestern Universities Research Association in the 1950s

DONALD W. KERST

Born 1911, Galena, Illinois; Ph.D., University of Wisconsin, 1937; nuclear physics and accelerators; University of Wisconsin (emeritus)

The possibilities of a great increase in the energy of particles opened up by the invention of alternating-gradient focusing aroused great interest in the Midwest. One result was the establishment of the Midwestern Universities Research Association (MURA) to build a large accelerator for the midwestern region. It was a time when there was much to be learned about accelerator design. To achieve the best results from machines, which were becoming conspicuously expensive, it was important to understand orbit behavior fully. The application of the digital computer after 1952, particularly at Brookhaven and at the University of Illinois, revealed the necessary details of manipulations of synchrotron and betatron phase space in the presence of realistic nonlinearities and perturbations.

At the dedication of the Brookhaven Cosmotron (on 15 December 1952), Samuel K. Allison of the University of Chicago and P. Gerald Kruger of the University of Illinois suggested that a meeting of the scientists in the midwestern region of the United States should be called to consider ways of providing a high-energy facility for that part of the country. The meeting, held on 17 and 18 April 1953, at the University of Chicago, was attended by a large number of midwestern scientists interested in high-energy research. The group invited Ernest Courant, John Blewett, and Robert R. Wilson, who were already working with the design problems of alternating-gradient accelerators, to give talks. Courant explained the problems of misalignment and the choice of optimum parameters for an alternating-gradient synchrotron. At the time, Brookhaven was considering ν (the number of betatron oscilla-

I wish to thank F. T. Cole, F. E. Mills, and K. R. Symon for their help in reviewing this chapter and for their suggestions for improvements and additions.

tions per revolution) in the vicinity of 10 and a three- to four-inch aperture. Blewett discussed magnet problems, injection problems, and buildings and costs. Wilson brought samples of a tube section and pole pieces that he was planning to use in the Cornell strong-focusing electron-synchrotron. This meeting served as a start for the formation of a working group that was to continue and to grow into MURA in the years ahead. Attendees who did not ultimately participate at the working level in MURA helped to provide the encouragement within their universities and institutions for others who ulti-mately joined the working group.

This first meeting brought out thoughts that various physicists had, including technical points and questions of organization. The possibility was discussed of using the squash court in the west stands of Stagg Field for laboratory space, using the area of Stagg Field itself for an accelerator. The representatives of physics departments formed an "Organizing Committee" and took steps to provide continuation of activity, asking me to be the technical director and to explore the possibility that some of the midwesterners could spend several weeks in the summer at Brookhaven to educate themselves about the new problems of alternating-gradient accelerators. Brookhaven made it possible for a group to visit there for the period 7–21 July 1953. The group was helped by lectures from Courant, Hartland Snyder, Robert Serber, and Ken Green. Milton White and M. Stanley Livingston also met with the group.

The physicists who participated in this first period of concentrated effort and maintained their connection with the working group in subsequent years were Lawrence Johnson of the University of Minnesota, Daniel Zaffarano and Jackson Laslett of Iowa State University, Lawrence W. Jones and Kent Terwilliger of the University of Michigan, Francis T. Cole of the University of Iowa, S. Courtney Wright of the University of Chicago, Norman Francis of Indiana University, John Powell of the University of Wisconsin, and myself from the University of Illinois.

A month after the Brookhaven meeting, a larger group assembled at the University of Wisconsin during the period 7 August through 5 September. Among those who became interested in the development of these studies and who joined the group at Madison was Keith R. Symon of Wayne University. Brookhaven continued to help, since Courant came to Madison for this period. The effort was directed toward orbit problems, with much attention immedi-ately going to the effects of nonlinearities on the particle motion. Coupling between vertical and horizontal oscillations was one of the orbit topics, and the damping of large phase oscillations was studied.

During those sessions in Madison, H. Richard Crane pointed out the exist-ence of what came to be called "locked-in motion," resulting from nonlineari-ties in the focusing forces. Crane's observation gave insight into characteristics of the motion that were soon to become more evident when digital-computer results were at hand. It was not then known whether or not this effect of non-linearities indicated greater stability with nonlinear forces. This locked-in motion turned out to be the motion about a second equilibrium position for an orbit (i.e., circulation about another fixed point in the phase plane). Soon, both digital and analog computers were used to determine the nonlinear motions, and it was not long before the Illiac at the University of Illinois was providing much information on coupled nonlinear differential equations with periodic coefficients.

Some structural problems and machines were considered in this summer period; for example, Laslett and Jones designed an alternating-gradient electron-synchrotron for 10 GeV and thought that with great effort on the radio-frequency (rf) problem one might be able to go to 15 or 20 GeV. It was to have a 15-msec rise time for the field.

In this period, the group called itself the Midwest Accelerator Conference and participated in the exchange of internal reports, especially with CERN and with Brookhaven, the laboratories that were especially active in the study of alternating-gradient accelerators. The decision was made to keep the activity proceeding throughout the year and to hold meetings every month or two at different institutions in the Midwest (two days in October at the University of Illinois, two days in November at the Institute of Nuclear Studies in Chicago, two days in January 1954 at the University of Minnesota, two days in February at the University of Indiana in Bloomington, two days in April at the University of Iowa, and two days in May at Purdue). The meetings were typically attended by approximately 20 people, usually including Cole, Robert Haxby, Jones, Johnson, Laslett, Morton Hamermesh, Powell, Fritz Rohrlich, Symon, Terwilliger, Wright, Zaffarano, John Livingood, James M. Snyder, and myself. Occasional visitors to the meetings were A. Taub, Enrico Fermi, Herbert Anderson, Keith Brueckner, K. Watson, Josef M. Jauch, R. Rollefson, Crane, A. Mitchell, and Kruger.

The work from the middle of 1953 to the middle of 1954 was aided by a National Science Foundation grant of $21,800 and generous leaves of absence, travel, and facility support from the universities. Topics studied and reported on in this period were effects of power-supply ripple on the Brookhaven Cosmotron, linac injector design, analog orbit studies made at the University of Michigan, and digital-computer studies of orbits. After a visit to CERN, problems of instrumentation became an active topic. In that year, experimental work with Hall-effect detectors, magnetic remnant-field effects, rf modulation of cavities, and strong-focusing lenses for linear accelerators was done at Iowa State College and at the University of Minnesota. An orbit analog device at the University of Michigan gave initial experience with physical strong-focusing, nonlinear systems that we were trying to compute.

The biggest effort went into theoretical work, both analytical and computational, on orbits. The question of existence of invariant curves in the phase plane for nonlinear motions was ever present. The digital computer demonstrated what appeared to be such invariants, but the topic was destined to have continual attention in the years ahead. Systematic surveys of the phase plane with nonlinearities and periodic coefficients was carried out in several ways. The effects of secular changes, misalignments, and errors in pole shape were included. These initial steps ultimately led to the ability to design and to run an accelerator on a computer in advance of any physical modeling; in fact, in the course of MURA's model program, the time arrived when the construction of iron models was suspended in a partially completed state, and the work was carried forward by the use of the digital computer for the determination of magnetic-pole shapes and the behavior of orbits focused by such poles in the presence of a variety of errors.

It was discovered that the nonlinear problem could be handled very rapidly by algebraic transformations, just as the linear problem is handled by transfer matrices. These transformations modeled the topology of phase space as given by the actual nonlinear differential equations, but allowed much more rapid calculation than with the differential equations. Symon pointed out that the physical interpretation of these algebraic transformations is equivalent to the concentration of the nonlinear part of focusing for a sector into an infinitesimal region of nonlinear forces where their effect is impulsive. Work using the computer for nonlinear forces commenced at Brookhaven, and analytic work began at CERN at about the same time.

Other topics studied in 1953–4 included the following: space charge effects for certain charge distributions in the beam; the rf "knockout" diagnostic technique, developed by the University of Michigan participants (to be widely used later in other machines) for determining orbit oscillation frequencies; and the cost of fabricating magnets with various lamination sizes, showing very little variation of cost over the range from $\frac{1}{16}$- to 1-inch-thick laminations.

A second summer session at Madison, from mid-June to 14 August 1954, found the group immersed in space charge studies, wide application of Liouville's theorem, coupling between the rf gap and the orbital motion (by participants from the Argonne Laboratory), and development of the smooth approximation for a general focusing field (by Symon). The group was kept informed on high-energy particle physics by Brueckner, Robert Sachs, and Watson and again benefited from the participation of Courant and Snyder from Brookhaven. Laslett was sent to Brookhaven for the summer to work with their accelerator development group; Haxby of Purdue went to Brookhaven to study magnetic measurement devices; and Wright of the University of Chicago was sent to Berkeley to work closely with the Bevatron group. Toward the end of the summer of 1954, some members of the group were examining the Ubangi-lip type of accelerator, in which the thin iron-pole lips saturate before high field is reached at the orbit, thus providing a large aperture at low field and at injection time, but no more than the necessary space for the beam at the time of high field. Carl E. Nielsen of Ohio State noted what came to be called the negative-mass instability. He and Andrew Sessler of Ohio State established the theory of that instability that involved Landau damping. The experimental confirmation was accomplished on the Cosmotron by Mark Barton. During this study, Symon discovered what was subsequently called the fixed-field alternating-gradient (FFAG) magnet for direct-current operation. The idea had only a few days of consideration at the end of the summer session, but it was to occupy a large proportion of MURA's future activities.

At the end of that summer in Madison, it was not at all clear how the Midwest Accelerator Conference could proceed. Funds from the Atomic Energy Commission (AEC) had not been forthcoming, and it became necessary to turn quickly to the National Science Foundation and to the Office of Naval Research for help. Assistance costs for the experimental programs and for the computer and travel expenses of the group were borne by the universities. In the fall, the Organizing Committee brought about the incorporation of the group into MURA, an Illinois corporation. At this low point in financial support, the group wanted to establish a permanent staff at some central location, to be based on support available from just the universities. On this basis, a small group was formed at the University of Michigan, but it was not possible to do it on a full-time basis. Every week in the fall and spring of 1954–5, this group met for two or three days. The group was composed of Jones, Laslett, Symon, Terwilliger, and myself, with participation from time to time by visitors who were continuing to work at their home institutions. Such visitors were Crane, Haxby, Mel Ferentz of the Argonne Laboratory, Cole, and Snyder. The five regular members of the group commuted to Ann Arbor for two or three days every week.

The original plan for the work, as described to the MURA board of directors, was to commence some engineering studies for particular accelerators and especially to examine the next higher energy for accelerators beyond the Brookhaven 30-GeV range. Studies were to be undertaken on means of automatic or servo control of magnet misalignment determined by sensing-beam behavior. Because an accelerator an order of magnitude greater than the Brookhaven or CERN accelerators might require very careful alignment in order to have a small aperture, these servo studies became the first topic when the group met in the fall. The group acquired tables of parameters and resulting focusing characteristics calculated by the University of Illinois computer.

The minutes of these meetings show that the interests of the group were so fundamental and general at this stage that engineering occupied very little of the group's time. The large number of questions that had been uncovered about alternating-gradient accelerators in general and the FFAG accelerator of Symon gradually became the group's main effort. In a period of two months, a half dozen varieties of FFAG accelerators had been invented using direct-current ring magnets with the possibility of greatly increased beam intensities. These accelerators were named Mark IA, IB, II, III, IV, and V.

Mark IA was Symon's original idea, in which the circumference was composed of identical magnet segments with adjacent segments having reverse polarity, but with the magnets of one polarity having a greater field strength

Figure 13.1 Mark IB FFAG accelerator by MURA [*Source*: *Proc. CERN Symp. High Energy Accel. Pion Phys.*, 11–23 June 1956, vol. 1, p. 362b].

in them than the magnets of the other polarity. In this way, particles could go completely around the accelerator with more bending occurring in the strong magnets than in the weaker magnets.

In the Mark IB (Figure 13.1) accelerator, the field strength has the same magnitude in all segments, but adjacent segments have reverse polarity, and one polarity has shorter circumferential width, so there can be a net bending around the accelerator. The circumference factor for Mark IB was originally thought to be about five, which was a little smaller than that for Mark IA.

For Mark II, Jones pointed out that the circumference factor might be reduced further by using a "bad" circumference factor at the injection energy, and by reducing the length of the reverse field segment for the high-energy orbit. The Mark III was the same as Mark IB, except that it was to operate on what was called the first patch of the necktie diagram, where a good circumference factor was achieved, but where vertical focusing was very weak. In the Mark IV, high-energy orbits were to cross low-energy orbits between the magnets. Although the magnetic-field direction was the same in all magnets, the field gradient was to alternate. Near injection, the orbits were to sample the weak-field regions of each magnet, and at high energy the orbits were to shift and pass through the high-field regions of each magnet. Closed orbits and radial focusing were to be achieved by tapering the magnet boundaries. For the Mark V (Figure 13.2), I proposed the spiral-ridge alternating-gradient structure in which the field was always in the same direction, but the spiral

Figure 13.2 Mark V spiral-ridge FFAG accelerator by MURA [*Source*: MURA Photo Collection].

ridges on the pole faces, or the edges of spiral magnets, gave alternating positive and negative gradients. This focusing, when applied to a cyclotron, as has been ably done by Henry Blosser and many others, allows intense beams at energies approaching the rest mass. The analysis of this variety of accelerator, and of many other complicated possibilities that it was hoped would have some merit, was greatly facilitated by Symon's elaboration of the smooth approximation and by Laslett's persistence in added rigor for more detailed calculations.

While this small group was busily exploring many new possibilities of acceleration, activity was continuing in various universities, and meetings were held at somewhat less frequent intervals than before (October 1954, at Iowa State College; November 1954, at the University of Chicago and at Argonne; January 1955, at Northwestern University; February 1955, at the University of Indiana; April 1955, at the University of Minnesota). They were attended by the same group of approximately twenty physicists, on the average.

In this period, plans were developing for a busy summer from 15 June to 15 August 1955 at the University of Michigan in Ann Arbor. Physicists from abroad and from the East Coast and West Coast attended, and Tihiro Ohkawa, who had independently invented the Mark I FFAG accelerator, came from Japan to join the group. Otto Frisch from England was also present. Topics studied at Ann Arbor were as follows:

1. Tuning methods for, and observations of, orbital and betatron frequencies in the new Michigan IB model (rf knockout was ultimately found to be the most useful method).
2. Spiral-ridge (Mark V) magnet design and methods of introducing straight field-free sections into the spiral magnet.
3. Radio-frequency beam handling – what happens to particles lost out of the phase stable region? This problem was being coded for the computer and was soon to be yielding nicely to computer experiments that discovered phase-displacement acceleration. Can the rf phase of a cavity be automatically controlled by sensing the beam bunch as it rotates? What happens to betatron oscillations excited by the synchrotron gap?
4. Computer problems for nonlinear forces in two dimensions for all types of accelerators. It was necessary to develop ways of displaying the influence of bumps in four-dimensional phase space. The introduction of magnet errors into the codes was also in process. The problem of coding nonscaling machines was studied.
5. Beam extraction.

During the first half of 1955, model construction of a Mark IB had been moving swiftly, with the help of people located at their different universities, each making contributions to design, theory, and hardware. For example, N. Vogt-Nilson had come to the University of Illinois from Norway, and Cole took a leave of absence from the University of Iowa to come to the University of Illinois for settling the design. While this design and theoretical work was going on close to the computer, iron magnets were being made at Purdue following the design developed at Illinois. I transported the magnets to Ann Arbor. By the time the summer study group assembled at Ann Arbor in 1955, the hardware was being assembled by Jones and Terwilliger at Michigan, and the plans for operation and tests of the first FFAG accelerator were under way. Although the machine was not quite finished during the summer, it operated successfully soon afterward with the betatron acceleration. Subsequently it was the device on which a wide variety of rf beam manipulations were tested. This accelerator gave 500 keV of electron energy.

In order to be closer to the computer, the group moved to the University of Illinois in the fall of 1955. Laslett, Sessler, Vogt-Nilson, Kerst, Snyder, Cole, Lloyd Fosdick, Symon, and Edward Akeley (who commuted from Purdue University) were at Illinois.

The first general meeting in the fall of 1955 attracted approximately seventy high-energy physicists, and others interested in accelerators, from all parts of the United States. The topics covered were the problems of high-energy physics, the East Coast and West Coast accelerators (discussed by representatives from each), and the new FFAG accelerator possibilities. In the course of this meeting, the possibilities of FFAG accelerators, with their direct-current ring magnets, which made possible the accumulation of coasting beams and which allowed the beams in two magnet systems to collide, were described in some detail. This colliding-beam scheme, patented by Rolf Wideröe in 1943 and mentioned by Wilson of Cornell, and brought up from time to time after he suggested it in a jocular vein, now became a real possibility. One could see how to stack or to accumulate a large enough beam of particles by carefully manipulating phase space to give observable interactions on collision of the circulating beams, so that the center of mass of the reacting particles is at rest in the laboratory. It was pointed out that 22-GeV beams colliding would give a center-of-mass energy equivalent to that from a beam of 1,000 GeV striking a stationary target. The scheme was sometimes called "synchroclash," because synchronism between colliding bunches was necessary to have a collision in the interaction region.

In 1956, a proposal was made to the AEC for a pair of 25-GeV high-current FFAG accelerators tangent to each other. A colliding-beam section for reactions having the center of mass at rest in the laboratory was to be incorporated. In 1958, MURA proposed an improved machine, a 15-GeV two-way FFAG accelerator that could accommodate several colliding-beam experimental areas. In the period that followed, the merits of high-intensity energy and/or high-center-of-mass energy were debated by the community.

In 1958, a special panel of GAC/PSAC (General Advisory Committee to the AEC/President's Science Advisory Committee) issued a report concluding that the most important priority in high-energy physics was the Stanford two-mile electron linac and that second in priority was a new high-intensity proton accelerator in the 10–20-GeV range. The rationale was that much higher energies probably would not be interesting, inasmuch as no sharp resonances could be expected at energies above a few GeV. The only machine proposed at that time for reaching much higher energies was the MURA colliding-beam FFAG accelerator. In view of the host of particles discovered since then, this position seems remarkably short-sighted. It reminds us that even experts in the field are not very good at predicting the future. As a result of this recommendation, MURA proposed a 12.5-GeV high-intensity FFAG signal-beam accelerator, but the proposal was turned down by President Lyndon Johnson at the end of 1963.

Starting in 1957–8, the MURA group designed and built a 50-MeV electron model of a colliding-beam FFAG synchrotron in which two beams could move in opposite directions, and beam collisions would occur between each magnet (Figure 13.3). The wide radial excursions of the equilibrium orbits in a field gradient made reverse orbits possible. This accelerator was used for about fifteen years for an injector into an electron storage ring called Tantalus. It was the first synchrotron light source, and it opened a new branch of research in condensed-matter physics.

In 1956, after it was pointed out by MURA scientists that colliding-beam experiments were now feasible, in view of the possibility of building up intense

Figure 13.3 The 50-MeV electron model of a colliding-beam FFAG synchrotron by MURA [*Source*: MURA Photo Collection].

beams by stacking in an FFAG accelerator, Roger Newton, Marc Ross, and Donald Lichtenberg of Indiana University (an active participant in MURA) and, independently, Gerard O'Neill of Princeton University suggested that similar methods might allow the storage of intense beams in small-aperture storage rings. This could be a less costly way to achieve colliding beams.

In the summer of 1959, MURA held a workshop for accelerator and high-energy physicists. The primary purpose was to discuss a high-intensity FFAG machine and the physics that might be done with such a machine, in accord with the recommendations of the GAC/PSAC panel. At the workshop, Matthew Sands of Caltech organized a group to consider the possibility of a synchrotron in the range of several hundred GeV. Sands proposed that the new concepts of beam stacking and beam handling and the new insights into accelerator physics would make it possible to use an intermediate-energy booster as an injector into a very high energy accelerator. It was not a new idea. It had previously been suggested by many people, but had not had the background of understanding of accelerator physics reached by 1959. By increasing the injection energy, it was possible to increase the space-charge limit. This and the stacking of many booster pulses during injection would give intensities almost as high as could be achieved with FFAG accelerators, and at much lower cost. Thus, high energy and high intensity became practical together.

The Sands proposal triggered a new wave of enthusiasm for a higher-energy accelerator. After a number of working groups and more advisory panels, it led to the construction of the 500-GeV accelerator at Fermilab.

The only practical FFAG accelerators constructed were the spiral-sector cyclotrons used in nuclear physics and in isotope production for medicine. The booster and storage-ring concepts provide a cheaper route to high energy and intensity. The lasting contributions of the MURA group have been to accelerator physics, especially to beam stacking and colliding beams. All later developments in high-energy accelerators and storage rings are based, at least in part, on those contributions.

From the vantage point of more than twenty years later, one can see that the importance of the MURA experience to the development of accelerators for science lies not so much in the direct application of FFAG (although FFAG cyclotrons are extremely important in nuclear physics), but in the many new accelerator ideas that came from the group. Nonlinearities, phase space, space charge, collective instabilities, beam stacking, and colliding beams probably all would have arisen and been worked on eventually, but they all came to fruition much earlier because of the work at MURA. It was a good time for accelerators and for the people of MURA.

14   Bubbles, sparks, and the postwar laboratory

PETER GALISON

Born 1955, New York City; Ph.D., Harvard University, 1983; history of
science and theoretical particle physics; Stanford University

Instruments and history

When we think about the history of physics, we usually think about
theories – the magisterial sweep of relativity, quantum mechanics, current
algebra, SU(3), and gauge theories seems to subsume and dwarf the details of
experiments. And if experiments appear small, how much less significant are
instruments? The great breaks and continuities of physics are typically defined by the epochal accomplishments of Maxwell's equations, Einstein's
special relativity, nonrelativistic quantum mechanics, and quantum electrodynamics. Such a "theorocentric" view has become ingrained in our pedagogical, physical, and historical literature. Most textbooks attach only a passing
section on the devices of experimentation. The subfields of physics are classified by the theories they lead to – not by all the results one can get from
instruments.

But it is hopelessly one-sided to let theory stand in for all of physics when
we want to understand the past sympathetically, to capture the felt continuities of physicists in the day-to-day practice of their discipline. Experimentalists
frequently move between theoretical areas linked only by a continuity in the
experimental skills and instruments they have at hand. Here I want to examine some of the "instrumental" continuities – and discontinuities – that

I would like to thank the following for helpful discussions and documentation: O. C.
Allkofer, L. Alvarez, G. Charpak, M. Conversi, B. Cork, J. Cronin, S. Fukui, D.
Glaser, A. Gozzini, P.-G. Henning, C. A. Jones, B. Maglich, S. Miyamoto, D. Nagle,
and W. Wenzel. Work conducted with support of the Howard Foundation, NSF SES
85-11076, and the Presidential Young Investigator Award.

existed in the 1950s, for that decade saw a proliferation of instruments unparalleled in the twentieth century. Necessarily, only a few can be discussed here, as it is hardly possible to cover the multitude of detectors that arose in that decisive decade. Regrettably, I shall not discuss developments in nuclear emulsions or scintillators. But my choice is not arbitrary, for during the 1950s two classes of instruments came into existence – the spark chamber and the bubble chamber – that held remarkable sway over elementary particle physics for almost a quarter of a century. Their respective histories, where they came from, and their relations to other classes of instruments will illustrate several important themes that may well hold for other disciplines as well.[1]

Spark chambers and bubble chambers offer the opportunity to address several important questions. What are the effects of the increased scale of instruments from the desktop devices that dominated physics through the 1930s to the airplane-hangar-sized experiments that took over after World War II? How did the computer revolution of the 1950s affect instrument development? How did the shift from cosmic-ray to accelerator physics condition not just the use, but also the design and invention of instruments? What role did theory play in the deployment of various new instruments?

In addition to exploring the origins of bubble and spark chambers, it is possible to set these two instruments in a broader context of experimental physics. In particular, the two types of chambers must be understood as participating in two competing experimental traditions – the spark chamber tied to a tradition of what I shall refer to as "logic" devices, and the bubble chamber linked to the tradition of "image-producing" instruments.

The idea of an experimental or instrumental "tradition" is itself in need of explanation. In what follows, the image and logic traditions designate continuity at three levels: a continuity of skills and technology, a continuity of personnel, and a continuity of the kind of evidence produced with the machinery at hand. Both traditions began near the turn of the century, and each was inaugurated with the invention of a new kind of instrument for atomic physics. The Geiger–Müller counter initiated the logic tradition, and the Wilson cloud chamber, the image tradition.

The cloud chamber rendered the paths of particles visible and so made subatomic processes "real" for generations of physicists. So well resolved were these trajectories that individual pictures could, and did, serve as evidence for novel phenomena in the 1930s and 1940s. Whereas the use of statistical arguments, based on many photographed events, was of course an option for the cloud-chamber (image) physicist, for the counter (logic) physicist of the 1930s it was a necessity. No individual click of a Geiger–Müller could serve as evidence; the electronic devices could persuade only by accumulating coincidences or anticoincidences. For example, to show that charged particles could penetrate thick lead plates, the counter physicist had to demonstrate that the joint counting rate of a device above the plate and below the plate was higher than that expected on the basis of chance.

Over the years that followed the first coincidence and anticoincidence experiments, the two traditions competed. Each had advantages, and each had weaknesses. On the one hand, the image devices provided detail, but often were vulnerable to the charge that "anything can happen once"; some unexplained fluke might have occurred in the device that could have fooled the experimenter. On the other hand, the electronic logic tradition consisted of experiments that typically had ample statistics, but remained open to the objection that they recorded only a very partial description of any single subatomic process. In contrast to the detailed evidence marshaled by the image devices, the electronic experiment might miss some crucial feature. It is not a story in which one side is right and the other wrong. The two ideals of experimentation, while frequently in competition, even tried to incorporate the best features of the other. The full story of the evolution of modern detectors, the long competition and eventual fusion of the image and logic traditions, cannot be given here. Nonetheless, the development of the bubble and spark chambers constitutes its central chapter and illustrates its major themes.

Images: clouds, films, and bubbles

The cloud chamber, with its vivid photographs of tracks curving, splitting, appearing, and disappearing, held a special charm for physicists and the public alike. C. T. R. Wilson built his chambers in quiet solitude, expertly blowing glass in the Cavendish Laboratory. In many ways he exemplified the small-scale craft tradition of that illustrious laboratory. From the Cavendish, precision cloud-chamber work spread to laboratories around the world. One of the most esteemed members of the corps was Carl Anderson, who, in the 1930s, used the device in his remarkable demonstrations of the existence of the positron and muon.[2] After his war work ended, Anderson returned to cloud-chamber work, applying the method to the new "strange" particles that George Rochester and Clifford Butler had discovered in 1947.[3] Meanwhile, some of Anderson's students, frustrated by the paucity of events to be found in the rarefied, sensitive volume of the cloud chamber, began thinking about alternative detectors that would be more useful to the new particle discoveries, but would remain faithful to the visual tradition.

Donald Glaser, in particular, was desperate to find a new kind of instrument within the image tradition.[4] For his thesis project he had used a double cloud chamber to measure the energy distribution of sea-level muons. Finishing his doctorate in 1950, the young physicist had to choose among laboratories, alternatives he winnowed first by his desire to work on his own and second by his interest in the design of a new visual detector. Glaser himself has stressed the importance to him of the nature of the experimental workplace:

> There was a psychological side to this. I knew that large accelerators were going to be built and they were going to make gobs of strange particles. But I didn't want to join an army of people working at the big machines. . . .

> I decided that if I were clever enough I could invent something that could extract the information from cosmic rays and you could work in a nice peaceful environment rather than in the factory environment of big machines.[5]

The University of Michigan offered the promised environment, and Glaser set to work with a $750 grant. At first he tried to continue the objectives of his old physics project; he even hoped to preserve its broad technique by designing an apparatus with two visual detectors separated by a strong electromagnet.

Reasoning that all known detectors worked by the catalysis of instabilities, Glaser systematically explored all the kinds he knew. Chemical instabilities lay behind nuclear emulsions – Glaser tried using a substance that was soluble in monomer form, but when polymerized was not. He hoped that when cosmic rays passed through (and polymerized) the dissolved monomer, it would create a solid track that he could extract and measure with calipers. The method failed, leaving only a darkened liquid. Turning to the next instability, Glaser noted that Geiger tubes worked on the principle of electrical instability. Building by analogy, Glaser sought to make a visual device by using glass sheets coated with a clear conducting substance to make big capacitors. In principle, when a charged particle sailed through, sparks would fly along the tracks, and Glaser would snap their picture. Again no luck. In exasperation, Glaser returned to the familiar theory and practice of cloud chambers as a technological staging point for his new device.

Cloud chambers work on a thermodynamic instability. When the pressure is suddenly decreased on water vapor in a gas, the temperature drops, and the vapor supersaturates. Where ions are present, they become nucleation sites for droplet formation, and it is these drops that can be photographed. Glaser hoped to be able to use a different thermodynamic instability, the formation of bubbles in a liquid, as the basis of a new image-producing device. One obvious problem was that if the electrical charge of an ion caused water to condense around it, an ion in an incipient bubble would tend to collapse the bubble, instead of making it grow. Now improvising, Glaser speculated that while this was true for a single ion, the presence of several ions might drive the bubble's growth by their mutual electrical repulsion. Under that (later abandoned) theory, Glaser found the thermodynamic conditions under which a superheated liquid might best be expected to work.

After many trials, Glaser and his students succeeded first in demonstrating that radioactive materials could induce generalized boiling in a superheated liquid. Then on 14 October 1952, using a movie camera, he took the first pictures of a 1-cc "bubble chamber" (Figure 14.1), concluding as follows:

1) Bubbles can grow in ~1 msec for sizes invisible with the optics used to diameters ~1 mm. . . .
4) Twice apparently very faint tracks of 4 or more bubbles were seen to precede

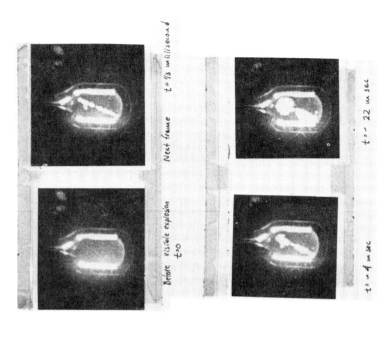

7

5) More light is needed to see smaller bubbles.

6) Consideration should be given to designing a
system in which the low pressure can be
maintained for a longer time than in the
present system.

Movies Taken October 18, at 8:30 p.m. (Saturday
300 feet of Faster 8mm super XX shot with two 1000 watt
projector lamps for illumination with full-size image of
their filaments projected on the bulb. T = 135°C

Results

Out of 8 events three had more than two bubbles and
one was a magnificent straight track of around ten
globes. It occurred in the 20th foot with the Variac
at 70 v ac — corresponding to about 3000 frames/second.
The prints here are about 1½ times the bulb size
and show the event developing.

Conclusions

Tracks can be photographed.

Plans

Next time must attempt to observe counter-controlled
expansions.

Before visible explosion   Next frame   t = 1/3 millisecond
t = 0

t = 1 m sec   t = 22 m sec

Figure 14.2 The seventy-two-inch bubble chamber. The actual chamber is obscured from view by the instrumentation for refrigeration. Operation began in March 1959 [*Source*: Lawrence Berkeley Laboratory, University of California].

a grown globe, but their faintness makes it possible to assume that these were dirt effects.
5) More light is needed to see smaller bubbles.[6]

Four days later, on 18 October, Glaser recorded the following under the photographs of the expanding bubbles:

> *Conclusions*: Tracks *can* be photographed.
> *Plans*: Next one must attempt to observe counter-controlled expansions.[7]

The immediate juxtaposition of recorded success with a self-admonition to make a counter control is highly significant. Cloud-chamber physicists had successfully borrowed simple logic circuits developed within the counter tradition and used them to expand the cloud chambers only on the occasion of an interesting event. Without a counter control, the nascent bubble chamber would be useless for cosmic-ray physics, because interesting events – especially of the new strange particles – occurred so rarely that one had to select for them. And because Glaser considered it one of his primary goals to "save cosmic rays," the counter control was part and parcel of the device he had to build. As physicists later realized, the goal of tying the bubble chamber to cosmic rays remained unattainable – in the decades since Glaser's first paper, and despite numerous attempts, no one has succeeded in making a triggerable bubble chamber.[8] Ironically, Glaser had created a device perfectly suited for the large-scale accelerator physics he had been avoiding, because with a controlled beam the operators knew when the beam would reach the detector; therefore, the chamber could be expanded just when interactions were likely to take place.

Though Glaser failed in his attempts to find a trigger, the bubble chamber was ready for presentation. On Saturday, 2 May 1953, Glaser offered a 10-min account of his results to a somewhat depleted audience at the Washington American Physical Society meeting, it being the last day of the conference. "My first paper," Glaser later wrote, "...was scheduled by the secretary of the American Physical Society, Karl Darrow, in the Saturday afternoon 'crackpot session'...." Shortly afterward, Glaser submitted a letter to *Physical Review*, which promptly returned it for the crime of using the word "bubblet" not found in *Webster's*.[9] In retrospect, it is surprising that two important listeners did hear Glaser's ideas – though they had very different ideas than Glaser about how the new detector ought to be used.[10]

Luis Alvarez, from Berkeley, spoke with Glaser during the conference banquet. Although enormously impressed with the new device, Alvarez's intention was to bury cosmic rays, not save them. Having matured scientifically under the tutelage of Ernest Lawrence, Alvarez had seen firsthand how a scientific entrepreneur was capable of building accelerators at a previously unimagined scale. During World War II, he witnessed the combined efforts of science and engineering, first in the MIT Radiation Laboratory, then, beginning in 1943, in the Manhattan Project. The stunning success of both multi-billion-dollar projects left an indelible imprint on Alvarez's conception of what science could do when linked in a massively funded hierarchical organization with the full resources of trained engineers. Soon after the war, Alvarez directed a project that exploited all of his experience to build a proton linear accelerator; the microwave techniques learned from the radar project, his new familiarity with large-scale engineering/scientific projects, and even $250,000 worth of war-surplus equipment he obtained from the government after demobilization all played a part.[11]

A basic infrastructure for large-scale physics was thus in place at Berkeley before the bubble chamber existed. Although Glaser's invention looked promising, Alvarez wanted to make a hydrogen chamber to exploit the device for the study of particles. Thus, the very first page of the Berkeley group's notebook, dated 5 May 1953 (just three days after Glaser's presentation), bore the word *hydrogen*, twice underlined. After some preliminary attempts to re-create a simple hydrocarbon chamber, one accelerator technician succeeded where his physicist contemporaries had failed: He built a track-producing two-inch chamber. The tiny device was "filthy" by the standards of the time; tracks were visible at the center of the photograph, but spurious bubbles were forming all around the edges. It had been a basic precept of chamber building for Glaser and for the Berkeley group that chambers had to be absolutely cleansed of impurities and rough edges.

In that "dirty" Polaroid photograph lay a key to the industrialization of the bubble chamber. Despite the dirty edges, the track was plainly visible. Suddenly, Alvarez and his colleagues realized what had eluded everyone – the chamber walls did not have to be clean. At once it became apparent that the chamber could function with the most contaminated of walls, walls that could be built out of thick reinforced steel as easily as of polished glass. Now the question was how to make, transport, and manipulate great vats of liquid hydrogen. Answers came from a source outside the purely academic world.

Late in the 1940s, as the cold war heated up, the debate whether or not to build the hydrogen bomb came to a head. At first the General Advisory Committee recommended that the Atomic Energy Commission (AEC) not give priority to the thermonuclear weapon, but the government overrode their objections, after a vigorous, secret dispute. On 31 January 1950, President Truman authorized a crash program. Lawrence surmised that the project would need large amounts of tritium (for the fusion bomb) and more fissionable material to stockpile for the fission arsenal. Gambling that Alvarez's proton linear accelerator could be scaled up to weapons-production size, Lawrence foresaw an enormous neutron foundry producing weapons material. Estimated to cost some $5 billion, the project, had it been completed, would have been more than two and a half times the size of the entire Manhattan Project.[12]

The first hydrogen bomb was "wet" – it used liquid hydrogen and its isotopes. To produce the material, the AEC and the National Bureau of

Standards (NBS) established a classified facility in Boulder, Colorado. There, physicists and engineers worked to understand the properties of the cryogenic substance: what its thermodynamic characteristics were, and how it could be safely produced, transported, and maintained. After the development of the Teller–Ulam idea and the idea that "dry" hydrogen bombs could be built with solid Li^{6}D (instead of cryogenic liquids), the lab's primary mission was finished, and the staff turned their attention to other applications, including propulsion, aerodynamic control, and communications equipment. Because none of these projects had the priority of the defunct "wet" bomb, the physicists and engineers could be recruited to other tasks.[13]

Alvarez, accustomed to working with the AEC, succeeded in bringing some of the AEC experts and their equipment to Berkeley. Beginning in 1955, the cryogenic engineers Dudley Chelton, Bascom Birmingham, and Douglas Mann all made the pilgrimage to help in building the ten-inch and larger devices, culminating in the mammoth seventy-two-inch chamber (Figure 14.2). An important part of their mission was to help in the establishment of a safety program for the laboratory. It is essential to remember how frightening safety problems were. There was at least one serious scare at Berkeley: In 1956, four liters of hydrogen flashed, reaching supersonic speeds as it literally "screamed" down the vent line and shot a flame between ten and twenty feet into the air. The safety system had worked. Infinitely worse was the fire and explosion in the liquid-hydrogen chamber at the Cambridge Electron Accelerator in the early morning hours of 5 July 1965. Hydrogen burned, igniting many secondary fires. In the conflagration, one man died, six were injured, and the plant sustained over a million dollars of damage.[14]

In addition to solving safety, construction, and operation problems, the bubble-chamber team rapidly expanded their data-processing ability. Already in 1955, Alvarez made it clear that the novel detector demanded a new form of data analysis.[15] Soon, experimentalists filled their instrumentation conferences with debates surrounding innovations in the role of computers in processing data; 1957 was the year of the "Franckenstein," a machine that Hugh Bradner and James Franck designed to follow and measure tracks automatically. As the chairman of one session in a 1960 conference put it, "the evolution is towards the elimination of humans, function by function."[16]

The transition to corporate practice was painful. As Alvarez reminded an audience of physicists, it was well known among industrialists that little companies often go bankrupt trying to make the transition from organizations in which individuals handle many jobs to a situation in which high efficiency comes from "production line operation with very expensive production tooling."[17] The main difficulty – so Alvarez argued – arises with the initially inefficient use of supervisory personnel; once this integration is completed, the company operates at a vastly higher level of productivity. As in business, so in the analysis of physical data. Aware that his business methods for accelerating laboratory work would startle his audience of physicists, Alvarez

Figure 14.2. The seventy-two-inch bubble chamber. The actual chamber is obscured from view by the instrumentation for refrigeration. Operation began in March 1959 [*Source*: Lawrence Berkeley Laboratory, University of California].

chided them that a pragmatic approach was the only reasonable one to take: "For those of you who may be horrified to hear a scientist setting such unscientific goals, let me remind you that I have my 'engineering hat' on at the moment so I have no apologies."[18] Although the value of physics goals could not be defined in production terms, clearly the preconditions for success could.

The combination of the hydrogen chamber and its associated data-analysis techniques (Figure 14.3) produced two kinds of physics discoveries. First, the large size of the chamber, along with the simplicity and low charge of the

Figure 14.3. Data reduction with spiral reader. After the operator (usually a woman) placed a spiral guide over the event vertex, an electrooptical reader labeled, measured, and computed the paths of tracks leading out from the vertex. At upper right and left is the computer that encoded the data onto tape. On the desk, one monitor showed the encoded image, the other an enlargement of the photographic image. At lower left is the interactive teletype that printed out errors as the operator proceeded [*Source*: Lawrence Berkeley Laboratory, University of California].

hydrogen nuclei, allowed physicists to follow complicated decay problems with remarkable accuracy. It was therefore possible to extract individual instances of particular processes in precisely the same way that cloud-chamber and emulsion physicists had. Such "golden events" could stand in and of themselves as demonstrations of the existence of an entity or inter-action. At Berkeley, such a discovery came with the identification of the cascade-zero, produced in tandem with a kaon in $K^- - p$ collisions. The puta-tive cascade-zero fell into a Λ^0 and a neutral pion, with the Λ^0 subsequently splitting into a negative pion and a proton. Though the Berkeley team found only one such event, its identification was so clear that it could be published as a sufficient demonstration in the *Physical Review Letters*.[19]

But the holy grail of bubble-chamber physics eluded the Berkeley team; ultimately it was brought home by their competitors at Brookhaven National Laboratory. The quest had begun when, from the floor of the 1962 international conference on high-energy physics, Murray Gell-Mann predicted the existence of a particle he expected to be the last of a group-symmetric set. He dubbed it the Ω^-. Two years later the Brookhaven National Laboratory prized a photograph of the particle from their big hydrogen chamber.[20] The salient event, persuasive to the community as a whole, was a classic success for the image tradition.

A democratic detector

Out of the established electronic logic tradition came a technology that struggled to compete with the bubble chamber, and the cloud chamber and nuclear emulsion that were its imaging predecessors. The new devices, spark chambers, are both easier and harder to analyze historically. They are easier to study because, at least at first, the spark chamber was child's play to build as compared with the imposing cryogenic facilities surrounding the larger hydrogen bubble chambers. Yet precisely this ease of construction makes the historian's task much more difficult. Where there was only a handful of costly, centralized bubble-chamber installations, the spark chambers soon proliferated everywhere. Where the bubble chambers de-manded teams of 100 or 200 people to service and run them and to analyze data, the spark-chamber groups rarely comprised more than ten physicists. Whereas bubble-chamber development can be recounted in a fairly linear fashion, the spark chamber grew up sporadically, in multiple versions and in parallel among laboratories scattered across the United States, England, Germany, Japan, and the Soviet Union. Exactly this democratic aspect of the spark chamber will prove important in understanding its cultivation as an instrument of physics.

The spark chamber is related to counters in many of the same ways that the bubble chamber is a descendant of the cloud chamber. As in the cloud-to-bubble-chamber transition, there is continuity of personnel, there is continu-ity of technique, and there is a continuity in the form of argumentation associated with experimental demonstration. However, because the spark chamber sprouted from at least three (probably more) independent roots, a narration of its evolution must reflect its ramified branches, where a few growths abruptly end, and some split into a multitude of smaller stems. As with so many of the new instruments of the postwar period, to grasp the origin of the spark chamber it is essential to understand the new reservoir of technical knowledge available after the war came to an end.

Both of the big American World War II weapons projects, the atomic bomb and radar, had to produce a new breed of fast electronic equipment to accomplish their tasks. At Los Alamos, physicists needed timing circuits for many purposes. Among other applications, Manhattan Project scientists required them to instrument the implosion process in the plutonium bomb and to analyze studies of neutron production. Seth Neddermeyer led one of the most sophisticated timing projects, the so-called chronotron, which determined the region of superposition of two pulses traveling along transmission lines in opposite directions.[21] Ultimately, Neddermeyer's goal was a determination of the time between the firing of counters and therefore a measurement of the velocity of fast charged particles. By the summer of 1945, Neddermeyer's team had in hand a device that could measure time differences to an accuracy of 3×10^{-10} sec. Potentially, it was a remarkable contribution to the logic tradition – but it suffered from a troublesome gap.

Missing from Neddermeyer's system was a counter with an accuracy commensurate with the timing system.[22] Lamentably, the conventional Geiger–Müller tubes that would have to be used with the chronotron had an inherent uncertainty of about 10^{-7} sec.[23] At root, the counter's precision was limited because electrons, liberated by the passage of the charged particle, had to drift from somewhere inside the tube toward the central wire before they encountered a field strong enough to initiate an avalanche. Experimentalists had to overcome this weak link in the instrumental chain if they were going to determine the lifetimes of the new particles.

Jack Warren Keuffel, a graduate student working with H. Victor Neher at Caltech, seized the opportunity presented by Neddermeyer's "timer in search of a detector." Keuffel realized that because charged parallel plates create a constant field throughout the volume between them, *spark counters* – essentially flattened Geiger counters – would allow avalanches to begin *anywhere* in sensitive volume. Technically, Keuffel's problem was to modify the known parallel-plate counters so that he could rapidly eliminate ions left over from earlier discharges. A perfected spark counter that could sweep the old ions away would therefore complete Neddermeyer's otherwise powerful measuring system.[24]

Keuffel's thesis advisor, Neher, was an expert on counters, and from him Keuffel must have learned the magic tricks of counter preparation: how to bake and bevel electrodes, how to seal and outgas containers, and how to handle the vagaries of electronic logic circuits and high-voltage apparatus. For years, counter building had held the reputation of "a kind of witchcraft," as one of its practicing magicians, Bruno Rossi, later put it.[25] Even the dry prose of physics textbooks could not hide the mystery behind counter manufacturing. In one of the most widely used instructional texts in prewar experimental physics, *Procedures in Experimental Physics*, Neher described the rudiments of the trade: Coat the copper tube with 6-*N* nitric acid; then cleanse with a 0.1-*N* nitric acid bath. The acolyte should then rinse the assembly (at least) ten times with distilled water, dry the electrodes, heat them over a flame until the copper turns brownish black, seal the counter, heat the equipment for several hours until the copper turns bright red with cupric oxide, evacuate, admit dry NO_2 until the copper turns dark and velvety, pump out the gas, and then admit argon bubbled through xylene. Then seal it off. "All the above steps may not be necessary in all cases," Neher confessed, "yet this procedure has been found to give very satisfactory counters" with reaction times of less than 10^{-5} sec.[26] If it works, use it.

Just as Glaser was able to move from the skills he had garnered with the cloud chamber to the bubble chamber, Keuffel could transfer the skills and techniques he learned as Neher's student to a novel instrument. But while the tradition of counter physics could open possibilities, it served at the same time to constrict the options Keuffel anticipated for the new instrument. Most important, Keuffel gave only passing mention in his 1948 thesis to the possibility of using the spark counter to *follow* a charged particle: "the discharge is localized, presumably, in the neighborhood of the initiating ion, with the streamer channel plainly visible. This has obvious possibilities as a means of determining the path of a particle...."[27] With that "obvious" remark, Keuffel let the subject drop.

Though spatial localization later proved of immense importance, we can understand why Keuffel was not particularly intrigued: The whole logic tradition from which he emerged, a tradition of fast timers and logic circuits, was oriented not toward the design of image-producing track-following devices, but toward the binary yes/no answers of the counter array. Keuffel wanted to fill a technological gap by providing a fast detector to a potentially successful timing circuit. The technical tradition and accepted experimental context defined a "need" for temporal, not spatial, localization.

Physicists read Keuffel's work, abroad as well as in the United States. One avid reader was Erich Bagge, who had also worked on a nuclear fission project during the war – but for the Germans. Enlisted to the task in September 1939, his efforts had been primarily directed toward the development of an isotope separation device on the principle of thermal diffusion. Heating the uranium source, he set in motion spinning disks, timed so that the faster-moving, lighter nuclei of ^{235}U would fly through the slits, while the ^{238}U would be caught.[28] Like several of his colleagues on the German project, Bagge also found time to pursue cosmic-ray physics, researching the problem of nuclear fragmentation in the lower atmosphere.[29]

As the Allied occupation forces advanced through Germany, Samuel Goudsmit's team (code-named Alsos) swept up Bagge, along with others, including Otto Hahn, Werner Heisenberg, and C. F. von Weizsäcker, and brought them as privileged prisoners to Farmhall in England. There the Anglo-American intelligence teams held them to ascertain how far the Germans had advanced in their quest for the bomb. Released after the Nazi surrender, Bagge explored issues related both to fission and to cosmic rays. In 1947, for example, he published a theoretical paper intended to account for the asymmetric fragments released in spontaneous nuclear fission.[30] Both nuclear and cosmic-ray physics demanded more precise and sophisticated measuring tools; Bagge turned to an examination of the rapidity of droplet growth in cloud chambers.[31] Responding to Keuffel's paper in the *Physical Review*, Bagge set his student, Jens Christiansen, the task of reproducing and refining the American work on parallel-plate spark counters.[32]

Bagge and Christiansen went to press with their improved spark counter in 1952, whereupon Bagge promptly presented another student, Paul-Gerhard Henning, the thesis problem of investigating the space behavior of the spark in counters like Christiansen's. What prompted Bagge to press the possibility of track localization where Keuffel had not? Perhaps Bagge's prior interest in cloud chambers catalyzed his attempt to transform the spark counter into a track-following device. But whatever the cause, over the next three years Bagge had Henning investigate the properties of a system of three spark counters, set one on top of another. To Keuffel's design Henning added a light intensifier (*Aufhellverstärker*) that discharged through the spark after the chamber had fired. The added current brightened the sparks of just those events where a charged particle plunged through the whole apparatus. Thus, by setting the camera so that only bright events would register, Henning effectively recorded only events in which coincidences occurred.

From March until October 1953, Henning designed and redesigned his electronics, calculating pulsing circuits, time constants, and coincidence circuits. Finally, on Saturday, 17 October 1953, he began to analyze film recording the frames one by one: "künstl. Funken" (spurious sparks), then "*nichts gesehen*" (nothing seen) for picture after picture. More spurious sparking...then something: "Koinzidenz gesehen?" – another spurious spark, whereupon the film switched to exposures of 1 min instead of 15 sec. In rapid succession, the excited comments follow: "Mit Sicherheit Koinz. gesehen!" (coincidences seen with certainty!), "vermutlich zwei Koinziden-zen," and finally, "mit Sicherheit Koinz. gesehen!" From these remarks and the ones that follow in the notebook it is clear that for Henning the persuasive experiments were these first double coincidences.

Simply showing double coincidences would not, however, persuade other physicists that the sparks delineated a particle track: It is always possible to draw a line between two points. Henning was obliged to extend the method to *threefold coincidence*, which was recorded in his notebook, without fanfare, almost half a year later (28 April 1954).[33] It was this demonstration that the Bagge group took as the conclusive exhibit for their case that spark counters could be used to follow the path of a penetrating particle. For the first time it was clear that an electronic detector could follow a single track.

Henning set his detector to photograph sparks left by the passage of multiple scatterings of cosmic-ray muons, and he reported his results to the Deutsche Physikalische Gesellschaft in Wiesbaden. Soon after completing his dissertation in 1955, Henning left physics for industry, and his work languished on the shelf. Bagge himself took his group to Kiel in 1957 and soon thereafter turned his attention to the installation of a nuclear reactor in Geesthact (Schleswig-Holstein). With both Henning and Bagge out of the picture, Henning's dissertation (and its published, single-paragraph summary) passed into obscurity – even among his Kiel colleagues "nobody felt responsi-ble to push the publication."[34] And when the work was finally ready, Bagge insisted that it appear in a largely ignored new journal, *Atomkernenergie*, for which he was an advisor.[35] Thus, the Kiel team's work, though innovative in its use of stereophotography, track-following sparks, multiple plates, selective recording, and cosmic-ray application, had no immediate effect outside the confines of that port city.[36]

A second example of an evanescent discovery occurred in Paris, where Georges Charpak settled after being liberated from a German concentration camp. Upon finishing his studies at the Ecole des Mines de Paris, he entered the Nuclear Chemistry Laboratory at the Collège de France under the supervision of F. Joliot-Curie. Virtually all of their equipment had to be built from scratch, and Charpak turned his attention to measurement of the excitation of atomic shells following β decay, using especially large angle Geiger counters.[37] This not only familiarized Charpak with the habit of building his own electronic equipment but also gave him a detailed knowledge about the tools of the logic tradition's trade: the choice of gases used in counters, the theory of sparks, the techniques of counter construction, and the assembly of electronic filters, coincidence circuits, and amplifiers.

By July 1957, Charpak had shifted his attention to the sparking process itself in order to create a device that would follow the tracks of charged particles. His idea was to provide a very fast high-voltage pulse in order to precipitate small ($\sim$1 mm) but visible avalanches around each electron liberated by the charged particle.[38] For that task he could use a gas combination in his new counter similar to that exploited in his thesis work (9 : 1 argon-to-alcohol; 8 : 1 argon-to-alcohol in his thesis work) and adapt his skills with high-voltage pulses and fast electronics. Reporting his results to Joliot-Curie in late 1957 gained him a tentative nod from his old thesis advisor: "I see that you have worked a lot, and already obtained some interesting results with your spark chamber. I think, like you, that this work ought be of substantial interest for physicists working on theories of the spark." Concurring with Charpak's own assessment, Joliot-Curie hoped to set him on a more productive path: "It would be desirable to continue this kind of research but I completely agree that it would be more worthwhile to return to your first work consisting of using the avalanche effect to amplify the effect of the liberation of electrons by particles in the gas."[39] As in the Keuffel work, Joliot-Curie's comment suggests that sparks and electronic devices fit more easily under the rubric of discharge physics than they did in the enterprise of following particle tracks.

Like Charpak, Adriano Gozzini had been imprisoned by the Nazis during World War II. When he returned to his laboratory in Pisa, he found a setting far more destitute than the privations of postwar Paris. His laboratory building had been ransacked by the German occupation troops. While holding the building, they had destroyed or stolen all the important equipment, as well as the library collection, and the laboratory's prize possession, a signed Roland grating. Research opportunities at the end of the war seemed almost nil, restricted essentially to experiments that used improvised materials. One good source of electronic hardware was the contingent of American troops pulling out from the woods outside the city and divesting themselves of unneeded gear. From these troops, in 1950, Gozzini purchased a magnetron that had been designed for radar use, and he set about converting the war relic into an instrument of microwave physics.[40]

With his refurbished microwave source, late in 1954 Gozzini devised a technique for detecting small impurities in chemical samples. His idea was this: If the molecules of the sample had electric moments different from those of the impurities in it, then the microwave should heat the two differently. Hotter and cooler liquid regions refract light differently; so a shining light could produce a visual display of the liquid's purity. To accentuate the effect, Gozzini chose for his liquid cyclohexane, a liquid that has almost no electric moment. Nothing happened.[41]

Thinking that his surplus magnetron was probably broken, Gozzini tested it – in a routine way – by holding the magnetron next to a tube of neon gas. When working properly, a magnetron causes such tubes to emit light. To see the tube more clearly, the Italian physicist brought the whole apparatus into a darkened room. The tube remained dark. When the room lights were turned on, however, the neon tube clearly, if faintly, glowed. After discussing this recalcitrant behavior with Marcello Conversi, the two physicists concluded that only free electrons, accelerated by the microwave's electromagnetic pulse, could ionize and excite atoms, producing a breakdown of the gas. Ionizing particles excite atoms; when the atoms deexcite, they radiate light. (Later it became clear that photons emitted in deexcitation played an important role in propagating the discharge throughout the tube.) In the dark, there were no free electrons with which the microwave could start an avalanche. Consequently, the neon tube would not glow.[42]

The two physicists deduced that they could build a detector by exploiting the fact that only gases with free electrons would glow, for if a charged particle traversed the tube, it would leave a trail of ions and free electrons. When a microwave pulse bathed the apparatus, those (and only those) tubes would glow that had been penetrated by a charged particle. As a precaution against one tube's light triggering another tube, the experimenters carefully wrapped each in a protective jacket of black paper.[43] On 25 March 1955, a pulsed electric field activated soda glass tubes of pure argon and, for the first time, produced a straight line of flashes in coincidence with counters. Soon, the Italian team switched to narrower, neon-filled tubes and began recording events (Figure 14.4).[44]

An essential feature of the flash tubes was that they were sensitive for only 10^{-5} sec prior to the pulse – any track older than that would no longer be made visible, because the free electrons would have diffused to the glass walls, where they would have recombined with ions. The Conversi–Gozzini tubes thus offered the possibility of tracking cosmic-ray particles as they

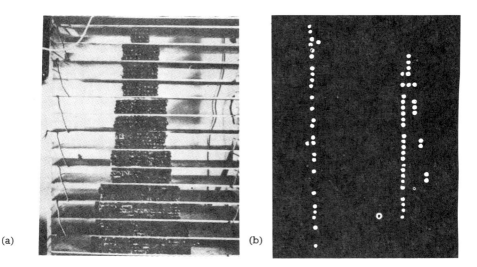

(a)

(b)

Figure 14.4. Conversi and Gozzini's "hodoscope chamber," which first operated on 25 March 1955: (a) the glass tubes wrapped in black paper; (b) representative photographs of track pictures taken in the hodoscope chamber showing single- and double-track events [*Source*: Conversi and Gozzini, "Track"[40]].

scattered and showered within the chamber. Just like the Geiger–Müller tubes on which they were patterned, the tubes provided a two-dimensional projection of the particles' true trajectory, though, as in the case of crossed Geiger–Müller tubes, it was possible to recover some of the track's three-dimensionality by alternating layers of tubes pointing one way and layers aligned perpendicularly.[45]

Unlike some of the other moves toward a spark-based track detector, the Italian work caught the logic community's attention. Recognizing the value of the triggered pulse system used by Conversi and Gozzini, two physicists working in Harwell (in England) sought to combine the spark counter (of Keuffel, R. W. Pidd, and L. Madansky) with the Italians' pulsing technique. That is, they tried to alleviate the problem of spurious discharges in the spark counter by pulsing the plates with high voltage only when a particle had entered the system. Not following the Italians, the Harwell physicists used air, not a noble gas; later, other physicists speculated that it might have been for this reason that the Harwell device never worked well.[46]

Once again, a temporary halt in one part of the world was the starting point for new modifications in another. A Japanese physicist, Shuji Fukui, was well prepared to reproduce the Pisa physicists' neon-filled device and, after 1957, to exploit T. E. Cranshaw and J. F. de Beer's more sophisticated triggering electronics. As an undergraduate, Fukui had had extensive training in glasswork for vacuum experiments, gaseous discharges, and classical atomic spectra. At the same time, the financial constraints of Japanese physics pointed to cosmic rays as the only economically viable way to participate in particle physics. War damage left only limited research funds, and although the emergency demands of the Korean War had revived industrial resources, the country was in no position to field an accelerator competitive with the Berkeley Bevatron. Moreover, in late 1945, American occupation forces had destroyed several Japanese cyclotrons: in the Institute for Physical and Chemical Research in Tokyo, at Osaka Imperial University, and at Kyoto University.[47]

Acting within these limits, in February of 1956 a workshop on "Physics of Super-High-Energy Interactions" set a twofold program for the cosmic-ray division of the recently completed Institute for Nuclear Study (INS) in Tokyo. First, cosmic-ray physicists would expose nuclear emulsions on mountain peaks and send films into the upper atmosphere in high-altitude balloons to probe elementary interactions directly. Second, experimenters would examine particle interactions indirectly by exploiting Conversi and Gozzini's recently invented detector to examine the particle distribution in the cores of extensive cosmic-ray showers. In pursuit of this second goal, Fukui, by then a research associate (*joshu*), wrote Conversi, requesting reprints of their *Nuovo Cimento* articles.[48]

Fukui and a doctoral student, Sigenori Miyamoto,[49] planned to assemble 5,000 small glass balls filled with neon. They would then stack the delicate containers on the laboratory floor in 7-m^2 array to detect the particle density and lateral distribution of extensive air showers.[50] At the beginning of their study, little was available from commercial sources. The hydrogen thyratron tubes (to generate the pulse), the old spring-driven Leica, and miscellaneous high-voltage equipment all were imported. Gozzini and Conversi had used their American war-surplus pulse-forming network, whereas Fukui and Miyamoto had to build their own pulser because they could not find a manufacturer capable of producing a square-pulse generator. Consequently, they had a cruder circuit, in which a capacitor, discharging through a resistor, produced a pulse with a steep rise and exponential decay – rather than the more elegant square wave. Their resistor was just a straight tungsten wire, and their first capacitor was an old-fashioned paper condenser in an oil container made in a small shop near Tokyo. In a moment we shall see how the (even more) homegrown character of Fukui and Miyamoto's apparatus was central to their success.

On receiving the Italian work, Fukui and Miyamoto began tests to determine if the flash tubes could be fired by their steeply rising, exponentially decaying pulse. During the fall of 1956 they obtained their first positive results with a few tubes, and they immediately turned to the task of assembling the extensive device foreseen at the INS workshop. By December of 1957 the Japanese experimenters had amassed a significant collection of gas-filled glass tubes. According to Conversi's instructions, the next step was to cover the tubes with dark paper, leaving one side open. Stuck in the laboratory without any technical assistance, they foresaw nothing but tedium in the prospect of wrapping five thousand little glass bulbs in black paper. Thus, before beginning mass production of tubes, the two physicists wanted to reflect on the best design: By what mechanism are adjacent tubes fired? How can gas pressure be lowered without losing too much light intensity? How is the efficiency changed by the pulse time constant? How can the high-voltage peak be lowered without changing light intensity? Could the chemical treatment of tubes be simplified to avoid the lengthy process of washing with chromic acid and then with water and then baking?

To lighten their burden and to explore the gas discharge, the physicists piled a few bulbs, unwrapped, with gas of varying pressures, on top of one another. They turned the pulser on and the lights off. Out of the darkness came a remarkable image that Conversi and Gozzini could not have seen because of the dark paper: Inside at least some of the tubes, *tracks* were forming (see Figure 15.1 on p. 255). Instead of the crude two-dimensional mosaic that had appeared to Conversi and Gozzini, the Japanese physicists observed a well-defined three-dimensional image.[51] Though limited in precision, it was a tentative step by the logic tradition to appropriate the image-producing capability of its competitors.

On the basis of the image recorded in Figure 15.1, Fukui and Miyamoto realized that they could build a new kind of detection chamber with much greater spatial resolution than had been possible using counters. But clear tracks appeared in only some of the tubes – those with higher pressure. This puzzled them at first, because they anticipated that the discharges in gases would be the same for two tubes at different pressures as long as the pulse height was greater than a critical minimum level. Because they thought that they were exposing all of the tubes to a field well above this threshold, the tubes' uneven responses was surprising.

Finally, they understood. Fortune had smiled on their poverty: The differing responses of the tubes with different pressures were traceable to their jury-rigged pulsing device, which produced pulses that were not square. For a bell-shaped pulse, the time t_1 during which the electric field was above a level E_1 was greater than the time t_2 during which the pulse had a field above some higher field value, $E_2 > E_1$ (Figure 14.5). For a truly square wave of the same peak value, the time during which the field was above the critical minimum was the same for all critical values between zero and the maximum height. Therefore, Fukui and Miyamoto were exposing the high-pressure tubes to a shorter effective pulse than the low-pressure tubes. As it turned out, short pulses on high-pressure tubes are optimal for producing the track-following discharges.[52]

Above all, Fukui and Miyamoto's observation of tracks meant that they no longer needed to model the detector on the *binary* response of Geiger

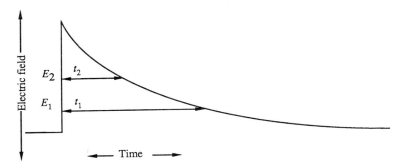

Figure 14.5. Consequences of a non-square-wave pulse. With Fukui and Miyamoto's non-square-wave pulses, the higher the voltage threshold, the shorter the pulse time. Thus, accidentally, they discovered the advantage of using steeply rising, short pulses on higher-pressure tubes.

counters.[53] Suddenly they were free to alter the geometry from hundreds of stacked cylinders to a few glass boxes made with electrically conductive surfaces. Within months they had used the new apparatus to produce clear pictures of particle tracks, and by September 1959 the two physicists had penned their contribution to *Nuovo Cimento*[54] (Figure 14.6).

The hegemony of bubble chambers

Meanwhile, the triumphant success of the bubble chamber at American accelerator laboratories would make the logic experimenters a ready audience for the Japanese achievement. Two Bevatron physicists, William A. Wenzel and Bruce Cork, along with two visiting colleagues, James Cronin from Princeton and Rodney L. Cool, had pushed counter experiments to the limit of complexity in an effort to capture the complicated topology of multiple interactions and decays. The prognosis for the logic tradition appeared grim. With each passing day it seemed more obvious that bubble chambers in general, and Alvarez's group in particular, would out-distance all electronic competition. Laboratory policy and scientific planning were intertwined; soon the sentiment that electronic apparatus was obsolete burst into the open.

During 1957, a decision had to be reached about future accelerator building. Alvarez, a leading advocate of big bubble-chamber physics, pressed for a high-energy low-intensity accelerator (good for bubble-chamber physics), while offering a dim estimate of the future contributions of counter experiments using high-intensity machines. In a letter of 17 September 1957 to the laboratory director, Edwin McMillan, Alvarez acknowledged that whereas high-intensity beams and counters may have made contributions in nuclear and medium-energy pion physics, these were domains where elastic scattering was common (i.e., only a few particles came out of the interaction).[55] For

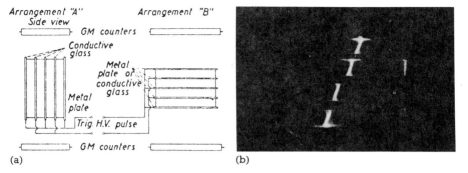

Figure 14.6. Fukui and Miyamoto's "discharge chamber" and early pictures (1959). Once they realized that the discharge was localized in the Conversi tubes, Fukui and Miyamoto could turn to a system of glass boxes (a) and photograph tracks, even somewhat oblique tracks (b) [*Source*: Fukui and Miyamoto (1959)[54]].

such experiments, visual detectors, cloud chambers and emulsions, were truly only "exploratory" devices that led to inadequate statistics.

High-energy physics, Alvarez contested, was in no way analogous. When particles interact inelastically, as is the rule in many high-energy events, the counter physicist may be able to measure total cross sections by simply seeing how many events take place. But real interest lay elsewhere, and the counter physicist "can hardly get started in a serious investigation of the much more common inelastic processes if he does not have a visual detector, and preferably a bubble chamber." If a high-energy negative pion interacts with a proton, various particles are produced, including many neutrals. "In order to find out what happened, one must measure the angles and momenta of the charged prongs and then calculate by relativistic kinematics the number, the nature, the energies, and the directions of the invisible prongs."[56] Counters, Alvarez concluded, were simply not up to the task.

Some of Cork and Wenzel's 1957 work figured in Alvarez's letter as an example of the inadequacy of counter work at high energies. The two counter physicists had devised an electronic experiment to measure the scattering of protons from protons at 2, 4, and 6 GeV. From Alvarez's perspective, the elastic cross section that they measured was not of primary theoretical concern because "the elastic part of the cross section is a small part of the whole affair. Some day the whole job will be done in a hydrogen filled bubble chamber, and in the process the elastic cross section will be measured again, so the theorists will have the real details of the interaction." With the exception of work on low-energy K mesons, Alvarez contended that "counters appear to me to be on the way out as precision instruments in high energy physics."[57] The logic tradition appeared squarely beaten by the image physicists.

With Alvarez ringing the death knell for their electronic livelihood, it was small wonder that Cork and Wenzel felt under siege. Appealing to McMillan just a few days after Alvarez, Wenzel conceded that he agreed with "most of Luis' observations," yet hastened to add that "counters are not on the way out as precision instruments."[58] First, he suggested, the counter is *flexible*, "it comes in small inexpensive units, easily moved and altered as to size and type on short notice." In high-energy experiments, the counter's fast time resolution could be used for particle velocities above those where "emulsions, cloud chambers and (potentially) bubble chambers" were useful. Cerenkov counters and devices that selected according to a particle's time of flight offered an extremely efficient way of separating K^- mesons from antiprotons.

Making use of the counters' efficiency in the study of antiprotons, Wenzel and Cork could plan experiments that would determine the angular distribution and total cross section of proton–antiproton scattering at low energies. Theorists had made predictions about this process even though they had no detailed account to offer about the annihilation process itself. For such an experiment, counters could generate much more data and do it faster than the bubble chamber. In short, Wenzel granted Alvarez's claim that bubble chambers would probably be the single most useful tool of high-energy physics.

Nonetheless, Wenzel continued, there was "no reason to regard it as the universal detector. While it may be true, as Luis says, that the bubble chamber does complete experiments, it is also true that completion may take a long time."[59] By offering rapid answers to theoretical predictions that dealt with "a few details of the complete experiment," counters might stake a provisional claim in territory that might eventually be deeded to the big bubble chambers.

Under the circumstances, Wenzel's qualified defense was probably as strong an endorsement of the electronic tradition as could be made. Caught in such straits, Wenzel, Cork, and Cronin greeted Fukui and Miyamoto's reprint with understandable enthusiasm. The three physicists submitted a counter paper on 26 May 1960 using their old techniques. But by 6 June 1960, Wenzel and Cork had drafted a proposal to build a spark chamber and within weeks had a machine yielding clear track photographs.[60] Immediately upon his arrival back East, Cronin, too, began constructing a modified Fukui–Miyamoto chamber, cannibalizing Keuffel's old spark counters for spark-chamber plates and exploiting Keuffel's old data for a design.[61] The new detector proved itself an ideal match to high-energy proton accelerators: The sensitive time of the chamber was of the order of several microseconds, and by applying a clearing field between the positive and negative spark-chamber plates, the experimenters could remove unwanted tracks.[62] Soon Cronin could report to Cork that "I am finally building a [high-voltage] pulser! Have you done more with your spark chamber?"[63]

Sparks fly

High-statistics electronic machines were back in the running. Cronin, at Princeton, consulted with engineers to find the best thyratrons[64] and by March 1960 could already report the successful test of an eight-gap chamber with plates four inches by five inches. Enthusiasm for the new device leapt ahead of specific projects: "We are now building a larger chamber with decent optics. It will have a volume of [6 × 6 × 12 inches] with 18 gaps. The purpose will be to test larger spark chambers. No particular experiment is in mind."[65] Unlike the bubble chambers, the new spark chambers could be prepared in short order. Within a few weeks the Princetonians hoped to test the device at the Cosmotron and then the following summer at Berkeley.

Word spread quickly. Letters of inquiry arrived from laboratories around the world – all the way from CERN to the University of Maryland[66] – for here was an affordable device for institutions that could not dream of competing with the great hydrogen bubble-chamber efforts at Berkeley or Brookhaven. As Cork put it, echoing Wenzel's defense against the Alvarez onslaught, "The first impressive thing about... spark chamber[s] is the fact that they are extremely easy to build, are inexpensive, and there seems to be no difficulty in making them go. Anyone can build one in the basement and make it work."[67] This was no exaggeration. At Princeton, the first model used by Gerard O'Neill's group "was largely the work of college sophomores majoring in physics,"[68] a far cry from the massive engineering and scientific efforts that supported the dangerous construction of bubble chambers.

As the new device entered the consciousness of the experimental community, applications proliferated. Cork, Wenzel, Cronin, and George Renninger (a graduate student) brought the device to accelerators for the first time to study the scattering of polarized protons and spin–spin correlations in proton–proton scattering[69] (Figure 14.7). In November of 1959, just a few months after the first of these experiments, students and faculty gathered around T. D. Lee at Columbia to debate how to test weak-interaction theory at high energies. Melvin Schwartz realized that it might be done with neutrinos, a thought that occurred almost simultaneously to Bruno Pontecorvo, who by then was in the Soviet Union.[70] But as Schwartz soon discovered, the difficulty was that tanks of scintillator, banks of Geiger–Müller counters, or stacks of neon-filled tubes all had poor spatial resolution. The bubble chamber was not a massive enough target to use to study neutrino interactions. During the early summer of 1960, Irwin Pless from MIT reported to Schwartz and Leon Lederman about Cronin's working desktop spark chamber. In pursuit of a better neutrino target, the two physicists raced to Princeton to look at it.[71] Again, as in the early stages of the "dirty" bubble chamber, the key was to transform a desktop device into an engineering-grade large-scale structure.

With money from the U.S. Navy, the AEC, Columbia, and Brookhaven,

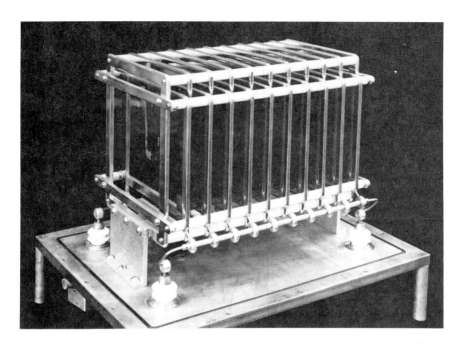

Figure 14.7. Cronin and Renninger's spark chamber (summer 1960). This device and pictures taken with it of scattering protons at the Berkeley Bevatron received wide attention at the 1960 conference on instrumentation and served to encourage many other groups to deploy spark chambers at accelerators [*Source*: Cronin and Renninger (1961)[65]].

the hunt for a second neutrino was set in motion. Engineering salaries came from Brookhaven, and electronics from Brookhaven, with money from Columbia's Nevis laboratory. The navy supplied gratis surplus cruiser deck plates, weighing between two and three thousand tons. By setting an example, the Columbia–Brookhaven machine yanked the spark chamber from the tabletop to the engineering age of physics. Only muons emerged from the neutrino events, rapidly convincing almost everyone that the two-neutrino hypothesis was correct.[72]

Spark-chamber work prospered during the next few years. Chambers bloomed in every conceivable size and shape. Experimentalists devoted specialized meetings and major sessions at more general gatherings to the noisy detectors,[73] drawing participants from laboratories around the world. By its ease of construction, the device satisfied a number of perceived needs, offering a fast, triggerable detector for accelerator physics and restoring to small groups and less wealthy laboratories the opportunity to experiment in particle physics without using the monster bubble chambers that were big enough to be working at the frontiers of the science. Simultaneously, the spark chamber infused the logic tradition with a vigor it had lost.

Excitement flashed through the electronic community at the possibility of competing again on something like equal footing with the image tradition. Cronin drafted a letter to Thomas J. Watson, Jr., president of IBM, pleading for faster delivery of his company's card punch so that the spark group could link it to their encoding system and beat the competition to a resolution of certain ambiguities in nuclear forces: "We have employed the newly-developed spark chamber technique and it will undoubtedly be a triumph for this technique if we are able to arrive at decisive results before groups in Russia and England, doing similar experiments, succeed with their more antiquated techniques."[74]

As the logic machines grew, both in size and sophistication, they produced ever more pictures. Soon the counter physicists clamored for better optics, faster film-advance mechanisms, more personnel (mostly women)[75] to analyze events, and subtler computer programs written to perform kinematic and particle-identification analyses. As Cronin put it, in many spark-chamber collaborations, experimenters were now finding themselves "in a situation similar to that of our bubble chamber colleagues."[76] The flexibility and small scale of the traditional counter system were yielding to the large scale of the contemporary image-producing devices.

Computers, physics, and the nature of the laboratory

Paradoxically, success put the counter physicists in exactly the situation that they had earlier shunned – large-scale image techniques were anathema to them. But as the spark chambers had grown, they had been forced to supplement the technical core of their tradition, electronics, with the skills and procedures of the "bubblers." They soon yearned for a return to the purity of circuits and counters. Wenzel: "I'd always been interested in electronics and after a while I got tired of the business of scanning and reconstructing the film...[the experiments were] a little tedious because I'd always done electronic experiments before. I started to think: what were other ways to record the information?"[77] By 1964, Wenzel's discontent was shared widely among the displaced counter physicists. In response, like-minded counter physicists convoked in March 1964 what they thought would be a small "informal meeting" at CERN on "Filmless Spark Chamber Techniques and Associated Computer Use."[78]

Originally, the organizers planned the assembly as a small colloquium for twenty people to set the future direction of the CERN program. But the idea resonated so forcefully with the concerns of the physics community that when the opening session began, over two hundred physicists had arrived to participate.[79] Introducing the meeting, P. Preiswerk welcomed the larger-than-expected crowd, expressing a hope he obviously shared with many in the audience that "with computers on-line the physicist will receive certain answers during the running of the experiment in the experimental halls, [which] might give him back the pleasure of being an experimenter and not only an operator, who is able to act and to put new questions on the grounds of this information during the running of the experiment." Prophesying the eventual triumph of the imageless devices, Preiswerk concluded, "The high cost of the new techniques might damp the speed of the development you have initiated, but not withhold it."[80]

G. R. Macleod then outlined the field, stressing that the technical basis for the antiimage movement came from rapid advances in electronics and computers. Just as the bomb work had proved important in laying technical prerequisites for the first large-scale bubble-chamber work, other military developments proved useful to the counter scientists. Macleod noted that the on-line use of computers was a "fairly widely used technique" in the guidance of satellites and missiles and the deployment of advanced radar systems.[81] Indeed, even a cursory perusal of military literature from the 1950s reveals the importance of computer-aided machine feedback for military applications. Turning, to take one example, to the U.S. Air Force's 1960 edition of its Air Training Command textbook, *Fundamentals of Guided Missiles*, there are several chapters that stress the importance of on-line computers for telemetric feedback. In missile delivery, many channels of information need to be recorded quickly, including the positions of control surfaces, airspeed, pitch, yaw, roll, temperature, acceleration, altitude, and ordnance functions. Typically, each such on-board instrument was connected to a transducer that produced an audio-frequency signal. This signal then modulated an FM carrier. Using a distributor, the system sampled each of these transducers and sent the data back to the tracking station. If guidance information was to be fed back to the missile in time for its in-flight correction, computers had to process and retransmit coded information quickly.[82]

Many of the problems that arose later in elementary particle physics were present earlier in such telemetry applications. For example, the information often came in faster than it could be processed; so systems for rapid recording on magnetic tape had to be developed. New technology was needed to separate the various channels, and interference between channels avoided. Similar technological work continued in other reseach areas, especially in advanced radar technology, and in the oil, chemical, and communications industries. Closer to home, particle physicists could borrow from the multichannel pulse-height analyzers used frequently by experimental low-energy nuclear physicists.[83] Together these various arenas of electronic applications provided a ready reservoir of techniques from which the counter physicists could draw.

For the ten years preceding the 1964 CERN meeting, computers had performed vital tasks in bubble-chamber research. In that context the computers were designed to take over ever more of the tedious and repetitive work involved in sorting and analyzing film, and then in collating the data into useful forms. Perhaps, one spark-chamber physicist ruminated, the "well known fact that all elementary particle physics in the last few years has been done in bubble chambers may not be unconnected with the fact that the bubble chamber physicists have had well developed data handling facilities." With a new class of instrument ready to defend the cause, "we should listen to the reports [of those] experimental groups who have been working very hard to try and change this lamentable situation."[84]

Spark-chamber workers had additional tasks in mind for the electronic brains, tasks that went beyond reducing photographic tracks. First, they hoped to use computers for data acquisition. As the data were produced by the detector, the electronic signals could be recorded on magnetic tape by a computer for processing at a later time. Second, computers could perform a check-and-control function by monitoring experimental parameters, magnet currents, counting rates, voltage levels, beam intensities, and so forth. By doing so, the computer could partially obviate gross errors by alerting the experimentalists to malfunctions in the apparatus. Such work was comparable to the upkeep of a logbook; recording "the sort of things which nowadays tend to be written (or worse, not written) in the notebooks in illegible pencil writing at three o'clock in the morning by harassed physicists [now can be] done automatically by the computer with rather more consistency and reliability."[85]

A third function of the computer was to perform sample computations. Suppose one knew that the beam particles entered at a certain point in the chamber. One might have the computer check whether many particles were produced or only a few. Thus, the computer functioned as a variable logic element in the detecting system and could be used to determine which events to record. Unlike the bubble chamber's post hoc use of the computer, this online use brought the computer into a role entirely analogous to Rossi's logical discriminators. They became part of the instrument itself.

After comparing notes, the various groups agreed that for a typical particle physics experiment, programming took "several man-months." All of this work was not equal. Macleod noted that "the thinking is part of the programmer's job." Once the physicists in charge of the project did that thinking, they could "give [the detailed computer work] to a very junior coder for example."[86] In all, by 1964, programming was occupying 10 to 20 percent of a spark-chamber group's work. To M. G. N. Hine, such an investment of time seemed quite reasonable: "The design of the programmes has quite properly occupied most of the physicists in the groups concerned as this is nowadays really another name for the design of the experiment itself."[87]

As always, bubble chambers were the competition. At each stage of the development of the electronic techniques, workers compared their machines to the great liquid detector that had produced so much physics during the 1950s. One obvious advantage of the bubble chamber was its ability to record immensely complicated interactions and decay patterns. Often these decays would be quite rare. Nonetheless, because of the detail of the bubble pictures, evidence from just a few events would be persuasive. For the electronic experimentalists, persuasive evidence came more from high statistics than from "golden events." The rarity of the fully analyzed bubble events seemed to Arthur Roberts

> one of the difficulties with the bubble chamber. Consider weak interactions: if we talk about decays of various particles there is nothing inherently different about the leptonic decay of the omega and the leptonic decay of the neutron, and the number of neutron decays that have been observed is numbered in the hundreds of thousands. To do an equivalent job on the decay of the omega which is equally interesting would require the same number of events. We are used to thinking of 50 [lambda] decays as a large sample not because there is anything in the physics which says that the [lambda] decay is any different from the neutron decay, but just because they are expensive.[88]

It was becoming clear that experimentation would be computer-limited, not simply because one lacked data-analysis computers, but because computers now constituted part of the detectors themselves. Roberts's comment about the Λ's captured the long-standing split between the image tradition, which would sacrifice number for detail, and the logic tradition's reluctance to settle for the rare event when high statistics could be extracted electronically. In the search for the power to sort and extract more data, S. J. Lindenbaum only half-jokingly contended that his group "could easily use up all the computers at CERN, Brookhaven and the whole East coast."[89]

For many at the conference, the spark chamber's fast (and inexpensive) response to complicated phenomena was its primary advantage; indeed, the possibility of eliciting meaningful data on the spot from the hardware was a salient feature of the "logic" physicist's mode of work. Computers coupled to imageless detectors allowed the counter physicist to regain this capacity, which had been lost as the spark chamber had drifted toward the technology and work organization of the bubble chamber. Even among the bubble-chamber physicists themselves there was a growing resistance to the reliance on large groups to sort, scan, and measure the photographs; Glaser left physics for biology in the early 1960s in large measure because the structure of the bubble-chamber workplace seemed to leave no room for solitary or small-group experimentation. By the mid-1960s, the malaise reached Alvarez, who began having second thoughts about the structure of laboratory life. In an interview in 1967 he explained:

> so much [work in the bubble-chamber laboratory] can be done by technicians. . . . You have technicians who run alpha particle spectrometers and beta ray spectrometers and gamma ray coincidence circuits. And the people working in the field are doing very much what our graduate students are doing, they are putting things into computers and analyzing the print-out,

and they are pretty disconnected from the experimental side of it, in the same way that we are. I can't complain because our people don't go down and look at the bubble chamber very often or at the Bevatron. They ask the bubble chamber operators to expose a certain number of millions of frames of film, and then they ask somebody else to measure them, and then run them through computer programs, and then they start with computer program output and process data.[90]

Alvarez turned away from high-energy accelerator physics and increasingly devoted his attention to cosmic-ray balloon work.

The logic experimentalists also wanted a style of experimental work in which they could respond to experimental conditions during the run of the experiment, indeed within a few minutes of the run. It was a desire that motivated a great deal of the instrumental design then under consideration. As one advocate of electronic experimentation reminded his audience, with the computer's aid,

one can see rather rapidly the effect of changing experimental parameters; one can make the setting up of an experiment a rather less laborious and less random affair; one can actually see that the data being recorded is [*sic*] the data one wants to record; and one can plan the experiment on the basis of analyzed data rather than having to make intuitive guesses based on what we did last time. All this I think makes for the physicist having a much better *control* over what is going on and gives him the information on which to base considered decisions on changes to be made during the experiment.[91]

Macleod's hope for "control over what is going on" is a refrain heard over and over again within what I have called the logic tradition. They were acutely aware of the changing organization of the experimental workplace that often forced experimentalists into supervisory jobs, unable to interact directly with their tools and the objects of their investigation. Concerns about the bubble chamber's noninteractive nature drove many of the electronic logic experimenters toward the development of imageless, on-line devices. In addition, the design of the instruments was intimately connected to the type of experimental problem that could be investigated. One can see this in one of the discussion sessions at the CERN conference, where an experimentalist commented that computers ought to be used to *exclude* certain events (or parameters describing events) from consideration. Otherwise the primary purpose of spark chambers would be lost – that is, the experimenter would sacrifice the possibility of presetting the logic for the type of event desired. If the computer were to record *everything*, the physicist would do better with bubble chambers.[92]

Conclusion: instruments and experimental work

The renaissance in instrumentation between 1952 and 1962 matches any in the entire history of the physical sciences. During this period, physicists invented or improved high-pressure cloud chambers, bubble chambers (hydrogen, hydrocarbon, and xenon), discharge chambers, spark chambers, flash chambers, acoustic chambers, current-division chambers, video chambers, and new kinds of scintillators. There were new timers, pulse-height analyzers, oscilloscopes, new computers, and the myriad of combinations and variations of these devices.

To understand this renaissance in the material culture of physics, our discussion had to proceed on several levels. On a first plane there was new physics to be done. When Glaser set about building a new chamber, the problem of strange particles and the origin of cosmic rays had the physics world puzzled. The old generation of detectors was neither dense enough to produce a sufficient number of interactions nor sufficiently well resolved in space and time to yield the detail needed to advance the field. For Fukui and Miyamoto, cosmic-ray air showers were similarly inaccessible – the cloud chamber simply could not handle phenomena that took place over several square meters. Cork, Cronin, and Wenzel also had a physics agenda – studying the decays and interactions of the strange particles first seen in cloud chambers and emulsions and then, more recently, produced in accelerators. They asked, for example, about the polarization properties of the Λ^0.

New instruments thus were partially motivated by the needs of various programmatic goals within physics. I say *programmatic* deliberately, to distinguish the physics ambitions of the experimental tradition from the suggestion that they were "simply" trying to resolve specific theoretical issues. Of course, there were instances where theorists "ordered" a particle – notoriously in the case of the Ω^-. In general, though, the physics questions that precipitated instrumental innovation were not nearly as specific as the properties of a single particle. More commonly, the experimenters responded to classes of phenomena; weak decays, nuclear lifetimes, and extensive air showers were typical of the broad-based physics concerns that drove the reformulation of equipment.

Closely coupled to such strategic, rather than tactical, physical goals were even broader issues about the direction that experimental physics should take. One such cut separated what I have called the image tradition from the logic tradition. We saw how Glaser drew on the skills developed within the visual-detector tradition as he built prototype instruments culminating in the bubble chamber. Once laboratories around the world established bubble-chamber programs, they rapidly drew together groups of physicists experienced in other image-type experiments, either with cloud chambers or with nuclear emulsions, for it was these scientists particularly who could most easily transfer their skills to the new technology. Similarly, it was from the resources of the logic tradition that physicists like Charpak, F. Bella, Gozzini, O. C. Allkofer, Henning, Bagge, Carlo Franzinetti, Keuffel, and others primarily drew.

There are several different levels of continuity hiding under the widely used term "technological transfer," and it is useful to distinguish among them. One strand of historical continuity embraces the laboratory skills of the experimenters themselves. As Neher described in his rough textbook sketch, there were myriad craft operations, many of them poorly understood, that were needed to build a working counter. Keuffel knew those skills and applied them in his important contribution to the spark counter. It is, obviously, not always possible to work by first principles in experimentation; often one has to *imitate* what worked if the device and experiment are to function properly. Charpak put it eloquently in 1962:

> You will see one physicist state that after filling his chamber he waits several days for the gas to get dirty – he says it works better then. Some other physicist will scrupulously purify his gas with a calcium oven to remove oxygen from it. Yet another adds a "quenching gas" while another adds nothing at all. In the end people do what's wise – they do whatever those are doing whose chambers work well – without trying to understand why![93]

Thus, part of the experimentalist's commitment to one or another of the traditions is practical – in one domain the physicist knows how to make devices work reliably.

Sometimes the technological transfer can involve not just skills but a material connection. Not only did the personnel from the NBS/AEC facility bring their knowledge, the government handed over the hydrogen compressor itself. On a much smaller scale, Gozzini's work was greatly facilitated by the surplus pulse-forming network he extracted from radar equipment bought from the American army. Frequently, as in Cronin's use of Keuffel's Princeton spark plates, one experiment is actually cannibalized for another. The spark counter is thus contained in the spark chamber in this most literal sense.

There is a third category of technological transfer that encompasses a carryover of whole structural systems from one device to the next. In the 1930s, physicists in the logic tradition designed electronic circuits to record selectively the firings of their Geiger–Müller counters. Two decades later, Conversi and Gozzini adapted such logic circuits to activate their flash tubes. And spark- and wire-chamber experimentalists later took analogous, though more sophisticated, circuits to sensitize their detectors at precisely the time when particles were to arrive from the accelerator. Such systemic similarities in the generation and analysis of data in the logic tradition are paralleled on the image side. There, techniques and skills developed around the processing of pictures are handed back and forth among cloud chambers, bubble chambers, and nuclear emulsions.

Along with the transfer of skills and hardware that link devices within a tradition, there are broader issues about the nature of experimental work that conditioned the development of modern instrumentation. Competition was stiff in the mid-1950s between accelerator and cosmic-ray physics. As Glaser himself put it, he wanted to "save" cosmic-ray physics as a viable enterprise within physics. Alvarez, by contrast, believed that the huge investment the government had put into the Bevatron and other big machines would be justified by producing a new and significantly better kind of physics using the bubble chamber. Thus, paradoxically, whereas the bubble chamber issued, in part, from Glaser's attempt to preserve cosmic-ray physics, Alvarez's success in merging the bubble chamber with the engineering and industrial/military organization of large-scale projects effectively killed cosmic-ray research as central to high-energy physics. Similarly, for the electronic family of detectors, many of the physicists, including the Japanese, the Italian, and the German groups, intended their small-scale new detectors for cosmic-ray physics. The Americans, by contrast, from the very start found the new flash chamber interesting precisely because it was adaptable to accelerators.

A similarly broad issue impinging on what I have called the "outer laboratory" occurred in the debate in the late 1950s over whether to build for high intensity or for high energy. And just as the cloud-chamber and emulsion techniques were suited to an outer laboratory of the cosmos, the bubble chamber was particularly suited to low-intensity, high-energy machines. The fast and flexible spark chambers gave their best performance when coupled with a high-intensity machine.

Concurrent with the debate over the appropriate site for the emerging field of high-energy physics lay another division, for not far beneath the surface of the rift between accelerator work and cosmic-ray work was a very human concern with the nature of laboratory work. Glaser's choice for cosmic-ray physics was not just a selection of a branch of the natural world to investigate; it was simultaneously and self-consciously a choice for a style of work, one removed from "the factory environment of the big machines." When Alvarez chose to develop the engineering/scientific team, he did so intentionally – his was a decision to further the integration of the two cultures he had witnessed during the war. Hierarchically structured, specialized "line work," he hoped, would prove as successful in advancing high-energy physics as it had in expediting the military enterprises on which it was modeled.

The effect of Alvarez's transformation of the experimental workplace was not lost on his contemporaries. Even before the invention of the spark chamber, physicists defended the electronic-counter tradition by citing its flexibility. Just a few years after the construction of the first spark chambers, experimental workers around the world tried to build into the device their vision of the ideal workplace itself. This is the sense of Macleod's remarks at the CERN conference to the effect that experimental physicists were going to build devices that would "give...back the pleasure of being an experimenter and not only an operator," or that physicists needed instruments that would give the experimenters back "control" over the apparatus, and so provide them with answers to their questions "in the experimental halls."

Throughout the fifties, experimentalists often expressed their optimism that new techniques would transform the life of their laboratories. There was a sense that almost any of the new technological systems could be turned to use in the search for the very small: hydrogen liquefiers, radar, computer cores, television, or stored-program computers. Not only would these techno-logical marvels serve as useful prototypes, but physicists could draw on the industrial and military experience of the war and cold-war years for models of the successful integration of engineering and physics. At the same time, there were real worries about how education and collaboration would proceed. Such issues even brought into question the nature of the experiment and the experimenter's place in the laboratory.

Thus, in addition to being seen as filling missing links between elements of experimental or theoretical systems, we must treat the organization of the laboratory workplace as a dynamic variable shaping the construction of in-struments. Just as the big bubble chambers embodied one form of work, so the early on-line spark chambers instantiated another.

Doubts arose, too, as to the style of technology being employed. To some observers the systems of the 1950s seemed awkwardly juxtaposed, a high-technology bricolage: Oscilloscopes scanned photographs; in one popular device, the acoustic chamber, transducers converted sound waves into electrical pulses, and then the computer sorted the signals. In another de-tector, a television camera swept its eye across the spark chamber, digitizing the location of discharges. Bubble-chamber physicists invented a myriad of new instruments to photograph and reduce the tracks of boiling hydrogen with the help of hired hands. A bit of unease with the kludgelike quality of some of the instruments emerged in Hine's concluding words to the CERN conference of 1964:

> I do not put my money on the acoustic chamber. I think it is too like the bubble chamber in having a 19th century steam-age feel about it. Especially I think that, apart from this esthetic disadvantage, its inability to cope with multiple tracks is going to be a practical reason why it will not be able to keep its present predominance. I would also vote against any photographic system or any vidicon system mainly because I think the difficulty of actually opti-cally looking into a spark chamber is going to become more and more of a nuisance as time goes on.[94]

Hine preferred Charpak's current-division method, which used the current produced by the spark to fix the spark's location. Solid-state detectors, too, might satisfy Hine. Here we catch a glimpse of the esthetics of machine building. It is a subject discussed much less than the "beauty" of group-theoretic symmetries in quantum field theory – but perhaps that is only because experimentalists have written fewer rhapsodic reflections on their discipline.

The origins of the renaissance of instrumentation in the postwar years is thus a story of many beginnings. It is a story that can be told in terms of theory – "explained," in a certain sense, by the necessity of clearing up the dynamics of strong resonances and weak interactions. It is equally a story about the extension of clusters of experimental skills propagated from one instrument to the next: from the cloud chamber to the bubble chamber, from the Geiger–Müller counter to the hodoscope chamber or the spark counter; from the spark counter to the spark chamber; and from the spark chamber to the many species of imageless electronic detectors. In a related sense, it is the story to be told as part of a historical epistemology – an account of competing ways of constructing demonstrations about the natural world. So, too, is the instrument renaissance linked inextricably to the history of wartime and post-war technology. Without the fast timing devices, cryogenics, pulse-forming circuitry, and computer systems developed for military and industrial uses, neither the bubble chamber nor the more advanced spark chambers would have amounted in the 1950s to more than curious prototypes for accelerator physics. Finally, the explosion of new instruments shaped a revolution in the laboratory. For experimentalists, instruments could hardly be more important: They constitute the stage for the entirety of laboratory life. Through the design and use of instruments, experimentalists continuously struggle to set the pattern in which they will interact with machines, collaborators, technicians, scanners, computers, and, finally, nature.

The quarter century following the acoustic chamber did not end the debate over the nature of experimental work. Among many other examples, it would be instructive to investigate the early stages of planning for the Superconducting Supercollider (SSC). Some physicists pressed for inclusion of a fixed-target facility within the SSC, in large part to preserve small collaborative enterprises within a project that would ordinarily function with groups of three or four hundred physicists. But if attempts to contain the ever-growing scale of physical experimentation made little headway, the effort to draw together the image and logic traditions fared well. The 1960s, 1970s, and 1980s brought a profound technological shift in the domain of microelectronics, one that welded together diverse tools of high-energy physics, for with the introduction of integrated circuitry that could time, sort, and amplify tens of thousands of independent channels, there emerged a system that could synthesize images electronically. By the 1980s, for the first time, the old dream came true – golden events like the one that certified the Ω^- could be produced by electronics, fished by a computational net out of the ocean of microphysical debris.

Notes

1 This material is excerpted from Peter Galison, *Image and Logic: Traditions of Experimentation in High Energy Physics* (in preparation).
2 P. Galison, "The Discovery of the Muon and the Failed Revolution against Quantum Electrodynamics," *Centaurus 26* (1983), 262–316; P. Galison, *How Experiments End* (University of Chicago Press, 1987), Chapter 3.
3 For a historical account, see G. D. Rochester, "The Early History of the Strange Particles," in *Early History of Cosmic Ray Studies, Astrophysics and Space Science Library*, Vol. 118, edited by Y. Sekido and H. Elliot (Boston: Reidel, 1985), pp. 299–321.

4 The following discussion of the invention and exploitation of the bubble chamber compresses a much fuller account given with full published and archival references in P. Galison, "Bubble Chambers and the Experimental Workplace," in *Observation, Experiment and Hypothesis in Modern Physical Science*, edited by P. Achinstein and O. Hannaway (Cambridge, Mass.: MIT-Bradford, 1985), pp. 309–73.

5 Cited in Galison, note 4 (p. 316).

6 Two of Glaser's notebooks survive (D. Glaser, personal papers, Department of Physics, University of California, Berkeley) and will be referred to by their original numbers GNB 1 and GNB 2. This reference: GNB 2, p. 5.

7 GNB 2, p. 7.

8 By the time the expansion can take place, the heat has diffused from the trajectory of the charged particles.

9 D. Glaser, interview, 4 March 1983; on the referee report from *Physical Review*, see D. Glaser to R. L. Weber, 17 August 1970, D. Glaser, private papers.

10 Darragh Nagle from Chicago also heard the talk. For more on the important work at Chicago leading to pion tracks in hydrocarbon chambers and the demonstration of hydrogen radiation sensitivity, see Galison, note 4 (pp. 325ff.).

11 R. Seidel, "Accelerating Science: The Postwar Transformation of the Lawrence Radiation Laboratory," *Historical Studies in the Physical Sciences 13* (1983), 375–400, especially 398–9.

12 P. Galison, note 4 (p. 331). On the MTA project, see J. L. Heilbron, R. W. Seidel, and B. R. Wheaton, *Lawrence and His Laboratory: Nuclear Science at Berkeley 1931–1961* (Berkeley: OHST, 1981), especially pp. 66–71.

13 See, for example, P. Lieberman, "E.R.E.T.S. Lox Losses and Preventive Measures," in *Advances in Cryogenic Engineering*, Vol. 2, edited by K. D. Timmerhaus (New York: Plenum, 1960), pp. 225–42; C. Hohmann and W. Patterson, "Cryogenic Systems as Auxiliary Power Sources for Aircraft and Missile Applications," in *Advances in Cryogenic Engineering*, Vol. 4, edited by K. D. Timmerhaus (New York: Plenum, 1960), pp. 184–95.

14 See, on the Berkeley incident, H. P. Hernandez, "Designing for Safety in Hydrogen Bubble Chambers," *Advances in Cryogenic Engineering*, Vol. 2, edited by K. D. Timmerhaus (New York: Plenum, 1960, pp. 336–56; on the Cambridge Electron Accelerator, see M. Stanley Livingston, "Semi-Annual Report," 1 July 1965 through 31 December 1965, CEA-1031, pp. 2–3.

15 L. W. Alvarez, "The Bubble Chamber Program at UCRL," typescript dated 18 April 1955, unpublished but widely circulated.

16 L. Kowarski, "VI a. Introduction," in *Proceedings of an International Conference on Instrumentation for High Energy Physics* (New York: Interscience, 1961), p. 223.

17 L. W. Alvarez, "Round Table Discussion on Bubble Chambers," in "Round Table Discussion. E. Data Analysis," in *Proceedings of the 1966 International Conference on Instrumentation for High Energy Physics* (Stanford: SLAC, 1966), p. 289.

18 Ibid.

19 L. W. Alvarez, P. Eberhard, M. L. Good, W. Graziano, H. K. Ticho, and S. G. Wojcicki, "Neutral Cascade Hyperon Event," *Phys. Rev. Lett. 2* (1959), 215–19.

20 M. Gell-Mann, "Comment," in *International Conference on High-Energy Physics*, edited by J. Prentki (Geneva: CERN, 1962), p. 805; V. E. Barnes, P. L. Connolly, D. J. Crennell, B. B. Culwick, W. C. Delaney, W. B. Fowler, P. E. Hagerty, E. L. Hart, N. Horwitz, P. V. C. Hough, J. E. Jensen, J. K. Kopp, K. W. Lai, J. Leitner, J. L. Lloyd, G. W. London, T. W. Morris, Y. Oren, R. B. Palmer, A. G. Prodell, D. Radojičić, D. C. Rahm, C. R. Richardson, N. P. Samios, J. R. Sanford, R. P. Shutt, J. R. Smith, D. L. Stonehill, R. C. Strand, A. M. Thorndike, M. S. Webster, W. J. Willis, and S. S. Yamamoto, "Observation of a Hyperon with Strangeness Minus Three," *Phys. Rev. Lett. 12* (1964), 204–6.

21 Upon leaving Los Alamos, Neddermeyer had the equipment transferred to him at Washington State University, where he continued to refine the technique. D. Hawkins, E. C. Truslow, and R. C. Smith, *Project Y: The Los Alamos Story* (Los Angeles/San Francisco: Tomash, 1983), pp. 320–1. S. H. Neddermeyer, E. J. Althaus, W. Allison, and E. R. Schatz,

"The Measurement of Ultra-Short Time Intervals," *Rev. Sci. Instrum. 18* (1947), 488–96.

22 "The difficulties in the achievement of [building a fast counter] are probably great but may not be fundamental." S. H. Neddermeyer, E. J. Althaus, W. Allison, and E. R. Schatz, "The Measurement of Ultra-Short Time Intervals," *Rev. Sci. Instrum. 18* (1947), 488–96, especially p. 488.

23 For example: C. W. Sherwin, "Short Time Delays in Geiger Counters," *Rev. Sci. Instrum. 19* (1948), 111–15. For an example of the difficulty of using Geiger counters for microsecond physics, see B. Rossi and N. Nerenson, "Experimental Determination of the Disintegration Curve of Mesontrons," *Phys. Rev. 62* (1942), 417–22.

24 Workers at the time *thought* of the parallel counter as a material analogue of the Geiger counter; this is not a retrospective connection. For example, R. W. Pidd and Leon Madansky, "Some Properties of the Parallel Plate Spark Counter I," *Phys. Rev. 75* (1949), 1175–80; on p. 1175: "The parallel plate counter is an adaption of a gas counter in which the electrodes have plane parallel symmetry." Pidd and Madansky's work is further developed in a second part to the preceding article: "Some Properties of the Parallel Plate Spark Counter II," *Rev. Sci. Instrum. 21* (1950), 407–10.

25 B. Rossi, "Arcetri, 1928–1932," in Sekido and Elliot, note 3 (pp. 53–73, especially p. 56).

26 J. Strong, *Procedures in Experimental Physics* (New York: Prentice-Hall, 1938), pp. 270–1.

27 J. W. Keuffel, unpublished Ph.D. dissertation, California Institute of Technology, 1948, p. 8; J. Keuffel, "Parallel-Plate Counters and the Measurement of Very Small Time Intervals," *Phys. Rev. 73* (1948), 531; J. Keuffel, "Parallel-Plate Counters," *Rev. Sci. Instrum. 20* (1949), 202–8. Similarly, other counter workers, F. Bella, Carlo Franzinetti, and D. W. Lee, maintained that "the main use of this type of counter is in the measurements of very short time intervals" of the order of one billionth of a second. See "On Spark Counters," *Nuovo Cimento 10* (1953), 1338–40; also see F. Bella and C. Franzinetti, "Spark Counters," *Nuovo Cimento 10* (1953), 1461–79.

28 David Irving, *The German Atomic Bomb. The History of Nuclear Research in Nazi Germany* (New York: Da Capo Press, 1967), pp. 43ff.

29 E. Bagge, "Nuclear Disruptions and Heavy Particles in Cosmic Radiation," in *Cosmic Radiation*, edited by Werner Heisenberg (New York: Dover, 1946), pp. 128–43.

30 E. Bagge, "Zur Theorie der Massen-Haüfigkeitsverteilung der Bruchstücke bei der spontanen Kernspaltung," *Z. Naturforsch. 2a* (1947), 565–8.

31 E. Bagge, F. Becker, and G. Bekow, "Die Bildungsgeschwindigkeit von Nebeltropfen in der Wilsonkammer," *Z. Angew. Phys. 3* (1951), 201–9.

32 E. Bagge and J. Christiansen, "Der Parallelplattenzähler als selbstlöschendes Teilchenmess-gerät," *Naturwissenschaften 39* (1952), 298.

33 P.-G. Henning, "Labor-Tagebuch," unpublished, Henning private papers; first double coincidence, 17 October 1953 (see notebook, pp. 48–50); first threefold coincidence, 28 April 1954 (p. 82).

34 See O. C. Allkofer, E. Bagge, P.-G. Henning, and L. Schmieder, "Die Ortsbestimmung geladener Teilchen mit Hilfe von Funkenzählern und ihre Anwendung auf die Messung der Vielfachstreuung von Mesonen in Blei," *Phys. Verh. 6* (1955), 166. A paper that is almost identical with the thesis was published by Henning in March 1957: "Die Ortsbestimmung geladener Teilchen mit Hilfe von Funkenzählern," *Atomkernenergie 2* (1957), 81–8; cf. O. C. Allkofer, *Spark Chambers* (Munich: Verlag Karl Thiemig, 1969), especially pp. 1–5, a brief history of the Kiel work; O. C. Allkofer, interview, 24 April 1983.

35 O. C. Allkofer to M. Conversi, 11 October 1972. Allkofer private papers, Institute für Reine und Angewandte Kernphysik der Christian-Albrechts-Universität Kiel.

36 O. C. Allkofer and colleagues continued research on cosmic radiation using the new detector. See, e.g., O. C. Allkofer, "Das Ansprechvermögen von Parallel-Platten-Funkenzählern für die harte Komponente der Kosmischen Ultrastrahlung," Diplomarbeit, Hamburg, 1956. The published version of this work and additional references from the Kiel group may be found in O. C. Allkofer, *Spark Chambers*, note 34.

37 "Thèses présentées à la faculté des sciences de l'université de Paris: i) Etude de phénomènes

atomiques de basse énergie liés à des désintégrations nucléaires. ii) La diffusion élastique des rayons [gamma] par les noyaux," 1954, Georges Charpak, *curriculum vitae*, 9 December 1981, Charpak private papers, CERN.

38 G. Charpak, "Principe et essais préliminaires d'un nouveau détecteur permettant de photographier la trajectoire de particules ionisantes dans un gaz," *J. Phys. Radium 18* (1957), 539–40. A longer version was presented at the International Conference on Mesons and Recently Discovered Particles, Venezzia, 28 September 1957: G. Charpak, "Principe d'un détecteur de temps d'efficacité très court, permettant de photographier la trajectoire de particules ionisantes dans un gaz," typescript, Charpak private papers, CERN.

39 F. Joliot-Curie to G. Charpak, 29 November 1957. Charpak papers, CERN.

40 A. Gozzini, interview 12 July 1984. See, e.g., A. Gozzini, "La costante dielettrica dei gas nella regione delle microonde," *Nuovo Cimento 8* (1951), 361–8; A. Gozzini, "Sull'effetto Faraday di sostanze paramagnetiche nella regione delle microonde," *Nuovo Cimento 8* (1951), 928–35. For historical recollections, see M. Conversi, "The Development of the Flash and Spark Chambers in the 1950's," CERN-EP/82-167, and the especially useful unpublished typescript, M. Conversi and A. Gozzini, "Electrically Pulsed Track Chambers and the Origin of the Spark Chamber Technique," dated July 1971, Gozzini private papers, Pisa. Note that, independent of the Pisa work, A. A. Tyapkin and his group developed a pulsed-fed hodoscope system of Geiger–Müller counters for use with accelerators. See, for example, V. V. Vishnyakov and A. A. Tyapkin, "Investigations of the Performance of Gas-Discharge Counters with a Controlled Pulsed Power Supply," *Sov. J. At. En. 3* (1957), 1103–13.

41 The original chemical sample that Gozzini used was still in Pisa in Gozzini's possession in 1983; it was originally given to Gozzini by A. Kastler from Paris (Gozzini to Galison, 17 Nov. 1986).

42 M. Conversi, "Flash," note 40.

43 M. Conversi and A. Gozzini, "The 'Hodoscope Chamber': A New Instrument for Nuclear Research," *Nuovo Cimento 2* (1955), 189–91. A longer paper was given at the 1955 Pisa International Conference on Elementary Particles: M. Conversi, S. Focardi, C. Franzinetti, A. Gozzini, and P. Murtas, "A New Type of Hodoscope of High Spatial Resolution," *Nuovo Cimento (Suppl.) 4* (1956), 234–7. Franzinetti brought to the collaboration his expertise with the theory and operation of spark counters. See F. Bella and C. Franzinetti, "Spark Counters," *Nuovo Cimento 10* (1953), 1461–79; F. Bella and C. Franzinetti, "On the Theory of the Spark Counter," *Nuovo Cimento 10* (1953), 1335–7; F. Bella, C. Franzinetti, and D. W. Lee, "On Spark Counters," *Nuovo Cimento 10* (1953), 1338–40. Note that Franzinetti and his collaborators built on and cited the earlier work by Keuffel and by Pidd and Madansky discussed in note 24. A. W. Wolfendale and his colleagues at Durham were the first to incorporate flash tubes in a usable detector; see, e.g., M. Gardener, S. Kisdnasamy, E. Rössle, and A. W. Wolfendale, "The Neon Flash Tube as a Detector of Ionizing Particles," *Proc. Phys. Soc. London, Sect. B 70* (1957), 687–99.

44 M. Conversi, "The Development of Flash and Spark Chambers in the 1950's," *J. Phys. (Paris) (Colloque C-8 Suppl.) 12* (1982), 91–99, especially 92ff.

45 Indeed, a triggering system altogether analogous to the Conversi–Gozzini system was used with Geiger tubes by Piccioni in 1948: O. Piccioni, "Search for Photons from Meson-Capture," *Phys. Rev. 74* (1948), 1754–8. This is not to imply that the flash tubes were always replaceable by Geiger–Müller tubes: The GM tubes notoriously have a strong field gradient around their central wire, and the flash tubes operate in a uniform electric field. This makes their response uniform.

46 T. E. Cranshaw and J. F. de Beer, "A Triggered Spark Counter," *Nuovo Cimento 5* (1957), 1107–16. On Pidd and Madansky's work, see note 24.

47 L. M. Groves, *Now It Can Be Told* (New York: Da Capo Press, 1983), pp. 367ff.

48 S. Fukui, "Chronological Review on Development of Spark Chamber in Japan," unpublished typescript, March 1983, courtesy S. Fukui. The INS's nuclear physics division had already completed reconstruction of their cyclotron by 1955 (S. Miyamoto to author, 21 October 1986).

49 Miyamoto was a doctoral student of K. Hushimi "for form's sake," but "actually" was a graduate student of Yuzuru Watase. Similarly, Fukui was formally Professor Ogata's *joshu*, but his actual affiliation was with Watase's laboratory. The Osaka laboratory was effectively a small branch of Professor Watase's laboratory (S. Miyamoto to author, 21 October 1986).

50 S. Fukui and S. Miyamoto, "A Study of the Hodoscope Chamber," INS-TCA 10 (14 December 1957), p. 1 (internal laboratory report).

51 Ibid. Also see S. Fukui, "Chronological Review," note 48. Miyamoto remembers the pressure tests as independent of the tube-wrapping problem (S. Miyamoto to author, 21 October 1986). In the Extensive Air-Shower experiment itself, Fukui and Miyamoto had better equipment: ceramic condensers by Nippon Condenser Co., an automatic film-advancing camera by Canon, a thyratron by Toshiba, etc. (Fukui to author, 8 Dec. 1986).

52 The discharge could be improved in quality by reducing the high-voltage pulse to 10^{-7} sec from the 10^{-3} sec used previously. INS-TCA 11 (10 April 1958), p. 4 (internal laboratory report, in Japanese, translation in typescript courtesy S. Fukui).

53 S. Fukui, "Chronological Review," note 48 (p. 5).

54 S. Fukui and S. Miyamoto, "A New Type of Particle Detector: the 'Discharge Chamber'," *Nuovo Cimento 11* (1959), 113–15.

55 L. Alvarez to E. McMillan, 17 September 1957.

56 Ibid.

57 Ibid.

58 W. Wenzel to E. McMillan, 27 September 1957.

59 Ibid.

60 B. Cork, "Charged-Particle Detector," draft typescript, 6 June 1960; subsequently revised by E. Beall, 13 June 1960, then by P. Murphy, 14 June 1960, and put in revised form as Bev-527, 1 July 1960. Some months later, published as "Properties of a Spark Chamber," *Nuovo Cimento 20* (1961), 502–8.

61 Interview, James Cronin, 4 January 1983.

62 B. Cork, "Charged Particle Detector," note 60.

63 J. Cronin to B. Cork, 24 November 1959. James Cronin, private papers, Physics Department, University of Chicago (hereinafter: JCP).

64 Howard M. Brady to J. Cronin, 17 February 1960 (JCP).

65 J. Cronin to B. Cork, 22 March 1960 (JCP). The first public report of the Cronin–Renninger accelerator work was made in J. Cronin and G. Renninger, "Studies of a Neon-Filled Spark Chamber," in *Proceedings of an International Conference on Instrumentation for High Energy Physics*, Berkeley 12–14 September 1960 (New York: Interscience, 1961), pp. 271–5.

66 For example, G. Bernardini to J. Cronin, 1 April 1960; R. A. Burnstein to J. Cronin, 18 January 1961 (JCP).

67 Remark by B. Cork in discussion at the Argonne National Laboratory, 7 February 1961, in "Spark Chamber Symposium," *Rev. Sci. Instrum. 32* (1961), 480–531, quotation on 486.

68 G. K. O'Neill, "The Spark Chamber," *Sci. Am. 207*: 2 (1962), 36–43, quotation on 43.

69 See discussion at Argonne conference, note 67 (pp. 487–9).

70 M. Schwartz, "Feasibility of Using High-Energy Neutrinos to Study the Weak Interactions," *Phys. Rev. Lett. 4* (1960), 306–7; B. Pontecorvo, "Electron and Muon Neutrinos," *Sov. Phys.–JETP 10* (1960), 1236–40. M. Schwartz, "Discovery of Two Kinds of Neutrinos," *Adventures in Experimental Physics [alpha]* (1972), pp. 82–100. Interview, Melvin Schwartz, 20 October 1983.

71 Interview, Melvin Schwartz, 20 October 1983.

72 G. Danby, J.-M. Gaillard, K. Goulianos, L. M. Lederman, N. Mistry, M. Schwartz, and J. Steinberger, "Observation of High-Energy Neutrino Reactions and the Existence of Two Kinds of Neutrinos," *Phys. Rev. Lett. 9* (1962), 36–44.

73 For example: the symposium on spark chambers, held at Argonne National Laboratory on 7 February 1961, note 67; session V of the proceedings of the 1962 conference on instrumentation for high-energy physics, held at CERN, 16–18 July 1962, proceedings in *Nucl. Instrum. Methods 20* (1963); session entitled "The Latest Advances in Spark and

Luminescent Chambers and Counters Method [*sic*]," in *Twelfth International Conference on High-Energy Physics*. Dubna, 1964, Vol. 2, edited by Y. A. Smorodinskii (Moscow: Atomizdat, 1966), pp. 301–77; translated from Russian by Israel Program for Scientific Translations, Jerusalem, 1969.

74 J. Cronin and E. Engels to T. J. Watson, 27 February 1961 (JCP).

75 A. Roberts, "Properties of Conventional Camera-Film Data Acquisition Systems with Narrow-Gap Spark Chambers," CERN (1964), note 78 (pp. 367–9); on p. 368: "Need for film scanning may require manpower (womanpower)"; Arthur Rosenfeld, "Current Performance of the Alvarez-Group Data Processing System," *Nucl. Instrum. Methods 20* (1963), 422–34; on p. 422: "Next, the scan cards are keypunched and verified by two full-time girls (currently we are scanning at about 200,000 events per year)."

76 J. Cronin, "Present Status of Spark Chambers," *Nucl. Instrum. Methods 20* (1963), 143–50, quotation on 150.

77 William A. Wenzel, interview, 7 March 1983.

78 Data Handling Division, CERN, *Proceedings of the Informal Meeting on Film-less Spark Chamber Techniques and Asociated Computer Use*, Yellow Report 64–30 (Geneva: CERN, 1964).

79 P. Preiswerk, "Introduction," in CERN, note 78 (p. 1).

80 Ibid. (pp. 1–2).

81 G. R. Macleod, "On-Line Computers in Data Analysis Systems for High-Energy Physics Experiments," in CERN, note 78 (pp. 3–9, especially p. 4).

82 Air Training Command, USAF, and Technical Staff Aero Publishers, *Fundamentals of Guided Missiles* (Los Angeles: Aero Publishers, 1960), especially pp. 497–569.

83 G. R. Macleod, "On-Line Computers in Data Analysis Systems for High-Energy Physics Experiments," in CERN, note 78 (p. 5).

84 Ibid. (p. 9).

85 Ibid. (p. 5).

86 G. R. Macleod, from transcription in "General Discussion on On-Line Computer Use," in CERN, note 78 (p. 310).

87 M. G. N. Hine, "Concluding Remarks," in CERN, note 78 (p. 374).

88 Arthur Roberts, transcription of discussion given in CERN, note 78 (p. 298).

89 S. J. Lindenbaum, transcription of discussion given in CERN, note 78 (p. 298).

90 L. Alvarez, transcript of interview by Charles Weiner and Barry Richman, 14–15 February 1967, American Institute of Physics.

91 G. R. Macleod, "On-Line Computers," in CERN, note 78 (p. 6) (emphasis added).

92 B. Maglich, transcription from discussion in CERN, note 78 (p. 305).

93 G. Charpak, "La Chambre à Etincelles," CERN reprint from *Industries Atomiques 6* (1962), 63–71, quotation on 68 (author's translation).

94 M. G. N. Hine, "Concluding Remarks," in CERN, note 78 (pp. 374–5).

15   Development of the discharge (spark) chamber in Japan in the 1950s

SHUJI FUKUI

Born 1923, Osaka, Japan; Ph.D., Osaka University, 1961; high-energy physics; Nagoya University, Sugiyama Jyogaku-en University

Introduction

Peter Galison's chapter in this volume correctly summarizes the history of the development of the flash-tube hodoscope and the spark chamber. However, books and survey articles that describe the development of the flash-tube hodoscope and also the spark chamber[1] refer to the historically incorrect review by Arthur Roberts presented at the spark-chamber symposium held 7 February 1961 at the Argonne National Laboratory.[2,3] Because, for some reason, the article I wrote in 1959 with Sigenori Miyamoto[4] was completely ignored by the researchers of cosmic-ray and particle physics outside of Japan, I would like to present here the precise history of that phase of the development of the discharge (spark) chamber.

In our work, Miyamoto and I recognized that the localization of discharges in the flash tubes depended on the values of E/p and on the time duration of applied high-voltage pulses, and we raised the possibility of a new triggerable track detector, using gaseous discharges illustrated with photographs that would show the discharge columns along the cosmic-ray trajectory. I should emphasize that this observation was the starting point of the discharge (spark) chamber and the transition from flash tubes. Copies of the article were mailed

It is a pleasure to express sincere gratitude to Professors Yuzuru Watase (who died in 1978) and Takanori Oshio, then of Osaka City University, and to Professor Minoru Oda, then of INS, for their kind assistance in aiding our work and for their encouragement. We are deeply indebted to Professor Kôdi Husimi, who showed interest in the localization of the discharges and strongly supported our work by his kind assistance in arranging financing. We are very much indebted to Professor Tsunesaburo Asada (who died in 1984) for showing interest and encouraging us in our work and for his kind introduction to the Mitsubishi Company, who made Nesa at no cost to us.

to the relevant institutes and laboratories outside of Japan from the Institute for Nuclear Study (INS), University of Tokyo. The 1957 report[5] was cited in the reference list of our article on the discharge chamber.[4]

Introduction of flash-tube apparatus in extensive air-shower experiments at INS

The INS, at the University of Tokyo, was founded in 1955 as the interuniversity research center for nuclear physics and cosmic-ray physics. One year later, in 1956, the high-energy physics division was added. At the workshop on super-high-energy interactions held at the Research Institute for Fundamental Physics, Kyoto University, in February 1956, we discussed what kind of physics the cosmic-ray division of INS should aim for and what kinds of instruments should be employed and developed. At the end of a week's discussion we decided to begin two projects on super-high-energy experiments: (1) a program of experiments on extensive air showers (EAS) to be carried out on the campus of the INS by Division A, and (2) experiments using nuclear emulsions that would be exposed at high altitudes on mountains and by balloons by Division B.

In the discussion of instruments to be employed in the EAS experiment, Jun Nishimura proposed that Marcello Conversi's flash tubes would be useful as a visual apparatus by which the lateral distributions of electrons in EAS could be observed precisely.[6] It was thought that the particle structures in the longitudinal and lateral directions from the shower axis would reveal the elementary mechanisms of the super-high-energy interactions. The flash-tube apparatus, which we named the "neon hodoscope," would cover the area as widely as possible and have a reasonable spatial resolution. I was asked to build it.

The neon hodoscope, which was a one-layer array of 5,040 neon tubes 2 cm in diameter, covering an area 2 × 3.5 m, was placed near the center of the EAS detector array at the end of 1958. Measurements using all the detectors began early in 1959. The first report of the overall experimental results was published in *Progress of Theoretical Physics.*[7] The structure of EAS observed by the neon hodoscope was discussed in the *Journal of the Physical Society of Japan.* The results of the measurements with the respective detectors were reported separately.[8]

Preparation of the neon hodoscope at Osaka University

I began this work in the Department of Physics of Osaka University with Miyamoto, who at that time was a graduate student. Except for us, all the members of the cosmic-ray research group in Osaka University had moved to INS or to Osaka City University, which was founded in 1949. My first act was to ask Conversi to send us the reprints and preprints of his works on the flash chamber. I had to import almost all the main materials for making flash tubes: a camera having an automatically advancing film mechanism, and even high-speed film. The industrial productivity of Japan in the years 1956 and 1957 had not yet improved, although economic conditions had gradually recovered after the Korean War ended in 1953.

Conversi had employed a pulse-forming network (PFN), obtained as war surplus, to generate a rectangular high-voltage pulse. We, in Japan, did not have any such PFN used in radar systems, and I could not find any manufacturer who could build such a PFN. Even the capacitors for high-voltage use had to be specially ordered. So we had to operate flash tubes without a PFN. Instead of a rectangular shape, we had to apply a pulse with an exponentially decreasing shape, obtained by condenser discharge through a resistance. Although it was not recognized until later, the fact that I did not have a PFN gave me the opportunity to succeed with the discharge chamber, as I shall describe in the following section.

Observation of discharges along the cosmic-ray trajectory

Our first task was to determine if the flash tubes could be fired by pulses of exponential shape. Before the fall of 1956 we had obtained good results. The second task was to learn what phenomena of gaseous discharges occurred in the flash tubes.

Conversi had reported that each tube was wrapped in black paper, to prevent extra discharges in the adjacent tubes induced by photons from the fired tubes. He observed the discharge light along the direction of the tube axis. We observed the fired tubes from a direction perpendicular to the tube axis in order to investigate what kind of discharge occurred. The arrangement of flash tubes for this observation is shown in Figure 15.1. Three 2-cm-diameter tubes were placed between Al-plate electrodes. The upper tube was filled with the gas mixture (Ne + 0.2 percent Ar) at the pressure of 500 Torr, the middle one at 350 Torr, and the lower one at 100 Torr. The array of 1-cm-diameter flash tubes was put over and under these three tubes. With this array, the trajectory of a cosmic-ray particle was determined. Triggered by a coincidence pulse of Geiger–Müller (GM) counters, a single high-voltage pulse was applied to the whole array of flash tubes. Therefore, the effective time durations of the pulses on 2-cm-diameter tubes were varied because of the different gas pressures. The higher the pressure, the shorter the time duration.

The photographs in Figure 15.1 (top to bottom) were taken at peak voltages of pulses of 5 kV, 8 kV, and 10 kV, respectively. From observation of these discharges, the optimum operational condition of flash tubes for emitting enough light was easily obtained. It was also noticed that the brighter discharge columns occurred in each tube at the location where the cosmic-ray particle passed. The appearance of discharges in the upper tube, compared with those in the other tubes, suggested the possibility of producing a single discharge column along the path of a particle. That was the starting point of the discharge chamber. Further investigation of the discharges in the flash tubes was performed in detail.[9]

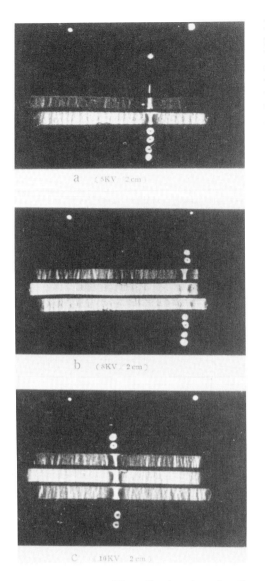

Figure 15.1. Typical examples of appearances of discharges in the flash tubes [*Source*: Fukui and Miyamoto (1957)[4]]. Each set of photographs was taken in a direction perpendicular to the axis of 2-cm-diameter tubes. The arrays of 1-cm-diameter flash tubes above and three 2-cm-diameter tubes beneath were set with their tubes' axes pointing at the camera to show the trajectory of the cosmic-ray particle. The high-voltage pulse of the condenser discharge was generated by a switching hydrogen thyratron triggered by the coincidence pulse between GM counters placed above and beneath the whole array of the flash tubes. The time constant used was 1 μsec. Because of the exponentially decreasing shape of the pulse, the effective time duration on each 2-cm-diameter tube varied with the gas pressure; that is, the higher the gas pressure, the shorter the time duration became, because the applied voltage was common to all three 2-cm-diameter tubes, and so the effective voltage above the threshold for starting discharges went up, depending on the pressure. It was easily recognized that there occurred many discharges forming the bright columns in each flash tube, and, remarkably, the brightest discharge columns appeared at the location of the cosmic-ray path that was shown by the 1-cm-diameter flash-tube array. Especially in photograph (a) there was a single column in the upper tube that was filled at 50 cm Hg pressure. The effective time duration of the high-voltage pulse on this tube was the shortest, as compared with those on other flash tubes. These observations gave us the idea that in order to produce only a single discharge column at the location of the charged particle's path in the flash tube, the high-voltage pulse should have a peak voltage not much over the threshold, and a time width as short as possible. It was a natural extension from this idea that applying this technique, that is, a triggered gaseous discharge in a relatively high pressure medium, we could develop a new instrument by which the trajectory of a charged particle could be determined more precisely than by the flash-tube hodoscope, perhaps as accurate as 1 mm or less. That was the starting point for developing our discharge chamber, which was the transition from the flash-tube hodoscope.

Development of the "discharge chamber"

We discovered how to localize a single discharge column along the path of an ionizing particle. In our case, the discharge current was restricted by the glass; so the voltage on the electrodes did not decrease quickly. That gave us the ability to produce a single-discharge or multidischarge column, even with our poor electronics. Kôdi Husimi showed interest in these ideas. We were also interested in the possibility of developing a visual track detector in a gaseous medium, as in the bubble or cloud chamber.[10] We now stood at the entrance to the "discharge chamber."

We made small rectangular boxes (8.5 cm wide × 13 cm long × 1 cm and 2 cm deep) of glass plates glued with thermoadhesive Araldite. To serve as the electrodes for these boxes, we employed the "Nesa," which had an electrically conducting surface and was optically transparent. We planned to observe the discharges projected on the electrode plane. Our paper described the experimental results, and we arrived at the discharge chamber in which beautiful tracks were shown in the multigaps, and multitracks in a single gap. We submitted the report with these photographs to *Nuovo Cimento.*[11]

In the course of investigating the discharges in both flash tubes and discharge chambers, Miyamoto determined their detailed characteristics by making ingenious trials with various combinations of operational conditions. Those results were submitted as his thesis to Osaka University and published in the *Journal of the Physical Society of Japan.*[10] Later we reported on the Penning effect in the pulsed field that showed a different mixing ratio from that in the dc field.[12]

The turning point of a new field

Conversi has stated that the starting point for his flash chamber was A. Gozzini's observation of the glowing properties of neon bulbs.[13] A neon bulb placed in the light glows by rf radiation, even with a short (1-μsec), low-power rf pulse, but when kept in the dark it does not glow even when irradiated by a high-power (1-MW) pulse. Conversi and Gozzini reached the conclusion that in the absence of light, and therefore of photoelectric emission from the body of the bulb, no free electron was present among the neon atoms in the bulb, unless some ionizing particle had crossed the bulb just before or during the rf pulse. That conclusion was the turning point for initiating the development of a new instrument that could show the trajectories of charged particles and be selectively triggered by their passages.

In the early 1950s, GM counters were commonly used in cosmic-ray experiments. To observe the trajectories of particles, an array of neon bulbs connected to the respective GM counters by an electronic circuit was arranged in a reduced but similar geometry to that of the counter arrangement. This apparatus was called the GM counter hodoscope. Applying the method used by Oreste Piccioni, the respective neon bulbs were fired when the GM counters were fired by the particles.[14] The pattern of fired neon bulbs was photographed.

We noticed that the firing efficiency of neon bulbs was 100 percent perfect in the room light, but became much less so in the dark box, even when higher voltage was applied. Also, we noticed that when the sunlight crept in through the chinks at the corner of the box and illuminated the array, the firing efficiency increased to 100 percent. So we put a small lamp in to illuminate the array, and then we had 100 percent firing of the neon bulbs.

Here, I should emphasize that we also observed a tendency for the neon bulb to glow, similar to Gozzini's observation, but through those observations we did not begin to develop a new technique, because our attention and efforts were concentrated on making 100 percent firing efficiency, in order to perfect data taking. We were not devoted to any other work, apart from that of the cosmic-ray experiment.

In 1962 I moved to CERN and joined the neutrino experiment. Carlo Franzinetti told me that he noticed a localization of sparks in the spark counter as J. Warren Keuffel did, but his interest at that time was to get very fast timing of pulses; therefore he did not try to establish the relationship between the localization of sparks and the places where the cosmic-ray particles passed.

Our turning point was the observation of localized discharges along the trajectory of the cosmic-ray particle and the use of an exponential shape of high-voltage pulse whose effective voltage and time duration on flash tubes varied naturally because of the gas pressure. The latter variation does not occur for the case of a rectangular pulse.

Concluding remarks

Even in 1956 we still suffered from the unsatisfactory circulation of scientific journals outside Japan. We received the copy of Georges Charpak's paper after our report was published in *Nuovo Cimento*.[15] I noticed A. A. Tyapkin's work at the International Conference on Instrumentation for High Energy Physics held at Berkeley, 12–14 September 1960.[16] E. Bagge gave me reprints of his work at the International Conference on Cosmic Rays and the Earth Storm, held at Kyoto, 4–15 September 1961.[17] The libraries of Osaka University and Osaka City University did not have those journals at that time. Their work and that of A. R. Bevan[18] should be mentioned as the relevant works in the history of the development of the detectors.

Notes

1 O. C. Allkofer, *Spark Chambers* (Munich: Verlag Karl Thiemig, 1969); P. Rice-Evans, *Spark, Streamer, Proportional and Drift Chambers* (London: Richelieu, 1974); J. A. Newth, "Devices for the Detection of Energetic Particles," *Rep. Prog. Phys. 27* (1964), 93–159; J. G. Rutherglen, "Spark Chambers," *Prog. Nucl. Phys. 9* (1964), 3–26; W. A. Wenzel, "Spark Chambers," *Annu. Rev. Nucl. Sci. 14* (1964), 205–38.

2 "Spark Chamber Symposium," *Rev. Sci. Instrum. 32* (1961), 480–531, especially A. Roberts, "Development of the Spark Chamber: A Review."

3 In Marcello Conversi's article in the proceedings of the International Colloquium on the

History of Particle Physics, 21–23 July 1982, Paris, France [M. Conversi, "The Development of the Flash and Spark Chambers in the 1950s," *Phys. (Paris) (Colloque C-8 Suppl.) 43 : 12* (1982), 91–9], he says that the starting point of the hodoscope chamber was Gozzini's observation in 1954 of rf-induced glow in neon bulbs in the dark or in the light and also in the irradiation of γ rays [M. Conversi and A. Gozzini, "The Hodoscope Chamber: A New Instrument for Nuclear Research," *Nuovo Cimento 2* (1955), 189–91]. Conversi states that the development of the flash-tube technique was done independently of the preexisting parallel-plate spark counter [J. W. Keuffel, "Parallel-Plate Counters and the Measurement of Very Small Time Intervals," *Phys. Rev. 73* (1948), A531; J. W. Keuffel, "Parallel-Plate Counters," *Rev. Sci. Instrum. 20* (1949), 202–8; L. Madansky and R. W. Pidd, "Characteristics of the Parallel-Plate Counter," *Phys. Rev. 73* (1948), 1215–16; L. Madansky and R. W. Pidd, "Some Properties of the Parallel Plate Spark Counter. I," *Phys. Rev. 75* (1949), 1175–80; L. Madansky and R. W. Pidd, "Parallel Plate Spark Counter. II," *Rev. Sci. Instrum. 21* (1950), 407–10; F. Bella and C. Franzinetti, "On the Theory of the Spark Counter," *Nuovo Cimento 10* (1953), 1335–7; F. Bella, C. Franzinetti, and D. W. Lee, "On Spark Counters," *Nuovo Cimento 10* (1953), 1338–40; F. Bella and C. Franzinetti, "Spark Counters," *Nuovo Cimento 10* (1953), 1461–79]. Conversi also mentions the transition from the flash to the discharge (spark) chamber (see note 5). Roberts (see note 2) states that the discharge chamber continued the work of the parallel-plate spark counter and the introduction of noble-gas filling, instead of air [T. E. Cranshaw and J. F. de Beer, "A Triggered Spark Counter," *Nuovo Cimento 5* (1957), 1107–16; P.-G. Henning, "Die Ortsbestimmung geladener Teilchen mit Hilfe von Funkenzählern und ihre Anwendung auf die Messung der Vielfach-Streuung von Mesonen in Blei," Ph.D. thesis, Hamburg, 1955; E. Bagge and O. C. Allkofer, "Das Ansprechvermögen von Parallel-Platten-Fundkenzählern für schwach ionisierende Teilchen," *Atomkernenergie 1* (1957), 1–11; P.-G. Henning, "Die Ortsbestimmung geladener Teilchen mit Hilfe von Funkenzählern," *Atomkernenergie 2* (1957), 81–8; E. Bagge and L. Schmieder, "Untersuchunger an Funken-plattenzählern," *Atomkernenergie 4* (1959), 169–81; O. C. Allkofer, "Ein neuartiger Ultrastrahlungs-Impulsspektrograph," *Atomkernenergie 4* (1959), 389–95; J. Trumper, "Parallel-plattenzähler mit elektronisch gesteuerter Funkenzündung für Teilchenbahnbestimmungen," *Atomkernenergie 5* (1960), 121–8; also see Fukui and Miyamoto, notes 4 and 5, and the works of Bella, Franzinetti, and Lee cited earlier in this note]. He claims that historically the discharge chamber was not the start. In the discussion after Roberts's review, Herbert L. Anderson made a comment recognizing my work with Miyamoto as the important contribution that originated the idea of the multiplate chamber, and he stated that the chamber looked in all outward appearances like the one of the Japanese, not like any of the previous ones, which really were quite different. Roberts said that the first observation of the discharge between parallel plates caused by a fast particle was apparently made by J. Warren Keuffel (1948, cited earlier in this note). However, Keuffel stated, without photographs and without confirming experiments, that the discharge was localized in a fine, plainly visible streamer channel, presumably in the neighborhood of the initiating ion. As described here, we did not work on the parallel-plate counter device, but on the hodoscope chamber.

4 S. Fukui and S. Miyamoto, "A New Type of Particle Detector: The Discharge Chamber," *Nuovo Cimento 11* (1959), 113–15.

5 S. Fukui and S. Miyamoto, "A Study of the Hodoscope Chamber," INS-TCA-10 (1957).

6 Jun Nishimura was, in 1956, a research associate of the Emulsion Group at Kobe University. He moved to INS as an associate professor and is now professor at the Institute of Space and Astronautical Science.

7 S. Fukui, H. Hasegawa, T. Matano, I. Miura, M. Oda, K. Suga, G. Tanahashi, and Y. Tanaka, "A Study on the Structure of the Extensive Air Shower," *Prog. Theor. Phys. (Suppl.) 16* (1960), 1–53.

8 S. Fukui, "The Structure of Extensive Air Showers near the Axes," *J. Phys. Soc. Jpn. 16* (1961), 604–15.

9 S. Fukui and S. Miyamoto, "A Study of the Hodoscope Chamber. II. A Preliminary Study of a New Device of a Particle Detector," INS-TCA-11 (1958) (in Japanese).

10 S. Fukui and S. Miyamoto, "New Type of Particle Detector: The Discharge Chamber," unpublished (1959); Sigenori Miyamoto, "The Discharge Chamber and Its Characteristics," Ph.D. thesis, Osaka University, 1961; S. Fukui and S. Miyamoto, "The Discharge Chamber and Its Characteristics," *J. Phys. Soc. Jpn. 16* (1961), 2574–5.

11 See note 5.

12 S. Fukui, S. Hayakawa, H. Nukushina, and T. Tsukishima, "Formation of Localized Plasmas in Pulsed Microwave Discharge," *J. Phys. Soc. Jpn. 17* (1962), 250–1; T. Tsukishima, "Formative Processes in Pulsed Microwave Discharges in Mixtures of Neon and Argon," *J. Phys. Soc. Jpn. 18* (1963), 558–71.

13 See Conversi (1982), in note 3.

14 O. Piccioni, "Search for Photons from Meson-Capture," *Phys. Rev. 74* (1948), 1754–8.

15 G. Charpak, "Principe et essais préliminaires d'un nouveau detecteur permettant de photographier la trajectoire de particules ionisantes dans un gaz," *J. Phys. Radium 18* (1957), 539–40.

16 A. A. Tyapkin, "Hodoscopic System for the Pulsed Feeding of Counters" (in Russian), *Prib. Tekh. Eksp. 3* (1956), 51–3; V. V. Vishnyakov and A. A. Tyapkin, "The Operation of Gas Counters under Pulsed Voltage Conditions" (in Russian), *At. Energ. 3* (1957), 298–307; I. M. Vasilevskij, V. V. Vishnyakov, E. Iliescu, and A. A. Tyapkin, "Hodoscopic System of Gas-Discharge Counters Used in Experiments with Accelerators," in *Proceedings of an International Conference on High Energy Accelerators and Instrumentation*, edited by L. Kowarski (Geneva: CERN, 1959), pp. 589–91.

17 See Bagge and Allkofer and Bagge and Schmieder in note 3.

18 A. R. Bevan, "High-Frequency Discharges Localized along Tracks of Ionizing Particles," *Nature (London) 164* (1949), 454–5.

16   Early work at the Bevatron: a personal account

GERSON GOLDHABER

Born 1924, Chemnitz, Germany; Ph.D., University of Wisconsin, 1950;
experimental particle physics; University of California at Berkeley and
Lawrence Berkeley Laboratory

The Bevatron started operating in early 1954 at what was then the Radiation
Laboratory and is now known as the Lawrence Berkeley Laboratory.

Some personal background

Sula (Sulamith Goldhaber) and I came to Berkeley from Columbia
University in 1953. I joined the Physics Department and Emilio Segrè's group
at the Radiation Laboratory. She joined Walter Barkas's group and later
Edward Lofgren's group. We had been working with photographic emulsions
at Columbia's cyclotron located at Nevis, with the help and encouragement
of Gilberto Bernardini. Before that, I used emulsions loaded with D_2O as a
γ-ray spectrometer for my Ph.D. thesis under Hugh Richards at the University of Wisconsin in Madison.

Setting up with photographic emulsions

Whereas in my earlier work I had used $100-600$-μm single small
emulsions on glass, this was the period in which emulsion stacks began to
be used in cosmic-ray work at Bristol and elsewhere; the electron-sensitive
emulsions had recently been introduced by Kodak Ltd. of England, followed
by C. Waller at Ilford, in close consultation with Cecil F. Powell and Giuseppe
P. S. Occhialini.

Thus, I started out at Berkeley to build up an emulsion-processing plant in
the Physics Department – the photographic-emulsion arm of the Segrè group.

Work supported in part by the U.S. Department of Energy under contract DE-
AC0376SF00098.

That involved new techniques for marking emulsion sheets, to allow easy track following from sheet to sheet, the modification of microscopes with special stages to hold and manipulate those large emulsion sheets after they were mounted precisely on glass, and the construction of accurate microscope stages for multiple scattering measurements.

I was very lucky to find that the shop foreman in the Physics Department, William Brower, loved to build precision equipment; his advice and consultation were invaluable to me. In addition, Stephen Goldsack, from England, visited the Brode–Fretter group for a year in 1954 and spent a good deal of time working with me and helped in the design of the multiple scattering equipment.

The start-up of the Bevatron

From the first day – I should actually say night – the Bevatron started accelerating proton beams, Sula and I were there to place emulsions into the beam. We were soon joined in those nightly vigils by Warren Chupp, who was working in the Bevatron group headed by Lofgren. At first we placed a few emulsions on an arm, which carried the target and was introduced into the Bevatron through a vacuum seal. The target carried a small polyethylene "lip" designed by Edwin McMillan to introduce a small energy loss and scrape off a small portion of the beam. (A similar device was introduced by Rodney L. Cool and Oreste Piccioni at the Cosmotron as a starting point for external beams.) As a result of this energy loss, the proton trajectories moved to a lower radius and hit the emulsions on the next pass. With these exposures we helped establish that we were indeed dealing with energetic protons and that we could get emulsion exposure, of sorts, inside the Bevatron vacuum tank.

I remember in particular one episode when Luis Alvarez was also spending the evening at the Bevatron and observed our procedures. That night the Bevatron operator charged with pulling the target probe holding our emulsions out through the vacuum lock gave a particularly vigorous pull and managed to yank the probe completely out, so that air started rushing into the vacuum tank. Alvarez, who was standing nearby, rushed over and placed the palm of his hand over the hole! This allowed the crew to close the vacuum lock without the entire Bevatron coming up to air. I must admit that I would not have thought of doing this – and furthermore probably would not have done it! Alvarez saved the day, and the Bevatron was able to pump back down without excessive loss of time, while Alvarez was rubbing the sore spot on his hand.

Our first interest was to study K mesons as well as any other new particle that might show up. One of the goals was to understand the $\tau-\theta$ puzzle. Little was known about the lifetimes of all the different charged K mesons (or were they possibly different decay modes?), and there was no reason to suppose

(a)

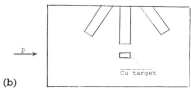

(b)

Figure 16.1. (a) The Bevatron vacuum tank [*Source*: LBL Photo Services]. (b) Sketch of the reentrant wells.

that some of these lifetimes could not be quite short, for example, as short as the K^0 or Λ lifetimes. It was clear to me that the emulsion exposures in which the emulsions were mounted on a target holder could not be well enough controlled for accurate experiments.

The vacuum tank of the Bevatron was enormous (Figure 16.1) because the machine had been designed before the invention of strong focusing (by Ernest Courant, Stanley Livingston, and Hartland Snyder and, independently, Nick Christofilos, 1952). This meant that if emulsions were exposed in an external beam, the kaons would have to travel at least 1–2 m, and hence any short-lived component would decay away.

I discussed this point with Lofgren and suggested that we might introduce reentrant wells into the vacuum tank to allow a close approach to the target from above. I also gave him a very crude sketch (for wells corresponding

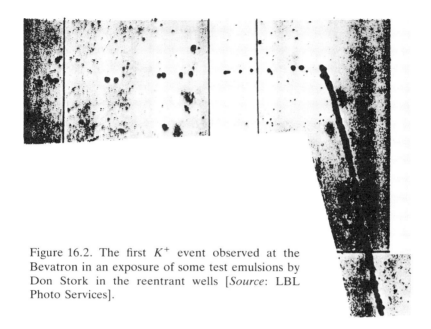

Figure 16.2. The first K^+ event observed at the Bevatron in an exposure of some test emulsions by Don Stork in the reentrant wells [*Source*: LBL Photo Services].

roughly to 45°, 90°, and 135° in the center-of-mass system). To my surprise and delight, when I saw Lofgren some ten days later he mentioned by the way that the reentrant wells were already installed! Thus, we were then able to expose emulsion stacks within a few centimeters from the target and could start looking for very short lived particles. The first K^+ decay event at the Bevatron was found by Don Stork in a test exposure in our reentrant wells (Figure 16.2).

During this period we also helped in the exposure of emulsion stacks from all over the world. Frequently we also processed the stacks in Berkeley using the techniques for stack alignment we had worked out. In particular, I remember an enormous stack brought over by Louis Leprince-Ringuet from Paris that we exposed and processed.

The next step was to expose emulsions in momentum-analyzed external beams originating at an internal target. This had the advantage that the three types of particles, π^+, K^+, and p, all of the same momentum, had different well-defined ranges in the emulsion, so that one could proceed directly to the region where the K^+ mesons come to rest without scanning the entire emulsion volume.

After consultation with my colleagues, we decided to introduce a 90° wedge magnet into the external beam to improve the intensity by focusing the beam. Although this device worked, Roy Kerth and Stork of the Richman group

STRONG - FOCUSING SPECTROMETER

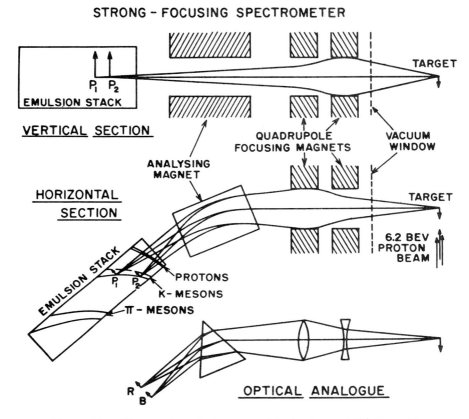

Figure 16.3. The quadrupole lenses used in the focused K^+ beam [*Source*: R. W. Birge].

came up with a better idea at about the same time. They used a set of strong-focusing quadrupoles – of the type built by Bruce Cork for focusing the proton beam at the linear accelerator – the injector to the Bevatron (Figure 16.3). With this improvement, relatively clean and easily studied K^+ as well as K^- beams became available. On some of this work we shared our stacks with Aihud Pevsner and David Ritson, who were both at MIT at that time, as well as with Mildred Widgoff and Gus Zorn. My first two students, Joe Lannutti and Ed Iloff, were also involved in this work.

We concentrated on interactions in flight (for cross-section determinations),[1] decays in flight (for lifetime determinations),[2] decays at rest of K^+ mesons (for study of the different particles, or decay modes),[3] and, later, interactions at rest of K^- mesons.[4] R. W. Birge, R. P. Haddock, Kerth, J. R. Peterson, J. Sandweiss, Stork, and Marion Whitehead of the Richman group concentrated on precision range measurements of θ^+ and τ^+ mesons yielding accurate mass measurements.[5] Alvarez, together with Sula, did the first τ^+ lifetime measurement by comparing τ production rates as observed close to the target (in the reentrant wells) and far away (in the external beams).[6] Harry Heckman, in the Barkas group, collected τ mesons for inclusion in the worldwide Dalitz plot.[7] We found a K^+H scattering event in our emulsion stacks that allowed a precision mass measurement of a single θ^+ event.[8] All this work was reported by Stork at the 1955 Pisa conference. This was clearly a milestone. In less than a year the Bevatron had begun to contribute significantly to what had largely been the domain of cosmic-ray physics.

What did we learn from all this work at the Bevatron?

We established that K^+ cross sections were significantly lower than K^- cross sections and that low-energy K^+ interactions did not produce pions, but only underwent either scattering or charge exchange – a clear confirmation of the Gell-Mann–Nishijima strangeness scheme. Furthermore, we confirmed the observations at the Cosmotron that K^- interactions produce Σ^+ and Σ^- hyperons, and we noted in particular from a few capture events on hydrogen in the emulsions that $M(\Sigma^-)$ was 14 m_e larger than $M(\Sigma^+)$, a surprising result at first.[4] The θ^+ and τ^+ mass measurements,[5,8] coupled with lifetime measurement[2,6] and particularly later lifetime measurement with counters (Alvarez, Frank S. Crawford, Myron L. Good, and M. Lynn Stevenson,[9] as well as those of Val Fitch and R. Motley[10] at the Cosmotron; see Chapter 31 by Fitch), pointed clearly to the puzzle that the θ and τ had nearly indistinguishable masses *and* lifetimes!

The stage was set, and the culmination of the cosmic-ray, Cosmotron, and Bevatron work, coupled with Richard Dalitz's analysis, led T. D. Lee and C. N. Yang to postulate two alternative possibilities: (a) either there is a parity doubling of particles or (b) parity is violated in weak decays. Furthermore, they suggested how this could be tested. As is well known, the experiment gave a resounding confirmation to hypothesis b!

The hunt for the antiproton

The Bevatron was designed to have enough energy for antiproton production in a $\bar{p}p$ collision. To search for the $\bar{p}$ was thus clearly on many people's minds. In the Segrè group, we decided to attempt a double-barreled attack on the antiproton. On the one hand, Owen Chamberlain, Clyde Wiegand, and Thomas Ypsilantis went ahead with the preparation of a beam for a counter experiment (the details are given in Chapter 17); on the other hand, Segrè and I went ahead to plan for an emulsion experiment in collaboration with Edoardo Amaldi and his group in Rome. When the $\bar{p}$ beam under construction by Chamberlain and associates[11] reached the first focus (i.e., it was about half done), we exposed our emulsion stack (Figure 16.4), pro-

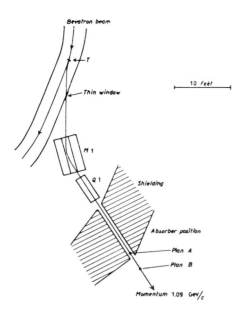

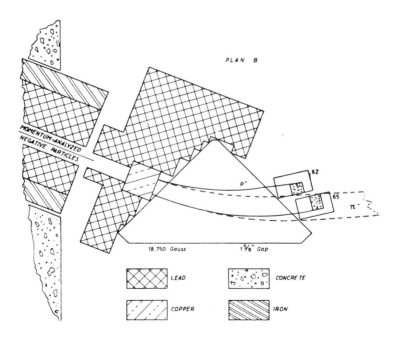

Figure 16.4. The first emulsion exposure to $\bar{p}$'s at 1,090 MeV/c. A plan view of the magnetic selection and focusing magnets is shown at top [*Source*: Chamberlain et al. "Antiproton Star Observed ..." (1956)[13]].

cessed it in Berkeley, divided it into two parts, and started scanning it at both Berkeley and Rome.

As it turned out, in this emulsion experiment we outsmarted ourselves. We calculated the effect of the Fermi motion and concluded that in order to get a reasonable $\bar{p}$ flux, we had to run at a momentum of 1,090 MeV/c, rather than 700 MeV/c. At the latter momentum, the $\bar{p}$'s could reach the end of their range in the stack. This meant that in order to stop $\bar{p}$'s in our emulsion stack, we had to place a sizable Cu absorber (132 g/cm^2) ahead of our emulsion stack. This had two deleterious effects. First of all, interactions of the beam particles in the absorber gave rise to a large number of protons, which managed to enter our stack together with the negative particles. This made track following of about 1.5 minimum ionizing tracks very difficult and meant that we had to rely in part on the very slow and laborious method of area scanning. Second, $\bar{p}$'s have a cross section that is considerably larger than the proton cross section. This fact, which we did not anticipate, reduced our $\bar{p}$ flux by more than a factor of 2 from what we expected.[12]

By October 1955, the counter experiment had clearly demonstrated the following:

1. There were negative particles of protonic mass within an accuracy of ±5 percent.
2. There was a threshold for the production of these particles at about 4 GeV of incident-proton-beam kinetic energy.

These were necessary conditions for the identification of $\bar{p}$'s.

Then, in November 1955, our efforts in the emulsion experiment, despite the handicaps mentioned earlier, yielded one event, found in Rome, that came to rest and produced a star with a visible energy release of about 826 MeV (Figure 16.5).[13] Again a necessary condition for $\bar{p}$'s. About the same time, John M. Brabant, Cork, Nahmin Horowitz, B. J. Moyer, Joseph J. Murray, Roger Wallace, and William A. Wenzel of the Lofgren and Moyer groups placed their special lead-glass Cerenkov counter behind the beam of Chamberlain and associates and observed "large pulses" consistent with the properties expected for $\bar{p}$'s.[14]

In December 1955 we decided to try another emulsion exposure – this time at 700 MeV/c, so that the $\bar{p}$'s could enter the emulsion stack and come to rest in it. Furthermore, I introduced a special sweeping magnet this time to guard against stray protons entering our stack. On this occasion, all emulsion groups at the laboratory participated in the exposure: Birge and associates of the Richman group, and Barkas and associates, who supplied their own emulsion stacks, as well as Amaldi's group in Rome, who shared our stack, together with Sula Goldhaber and Chupp of the Lofgren group. Also, in September 1955, Gösta Ekspong came to visit from Sweden and joined me in my efforts to find more $\bar{p}$'s in emulsions.

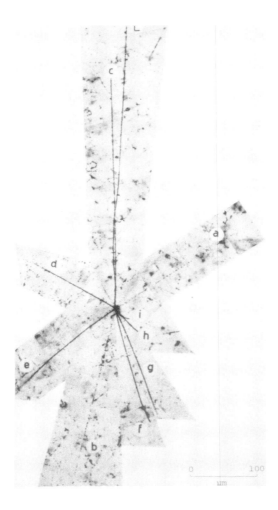

Figure 16.5. Photomicrograph of first event in our emulsion exposure found in Rome. *L* indicates the incoming antiproton track. Tracks *a* and *b* are pions, and *c* is a proton. The remaining tracks could be protons or *α* particles. [*Source*: Chamberlain et al., "On the Observation ..." (1956)[13]].

Just before we started the exposure, we went through the usual period of doubt – had all the magnets been connected up correctly? As a last check, we brought out a battery and connected a piece of thin wire and checked the direction of the forces on it in the various magnets. This was followed by all present holding up either their right hand or their left hand with three fingers held perpendicular to each other to ascertain that negative particles would be bent correctly by the magnets.

This exposure was extremely successful. As soon as the emulsions were developed, we could see $\bar{p}$ candidates entering the emulsion stack. Our procedure was to scan along the upstream edge of each emulsion and look for about twice-minimum ionization tracks – which were easily distinguishable from the large background of 700-MeV/c pions that were at minimum ionization.

The emulsion processing was started over the New Year holiday, and early in January 1956, as soon as the emulsions were dry from the developing, fixing, and washing cycle, Ekspong would scan the leading edge and look for $\bar{p}$ candidates. We found a few twice-minimum tracks, and Ekspong started to follow along the track through a series of plates. In the morning of 11 January 1956, he followed a track to the end of its range, where it came to rest and formed a large star! That same afternoon a scanner working with Sula found another star! The first star occurred at the interface between two emulsion sheets, with half the tracks going upward, and the other half downward. We had to wait another week or so, until the rest of the stack was developed, before we could follow all the tracks from this star.

After Ekspong and I developed a new method for the multiple scattering measurements of steep tracks – and here our precision placement of the emulsion sheets was of crucial importance – we evaluated the total visible energy. This event turned out to be particularly important because it gave the conclusive proof ("sufficient condition" for those who were still in doubt) of the annihilation process. The visible energy released in this star was $1,300 \pm 50$ MeV (Figure 16.6).[15] Clearly greater than the mass of the incident negative particle!

I remember two amusing consequences of our discovery. First, the day after the annihilation event was found, Segrè saw to it that a telephone was installed in my laboratory in LeConte Hall. Second, Chamberlain gave an invited talk at the 1956 New York meeting of the American Physical Society. There he reported on both the counter experiment and our annihilation event. He told me afterward that the proof supplied by the annihilation event was an important ingredient in the minds of the audience. In fact, in a subsequent interview with the press, my hand drawing of the first annihilation event was reproduced in *Time* magazine.[16]

Subsequently, all the groups participating in this exposure found $\bar{p}$ events in their emulsion searches. We pooled our data and published our results as the "Antiproton Collaboration Experiment" – thirty-five events and eighteen authors![17] Figure 16.7 shows the visible energy distribution for these thirty-five events in units of $2\,M_p$. About two-thirds of the events showed a visible energy release above 0.5 (i.e., above M_p). Aside from proving that $\bar{p}$ annihilation occurs, we found many interesting properties of the annihilation process. When we were first looking for $\bar{p}$ events in emulsion, some expectations were that we would see $\bar{p}p \rightarrow e^+e^-$ or $\bar{p}p \rightarrow \pi^+\pi^-$, so-called T events. This

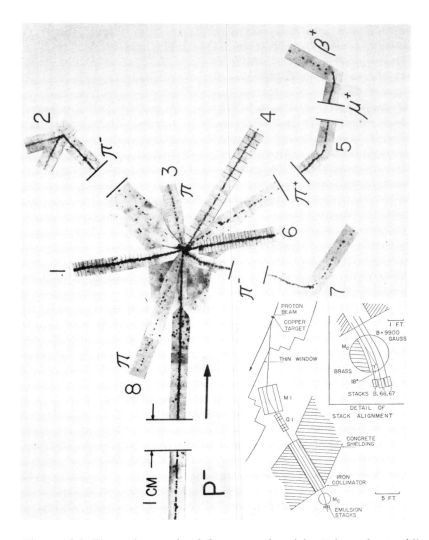

Figure 16.6. Photomicrograph of first event found by "along the track" scanning in the second exposure. This event, which released 1,300 ± 50 MeV of visible energy, gave the conclusive proof for the annihilation process [*Source*: Chamberlain et al. (1956)[15]].

was certainly not the case. We found a surprisingly large pion multiplicity, $\bar{N}$ = 5.3 ± 0.4, which, if one took Fermi's statistical model seriously, implied a rather large interaction volume of radius over two times the expected radius $(\hbar/m_\pi c) \approx 1$ fermi.

Actually, the high multiplicity probably is the result of the fact that meson

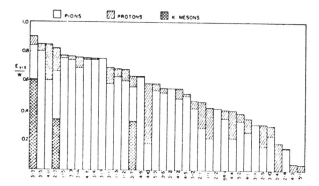

Figure 16.7. Visible energy release for thirty-five $\bar{p}$ events observed in the Antiproton Collaboration Experiment. Energy is given in units of the total available energy: $2\,M_p$. The star reference number is given for each entry [*Source*: Barkas et al. (1957)[17]].

resonances, rather than individual particles, are produced in the $\bar{p}$ annihilation process. But the discovery of meson resonances at the Bevatron came nearly five years later, as discussed by Alvarez in Chapter 19.

Notes

1 W. W. Chupp, G. Goldhaber, S. Goldhaber, E. L. Iloff, J. E. Lannutti, A. Pevsner, and D. Ritson, "Interactions and Decay of Positive *K*-Particles in Flight," *Nuovo Cimento (Suppl. 10) 4 : 2* (1956), 361–6.

2 E. L. Iloff, W. W. Chupp, G. Goldhaber, S. Goldhaber, J. E. Lannutti, A. Pevsner, and D. Ritson, "Mean Lifetime of Positive *K* Mesons," *Phys. Rev. 99* (1955), 1617–18.

3 D. M. Ritson, A. Pevsner, S. C. Fung, M. Widgoff, G. T. Zorn, S. Goldhaber, and G. Goldhaber, "The Characteristics of *K*-Particles Decay," *Nuovo Cimento (Suppl. 10) 4 : 2* (1956), 390–3.

4 G. Goldhaber, S. Goldhaber, E. L. Iloff, J. E. Lannutti, F. H. Webb, M. Widgoff, A. Pevsner, and D. Ritson, "Interaction and Decay of Negative *K*-Particles in Flight," *Nuovo Cimento (Suppl. 10) 4 : 2* (1956), 367–70.

5 R. W. Birge, R. P. Haddock, L. T. Kerth, J. R. Peterson, J. Sandweiss, D. H. Stork, and M. N. Whitehead, "Positive Heavy Mesons Produced at the Bevatron," *Nuovo Cimento (Suppl. 10) 4 : 2* (1956), 351–8.

6 W. Alvarez and S. Goldhaber, "The Lifetime of the π-Meson," *Nuovo Cimento (Suppl. 10) 4 : 2* (1956), 33.

7 H. H. Heckman, "Analysis of the π-Meson Decay," *Nuovo Cimento (Suppl. 10) 4 : 2* (1956), 230–2.

8 W. W. Chupp. G. Goldhaber, S. Goldhaber, W. R. Johnson, and J. E. Lannutti, "*K*-Meson Mass from a *K*-Hydrogen Scattering Event," *Nuovo Cimento (Suppl. 10) 4 : 2* (1956), 379–80.

9 Luis W. Alvarez, Frank S. Crawford, Myron L. Good, and M. Lynn Stevenson, "Lifetime of *K* Mesons," *Phys. Rev. 101* (1956), 503–5.

10 V. Fitch and R. Motley, "Mean Life of K^+ Mesons," *Phys. Rev. 101* (1956), 496–8.

11 Owen Chamberlain, Emilio Segrè, Clyde Wiegand, and Thomas Ypsilantis, "Observation of Antiprotons," *Phys. Rev. 100* (1955), 947–50.

12 Gösta Ekspong tells me that one day Edward Teller came rushing into my lab looking for me. Edward was all excited – he had the explanation for why we were not seeing any events in our emulsions – the large cross section was the cause! Hans-Peter Duerr and Edward Teller, "Interaction of Antiprotons with Nuclear Fields," *Phys. Rev. 101* (1956), 494–5.

13 O. Chamberlain, W. W. Chupp, G. Goldhaber, E. Segrè, C. Wiegand, E. Amaldi, G. Baroni, C. Castagnoli, C. Franzinetti, and A. Manfredini, "Antiproton Star Observed in Emulsion," *Phys. Rev. 101* (1956), 909–10; O. Chamberlain, W. W. Chupp, G. Goldhaber, E, Segrè, C. Wiegand, E. Amaldi, G. Baroni, C. Castagnoli, C. Franzinetti, and A. Manfredini, "On the Observation of an Antiproton Star in Emulsion Exposed at the Bevatron," *Nuovo Cimento 3* (1956), 447–67.

14 John M. Brabant, Bruce Cork, Nahmin Horowitz, Burton J. Mayer, Joseph J. Murray, Roger Wallace, and William A. Wenzel, "Terminal Observations on Antiprotons," *Phys. Rev. 101* (1956), 498–501.

15 O. Chamberlain, W. W. Chupp, A. G. Ekspong, G. Goldhaber, S. Goldhaber, E. J. Lofgren, E. Segrè, C. Wiegand, E. Amaldi, G. Baroni, C. Castagnoli, C. Franzinetti, and A. Manfredini, "Example of an Antiproton–Nucleon Annihilation," *Phys. Rev. 102* (1956), 921–3.

16 *Time*, 13 February 1956, p. 34.

17 W. H. Barkas, R. W. Birge, W. W. Chupp, A. G. Ekspong, G. Goldhaber, S. Goldhaber, H. H. Heckman, D. H. Perkins, J. Sandweiss, E. Segrè, F. M. Smith, D. H. Stork, L. van Rossum, E. Amaldi, G. Baroni, C. Castagnoli, G. Franzinetti, and A. Manfredini, "Antiproton–Nucleon Annihilation Process (Antiproton Collaboration Experiment)," *Phys. Rev. 105* (1957), 1037–58.

17   The discovery of the antiproton

OWEN CHAMBERLAIN

Born 1920, San Francisco, California; Ph.D., University of Chicago, 1949;
experimental physics; Nobel Prize in physics, 1959, for discovery of the
antiproton; University of California, Berkeley

I believe the antiproton story starts with P. A. M. Dirac, who in
1930 published his paper "A Theory of Electrons and Protons." Later, the
positive particles in the theory were identified as antielectrons (we call them
positrons).[1] Dirac's theory started out with electron energies that were both
positive and negative. He considered the possibility that the negative-energy
states were completely full, or were full except for an occasional missing
electron, understanding very well that the missing electron would appear to
be a positive charge. Thus, the theory called for positive particles of electron
mass, namely, the positron.

The positron was promptly (in three years) found by Carl D. Anderson.[2]
His finding greatly strengthened belief in the importance of the Dirac equation. The question then on people's minds was, Does the proton have an
antiparticle? I think most physicists believed that the proton did have an
antiparticle, but there was some doubt. I remember someone saying: The
proton has a large anomalous magnetic moment. That may be the signal that
the proton is very different from the electron.

A number of cosmic-ray papers did mention the antiproton as a possible
explanation of certain hard-to-explain events. For example, Evans Hayward
published a paper in 1947 that included three singular events observed in her
cloud chamber.[3] One of them (Figure 17.1) was discussed in her paper as a
possible antiproton. It showed considerable shower activity attributable to
a particle that might have been of proton mass entering her cloud chamber.
As she said in her publication, "Other possible explanations are that it is an
extremely high-energy electron, if no dielectric absorption effect exists, or
that it is several coincident electrons, or that it is a negative proton giving up
all its energy in interacting with the lead plate."

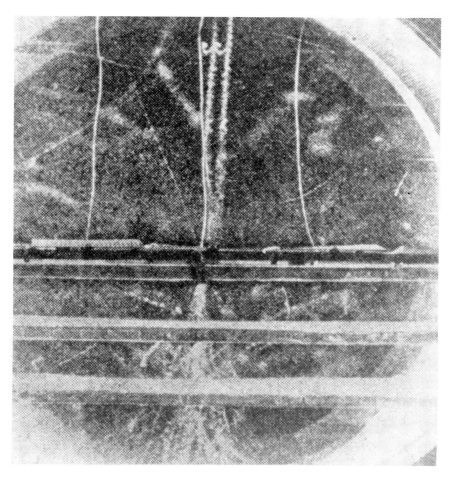

Figure 17.1. Singular event observed in a cloud chamber by Evans Hayward. The negative proton is mentioned as a possible identification for the incoming particle [*Source*: Hayward (1947)[3]].

In 1954, H. S. Bridge, Hans Courant, H. De Staebler, and Bruno Rossi observed the "possible annihilation of a heavy particle"[4] (Figure 17.2.) Their cloud chamber had multiple plates of brass, $\frac{1}{2}$ inch in thickness. The figure shows the incoming particle (a) from above; there are a backward shower (b), an upward shower, and two small showers below.

Analysis showed that the visible energy was about 1,000 MeV, of the order of one nucleon rest energy. In their report they say: "In view of the difficulties of interpreting the event as a decay or an absorption process, one should consider the possibility that the event represents the annihilation process of two

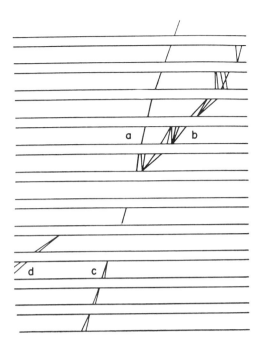

Figure 17.2. Sketch of the cosmic-ray event observed by Bridge, Courant, De Staebler, and Rossi [*Source*: Bridge et al. (1954)[4]].

heavy fermions. For example, the incident particle might be an antiproton (or an antihyperon) that undergoes annihilation with an ordinary proton."

In early 1955, Edoardo Amaldi, C. Castagnoli, G. Cortini, C. Franzinetti, and A. Manfredini published an article in *Nuovo Cimento* in which, referring to Figure 17.3, they say:[5]

> We are left to consider the star *B* as produced by the track *p*. Then the corresponding particle either has a rest energy of the order of $1.5 \div 2$ GeV, or, being an antiproton, it has been annihilated by a nucleon, releasing $2m_p c^2 = 1876$ MeV.
>
> We do not have any argument in favor of one or the other of these two possibilities apart from the fact that unstable particles of rest energy of the order of $1.5 \div 2$ GeV have never been observed; nor has the antiproton, but this, at least, is expected to exist as a consequence of very general arguments based on symmetry with respect to the sign of the electric charge.

It is small wonder that Amaldi and his colleagues were anxious to have emulsion stacks exposed to secondary-particle beams at the Bevatron.

Obviously, a fair number of cosmic-ray physicists were thinking about antiprotons. Indeed, years earlier, when proposals were being contemplated for accelerators at Brookhaven and at Berkeley, the energy at Berkeley was

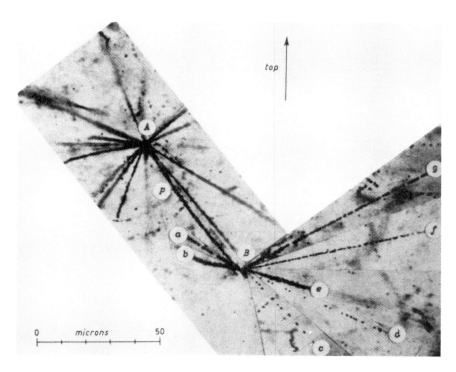

Figure 17.3. The authors believe that a cosmic-ray interaction may have produced an antiproton that underwent annihilation at B [*Source*: Amaldi et al. (1955)[5]].

varying between 1.8 GeV and 10 GeV. The decision for 6 GeV was based on calculations by Edwin McMillan and Pief Panofsky that determined that this energy was needed to produce nucleon pairs. Of course, since the proton appeared stable, it was believed that antiprotons would not be created singly.

As the Bevatron came into fairly stable operation, a number of groups were trying to observe antiprotons. It was my belief that we should not discuss our plans, other than among ourselves, during the planning stages. I felt I did not want to try to establish a patent on what we were doing (meaning other physicists should please stay away and leave the field to us). Nor did I want to encourage other physicists to start thinking about what we were thinking about. Of course, once we got beyond the planning stage and wanted to have apparatus built in the shop, we had to give a justification and a purpose.

B. Cork, N. Horowitz, J. J. Murray, and W. A. Wenzel used three Ceren-kov counters with indices of refraction of about 1.3, 1.4, and 1.5 to try to pick out a promising group of particles that might be antiprotons. Wenzel also had a time-of-flight system in mind that I believe was never fully tried, at least not in the earlier days. He hoped to use large counters, say 30 × 30 cm, in a long flight path that would allow at least two time-of-flight measurements.

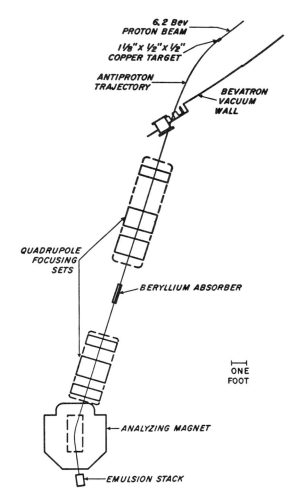

Figure 17.4. Arrangement of quadrupole magnets, bending magnets, and beryllium absorber, used by Stork, Birge, Haddock, Kerth, Sandweiss, and Whitehead [*Source*: Stork et al. (1957)[6]].

D. H. Stork, R. W. Birge, R. P. Haddock, L. T. Kerth, Jack Sandweiss, and M. N. Whitehead attempted a partial separation of antiprotons from pions by placing a beryllium absorber between two magnetic analyzers.[6] The idea was that at the same momentum, antiprotons had less speed than negative pions. They would lose energy faster than the pions, and after the absorber the antiprotons would have less momentum than the pions and could be separated in a final magnetic bending.

At the top of Figure 17.4, which shows their apparatus, is the Bevatron beam hitting a copper target. The desired antiproton trajectory comes through focusing quadrupoles, a beryllium absorber, more focusing quadrupoles, and the final magnet, with an emulsion stack as the detector at the bottom of the figure. Incidentally, they later reported that the cross section of antiprotons in beryllium was about 1 b (barn). This turned out to have a great effect on the experiment. This high cross section, especially for annihilation of the anti-protons, put experiments with thick absorbers at a great disadvantage, it was later recognized.

J. M. Brabant, Cork, Horowitz, B. J. Moyer, Murray, R. Wallace, and Wenzel had a large lead-glass Cerenkov counter.[7] They were not able to recognize antiprotons in a large background of negative pions, but the equipment was used to study stopping antiprotons, once a satisfactory trigger from our apparatus could be had. In fact, their apparatus gave the first indica-tion that large energy releases, commensurate with proton–antiproton annihi-lation, were seen.

Our group (Chamberlain, Emilio Segrè, Clyde Wiegand, and Thomas Ypsilantis), a subgroup of the Segrè group, had planned to set up a pair of magnetic analyses in tandem, each with focusing by quadrupole triplet. As soon as the equipment for the first magnetic analysis was in place, it became a suitable place to irradiate emulsion stacks. Amaldi and his collaborators, Gerson Goldhaber and his collaborators (the part of the Segrè group that was using emulsion techniques), and Birge and his collaborators all had irradia-tions using the first leg of the two-leg magnetic analysis.

In Figure 17.5, the incoming particles (incoming from the upper left) have already been magnetically analyzed. A flaw in this arrangement was that an absorber was built into the setup. Negative particles in that absorber could make protons and other particles that would serve to confuse the person scan-ning the emulsions. So the emulsion exposure in this arrangement was not as clean as those we could make later. There came from that stack, however, one event that was found in the part of the emulsion that went to the Univer-sity of Rome. I believe it was known for a while as "Leticia." The energy released was considerable, but the visible energy was not above 938 MeV.

Later we made an exposure at lower momentum. That seemed to be the secret, because by going to lower momentum we could arrange that the whole range of the antiprotons was spent in the emulsion stack. The antiprotons could be identified at the entrance to the stack by observing their higher specific ionization. Very soon that exposure produced an important event, which I shall mention later.

I want now to return to our counter subgroup. We built the rest of the mag-netic analysis system, as shown in Figure 17.6. Magnetic analysis and focusing gave a first focus near the center of the figure, approximately where the emul-sions had been exposed. Then another quadrupole triplet and magnetic bend-ing gave a final focus. The claim was that with two magnetic analyses, one right after the other, we were twice measuring the momentum of each can-didate particle. We also had two velocity measurements. One was a signal

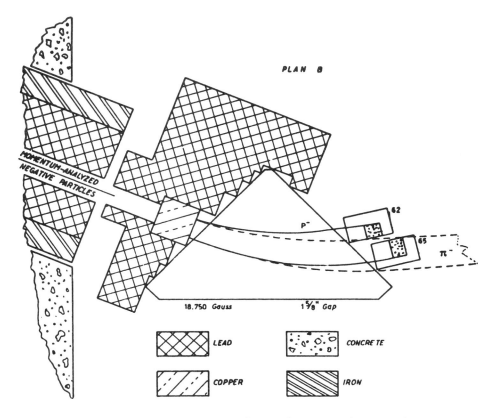

Figure 17.5. Setup designed to allow antiprotons to be stopped in nuclear-emulsion stacks. Close to the center of the figure is a copper absorber.

from our velocity-selecting Cerenkov counter, here called C2, and the other a time-of-flight measurement between the counters labeled S1 and S2. We declared that double measurement of momentum and of velocity gave us reliable particle masses. We made our announcement based on the observation of negative particles of protonic mass.[8]

Figure 17.7 shows time-of-flight information for a few cases. Signals from the two counters S1 and S2 were superposed on the same oscilloscope trace, as well as C1, with opposite sign. The first part of the figure (a) shows a fast particle (a pion), the two signals being somewhat close together. Part (b) shows an antiproton, and (c) is an accidental coincidence. The signal cables from S1 to the oscilloscope and from S2 to the oscilloscope were of equal length, so the separation of pulses S1 and S2 is (approximately) the actual flight time. The presence of a downward pulse (from C1) indicates a particle moving faster than would an antiproton of the chosen momentum. Plotting

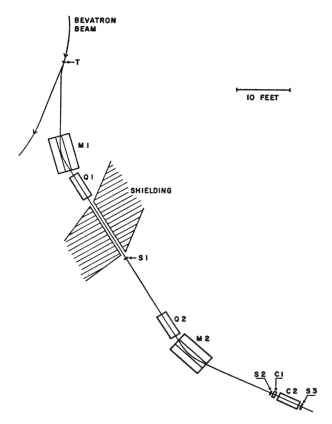

Figure 17.6. Experimental arrangement of Chamberlain, Segrè, Wiegand, and Ypsilantis. T represents the production target, Q stands for quadrupole triplet, M for bending magnet. Scintillation counters are designated by S, and Cerenkov counters by C [*Source*: Chamberlain et al. (1955)[8]].

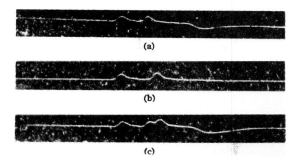

Figure 17.7. Oscillograph traces showing (from left to right) protons from scintillation counters S1 and S2 and from the threshold Cerenkov counter C1; (a) pion, (b) proton, (c) accidental coincidence [*Source*: Chamberlain et al. (1955)[8]].

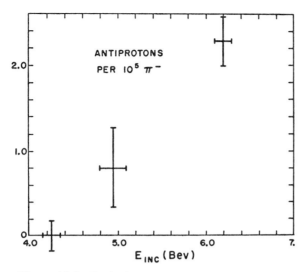

Figure 17.8. Excitation curve from the production of antiprotons relative to meson production as a function of Bevatron beam energy [*Source*: Chamberlain et al. (1955)[8]].

up histograms of the time of flight, as measured from the film, we found the mesons at 40 nsec, the antiprotons near 52 nsec, and the accidental coincidences making a general smear.

We wanted also to show that we were dealing with something that had a reasonable requirement of energy at production. So we studied the rate of these antiproton signals as we changed the energy of the Bevatron. As far as we could tell at that time, it was more or less zero at 4 GeV; it was visible at 5 GeV and gave full production at 6 GeV. Full production was one antiproton counter every 15 min. Figure 17.8 shows the energy dependence of the counting rate for antiprotons.

When we first turned this detection system on, as I remember it, we got a not-very-clear signal. We were not as sure as we should have been that we were turned up correctly. (Here we are at one of those points where our memories differ on just what happened.) I remember Wenzel coming by and saying, "Don't forget you can reverse all the magnets and see protons come through your setup." Anyway, when we did check our system with protons, we were further off than we hoped we were, and it helped a great deal to tune up the apparatus on protons. Then we knew we were most sensitive to particles of protonic mass, and our rate of counting antiproton candidates increased appreciably.

In the further testing of the way the detector operated, we studied the response of the system to the mass value to which the system was tuned. Different mass values could be chosen by varying the momentum to which the system was tuned, while keeping all velocity-selecting components un-

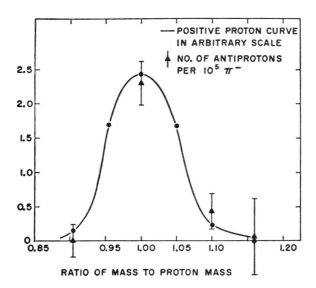

Figure 17.9. Counting rates as a function of the mass to which the system was tuned. The solid curve is for protons, obtained with all bending magnets and quadrupole triplets operated with reversed currents. Also shown are the experimental points obtained with antiprotons [*Source*: Chamberlain et al. (1955)[8]].

changed. Thus, we kept all counter timing constant and kept our velocity-selecting Cerenkov counter at a constant setting while varying the magnetic fields in M1, M2, Q1, and Q2. Being uncertain how efficiently the Bevatron proton beam was hitting the internal copper target, we chose to normalize our counting rates for antiprotons to a standard number of pions coming through the magnet system.

Figure 17.9 shows the results. The solid curve is proportional to the response to protons (magnets having reversed currents). The experimental points obtained with antiprotons (magnet of normal polarity) are as shown. The excellent fit between the points and the curve gave confidence that the counts we were getting were indeed due to particles whose mass was very close to the proton mass.

Soon after we had announced that we had observed antiprotons, this event was found in emulsion, from the better irradiation at lower momentum (Figure 17.10). It was found in Berkeley by A. G. Ekspong. It had the magic property that the observable energy was well above 938 MeV, so that besides providing the energy of a proton-mass particle coming in, we must have annihilated a preexisting proton-mass particle in the emulsion stack. That is the one that put us over the top. We said this tied the knot on the whole thing.

Looking back on the antiproton discovery, one sees that quite a number

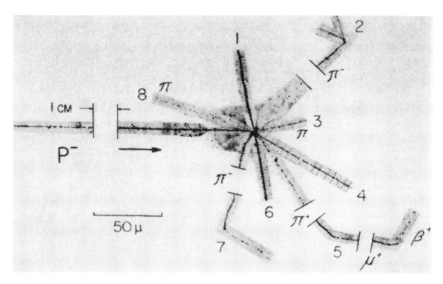

Figure 17.10. Reproduction of the antiproton star that showed visible energy well above 938 MeV. This showed that annihilation was involved [*Source*: Chamberlain et al. (1956)[10]].

of physicists were looking for antiprotons; so it is clear that if our experiment had failed, someone else would soon have found them. If the development and scanning of emulsions had not been so time-consuming, the discovery might have come from emulsion work. Later work shows that at least one-third of the antiproton stars exhibit visible energy convincingly above 938 MeV; so the emulsion method was quite powerful.[9] Our method was sound in that it had effectively very low background. However, it was not a high-intensity experiment. Soon after the discovery, antiproton beams of much greater intensity were designed and built.

In assessing the impact of the discovery on physics, I would say it was certainly no surprise. Most theorists predicted that the antiproton was there to be found when conditions were right. Still, the discovery cleared the air: It allowed people to proceed more confidently into a rewarding future.

Notes

1 P. A. M. Dirac, "A Theory of Electrons and Protons," *Proc. R. Soc. London 126* (1930), 360–5.
2 C. D. Anderson, "The Positive Electron," *Phys. Rev. 43* (1933), 491–4.
3 Evans Hayward, "Ionization of High Energy Cosmic-Ray Electrons," *Phys. Rev. 72* (1947), 937–42.
4 H. S. Bridge, H. Courant, H. De Staebler, Jr., and B. Rossi, "Possible Example of the Annihilation of a Heavy Particle," *Phys. Rev. 95* (1954), 1101–3.

5 E. Amaldi, C. Castagnoli, G. Cortini, C. Franzinetti, and A. Manfredini, "Unusual Event Produced by Cosmic Rays," *Nuovo Cimento (Sec. 10) 1* (1955), 492–500.

6 D. H. Stork, R. W. Birge, R. P. Haddock, L. T. Kerth, J. Sandweiss, and M. N. Whitehead, "Search for Antiprotons," *Phys. Rev. 105* (1957), 729–30.

7 J. M. Brabant, B. Cork, N. Horowitz, B. J. Moyer, J. J. Murray, R. Wallace, and W. A. Wenzel, "Terminal Observations on Antiprotons," *Phys. Rev. 101* (1956), 498–501; "Interactions of Antiprotons in Lead Glass," *Phys. Rev. 102* (1956), 1622–5.

8 O. Chamberlain, E. Segrè, C. Wiegand, and T. Ypsilantis, "Observations of Antiprotons," *Phys. Rev. 100* (1955), 947–50.

9 W. H. Barkas, R. W. Birge, W. W. Chupp, A. G. Ekspong, G. Goldhaber, S. Goldhaber, H. H. Heckman, D. H. Perkins, J. Sandweiss, E. Segrè, F. M. Smith, D. H. Stork, L. van Rossum, E. Amaldi, G. Baroni, C. Castagnoli, C. Franzinetti, and A. Manfredini, "Antiproton–Nucleon Annihilation Process (Antiproton Collaboration Experiment)," *Phys. Rev. 105* (1957), 1037–58.

10 O. Chamberlain, W. W. Chupp, A. G. Ekspong, G. Goldhaber, S. Goldhaber, E. J. Lofgren, E. Segrè, C. Wiegand, E. Amaldi, G. Baroni, C. Castagnoli, C. Franzinetti, and A. Manfredini, "Example of an Antiproton–Nucleon Annihilation," *Phys. Rev. 102* (1956), 921–3.

18   On the antiproton discovery

ORESTE PICCIONI

Born 1915, Siena, Italy; *laurea*, University of Rome, 1938; elementary
particle physics; University of California at San Diego

My plans at Brookhaven

Cosmic-ray physicists had been dreaming of observing antiprotons
long before the construction of the Bevatron. In fact, some events that could
be interpreted as revealing antiprotons had already been observed in cloud
chambers and in photographic emulsions, but the indications were not compelling. Being a cosmic-ray physicist, I shared that "antimatter rush" when
I was at Brookhaven working with the Cosmotron. At the Brookhaven cafeteria, discussing antiprotons was the order of the day. Hartland Snyder and
Maurice Goldhaber bet $400 on whether or not the antiproton existed. Gold-
haber lost.

In contrast to this, physicists from Berkeley had hardly any interest in antiprotons. They considered the wish to discover the antiprotons unoriginal, if
not naïve. After all, the energy of the Bevatron was the bare minimum for the
production of antiprotons, and no one could be confident about it. Moreover,
the hot subjects at the time were the strange particles (copiously produced by
the Bevatron), which had opened a field of physics monopolizing the atten-
tion of the theorists, with the K^0, $\bar{K}^0$, K_1, and K_2 quartet, and the experiments
that enabled T. D. Lee and C. N. Yang to perceive the parity violation. Even
as late as December 1954, there was so little interest in the antiproton that
at the American Physical Society (APS) meeting in Berkeley no one presented a paper speculating on its discovery. However, I thought that even a
slight chance of finding antimatter justified a great effort.

The issue for me was the purely experimental question: Which instrumental
setup was best? Because the production cross section was not known, I tried
to invent a method of detection that would make possible the observation of a
minimal number of antiprotons amidst an expected large number of negative

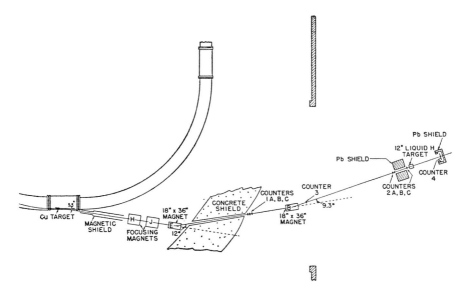

Figure 18.1. The BNL experiment on total cross section of pions, where the time-of-flight method was used to distinguish pions from protons [*Source*: Cool et al. (1956)[1]].

pions. Because antiprotons would be produced almost at rest in the center of mass of the incoming and the target nucleon, the best angle of production was clearly zero, and a simple computation showed the best momentum to be about 2 GeV/*c*. For instrumental reasons, a somewhat lower value was advisable.

I was working at the Cosmotron with Rodney Cool, Leon Madansky,[1] James Cronin, and Al Abashian, and we were using counters (Figure 18.1) and the method of "time of flight" to select a clear sample of positive pions in a beam with a greatly prevalent component of protons. That job was similar to fishing out a few antiprotons from a crowd of negative pions.

The distance between the counters that determined the time of flight of the particles should have been as large as possible, but it was limited by the wall of the building. So I had the idea of making a hole in that wall through which the beam of particles could pass. George Collins, the director of the Cosmotron, laughed and approved. It worked well, which is relevant to the antiproton project, because both the limitation of distance and the possibility of drilling a hole in the wall of the building existed also at the Berkeley Bevatron.

Another feature that proved to be very useful at Brookhaven National Laboratory (BNL) was the repeated determination of the time of flight by independent counters. The idea was based on the recognition that the error in

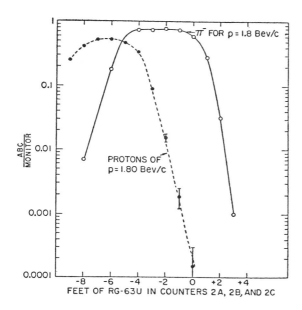

Figure 18.2. The discrimination obtained with time-of-flight measurement[3] in the experiment of Figure 18.1 [*Source*: Cool et al. (1956)[1]].

measuring the velocity was virtually always produced by a fluctuation in the scintillation counters themselves, not by variations in the path lengths of the particles, which could hardly have been remedied.

With larger distance between the counters and the repetition of the time measurement with three pairs of counters, we achieved the discrimination represented in Figure 18.2. It was not only more than adequate for our purpose at Brookhaven but also indicated that the method could be pushed much further to fish for a few antiprotons at Berkeley. Figure 18.3, taken during the antineutron experiment, which I later did at Lawrence Berkeley Laboratory (LBL) together with Bruce Cork, Glen Lambertson, and William Wenzel,[2] shows the benefit of repeated determinations of the time of flight.

Finally, to have a long distance to the extreme counters without losing quality and intensity of the beam, I conceived of the use of a magnetic double-focusing system. A first set of lenses would make the image of the target on an intermediary focus, and a second set would make the image of that focus on the final counter of the system. The solid angle accepting the particle was thus as large as it would be for a beam half as long. Moreover, a collimator placed at the first image would conveniently determine the maximum momentum difference among the particles. Such a scheme had never been proposed at Brookhaven, which was at the time the only laboratory where strong-focusing magnets had been used on experimental beams. It became almost standard

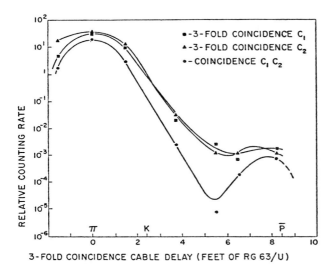

Figure 18.3. The selection of antiprotons with the method of time of flight, without any Cerenkov counters, obtained in the antineutron experiment [*Source*: Cork et al. (1957)[2]].

procedure in subsequent years. It was easy to see that the scheme was advantageous in the ideal case of a target of no dimensions and with particles of exactly the same momentum. However, once one took into account the size of the images and the difference in the momenta of the particles, it was no longer so obvious that the system was worth the greater expenditure in money and fabrication time.

It was very tedious to compute by hand, with the help of drawings, the trajectories and magnifications, but eventually I convinced myself that the experiment should make use of that principle, which would also improve the quality of the beam from other points of view. I received encouragement from Ernest Courant, one of the inventors of the strong-focusing principle.

I was very satisfied with my plan and had confidence that, contrary to the popular opinion, neither the cloud chambers nor the photographic emulsions were the most promising tools to make the first observation of antimatter. I thought that the common opinion was based on the naïve preference for seeing a "picture." I remembered the success that we had had in Rome when we obtained clear proof of the muon decay with the method of counting events versus time delays. That had proved to be a stable and efficient tool, allowing us to discover the leptonic nature of the muons. Similarly, in Berkeley, the quadrupoles, the counters, and the time of flight allowed us not only to discover the antiproton but also to proceed with the observation of antineutrons, which would have been impossible by other methods. The points of my plan that I did not work out were that I needed funds to go to Berkeley for a long visit, that I probably needed security clearance (I was not a U.S. citizen), and that several physicists would be needed for the project.

Visiting LBL

A meeting in Berkeley had been organized by the APS for the last week in December 1954. I submitted a paper to describe the external beam of the Cosmotron and another idea of mine that had led to work done with my Brookhaven collaborators, Cool, Alan Thorndike, and others. The idea was to use large quadrupole magnets to focus a thousand high-energy protons in a very narrow beam so that when the beam pulse crossed a cloud chamber it would allow clear observation of the secondaries even in the vicinity of the interactions. The probability of producing an interaction in the gas of the cloud chamber was appreciable with a thousand protons. The individual proton that interacted could not be distinguished, but that did not matter, because all protons had almost the same momentum and direction. I was proud that Enrico Fermi had told me he liked "the idea of making use of the large intensity of the accelerator to select a beam of very few particles in a very narrow range of momenta."

During the APS meetings, Luis W. Alvarez was very friendly to me. He was chairing the session where I gave the paper, and as a compliment he let me chair half of the session. Later, having lunch with him and Arthur Rosenfeld, I talked to him about the idea of searching for the antiprotons with counters, but he did not think that was the way to do the search. In fact, the discovery of antiprotons was not at all his top priority, just as it was not in the mind of any experimental or theoretical physicist to whom I talked.

Emilio Segrè and Owen Chamberlain were deeply rooted in work at the synchrocyclotron studying the scattering of protons in their various spin states. That subject did not appear exciting to me. Moreover, neither Segrè, Chamberlain, Clyde Wiegand, nor Thomas Ypsilantis had ever done anything with the Bevatron, nor had they taken any part in its design or construction. I concluded that even if I wanted to involve Segrè in the antiproton experiment, which I had no intention of doing, he would not be interested. However, one morning I told him that I had a plan for an important experiment with the Bevatron, and I was looking for a group working with that accelerator to start a collaboration on my plan. Segrè's answer was remarkable: first, because he did not even ask if the experiment was about the antiproton; second, because he answered with the words (which I remember very clearly), "Describe the experiment to us and if we like it we will do it together." Though surprised, I felt that Segrè's proposal represented a viable solution. I knew that Chamberlain was responsible for the success of Segrè's group at the cyclotron, and I liked the prospect of working together with him and Wiegand.

I started talking about the experiment, but Segrè followed my descriptions with difficulty and pretty soon wanted to have Chamberlain at his side. The three of us had several meetings in Segrè's office in LeConte Hall on the campus and at the LBL. Wiegand was sometimes present, and at least once Ypsilantis (then a student) was also around.

I described the experiment in much detail. I was talking at the blackboard, and they were sitting. Chamberlain was writing in notebooks. Segrè was sleeping part of the time. I talked about the criterion of constructing an apparatus that could analyze the largest number of pions, that there was no point in asking the theorists to compute the rate of production, which could not be reliable. I emphasized that in a few months the Bevatron could be expected to be ready for the project and that no time should be wasted in waiting for a larger intensity, because the apparatus could not be flooded with too many pions. Then I talked at length about the discrimination obtainable by repeating the time-of-flight measurement and the need for having a large distance between the counters, and I emphasized the advantage of having a double set of focusing lenses with an intermediate focus, which would provide a more intense and better beam, with particles of the same momentum and with little background from scattering in the walls of the lenses. Such a beam would also be advantageous if we wanted to add a Cerenkov counter.

Segrè objected that the focusing magnets "would cost as much as $5,000 each." I replied that they would cost more than that, but a search for the anti-proton was well worth the price. I also tried to convince Segrè and Chamberlain that the experiment of the antiproton was much more interesting than their study of proton scattering in various spin states. If successful, I emphasized, it would give a much greater reward than deserved, and greater than any other experiment. Notably, I did not think of a Nobel Prize, because I thought that the effort in building the Bevatron dwarfed that of doing a single experiment.

I talked at length of the fortunate circumstance that, due more to pre-conceptions than to any profound analysis, everyone who had thought even vaguely of an antiproton experiment felt that surely, without a picture of the annihilation, the experiment could not be meaningful. To make my points, I stayed several days after the APS meeting, postponing my return to Long Island until after the end of the year. There were other meetings in which Wiegand participated, because I wanted to start an approximate design of the focusing lenses. Despite the well-known competence of the Berkeley engineers, I did not want to turn that job over to them completely, without any guidelines. I sat at their drawing table and drew a full-scale cross section of the quadrupole lenses I had computed at Brookhaven, taking the Cosmotron flux of pions as a reference, and the flux expected at Berkeley within a given solid angle of acceptance of the magnetic lenses. I decided that the aperture of our magnetic lenses should be four inches, three times less than the twelve inches of the BNL lenses that I used. It was a decision made with good judgment, and it took into account the time to build the magnets. If I had chosen twelve or even eight inches, it would have taken so much more time that we would have lost jurisdiction of the experiment. When I talked about the diameter, no comment at all came from Chamberlain or Wiegand, who were not familiar with the Cosmotron's fluxes and obviously had given no previous thought to the point. I was unquestionably the leading spirit of the group at those meetings. In describing the magnets to the engineers, Wiegand or Chamberlain split one of the two quadrupoles into two pieces, which increased the labor and cost, but did not much improve the quality of the devices.

I expected the double-lens spectrometer to be too long to be accommodated inside the building of the Bevatron, so I insisted on drilling a hole in the building, emphasizing the advantage of the climate at Berkeley. I even indicated approximately where the hole should be made, and the hole was made there, several months afterward.

After hours and hours of talking, I rested my case. Chamberlain raised no questions. He did *not* say "I have also given some thoughts to a project like this," as he claimed at this symposium, thirty years later. At the end of my presentation, on our way from his office to his house, Segrè told me in Italian: "l'esperimento e' ben pensato" (the experiment has been well thought out). He was not a person prone to paying compliments.

I finally came to the point of finding financing for my visit to Berkeley. I said that I would write a letter to Ernest Lawrence to describe the plans and ask for financing. Segrè said no, because Lawrence would take the funds from his own group. Knowing of the adequate funds of the Berkeley laboratory, I was surprised. "What should I do?" I asked. "Go to Alvarez, who likes you, and ask him to invite you to Berkeley and pay for your expenses." "All right," I said, "then I'll describe to him what we want to do." "Certainly not," said Segrè, adding that the idea and the plans for this project should be kept secret because people with power could steal them. I turned to Chamberlain for support, but he said, "I am afraid Emilio is right." So I went to Alvarez, to ask him to pay for my visit to do an experiment that I should not describe to him.

I had another difficulty that probably would have disappeared if I could have described my plans in a letter to Lawrence. Someone in Italy had, by error or by malice, informed the FBI that I was a member of the Italian Communist Party, and my U.S. citizenship was being withheld. This fact was expected to make it difficult for me to obtain a clearance for the Berkeley laboratory. Thus, a letter from Lawrence or Segrè or Alvarez describing the importance of the experiment might have helped solve that problem. However, solving that problem was not necessarily in the best interest of Chamberlain and Segrè, because they knew that after I had described all my ideas, they could proceed to perform the experiment, which presented less complexity than an ordinary high-energy experiment. Books on patents and copyrights are full of examples illustrating the difficulty of protecting ideas.

My "default" by returning to Brookhaven

After returning to Brookhaven, I continued to work on the plans for the antiproton project. I was moved by the urgency of the project and tried to think of ways to simplify the construction of the strong-focusing magnets. To study the magnets, I was using a method I had learned from Hildred Blewett. It consisted of modeling the cross section of a magnet with carbon paper and injecting into it electric currents that simulated the real currents in the coil of the magnet. Playing with that paper suggested to me that two separate parallel plates, with coils around each, had a sizable region in the gap between them where the magnetic field had a constant gradient, which is the essential property of a lens. Unfortunately, the magnet was wasteful of electric power, and when I proposed it to Wiegand, he and Chamberlain correctly objected on that score.

I talked about the idea to Wolfgang Panofsky at the Rochester meeting. He liked it, and later on he made an important improvement that decreased the power consumption. He constructed one example of such a "rectangular quadrupole" and published its description in the *Review of Scientific Instruments*, acknowledging my idea.

I have recalled this episode because during a telephone conversation in 1959, Chamberlain said that I "defaulted" by not being present at Berkeley. Though the plates were never constructed, that study shows that I was doing all I could for the antiproton, considering the difficulties of a three-thousand-mile distance and not having citizenship nor the funds nor the approval to move to Berkeley.

Moreover, Segrè on two occasions communicated to me that there was no urgency that I go to Berkeley. At Rochester (March 1955), I told him I felt frustrated that Alvarez had not found funds for my visit and asked him to write to Lawrence. I also expressed my feeling to Gilberto Bernardini, a good friend of mine and Segrè's. Bernardini told Segrè that if he could not invite me to Berkeley, in all fairness he should not contemplate doing the antiproton experiment. Segrè then told me, "I will not write to Lawrence about you and the antiproton project, but if you prefer we will wash our hands of that project." With the hope that fairness would eventually prevail with Segrè, and above all with my enthusiasm about my plan, I could not accept the offer of abandoning the experiment. Segrè also said, "I think this experiment should proceed not too fast, not too slow." The message to me clearly was that no great progress had been made in the preparations by Segrè's group; thus, I could not feel "in default" for not being in Berkeley. Actually, the preparation for the experiment, as I learned later, was well under way.

Shortly before the Washington meeting, Alvarez phoned and with his warm style told me he had obtained funds for my visit. In Washington, I saw Segrè again at a party given by Ugo Fano. In the presence of several friends (Fano, Giancarlo Wick, Giuseppe Cocconi, and others), he repeated to me: "You are crazy if you hope that Alvarez will let you do that experiment and discover the antiproton." Thus, as late as May 1955, Segrè still fully recognized that the idea and the initiative of the experiment were mine. At the same time, he again implicitly communicated to me the clear message that the preparation was proceeding very slowly.

With hindsight, I should not have taken Segrè's word for granted, but I strongly felt that the use of my plans gave me the unquestionable right to be part of the experiment, and I trusted him. We had in common our nationality and the school of Enrico Fermi. I should also note that at no time did Chamberlain, who claimed that I "defaulted," contact me by telephone or by letter to advise me that my absence from Berkeley could be interpreted as a "default."

The ups and downs of the acknowledgments to me

After a month, and without telling me, Segrè relaxed his rule of secrecy in order to have the magnets built. His first communication was in the quarterly LBL report, January 1955 (UCRL 2920). Not wanting to arouse jealousies or competition, he gave it the cryptic, nondescriptive title of "Bevatron," instead of antiprotons. It said (emphasis mine): "We are preparing an experiment to detect *negative protons*, if they are produced in the Bevatron. *Parts* of this project were *discussed* with O. Piccioni of Brookhaven National Laboratory." It had been a peculiar discussion during which Segrè, Chamberlain, Wiegand, and Ypsilantis kept silent and took notes. I never received a copy of that report.

The succession of the quarterly LBL reports shows that Segrè's property rights to the antiproton beam resulted not only from the quality of my plans (particularly the good compromise of four inches in diameter) but also from my urging to propose the quadrupole immediately in January 1955. In the mentioned November 1954–January 1955 report, the progress in *thirty-two* experimental physics projects was reported, but the "Bevatron" of Segrè was the only one about antiprotons. It communicated the construction of the quadrupoles and thus established his title to them.

In the next report, February–April 1955 (UCRL 3014), *twenty-eight* projects are mentioned. Again, none on antiprotons. Only in the May–July 1955 report (UCRL 3115) did the antiproton plans appear: one by the excellent Lofgren group (Wenzel, Cork), another, with a cloud chamber, by the Powell group, and another with photographic plates by the Richman group. By then, however, Segrè's ownership of the beam was well established. It is clear that *even if* Chamberlain had given some thought to the antiproton experiment, he certainly would have needed a couple of months to make concrete proposals, and that would have been enough for him to lose the front seat.

Despite a feeling of urgency, I could arrive in Berkeley with my family only late in July. I first saw Ypsilantis, who told me that the magnets and the counters were assembled, but the execution of the experiment had not started. I went to see Segrè to talk about the experiment. He said they were working

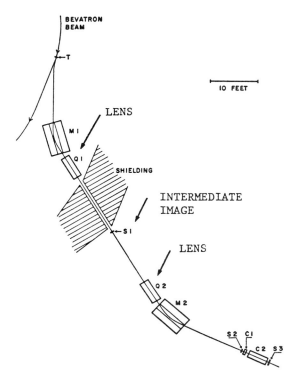

Figure 18.4. Segrè's double-image setup (constructed according to my plans) [*Source*: Chamberlain et al. (1955)[4]].

at it. I asked if the work was along the lines of my plans. Incredibly, he simply said he did not remember the plans I had described to him seven months before. I reminded him that they had taken notes. He casually grabbed a notebook, shaking his head as if he did not believe me, and started reading the first page. He read: "Piccioni describes the antiproton experiment etc." He did not even laugh. Also hard to believe, Wiegand, who was present, did not say a word on my behalf. Chamberlain did not try to get in touch with me. Both chose to share in the onus of Segrè in order to share in the benefits. In 1959, Segrè and Chamberlain received the Nobel Prize for the discovery of the antiproton, with specific recognition "for the ingenious method" (Figure 18.4).

Two important facts must be noted at this point. At the Berkeley meeting in 1985 for the celebration of the discovery of the antiproton, Wiegand, Robert Birge, and Ypsilantis opposed my description on the grounds that some very little quadrupoles, by far inadequate for the antiprotons, had been proposed by someone shortly after my visit to Berkeley. The occasion of the meeting with Segrè and Wiegand, described earlier, would have been a wonderful opportunity to tell me that the quadrupoles of the antiprotons stemmed from those small devices. Obviously, Segrè did not know about them even at the time. The Nobel speeches also prove my point, and so does an interview with Wiegand and Ypsilantis published in *Physics Today*,[3] as well as the transcription of a conversation I had with Chamberlain in 1959. Of course, I never claimed having invented the principle of quadrupoles.

The second fact, widely known around the Berkeley laboratory and told to me by Wiegand, is that none of the books containing the notes of the plans of the experiments were seen after I filed a lawsuit in 1972. This is a most eloquent fact, as Segrè is known to take meticulous care of the notes of his experiments. It is hard to see why the books leading to his Nobel Prize have been so poorly taken care of. They could show that I am wrong and that their plans had nothing to do with me.

Chamberlain and Wiegand added to my plans a well-built Cerenkov counter, similar to one built at Brookhaven by Sam Lindenbaum. However, its usefulness was debatable because it reduced the rate of antiprotons. Our group, which observed the antineutron shortly after the antiproton experiment, obtained an efficiency about ten times larger than Segrè's.

In 1972, after seeing that Segrè and Chamberlain did not give me any recognition, I consulted Melville Nimmer of UCLA, a foremost authority on copyright, and he advised me to initiate a lawsuit, which eventually was heard by the Court of Appeals. The only point upheld by the court was the statute of limitations. The case was never heard on its merits.

Notes

1 R. L. Cool, L. Madansky, and O. Piccioni, "Total Cross Section of Pions at 1.5 BeV," *Phys. Rev. 93* (1954), 249–50; R. L. Cool, L. Madansky, and O. Piccioni, "Total Interaction Cross Section of Pions with Protons and Deuterons at 1.0 BeV," *Phys. Rev. 93* (1954), 637–8; R. L. Cool, L. Madansky, and O. Piccioni, "Total Interaction Cross Section of Negative Pions at Kinetic Energies of 1.0 and 1.45 BeV," *Phys. Rev. 93* (1954), 918; R. Cool, O. Piccioni, and D. Clark, "Pion–Proton Total Cross Sections from 0.45 to 1.9 BeV," *Phys. Rev. 103* (1956), 1082–97.

2 B. Cork, G. R. Lambertson, O. Piccioni, and W. A. Wenzel, "Cross Sections for Antiprotons in Hydrogen, Beryllium, Carbon and Lead," *Phys. Rev. 107* (1957), 248–56.

3 Gloria Lubkin, "Piccioni Sues for Share of Antiproton Credit," *Phys. Today 25 : 9* (1972), 69–71.

4 O. Chamberlain, E. Segrè, C. Wiegand, and T. Ypsilantis, "Observations of Antiprotons," *Phys. Rev. 100* (1955), 947–50.

PART V

THE STRANGE PARTICLES

There's no clarity or regularity
Such profusion's mere vulgarity,
We're guided by faith, hope and parity –
   Some people don't know where to stop.
   – © Arthur Roberts', "Some People
   Don't Know Where to Stop" (1952)

19 The hydrogen bubble chamber and the strange resonances

LUIS W. ALVAREZ

Born 1911, San Francisco, California; Ph.D., University of Chicago, 1936; experimental physics; Nobel Prize, 1968, for decisive contributions to elementary particle physics, in particular the discovery of a large number of resonance states, made possible through his development of the technique of using hydrogen bubble chambers and its data analysis; Lawrence Berkeley Laboratory [deceased]

I have been out of high-energy physics for some twenty years, and to get myself back into the mood of a particle physicist, I would like to quote some recent remarks by Carlo Rubbia:

> Detectors are really the way to express yourself. To say somehow what you have in your guts. In the case of painters, it's painting. In the case of sculptors, it's sculpture. In the case of experimental physicists, it's detectors. The detector is the image of the guy who designed it.

I've never heard it expressed so well, but I'd like to add that particle physics has always been done by a triad of equally important professionals: accelerator builders, experimental physicists, and theoretical physicists. I'll have some comments at the end on how I hope the members of the triad will interact in the future. I have been a member of the first two categories, but never of the third.

My ten years in the bubble-chamber trenches (also discussed by Peter Galison in Chapter 14), the most exciting period in my life, started at the 1953 Washington meeting of the American Physical Society (APS), when I met Donald Glaser.[1] He showed me his first cosmic-ray tracks in a tiny bubble chamber (1 × 2 cm) filled with ether. I had been unsuccessfully racking my brain to find an appropriate detector for the Bevatron, which was about to be turned on. It was immediately clear to me that Glaser's chamber filled the bill exactly – if it could be made to work with liquid hydrogen, and if it would operate in large enough sizes. I wanted one big enough to see the production and decay of the strange particles that had first been observed in cosmic rays by George Rochester and Clifford Butler, in a Wilson cloud chamber (see Chapter 4 by Rochester), and had recently been seen by Ralph Shutt's group at the Brookhaven Cosmotron, in a hydrogen diffusion cloud chamber (see Chapter 21 by Chinowsky and Chapter 22 by Fowler). The properties of these chambers were well suited to the discovery of the particles and of their production mode, respectively, but not for systematic studies of their properties. I should add also that on the theoretical side, Abraham Pais had predicted the phenomenon of associated production, and Murray Gell-Mann had invented the strangeness rules that tied the few available experimental facts together and that predicted many of the reactions we would later observe, a good fraction of them for the first time (see Chapter 23, by Pais and Chapter 47 by Gell-Mann).

As soon as I returned to Berkeley, my colleagues, Lynn Stevenson and Frank Crawford, started to repeat Glaser's experiments, with the explicit aim of seeing tracks in liquid hydrogen. They fired up two technicians in the synchrotron shop where we all were working, and these two men, John Wood and A. J. (Pete) Schwemin, collaborated with us in building the first hydrogen chamber to show tracks. Wood sent his letter to the editor of *Physical Review*, with pictures of his first tracks. I ghost-wrote the letter, since John had never published anything before. John's pictures showed an unexpected effect that was the key to the successful operation of large bubble chambers. One could see bubbles forming at the glass walls, while sharp tracks were forming in the central region. This was contrary to Glaser's feeling that bubble chambers had to have such clean, smooth walls that bubbles wouldn't form there, but only on the tracks. As soon as I pointed out the importance of John's discovery, Schwemin, together with Doug Parmentier, started to build a two-inch-diameter metal chamber with gasketed glass windows – the first purposely "dirty" bubble chamber. They had it working very quickly, and at Schwemin's request I ghost-wrote their article for the *Review of Scientific Instruments*. Schwemin and Parmentier then built a four-inch-diameter chamber, which was the first bubble chamber of any kind to be fitted with a magnetic field, and in which we saw the first V particles during a short exposure to a negative-pion beam at the Bevatron.

We now felt we were on the right track, and we enlisted the help of Dick Blumberg, a mechanical engineer, to design a ten-inch-diameter hydrogen chamber to fit in the well of a wonderful magnet that Wilson Powell very kindly let us use. Wilson had two nearly identical magnets for the beautiful cloud chambers he built and used, and he simply let us have one on an indefinite loan. The ten-inch chamber was our first to be designed; the previous ones had been fashioned on a lathe by Schwemin, who would say to himself, "The flange should be about this wide, and it should have a groove about here, to take a solder wire gasket that I'll make to fit it."

We spent a lot of time becoming familiar with Gell-Mann's strangeness rules, and I decided (after all, a group leader has to do something) that we would do our first experiment with stopping K^- mesons in hydrogen. From the theoretical and experimental standpoints, it appeared to be a potential gold mine, and from the sociological standpoint, it was also a real winner.

Everyone else waited in line for high-energy negative pions, kaons, or anti-protons that came out of the one useful straight section of the Bevatron. But we were able to use a "private" target that could be flipped up in a curved section of the Bevatron and that sent its sharply curved low-momentum pions and kaons between the outside iron return yokes and into a very crude mass spectrometer. This separator consisted of a thin absorber that subtracted away almost all the momentum of the kaons, and much less of that of the pions. The cloud-chamber magnet then bent the negative kaons into the active volume of hydrogen, where a reasonable fraction of them came to rest. No one had ever before seen K^- particles stopping in hydrogen; so we had the pleasure of seeing the copious production of all the hyperons with strangeness equal to -1: $\Lambda, \Sigma^+, \Sigma^0, \Sigma^-$. We very accurately measured the masses and the lifetimes of all these particles. We saw Σ^- hyperons interact in the hydrogen. (Anyone who wants to experience the impact of this experiment on the particle physics community should read the enthusiastic summary in the *Supplement to Nuovo Cimento*, Vol. 2, 1957, pp. 773–5, with three photographs.)

We were fortunate that the separation efficiency of our crude K^--beam system was so poor that it let in large numbers of negative pions, as well as some negative muons. That permitted us to be the first to see the now well-known muon-catalyzed fusion reactions. We thought at first that we had discovered a new particle that decayed into a muon, but it was soon apparent that the negative muon was forming a tightly bound $p–d–\mu$ molecular ion, in which the p and d quickly fused to make ^{3}He plus energy, which internally converted to eject the muon, so the process could be repeated. (We found two cases in which successive fusions were catalyzed by the same muon.)

We were surprised to see the reaction happen so very often in our "pure" hydrogen, with only one part deuterium per 5,000 ordinary hydrogen atoms. The reason was that the reduced mass effect had the proper sign to make muons captured on protons become neutronlike objects that quickly transfer muons to deuterons at their first collision. The $\mu–d$ system then quickly captured another proton to form the molecular ion. We thought for an exciting hour that we might have solved the energy problems of the world by going to very low temperatures, where conventional wisdom said one had to go to many millions of degrees. Although we were quickly disabused of that notion, a recent resurgence in interest in catalyzed fusion has centered on experiments with $d–t$ fusion, and several groups have found surprisingly high yields of 14-MeV neutrons per negative muon. Steve Jones, at Los Alamos, finds an average number of catalyzed reactions of the order of 150 per muon. Because each reaction yields 17 MeV, the energy released is thus about 2.5 GeV, which is within a factor of about 10 of what it takes to produce a muon. So we were originally overoptimistic, but not by such a large factor as we've thought for the past twenty-nine years.

Before the ten-inch chamber was operated, I became convinced that if we were ever to do the kind of strange-particle physics that liquid-hydrogen chambers should permit, we'd need a very large chamber. My first guess was fifty inches by twenty inches by twenty inches, but I soon realized that because of the magnetic field, the particles would fan out more in the horizontal direction than the vertical. So we could exchange some unneeded depth for extra length, and the chamber became the well-known seventy-two-inch chamber; later, when it was moved to Stanford, it became the eighty-two-inch chamber.

We now needed a special appropriation from the Atomic Energy Commission (AEC), and after thinking about it for a few hours, the commissioners voted us 2.5 million of the 1955 variety, or 10 million in 1985 terms. It couldn't have taken them very long, because I first briefed them one morning, and that same evening at a cocktail party John von Neumann told me that he and the four other commissioners had given my proposal their stamp of approval. They didn't bother to ask for any peer review – that dismal procedure hadn't yet been invented! I didn't use much of the time in my presentation to remind them that the largest operating liquid-hydrogen bubble chamber anywhere in the world was our four-inch device.

Ernest Lawrence, from whom I learned how to make such a large extrapolation, thought I was sticking my neck out a bit too far, and one of my greatest disappointments is that he died a few months before the seventy-two-inch chamber showed its first tracks. But I did have the pleasure of giving him some escorted tours of the bubble chamber and its new building as the construction proceeded. Someone asked me why it took longer to build the building than it did to design the bubble chamber and make it operational. My reply was that people had been putting up buildings for thousands of years, so there were long shelves of regulations that had to be met, but bubble chambers were too new to be so encumbered. In fact, the laboratory's safety department, which one might have thought would get involved in such a potentially dangerous project, left us totally alone as we did our own tests on hydrogen safety.

The seventy-two-inch chamber was a major engineering effort, and we assembled a very strong design team under the leadership of Paul Hernandez and an equally strong operational group under the direction of my closest associate, J. Donald Gow. We decided to test several novel features of the seventy-two-inch design in a smaller fifteen-inch chamber. One new feature was the single-window design, which increased the safety and, more important, the strength of the magnetic field. We went full speed ahead with the single-window design, even though we didn't know how to illuminate the bubbles and photograph them through the same window until shortly before the fifteen-inch chamber became operational. The seventy-two-inch chamber worked for the first time on 24 March 1959, and it had a long and very useful life.

I am pleased that we decided, early in our bubble-chamber program, to share all our technical information with anyone interested in hydrogen chambers. As physicists, we thought of ourselves as competitors, trying to do the best experiments before our friends in other parts of the world could get around to them. But as engineers, we considered ourselves as "members of a club" and custodians of a lot of government-funded development work; so we sent copies of our voluminous unpublished "engineering notes" and "physics notes" to everyone else in the club. Very quickly, all of our potential physics competitors knew everything that we did about how to build chambers and how to use our rapidly increasing volume of software to analyze the bubble-chamber pictures. The leaders in this important phase of our work were Frank Solmitz and Art Rosenfeld.

In my 1955 proposal to the AEC for money to build the big chamber, I pointed out that unless we could greatly increase our ability to analyze bubble-chamber film – compared with cloud-chamber film – the big chamber would simply be a very expensive toy that would produce enough interesting events in a single day to keep all of the world's cloud-chamber experts busy for a year. Cloud-chamber events usually were "solved" by reprojecting the two stereo views of each track onto tiltable and rotatable "space tables" until the two images coincided everywhere. Then the orientation of the tracks in space could be read from angular scales, and the curvature of the tracks could be measured using sets of circles with varying diameters. It was very time-consuming – one might solve two events per day, but it fitted the production rate in the low-density gas. But in going to liquid density, plus very long path lengths, the event rate would rise by about three orders of magnitude.

I proposed that we use what later became known as "Franckensteins" (for their designer, Jack Franck), which would quickly measure the track coordinates on the film itself, in a semiautomatic track-following mode. These coordinates would then be subjected to computer analysis, which would give us the coordinates in real space. That was a very successful program, involving several generations of more and more automatic devices, culminating in the "spiral readers," of which we had two, each capable of measuring very nearly 1 million events per year. Our group's scanning and measuring department eventually employed about a hundred people, most of whom were undergraduate students working part time. From the earliest days, we always measured more events per year than any other group, and we (almost) always had the largest hydrogen bubble chamber, from our first one, in 1953, until the Brookhaven eight-inch chamber became operational in 1964. (The exception was a period of a few months when Jack Steinberger's twelve-inch chamber came on shortly before our fifteen-inch chamber was operational.)

I shall now take off my detector designer's hat and exchange it for my user's hat. The first experiment we did with the fifteen-inch chamber was designed as a test of the Gell-Mann–Nishijima strangeness rules that predicted the existence of a neutral cascade particle, the Ξ^0. We set out to measure the mass of a new neutral particle, the Ξ^0, which should decay into two other neutral particles, the K^0 and the Λ. (Victor Weisskopf had recently given a humorous Rochester conference banquet talk, the high point of which had him showing a perfectly blank cloud-chamber picture and claiming that it showed the existence of a new neutral particle decaying into two other neutral particles. So that was the unlikely task we set for ourselves.) We found one excellent event that let us measure the mass of the Ξ^0, thereby proving it existed.

But the most important result of that first fifteen-inch-bubble-chamber exposure to a medium-energy well-separated K^- beam was the discovery of the first three "strange resonances" that set off the population explosion of what were for a time called fundamental particles, but which couldn't hold that title for long, in view of their rapidly increasing number. We have all been brought up with Eddington's concept of an "experimental fishnet." A fisherman who throws out a net will catch no fish with dimensions less than those of his mesh. But our discovery of the strange resonances violated Eddington's rule by an enormous factor. We had designed our chambers with a fishnet to match the decay lengths of the strange particles, in the range from millimeters to perhaps 20 cm. But the most important "fish" we caught had decay lengths shorter by factors of about 10^{12} – just the factor by which the lifetimes of the strange particles had been increased over typical nuclear times to make physicists call their behavior "strange."

What made the discovery possible was, of course, our extensive set of computer software that came from the Solmitz–Rosenfeld collaboration and the many talented associates they had recruited to work with them. And it also took the perseverance of two dedicated graduate students, Stan Wojcicki and Bill Graziano. We all know that Jocelyn Bell discovered the pulsars for which her graduate advisor, Antony Hewish, was subsequently honored. In the resonance business, Stan and Bill were my equivalents of Bell, and I am pleased to acknowledge their discovery. Bill is now doing other things, but Stan is well known to all of you as a leader in the plan to build the Superconducting Supercollider (SSC). Stan and Bill accidentally discovered the now standard method of finding new particles by looking for bumps in invariant mass plots. (For many years, I thought of myself as a "professional bump-hunter," and I've found that that is still a pretty good job description, now that I'm working in geology and paleontology.)

The discoveries of the first three strange resonances were published by a group of seven of us, known collectively as (Margaret) "Alston et al." The first resonance is now known as the Σ 1385, the second as the Λ 1405, and the third as the K^* 892. Bogdan Maglich soon found the ω meson in a seventy-two-inch exposure, using the bump-hunting technique; Harold Ticho led a group at UCLA that used seventy-two-inch film to find the Ξ 1530, and Aihud Pevsner led a group at Johns Hopkins that found the η meson in seventy-two-inch film. By giving our precious film to other laboratories, we were following the example set by Lawrence. (The first of the four "missing elements," technetium, was discovered in Palermo, Sicily, by Emilio Segrè and C. Perrier in a molybdenum deflector strip that Lawrence sent them from the twenty-eight-inch cyclotron that had been bombarded by 6-MeV deuterons for two years.) Bump-hunting soon became a popular activity, and physicists with access to other bubble-chamber film reported the discovery of other resonances, or particles as they now are called; the most important of these was the ϱ meson, discovered by William Walker at Wisconsin.

Until the population explosion started in the fifteen-inch chamber, there was only one resonance known in particle physics, and that was Enrico Fermi's famous "3-3 resonance" in the pion–nucleon system. Although I went to all the Rochester conferences in this period, I never heard anyone call the 3-3 resonance a particle; it was always thought of simply as a resonance or bump in a production cross-section curve as a function of energy. But it was clear that the objects we found as bumps in mass plots were really particles; they stayed together long enough for other particles to recoil against them, and then they came apart in times of the order of 10^{-22} sec, as one could measure from the energy widths of the bumps, using the uncertainty principle. It was soon apparent that the 3-3 resonance was the first of the "new particles," and Walter Barkas and Rosenfeld started a new cottage industry to keep everyone abreast of the best values of masses, lifetimes, spins, and so forth, for *all* the particles that gave our profession its name. If the proton is found to decay, the lifetime range will span sixty decades!

It wouldn't be fair to say that as soon as all the new particles were found, the theorists came into the picture, and explained their taxonomy – the theorists had various frameworks to codify the particles known before the population explosion started, the most notable being the "eightfold way" of Gell-Mann and Yuval Ne'eman. They extended these ideas to embrace the newly discovered very short-lived particles. Their most famous prediction was the existence and mass of the Ω^- hyperon. That prediction came from the equality of the mass spacings of the 3-3 resonance (now known as the Δ), the Σ 1385, and the Σ 1530. The Bevatron didn't have enough energy to make Ω^- hyperons, which was a big disappointment to my group; we had the right detector, but the wrong accelerator. So we had to wait a few years until the eighty-inch chamber came into operation, when we sent our congratulations to the Brookhaven group. But we did have the satisfaction of knowing that the important equality in the mass spacings came out of measurements made in our hydrogen chambers, plus, of course, Fermi and Anderson's old mass value for the Δ. So we could feel, in the language of an official baseball scorer, that we had "an assist" in the important discovery of the Ω^-; Gell-Mann used our data to tell the Brookhaven group where to look for the Ω^-.

I would like to close by distinguishing between two classes of discoveries. If I had been born a few hundred years ago, I probably would have been an explorer. My heroes in the world of exploration are James Cook and Roald Amundsen (and Amundsen's unlucky rival, Robert Scott). They made great geographical discoveries that are correctly acclaimed by everyone. But just think how different they were; Amundsen found the South Pole, whose existence could be questioned only by members of the flat-earth society, whereas

Cook found the Hawaiian Islands, whose existence was a surprise to everyone. One can't decide which discovery was more praiseworthy; we need both kinds. I have described our satisfaction in finding the predicted Ξ^0 and in watching our friends at Brookhaven find the Ω^-. We all have enormous admiration for the discovery of the W's and the Z, which are in the same category, but much harder. But I wonder if in the future anyone will be able to find something completely unexpected, such as the J/Ψ, or the strange resonances. As I look ahead in particle physics, I see support only for the enormously expensive detectors to find particles whose existence has been predicted by theorists. I think that is a very unhealthy situation, and I hope that those who are pressing for construction of the SSC will turn some of their attention to this dilemma; if one is allowed to look only for things that are predicted from earlier knowledge, both experimental and theoretical, how can we use our powerful new accelerator to find something *really* new, such as the examples I've just mentioned, both in geography and in particle physics?

Notes

1 I refer the reader to my Nobel lecture: Luis W. Alvarez, "Recent Developments in Particle Physics," *Science 165* (1969), 1071–91, which covers the same ground as the present chapter, but in more detail, and contains full referencing to the developments.

20   A particular view of particle physics in the fifties

JACK STEINBERGER

Born 1921, Bad Kissingen, Germany; Ph.D., University of Chicago, 1948;
high-energy particle physics; Nobel Prize, 1988, for the discovery of the
neutrino beam method and the demonstration of the doublet structure of the
leptons through the discovery of the muon neutrino; European Center for
Nuclear Research, Geneva

The neutrino, neutron, positron, and muon were discovered in the
1930s. The pion and strange particles had their turn in the late forties. During
the fifties, which saw the change from dependence on natural sources to
accelerators in the study of particles, the particles and their properties were
sorted out, the strange particles were systematized, the first particle resonances were seen, and parity was found to be violated, with a concomitant clarification of the weak interaction. It was an exciting time. The field advanced
rapidly. I was young and played my part in this adventure.

Cosmic rays and muon decay

For me, particle physics began in 1947, when I was a graduate student
at the University of Chicago. Enrico Fermi gave a seminar on the results of
the Conversi, Pancini, and Piccioni experiment in which he explained that
because the negative mesotrons rapidly find their way to the atomic K shell,
the observed long lifetimes were incompatible with the role of these mesons
as Yukawa particles.[1,2] It was a beautiful and important experiment, and
Fermi's explanation was extraordinarily lucid, as well as stimulating and
exciting. The puzzle was at least in part cleared up a few months later when
the Bristol group discovered the pion in nuclear emulsions.[3]

It was a great privilege to be a student in that department at that time.
There were excellent teachers: Fermi, Edward Teller, Willie Zachariasen,
and Maria Mayer, to name a few. Fermi's courses, in particular, were models
of transparent and simple organization of the most important concepts. He
went to great length to show those of us who had finished the courses and
were working on our Ph.D. theses how to attack a variety of simple, general

problems in different branches of physics, by gathering us together one or two evenings a week (I don't remember just how often), proposing a problem, and then, perhaps later, going through the solution.

My fellow students also had a great impact on my formation. They were a remarkable group: Chen Ning Yang, Tsung Dao Lee, Marvin L. Goldberger, Marshal Rosenbluth, Lincoln Wolfenstein, Owen Chamberlain, Geoffrey Chew, Richard Garwin, and others. We often worked together on some problem, perhaps posed by Fermi. The most impressive student-teacher in that group was Yang, who had come from China after the war and at the age of twenty-four, despite the limited wartime possibilities in China, knew all of modern physics fluently by the time he entered graduate school in Chicago.

My thesis topic has an interesting history. Fermi asked me to look into a peculiar result on stopping cosmic-ray muons, obtained by Matthew Sands, a student of Bruno Rossi.[4] The observed rate was several times less than expected on the basis of the known cosmic-ray muon flux. Part of the discrepancy could be understood in terms of an overestimate of the acceptance of the apparatus, but still a factor of two was missing. Perhaps the energy of the decay electrons was not half of the muon rest energy as the authors had assumed. At the time, I was an aspiring theoretician, but Fermi suggested that I measure the energy of those electrons for my Ph.D. thesis. That was the fall of 1947.

By the spring of 1948, the apparatus was ready. It registered the delayed coincidence between an incident meson that had stopped in an absorber and the outgoing decay electron, as a function of the thickness of a second absorber. The experiment used eighty Geiger tubes and was at the time a comparatively large experiment. The results showed the average decay energy to be clearly less than the 50 MeV that would be expected for a decay into an electron and a massless neutrino.[5] Improved statistics gained by mounting the experiment in a truck and exposing it at the top of Mt. Evans, Colorado, in the summer of 1948 showed the spectrum to be continuous and therefore showed that the muon decays into an electron and two other light, neutral, undetected particles, perhaps neutrinos.[6] That was the third process, after β decay and muon capture, to be describable as a four-fermion interaction. Giovanni Puppi and J. Tiomno and J. A. Wheeler discovered that the same coupling constant would account for all three processes. This was the birth of the idea of a universal weak interaction.[7,8] The experimental result was quickly verified by Robert B. Leighton, Carl D. Anderson, and A. J. Seriff,[9] and by E. P. Hincks and Bruno Pontecorvo.[10]

Accelerators and the properties of pions

After a postdoctoral year at the Institute for Advanced Study, I came to Berkeley in the fall of 1949 at the invitation of Professor Gian Carlo Wick to be his assistant. Once there, I was attracted by the possibilities of experimentation at the Radiation Laboratory, and Wick very generously left me free to do this. Artificial mesons had only recently been seen at the 184-inch cyclotron.[11]

My first effort was to try to trigger Wilson Powell's magnetic cloud chamber on stopping pions, to measure the muon-decay β spectrum more precisely. I remember the kindness and hospitality with which I was accepted in Powell's lab, and also the fact that I had trouble with leaks in the little liquid scintillation counter inside the cloud chamber in which the pions were supposed to stop and decay.

However, the project never got off the ground. Somehow I was diverted, probably by Edwin McMillan, to looking for mesons at the 330-MeV electron-synchrotron. McMillan was codiscoverer with V. I. Veksler of the synchrotron phase-stability principle. The accelerator, based on this principle, was just finished and waiting for customers. There was surprisingly little interest in this new machine; the 335-MeV cyclotron had attracted most of the physicists.

The external beam of the synchrotron was a bremsstrahlung spectrum of photons; the electrons themselves could not be extracted. The detector consisted of three anthracene scintillation counters, each about two inches square and half an inch thick, with a variable thickness of absorber in between them, which determined the meson energy. The mesons that stopped in an absorber following the second counter were detected by means of a delayed coincidence in the third counter. It was an early use of scintillators, and I am much indebted to Robert Hofstadter for initiating me into this new technique.

The mounting of this experiment did not take more than a few weeks. I remember soldering up the circuits myself in a few nights. The experiment enjoyed the enthusiastic support of the synchrotron operators and, above all, McMillan, who was always at hand with advice and encouragement. Only positive mesons could be detected. The hydrogen process $\gamma + p \to n + \pi^+$ was measured by comparing productions in polyethylene and carbon. The angular distribution (Figure 20.1), the first detailed measurements for the production of mesons in any process, gave some initial insight into the meson–nucleon coupling.[12] It was clearly in disagreement with the electric dipole distribution expected for scalar coupling, but compatible with the flat distribution expected for pseudoscalar coupling. The same apparatus, equipped with an oscilloscope for the measurement of the time delay between the pion and muon, produced the first precise pion lifetime,[13] $2.60 \pm 0.12 \times 10^{-8}$ sec (the presently accepted lifetime is $2.603 \pm 0.002 \times 10^{-8}$ sec).

In the winter of that year, R. Bjorklund, W. E. Crandall, B. J. Moyer, and H. F. York observed the photon spectrum from an internal target at the cyclotron and found the energy and angular dependences to be very different from the expectations of any currently known process, but in striking agreement with what would be expected if neutral mesons with mass similar to the $\pi^\pm$ mesons were produced and then decayed into photons.[14] The technique that I had been using at the synchrotron could easily be adapted to look for the decay of a neutral meson into two photons. Observation of that decay

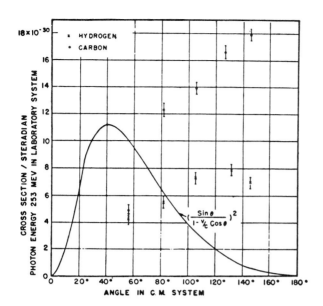

Figure 20.1. Angular distribution for photoproduction of π^+ meson by a 330-MeV bremsstrahlung beam. The upper set of points are on carbon, the lower on hydrogen. The results are in clear contradiction to the dipole production expected for scalar mesons (continuous curve), but compatible with those for pseudoscalar mesons. [*Source*: Steinberger and Bishop (1950)[12]].

would constitute proof of the existence of the neutral meson. In this search I was joined by Pief Panofsky, much more experienced than I, and J. Steller, a graduate student.[15] A second counter telescope was added, and $\frac{1}{4}$-inch-thick pieces of lead between the first two counters served as a converter, so that two γ's could be detected and the angle between them roughly measured. The detector is shown in Figure 20.2. After a few days of effort to reduce the background, a signal was observed, and the angular correlation (Figure 20.3) gave clear evidence of the two-photon decay of a neutral meson of similar mass (produced with similar energies) as the charged pions previously observed. [May I recall here that when we first looked for, and finally found, real γ-ray coincidences, the two telescopes were at 180°, i.e., in a straight line, perpendicular to the beam. We then arranged the angle between the telescopes to be 90°, and the signal was expected to go away. Lo and behold (Figure 20.3), it was ten times bigger. It took a full day for us to understand the reason, the velocity of the produced meson. Of course, it was Panofsky who was the first to see this.]

It was an exciting year of experimentation. Particle physics was just opening up, and a great deal could be learned with simple means, provided, of course, that one was at Berkeley, the only place in the world where, at that

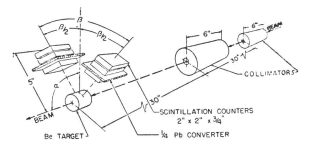

Figure 20.2. Apparatus used in the experiment that demonstrated the existence of the neutral pion decaying to two photons [*Source*: Steinberger et al. (1950)[15]].

time, such experiments were possible. It came to an abrupt end on 30 June 1950, when I was asked peremptorily to "hand in my badge" that same day. AEC clearance and a badge were required then to work in the Berkeley Radiation Laboratory. I had not signed the loyalty oath that the regents of the university had imposed on the staff, and this displeased the laboratory director, Ernest O. Lawrence. Wick was similarly treated at that time.

One of the consequences of the explosion of nuclear bombs at Hiroshima and Nagasaki in 1945 was a substantial expansion, immediately after the war, of university physics research, in particular the construction of five cyclotron laboratories all comparable in energy to the Berkeley machine. These were at the Harvard, Rochester, Columbia, and Chicago universities and at the Carnegie Institute in Pittsburgh, in order of increasing energy. From assistant in Berkeley I became assistant professor at Columbia in the fall of 1950. The 380-MeV cyclotron had already begun to function. The laboratory director was Eugene Booth. James Rainwater was responsible for the most difficult part of the cyclotron, the radio-frequency system. Gilberto Bernardini was guru-in-residence. Among the graduate students was Leon Lederman. Columbia's big strength was the molecular-beam laboratory of Isidor Isaac Rabi, with Willis Lamb, Polykarp Kusch (who was also department chairman), and Charles Townes. Chien Shung Wu, who had already demonstrated the correctness of the Fermi β spectrum and in a few years would discover parity violation, was installed in the nuclear laboratory.

The cyclotron was located twenty miles north along the Hudson River, on the beautiful old Nevis estate, built by Alexander Hamilton for his son, and bequeathed to Columbia University by the DuPont family. At the time I arrived, the laboratory had a unique facility, an external meson beam. The Chicago and Carnegie cyclotrons were not yet in operation; I have forgotten why external meson beams were not possible in Berkeley. The mesons were produced on an internal target, and momentum selected by the cyclotron magnet fringing field and a slot in the shielding. When I arrived, the cyclotron had been running for a year, but only emulsion experiments were in progress.

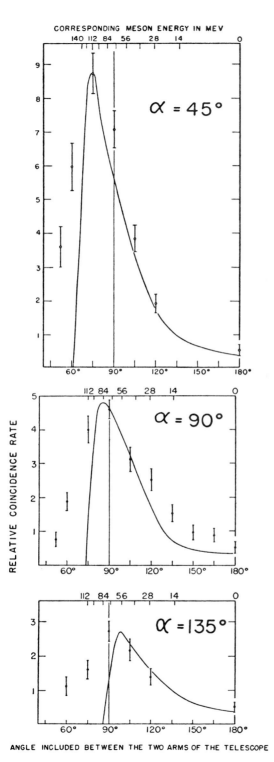

Figure 20.3 Angular correlation of the two photons in π^0 decay [*Source*: Steinberger et al. (1950)[15]].

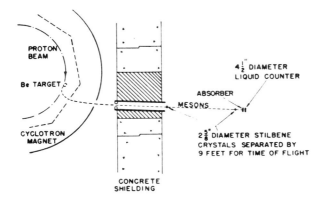

Figure 20.4. Experimental layout for the first determination of pion total cross sections [*Source*: Chedester et al. (1951)[16]].

The electronic detection techniques with which I was familiar offered new possibilities to investigate the properties of the new mesons.

In the next years we exploited those beams to determine the spins and parities of charged and neutral pions and to study the nuclear interaction of the charged pions. The first of the experiments measured the total cross sections of π^- mesons in several nuclei by a simple absorption technique.[16] The layout is shown in Figure 20.4. Another very early experiment was the measurement of the differential cross section of the pion absorption process $\pi^+ + d \rightarrow p + p$.[17] Comparison with the inverse process allowed the definitive assignment of zero spin to the pion on the basis of detailed balance.[18,19] The apparatus is reproduced in Figure 20.5.

The next series of experiments involved the use of liquid hydrogen and deuterium targets and the manufacture of liquid hydrogen. The construction of those targets was a challenge, not only in their design, because we had no engineering help, but also in their construction in the Columbia Physics Department shop. In these early years, Alan Sachs and I made the liquid hydrogen ourselves, after reactivating on campus a primitive and somewhat dangerous prewar hydrogen liquefier that could produce about 1.5 liters per hour by compressing the hydrogen and expanding it in a Joule–Thompson valve. This liquid, maybe six or seven liters at a time, we then drove (surely illegally) in our own car up the West Side Highway to Nevis.

In the study of the differential scattering of pions in hydrogen we were overtaken by the beautiful work of Anderson, Fermi, and collaborators.[20] Our own work was also quite beautiful (if I can say so), but it was restricted to lower energy by the more limited Nevis cyclotron energy.[21] Figure 20.6 shows our results at 60 MeV. The crowning achievement of these scattering experiments, the discovery of the $\frac{3}{2}$-$\frac{3}{2}$ resonance at ~180 MeV, was due to J. Ashkin

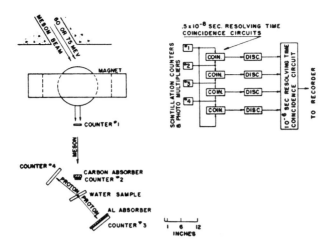

Figure 20.5. Experimental layout used in the determination of the differential cross section for the absorption of positive pions in deuterium [*Source*: Durbin et al. (1951)[17]]. Together with the measurements on the inverse process,[19] this determined the pion spin to be zero.

and collaborators at the Carnegie cyclotron.[22] That was possible because the energy of that machine was high enough to permit passage through resonance, as shown in Figure 20.7. Figure 20.8 shows the differential cross sections near resonance, given in a later paper.[23]

Another experiment characteristic of our work on the properties of pions was the observation of the capture reaction of π^- in deuterium: $\pi^- + d \rightarrow n + n$.[24] The reaction had been surmised on the basis of the lower yield of γ rays in deuterium capture relative to that in hydrogen, but we set out to observe the reaction itself.[25] The same target was used, but filled with deuterium. The two neutrons were detected, and their angular correlation measured (Figure 20.9). It could be convincingly argued that the capture must proceed from the *s* state and that therefore the π^- could only be pseudoscalar.[26]

Strange particles

The discovery in 1947 of the "*V* particles" by George Rochester and Clifford Butler did not immediately stir up a great deal of activity in the States, although in Europe important cosmic-ray results, obtained in emulsions and cloud chambers, continued to appear.[27] I remember in 1949, on a bulletin board at the Institute for Advanced Study in Princeton, a beautiful photomicrograph of a τ^+ meson decaying into three pions, obtained in an emulsion in C. F. Powell's lab. We all saw it, but did not react. It was too difficult to accept the notion of entirely new types of particles. The field began

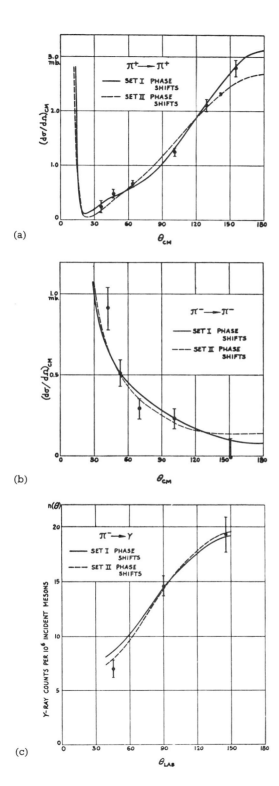

Figure 20.6. Differential cross sections observed at 60 MeV by Bodansky and associates: (a) $\pi^+ + p \to \pi^+ + p$, (b) $\pi^- + p \to \pi^- + p$, and (c) $\pi^- + p \to \pi^0 + n$, $\pi^0 \to \gamma$ [*Source*: Bodansky et al. (1954)[21]].

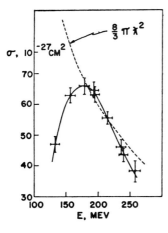

Figure 20.7 Excitation function of Ashkin and associates showing first hadron resonance $(\frac{3}{2}, \frac{3}{2})$ [*Source*: Ashkin et al. 1954)[22]].

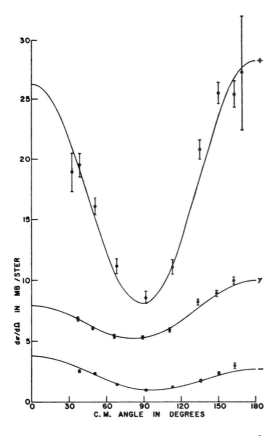

Figure 20.8. Angular distributions at the $\frac{3}{2} - \frac{3}{2}$ resonance of Ashkin and associates [*Source*: Ashkin et al. (1956)[23]].

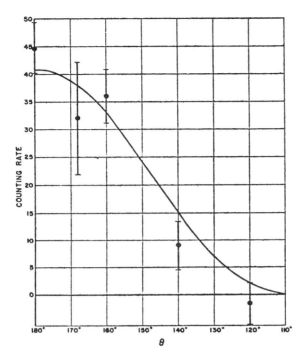

Figure 20.9. Angular correlation of the two neutrons in the capture reaction $\pi^- + d \rightarrow n + n$ used to argue the relative parity of the charged pion and nucleon [*Source*: Panofsky et al. (1951)[24]].

to come alive in America only after the publication in 1950 by Leighton and Anderson of a larger number (thirty-four) of cloud-chamber events.[28]

Because many new particles were being observed, the early experimental situation was most confused. I would like to recall here an incident at the 1952 Rochester conference, in which the puzzle of the neutral V's was instantly clarified. It was the session on the neutral V particles. Anderson was in the chair, but J. Robert Oppenheimer was dominant. He called on his old friends, Leighton from Caltech and W. B. Fretter from Berkeley, to present their results, but no one was much the wiser after that. Some in the audience, clearly better informed than I was, asked to hear from Robert W. Thompson from Indiana, but Oppenheimer did not know Thompson, and the call went unheeded. Finally there was an undeniable insistence by the audience, and reluctantly the lanky young midwesterner was called on. He started slowly and deliberately to describe his cloud chamber, which in fact was especially designed to have less convection than previous chambers, an improvement crucial to the quality of the measurements and the importance of the results. Oppenheimer was impatient with these details, and sallied forth from his

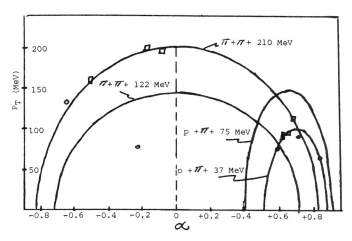

Figure 20.10. Distribution in the P_T–α plot of twelve V^0's $[\alpha(p_+^2 - p_-^2)]$ shown by Thompson at the 1952 Rochester conference, demonstrating that there are two different V^0 particles, the Λ^0 and θ^0.

corner to tell this unknown that we were not interested in details, that he should get on to the results. But Thompson was magnificently imperturbable: "Do you want to hear what I have to say, or not?" The audience wanted to hear, and he continued as if the great master had never been there. A few minutes later, Oppenheimer could again no longer restrain himself, and tried again, with the same effect. The young man went on, exhibited a dozen well-measured V^0's, and, with a beautiful and original analysis, showed that there were two different particles, the $\Lambda^0 \to p + \pi^-$ and the $\theta^0 \to \pi^+ + \pi^-$. The θ^0 (θ for Thompson) is the present K^0. The graph he showed us at that conference is reproduced in Figure 20.10.[29]*

The first accelerator experiment on the "new unstable particles" was performed at the Brookhaven National Laboratory Cosmotron, which could produce external pion beams of up to 1.5 GeV.[30] That experiment demonstrated the very important concept of "associated production," which had only some months before been proposed by Abraham Pais and others to account for the quasi-stability of the new particles. The group had developed the "diffusion chamber," a continuously sensitive cloud chamber in which supersaturated vapor was produced in a thermal gradient. The chamber was filled with 20-atm hydrogen gas and exposed to 1.5-GeV π^- mesons. Five cases of the associated-production reaction $\pi^- + p \to \Lambda^0 + \theta^0$ were found.

* *Ed. note:* For Thompson's own story, see L. M. Brown and L. Hoddeson (eds.), *The Birth of Particle Physics* (Cambridge University Press, 1983), pp. 251–7.

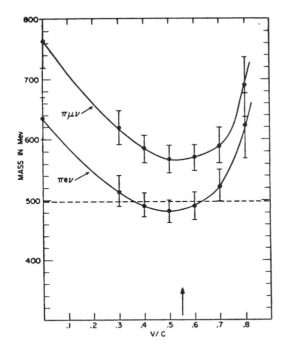

Figure 20.11. Mass of the decaying particle as a function of assumed velocity and decay mode in the experiment of Lande and associates in which the K^0_L was discovered [*Source*: Lande et al. (1956)[31]].

Another beautiful early accelerator experiment on strange particles that I want to mention here is that of Lederman and friends, in which the long-lived neutral kaon was discovered.[31] Murray Gell-Mann and Pais had noticed, in a penetrating analysis, that there should be two K^0's, one the antiparticle of the other, that the observed K^0 was one linear combination, and that the other combination should have a much longer life, comparable to that of the charged K particles.[32] To this end, an old but hardly used thirty-six-inch cloud chamber was refurbished and put 6 m from a target struck by the external proton beam at the Cosmotron, far enough from the target so that the known V^0's could not reach it.[33] It was by no means easy to find the twenty-three V^0 events, all of which had to be three-body decays, since they did not conserve transverse momentum. The observed momenta were consistent with the decay of a particle of 500 MeV mass into pion, electron, and neutrino (Figure 20.11).

My own participation in the study of these particles began in 1954, when three students, Jack Leitner, Nick Samios, and Mel Schwartz, and I started to experiment with the bubble chamber, which had been invented by Don Glaser two years before. We made a 10-cm-diameter propane chamber, with much advice from Glaser himself, but with one nonnegligible improvement: The liquid was recompressed some milliseconds after the expansion, to prevent the bubbles from rising and collecting on top. This made it possible to cycle at a reasonable rate (a few times per second). Our first publication, "Properties of Heavy Unstable Particles Produced by 1.3 GeV π^- Mesons," was the first application of Glaser's invention.[34] The chamber had grown to 15 cm in diameter and is shown in Figure 20.12.

In a two-day exposure, fifty-five associated-production events in hydrogen were obtained, many more than had been obtained in the previous two years by older techniques. Figure 20.13 reproduces one of these events. The paper gives evidence that the spins of Λ^0 and Σ^- hyperons are $\frac{1}{2}$ and broaches the subject of the possible observation of parity violation in hyperon decay, some six months before the demonstration of parity violation by Wu and associates. The relevant distribution is reproduced in Figure 20.14. The up–down asymmetry (the angle ϕ in the figure) is $7 \div 15$, a two-standard-deviation effect, but the authors (correctly) considered the statistics inadequate for such an important conclusion.

These early bubble chambers could be constructed rather quickly. Less than half a year after this first publication, we submitted results from the next chamber, with eight times the volume and in a magnetic field.[35] It described two events in which Σ^0's were produced on free protons, according to the reaction $\pi^- + p \to \Sigma^0 + \theta^0$, where both the Λ^0 and the γ from the Σ^0 decay are observed. One of the events is reproduced in Figure 20.15. These events demonstrated the existence of the Σ^0 and determined its mass to a few million electron volts. The existence of the Σ^0 was one of the first confirmations of SU(3), which had put the Σ^+ and Σ^- into an isospin triplet together with the as yet unobserved Σ^0.

Toward the end of that year, little more than a year after our first bubble-chamber publication, we published results on the angular correlation in Λ production and decay that were conclusive in showing parity violation.[36] There were now ten times the number of events, obtained mostly in the 30-cm propane chamber, but including also some events from our first liquid-hydrogen chamber, also 30 cm in diameter and in the same magnet, and events as well from a Michigan group that included Glaser and Martin Perl. The relevant angular distribution is shown in Figure 20.16. [Note that the up–down angle here called θ was labeled ϕ in the previous publication (Figure 20.14).]

The same exposures also provided the data for an analysis, based on an argument of Robert K. Adair, to determine the spins of the Λ^0 and Σ^- hyperon.[37,38] The argument shows that near threshold, and for forward and backward hyperon production angles, the angular correlation between incident pion and decay pion is a predictable function of the hyperon spin. The

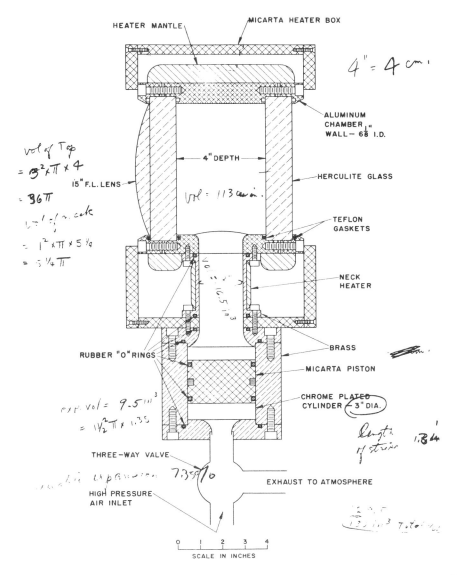

Figure 20.12. The chamber used in the first bubble-chamber experiment.[34] (I am indebted to Bill Fowler for this figure from an extinct Nevis report.)

prediction for spin $\frac{1}{2}$ is a flat distribution, which was found to be the case for both Λ^0 and Σ^- (Figure 20.17).

The twelve-inch hydrogen chamber, which provided some of the data for these last two experiments, was also exposed at the Nevis cyclotron in a

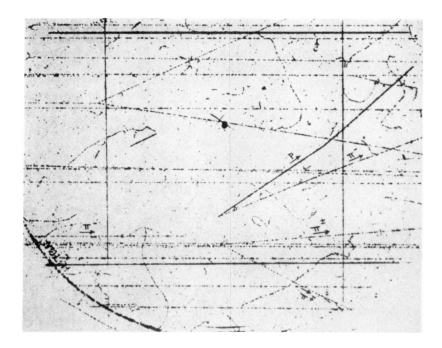

Figure 20.13. Example of the reaction $\pi^- + p \rightarrow \Lambda^0 + \theta^0$, $\theta^0 \rightarrow \pi^+ + \pi^-$, $\Lambda^0 \rightarrow p + \pi^-$, in the first bubble-chamber experiment [*Source*: Budde et al. $(1956)^{34}$].

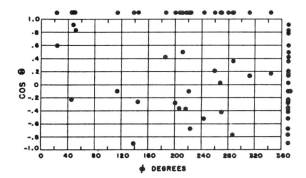

Figure 20.14. Angular correlation in the production and decay of twenty-two Λ^0's seen in the first bubble-chamber experiment. The angle ϕ is the up–down angle. The observed up–down asymmetry of $7 \div 15$ is a two-standard-deviation evidence in favor of parity violation, half a year before its discovery [*Source*: Budde et al. $(1956)^{34}$].

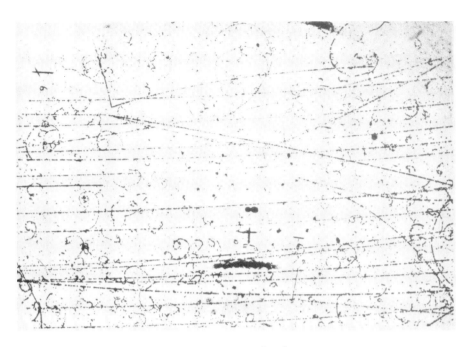

Figure 20.15. First observation of Σ^0, $\Sigma^0 \to \Lambda + \gamma$ [*Source*: Plano et al. (1957)[35]].

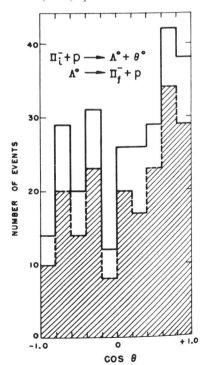

Figure 20.16. Up–down angular distribution in the correlation of production plane and decay angle (the angle called θ here is the same as that called ϕ in Figure 20.14). The parity violation is now statistically firm [*Source*: Eisler et al. (1957)[36]].

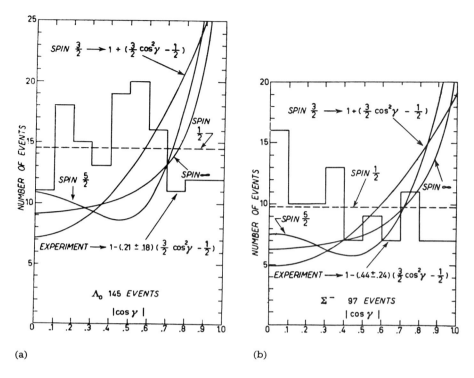

Figure 20.17. Distribution in the angle between the incident and the decay pion in the process $\pi^- + p \to \Lambda^0$ and $\pi^- + p \to \Sigma^- + K^+$ near threshold. The flat distributions observed both for the Λ^0 and the Σ^0 show that the spins of both hyperons are $\frac{1}{2}$ [*Source*: Eisler et al. (1958)[37]].

search for the rare decay $\pi^+ \to e^+ + \nu$ and in order to measure the π^0 parity. A drawing of the chamber in its magnet is shown in Figure 20.18. The pion β decay search was successful; six cases were found, at a rate consistent with the prediction of $\mu-e$ universality.[39] The results were published simultaneously with similar results from a CERN counter experiment.[40] A second experiment searched for the very rare double internal conversion of π^0 decay, $\pi^0 \to 2e^+ + 2e^-$, which was theoretically expected at a rate of $\sim 2\,\alpha^2/\pi$.[41] The angular correlations permit a direct test of the π^0 parity: If two pairs are chosen so that the distribution between the members of a pair is the smaller of the two possible choices, then the distribution in the angle between the two planes so defined is expected to be peaked at $0°$ for scalar mesons and at $90°$ for pseudoscalar mesons. Some 7×10^5 pictures were taken, with approximately ten stopping π^- in each. About half of these were converted to 4-MeV π^0's in the capture. Figure 20.19 shows one of the 103 double-internal-conversion events

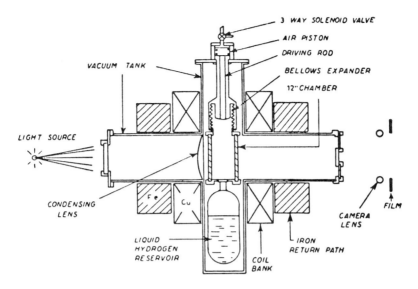

Figure 20.18. Our first hydrogen chamber, twelve inches in diameter, in a 13.4-kG field [*Source*: Eisler et al. (1958)[45]]. The chamber was used for numerous experiments.[36,37,39,41]

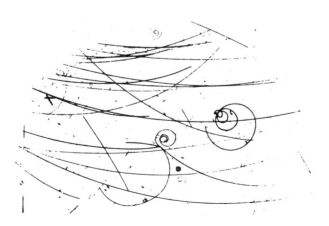

Figure 20.19. A double internal conversion observed in the stopping and capture of π^- mesons in hydrogen: $\pi^\pm + p \to \pi^0 + n$, $\pi^0 \to 2e^+ + 2e^-$ [*Source*: Plano et al. (1959)[41]].

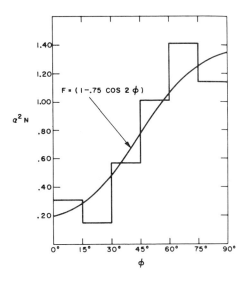

Figure 20.20. Angular correlation of the two decay planes in the double internal conversion of the π^0. The correlation is that expected for pseudoscalar mesons [*Source*: Plano et al. (1959)[41]].

found. The angular correlation (Figure 20.20) shows the pseudoscalar parity of the π^0.

The two-neutrino experiment

In the two preceding sections I have recalled some of the experiments performed at Nevis, or at the Brookhaven National Laboratory. This Columbia–BNL collaboration was very fruitful and contributed a great deal to the output of the Cosmotron and the early alternating-gradient synchrotron (AGS). The crowning achievement of the Columbia–BNL collaboration was probably the two-neutrino experiment of 1961–2.[42]

The experiment followed the suggestion of Pontecorvo and Schwartz that at the higher-energy accelerators then being constructed it should be possible to achieve neutrino "beams" from pion and kaon decay, with sufficient intensity that neutrino reactions could be observed in multiton detectors.[43,44] Such detectors had never been constructed. Luckily, optical multiplate spark chambers had just been invented, and these formed the basis of the detector. The ten modules were arranged in two stories of five. Each consisted of nine one-inch aluminum plates and had a mass of about one ton. The beam was taken at the smallest possible angle, 7.5°, from an internal target at the AGS. It had a decay length of 20 m, and a muon shield of 13.5 m of iron. The detector was protected against cosmic rays by anticoincidence counters and was triggered by means of counter planes between the modules. After some initial skirmishes with the shield, which had to be "neutron-proofed," fifty-six events were obtained in a run of about a month. These could be shown to be neutrino events, contaminated by about 10 percent stopping muons from cosmic rays and not more than 20 percent machine-produced background.

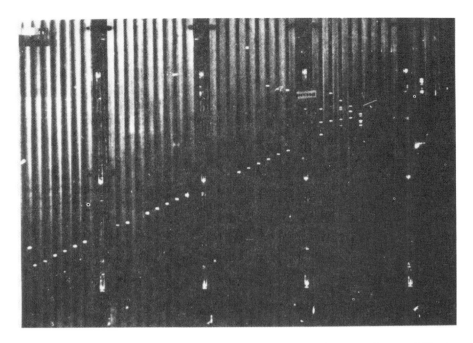

Figure 20.21. Typical event observed in the neutrino experiment of Danby and associates in multiplate spark chambers. It shows a penetrating track, probably a muon, and some hadronic debris [*Source*: Danby et al. (1962)[42]].

Most of the events contained a penetrating track, presumably a muon, such as the event shown in Figure 20.21. Six events were seen with irregular sparks indicating a shower, but these had fewer sparks than would be expected for electrons of the energy of the observed muons, as can be seen from Figure 20.22. This established that the neutrino emitted in the dominant pion and kaon decays is different from the neutrino of β decay.

We had discovered that there are two neutrinos, but we did not discover neutral currents. Why not? The reason may be twofold. On the one hand, and very important, we were not looking for it; at the time, no one imagined that neutral currents existed. On the other hand, the experiment was too marginal, both systematically and statistically. Now we know that neutral currents exist, but the rates are only about one-fifth of the charged-current rates, and the visible energy is only one-half, on the average, since the outgoing neutrino leaves no sparks. The six "shower" events referred to earlier may very well be neutral-current events, but given our uncertainty concerning possible neutron background and neutrino background with short-range muons, no positive conclusion could have been drawn.

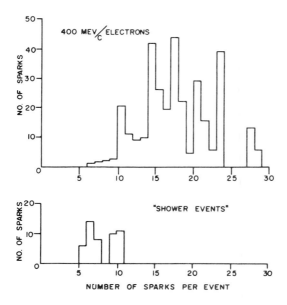

Figure 20.22. Spark distribution of the six muonless events (bottom), as well as the expected distribution (top) obtained in an auxiliary exposure to 400-MeV electrons, normalized to the expected number of showers, assuming equal electron and muon production. The missing electrons are the evidence that the muon neutrino is distinct from the electron neutrino [*Source*: Danby et al. (1962)[42]].

Notes

1 M. Conversi, E. Pancini, and O. Piccioni, "On the Disintegration of Negative Mesons," *Phys. Rev. 71* (1947), 209–10.
2 E. Fermi, E. Teller, and V. Weisskopf, "The Decay of Negative Mesotrons in Matter," *Phys. Rev. 71* (1947), 314–15.
3 C. M. G. Lattes, G. P. S. Occhialini, and C. F. Powell, "Observations on the Tracks of Slow Mesons in Photographic Emulsions. I. Existence of Mesons of Different Mass," *Nature (London) 160* (1947), 453–6; "II. Origin of the Slow Mesons," ibid., 486–92.
4 Bruno Rossi, Matthew Sands, and Robert F. Sard, "Measurement of the Slow Meson Intensity at Several Altitudes," *Phys. Rev. 72* (1947), 120–5.
5 J. Steinberger, "On the Range of the Electrons in Meson Decay," *Phys. Rev. 74* (1948), 500–1.
6 J. Steinberger, "On the Range of the Electrons in Meson Decay," *Phys. Rev. 75* (1949), 1136–43.
7 G. Puppi, "Cosmic Ray Mesons," *Nuovo Cimento 6* (1949), 194–9.
8 J. Tiomno and J. A. Wheeler, "Energy Spectrum of Electrons from Meson Decay," *Rev. Mod. Phys. 21* (1949), 144–52.
9 Robert B. Leighton, Carl D. Anderson, and Aaron J. Seriff, "The Energy Spectrum of the Decay Particles and the Mass and Spin of the Mesotron," *Phys. Rev. 75* (1949), 1432–7.
10 E. P. Hincks and B. Pontecorvo, "On the Disintegration Products of the 2.2-μSec. Meson," *Phys. Rev. 77* (1950), 102–20.

11 E. Gardner and C. M. G. Lattes, "Production of Mesons by the 184-Inch Berkeley Cyclotron," *Science 107* (1948), 270–1.

12 J. Steinberger and A. S. Bishop, "The Detection of Artificially Produced Photo-Mesons with Counters," *Phys. Rev. 78* (1950), 493–4.

13 O. Chamberlain, R. F. Mozley, J. Steinberger, and C. Wiegand, "A Measurement of the Positive π-μ-Decay Lifetime," *Phys. Rev. 79* (1950), 394–5.

14 R. Bjorklund, W. E. Crandall, B. J. Moyer, and H. F. York, "High Energy Photons from Proton–Nucleon Collisions," *Phys. Rev. 77* (1950), 213–18.

15 J. Steinberger, W. K. H. Panofsky, and J. Steller, "Evidence for the Production of Neutral Mesons by Photons," *Phys. Rev. 78* (1950), 802–5.

16 C. Chedester, P. Isaacs, A. Sachs, and J. Steinberger, "Total Cross Sections of π-Mesons on Protons and Several Other Nuclei," *Phys. Rev. 82* (1951), 958–9.

17 R. Durbin, H. Loar, and J. Steinberger, "The Spin of the Pion via the Reaction $\pi^+ + d \rightleftarrows p + p$," *Phys. Rev. 83* (1951), 646–8.

18 Vincent Z. Peterson, "Mesons Produced in Proton-Proton Collisions," *Phys. Rev. 79* (1950), 407–9; W. F. Cartwright, C. Richman, M. H. Whitehead, and H. A. Wilcox, "The Production of π^+ Mesons by Protons on Protons in the Direction of the Beam," *Phys. Rev. 81* (1951), 652–3.

19 R. E. Marshak, "Meson Reactions in Hydrogen and Deuterium," *Phys. Rev. 82* (1951), 313.

20 H. L. Anderson, E. Fermi, R. Martin, and D. E. Nagle, "Angular Distribution of Pions Scattered by Hydrogen," *Phys. Rev. 91* (1953), 155–68.

21 D. Bodansky, A. M. Sachs, and J. Steinberger, "Scattering of 65-MeV Pions in Hydrogen," *Phys. Rev. 93* (1954), 1367–77.

22 J. Ashkin, J. P. Blaser, F. Feiner, J. Gorman, and M. O. Stern, "Total Cross Sections of 135-MeV to 250-MeV Negative Pions in Hydrogen," *Phys. Rev. 93* (1954), 1129–30.

23 J. Ashkin, J. P. Blaser, F. Feiner, and M. O. Stern, "Pion–Proton Scattering at 150 and 170 MeV," *Phys. Rev. 101* (1956), 1149–58.

24 Wolfgang K. H. Panofsky, R. Lee Aamodt, and James Hadley, "The Gamma-Ray Spectrum Resulting from Capture of Negative π-Mesons in Hydrogen and Deuterium," *Phys. Rev. 81* (1951), 565–74.

25 K. Brueckner, R. Serber, and K. Watson, "The Capture of π-Mesons in Deuterium," *Phys. Rev. 81* (1951), 575–8.

26 W. Chinowsky and J. Steinberger, "Absorption of Negative Pions in Deuterium: Parity of the Pion," *Phys. Rev. 95* (1954), 1561–4.

27 G. D. Rochester and C. C. Butler, "Evidence for the Existence of New Unstable Elementary Particles," *Nature (London) 160* (1947), 855–7.

28 A. J. Seriff, R. B. Leighton, C. Hsiao, E. W. Cowan, and C. D. Anderson, "Cloud-Chamber Observations of the New Unstable Cosmic-Ray Particles," *Phys. Rev. 78* (1950), 290–1.

29 R. W. Thompson, A. V. Buskirk, L. R. Etter, C. J. Katzmack, and R. H. Rediker, "The Disintegration of V^0 Particles," *Phys. Rev. 90* (1953), 329.

30 W. B. Fowler, R. P. Shutt, A. M. Thorndike, and W. L. Whittemore, "Production of Heavy Unstable Particles by Negative Pions," *Phys. Rev. 93* (1954), 861–7.

31 K. Lande, E. T. Booth, J. Impeduglia, L. M. Lederman, and W. Chinowsky, "Observation of Long-lived Neutral V-Particles," *Phys. Rev. 103* (1956), 1901–3.

32 M. Gell-Mann and A. Pais, "Behavior of Neutral Particles under Charge Conjugation," *Phys. Rev. 97* (1955), 1387–9.

33 O. Piccioni, D. Clark, R. Cool, G. Friedlander, and D. Kassner, "External Proton Beam of the Cosmotron," *Rev. Sci. Instrum. 26* (1955), 232–3.

34 R. Budde, M. Chretian, J. Leitner, N. P. Samios, M. Schwartz, and J. Steinberger, "Properties of Heavy Unstable Particles Produced by 1.3 BeV π^- Mesons," *Phys. Rev. 103* (1956), 1827–36.

35 H. Plano, N. Samios, M. Schwartz, and J. Steinberger, "Demonstration of the Existence of the Σ^0 Hyperon and a Measurement of Its Mass," *Nuovo Cimento 5* (1957), 216–19.

36 F. Eisler, R. Plano, A. Prodell, N. Samios, M. Schwartz, J. Steinberger, P. Bassi, V. Borelli,

G. Puppi, G. Tanaka, P. Woloschek, V. Zoboli, M. Conversi, P. Franzini, I. Manelli, R. Santangelo, V. Silverstrini, D. A. Glaser, C. Graves, and M. L. Perl, "Demonstration of Parity Nonconservation in Hyperon Decay," *Phys. Rev. 108* (1957), 1353–5.

37 F. Eisler, R. Plano, A. Prodell, N. Samios. M. Schwartz, J. Steinberger, P. Bassi, V. Borelli, G. Puppi, H. Tanaka, P. Woloschek, V. Zaboli, M. Conversi, P. Franzini, I. Manelli, R. Santangelo, V. Silverstrini, G. L. Brown, D. A. Glaser, and C. Graves, "Experimental Determinations of the Λ^0 and Σ^- Spins," *Nuovo Cimento 7* (1958), 222–30.

38 R. K. Adair, "Angular Distribution of Λ^0 and θ^0 Decays," *Phys. Rev. 100* (1955), 1540–1.

39 G. Impeduglia, R. Plano, A. Prodell, N. Samios, M. Schwartz, and J. Steinberger, "β Decay of the Pion," *Phys. Rev. Lett. 1* (1958), 249–51.

40 T. Fazzini, G. Fidecaro, A. W. Merrison, H. Paul, and A. V. Tollestrup, "Electron Decay of the Pion," *Phys. Rev. Lett. 1* (1958), 247–9.

41 R. Plano, A. Prodell, N. Samios, M. Schwartz, and J. Steinberger, "Parity of the Neutral Pion," *Phys. Rev. Lett. 3* (1959), 525–7.

42 G. Danby, J. M. Gaillard, K. Goulianos, L. M. Lederman, N. Mistry, M. Schwartz, and J. Steinberger, "Observation of High-Energy Neutrino Reactions and the Existence of Two Kinds of Neutrinos," *Phys. Rev. Lett. 9* (1962), 36–44.

43 B. Pontecorvo, "Electron and Muon Neutrinos," *Zh. Eksp. Teor. Fiz. [Sov. Phys.–JETP] 37* (1959), 1751–7.

44 M. Schwartz, "Feasibility of Using High-Energy Neutrinos to Study the Weak Interactions," *Phys. Rev. Lett. 4* (1960), 306–7.

45 F. Eisler, R. Plano, A. Prodell, N. Samios. M. Schwartz, J. Steinberger, P. Bassi, V. Borelli, G. Puppi, H. Tanaka, P. Woloschek, V. Zaboli, M. Conversi, P. Franzini, I. Manelli, R. Santangelo, and V. Silverstrini, "Bubble Chamber Study of Unstable Particle Production in $\pi^- - p$ Collisions at 910, 960, 1200 and 1300 MeV," *Nuovo Cimento 10*(3) (1958), 468–89.

21   Strange particles

WILLIAM CHINOWSKY

Born 1929, New York City; Ph.D., Columbia University, 1955; experimental
high-energy physics; University of California, Berkeley

The earliest accelerator results involving a substantial number of
strange particles were obtained at Brookhaven National Laboratory in
Upton, Long Island, New York, at the Cosmotron, beginning in 1954. I was
fortunate to be a part of that work, which I shall now describe.

By the time the 1954 Rochester conference took place, there appeared
to be a number of distinct strange particles. Table 21.1 lists the strange particles as known in that year, my first at Brookhaven. Notice that there is an
entry for neutral τ and for an anomalous θ^0 that "could be a three-body
decay." Both Λ^0 and θ^0 were firmly established, and their properties were
somewhat clearer than they had been in the past. The next few years saw
things become clearer still, largely because of work done at the Brookhaven
Cosmotron.

Figure 21.1 shows the setup of experiments at the Cosmotron in that summer of 1954. There were pion beams for total-cross-section measurements
with counters. The visual detectors were emulsions and cloud chambers; a
few years later they had been displaced by bubble chambers. All those devices
were set up to study strange-particle production cross sections and properties.
Among other things, K^- mesons were put into emulsion. John Hornbostel
and Edward Salant showed that K^- produce hyperons and estimated a geometric value for the interaction mean free path.[1]

In the early 1950s the most important experimental results were those of
William B. Fowler, Ralph Shutt, Alan Thorndike, and William Whittemore
with π^- in a hydrogen diffusion chamber.[2] It was shown that strange particles
are made in associated production, and the conservation of strangeness was
confirmed. That work is discussed elsewhere in this volume (see Chapter 22 by
Fowler), so I shall concentrate on the *really* strange particles, the long-lived

Table 21.1. *Tabulated properties of strange particles shown at the 1954 Rochester conference*

Name	Daughters	Lifetime	Remarks
$\Lambda^0 \rightarrow p + \pi^-$	$+(35 \pm 2)$ MeV	$\sim 3 \times 10^{-10}$ sec	Some evidence also for $Q \sim 20$ and $Q \sim 75$
$\Lambda^+ \rightarrow n + \pi^+$	$+(?)135$ MeV		6–7 cases in emulsion
$[\Lambda^+ \rightarrow p + \pi^0$	Consistent with 135 MeV	$<3 \times 10^{-10}$ sec	Not universally accepted]
$Y^- \rightarrow \Lambda^0 + \pi^-$	$+(65 \pm 12)$ MeV		Explains some cascades (sometimes called Ω^-)
$[Y^0 \rightarrow (?)^- + \pi^+$	$+12$ MeV		$(?)^-$ is approximately of nucleonic mass, but no evidence it is p^-]
K particles			
$\tau^\pm \rightarrow \pi^\pm + \pi^+ + \pi^-$			
$\tau^\pm \rightarrow \pi^\pm + 2\pi^0$	$+74$ MeV	$\sim 10^{-9}$ sec	Only τ^+ has been observed to decay (interpretable as stars)?
$(\tau^0?)$			
$\theta^0 \rightarrow \pi^\pm + (\pi^\mp \text{ or } \mu^\mp)$	$+(214 \pm 5)$ MeV	$\sim 1.5 \times 10^{-10}$ sec	Evidence favors 2nd π over a μ
$?^0 \rightarrow (\pi^\pm \text{ or } \mu^\pm) + (\pi^\mp \text{ or } \mu^\mp)$	$+(41 \pm 5)$ MeV		(Other Q values 50–150); could be a 3-body decay
$[\chi^\pm \rightarrow \pi^\pm + ?^0$	$(P_\pi \sim 200$ MeV/c$)$	$\gtrsim 10^{-9}$ sec	(Existence sometimes questioned)]
$K^\pm \rightarrow \mu^\pm + ?^0 + ?^0$	$(P_\mu$ sometimes exceeds 220 MeV/c$)$		
$K_\mu^\pm \rightarrow \mu^\pm + ?^0$	$(P_\mu = 220 \pm 3$ MeV/c$)$	$\gtrsim 10^{-9}$ sec	May include many cases formerly called K, but not all

Source: Proceedings of the Fourth Annual Rochester Conference on High Energy Nuclear Physics, edited by H. P. Noyes, E. M. Hafner, J. Klarmann, and A. E. Woodruff (University of Rochester, 1954).

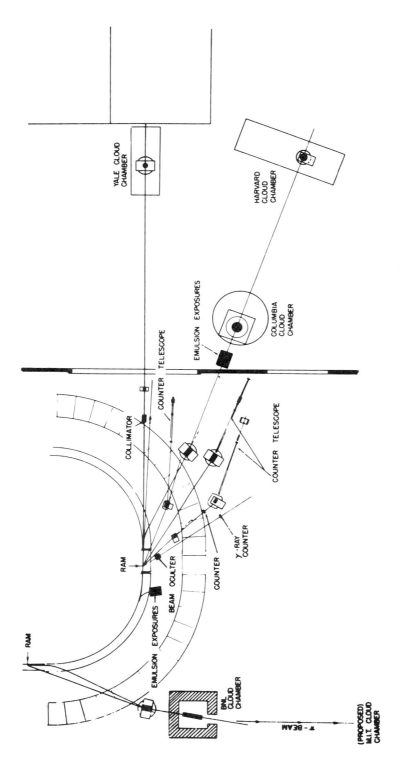

Figure 21.1. Arrangement of experimental apparatus at the Cosmotron in early 1954.

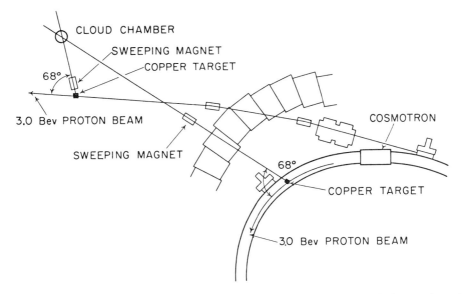

Figure 21.2. Experimental arrangements for production and observation of long-lived neutral strange particles [*Source*: Bardon et al. (1958)[8]].

neutral K, alias θ_2 or K_2 or K_L. (My terminology speaks to all generations.)

In that summer of 1954, I joined up with Leon Lederman, who promised fame and fortune if I would come and help with the Brookhaven piece of the soon-to-be-far-flung Lederman operations. That piece was the study of strange-particle production and decays in more detail than could be obtained from cosmic-ray studies. To the practiced eye, the floor plan of the Cosmotron's experimental area would indicate the outstanding problems in particle physics at that time. As shown in Figure 21.2, Lederman had placed a thirty-six-inch-diameter expansion cloud chamber, in a 10,000 magnetic field, in a 1.9-GeV negative-pion beam. About 100 Λ^0's and 50 θ^0's were produced in carbon and lead plates in the cloud chamber in that exposure. Some nice, if not profound, results came from that study.[3] The Λ^0 and θ^0 lifetimes were measured correctly – the result from cosmic-ray Λ^0 was too long, but only by one standard deviation. That experiment was becoming exhausted in 1955.

Fortunately, in the early part of that year, a remarkable seminar at Brookhaven got the θ_2 business started. Abraham Pais walked in on a meeting at the Cosmotron, and George Collins asked him to give us a talk. "I have nothing to say," said Pais. You know that cannot have been true; Pais always had something to say. This time it was about a "funny" paper that he and Murray Gell-Mann had written.[4] I thought that was its first mention in public, but in fact Pais had already said something in passing about "particle mixtures" at the Rochester conference in that year.

In the "funny" paper, it was shown that charge conjugation invariance required that θ^0 and $\bar{\theta}^0$, which have definite strangeness, be mixtures of particles θ_1 and θ_2, which have definite lifetimes or, equivalently,

$$\theta_1 = \frac{1}{\sqrt{2}} (\theta + \bar{\theta})$$

$$\theta_2 = \frac{1}{\sqrt{2}} (\theta - \bar{\theta})$$

Furthermore, only the θ_1 decays to $\pi^+\pi^-$, whereas θ_2 does not and has a longer lifetime, about 100 times longer, supposing the decay mode to be $\theta_2 \rightarrow \pi^+\pi^-\gamma$. In the discussion, just about all the bizarre behavior of those objects was pointed out to Pais, and soon became well known. A beam of particles that start as pure positive-strangeness θ^0's will, in a few θ_1 lifetimes, become essentially one-half negative-strangeness $\bar{\theta}^0$'s. Their interaction with nucleons will, for instance, yield hyperons. Also, as Oreste Piccioni stressed, the beam transmitted through an absorber will be mostly θ^0, and so $\theta_1 \rightarrow \pi^+\pi^-$ decays will be regenerated.[5] Pais was reluctant to believe all that. It seemed that he and Gell-Mann had not realized quite what kind of monster they had created.

When I talked with Lederman, not long after, about those weird predictions, he immediately suggested that we look for them with the cloud chamber. The setup we used is shown in Figure 21.2. The extracted proton beam was focused with a quadrupole-pair lens on a copper target sixteen feet from the entrance to the cloud chamber.[6] Charged particles were swept away in a magnet. Altogether there were six magnets in operation – a substantial and difficult technical feat. Photographs of the setup are reproduced in Figures 21.3 and 21.4. Visible in Figure 21.3 are the back wall of the Cosmotron building, the target position and shielding wall with the sweeping magnet buried inside. The corn crib doesn't hold Long Island corn, but the cloud chamber within its magnet. The mode of construction appears in Figure 21.4 with Kenneth Lande, myself, and helpers laboring at the shielding.

The first run gave twenty-three events in 1,200 pictures; an example is shown in Figure 21.5, which shows a particularly nice one, having a $\pi \rightarrow \mu$ decay. With those two dozen events we made convincing arguments, I think, first that they were decays of particles produced at the target and that the yield was sensible only if the mass were near 500 MeV and the lifetime greater than 3×10^{-9} sec, more than ten times longer than that of the θ_1^0's.[7] Running continued until 152 events were gotten.[8] Then another run was made with a target in the circulating beam – seventy feet from the cloud chamber. The yield from that was thirty-four events. The ratio gave a lifetime $\tau = 8^{+3}_{-2} \times 10^{-8}$ sec.[9] Today's value is 5×10^{-8} sec. Examples of the major decay modes were identified – $\pi\mu\nu$, $\pi e\nu$, and 3π were found – with branching ratios not terribly wrong.[10]

Confirmation of those results, then observations of interactions, and finally

Figure 21.3. The K_2^0 experimental arrangement.

Figure 21.4. The K_2^0 experimental setup during construction.

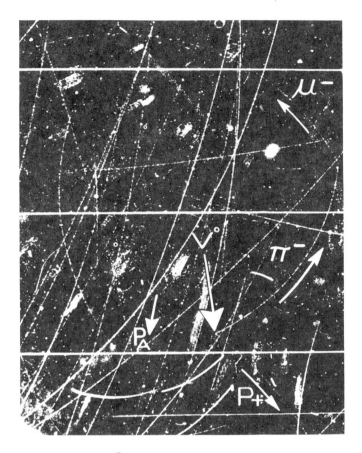

Figure 21.5. A K_2^0 decay with a pion decay product visible in the cloud chamber [*Source*: Lande et al. (1957)[10]].

regeneration came thick and fast. They started to appear everywhere. William F. Fry, Jack Schneps, and M. S. Swami found stars made by incident neutral particles in emulsions that had been exposed to a K^- beam.[11] They gave evidence that the long-lived neutrals carry negative strangeness. That was also in 1956. Raymond Ammar and associates found both decays and interactions in emulsions exposed in a neutral beam at the Bevatron in Berkeley.[12] Figure 21.6 shows three of their τ decay events. W. K. H. Panofsky, Val Fitch, and R. M. Motley found decays in a neutral beam using counters to detect the decay products.[13] Among events that appeared in bubble chambers were the Σ^0 and three-body θ_2 decay in hydrogen shown in Figure 21.7, from Frank S. Crawford and others at the Radiation Laboratory.[14] Another, in Figure 21.8, a nice one, shows a $\Sigma^0 \to \Lambda^0 + \gamma$ and a θ_2 decay far away, from Jack Steinberger's group.[15] An interaction of a K^0 in hydrogen from Craw-

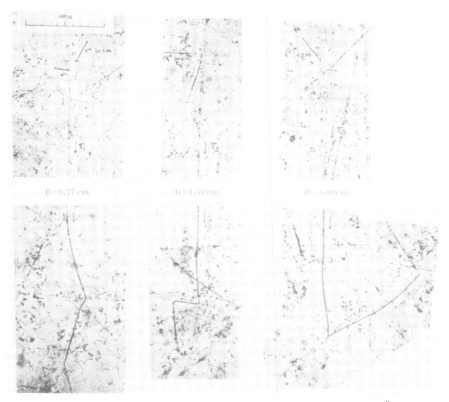

Figure 21.6. Production of τ mesons in interactions of K_2^0 mesons in emulsion, observed by Ammar and associates [*Source*: Ammar et al. (1957)[12]].

ford and associates[16] is shown in Figure 21.9. Wilson M. Powell and others, in 1957, directed θ_2's made by pions into a propane bubble chamber. They found θ_2 decays, θ_1 decays, and Λ's produced in θ_2 interactions.[17] Perhaps the most satisfying corroborating result was the report by Steinberger and associates, early in 1957, that of the θ^0's produced together with Λ^0's by π^- in propane, 49 percent of those decayed into $\pi^+\pi^-$ or $\pi^0\pi^0$, leaving the predicted 50 percent to decay far away, outside the chamber.[18]

Finally, toward the end of the decade, regeneration was demonstrated in abundance by the Powell group, together with Piccioni. Figure 21.10 shows their setup. There are many more than six magnets now – the technique has advanced. The event shown in Figure 21.11 is one of the 200 regenerated θ_1 decays they found. Their analysis first gave a value, $0.84\tau_1/\tau_1$, of the $\theta_1-\theta_2$ mass difference. That should be an important subject for the next symposium.

Figure 21.8. (*facing*). Associated production of a Σ^0 and K_2^0 in hydrogen, observed by Eisler and associates [*Source*: Eisler et al. (1958)[15]].

Figure 21.7. Associated production of a Σ^0 and K_2^0 in hydrogen, observed by Crawford and associates [*Source*: Crawford et al. (1959)[14]].

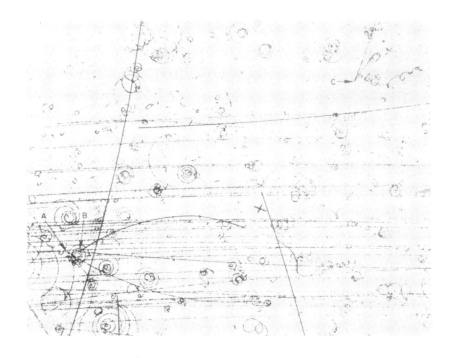

Figure 21.9. Associated production of a Λ^0 and K_2^0 in hydrogen. The K_2^0 interacts at A to produce a Σ^+ [*Source*: Crawford et al. (1959)[16]].

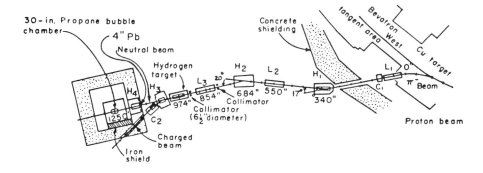

Figure 21.10. Experimental arrangement of Good and associates to produce K_2^0 mesons and observe the regeneration of K_1^0 decays in propane [*Source*: Good et al. (1961)[19]].

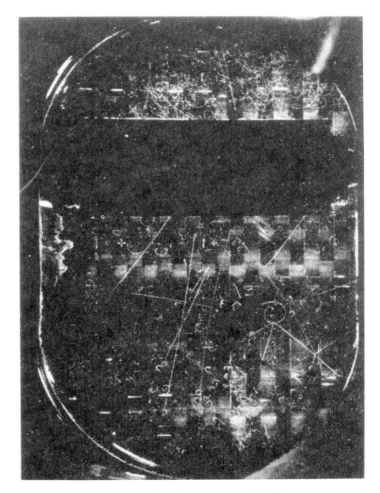

Figure 21.11. An example of a regenerated K_1^0 decay, observed by Good and associates [*Source*: Good et al. (1961)[19]].

Notes

1 J. Hornbostel and E. O. Salant, "Capture of Negative K Mesons," *Phys. Rev. 98* (1955), 218–19.
2 W. B. Fowler, R. P. Shutt, A. M. Thorndike, and W. L. Whittemore, "Production of Heavy Unstable Particles by Negative Pions," *Phys. Rev. 93* (1954), 861–7.
3 H. Blumenfeld, W. Chinowsky, and L. M. Lederman, "Observation of Production, Decay, and Interaction of Strange Particles," *Nuovo Cimento 8* (1958), 296–315.
4 M. Gell-Mann and A. Pais, "Behavior of Neutral Particles under Charge Conjugation," *Phys. Rev. 97* (1955), 1387–9.
5 A. Pais and O. Piccioni, "Note on the Decay and Absorption of the θ^0," *Phys. Rev. 100* (1955), 1487–9.

6 O. Piccioni, D. Clark, R. Cool, G. Friedlander, and D. Kassner, "External Proton Beam of the Cosmotron," *Rev. Sci. Instrum. 26* (1955), 232–3.

7 K. Lande, E. T. Booth, J. Impeduglia. L. M. Lederman, and W. Chinowsky, "Observation of Long-lived Neutral *V* Particles," *Phys. Rev. 103* (1956), 1901–4.

8 M. Bardon, K. Lande, L. M. Lederman, and W. Chinowsky, "Long-lived Neutral *K* Mesons," *Ann. Phys. (N.Y.) 5* (1958), 156–81.

9 M. Bardon, M. Fuchs, K. Lande, L. M. Lederman, W. Chinowsky, and J. Tinlot, "Lifetime and Decay of the K_2^0 Meson," *Phys. Rev. 110* (1958), 780–2.

10 K. Lande, L. M. Lederman, and W. Chinowsky, "Report on Long-lived K^0 Mesons," *Phys. Rev. 105* (1957), 1925–7.

11 W. F. Fry, J. Schneps, and M. S. Swami, "Evidence for a Long-lived Neutral Unstable Particle," *Phys. Rev. 103* (1956), 1904–5.

12 R. Ammar, J. I. Friedman, R. Levi-Setti, and V. L. Telegdi, "Nuclear Interactions of Long-lived Neutral Strange Particles," *Nuovo Cimento 5* (1957), 1801–7.

13 W. K. H. Panofsky, V. L. Fitch, R. M. Motley, and W. G. Chesnut, "Measurement of the Total Absorption Coefficient of Long-lived Neutral *K* Particles," *Phys. Rev. 109* (1958), 1353–7.

14 Frank S. Crawford, Jr., Marcello Cresti, Roger L. Douglass, Myron L. Good, George R. Kalbfleisch, and M. Lynn Stevenson, "Three-Body Decays of K_2^0 and K_1^0," *Phys. Rev. Lett. 2* (1959), 361–3.

15 F. Eisler, R. Plano, N. Samios, J. Steinberger, and M. Schwartz, "Associated Production of Σ^0 and θ_2^0; Mass of the Σ^0," *Phys. Rev. 110* (1958), 226–7.

16 Frank S. Crawford, Jr., Marcello Cresti, Myron L. Good, Klaus Gottstein, Ernest M. Lyman, Frank T. Solmitz, M. Lynn Stevenson, and Harold Ticho, "Evidence for the Transition of a K^0 into a $\bar{K}^0$ Meson," *Phys. Rev. 113* (1959), 1601–4.

17 William B. Fowler, Richard L. Lander, and Wilson M. Powell, "Neutral *K* Meson as a Particle Mixture," *Phys. Rev. 113* (1959), 928–34.

18 F. Eisler, R. Plano, N. Samios, M. Schwartz, and J. Steinberger, "Systematics of Λ^0 and θ^0 Decay," *Nuovo Cimento 5* (1957), 1700–15.

19 R. H. Good, R. P. Matsen, F. Muller, O. Piccioni, W. M. Powell, H. S. White, W. B. Fowler, and R. W. Birge, "Regeneration of Neutral *K* Mesons and Their Mass Difference," *Phys. Rev. 124* (1961), 1223–39.

22 Strange particles: production by Cosmotron beams as observed in diffusion cloud chambers

WILLIAM B. FOWLER

Born 1924, Owensboro, Kentucky; Ph.D., Washington University, St. Louis, Missouri, 1951; high-energy physics; Fermi National Accelerator Laboratory

The physicist's ability to study the properties of strange particles expanded greatly following the observation of machine-produced V particles. Protons were accelerated by the Cosmotron to 1 GeV in May 1952; however, by the time shielding was added and the target, beam, and detector equipment put in place, it was the spring of 1953. At this point, experimenters could begin to collect data. Among the first detectors to be used was the high-pressure hydrogen diffusion cloud chamber with an 11,000-G magnetic field. In the first 4,000 photographs, with the chamber exposed to a high-energy neutron beam, several V particles popped up. Two of these had tracks that gave momentum and angle measurements of sufficient accuracy to be convincing. That led to the first studies exploiting the advantages the accelerator had over cosmic rays.

In the summer of 1950, several groups were experimenting with continuously sensitive diffusion cloud chambers, and Ralph P. Shutt recognized their potential use for Cosmotron experiments, in particular if high-pressure hydrogen could be made to work. Great enthusiasm and a wide variety of skills existed at that time in Shutt's cloud-chamber group, and very shortly two 20-atm hydrogen diffusion chambers had been constructed.[1] Through the effort of Gilberto Bernardini this equipment was moved to the Nevis cyclotron, where exposures to pion beams demonstrated the usefulness of this new tool. With the addition of a new magnet, the high-pressure hydrogen diffusion chamber was on the floor of the Cosmotron when the first experiments began. Returning to the first observation of artificial production of V's, the maximum energy of an incident neutron was 2.2 GeV, because the Cosmotron energy was known. Various possible production mechanisms were investigated, and the following conclusion was stated:

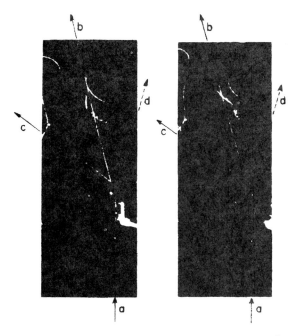

Figure 22.1. Stereoscopic photograph of a V_1^0 (tracks b and c) produced in hydrogen by a negative pion (track a) of an energy of 1.5 BeV. If one heavy meson of mass of about 1,350 m_e has been produced in addition, its line of flight is indicated by d [*Source*: Fowler et al. (1953)[3]].

Production of V_1^0 from pre-existing nucleons, singly or in pairs, or of (V_1^0, V_2^0) pairs, would... be consistent with the present observation.[2]

That is, the question of associated production was still open.

In those days, it was possible to move the detectors from one beam to another for different exposures. In fact, the high-pressure hydrogen diffusion chamber's magnet was on wheels, and the experimenters (with the help of a few strong technicians) crossed the Cosmotron floor and set up in a negative-pion beam of 1.5 GeV energy. A fantastic photograph appeared one afternoon on the scanning table being used by Alan Thorndike (Figure 22.1). Everyone knew at a glance that something significant was at hand. The pion–proton collision had produced a neutral V particle that was identified by its decay.[3] Track (b) was unambiguously a proton, and from momentum and angle measurements the decay was classified as a V_1^0 (Λ). Because momentum and energy are conserved in the production, particle (d) could be determined to be a heavy meson of 1,350 ± 70 m_e, if only one other particle was produced with the Λ. Another event yielded similar results.

This paper did not claim associated production of strange particles. Its carefully worded conclusion was as follows:

> Of course, instead of one heavy particle, several lighter ones (for instance two π^0's or a π^0 and a V_1^0) could originate from the events in addition to the V_1^0. However, the present results are consistent with the possibility of production of V_1^0 together with one other heavy unstable particle.[3]

More running and more scanning clarified the situation, so that by November 1953 the group submitted new data on the associated production of strange particles in $\pi^- - p$ collisions.[4] The reactions seen were

$$\pi^- + p \rightarrow \Lambda + \theta$$

as well as

$$\pi^- + p \rightarrow \Sigma^- + K^+$$

where Σ^- was called Λ^- (V_1^-). This paper, which reports a Σ^- decay into a π^- + neutron, with $Q = 130$ MeV, reports the initial discovery of the Σ^-. Lifetime information, although sparse, was consistent with the 10^{-10} to 3×10^{-10} sec quoted in the literature at that time.

The cross section for production was ~ 1 mb, which is what made the experiment possible. Even so, after the initial flurry of events, it was a year later, December 1954, before five additional events had been added.[5] More precise data were available on masses, and the new events plus a reanalysis of the older events led to the conclusion that "these and earlier observations can be made consistent with known hyperon and K-meson masses if the reaction $\pi^- + p \rightarrow \Sigma^0 + \theta^0$ is assumed to be a possibility with an immediate γ decay $\Sigma^0 \rightarrow \Lambda^0 + \gamma$ as suggested by Murray Gell-Mann and Pais."

To summarize:

$$
\begin{aligned}
\pi^- + p &\rightarrow \Lambda^0 + \theta^0 && \text{2 events} \\
&\rightarrow \Sigma^0 + \theta^0 && \text{4 events} \\
&\rightarrow \Sigma^- + \theta^+ \ (K^+) && \text{3 events}
\end{aligned}
$$

The Brookhaven cloud-chamber group, which made the previously described work possible, consisted in 1952 of those listed in Table 22.1.

I would like to express my opinion of the impact of the work with the high-pressure diffusion cloud chamber in the period following the 1950s, and extending up to present-day high-energy-physics research. The desire to have available a better detector that could extend the capability of the hydrogen diffusion chamber (i.e., accelerator particle-beam interactions with protons and decays of strange particles produced in the collisions) helped to drive the discovery and development of the much superior hydrogen bubble chamber discussed by Luis Alvarez in Chapter 19. It is not accidental that some of the participants in the two developments are the same. The photo of the Brook-

Figure 22.2. The Brookhaven bubble-chamber group of 1963, photographed in front of the eighty-inch hydrogen bubble chamber [*Source*: Henry Margeneu et al. (1964), *The Scientist, Life* Science Library. © 1984 Time Inc. Print provided by BNL Photography Dept.].

haven bubble-chamber group of 1963 on the occasion of the Ω^- discovery (Figure 22.2) shows Nick Samios, Shutt, and William Fowler, along with the other physicists, engineers, technicians, film scanners and measurers, and various other support staff necessary for winning the friendly race with our European colleagues. The Brookhaven eighty-inch hydrogen bubble chamber designed, constructed, and used to take the photo of the Ω^- (featured on the

Table 22.1. *The 1952 Brookhaven cloud-chamber group*

Physicists	Engineers	Technicians
W. B. Fowler	A. Johnson	S. Hasselriis
R. P. Shutt	W. Tuttle	K. Kristoffersen
A. M. Thorndike		F. Pallas
W. L. Whittemore		J. Pokorney
		F. Ratheny
Research assistant	Scanners	A. Rosech
S. Cornish	M. Burns	O. Thomas
	B. Carpenter	T. Tuttle
Grad student	Secretary	E. Wohr
P. Kenney	R. Kelly	A. Wright

announcement for this meeting) was the major activity of the Shutt group from 1958 to 1962 and appears in the background.

A further connection from the days of widespread use of hydrogen bubble chambers to the present is that the cryogenic technology that the high-energy physicists learned about carried over into the application of superconductivity the necessary ingredient that made possible the first superconducting particle accelerator, the Fermilab Tevatron (or Energy Doubler). Superconducting-magnet technology used on all the large hydrogen bubble chambers, such as Fermilab's fifteen-foot chamber, helped in convincing skeptics like Robert R. Wilson that operating 1,000 six-meter-long pulsed-accelerator-quality magnets at liquid-helium temperature (4.5 °K) was a piece of cake. If I am not mistaken, Wilson sometimes still looks a little skeptical, but Leon Lederman is convinced. (Well, maybe?)

Notes

1 W. B. Fowler, R. P. Shutt, A. M. Thorndike, and W. L. Whittemore, "Diffusion Cloud Chambers for Cosmotron Experiments," *Rev. Sci. Instrum. 25* (1954), 996–1003.
2 W. B. Fowler, R. P. Shutt, A. M. Thorndike, and W. L. Whittemore, "Observation of V^0 Particles Produced at the Cosmotron," *Phys. Rev. 90* (1953), 1126–7.
3 W. B. Fowler, R. P. Shutt, A. M. Thorndike, and W. L. Whittemore, "Production of V_1^0 Particles by Negative Pions in Hydrogen," *Phys. Rev. 91* (1953), 1287.
4 W. B. Fowler, R. P. Shutt, A. M. Thorndike, and W. L. Whittemore, "Production of Heavy Unstable Particles by Negative Pions," *Phys. Rev. 93* (1954), 861–7.
5 W. B. Fowler, R. P. Shutt, A. M. Thorndike, and W. L. Whittemore, "Production of Heavy Unstable Particles by 1.37 BeV Pions," *Phys. Rev. 98* (1955), 121–30.

23   From the 1940s into the 1950s

ABRAHAM PAIS

Born 1918, Amsterdam; Ph.D., Utrecht University, 1941; theoretical physics; Rockefeller University.

This brief memoir deals with some subjects I was involved with or that otherwise came to my notice during the period from July 1946 through June 1953. A memoir, according to the *Oxford English Dictionary*, is a record of events "not purporting to be a complete history, but treating of such matters as come within the personal knowledge of the writer, or are obtained from certain particular sources of information." During the last few years I have had occasion to consult numerous memoirs, some old, some recent, because I have been writing a book on the history of particles and fields.[1] I was not surprised to find that I did not always agree with what those memoirs said (as, for example, in regard to the genesis of the neutral-K particle mixing idea[2]). Nor did they always agree with each other. Such variants can be revealing, and sometimes even interesting, but should, of course, always take second place compared with the published scientific record.

Now to July 1946. That was the time of the first international post–World War II physics conference, held at the Cavendish Laboratory. I was coauthor of two papers presented there. The first, written with Lamek Hulthen, dealt with calculations of $n-p$ scattering up to 25 MeV.[3] In this work, noninteger orbital angular momentum played the role of a variational parameter.[4] This paper is of some interest because it was stimulated by nucleon–nucleon scattering experiments at the highest energy then reached. In September 1945, Rubby Sherr had scattered 25-MeV neutrons, produced at the Harvard cyclotron, off protons.[5] Thus, 25 MeV is the benchmark on the scale of energies at which postwar high-energy physics began.

Work supported in part under U.S. Department of Energy grant DE-AC02-81ER40033B.

The other paper, jointly with Christian Møller, deals with the possible existence of mass spectra of fundamental particles.[6] We wrote:

> Measurements of the mass of the particles in the hard component of the cosmic radiation give a spread of mass values between, say, 100 and 400 electron masses. Meson masses smaller than, say, 150 are very difficult to reconcile with the range of nuclear forces as determined from the proton–proton scattering. This difficulty may, however, be avoided by assuming that the particles with masses smaller than 150 are not a kind of mesons, but heavy electrons.

Why would we be thinking of cosmic-ray mesons as heavy electrons, at a time preceding the emission-versus-absorption paradoxes of cosmic-ray mesons?[7] Because we had been studying the Dirac equation of the electron in a de Sitter space pictured as a five-dimensional spherical shell of finite thickness. That, as is now well known, gives rise to mass spectra, now called towers. Such spectra, we wrote, should appear for the three types of fundamental particles: the nucleons, the mesons, and the light particles. For the latter, we wanted to have an expressive generic name. So after consulting with Danish classicists, we added: "For the 'light' particles we propose the name 'leptons'." From that time dates my not quite rationally founded interest in particles as families.

In September 1946, there took place an American Physical Society meeting with international participation. Three aspects of that meeting are of particular interest. First, there was a panel discussion on "relative advantages of proton and electron accelerators," in which Luis Alvarez, Ernest Lawrence, Edwin McMillan, and Robert Wilson participated. Second, Louis Leprince-Ringuet presented a short paper suggesting the "probable existence of a heavy meson $(1,000\ m_e)$."[8] I do not think that paper, the first presented in the United States related to the new particles, made a particularly strong impression at that time. Third, and once again setting a benchmark, there is the following statement in the minutes:

> [The meeting] was confined to papers on three topics: cosmic-ray phenomena, theories of elementary particles and the design and operation of accelerators of nuclear particles and electrons. Disparate as these subjects may appear to be, the trend of physics is rapidly uniting them.[9]

I turn next to a meeting in Copenhagen, in September 1947, where I met Cecil F. Powell for the first time. Powell gave a report on his new discovery of the π–μ–e chain. After that meeting, I went back to the United States by slow freighter. I remember pacing the deck, wondering if there was a lepton family after all. But then, I said to myself, why don't we see nucleons (as I called them then) heavier than the proton and neutron? Since those do not exist, it must all be nonsense. So I forgot about families and turned to other pursuits.

Then, in December 1947, the paper by George Rochester and Clifford C.

Butler appeared with photographs of one kink, interpreted as the spontaneous decay of a new kind of charged particle, and one fork, the decay of a new kind of neutral particle.[10] In their paper, Rochester and Butler drew attention to "peculiar cloud chamber photographs" taken earlier by others, suggesting that these might perhaps be akin to their own pictures.[11] Others, however, did not at once reexamine earlier cloud-chamber pictures for *V*-particle evidence. Years later I raised this issue with Wilson Powell, the cosmic-ray expert from Berkeley, asking him in particular why *V* particles had not been discovered earlier, as well they might have been. He promised an answer by the following day. When next we met, he had with him a handful of cloud-chamber pictures he had taken in the 1930s, each one showing either a fork or a kink. He told me that at some time in the early fifties he had convinced himself that these were *V* particles, adding with a smile that because of the great interest in showers in the thirties, experimentalists would load their cloud chambers with many metal plates in order to obtain high particle multiplicities and would pay little attention to individual tracks. Rochester and Butler had noticed their two events in a chamber containing only a single plate.

In 1948 there was hardly any reaction to this news. The report of the discussion on cosmic rays at the Pocono conference (20 March–1 April 1948), led by Bruno Rossi, does not contain any mention of *V* particles.[12] Rochester's own report on cosmic-ray work to the symposium in honor of Robert A. Millikan's eightieth birthday (June 1948) does not include *V* particles in the list of main results.[13] When I recently queried Rochester about this, he replied: "It is true that there is meager reference in the printed version...I did, however, say more about the *V* particles in my talk and I spoke privately with a number of participants, notably, Carl Anderson, R. B. Brode, and Rossi."[14]

At the same symposium, Brode from Berkeley mentioned, almost as an aside, the observation in cosmic rays of eight particles with masses between 500 and 800 m.[15] Leprince-Ringuet was more emphatic in announcing an example of "very heavy mesons which we call τ-mesons."[16] This event, found in emulsions, was a nuclear disintegration caused by a particle with mass of at least 700 m_e. As the year drew to a close, the Bristol group found the first example of a charged particle with mass between 870 and 985 m decaying into three particles believed to be probably all pions, although $\pi + 2\mu$ or $2\pi + \mu$ "cannot be excluded."[17] This is the first example of τ decay:

$$\tau^+ \to \pi^+ + \pi^+ + \pi^-$$

Nothing worthy of note happened in the first few months of 1950. In April, Enrico Fermi, lecturing at Yale on "elementary particles," made no mention of *V*'s or τ's.[18] Then, later that spring, I read a Caltech report on results obtained with "the famous cloud chamber of Anderson that discovers all the new particles."[19] Thirty forks and four kinks had been observed. "One must come to the same remarkable conclusion as that drawn by Rochester and Butler...these events...represent...the spontaneous decay of neutral and charged unstable particles of a new kind."[20] No mass values were quoted. The decay products of the neutrals could be two π's, a π and a μ, or a proton and a meson (π or μ). I was startled. Were there meson and nucleon families after all?

Now, slowly, the new particles began to gain attention. During the Harwell conference (September 1950), P. M. S. Blackett reported that the Manchester group, having moved their cloud chamber to the observatory at the Pic-du-Midi, had found ten more events.[21] In October, the first case was found of a V particle, now called Λ, with a proton as one identified decay product.[22] During the first Rochester conference (16 December 1950), Robert Oppenheimer suggested a discussion of τ-mesons.[23] Nothing came of that, and the day was spent on nucleon and muon physics. In a public lecture given in October 1951, Fermi was now taking the new particles quite seriously.[24] During the two-day-long second Rochester conference (11–12 January 1952), one full day was devoted to that subject. Robert Marshak's textbook of that year, *Meson Physics*, was the first to contain a chapter on the new particles.[25]

As all of us recall, the coming of the new particles raised the problem of how to reconcile their copious production with their slow decay. Some time early in 1951 it was suggested by Willy Fowler[26] that a steep potential barrier between nucleon and pion might suppress Λ decay, a low-energy process, without inhibiting Λ production, a high-energy process, whereupon Fermi noted that this would be the case if the Λ had a spin of, say, $\frac{13}{2}$. I did not much care for that idea. Because of my thoughts about particle families, I had the vague notion that the Λ should be related to the nucleon as the muon is to the electron. If so, one should seek a solution in which "the heavy fermion [Λ] is as elementary as the nucleon."[27] Thus, I began to look for a model in which "the emphasis will be...on the role of selection rules," rules that had to be of a new kind.

Being familiar with the difficulties of theoretical pion physics, I knew that perturbation theory would be of no use. In addition, there was the further problem of electromagnetic decay. Further prohibition was demanded to slow down processes like $K^0 \to 2\gamma$ by at least six orders of magnitude. I looked for *selection rules that would hold for strong and electromagnetic processes, but not for weak processes*, meaning by "weak processes" reactions "similar to... neutrino processes where a coupling constant occurs which is very small."[27] I proposed such a rule (details are uninteresting) according to which the new particles, loosely speaking, come in pairs, a mechanism later named associated production. I do not know who coined that term. That work was done in the summer of 1951, long before the discovery of the cascade particles, which would not have fitted in that simple-minded scheme. Later I learned that several Japanese colleagues had been considering a series of options that also included the idea of a strong-interaction selection rule.[28]

I noted at that time:[27] "There is perhaps an indication of the existence of families of elementary particles like a nucleon and an electron family in which, not unlike the levels in a given kind of atom, the members of a given family are distinguished from each other through a quantization process, but one of a new kind," adding that "the search for ordering principles at this moment may ultimately have to be likened to a chemist's attempt to build up the periodic system if he were only given a dozen odd elements."[29] I lectured on these ideas in the autumn of 1951 at the Institute for Advanced Study and again at the second Rochester conference. I am not responsible for the term "megalomorphs" in the title of that last talk. That was an invention of Oppenheimer, made because, in his words, "Fermi had become bored with the name 'elementary particles.'" Toward the end of that talk, I suggested "to look at this schematization as the unfolding of an ordering in which one talks about families of elementary particles rather than elementary particles themselves."

Early in 1953 I made a try, the first of its kind, at enlarging the isospin group so as to find room for an additional quantum number that should obey an additional selection rule. In other words, the plan was to give the additional selection rule a group-theoretical foundation. The new particles should be grouped into isospin multiplets, like the old ones. Isospin should be conserved in *all* strong interactions, but not in electromagnetic interactions, while the new rule should hold for both these forces, but not for weak interactions. As I said then: "The present picture seems to involve a hierarchy of interactions corresponding to the symmetry classes of the (intrinsic) variables."[30] That picture, presented at the Lorentz–Kamerlingh–Onnes conference in Leiden in June 1953, has survived. This first concrete example of an enlarged group was richly premature, however. For one thing, there still was no cascade particle!

As to associated production, cosmic-ray evidence seemed at first against it.[31] Cyclotron experiments at energies too low for production of the new particles in pairs gave no evidence for their single production.[32] The issue was settled when accelerators in the giga-electron-volt region became available. A Cosmotron experiment (November 1953) yielded the first convincing results.[33]

In concluding this memoir, I should like to recall my first encounter with non-Abelian gauges (a term I had never heard of then), which dates from the Leiden conference just mentioned.[34] In the paper presented there, the following assumption was introduced: "The element of space time is not a point but a manifold" (a two-dimensional sphere, as it happened).[35] I was unaware that I was dealing with a fiber bundle of the most trivial kind. Wolfgang Pauli, in the audience, was intrigued, and in the subsequent discussion raised the following question. I now quote from the proceedings of the conference:

> I have a particular question regarding the interaction between mesons and nucleons.... I would like to ask...whether the transformation group with constant phases can be amplified in a way analogous to the gauge group for the electromagnetic potentials in such a way that the meson–nucleon interaction is connected with the amplified group.[36]

Pauli kept pondering that question. (I believe his interest may have stemmed from his earlier work on a unified theory of gravitation and electromagnetism in which a five-dimensional manifold is hung at every space–time point.)[37] Right after the Leiden conference he wrote me about a possible answer in terms of a generalized Kaluza–Klein theory.[38] A few weeks later he sent me a manuscript, entitled "Meson Nucleon Interaction and Differential Geometry," which begins: "Written down July 21 till 25 (1953) in order to see how it is looking."[39] In this manuscript, Pauli noted that local isospin gauge invariance demands the introduction of a compensating isotriplet of gauge field vector potentials and found, what he called his "main result," the correct expression for the corresponding field strengths, the gauge-covariant derivatives. He did not, however, write down the associated dynamical field equations.

Later in 1953, Pauli's enthusiasm began to wane. "If one tries to formulate field equations...one will always obtain *vector mesons with reset mass zero* [his italics]. One could try to get other meson fields – pseudoscalars with positive rest mass – ... But I feel that is too artificial."[40]

Meanwhile, C. N. Yang and Robert L. Mills had independently tackled the same problem. In February 1954, Yang reported on their results in a seminar at the Institute for Advanced Study at Princeton. Pauli was in the audience, and I well recall his critical and negative reaction; see also Yang's own recollection of that event.[41] Pauli never published (but notes of his lectures on this subject, given in the autumn of 1953, later appeared in print);[42] Yang and Mills did. In two short brilliant papers they founded modern gauge theory.[43]

As a final remark on the origins of non-Abelian gauge theory, I note that the Yang–Mills equations were independently derived by Ronald Shaw in his unpublished Cambridge thesis, completed in August 1955.[44] This work contains the following footnote: "The work described in this chapter was completed, except for its extension [to a four-dimensional case] in Section 3, in January 1954, but was not published. In October 1954, Yang and Mills adopted independently the same postulate and derived similar consequences."

Notes

1 A. Pais, *Inward Bound* (Oxford University Press, 1986).
2 Compare Pais, note 1 (Chapter 20) with M. Gell-Mann, "Strangeness," in International Colloquium on the History of Particle Physics, *J. Phys. (Paris) (Suppl.) 43 : 12* (1982), 395–402.
3 *Conference on Fundamental Particles and Low Temperatures*, Vol. 1, 22–27 July 1946 (London: Taylor & Francis, 1947), p. 177.
4 A. Pais, "On the Scattering of Fast Neutrons by Protons," *Proc. Cambridge Philos. Soc. 42* (1946), 45–54.
5 R. Sherr, "Collision Cross Sections for 25-MeV Neutrons," *Phys. Rev. 68* (1945), 240–5.
6 Note 3 (p. 184).
7 M. Conversi, E. Pancini, and O. Piccioni, "On the Disintegration of Negative Mesons," *Phys. Rev. 71* (1947), 209–10.
8 L. Leprince-Ringuet, "Measurement of Meson Masses by the Method of Elastic Collision. Probable Existence of a Heavy Meson (1000 m_0) in the Cosmic Radiation," *Phys. Rev. 70* (1946), 791–2.

9 Minutes of the American Physical Society meeting of 19–21 September 1946 at New York, *Phys. Rev. 70* (1946), 784.

10 G. D. Rochester and C. C. Butler, "Evidence for the Existence of New Unstable Elementary Particles," *Nature (London) 160* (1947), 855–7.

11 Ibid. (footnotes 3 and 4).

12 Notes of the Pocono conference, unpublished.

13 G. D. Rochester, "The Penetrating Particles in Cosmic-Ray Showers," *Rev. Mod. Phys. 21* (1949), 20–6.

14 G. D. Rochester, letter to A. Pais, 25 November 1984. I am further indebted to Professor Rochester for sending me copies of letters (all from 1947) to him by Enrico Fermi (3 December), Walter Heitler (23 November), Powell (24 November), Rossi (28 November), and John Wheeler (9 December), all expressing great interest in the preprint they had received of the Rochester–Butler paper. A discussion on these particles did take place during the eighth Solvay conference (Brussels, October 1948), but its proceedings were not published until 1950.

15 R. B. Brode, "The Mass of the Mesotron," *Rev. Mod. Phys. 21* (1949), 37–41.

16 L. Leprince–Ringuet, "Photographic Evidence for the Existence of a Very Heavy Meson," *Rev. Mod. Phys. 21* (1949). 42–3.

17 R. Brown, U. Camerini, P. H. Fowler, H. Muirhead, C. F. Powell, and D. M. Ritson, "Observations with Electron-Sensitive Plates Exposed to Cosmic Radiation. II. Further Evidence for the Existence of Unstable Charged Particles of Mass $\sim 1000\ m_e$ and Observations on their Mode of Decay," *Nature (London) 163* (1949), 82–7.

18 Enrico Fermi, *Elementary Particles* (New Haven: Yale University Press, 1951).

19 W. B. Fretter, Proceedings of the 2nd Rochester conference (University of Rochester Report NYO-3046), p. 56.

20 A. J. Seriff, R. B. Leighton, C. Hsiao, E. W. Cowan, and C. D. Anderson, "Cloud-Chamber Observations of the New Unstable Cosmic-Ray Particles," *Phys. Rev. 78* (1950), 290–1.

21 P. M. S. Blackett, "The V-Meson," in *Proceedings of the Harwell Nuclear Physics Conference*, edited by E. W. Titterton (Harwell, Berks.: Ministry of Supply, 1950), pp. 20–1.

22 V. D. Hopper and S. Biswas, "Evidence Concerning the Existence of the New Unstable Elementary Neutral Particle," *Phys. Rev. 80* (1950), 1099–100.

23 Unedited and unpublished transcript.

24 Enrico Fermi, "The Nucleus," *Phys. Today 5* (1952), 6–9.

25 Robert Marshak, *Meson Physics* (New York: McGraw-Hill, 1952).

26 Cf. R. P. Feynman, "High Energy Phenomena and Meson Theories," unpublished notes of Caltech lectures, January–March 1951, p. 76.

27 A. Pais, "Some Remarks on the V-Particles," *Phys. Rev. 86* (1952), 663–72.

28 Y. Nambu, K. Nishijima, and Y. Yamaguchi, *Prog. Theor. Phys. 6* (1951), 615–19; K. Aizu and T. Kinoshita, ibid., 630; H. Miyazawa, ibid., 631; S. Oneida, ibid., 633.

29 In a table I drew on the board during the Rochester conference, I made parallels between three families, then called nucleons, bosons, and leptons. That table was recently reproduced by Robert E. Marshak in "Particle Physics in Rapid Transition: 1947–1952," in *The Birth of Particle Physics*, edited by Laurie M. Brown and Lillian Hoddeson (Cambridge University Press, 1983), pp. 376–401.

30 A. Pais, "Isotopic Spin and Mass Quantization," *Physica 19* (1953), 869–87.

31 R. B. Leighton, S. D. Wanlass, and C. D. Anderson, "The Decay of V^0 Particles," *Phys. Rev. 89* (1953), 148–67; W. B. Fretter, M. M. May, and M. P. Nakada, "A Study of Neutral V Particles," *Phys. Rev. 89* (1953), 168–80.

32 R. L. Garwin, "A Search for V^0 Particles Produced by 450-MeV Protons," *Phys. Rev. 90* (1953), 274–8; A. H. Rosenfeld and S. B. Treiman," Search for V Particles Produced by 430-MeV Protons," *Phys. Rev. 92* (1953), 727–9.

33 W. B. Fowler, R. P. Shutt, A. M. Thorndike, and W. L. Whittemore, "Production of Heavy Unstable Particles by Negative Pions," *Phys. Rev. 93* (1954), 861–7.

34 I did not know at that time of a paper by Oskar Klein, "New Theories in Physics" (Warsaw

Conference, May 30–June 3, 1938) (The Hague: Nyhoff, 1939), p. 77. This article does not contain non-Abelian gauge transformations, but nevertheless bears on the subject at hand.

35 See note 30.

36 W. Pauli, "Discussion Remarks at Lorentz Kamerlingh Onnes Conference," *Physica 19* (1953), 887.

37 W. Pauli and J. Solomon, "Dirac's Equations and the Unitary Theory of Einstein and Mayer," *J. Phys. Radium 3* (1932), 452–63; W. Pauli and J. Solomon, "Unitary Field Theory and Dirac's Equations," *J. Phys. Radium 3* (1932), 582–9.

38 W. Pauli, letter to A. Pais, 3 July 1953.

39 A copy of this manuscript as well as an important "Mathematischer Anhang," written in December 1953, are in the Pauli Collection, CERN Archives.

40 W. Pauli, letter to A. Pais, 6 December 1953.

41 Chen Ning Yang, *Selected Papers 1945–1980 with Commentary* (New York: W. H. Freeman, 1983), pp. 19–21.

42 P. Gulmanelli, "Su una teoria dello spin isotopico," Pubblicazione sezione di Milano dell' Istituto Nazionale di Fisica Nucleare, Casa Editrice Pleion, Milano, undated (probably 1954).

43 C. N. Yang and R. L. Mills, "Isotopic Spin Conservation and a Generalized Gauge Invariance," *Phys. Rev. 95* (1954), 631; C. N. Yang and R. L. Mills, "Conservation of Isotopic Spin and Isotopic Gauge Invariance," *Phys. Rev. 96* (1954), 191–5.

44 Ronald Shaw, "The Problem of Particle Types and Other Contributions to the Theory of Elementary Particles," Dissertation submitted for the annual election of Fellows at Trinity College, Cambridge, August 1955 (see part 2, p. 37 for the footnote quoted in the text).

PART VI

WEAK INTERACTIONS

Once upon a time, the world was less complex,
With fewer nervous wrecks, with lots more time for sex.
  Electrons and protons were all we could afford,
  They were good enough for dear old Rutherford.
Then the troubles started; they gave us, just as we know
The positron, the neutron, the mysterious neutrino.
– © Arthur Roberts, "Some People Don't Know
Where to Stop" (1952)

24   Detection of the neutrino

FREDERICK REINES

Born 1918, Paterson, New Jersey; Ph.D., New York University, 1944;
cosmic rays, experimental neutrino physics; University of California, Irvine

The efforts that came to fruition with the detection of the free neutrino started at Los Alamos in 1950–1, when several of us who were engaged in the testing of nuclear bombs wondered if that man-made star could be used to advance our knowledge of physics.[1] For one thing, the unusual object featured lots of fissions, followed by β decays; so why not use it as a source of neutrinos? After all, it is an extraordinarily intense pulse, and signals produced by such a source might be sufficiently copious and distinguishable from background. Some hand-waving and rough estimates led me to conclude that the bomb was, in fact, the best source. All that was required was a massive detector, measuring a cubic meter or so.

As it happened, Enrico Fermi was at Los Alamos during the summer of 1951, and I took the opportunity to talk with him about these matters. The conversation went something as follows: "I would like to talk with you a few minutes about the possibility of neutrino detection." He was very pleasant, and said, "Well, tell me what is on your mind." "First off as to the source," I said, "I think that the bomb is best." After a few moments thought he replied, "Yes, the bomb is the best source." So far so good. And then I said, "But one needs a detector which is so big and I don't know how to make such a detector." He thought about it further and said he didn't know either.

Coming from the master, that was very discouraging. I put it on the back burner until a chance conversation with Clyde Cowan some months later. "Let's do a real challenging problem," I said. He said, "Let's work on positronium." I said, "No, positronium is a very good thing, but Martin Deutsch has that sewed up. So let's not work on positronium; let's work on neutrinos." His immediate response was, "Great idea!" So we shook hands and got off to working on the neutrinos. At that time we had not the vaguest clue how to build a suitable detector or even what reaction to explore.

Why detect the free neutrino? By 1951, essentially everybody had accepted the idea of the neutrino's existence. The Pauli–Fermi theory was beautiful. It explained an enormous number of phenomena, and who needed to *see* this thing? What did that mean, anyway? Why were we interested in the neutrino? Because everybody said one could not see it. Not very sensible, but we were attracted by the challenge. Besides, we had a bomb that constituted a unique intense source; so maybe we had an edge on other people. But there is a more profound reason that occurred to us as we worked on the problem. In order to determine that the neutrino existed, one had to see it do something at a point remote from its place of origin.

In rebuttal to this point of view, people said that after all, the observation of recoils during K capture would accomplish the equivalent. Clearly, if one did not see a recoil, the neutrino hypothesis would be in trouble. But even if one did see a recoil, that did not prove the existence of the neutrino. It would not be proved because it might be that energy/momentum is not conserved in the act of β decay. In fact, that was a possibility pointed out by Niels Bohr. Therefore, if one did not have conservation at the "scene of the crime," so to speak, then it was inappropriate to use that lack to attribute to the neutrino the responsibility for the recoil in K capture. That simply was an exercise in circular reasoning. Well, being brash, but having respect for certain authorities, I commented in this vein to Fermi, who agreed.

If we demonstrate the existence of the neutrino in the free state, that is, by an observation at a location remote from its origin, we extend the range of applicability of these fundamental conservation laws to the nuclear realm. On the other hand, if we do not see this particle in the predicted range, then we have a very real problem. As Bohr is reputed to have said, a deep question is one for which either a yes or no answer is interesting. So I guess this question of the existence of the free neutrino might be construed to be deep.

What about the problem of detection? We fumbled around a great deal before we got to it. Finally we chose to look for the following reaction (in modern parlance, this reaction is written $\bar{\nu}_e + p \rightarrow n + e^+$, where $\bar{\nu}_e$ is the electron antineutrino produced by a neutron-rich β emitter):

$$\nu + p \rightarrow n + e^+$$

If the neutrino exists in the free state, this inversion of β decay must occur. We chose to consider this reaction in particular because if we believe in detailed balancing and use the measured value of the neutron half-life, we know what the cross section must be – a nice clean result. (In fact, as we learned some years later from Tsung Dao Lee and Chen Ning Yang, the cross section is greater by a factor of 2 because of parity nonconservation.)

How big a detector is required? How many counts do we expect? What features of the interaction do we use for signals? Hans Bethe and Rudolph Peierls, in 1934, almost immediately after the Fermi paper, estimated that in the range of a few million electron volts the cross section would be of the order of 10^{-44} cm^2. That is about a thousand light-years of liquid hydrogen. No wonder they concluded in the article that appeared in *Nature* in April 1934 that "there is no practically possible way of observing the neutrino." I confronted Bethe with this pronouncement some twenty years later, and with his characteristic good humor he said: "Well, you should not believe everything you read in the papers."[2]

It may have been this paper that led to the neutrino being labeled as elusive. Despite these considerations, the situation was reviewed to predict whether or not the intense neutrino flux expected from a nuclear reactor would be harmful. I recall one such review, and in a recent conversation John A. Wheeler did not disclaim having considered the neutrino radiation problem, either by himself at DuPont or jointly with Eugene Wigner in 1943, before the Hanford piles were turned on.

What, in the face of this extraordinarily small interaction, changed the prospects for neutrino detection? Two developments: the discovery, which was old, of fission, and the discovery by Hartmut Kallmann and others of organic liquid scintillators. The latter development meant to us that a large detector was possible, although at the time Cowan and I started our quest, a "big" detector was only a liter or so in volume. Despite the three-orders-of-magnitude extrapolation in detector size required, it seemed to us an approach worth pursuing. The detection idea then was to use the $\nu + p$ reaction with protons in the scintillator as the target. The bomb neutrinos would be emitted in a great pulse lasting for a second or so until the mushroom cloud left the vicinity of the detector. But how does one identify a signal as due to neutrinos? Bethe asked me that question when I was describing our plans to him, and I said, "That's easy – you use a delayed coincidence and that identifies it." Cowan and I were well aware of the delayed coincidence between the positron and neutron capture pulses, but it had not yet occurred to us that we were describing a way to cut the background and, more important, that such a distinctive signature might mean one did not have to use a bomb. We realized its significance somewhat later, fortunately before we actually carried out the planned bomb experiment.

But how sensitive would the bomb experiment have been? It is interesting that when we presented this idea for approval by the Los Alamos director, Norris Bradbury, we could not see how to detect the predicted cross section for bomb neutrinos of $\sim 10^{-43}$ cm^2. In fact, we figured that the best we could do was a cross section of 10^{-39} cm^2, four orders of magnitude short of our goal. Yet it was an improvement by a factor of a thousand or so from the previous experimental limits, and the director, in his wisdom, and with an assist from Bethe and Fermi, decided to let us try it. So we worked, and while we were working, it occurred to us that we could use the delayed coincidence and search in the quieter circumstances associated with a nuclear pile. We wrote to Fermi, and he sent us the letter shown in Figure 24.1.

After some preliminary tests, we built the detector (Figure 24.2) and brought

THE UNIVERSITY OF CHICAGO
CHICAGO 37 · ILLINOIS

INSTITUTE FOR NUCLEAR STUDIES

October 8, 1952

Dr. Fred Reines
Los Alamos Scientific Laboratory
P.O. Box 1663
Los Alamos, New Mexico

Dear Fred:

Thank you for your letter of October 4th by Clyde Cowan and yourself. I was very much interested in your new plan for the detection of the neutrino. Certainly your new method should be much simpler to carry out and have the great advantage that the measurement can be repeated any number of times. I shall be very interested in seeing how your 10 cubic foot scintillation counter is going to work, but I do not know of any reason why it should not.

Good luck.

Sincerely yours,

Enrico Fermi

EF:vr

Figure 24.1. Letter from Fermi on hearing about our plan to use the Hanford reactor to attempt to observe the neutrino [*Source*: Reines, *Science 203* (1979),[1] 11–16; copyright © 1979 by the AAAS].

it to a Hanford plutonium production reactor. It was a terribly difficult business; the Tygon reflective-paint coating in the detector hung loose from the walls, and we encountered electrical noise during reactor-off periods as a maintenance elevator ran up and down the reactor face. We stacked and restacked hundreds of tons of lead bricks and borated-paraffin shielding blocks, trying to make the best shield against reactor γ's and neutrons. The backgrounds from cosmic-ray μ capture in our detector were some ten times the predicted signal, and we finally decided we could not do any better. Exhausted, we gathered our data, got on the train, and went back to Los Alamos.

Figure 24.2. First large (0.3-m³) liquid scintillation detector in shield. The liquid was viewed by ninety two-inch photomultiplier tubes. Before development of this detector, a 0.01-m³ volume was considered large [*Source*: Reines, *Science 203*, (1979),[1] 11–16; copyright © 1979 by the AAAS].

When we analyzed the data, we found we had a two-standard-deviation effect, which is just a hint at best. Although tests showed that this reactor-associated effect could not be due to neutrons, these marginal results merely served to whet our appetites. Accordingly, we designed a detector that was rather more sophisticated (Figures 24.3 and 24.4). In this approach, the neutrino came from the reactor and interacted in the water, producing a positron that annihilated, throwing γ's into the counters. The neutron was moderated, and then, in ~5.5 μsec, was captured in the cadmium compound dissolved in the water target, emitting γ rays, which were again seen in the counters above and below. That sequence of coincidences was extraordinarily distinctive and militated against backgrounds of various kinds, and most particularly against neutrons. It is noteworthy that this setup, a composite of 330 five-inch photomultiplier tubes viewing several tons of liquid scintillators, heralded an early use in particle physics of very large detectors.

Wheeler, hearing of our experiment, suggested that the Savannah River plant, with its well-shielded 700 MW (at that time), would be a good place to go. We located our detector system ~11 m from the reactor core, and the whole setup was located 12 m underground to shield against cosmic rays. Several redundant tests, which we devised, showed the reactor-associated

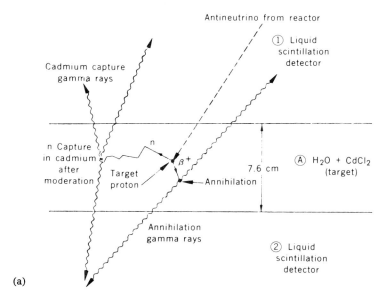

(a)

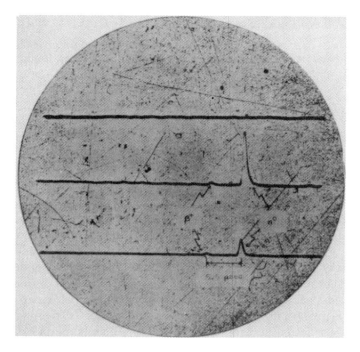

(b)

Figure 24.3. (a) Schematic of neutrino experiment showing one triad; (b) a characteristic record [*Source*: Reines, *Science 203* (1979),[1] 11–16; copyright © 1979 by the AAAS]. Each of the three oscilloscope traces corresponds to a detector tank. The event recorded occurred in the bottom triad. First seen in coincidence are the positron-annihilation γ-ray pulses in each tank, followed in 5.5 μsec by the larger "neutron" pulses. The positron is denoted by B^+, and the neutron by n^0.

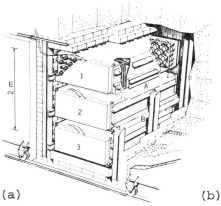

(a)   (b)

Figure 24.4. (a) Sketch of detectors inside their lead shield [*Source*: Reines (1979)[1]]. The tanks marked 1, 2, and 3 contained 1,400 liters of triethyl-benzene (TEB) based liquid scintillator solution, which was viewed in each tank by 110 five-inch photomultiplier tubes. The tubes were immersed in pure nonscintillating TEB to make light collection more uniform. Tanks A and B were polystyrene and contained 200 liters of water, which provided the target protons and contained as much as 40 kg of dissolved $CdCl_2$ to capture the product neutrons. (b) Inside view of electronics van showing equipment required to select and record neutrino signals [*Source:* Reines, *Science 203* (1979),[1] 11–16; copyright © 1979 by the AAAS].

delayed coincidence rate to be explicable only in terms of neutrino interactions via the $\bar{\nu}_e p$ reaction.[3]

The results were, we felt, rather convincing, and so we thought we should tell Wolfgang Pauli. It was time to tell the man who started it all when, as a young fellow, he wrote his famous letter to Tübingen (1930) in which he postulated what Fermi later called the neutrino, saying something to the effect that he could not come to the meeting and tell about it in person because he had to go to a dance! We understand that on receiving our message, he interrupted Gilberto Bernardini, during a meeting they were attending at CERN, to tell about all this.

It took many years before anyone checked our findings. I have never been able to figure out why, because science is not made by one person or one small group doing a measurement. It has to be done by separate groups so that they can either reinforce or destroy each other. My guess is that the reason it was not expeditiously checked by others was that almost everyone believed that the neutrino was there anyway, and because it appeared to behave as expected, there was no need to check. Another possible reason is the enormous effort required relative to the standards of the early and middle fifties – many tons of shielding, big trailers carrying tons of liquid for the detectors, hundreds of photomultiplier tubes, a large trailer full of electronics, and so forth.

In view of subsequent developments, that inactivity was unfortunate. Had reactor neutrinos been more vigorously pursued, it is possible that neutral currents would have been detected earlier (the neutral-current reaction $\bar{\nu}_e + d \rightarrow n + p + \bar{\nu}_e$ was seen at a reactor in 1979, several years after it was observed at accelerators). In any event, observation of the free neutrino probably served to reduce the credibility gap that might have been associated with an already unusual unobserved particle that in addition did not conserve parity.

As a postscript, I would like to remark on the question of the identity of ν_e with the "neutretto," as the neutrino ν_μ associated with $\pi - \mu$ decay was sometimes referred to in the early literature. Having just detected $\bar{\nu}_e$, it seemed to us appropriate to investigate the matter experimentally. Accordingly, in early 1957 we said to the laboratory management that "we would next like very much to test the identity of ν_e and ν_μ." The response we received was that "you fellows have had enough fun playing around, why don't you go back to work." Fortunately for physics, Melvin Schwartz and his colleagues at Brookhaven settled the two-neutrino question.[4]

Notes

1 Frederick Reines, "The Early Days of Experimental Neutrino Physics," *Science 203* (1979), 11–16.

2 H. Bethe and R. Peierls, "The 'Neutrino'," *Nature (London) 133* (1934), 532. However, in May 1934, they wrote that "it is not impossible in principle to decide experimentally whether they exist," and they proposed two possible experiments. See Bethe and Peierls, "The Neutrino," *Nature (London) 133* (1934), 689–90.

3 The tests consisted of measuring a reactor-associated delayed coincidence signal ($\sim$3/hour), the first pulse of which was shown to be a positron and not a neutron, and the second pulse to be due to capture of a neutron. The signal varied as required with the number of target protons and was unaffected by bulk shielding, which reduced neutrons and γ's by more than an order of magnitude.

4 G. Danby, J. M. Gaillard, K. Goulianos, L. M. Lederman, N. Mistry, M. Schwartz, and J. Steinberger, "Observation of High-Energy Neutrino Reactions and the Existence of Two Kinds of Neutrinos," *Phys. Rev. Lett. 9* (1962), 36–44.

25 Recollections on the establishment of the weak-interaction notion

BRUNO M. PONTECORVO

Born 1913, Italy; *laurea*, University of Rome, 1933; Lenin Prize, 1963; experimental nuclear and high-energy physics; Academy of Sciences, Moscow

Introduction

I am very glad to have the opportunity to describe for this symposium some early, practically unknown, Dubna work on strange particles. It is quite natural that I would like people to be informed about some of my work, significant in my opinion, performed a long time ago, and the only way of fulfilling such a desire decently is to be invited to take part in a symposium. True, in his delightful talk on strange particles at the 1982 Paris colloquium on the history of particle physics, Murray Gell-Mann mentioned my work.[1] I shall cover mainly Dubna work on new particles performed in 1951–5[2,3] in the context of the notion of weak interaction, a notion that was certainly not taken for granted in the early 1950s, but that had become one of my pet ideas as early as 1947.[4]

Nuclear β decay, the first known weak process, was discovered by Ernest Rutherford about eighty-five years ago. However, not every physicist knows that the notion of weak interaction, a conception much wider than that of the single process of β decay, came to be well established only in the 1950s, that is, about fifty years after the discovery of β rays and about forty years after James Chadwick's discovery of the continuous β spectrum.[5] Here I shall present personal recollections about the way the notion of weak interaction was born and then became well established. Of course, my story is going to be neither objective nor complete. I shall talk about some episodes that I saw with my own eyes or in which I directly took part. Naturally, the reader must keep in mind that I am writing the present note in June 1985. Furthermore, I have been relying mostly on my memory, not on the literature.

I am very grateful to S. M. Bilenky for discussions.

There is no need to recall here the most decisive ideas and experiments, I would say the "final" contributions to the creation of the universal electroweak-interaction theory. I shall limit myself to the evidence in favor of my 1947 idea, that the β decay "is not alone." The processes, other than β decay, that pointed to some kind of universal behavior concerned first the muon and then strange particles. This story begins in 1947 and terminates in 1955.

Muon capture by nucleons and muon decay

Marcello Conversi, Ettore Pancini, and Oreste Piccioni[6] in 1947 demonstrated that the (cosmic) 2.2-μsec-lifetime mesotrons, that is, the muons, do not have the properties postulated for the Yukawa particles: The muon interacts much more weakly with nucleons than the Yukawa particle should.[7] I have already described in detail elsewhere how the experiment of Conversi and others personally influenced my way of thinking. Briefly, because the muon was not the Yukawa particle, there were no compelling reasons to believe that the muon had the properties that were being postulated for the Yukawa particle. Thus, in my opinion, the following questions were entirely open:

1. Why should the spin of the muon be integral?
2. Who said that the muon must decay into an electron and a neutrino, rather than into an electron and two neutrinos, or into an electron and a photon?
3. Is the charged particle emitted in the muon decay an electron?
4. Are particles other than electrons and neutrinos emitted in muon decay?
5. In what form is the nuclear muon capture energy mainly released?

Some of these questions were answered experimentally by E. P. Hincks and myself, and by other groups. I wish to discuss question 5 here in some detail.

The nuclear muon capture energy, I thought in 1947, must be released mainly in the form of neutrinos. The relevant reaction is then $\mu^- + Z \rightarrow (Z - 1) + $ neutrino, very similar to the process of nuclear K capture $e^- + Z \rightarrow (Z - 1) + $ neutrino. I interpreted the similarity of these two processes as a very significant and deep effect, because, as a matter of fact, the rate of nuclear electron capture and that of muon capture are quite close (when proper account is taken of phase-space effects and of the different electron and muon orbit volumes).[4] I excluded the possibility of a chance coincidence and reached the following conclusions:

(i) The muon capture must be a process in some way identical with the β process, proceeding according to the reaction $\mu^- + p \rightarrow $ neutrino + n.[5]

(ii) In muon capture, most of the released energy is "invisible," because

it is carried away in the form of neutrinos, a conjecture that is supported by experiments and agrees with (i).
(iii) The muon spin must be $\frac{1}{2}$.

Thus, in 1947, I started to think in terms of weak-interaction processes[4] and understood first that both the muon capture by nuclei and the β decay are processes due to a definite weak interaction existing in nature. It was clear to me that the muon is a sort of heavy electron and that the muon–electron symmetry is taking place under a type of interaction that is properly called weak, thanks to the smallness of the corresponding constant G – the Fermi β decay constant. A similar point of view – namely, to include the muon decay among weak processes – was adopted later[8] by others: Oskar Klein; Giovanni Puppi; T. D. Lee, M. Rosenbluth, and C. N. Yang; Jayme Tiomno and John A. Wheeler.

The original 1947 idea that there exists a muon–electron symmetry in nature was the first hint of a universal weak interaction. (But how far this was still from the 1957 form of such interaction: the $V-A$ theory of Marshak–Sudarshan and Feynman–Gell-Mann, implemented later by the Cabibbo hadron mixing, the Glashow–Weinberg–Salam "final" electroweak interaction with the Higgs mechanism, and the discoveries of neutral currents and of W^+, W^-, Z^0!)

The main physical content of my 1947 idea is still not understood today; it concerns the existence of families of leptons (and families of quarks). Why do such families exist in nature? I must say that the existence of several weak processes, in addition to the β decay process, seemed clear to me in 1947 (much clearer than today). Anyway, my credo in 1947 led to my expectation that there must exist a number of weak-interaction processes in addition to the β decay. Herein I shall be concerned only with processes of the "charged-current" type, although neutral-current processes later turned out to be quite relevant.

Because the weak-interaction conception was first formulated for the capture of muons and electrons, I believed for some time that every weak process must imply the participation of neutrinos. That wrong idea may have slowed down the development of the notion of weak interaction, but the discovery of new unstable particles unmistakably widened the weak-interaction conception to include hadrons.

Strange particles and the weak interaction

I shall not give details about the very important investigations and discoveries of new particles.[9] I am limiting myself to a few particle discoveries, sufficient to illustrate the question about the weak interaction being responsible for the particle decay. In a short period, starting in 1947, a number of unstable new particles were discovered, some electrically neutral and some electrically charged. Among the neutral particles, one could definitely recognize in a cloud chamber those having baryonic charge, later called Λ^0, and decaying slowly according to the scheme $\Lambda^0 \to p + \pi^-$.[9] Besides, it was shown in a very clean way[8] that some charged mesons, now called K mesons, decay into pions: $K^+ \to \pi^+\pi^+\pi^-$. Here, too, the decay is slow, as indicated by the very fact that the meson has enough time to stop in a thick photographic plate before its decay. The properties of Λ^0 and K^+ were, in my opinion, an indication that the decays $\Lambda^0 \to p + \pi^-$ and $K^+ \to \pi^+\pi^+\pi^-$ are due to a weak interaction, probably the same weak interaction that is responsible for the β decay and muon processes. A similar point of view was expressed independently by N. Dallaporta.[10]

At the time, physicists usually reasoned in terms of the Yukawa process, and at high energy only strong processes were considered. However, such a picture would fail to explain the generation and the decay of such (strange) particles as Λ^0 and K. These particles are copiously produced in cosmic rays, but have quite long lifetimes; they demonstrate strange behavior, if one assumes that the process of particle generation is fundamentally the same as the decay process. However, if we assume that the strange particles are generated in strong processes, but decay in weak-interaction processes, then there are no more difficulties. If we assume that Λ^0 and K are generated together, the difficulties connected with the long mean lives of baryons such as Λ^0 and of mesons such as K are resolved together.[2]

In the early 1950s, several theoreticians came once a week from Moscow to Dubna to conduct seminars on a high level, often among them Isaak Pomeranchuk. I presented my arguments briefly to Pomeranchuk, who liked them very much and right away organized a seminar to illustrate the curious properties of hyperons and kaons along the lines I had suggested; that is, the (weak) decays of hyperons and kaons are not due to the (strong) interaction that generates them together. Since 1947 I had been expecting new weak processes; so I was very happy about all this. I felt that the notion of weak interaction became wider once again, but in new processes. Thus, at the time, the weak interaction appeared to me to be a universal interaction acting between any group of four fermions. That was not very far from today's point of view: W decays into elementary particles, leptons and quarks (and only that way), there being some choice of flavor (masses) for the decay products.

On the basis of these simple arguments, I introduced,[2] independently of Abraham Pais,[11] the idea of pair production of the new particles, more exactly, pair production of hyperons and kaons.

The reactions $N + N \to N + \Lambda^0$ and $n + n \to \Lambda^0 + \Lambda^0$

The question of strange-particle generation can be investigated effectively in experiments performed near the production threshold. We investigated experimentally the question of a possible generation of Λ^0 particles in nucleon–nucleon collisions.[3] The method we used was due to a brilliant suggestion of Richard Garwin,[12] who also was investigating the production of

Λ^0 particles. Garwin's idea was that in some experiments it is convenient to register Λ^0 particles by detecting photons from π^0's emitted in the channel $\Lambda^0 \to n + \pi^0$. Our experiment at the time was interesting, because Marcel Schein and associates[13] claimed to have detected the production of Λ^0 particles, and the question whether or not Λ^0 is produced singly is one of principle.[1,2,3]

In our experiment, in which 670-MeV protons from the Dubna synchro-cyclotron impinged on the accelerator's internal carbon target, we reached the conclusion that Λ^0's are not produced, either in the reaction $N + N \to N + \Lambda^0$ or in the reaction $n + n \to \Lambda^0 + \Lambda^0$. The absence of the reaction $N + N \to N + \Lambda^0$ agreed well with the idea[2,10] of the generation of two new particles together.

As for the vanishing value of the cross section for the reaction $n + n \to \Lambda^0 + \Lambda^0$, this was just the expectation of Gell-Mann (and of K. Nishijima), for reasons that today are obvious to everybody. Two words about our interpretation of this vanishing value, which we were able to give correctly, even without possessing the notion of strangeness.

I had figured out a scheme based on the assumption that there is a strong interaction responsible for the generation of new particles (two at the same time), and conserving the isotopic spin, and the weak interaction responsible for the decays of particles, and not conserving the isotopic spin. The isotopic spin has a meaning only in strong interactions and cannot be determined by weak decays. There arises the possibility of the existence of fermions with integral isotopic spin (e.g., Λ^0) and of bosons with half-integral isotopic spin (e.g., kaons). The scheme allowed one to interpret the failure to observe the reaction $n + n \to \Lambda^0 + \Lambda^0$ through the assumption that the isotopic spin of the kaon is $\frac{1}{2}$ (that is, $K^0 \neq \bar{K}^0$) and to make a number of predictions. Neverthe-less, the notion of strangeness was a very powerful tool without which physics could not have made the great advances it did. As we know now, the physical content of strangeness is that charge multiplets of hadrons are classified by the number $(0, 1, 2, \ldots)$ of some material particles – the number of s quarks – they contain.

In conclusion, I would like to say that at the Pisa conference of 1955, mainly as a result of Gell-Mann's wonderful talk, the notion of weak interaction, which was introduced in 1947,[4] finally became established.

Notes

1 M. Gell-Mann, "Strangeness," in Colloque International sur l'Histoire de la Physique des Particules, *J. Phys. (Paris) (Suppl.)* 43:12 (1982), 395–402.

2 B. M. Pontecorvo, "On the Processes of Production of Heavy Mesons and V_1^0 Particles," *Zh. Eksp. Teor. Fiz.* [*Sov. Phys.–JETP*] 29 (1955), 140–6.

3 M. P. Balandin, B. D. Balashov, V. A. Zhukov, B. M. Pontecorvo, and G. I. Selivanov, "On the Possibility of Production of Λ^0-Particles by Protons of Energy Up to 700 MeV," *Zh. Eksp. Teor. Fiz.* [*Sov. Phys.–JETP*] 29 (1955), 265–73.

4 B. Pontecorvo, "Nuclear Capture of Mesons and the Meson Decay," *Phys. Rev. 72* (1947), 246–7.

5 It took 15 years before the two particle reactions $\mu^- + p \rightarrow n + \nu_\mu$, $\mu^- + {}^3\mathrm{He} \rightarrow {}^3\mathrm{H} + \nu_\mu$ were directly observed in the experiments of R. Hildebrand and in our own experiments (together with R. Sulyaev et al.).

6 M. Conversi, E. Pancini, and O. Piccioni, "On the Disintegration of Negative Mesons," *Phys. Rev. 71* (1947), 209–10.

7 E. Fermi, E. Teller, and V. Weisskopf, "The Decay of Negative Mesotrons in Matter," *Phys. Rev. 71* (1947), 314–15.

8 O. Klein, "Mesons and Nucleons," *Nature (London) 161* (1948), 897–9; G. Puppi, "On Cosmic Ray Mesons," *Nuovo Cimento 5* (1948), 587–8; T. D. Lee, M. Rosenbluth, and C. N. Yang, "Interaction of Mesons with Nucleons and Light Particles," *Phys. Rev. 75* (1949), 905; J. Tiomno and J. A. Wheeler, "Energy Spectrum of Electrons from Meson Decay," *Rev. Mod. Phys. 21* (1949), 144–52.

9 See, for example, C. C. Butler, "Early Cloud Chamber Experiments at the Pic-du-Midi," in Colloque International sur l'Histoire de la Physique des Particules, *J. Phys. (Paris) (Suppl.) 43:12* (1982), 177–84; R. H. Dalitz, "Strange Particle Theory in the Cosmic Ray Period," ibid., 195–205; W. S. Fretter, "Cosmic Rays and Particle Physics at Berkeley," ibid., 191–4; L. Leprince-Ringuet, "Les Rayons Cosmiques et la Physique des Particules á l'Ecole Polytechnique," ibid., 165–8; C. O'Ceallaigh, "A Contribution to the History of C. F. Powell's Group in the University of Bristol 1949–65," ibid., 185–9; C. Peyrou, "The Role of Cosmic Rays in the Development of Particle Physics," ibid., 7–66; G. D. Rochester, "Observation on the Discovery of the Strange Particles," ibid., 169–75; J. Rösch, "La Venue au Pic-du-Midi du Groupe Blackett et du Groupe Leprince-Ringuet," ibid., 215–18; J. Six and X. Artru, "An Essay of Chronology of Particle Physics until 1965," ibid., 465–93.

10 N. Dallaporta, "On the Mean Lives of Heavy Unstable Particles," *Nuovo Cimento 1* (1955), 962–5.

11 A. Pais, "Some Remarks on the *V*-Particles," *Phys. Rev. 86* (1952), 663–72.

12 R. L. Garwin, "A Search for V^0 Particles Produced by 450-MeV Protons," *Phys. Rev. 90* (1953), 274–8.

13 M. Schein, D. Haskin, R. Glasser, F. Fainberg, and K. Brown (eds.), *Congrès International sur le Rayonnement Cosmique*, Bagnéres de Bigorre (University of Toulouse, 1953).

26   Symmetry and conservation laws in particle physics in the fifties

LOUIS MICHEL

Born 1923, Roanne (Loire), France; Ph.D., Sorbonne, 1953; theoretical physics; IHES, Bures-sur-Yvette, France

The "Gruppenpest" that so many physicists fought against before the war died out during the 1940s. Although some famous physicists made pioneering, fundamental contributions in the theory of infinite-dimensional representations of groups, the community of physicists ignored them.* However, in the late forties, with the discovery of new particles (π mesons and V particles), the need arose to know more about conservation laws. For instance, that a spin-1 boson cannot decay into two photons was known from a short, brilliant paper by Lev Landau and from Eugene P. Wigner's un-

---

* The first mathematical paper that gave the complete series of unitary irreducible representation of a noncompact, nonsemisimple Lie group was Wigner's paper on the Poincaré group. This paper, one of the most quoted of the century, according to a citation index, was refused by several journals before being accepted by *Annals of Mathematics*. E. Majorana, and Dirac later, implicitly considered unitary representations of the Lorentz group earlier.[1] The problem was solved completely in 1947, by V. Bargmann and J. M. Gelfand and M. A. Naimark, independently.[2] In his paper, Wigner used unpublished results of J. von Neumann, which appeared several years later. In fact, he solved an even much more difficult problem, that of the *projective* representations, since they are the ones exactly required by physics. This fundamental paper was used by few physicists in the fifties and was practically ignored by the community.[3] It is intriguing that Wigner, who had taught us that time reversal should be represented by an antiunitary operator, studied unitary time reversal in this famous paper.[4] He published the physical time-reversal version only in 1962, but taught it many years before.[5] Some of these irreducible corepresentations of the full Poincaré group correspond to parity doublets of particles. Wightman and I described them in a fat set of unpublished but widely distributed notes at Princeton around 1954–5.

I am grateful to the Department of Physics of Politecnico of Turin and to the Institute for Scientific Interchange, Villa Gualino, Turin, for their hospitality. The Appendix was written during my stay in Turin.

published findings, quoted by Jack Steinberger.[6] Using field-theory concepts, particle physicists were eager to establish the list of selection rules for particle decays; for example, C. N. Yang and D. C. Peaslee gave their proof of spin $1 \nrightarrow 2\gamma$.[7] In a long review I wrote in 1951, I carefully referenced all that was known.[8] A complete list of selection rules for angular momentum and parity conservation appeared only later.[9]

Then came the $\theta{-}\tau$ puzzle. Its history is recounted excellently by Richard Dalitz in Chapter 30. Apparently this was a typical case of the parity doublet; T. D. Lee and Yang described it in an elegant paper,[10] six months before they proposed parity violation as the explanation.[11] So much has been written on the history of P violation (and some contributions to this conference are devoted to it) that I wish here only to mention some "prehistoric" relevant publications. Pierre Curie, in a fundamental and classic paper (in which he formulated his famous symmetry principles), explained parity conservation and the difference between axial and polar vectors in classical physics.[12] Wigner showed the importance of parity in quantum mechanics.[13] For different particles, one may be able to define only relative parity. As explained by G. Racah, because Emil Konopinski and George Uhlenbeck made a different choice from Enrico Fermi of what is really "relative parity" among n, p, e, and ν, they found an opposite sign for some terms proportional to the neutrino mass in the β decay "allowed" spectrum.[14]

Mario Schoenberg did consider the possibility of violating parity by introducing both scalar and pseudoscalar coupling constants (as I did in my thesis).[15] Let me also mention B. Ferretti's spin-zero meson theory, in which the three members of an isospin triplet have different parity![16] (I used this idea in an unpublished manuscript on the $\theta{-}\tau$ puzzle.) Also, parity was sometimes violated implicitly – for example, by Bruno Touschek in his theory of double β decay, or by H. Enatsu, who explained the V and A weak couplings (before some wrong experiments excluded them!) through a unique charged pair of spin-1 intermediate bosons.[17] Nearly all these references were discussed in my 1952 review.[8]

In a parity-violating decay, the asymmetry in the angular distribution of the decay products becomes a powerful tool to measure the spin of the initial particle. This was used immediately by Lee and Yang in their beautiful paper showing that Λ^0 has spin $\frac{1}{2}$.[18] However, the systematic use of angular-momentum conservation and (for strong and electromagnetic interaction) of parity conservation was mainly done outside particle physics. The cornerstone is Racah's 1949 paper.[19] The first three papers in that series (in 1942 and 1943) barely mentioned group theory, probably as a result of the intolerant war against it. The fourth paper explains successfully hundreds of spectral lines from a few radial integrals; it not only introduces the seniority quantum number but also uses many results on the theory of irreducible unitary representation of the classical simple Lie groups (and also G_2). All this was completely unknown to particle physicists: they learned it more than ten years later when they badly needed it. However, Racah's work was immediately extended to nuclei by H. A. Jahn just before the birth of the nuclear-shell model by Maria Goeppert-Mayer and J. H. D. Jensen.[20] So this use of group theory among nuclear physicists culminated in the fifties.

In the same period, the general theory of angular correlation was developed, mainly by and for nuclear physicists (see the historical introduction in the chapter by Samuel Devons of the *Handbuch der Physik*).[21] Hence, it is not astonishing that the simple constraint on particle polarization due to parity conservation in a two-body reaction or decay, so useful in particle physics, was found by Aage Bohr.[22] Finally, in the last year of the fifties came the work of Maurice Jacob and Gian Carlo Wick.[23] So it was only in the following decade that particle physicists became interested in these problems.*

Parity violation is accompanied by *C* violation (*C* = charge conjugation). The story of the other discrete symmetries *C* and *T* has not been as thoroughly written as that of *P*. (I wrote on *C* before 1955.[27]) In 1950, some of us were asking if the neutrino was a Dirac or a Majorana particle. Meanwhile, several famous and basic papers (and *many more* less famous, such as some of mine) were violating *C* and *T*. Only in the beginning of the fifties were β decay phenomenologists beginning to understand a twenty-year-old Wigner paper and to include the effects of *T* conservation in their work.[4] Field theorists went deeper and unearthed the *CPT* theorem.[28] More generally, symmetry principles were systematically exploited by axiomatic field theory from the middle of the fifties, mainly under the influence of Arthur Wightman (see Chapter 42).

I shall be more sketchy about internal (i.e., nongeometrical) quantum numbers, as the history of isospin has been well considered by several authors; the most recent contribution contains all earlier references.[29] But I became really disturbed when I could not trace back the *basic law of conservation of nucleons* to the work of P. A. M. Dirac. Professional science historians should study the question. An Appendix to this chapter gives my recollections on this subject. It seems to me that the basic paper is from Wigner again, but as late as 1952.[30] He asks rightly whether this conservation law is due to a gauge invariance of the first or second kind (Wolfgang Pauli's

---

* When the ϱ and hundreds of other resonances were found, after 1960, their observation through their decay products yielded some information on their polarization. Angular momentum and parity conservation impose some boundary to the domain of polarization parameters, just as the boundary of the Dalitz plot gives the constraints due to energy–momentum conservation. For the usual experimental observations of spin-1 and spin-$\frac{3}{2}$ resonances (e.g., ϱ and $\varDelta$), the boundary of the polarization domain has been given by two of my students.[24] Together, and at times with other physicists, we studied these polarization domains systematically.[25] We even extended the study to sets of reactions related by internal symmetry (e.g., isospin).[26] These constraints due to angular momentum and parity conservation were often used by European physicists (namely, French); they have never been used in the United States outside the Allan Krisch group. I am still astounded by such a lack of intellectual curiosity.

terminology in his old paper on quantum mechanics), and he deduces non-trivial physical consequences. Happily for the reader of this chapter, Wigner asked himself about the origin of this conservation law in a long footnote that I reproduce in the Appendix. Then came Abraham Pais's associated production and Murray Gell-Mann's strangeness: Strange particles violated the rule known at that time, $(-1)^{2(j+t)} = 1$, where j is the spin and t the "isotopic" spin (as we used to say). Gell-Mann wrote this part of the history freshly and vividly for the Paris history conference.[31]

In that period there emerged the framework that we kept for the next twenty years. There is a hierarchy in increasing strength and increasing internal symmetry for four fundamental interactions: gravitational, weak, electromagnetic, and strong, the latter being already broken at the level of nuclear forces. Many efforts were made to guess the internal symmetry group of the strong interaction. As is well known, success came only at the beginning of the sixties. The story of that success has been given by its authors: Gell-Mann, Yuval Ne'eman, and the less publicized Michael Nauenberg and T. W. Speiser.[32] As is usual, the story of the previous failures has not been told, and it is one of those rare cases where that history may not be very instructive. There was also the question of lepton conservation; lepton charge attributions changed several times. At the end of the period, the Brookhaven experiment taught us a deep truth: ν_e and ν_μ are different!

Seen from the present perspective, there should have been two important questions for the end of the fifties:

1. Can local gauge invariance of electromagnetic theory be extended to the other fundamental interactions?
2. Why are some internal or discrete symmetries only approximate? Are they spontaneously broken?

Of course, it is ridiculously easy to ask questions twenty-five years too late, and one could not expect them to correspond to the fashion of the period: Fashion was exerting and still exerts too great an influence on research in physics. Alas, Jun John Sakurai is no longer with us to tell about the paper he wrote in a prophetic style about vector dominance. But he could not explain why the Yang–Mills field was so heavy; there were other prophets.

What strikes me now (and I say frankly that I did not really comprehend it then) is that two of the giants were studying these very questions. Up to his death, Pauli worked on the gauge invariance of the strong interactions. Feza Gürsey, whose work was then commented on by Pauli, should be invited to tell this story.

Before the war, Landau understood second-order phase transitions and gave a model for the spontaneous breaking of symmetry that might accompany them.[33] In the fifties, this became an active topic of research, but mainly in the USSR; it led to spontaneous gauge breaking in superconductivity. Thanks to Jeffrey Goldstone, and mostly Yoichiro Nambu, the high-energy physics community absorbed this idea into its culture, but only in the sixties. Although it was too early to have the possibility of success, Werner Heisenberg was working on unified theory and knew that the internal symmetry of his equation had to be spontaneously broken. The corpus of data had been considerably enriched when Gell-Mann and Zweig independently made the bold invention of the quarks, seven years later.

I know how dangerous oversimplifications are, but nevertheless I venture to say that since the seventies, symmetry plays a central role in our philosophy of unification of physics. Such an idea was alien to the physics literature in the late forties. But there was an irresistible ascent of the role of symmetries in the fifties, and the attitude had changed completely. Most of us then shared the enthusiasm that Heisenberg had at that time (although his theory was not fashionable). In his autobiography, *Der Teil und das Ganze*, he recalls his discussion with Pauli in 1958; with Genesis, and maybe Faust, in the background, he proclaimed: "*Am Anfang war die Symmetrie!*" It is interesting to note that two weeks after Lee and Yang had sent their famous nonconservation paper to *Physical Review*, they sent another one, very detailed, on their parity doublet model.[10,11]

Appendix

After enjoying the first two days of this symposium, it appeared to me that the oral lecture should be rather different from the "scholarly" written text I had prepared. The oral lecture should be more informal and also give some personal reminiscences, and it would also be a good occasion to raise questions. This appendix is based on my oral lecture.

I wish to thank the organizers of this interesting conference for the invitation to write a "scholarly contribution" on the history of particle physics in the fifties, which also included my coming to Fermilab to meet many of the particle physicists who were active in this period. That was a great pleasure, but because I have fewer opportunities to speak to science historians, I shall mainly address them.

During the meeting, we heard the history of a long battle waged by Man to unravel some secrets of Nature. But we heard only the accounts made by the generals and commanders-in-chief. I hope the historians are also interested in the eyewitness accounts of low-ranking officers. We form a large group, happy to do physics. We have to work, we have to sweat, but at least our life does not know a great danger: that of losing the friendship of the colleagues with whom we have made our discoveries. Of course, I had the happiness to make *some* discoveries in physics, but I had more occasions to make permanent friendships with colleagues, and this brought me even deeper joys.

At the meeting, we heard a lot on particle physics at the end of the 1940s. However, I have a strong feeling that this does not give a balanced account of how it really was. There were many more conferences than we were told about, and many of them left no proceedings. Some of those forgotten conferences were very important for communication among physicists; remember that there were no preprints then, and no long-distance phone calls. It seems to me very necessary for the historians to have a compilation on all scientific conferences in this domain between 1946 and 1950, to have a list of their participants, and to have an idea of what they said. My eyewitness report will mention the first scientific conferences I attended.

By a student exchange, in spring 1947 I was working in Patrick M. S. Blackett's laboratory in Manchester. There were two "V particles" (discussed by George Rochester in Chapter 4). I remember so well the seminar given there by Cecil F. Powell and G. P. S. Occhialini. This new particle – was it the Yukawa meson? F. C. Franck had another explanation of these $\pi-\mu$ tracks. He pointed out that the μ^-p^+ bound state, neutral and very small, could induce fusion with another hydrogen nucleus; the spectator μ^- would then receive some kinetic energy. This possibility was observed ten years later.[34]

I wished to understand this better, so I studied theory. Before going back to France, I went to see Léon Rosenfeld, who had just arrived in Manchester. He agreed to accept me in his group if I could find some French financial support.

I was back in Manchester in July 1948, working on particle decays. I soon discovered that I was not the only one doing it in the world! To save money, I hitchhiked to my first scientific conference; it was in Bristol, 20–24 September 1948. I remember very vividly Powell's formal night lecture, celebrating "nuclear emulsions" and the pageantry going with it. (Donald Perkins tells in Chapter 5 about the important experimental contributions presented.) The proceedings of that conference have been published in book form.[35] Under the title, it says: "COLSTON PAPERS based on a Symposium promoted by the COLSTON RESEARCH SOCIETY and the UNIVERSITY OF BRISTOL in September 1948, now published as a Special Supplement to Research a Journal of Science and its Applications." There were only four theoretical papers (I quote from the table of contents):

The week before, 14–18 September, there had been a conference in Birmingham. I think there were no proceedings. American physicists attended (e.g., Robert Oppenheimer and Pais). Indeed, Møller quotes their contributions in his report (and so he was probably there himself). I cannot refrain from quoting Oppenheimer: "An obvious way out of this difficulty would be to say that the nucleons are not correctly described by the Dirac equation in the region of pair production and that antinucleons do not really exist."

There is not a word about the so *famous* and *important* Pocono conference, which had occurred a few months before in the United States. Could physicists now believe this extraordinary lack of propagation of news? Historians have to check these facts if they want to understand how particle physics research went in this period. The only thing I can say about this situation, so hard to imagine today, is that it meant great luck to me! Otherwise I should not have written my first scientific paper, which made my name a label for a parameter characterizing the possible electron energy spectra in

μ decay. Indeed, Møller did not mention the extensive work of J. Tiomno and J. A. Wheeler (presented at the Pocono conference) on this subject. But he presented that of J. J. Horowitz, O. Kofoed-Hansen and J. Lindhard on the decay $\mu^+ \to \mu^0 + E^+ + \nu$ analogous to β decay.[36] However, since the μ^0 mass had probably dwindled to zero, the electron energy spectrum should be computed again. Møller gave two examples; in one of them there was a completely new feature: The spectrum was not going to zero at maximal energy.

I was very excited about "particle spectrum." Oskar Klein had already pointed out that μ was an excited electron, and Møller had explained in his talk that τ's were excited π's. However, back in Manchester, I first did the computation suggested by Møller on the electron spectrum from μ decay. Why did I not quote him in the letter sent to *Nature* a few months later? To this day, I do not know. Probably because there was no text to quote (no preprints in those times, at least in Europe). My revered teacher Rosenfeld, who had to correct my first writings in English a great deal, did not comment on this omission, although Møller was his personal friend. Together they had formulated what was then "the best" meson theory: a mixture of vector and pseudoscalar meson geared to suppress the r^{-3} singularity of the static potential [we would explain it today as the effect of its supermultiplet SU(4) symmetry].[37] This theory explained the deuteron tensor forces; the pseudoscalar lifetime was that of the cosmic-ray meson, while that of the vector meson was much shorter. At last, I can repair this omission today.

Some time after I had sent my first letter to *Nature*, we received in Manchester the 1949 issue of *Reviews of Modern Physics* containing the Tiomno and Wheeler paper.[38] There is no doubt of their priority, and no American physicist quoted my paper for five years, although Wheeler, very generously, propagandized for it. I say that for two reasons. First, I want to emphasize the generosity of the physicists of the generation previous to mine. To me, most of them were models that I tried to pass to my students. Second, one gets the impression from any textbook that the parameter was immediately used, but that was true only in Europe, and by French physicists.[39]

The second physics conference that I attended was at Edinburgh, on 14 November 1949, sitting beside Rosenfeld, while listening to the lecture of Norman Feather explaining that he had attempted to observe the spontaneous emission of antiprotons from nuclei. "Bien sur, c'est impossible!" said I to Rosenfeld. "Ah! Pourquoi?" he answered. As far as I know, there were no proceedings of this meeting; but there is a general photograph that shows Born, Darwin, Peierls, Møller, Fröhlich, Fierz, Proca, Pryce, Fuchs, Dyson, Matthews, Hamilton, Touscheck, Abragam, Bloch, Blackett, Powell, Wilson (J. G.), Rochester, Butler, Pontecorvo, Perkins, and others. (Niels Bohr is not in the photograph, but my wife and I remember that he and his wife Margrethe were at this meeting because we saw them there for the first time.) I cannot say why no one raised an objection. I was much too shy myself to do it, but I can say that later on I tried to convince Rosenfeld. When I succeeded, he advised me to write a letter to *Nature*. This was sent on 7 March 1950, with a copy for Feather.

I was at the Paris international conference in 1950, and I remember well the lectures by Källen and by Richard P. Feynman, whom I saw then for the first time. Of course, I was not invited to the UNESCO-sponsored conference held at the Tata Institute of Fundamental Research in Bombay, 14–22 December 1950. But I find it very astonishing that no physicist from the United States was invited. The list of invited participants is in the proceedings.

Also, on pages 117–23, in the contribution of Feather, "Recent Attempts to Detect Negative Protons and Dineutrons," the conclusion of the first half is that "thus all attempts to detect the p^- by experiments with low energy particles have failed. Michel has recently pointed out that just because the nuclei are stable and do not disappear spontaneously by the n, $n \rightarrow p^+$, p^- reaction, the spontaneous emission of negative proton from a nucleus is impossible." At the end of the lecture, Rosenfeld gave a two-page summary of my letter to *Nature*.[40] In it I spoke of "mesic charge." But this letter is complicated because it also considers the case of Majorana particles of spin $\frac{1}{2}$ (in his paper, Majorana thought wrongly that neutrons could be of that sort).[41] Feather was not the only one looking for antiprotons from nuclei. I also quoted a letter to the *Physical Review*, and I can add others.[42]

If I had been asked a few months ago about the conservation law for nucleons and its consequence for antiproton production, I would have answered that this was well understood from Dirac's 1931 paper. Thanks to this conference, I had to read, or recollect, much conflicting evidence. Wigner's short paper "On the Law of Conservation of Heavy Particles" is clear and basic; note that it does not speak of baryonic charge, but of "neutronic charge" in one paragraph and "mesonic charge" in the next one.[30] I reproduce here the first footnote, in which Wigner raises the question:

> It is difficult to trace the first statement of this principle. It is clearly contained in the writer's article in *Am. Philos. Soc. 93*, 521, (1949), but may have been recognized about that time also by others, cf. T. Okayama, *Phys. Rev. 75*, 308, (1949). C. N. Yang informs me that the purpose of introducing an imaginary character to the reflection properties of certain fermions in the paper of J. Tiomno and C. N. Yang (*Phys. Rev. 79*, 495, (1950)) was to explain this principle. Cf. also L. I. Schiff, *Phys. Rev. 85*, 374, (1952), and, in particular, P. Jordan, *Z. f. Naturf. 7a*, 78, (1952). [Remark: K. B. Jarkov is similar to Yang and Tiomno.[43]]

When I wrote my letter to *Nature*, I did not know the Wigner and Okayama references. Only footnote 9 of the former reference deals with the subject I am discussing: "It is conceivable, for instance, that a conservation law for the number of heavy particles (protons and neutrons) is responsible for the stability of the protons." Okayama's work, quoted by Wigner, is a letter to the *Physical Review*, "On the Mesic Charge."[44] For its author, "it is unlikely that a negative proton exists," but he did point out a new conservation law for nucleons if they obey the Dirac equation, and his remarks on charge conjugation in Dirac theory were new and right! There were also interesting papers on the conservation of heavy particles in Japan in 1951.[45]

I was very much interested to read that in 1947, when the proposal to build a 10-GeV accelerator (in Berkeley) was submitted to the AEC, to reduce the cost, the energy was first lowered to 5 GeV and immediately raised to 6 GeV, so that it was at least above the threshold of nucleon pair production.[46] This corresponds much better to what I believe happened in the history of particle physics, even if it conflicts with what I heard, read, and lived thirty-five years ago! I was only a low-ranking officer.

P.S.: Just after this manuscript was finished, I received a letter from Dalitz, to whom I am grateful for answering several questions I asked him. There seem to have been no proceedings of the Birmingham conference on "Fundamental Physical Theory" (23–26 July 1947), but there is a report by M. H. L. Pryce, in *Nature*, 11 October 1947, pp. 627–8. Similarly for the nuclear physics Harwell conference (18–19 September 1947), as *Nature* reports on 11 October 1947, pp. 492–4. Dalitz was at the 1948 Birmingham conference (14–18 September) and has from it a forty-page "Informal Report" due to M. L. Oliphant, Rudolph Peierls, and P. B. Moon. Oppenheimer and Hans Bethe attended this conference. The former gave a report on the work done in the United States on QED and radiative corrections. But he did not speak on μ decay or μ capture. This agrees with what I heard from Rosenfeld, who attended and spoke at this conference.

Notes

1 P. A. M. Dirac, "Unitary Representations of the Lorentz Group," *Proc. R. Soc. London, Ser. A 183* (1945), 284–95; E. Majorana, "Relativistic Theory of Particles with Arbitrary Intrinsic Momentum," *Nuovo Cimento 9* (1932), 335–44.

2 V. Bargmann, "Irreducible Unitary Representations of the Lorentz Group," *Ann. Math. 48* (1947), 568–640; J. M. Gelfand and M. A. Naimark, "Unitary Representations of the Lorentz Group" (in Russian), *Izv. Akad. Nauk. SSSR, Math. Ser. 11* (1947), 411–504.

3 E. P. Wigner, "On Unitary Representations of the Inhomogeneous Lorentz Group," *Ann. Math. 40* (1939), 149–204.

4 E. Wigner, "Ueber die Operation der Zeitumkehr in der quanten Mechanik," *Göttingen Nach.* (1932), 546–59.

5 E. P. Wigner, "Unitary Representations of the Inhomogeneous Lorentz Group Including Reflections," in *Group Theoretical Concepts and Methods in Elementary Particle Physics*, edited by Feza Gürsey (New York: Gordon & Breach, 1964), pp. 37–80.

6 L. D. Landau, "The Moment of a 2-Photon System," *Dok. Akad. Nauk SSSR 60* (1948), 207–9 (in Russian); J. Steinberger, "On the Use of Subtraction Fields and the Lifetimes of Some Types of Meson Decay," *Phys. Rev. 70* (1949), 1180–6.

7 C. N. Yang, "Selection Rules for the Dematerialization of a Particle into Two Photons," *Phys. Rev. 77* (1950), 242–5; D. C. Peaslee, "Absolute Selection Rules for Meson Decay," *Helv. Phys. Acta 23* (1950), 845–54.

8 L. Michel, "Coupling Properties of Nucleons, Mesons and Leptons," in *Progress in Cosmic Ray Physics*, edited by J. G. Wilson (Amsterdam: North Holland, 1957), pp. 127–85.

9 L. Michel, "Selection Rules Imposed by Charge Conjugation," *Nuovo Cimento 10* (1953), 319–39; L. Michel in *Proceedings Congrès International sur le Rayonnement Cosmique*, Bagnères de Bigorre, edited by M. Schein et al. (University of Toulouse, 1953), p. 272.

10 T. D. Lee and C. N. Yang, "Mass Degeneracy of the Heavy Mesons," *Phys. Rev. 102* (1956), 290–1; "Possible Interference Phenomenon Between Parity Doublets," *Phys. Rev. 104* (1956), 822–7.

11 T. D. Lee and C. N. Yang, "Question of Parity Conservation in Weak Interactions," *Phys. Rev. 104* (1956), 254–8.

12 P. Curie, "Sur la symètrie dans les phénomenes physiques, symetrie d'un champ électrique et d'un champ magnétique," *J. Phys. (Paris) 3* (1894), 393–415.

13 E. P. Wigner, "Einige Folgerungen aus der Schrödingerschen Theorie für die Termstrukturen," *Z. Phys. 43* (1927), 624–52.

14 E. Fermi, "Theory of β Rays," *Z. Phys. 88* (1934), 61–171; E. J. Konopinski and G. E. Uhlenbeck, "On the Fermi Theory of β-Radioactivity," *Phys. Rev. 48* (1935), 7–12; G. Racah, "Symmetry between Particles and Anti-Particles," *Nuovo Cimento 14* (1937), 322–8.

15 L. Michel, *Memorial des Poudres 35* (1953), Annexes; M. Schoenberg, "On the Theory of Integer Spin Mesons," *Phys. Rev. 60* (1941), 468.

16 B. Ferretti, "The Theory of the Collision between Protons and Rapid Neutrons. II," *Ricera Sci. 12* (1941), 993–1019.

17 H. Enatsu, "On the Nuclear Forces," *Prog. Theor. Phys. 5* (1950), 102–16; Bruno Touschek, "Zur Theorie des doppelten β-Zerfalls," *Z. Phys. 125* (1948), 108–32.

18 T. D. Lee and C. N. Yang, "Possible Determination of the Spin of Λ^0 from Its Large Decay Angular Symmetry," *Phys. Rev. 109* (1958), 1755–8.

19 G. Racah, "Theory of Complex Spectra. IV," *Phys. Rev. 76* (1949), 1352–65.

20 H. A. Jahn, "Theoretical Studies in Nuclear Structure. I. Enumeration and Classification of the States Arising from the Filling of the Nuclear d-Shell," *Proc. R. Soc. London, Ser. A 201* (1950), 517–44; H. A. Jahn, "Theoretical Studies in Nuclear Structure. II. Nuclear d^2, d^3 and d^4 Configurations. Fractional Parentage Coefficients and Central Force Matrix Elements," *Proc. R. Soc. London, Ser. A 205* (1951), 192–237.

21 Samuel Devons and L. J. B. Goldfarb, "Angular Correlations," in *Handbuch der Physik, Band 42, Nuclear Reactions III*, edited by S. Flügge (Berlin: Springer-Verlag, 1957), pp. 362–554, especially pp. 362–70.

22 A. Bohr, "Relation Between Intrinsic Parities and Polarizations in Collision and Decay Processes," *Nucl. Phys. 10* (1959), 486–91.

23 M. Jacob and G. C. Wick, "On the General Theory of Collisions for Particles with Spin," *Ann. Phys. (N.Y.) 7* (1959), 404–28.

24 M. G. Doncel, "Polarization of Resonances in a Simple Peripheral SU_{6w} Model," *Nuovo Cimento 52A* (1967), 617–27; P. Minnaert, "Positivity Conditions for the Density Matrix of Spin-One Particles," *Phys. Rev. Lett. 16* (1966), 672–4.

25 M. G. Doncel, P. Mery, L. Michel, P. Minnaert, and K. C. Wali, "Properties of Polarization Density Matrix in Regge-Pole Models," *Phys. Rev. D7* (1973), 815–35; L. Michel, "Analysis of Polarization Measurements and Test of Selection Rules and of Models," in *Proceedings of the Summer Studies on High-Energy Physics with Polarized Beams*, edited by J. B. Roberts (Argonne National Laboratory, 1975), pp. XXVII 1–18.

26 M. G. Doncel, L. Michel, and P. Minnaert, "Isospin Constraints Between Three Cross Sections and Two Polarization Density Matrices," *Phys. Lett. B42* (1972), 96–8.

27 L. Michel, *First International Congress on the History of Scientific Ideas*, San Feliu de Guixals, edited by M. Doncel and K. von Meyenn (in press).

28 R. Jost, "A Remark on the Relation between the Scattering Phases and the Potential" (in German), *Helv. Phys. Acta 29* (1956), 410–18; G. Lüders, "On the Equivalence of Invariance under Time Reversal and under Particle–Antiparticle Conjugation for Relativistic Field Theories," *K. Dan. Vidensk. Selsk., Mat.-Fys. Medd. 28* (1954), 17; "Exclusion Principle, Lorentz Group and Reflection of Space-Time and Charge," in Wolfgang Pauli (ed.), *Neils Bohr and the Development of Physics* (New York: Pergamon, 1955), pp. 30–51.

29 N. Kemmer, "Isospin," in Colloque International sur l'Histoire de la Physique des Particules, *J. Phys. (Paris) (Suppl.) 43:12* (1982), 359–90.

30 E. P. Wigner, "On the Law of Conservation of Heavy Particles," *Proc. Nat. Acad. Sci. U.S.A. 38* (1952), 449–51.

31 M. Gell-Mann, "Strangeness," in Colloque International sur l'Histoire de la Physique des Particules, *J. Phys. (Paris)(Suppl.) 43:12* (1982), 395–402.

32 See M. Gell-Mann, Y. Ne'eman, M. Nauenberg, and T. W. Speiser, in *First International Congress on the History of Scientific Ideas*, San Feliu de Guixals, edited by M. Doncel and K. von Meyenn (in press).

33 L. D. Landau, "Theory of Phase Transformations. Part II" (in German), *Phys. Z. Sowjetunion 11* (1937), 545–55. L. D. Landau and E. M. Lifschitz, *Statistical Physics*, translated from Russian by J. B. Sykes and M. J. Kearsley (New York: Pergamon, 1958).

34 L. W. Alvarez, H. Bradner, F. S. Crawford, Jr., J. A. Crawford, P. Falk-Variant, M. L. Good, J. D. Gow, A. H. Rosenfeld, F. Salmitz, M. L. Stevenson, H. K. Ticho, and R. D. Tripp, "Catalysis of Nuclear Reactors by μ Mesons," *Phys. Rev. 105* (1957), 1127–8; F. C. Franck, "Hypothetical Alternative Energy Sources for the 'Second Meson' Events," *Nature (London) 160* (1947), 525–7.

35 *Cosmic Radiation: The Colston Papers* (London: Butterworth, 1949).

36 J. J. Horowitz, O. Kofoed-Hansen, and J. Linhard, "On the β-Decay of Mesons," *Phys. Rev. 74* (1948), 713–17.

37 C. Møller and L. Rosenfeld, "On the Field Theory of Nuclear Forces," *K. Dan. Vidensk. Selsk., Mat.-Fys. Medd. 17:8* (1940), 1–72.
38 J. Tiomno and J. A. Wheeler, "Energy Spectrum of Electrons from Meson Decay," *Rev. Mod. Phys. 21* (1949), 144–52; J. Tiomno and J. A. Wheeler, "Charge-Exchange Reaction of the μ-Meson with the Nucleus," *Rev. Mod. Phys. 21* (1949), 153–65.
39 L. Leprince-Ringuet, in *Report of an International Conference on Elementary Particles* (Bombay: Tata Institute of Fundamental Research, 1950), pp. 25–9; Charles Peyrou and Andre Lagarrique, "Mesure des Masses sur les Particules du Rayonnement Cosmique," *J. Phys. (Paris) 11* (1950), 666–72.
40 L. Michel, "Reactions between Nucleons and Mesons, and the Search for Negative Protons," *Nature (London) 166* (1950), 654–5.
41 E. Majorana, "Symmetrical Theory of Electrons and Positrons," *Nuovo Cimento 14* (1937), 171–84; K. H. Sun, "On the Negative Proton," *Phys. Rev. 76* (1949), 1266.
42 F. J. Belinfante, "On the Current and the Density of the Electric Charge, the Energy, the Linear Momentum, and the Angular Momentum of Arbitrary Fields," *Physica 7* (1940), 449–74; E. Broda, N. Feather, and D. H. Wilkinson, "A Search for Negative Protons Emitted as a Result of Fission," in *International Conference on Fundamental Particles and Low Temperature, Vol. I* (London: The Physical Society, 1947), pp. 114–25; J. M. C. Scott and E. W. Titterton, "Search for the Negative Proton in Fission," *Philos. Mag. 41* (1950), 918–20.
43 K. B. Jarkov, *Zh. Eksp. Teor. Fiz. [Sov. Phys. – JETP] 20* (1950), 492.
44 T. Okayama, "On the Mesic Charge," *Phys. Rev. 75* (1949), 308; S. Oneda, "On Some Properties of the Interactions between Elementary Particles," *Prog. Theor. Phys. 9* (1953), 327–44.
45 Y. Nambu, K. Nishijima, and Y. Yamaguchi, "On the Nature of *V*-Particles, I–II," *Prog. Theor. Phys. 6* (1951), 615–22.
46 U. Mersits, in *Studies in CERN History, Vol. 17* (Geneva: CERN, 1985).

27　A connection between the strong and weak interactions

SAM B. TREIMAN

Born 1925, Chicago, Illinois: Ph.D., University of Chicago, 1952; theory of fundamental particles, field theory; Princeton University

---

　　　　In the mid-1950s, my colleague Marvin L. Goldberger had acquired a leading position in the burgeoning field of dispersion relations (i.e., the study and exploitation of analyticity properties of scattering amplitudes). To speak of scattering in that era was to speak of strong or perhaps electromagnetic reactions, not weak ones. The weak interactions were, of course, of great and growing interest; but the focus was mainly on *decay* of weakly unstable particles, rather than on weak *scattering* reactions. Even for the simplest kind of scattering process (e.g., forward pion–nucleon scattering) there *is* a continuous physical variable, the energy. It is a well-posed question, therefore, to ask if the amplitude function can be continued into the complex energy plane, to ask what singularities one encounters there, and so forth.

　　It turned out that these inquiries and others that grew out of them were illuminating and fruitful. In fact, they were among the dominant themes of particle physics throughout much of the 1960s. On the other hand, consider a simple one-body-to-two-body decay reaction (e.g., $\pi \to \mu + \nu$ decay). Here there is no continuous physical variable, no spectrum, just a fixed number for the amplitude – thus, seemingly, no role for analyticity thinking. Maybe this judgment is, in fact, correct. But let us not get ahead of our story; we return to Goldberger. While he was conducting his dispersion-relation pyrotechnics, I was busy pursuing various interests in weak-interaction physics. However, he managed, all the while, to keep a somewhat avuncular eye on my doings, and I tried to keep up with his. One day in 1957 he announced to me that the time had come to train the big guns of analyticity on the weak interactions. "What needs doing?" he asked me. What followed is a very minor footnote in the story of mankind. It happens, however, to be my topic today. Luckily for all of us, I never flattered myself that I should save the worksheets, correspondence, or other memorabilia and documentation. My remarks are therefore based only on a possibly faulty memory and, of course, on the published literature.

One thing I do remember well is that Goldberger entered into our project in a somewhat combative spirit. He had learned – I do not recall by what means and how reliably – that N. N. Bogoliubov had somewhere expressed the opinion already noted earlier: namely, that there was no role for analyticity thinking in the weak-interaction physics of the time. Goldberger was not prepared to concede such a thing. There were no limits to the power of his dispersion relations. Anyhow, we decided to launch a dispersion attack on $\pi \rightarrow \mu + \nu$ decay. Why did we choose that weak process rather than another? The answer, within the framework of the demented methods that we were going to employ, is that the empirical data we would need (e.g., the strength of the pion–nucleon coupling constant) were more readily available there than for other processes we could think of.

After rather lengthy and brutal calculation, we came to the following relation (expressed here in modern notation):[1]

$$f_\pi = \frac{m g_A}{g_{\pi NN}}$$

where f_π is the pion decay constant, m is the nucleon mass, g_A is the axial vector coupling constant in nucleon β decay, $g_{\pi NN}$ is the strong-interaction pion–nucleon coupling constant. Actually, the formula that we obtained was as quoted here, but with a multiplying factor that I shall here denote by $(1 - r)$. We had a definite expression for the quantity r, but it is much too complicated to reproduce here. It involved a dispersion integral whose integrand depended on, of all things, the complex phase shift for neutron–antiproton scattering in the $^1S_0$ state. This phase shift was not then, and is not now, known; nevertheless, we could convince ourselves that r must be very small compared with unity. We advocated the view, and others took our "theory" to imply, that r is essentially zero. Empirically, the quantity Δ defined by

$$\Delta = 1 - \frac{m g_A}{f_\pi g_{\pi NN}}$$

was already remarkably small at the time, and is still small now. With current data, $\Delta \simeq 0.06$.

The dispersion treatment and approximations that we adopted are not easily summarized. (Even less easily can they be justified.) Roughly, however, the dispersion variable was taken to be the square of the off-shell pion mass. The absorptive part of the pion decay "constant," regarded as a function of this variable, involves transitions from off-shell pion to a complete set of intermediate states, the latter coupling through the axial current to the leptons. In those prequark days, when nucleons were the preeminent funda-

mental particles, we deemed it reasonable to assume that the dominant intermediate state must be the nucleon–antinucleon pair state. This brings in
the pion–nucleon vertex function, parameterized in part by nucleon pair
scattering (hence the appearance of the pair-scattering phase shift in r). The
coupling of nucleon pair to the leptons brings in g_A and so on. As said, we
went through a lengthy sequence of steps. The final result, however, was a
simple formula that provided what was, for the time, a rather remarkable and
unexpected connection between strong- and weak-interaction quantities. The
formula itself was accorded quite a good reception right from the start. Even
the steps that led to it were given respectful attention, at least in some
quarters. After all, there was still a certain awe in those days about the magic
of dispersion relations. Moreover, the agreement with experiment was good –
surely, we told ourselves and others, sloppy reasoning could not have produced such a happy result.

Well, there were some people who thought it could. They set off in search
of more persuasive hypotheses. Two years after Goldberger and I published
our paper, Yoichiro Nambu came out with a different and much simpler
approach.[2] He was intrigued by an idea that had been in the air: namely, a
possible axial-current analogue of the conserved-vector-current hypothesis of
Richard Feynman and Murray Gell-Mann.[3,4] One knew that the axial current
A_μ could not be strictly conserved, but maybe, thought Nambu, it is *almost*
conserved. Operationally, here is what he meant by that (I alter the notation
somewhat). Consider the matrix element of A_μ between nucleon states:

$$\langle p'|A_\mu|p\rangle = \bar{U}(p')\Gamma_\mu^A U(p)$$
$$\Gamma_\mu^A = ig_A(q^2)\gamma_\mu\gamma_5 - G_A(q^2)q_\mu\gamma_5, \qquad q = p - p'$$

The form factor $G_A(q^2)$ has a pole at $q^2 = -m_\pi^2$, arising from exchange of a
virtual pion between nucleons and axial current. If only the pole-term contribution is kept, one has

$$G_A(q^2) = \frac{2f_\pi g_{\pi NN}}{q^2 + m_\pi^2}$$

Suppose, moreover, that the form factor $g_A(q^2)$ depends little on q^2 (over a
range large compared with m_π^2). The almost-conserved-axial-current hypothesis asserts that the matrix element of the divergence $\partial_\mu A_\mu$ essentially
vanishes once $|q^2|$ is large compared with m_π^2 (i.e., once one is well away from
the pole). These suppositions, taken together, immediately lead to the
equation that Goldberger and I had propounded.

Another approach, developed at about the same time, was proposed by
Gell-Mann and Maurice Levy.[5] The divergence of the axial vector current,
$\partial_\mu A_\mu$, has the quantum numbers of the pion. Even if this divergence were a
composite operator formed out of the "canonical" fields of the underlying
theory, one could take it (up to a suitable proportionality constant) to be the

operator that in these theories creates and destroys pions. However, Gell-Mann and Levy thought the divergence would be an especially *gentle* operator if the theory really contained a canonical pion field and if $\partial_\mu A_\mu$ were literally proportional to this field. They developed and discussed several model field theories that achieved this identification of axial divergence with pion field. In particular, they devised their celebrated σ model. It nicely implemented a Nambu–Goldstone realization of $SU(2) \times SU(2)$ chiral symmetry and symmetry breaking, and it took on quite a life of its own. For present purposes it is enough to record the PCAC (partially conserved axial vector current) relation between the axial current divergence and the canonical pion field ϕ_π:

$$\partial_\mu A_\mu = f_\pi m_\pi^2 \phi_\pi$$

Assuming that the pion field, hence the current divergence, is a gentle operator, Gell-Mann and Levy could quickly recover the equation that Goldberger and I found.

The axial current is said to be partially conserved in the sense that the coefficient on the right-hand side of the foregoing relation is proportional to the square of the notoriously small pion mass. If the axial current were exactly conserved, this, taken together with conservation of the vector current, would imply $SU(2) \times SU(2)$ symmetry of the underlying strong-interaction theory. Moreover, from the foregoing equation one would conclude that either $m_\pi = 0$ or $f_\pi = 0$. The latter would correspond to a conventional Wigner–Weyl realization of the symmetry, implying the existence of parity doublets. This does not remotely resemble the real world. The former corresponds to the favored Nambu–Goldstone realization, with the pion playing the role of Goldstone boson, and looks much more like the real world. Although not precisely massless, the pion *is* very light. There are deep issues still alive today in the context of quantum chromodynamics concerning chiral symmetry breaking. But those current issues are not part of my present story.

From the point of view of practical developments back in the 1960s, the historical point is that the PCAC identification, especially when adjoined to the ideas of current algebra, began to produce other successes going well beyond what Goldberger and I had done. There were the Adler consistency conditions and the Adler–Weisberger relation for pion–nucleon scattering – and then a flood of other authors and other applications.[6] There was and still is the question why PCAC works out quantitatively as well as it does – with accuracy, sometimes, in the 10–20 percent range. The lightness of the pion helps, because PCAC can be looked on in some sense as a kind of pion pole dominance principle. But the natural parameter of smallness for measuring departures from PCAC is, up to numerical factors, something like (m_π^2/f_π^2), and one has to face the fact that $f_\pi \approx 93$ MeV is also small. What may save the situation are numerical factors. From the experience of attempts at computing corrections to PCAC, Heinz Pagels argues that the parameter of smallness is really something like $(\frac{1}{32}\pi^2)(m_\pi^2/f_\pi^2)$.[7] Perhaps.

Back to the early history. When Goldberger and I set out to attack pion decay, we knew that the axial current could not be conserved. Indeed we said so in a publication of about that time, in response to suggestions to the contrary that had appeared in the literature.[8] We had no further reason, therefore, to focus on the divergence of the axial current; we had no reason to substitute the notion of "almost conservation." Had we thought of chiral symmetry, or "almost chiral symmetry," we would have dismissed it out of hand, on the grounds that the Wigner–Weyl realization is untenable. The idea of broken symmetry was not in the air in our circles. Dispersion relations were the thing, and if one had to make some drastic approximations along the way in using them, one did so. Goldberger and I were of course pleased with our nice final result for pion decay. It didn't occur to us that such a simple outcome should suggest that there are simpler explanations. What we *did* do, in order to check certain matters of principle in the application of dispersion relations to decay processes, was invent a problem analogous to pion decay in the framework of the soluble Lee model. We tested the dispersion methods there against an easily obtained direct answer. All was in agreement; we published a paper on this and turned to other things.[9]

But as I have already related, other people did not let it go at that. Nambu was very gentle with us. In his paper on the almost-conserved axial current, a paper that made a very deep impression on me, he merely noted softly that Goldberger and I had arrived at our formula from an entirely different approach than his own. He did not blast us, as he might have been tempted to do. Gell-Mann and Levy let us have it more forthrightly. Here is a quotation from their paper: "Their derivation. . .involves several violent approximations which are not really justified." Gell-Mann and Levy *do* go on to say that "the formula, however, is in excellent agreement with experiment."

Gell-Mann came to give a colloquium at Princeton early in the 1970s, as I recall, and there, still rankling after a decade, he was even more outspoken about the quality of our approximations. There were numerous other jabs by others, as well, over the years. It was interesting in a way to be such a center of attention, but frankly, I would have preferred being treated with a bit more respect. In the hope of recovering at least some self-respect, I remember one time asking myself: Would the accursed formula have been discovered by these newfangled hypotheses and reasonings if Goldberger and I had not stumbled onto it first? Remember, the relation we had found connecting strong and weak parameters was quite unusual for its time. The question is hardly a momentous one. Still, for what it's worth, I can tell you that my guess is that Nambu probably would have done what he did, and maybe others, too, would have done what they did. The one thing I *am* certain of is that no one but Goldberger and I would have had the effrontery to do what Goldberger and I did.

Notes

1 M. L. Goldberger and S. B. Treiman, "Form Factors in β Decay and μ Capture," *Phys. Rev. 111* (1958), 354–61.
2 Y. Nambu, "Axial Vector Current Conservation in Weak Interactions," *Phys. Rev. Lett. 4* (1960), 380–2.
3 J. C. Taylor, "Beta Decay of the Pion," *Phys. Rev. 110* (1958), 1216; J. C. Polkinghorne, "Renormalization of Axial Vector Coupling," *Nuovo Cimento 8* (1958), 179–80, 781.
4 R. P. Feynman and M. Gell-Mann, "Theory of the Fermi Interaction," *Phys. Rev. 109* (1958), 193–8.
5 M. Gell-Mann and M. Levy, "The Axial Vector Current in Beta Decay," *Nuovo Cimento 16* (1960), 705–26.
6 M. D. Scadron, "Current Algebra, PCAC, and the Quark Model," *Rep. Prog. Phys. 44* (1981), 213–92; Sam B. Treiman, Roman Jackiw, and David J. Gross, *Lectures on Current Algebra and Its Applications* (Princeton University Press, 1972).
7 H. Pagels, "Departures from Chiral Symmetry," *Phys. Rep. 16C* (1975), 219–311.
8 M. L. Goldberger and S. B. Treiman, "Conserved Currents in the Theory of Fermi Interactions," *Phys. Rev. 110* (1958), 1478–9.
9 M. L. Goldberger and S. B. Treiman, "A Soluble Problem in Dispersion Theory," *Phys. Rev. 113* (1959), 1663–9.

28   The weak interactions from 1950 to 1960: a quantitative bibliometric study of the formation of a field

D. HYWEL WHITE

Born 1931, Cardiff, Wales; Ph.D., University of Birmingham, 1956; experimental high-energy physics; Los Alamos National Laboratory

DANIEL SULLIVAN

Born 1944, Bethesda, Maryland; Ph.D., Columbia University, 1971; sociology and history of science; President, Allegheny College

Background and introduction

About ten years ago, a group of sociologists and scientists gathered at Cornell University in a new program of research designed to discover and validate new methods for studying the evolution and health of scientific fields using quantitative measures. The interests of the group's founders lay both in finding ways to obtain a better basic understanding of how scientific fields develop – how, for example, the social and intellectual aspects of a scientific community interact – and in producing work that might have an influence on the way in which science policy is formulated by showing policy analysts better ways of monitoring the outputs of science and relating them to the inputs.

We believed that the best way to proceed with this program was to undertake several case studies of scientific specialties using teams of collaborators made up of sociologists and scientists familiar with the specialties in question. We believed, further, that one case study should focus on a field with a strong and highly developed theoretical basis that also involved major experimental work. That meant a field like particle physics, and we thought that within particle physics the study of the weak interactions might be a manageable specialty on which to focus.

A large and comprehensive data base was constructed, comprised of all serial articles published in the specialty from 1950 to 1975 ($N = 5,765$), and our own weak-interactions citation index was constructed from the refer-

*   Work supported by NSF grant GS-41697 to the Research Program on Social Analyses of Science Systems (SASS), Cornell University, Ithaca, New York, and NSF grant SOC76-84482 to Carleton College, Northfield, Minnesota.

ences that appear at the end of the serial articles. Over 80,000 references, originating in one of the articles in our weak-interactions bibliography (citing articles both within our bibliography and in other fields), make up our citation index. We have therefore been able to supplement the reading of important papers, reviews, proceedings, and other documents with types of quantitative analysis of the development of the field of weak interactions that were developed especially for this purpose. These quantitative analyses have given us a view of trends and patterns for the specialty as a whole that is unavailable from traditional techniques of historical study.[1] We deal with the period from 1950 to 1960 in this chapter.

We shall first give an overview of what the physicists working in weak interactions during this period were doing, as indicated by an analysis of the subjects of their papers. The dominant problems and concerns are discussed. We shall then focus on the events surrounding the emergence of the $\tau-\theta$ puzzle, the discovery of parity nonconservation, and the resolution offered by the $V-A$ theory. For this latter analysis we take advantage of data from our citation index, displayed in some unusual ways that highlight the dominant issues of the period and the especially close relationship between theory and experiment in the latter half of the decade.

The formation of the field

The idea that weak interactions should form a separate field of study became well accepted only in the late 1940s, when it emerged that a number of weak reactions, including β decay and μ decay, could be described by approximately the same coupling constant. By the beginning of the 1950s, both the pion and muon were known, K mesons had been observed in cosmic rays, and it was appreciated that there existed a class of heavy strange particles, subsequently called hyperons, that also had been observed in cosmic rays. The attention of a rapidly increasing number of physicists began to be focused on this new field, though initially weak interactions drew many more experimentalists to work on its mysteries than theorists.[2] In Figures 28.1 and 28.2 we show the numbers of different theorists and experimentalists (that is, an author was counted only once in a year in which he published) from all over the world who published in weak interactions each year during 1951–60. The rapid growth observed was due, we believe, to a combination of growth in the scientific community as a whole during this period and the relatively attractive character of weak interactions as a field that drew physicists to it from other fields. By 1957, nearly 400 physicists had published at least once in weak interactions.

The patterns of theoretical and experimental article production are shown in Figures 28.3 and 28.4. The number of experimental articles published each year grew more rapidly than the number of theoretical articles until 1957 and 1958, when theorists published over 160 articles, almost four times the number published in 1956. The huge increase was due entirely to the

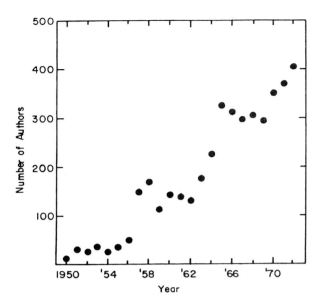

Figure 28.1. Number of different theoretical authors who published each year.

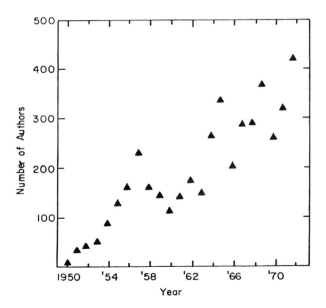

Figure 28.2. Number of different experimental authors who published each year.

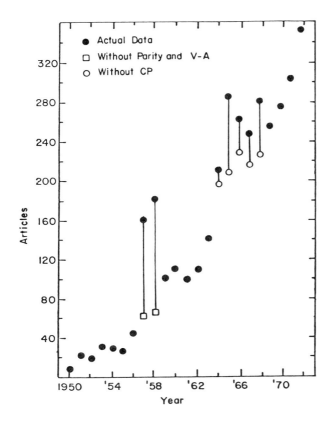

Figure 28.3. Number of theoretical articles per year.

discovery of parity nonconservation in 1956, about which we shall have much more to say later. The bottom points shown for 1957 and 1958 in Figure 28.3 represent the number of articles published in those years exclusive of those concerned with parity. Such an outpouring of papers by theorists could occur because they did not have to design experiments, construct them, take data, and analyze data before publishing, as do experimentalists. Their ability to change their research focus on short notice and get into print is much greater than in the case of experimentalists.

Figure 28.5 gives a capsule summary of what experimentalists were doing during this period. The analysis of $\pi-\mu$ decay dominates the early years, since examples were plentiful in cosmic-ray experiments, and pions and muons could be made plentifully in the synchrocyclotrons then available. With the construction of larger accelerators (the Cosmotron came on line in 1952, and the Bevatron in 1954), the decays and characteristics of heavier particles involved in weak interactions could be studied, and one saw a

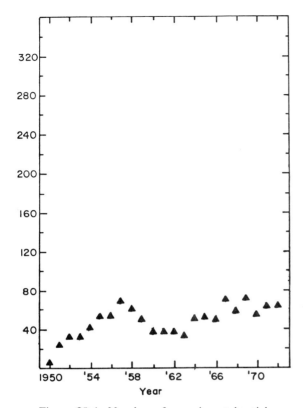

Figure 28.4. Number of experimental articles per year.

growing emphasis on experiments concerned with K decay and with the analysis of hyperons and their decays. The several experiments concerned with the $\tau-\theta$ puzzle, which began to capture the attention of more and more people in the field in the middle of the decade, are included here in the K decay numbers.

T. D. Lee and C. N. Yang's paper on parity nonconservation transformed the field in 1956,[3] and the effect is dramatically visible in Figure 28.5. Figure 28.6 is a comparable plot for theoretical papers, dominated almost entirely by the emergence of the $\tau-\theta$ puzzle, parity nonconservation, and $V-A$. (We assume that with this audience we do not have to describe in detail the questions of physics involved during those exciting years.) What is remarkable to us, however, is the extent to which, first, the $\tau-\theta$ puzzle, then parity nonconservation, and finally $V-A$ theory dominated the publications in the field. We know of no parallels in other scientific fields where such a high fraction of the players' attention became focused so closely on a single set of issues. We are tempted to say that such concentration is possible only in fields

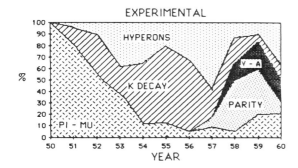

Figure 28.5. Percentage of experimental articles devoted to each topic.

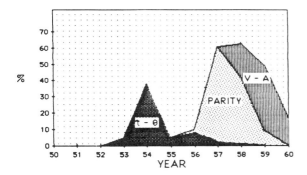

Figure 28.6. Percentage of theoretical articles devoted to each topic.

where the central theoretical structure is both coherent and basic to all of the work that can be done.

There is a technique of analysis (which we have used extensively in other contexts and which we have had a role in refining) that helps to illuminate the intellectual activities in a field and also gives a quantitative dimension.[4] It is called cocitation analysis. We shall use it to focus on the period 1956–60, which includes the discovery of parity nonconservation and the development and validation of $V-A$ theory.[5]

Parity and $V-A$

In a cocitation analysis, one examines the referencing patterns in the papers published in a given year y to see what research questions, *defined by the content of the papers being cited*, were of common concern to the researchers publishing in that year. A start is made by identifying the papers cited most frequently by the weak-interactions papers published in year y. In the analysis presented here, only papers cited ten or more times in papers

*Konopinski & Mahmoud   '53

Figure 28.7. Cocitation multidimensional scaling plot for weak interactions in 1956.

published in a given year are considered eligible for inclusion in a cocitation cluster. Then a count is made of the number of times these highly cited papers are cited together in papers published in year y. We infer that groups of earlier papers that are cited *together* frequently (in this case cocited seven or more times) by papers published in year y defined a common focus for the papers published in y that made reference to them. Then, using a fitting technique called multidimensional scaling, one can produce a two-dimensional plot that arranges the papers in a plane such that their planar distances correspond as closely as possible to their relative cocitation levels.[6]

Figure 28.7 shows the kind of result the multidimensional scaling program produces. Nine papers (see Appendix) were cited ten or more times by weak-interactions papers published in 1956 and were cocited seven or more times with at least one other paper.[7] The multidimensional scaling program clustered them into four groups. The lower center group is made up of four papers, two by Richard Dalitz, one by E. Fabri, and an experimental paper by G. Harris, Jay Orear, and S. Taylor. The Dalitz and Fabri papers defined the $\tau-\theta$ puzzle, and the paper of Harris and associates, concerned with the lifetimes of K^+ mesons in two- and three-body decays, presented data that those publishing in 1956 believed to be crucial to discussions of the $\tau-\theta$ puzzle. The lower left cluster contains two papers, both experimental and coming a little later, that presented data and analyses on the lifetimes of the K meson in two-body decay modes and that were only peripherally related to the $\tau-\theta$ puzzle. The lower right cluster has two papers devoted to the expected $K_1^0-K_2^0$ lifetime difference. Finally, the top center cluster is a single paper, by Emil Konopinski and Hormoz M. Mahmoud, in which the STP or VA character of the weak interaction is discussed, together with implications of the particle–antiparticle assignments for the μ^+.

We believe the contents of the four clusters of papers shown in Figure 28.7

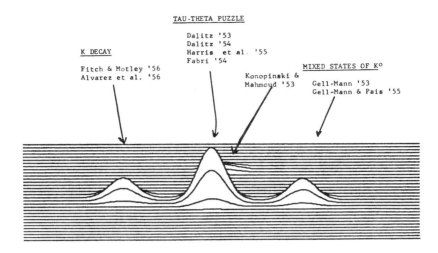

TAU-THETA PUZZLE

Dalitz '53
Dalitz '54
Harris et al. '55
Fabri '54

K DECAY

Fitch & Motley '56
Alvarez et al. '56

Konopinski &
Mahmoud '53

MIXED STATES OF K⁰

Gell-Mann '53
Gell-Mann & Pais '55

Figure 28.8. Cocitation hill diagram, weak interactions, 1956.

represent quite well the issues to which physicists working in weak interactions gave their attention during much of 1956. To better indicate the degree of emphasis those working in the field placed on each cluster of activity, we devised the plots shown in Figures 28.8–28.12. We begin with the papers clustered as in Figure 28.7. Now imagine that associated with each paper on the plane there is a third dimension perpendicular to the plane that corresponds to the number of times each paper was cited. Each paper is then treated as if it were a cylindrically symmetric Gaussian hill whose height is the number of times it was cited overall (its visibility) and whose standard deviation is *arbitrarily* set so as to produce pictures at different levels of resolution. The Gaussians are then added together, and the resulting plot of hills and valleys gives a sense of how much activity was going on in different parts of what might be called the "subject space." In some of these plots we present simultaneously a picture with small standard deviations (so that the location of specific papers can be seen) and one with larger standard deviations that give a good sense of the overall pattern.

The new information provided in Figure 28.8 is a clear indication of how important and dominant the concern with the $\tau-\theta$ puzzle was. The hill representing the papers of Dalitz, Fabri, and Harris and associates is large and central.

The year 1957 yields the cocitation hill diagram shown in Figure 28.9. It is quite different from the picture of the previous year. Lee and Yang had asserted that parity may not be conserved (center hill), explaining the $\tau-\theta$ puzzle, and a major experimental effort was launched to determine if they were right (left hill). The papers verifying parity violation form that peak

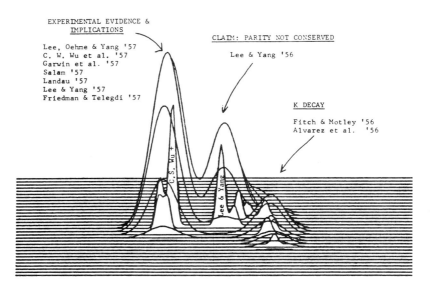

Figure 28.9. Cocitation hill diagram, weak interactions, 1957.

(C. S. Wu et al., Richard L. Garwin, Leon Lederman and M. Weinrich, and J. Friedman and Val Telegdi), along with several that extended the implications of parity violation (Salam, Landau, and a two-component neutrino theory by Lee and Yang). Lee, Oehme, and Yang discuss the $K^0-\bar{K}^0$ system in the light of parity violation.

To the right and right foreground are some of the older K-decay-lifetime measurements, together with the K_2^0 discovery by Lande and associates, discussion of the differential absorption of $K^0-\bar{K}^0$ by Abraham Pais and Oreste Piccioni, an older paper by G. Takeda discussing the two-body decay modes of the θ, and the paper by Murray Gell-Mann and Pais on the $K_1^0-K_2^0$ effect.

Figure 28.9 is a dramatic representation of the dominant concerns in weak interactions in 1957. It shows not only the clusters of activity but also how important they were, relative to each other, in the eyes of those working in the field at the time. Parity had been found not to be conserved, but a general explanation for that, one that could form the basis for an ongoing theoretical and experimental research program, had not yet been developed.

However, in late 1957, a general theory of the weak interactions – the universal $V-A$ theory – was proposed by E. C. G. Sudarshan and Robert Marshak and by Richard P. Feynman and Gell-Mann.[8] Because only serial articles are included in our data base, the Sudarshan and Marshak paper from the Padua–Venice conference is not included in the cocitation plot. However, its absence does not, we believe, alter the general structure that is

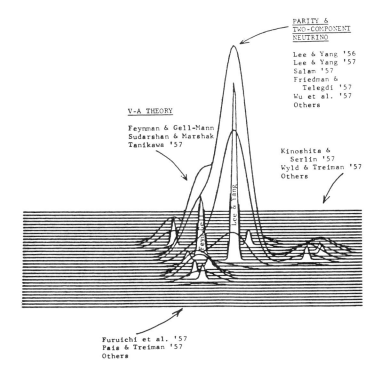

Figure 28.10. Cocitation hill diagram, weak interactions, 1958.

revealed in Figure 28.10. It takes some time for new ideas to make their way into the published literature. That is why the large center hill is still dominated by parity nonconservation. The 1956 Lee and Yang paper and their paper describing a two-component theory (which received the most citations in 1958) dominate the central region. They are joined there by the experimental papers verifying parity nonconservation, which are subsumed in the central region. Feynman and Gell-Mann on the theory of the Fermi inter-action $(V-A)$, with the *Physical Review* paper of Sudarshan and Marshak and the renormalization paper of Yasutaka Tanikawa, form a closely related hill to the left. On the other side and in the foreground, analysis of decay pro-cesses by Toichiro Kinoshita and A. Sirlin, H. W. Wyld and Sam Treiman, and Pais and Treiman form lower peaks.

With the 1959 plot, a further interpretation of the 1958 data becomes possible. By 1959, $V-A$ dominates the scene. One can think of the $V-A$ hill in the 1958 plot as moving toward center stage and about to displace the hill concerned primarily with parity nonconservation. It is remarkable how quickly the content of the literature in this field changed in those years.

We present Figure 28.11 without the fine structure of the internal detail.

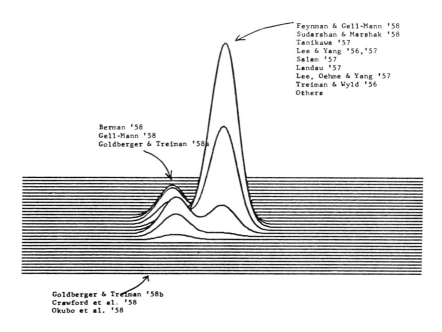

Feynman & Gell-Mann '58
Sudarshan & Marshak '58
Tanikawa '57
Lee & Yang '56,'57
Salam '57
Landau '57
Lee, Oehme & Yang '57
Treiman & Wyld '56
Others

Berman '58
Gell-Mann '58
Goldberger & Treiman '58

Goldberger & Treiman '58b
Crawford et al. '58
Okubo et al. '58

Figure 28.11. Cocitation hill diagram, weak interactions, 1959.

In the central peak are Feynman and Gell-Mann, Sudarshan and Marshak, Tanikawa, Lee and Yang, Salam, Landau, and others. In the satellite peak to the left front is an experimental article by Frank Crawford and associates on β decay of the Λ. All the other experiments are gone. The remaining satellites are theoretical articles exploiting $V-A$ theory by Jun John Sakurai, Marvin L. Goldberger, and Treiman, Susumu Okubo and associates, Gell-Mann, and S. M. Berman.

One still senses great excitement in this field in 1959. The structure of the cocitation plot conveys focus and sense of purpose, and the height of the central hill shows a high level of activity. By 1960, the field appears to pause intellectually (Figure 28.12). A paper by F. Zachariasen is the only new entry to the central hill, exploring K-meson decay as a test of the universality of μ and e, and the central hill is lower, which means that references in 1960 papers were less concentrated. Figures 28.13 and 28.14 provide dramatic indicators of this return to normal, and at the same time they show how stunning the discovery of parity nonconservation and the subsequent development of $V-A$ theory were to the physicists working in the field.

In Figure 28.13 we display what is called a concentration ratio – the percentage of references in theoretical weak-interactions articles published in a given year that go to the most highly cited 5 percent of all weak-interactions articles cited by theory articles in that year – for the entire

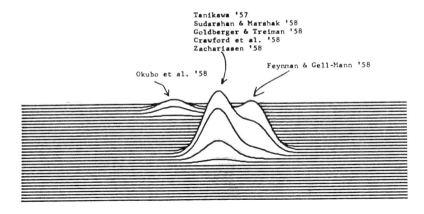

Tanikawa '57
Sudarshan & Marshak '58
Goldberger & Treiman '58
Crawford et al. '58
Zachariasen '58

Okubo et al. '58

Feynman & Gell-Mann '58

Figure 28.12. Cocitation hill diagram, weak interactions, 1960.

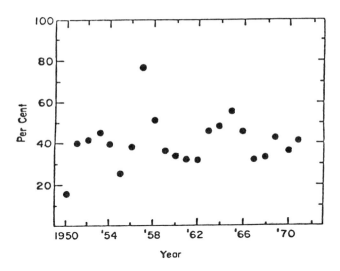

Figure 28.13. Concentration ratio: percentage of references in theoretical articles that go to the most highly cited 5 percent of all weak-interactions articles published that year.

period from 1950 to 1971. In 1957, the first full year in which theory articles could cite Lee and Yang, almost 80 percent of all the references in the theory articles published went to the most highly cited 5 percent. That is a remarkable level of concentration on a tiny fraction of the literature in the field. By 1959, the concentration ratio dropped below 40 percent and continued to decline until 1963. We know from other analyses that the increase in 1963, and the even higher peak in 1965, were due, respectively, to the interest

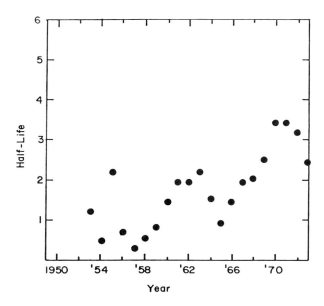

Figure 28.14. Reference half-life: theoretical weak-interactions articles citing other theoretical weak-interactions articles.

caused by the $\Delta S = \Delta Q$ selection-rule controversy and CP violation, which came on the scene in 1962 and 1964.

Another indication of the degree to which almost everyone was focused on parity and $V-A$ in the latter part of the decade may be had by examining Figure 28.14. There we show something we have called the reference half-life for theoretical weak-interactions articles citing other theoretical weak-interactions articles. The reference half-life is defined as the elapsed time (h), such that half the references made in a given year are to papers less than h years old. The reference half-life is a measure of how dependent today's papers are on results from the immediate past. The shorter the reference half-life, the faster the field is moving. In 1957, half the references from theory articles were to other theory articles that were only a few months old. By 1960, the reference half-life had grown to about two years, which meant that most theorists were publishing papers whose focus still depended heavily on work published during and immediately after the parity and $V-A$ papers. The data do seem to suggest that the field paused intellectually for a time. At least for the field of weak interactions, the fact that this conference focused on the decade from 1950 to 1960 was appropriate: 1960 was the end of one phase and the beginning of another.

From a situation at the start of the decade when new particles were being found and the first measurements of the weak interactions were being made, the field progressed through a major upset of the physical laws that were held to be appropriate to weak decays (parity nonconservation), arriving at a theory that formed the foundation of the field for some time. It was striking how rapidly, once the possibilities for a new direction were sensed (parity nonconservation again), theorists and experimenters moved to develop implications and verify predictions. In many ways this fits our preconceptions of the way science should proceed.

It should not be overlooked that the progress of the experiments demanded the new accelerators that began producing particle beams in this period. The $\tau-\theta$ puzzle was only a puzzle until the precision of the experimental data forced an initially unpalatable position on the theorists and demanded a basic review of the underlying concepts.

Notes

1 The following papers are reports of our earlier analyses: Daniel Sullivan, D. Hywel White, and Edward J. Barboni, "The State of a Science: Indicators in the Specialty of Weak Interactions," *Social Studies of Science 7* (May 1977), 167–200; Daniel Sullivan, D. Hywel White, and Edward J. Barboni, "Co-Citation Analyses of Science: An Evaluation," *Social Studies of Science 7* (May 1977), 223–40; D. Hywel White, Daniel Sullivan, and Edward J. Barboni, "The Interdependence of Theory and Experiment in Revolutionary Science: The Case of Parity Violation," *Social Studies of Science 9* (November 1979), 303–27; D. Hywel White and Daniel Sullivan, "Social Currents in the Weak Interactions," *Phys. Today* (April 1979), 40–7; Daniel Sullivan, David Koester, D. Hywel White, and Rainer Kern, "Understanding Rapid Theoretical Change in Particle Physics: A Month-by-Month Co-Citation Analysis," *Scientometrics 2:4* (1980), 309–19; Daniel Sullivan, David Koester, and D. Hywel White, "Survival in Particle Physics: An Analysis of Experimentalists' Persistence in Research," *Scientia Yugoslavica 6* (1980), 191–201; Daniel Sullivan, Edward J. Barboni, and D. Hywel White, "Problem Choice and the Sociology of Scientific Competition: An International Case Study in Particle Physics," *Knowledge and Society: Studies in the Sociology of Culture Past and Present 3* (1981), 163–97; David Koester, Daniel Sullivan, and D. Hywel White, "Theory Selection in Particle Physics: A Quantitative Case Study in the Evolution of Weak-Electromagnetic Unification Theory," *Social Studies of Science 12* (May 1982), 73–100.

2 All of the articles in our weak-interactions bibliography were classified by a postdoctoral student in particle physics as theoretical or experimental from a reading of *Nuclear Science Abstracts*. A sample was checked by one of us (White), and agreement was about 95%. Later we acquired copies of all of the articles in our bibliography, which allowed an additional check.

3 T. D. Lee and C. N. Yang, "Question of Parity Conservation in Weak Interactions," *Phys. Rev. 104* (1956), 254–8.

4 Sullivan, White, and Barboni, note 1 (1977), 223–40.

5 Credit for developing the technique of cocitation analysis must go primarily to Henry Small and Belver Griffith. See, for example, H. G. Small, "Co-Citation in the Scientific Literature: A New Measure of the Relationship Between Two Documents," *Journal of the American Society for Information Science 24* (July–August 1973), 265–9; H. G. Small and B. C. Griffith, "The Structure of Scientific Literatures. I: Identifying and Graphing Specialties," *Science Studies 4* (1974), 17–40.

6 The two-dimensional plots were produced by a multidimensional scaling program called TORSCA. See V. R. Rao, "Computer Programs for Multidimensional Scaling," Cornell University Graduate School of Business and Public Administration, 1972.

7 These thresholds, by the way, were used to reduce the papers in the analysis to a manageable number. Our experience and the experience of others who have used cocitation analysis

indicate that using such cutoffs, after suitable exploration of the effects on the plots of higher and lower cutoffs, highlights the core intellectual foci of the field. The less frequently cited papers that are excluded do not tend to represent new clusters of activity that ought to become part of the analysis.

8 E. C. G. Sudarshan and R. E. Marshak, in *Proceedings of the Padua–Venice Conference on Mesons and Recently Discovered Particles* (Padova: Societa Italiana di Fisica, 1958), and "Chirality Invariance and the Universal Fermi Interaction," *Phys. Rev. 109* (1958), 1860–2; R. P. Feynman and M. Gell-Mann, "Theory of the Fermi Interaction," *Phys. Rev. 109* (1958), 193–8.

Appendix

Articles in cocitation plot for 1956

L. Alvarez, F. S. Crawford, M. L. Good, and M. L. Stevenson, "Lifetimes of K Mesons," *Phys. Rev. 101* (1956), 503–5.

R. H. Dalitz, "On the Analysis of τ-Meson Data and the Nature of the T-Meson," *Philos. Mag 44* (1953), 1068–80.

"Decay of τ-Mesons of Known Charge," *Phys. Rev. 94* (1954), 1046–51.

E. Fabri, "A Study of τ-Meson Decay," *Nuovo Cimento 11* (1954) 479–91.

V. Fitch and R. Motley, "Mean Life and K^+ Mesons," *Phys. Rev. 101* (1956), 496–8.

M. Gell-Mann, "Isotopic Spin and New Unstable Particles," *Phys. Rev. 93* (1953), 833–4.

M. Gell-Mann and A. Pais, "Behavior of Neutral Particles Under Charge Conjugation," *Phys. Rev. 97* (1955), 1387–9.

G. Harris, J. Orear, and S. Taylor, "Lifetimes of T^+ and KL^+ Mesons," *Phys. Rev. 100* (1955), 932.

E. J. Konopinski and H. M. Mahmoud, "The Universal Fermi Interaction," *Phys. Rev. 92* (1953), 1045–9.

Articles in cocitation plot for 1957

L. Alvarez, F. S. Crawford, M. L. Good, and M. L. Stevenson, "Lifetimes of K Mesons," *Phys. Rev. 101* (1956), 503–5.

H. Blumenfeld, E. T. Booth, and L. M. Lederman, "Associated Production and the Lifetime of the $Lambda^0$ and $Theta^0$ Particles," *Phys. Rev. 102* (1956), 1184–5.

V. Fitch and R. Motley, "Mean Life and K^+ Mesons," *Phys. Rev. 101* (1956), 496–8.

J. I. Friedman and V. L. Telegdi, "Nuclear Emulsion Evidence for Parity Non-conservation in Decay Chain $\text{Pi}^+ \rightarrow \text{Mu}^+ \rightarrow e^+$," *Phys. Rev. 105* (1957), 1681–2.

R. L. Garwin, L. M. Lederman, and M. Weinrich, "Observations of the Failure of Conservation of Parity and Charge Conjugation in Meson Decays: The Magnetic Moment of the Free Muon," *Phys. Rev. 105* (1957), 1415–17.

R. Gatto, "Charge Properties of the Weak Decay Interactions of the New Particles," *Nuovo Cimento 3* (1956), 318–35.

M. Gell-Mann and A. Pais, "Behavior of Neutral Particles Under Charge Conjugation," *Phys. Rev. 97* (1955), 1387–9.

L. Landau, "On the Conservation Laws for Weak Interactions," *Nucl. Phys. 3* (1957), 127–31.

K. Lande, E. T. Booth, J. Impeduglia, and L. M. Lederman, "Observation of Long-Lived Neutral V Particles," *Phys. Rev. 103* (1956), 1901–4.

T. D. Lee and C. N. Yang, "Mass Degeneracy of the Heavy Mesons," *Phys. Rev. 102* (1956), 290–1.

"Question of Parity Conservation in Weak Interactions," *Phys. Rev. 104* (1956), 254–8.

"Parity Nonconservation and a Two-Component Theory of the Neutrino," *Phys. Rev. 105* (1957), 1671–5.

T. D. Lee, R. Oehme, and C. N. Yang, "Remarks on Possible Noninvariance Under Time Reversal and Charge Conjugation," *Phys. Rev. 106* (1957), 340–5.

A. Pais and O. Piccioni, "Note on the Decay and Absorption of the $Theta^0$," *Phys. Rev. 100* (1955), 1487–9.

A. Salam, "On Parity Conservation and Neutrino Mass," *Nuovo Cimento 5* (1957), 299–301.

G. Takeda, "Branching Ratio for Alternative Modes of Decay of Hyperons and $Theta^0$ Mesons," *Phys. Rev. 101* (1956), 1547–51.

C. W. Wu et al., "Experimental Test of Parity Conservation in Beta Decay," *Phys. Rev. 105* (1957), 1413–15.

Articles in cocitation plot for 1958

R. P. Feynman and M. Gell-Mann, "Theory of the Fermi Interaction," *Phys. Rev. 109* (1958), 193–8.

J. I. Friedman and V. L. Telegdi, "Nuclear Emulsion Evidence for Parity Non-conservation in Decay Chain $Pi^+ \rightarrow Mu^+ \rightarrow e^+$," *Phys. Rev. 105* (1957), 1681–2.

S. Furuichi et al., "On the Spins and Decay Interactions of K_{mu3} and K_{03}," *Progr. Theor. Phys. (Japan) 17* (1957), 89–106.

R. L. Garwin, L. M. Lederman, and M. Weinrich, "Observations of the Failure of Conservation of Parity and Charge Conjugation in Meson Decays: The Magnetic Moment of the Free Muon," *Phys. Rev. 105* (1957), 1415–17.

T. Kinoshita and A. Sirlin, "Muon Decay with Parity Nonconserving Interactions and Radiative Corrections in the Two-Component Theory," *Phys. Rev. 107* (1957), 593–9.

L. Landau, "On the Conservation Laws for Weak Interactions," *Nucl. Phys. 3* (1957), 127–31.

T. D. Lee and C. N. Yang, "Question of Parity Conservation in Weak Interactions," *Phys. Rev. 104* (1956), 254–8.

"Parity Nonconservation and a Two-Component Theory of the Neutrino," *Phys. Rev. 105* (1957), 1671–5.

A. Pais and S. B. Treiman, "Angular Correlations in K^0 Decay Processes," *Phys. Rev. 105* (1957), 1616–19.

A. Salam, "On Parity Conservation and Neutrino Mass," *Nuovo Cimento 5* (1957), 299–301.

E. C. G. Sudarshan and R. E. Marshak, "Chirality Invariance and the Universal Fermi Interaction," *Phys. Rev. 109* (1958), 1860–2.

Y. Tanikawa, "Renormalization Theory for Fermi Interactions," *Phys. Rev. 198* (1957), 1615–19.

C. W. Wu et al., "Experimental Test of Parity Conservation in Beta Decay," *Phys. Rev. 105* (1957), 1413–15.

H. W. Wyld and S. B. Treiman, "Charge Asymmetries in the Decay of Long-Lived Neutral K-Mesons," *Phys. Rev. 106* (1957), 169–70.

Articles in cocitation plot for 1959

S. M. Berman, "Radiative Corrections to Muon and Neutron Decay," *Phys. Rev. 112* (1958), 267–70.

F. Crawford, M. Cresti, et al., "Beta Decay of the Lambda," *Phys. Rev. Lett. 1* (1958), 377–80.

R. P. Feynman and M. Gell-Mann, "Theory of the Femi Interaction," *Phys. Rev. 109* (1958), 193–8.

M. Gell-Mann, "Test of the Nature of the Vector Interaction in Beta Decay," *Phys. Rev. 111* (1958), 362–5.

M. L. Goldberger and S. B. Treiman, "Decay of the Pi Meson," *Phys. Rev. 110* (1958), 1178–84.

"Form Factors in Beta Decay and Mu Capture," *Phys. Rev. 111* (1958), 354–61.

L. Landau, "On the Conservation Laws for Weak Interactions," *Nucl. Phys. 3* (1957), 127–31.

T. D. Lee and C. N. Yang, "Question of Parity Conservation in Weak Interactions," *Phys. Rev. 104* (1956), 254–8.

"Parity Nonconservation and a Two-Component Theory of the Neutrino," *Phys. Rev. 105* (1957), 1671–5.

T. D. Lee, R. Oehme, and C. N. Yang, "Remarks on Possible Noninvariance under Time Reversal and Charge Conjugation," *Phys. Rev. 106* (1957), 340–5.

S. Okubo. R. E. Marshak, E. C. G. Sudarshan, W. B. Teutsch, and S. Weinberg, "Interaction Current in Strangeness-Violating Decays," *Phys. Rev. 112* (1958), 665–8.

J. J. Sakarai, "Mass Reversal and Weak Interactions," *Nuovo Cimento 7* (1958), 649–60.

A. Salam, "On Parity Conservation and Neutrino Mass," *Nuovo Cimento 5* (1957), 299–301.

E. C. G. Sudarshan and R. E. Marshak, "Chirality Invariance and the Universal Fermi Interaction," *Phys. Rev. 109* (1958), 1860–2.

Y. Tanikawa, "Renormalizable Theory for Fermi Interactions," *Phys. Rev. 198* (1957), 1615–19.

S. B. Treiman and H. W. Wyld, Jr., "Decay of the Pi Meson," *Phys. Rev. 101* (1956), 1552–7.

Articles in cocitation plot for 1960

F. Crawford, M. Cresti, et al., "Beta Decay of the Lambda," *Phys. Rev. Lett. 1* (1958), 377–80.

R. P. Feynman and M. Gell-Mann, "Theory of the Fermi Interaction," *Phys. Rev. 109* (1958), 193–8.

M. L. Goldberger and S. B. Treiman, "Form Factors in Beta Decay and Mu Capture," *Phys. Rev. 111* (1958), 354–561.

T. Kinoshita and A. Sirlin, "Radiative Corrections to Fermi Interactions," *Phys. Rev. 113* (1959), 1652–60.

S. Okubo, R. E. Marshak, E. C. G. Sudarshan, W. B. Teutsch, and S. Weinberg, "Interaction Current in Strangeness-Violating Decays," *Phys. Rev. 112* (1958), 665–8.

E. C. G. Sudarshan and R. E. Marshak, "Chirality Invariance and the Universal Fermi Interaction," *Phys. Rev. 109* (1958), 1860–2.

Y. Tanikawa, "Renormalizable Theory for Fermi Interactions," *Phys. Rev. 198* (1957), 1615–19.

F. Zachariasen, "Decay of K Mesons as a Test of the Universal Fermi Interaction," *Phys. Rev. 110* (1958), 1481–2.

PART VII

WEAK INTERACTIONS AND PARITY NONCONSERVATION

Brush up on parity, start testing it now.
Brush up on parity, and your colleagues you will wow.
Just you mention a pi dot sigma,
And they'll think you've cleared up an enigma.
If the weak interaction's your sector,
You can sell 'em on axial plus vector,
And if you can observe a helicity
You'll attain to the utmost felicity.

– © Arthur Roberts, "Brush Up on Parity" (1957)

29   The nondiscovery of parity nonconservation

ALLAN FRANKLIN

Born 1938, Brooklyn, New York; Ph.D., Cornell, 1965; history and
philosophy of science; University of Colorado

Introduction

In 1928 and 1930, two experiments, performed by Richard Cox,
C. G. McIlwraith, and Bernard Kurrelmeyer, and by Carl Chase, provided,
at least in retrospect, evidence for the nonconservation of parity in the weak
interactions.[1-3] The anomalous nature of the experimental results seems to
have been fairly well known, although the exact nature of the anomaly was
not clear. One thing is certain, however; the relation of these results to the
conservation of parity was not recognized or understood by any contemporary
physicists, including the authors themselves. In this chapter, we examine these
experiments in some detail and discuss whether or not they really provided
evidence for the nonconservation of parity.

Did the experiments show the nonconservation of parity?

The intellectual context for these experiments was the desire to
demonstrate the vector nature of electron waves. Louis de Broglie[4] had
suggested in 1923 that just as light exhibited both particle and wave characteristics, so should what we normally consider particles exhibit wave characteristics. This wave nature had been brilliantly confirmed in 1927 in the
experiment on the diffraction of electrons by crystals performed by Clinton
J. Davisson and Lester H. Germer.[5] The idea of electron waves was then

This paper summarizes part of "The Discovery and Nondiscovery of Parity Nonconservation," *Studies in History and Philosophy of Science 10* (1979), 201–57, copyright
© 1979 Pergamon Press plc.; reprinted by permission. For more details, see that or my
book, *The Neglect of Experiment* (Cambridge University Press, 1986). Some of this
work was done while I was a fellow at the Center for Philosophy of Science, University
of Pittsburgh. I am grateful to the center for its support and hospitality.

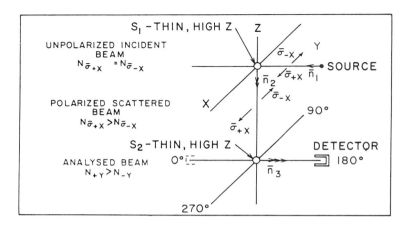

Figure 29.1. Mott double scattering for an unpolarized beam [*Source*: Grodzins (1959)[3]].

combined with the concept of electron spin, introduced by George Uhlenbeck and Samuel Goudsmit.[6] The mathematics of a vector electron were worked out by Charles G. Darwin.[7] Cox and his collaborators thought that an experiment in which electrons were double-scattered would provide evidence, in analogy to experiments with light and x rays, that the first scattering polarized the light and the second scattering acted as an analyzer.

Although the general nature of the effect to be observed in this experiment was known from the optical analogies, the detailed calculation of the effect was not made until 1929 by Nevill Mott.[8] He recognized that the double scattering of electrons was one of the few methods available to observe the spin of the free electron. The direct measurements would fail for technical reasons. Mott's discussion was not available to Cox and his collaborators, although it was known to Chase. Mott showed that for double scattering of relativistic electrons from high-Z nuclei, the number scattered at $180°$ is greater than that at $0°$ (Figure 29.1).

Let us consider the effects to be observed if the initial electron beam is longitudinally polarized (i.e., the spin is preferentially either along the direction of the electrons' motion or opposite to the motion). Although this situation was not treated by Mott, it is of crucial importance in understanding the experimental results of Cox and his collaborators and Chase. The very existence of a longitudinal polarization for electrons from β decay is evidence for the nonconservation of parity. This is made clear by reference to Figure 29.2. Let us assume that the electron spin is opposite to its momentum. A one-dimensional mirror reflection will change the spin direction, but not the momentum; so the mirror image will have its spin in the same direction as the momentum, clearly showing a difference. This longitudinal polarization

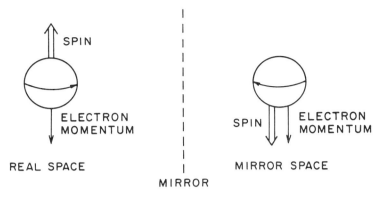

Figure 29.2. Spin and momentum in real space and mirror space.

can also be detected in a double-scattering experiment. In that case, a scattering asymmetry will exist for electrons scattered at 90° and at 270° (Figure 29.3).

Cox and his colleagues described their experiment as follows:

> In our experiment β-particles, twice scattered at right angles, enter a Geiger counter. The relative numbers entering are noted as the angle between the initial and final segments of the path is varied. For reasons to be mentioned later, the angles at which most of the observations have been made are indicated...as 270° and 90°. The difference between the configurations of the three segments of path at these two angles is the same as the difference *between right- and left-handed rectangular axes.*[9] [emphasis added]

The targets consisted of gold plugs, and a milligram of radium was used as the source of β particles. The scattered β particles were then detected by platinum-point Geiger counters, which were quite unreliable and were also sensitive to background electrons. Their experimental results are given in Table 29.1. The weighted average of the results gives the ratio of the count at 90° to the count at 270° as 0.91 ± 0.01. They noted, however, that their results varied considerably among the different runs.

They then examined the possible sources of systematic error in their experiment, such as the lack of proper centering of the radioactive source and Geiger counter, some asymmetry in target orientation, the possibility of some residual magnetic field in the apparatus, and the possibility that the electron was polarized in passing through some material in the apparatus; they rejected all of these as unlikely and concluded as follows:

> It should be remarked of several of these suggested explanations of the observations that their acceptance would offer greater difficulties in accounting for the discrepancies among the different results than would the acceptance

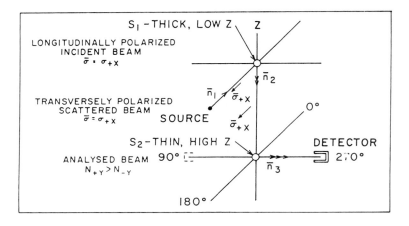

Figure 29.3. Double scattering for a longitudinally polarized incident beam [*Source*: Grodzins (1959)³].

Table 29.1. *Experimental results of R. T. Cox, C. G. McIlwraith, and B. Kurrelmeyer¹*

$\dfrac{\text{Count at }90°}{\text{Count at }270°}$	0.76	0.90	0.94	0.87	0.98	1.03	1.03	0.91
Probable error	0.01	0.07	0.01	0.02	0.01	0.03	0.02	0.02
$\dfrac{\text{Count at }90°}{\text{Count at }270°}$	0.95	0.99	1.01	1.06	1.05	0.55	0.91	
Probable error	0.05	0.03	0.04	0.05	0.02	0.05	0.03	

Note: Weighted average = 0.91 ± 0.01. (This average was not calculated in the original paper.)

of the hypothesis that we have here a true polarization due to the double scattering of asymmetrical electrons. This latter hypothesis seems the most tenable at the present time.[10]

The authors offered no theoretical explanation of their results, but did suggest that the discrepancies in their results might be attributable to a velocity dependence in their Geiger-counter efficiencies, or to a velocity dependence of the polarization.[11]

Before discussing whether or not these results are valid and, if so, what they mean, it is worth examining the continuation of the experiment performed by Chase, then a graduate student working under the supervision of Cox. Chase's first experiment started with an apparatus that was quite similar to that of Cox and associates, with the major difference being that the targets were made of lead rather than gold.[12] He obtained in the early data "an asymmetry of the same nature as had been observed by the earlier three, the effect being in the same direction and on the average of the same magnitude." There was, however, a large background due to β rays, which had also been present in Cox's experiment, and the apparatus was redesigned to eliminate it. His new results are as follows:

Azimuthal angle	Relative count
0°	1.000
90°	0.977
180°	0.958
270°	0.969

The error is less than 1 percent. He concluded that there was no evidence of an electron polarization.

By this time, Mott's first paper on electron scattering, which predicted an asymmetry between the 0° data and the 180° data, had already been published. Chase, who had observed such an asymmetry, attributed it to a difference in path rather than a polarization effect. Chase continued his work, including a survey of the literature, and noted that

> All experiments with slow electrons have given negative results, while some of the experiments with fast electrons have shown evidence of polarization effects. As far as any theory is available, the prediction is that the spin vector of the electron may possibly show itself as the analog of a transverse vector in the electron waves; but only when the electrons have high velocities.[13]

He then proceeded to examine the velocity sensitivity of the Geiger counters, an effect that had been noted earlier by Cox. He concluded that the sensitivity of the counters depended quite strongly on the counter windows and the voltage applied and that this cast some doubt on his own previous work. In view of the difficulties involved with the use of Geiger counters, Chase continued his work, but this time with an electroscope used as a detector. His results are given in Table 29.2. He obtained a result for the ratio of counter at 90°/count at 270° = 0.973 ± 0.004 and concluded:

> The following can be said of the present experiments: the asymmetry between the counts at 90° and 270° is always observed, which was in no sense true before. Not only every single run, but even all the readings in every run, with a few exceptions, show the effect. As an interesting sort of check, the apparatus that had previously given a negative result was set up again; with the counters used as they were before, at lower voltages, the results were negative as before, but with high voltages on the counter, high enough to ruin the point within an hour or two, the effect was very likely to appear. Making no changes except in the voltage on the counter, the effect could be accentuated or suppressed.[14]

Table 29.2. *Experimental results of C. Chase: relative counts*[2]

At 0°	At 90°	At 180°	At 270°	Weight
1.000	0.972	1.009	1.024	1
1.000	0.975	1.075	1.075	1
1.000	0.997	0.986	1.005	1
1.000	0.990	0.986	1.015	1
1.000	0.988	1.000	1.008	1
1.000	0.994	0.976	1.010	1
1.000	1.034	1.044	1.044	1
1.000		0.950		4
	1.000		1.030	3
	1.000		1.040	3
		1.000	1.020	2
1.000		0.933		1
	1.000		1.030	2
		1.000	0.969	2
1.000			1.003	1
1.000		1.037		2
	1.000	0.933		2
1.000	0.993	0.985	1.021	Weighted mean
±0.003	±0.003	±0.003	±0.003	

Note: Experimental error = 1%.

In the experiment, Chase also obtained the value 0.987 ± 0.003 for the ratio of the count at 180° to the count at 0°. This time he did attribute it to a Mott scattering effect.

We are now in a position to discuss the significance of these results. If the results of Cox and associates and Chase's last results are valid, we conclude that the experiments give evidence for a longitudinal polarization of electrons from β decay and thus for parity nonconservation in the weak interactions, as discussed earlier.

The first question is whether or not the results are statistically significant. We recall that Cox and associates obtained an asymmetry of 0.91 ± 0.01, which would normally be regarded as fairly strong statistical evidence. If one looks, however, at the data presented in Table 29.1, we see that there are wide fluctuations in the results and that the asymmetry is really determined by a few low results. Cox's paper also cast some doubt on the statistical reliability of the result by stating that "it must be admitted that the probable error in many cases is reckoned from too few values."[15] Despite these reservations, one should still regard the results as suggestive. Doubts concerning statistical reliability do not, however, apply to Chase's results. He obtained an asymmetry of 0.973 ± 0.004, which is statistically significant, and in addition, as we have seen earlier, Chase's results are very consistent, with the asymmetry appearing not only in almost every run but also in almost every reading. Thus, we would conclude, at least on statistical grounds, that the asymmetry was established.

There remains the possibility, however, of systematic effects that might have influenced the result, some of which we mentioned earlier. We consider first the experiment of Cox and associates. The first problem is the wide fluctuation in the behavior of the Geiger counters. This behavior, in which some of the counters were sensitive only to slow electrons, would, in fact, lessen the effect, because the polarization effects are larger for fast electrons. There is also the possibility of a misalignment of the source, the counter, or the targets. The experimenters reported that these items were removed and replaced repeatedly during the experiment and stated that "it seems unlikely that the accidental dislocations could be preponderantly in one direction as are the observations."[15] They also discussed the possibility that a stray magnetic field might orient the electron spins or that the electron was polarized by passing through other material in the apparatus. They pointed out that this was unlikely, and was absent in an auxiliary experiment, and that it would also require "a polarity in the electron as definite as that required to explain the observations as due to double scattering." There seems to be no reason to reject their statement that "the acceptance [of the explanations] would offer greater difficulties in accounting for the discrepancies among the different results than would the acceptance of the hypothesis that we have here a true polarization due to the double scattering of asymmetrical electrons."[16] Similar arguments apply to possible systematic effects in Chase's experiment.

There are still, however, several problems that can be examined only with the insights provided by later theoretical and experimental work. We recall the earlier discussion of the double scattering of a longitudinally polarized electron beam. The relativistic electrons should first be scattered by a thick, low-Z target and then by a thin, high-Z target, and this will give rise to a $90°-270°$ asymmetry. That experiment was done in the 1950s by several groups, all of whom observed definite asymmetries (6 percent and up). Lee Grodzins examined these results and concluded that a measurable effect should have been seen in the Cox and Chase experiments.[17]

There is one remaining and quite puzzling problem connected with the experimental results of Cox and Chase. In his earliest work, Grodzins concluded that those two experiments did show a $90-270°$ asymmetry and thus could have given evidence for parity nonconservation. In a later publication, Grodzins pointed out that his earlier analysis was incorrect because both experiments found fewer counts at $90°$ than at $270°$, whereas the modern theory predicts more counts at $90°$, and thus both Chase and Cox had an effect of the wrong sign.[18] Their error was confirmed in an experiment performed by Sidney Altman, a student at MIT working for Grodzins. Our own theoretical analysis as well as a careful comparison of the results of later experiments with those of Cox and Chase confirm that the sign of the asymmetry obtained by Chase and Cox is indeed wrong. Grodzins, however, concluded that although the published sign of the asymmetry is incorrect, Cox and Chase did a correct experiment:

> It has long been my view that Chase and Cox did correct experiments, but that between the investigation and the write-up the sign got changed. . . . Did Cox mislabel his angles? Did he use a right-handed coordinate system instead of the left-handed one shown in his figure? If, as I suspect, he did make some such slip then the error would undoubtedly have been retained in subsequent papers. Such errors are neither difficult to make nor particularly rare. Many a researcher and at least one former historian of science have erred similarly.[19]

I might point out that my own initial analysis was also wrong in the same way as that of Grodzins. I also believed that the asymmetry reported by Chase and Cox was correct. We do, however, have more evidence on this point. Cox was unaware of Grodzins's later analysis, and even though his own reminiscences are included in the same volume, he did not have a chance to respond to it. His own recollections of the problem are as follows:

> Grodzins in his article expressed the opinion that we (or I should say I, since I think our paper as published was mainly written by me) made a slip between the experimental observations and its published description. He supposes that the asymmetry we found was actually in the sense the theory predicts but that, in describing the experiment, I accidentally reversed it. At first sight, at least, this seems unlikely. But the alternative explanation, which assumes a persistent instrumental asymmetry, also seems unlikely when I consider how often we removed the Geiger counter to change electrodes (as was necessary in the early short-lived type of counter which McIlwraith, Kurrelmeyer and I used) and when I remember also other changes which Chase made in the very different equipment with which he replaced ours.
>
> I have thought about the matter off and on for a long time without coming to any conclusion either way.[20]

Although Cox draws no conclusions, I find his argument against a persistent instrumental asymmetry, both in his letter and also in the published paper, convincing. The correctness of the results is not, however, important for our further discussions. Because, as we shall see, no one ever performed a similar experiment or realized the significance of the results, their accuracy was never really questioned, although some later work did attribute their results to instrumental effects. Both experiments showed the velocity dependence of the polarization that is both predicted by modern theory and observed. In view of the foregoing argument and the discussions of the statistical validity and the systematic errors given earlier, we must agree with Professor Cox's conclusion with regard to his own experiments and those of Chase: "It appears now, in retrospect, that our experiments and those of Chase were the first to show evidence for parity nonconservation in weak interactions."[21]

That was not, however, the reaction of the 1930 physics community. Although some authors mention these results as an anomaly, there is absolutely no recognition by the authors or by anyone else of their significance for the question of parity conservation in electron scattering. Kurrelmeyer states that "as to our understanding of parity, it was nearly nil. Even the term had not been coined in 1927, and remember, this experiment was planned in 1925, and none of us were theoreticians."[22,23] Cox, in discussing the reaction of the physics community, stated, "I should say that the experiments were widely ignored."[24] Also, "our work was, prior to 1957, generally unaccepted, disbelieved, and poorly understood."[25] As we shall see in the next section, Cox's appraisal is not entirely accurate.

The community's reaction

In this section, I examine the question of why those experiments were almost completely ignored by the physics community. The standard textbook explanation for this is that when the experiments were redone with electrons from thermionic sources, rather than from decay sources, the effect did not occur, and so it was dismissed. Although there is an element of truth to this, it is by no means an accurate explanation.

Cox's own recollections are a useful starting point:

> As to the reaction of other physicists to the experiment of McIlwraith, Kurrelmeyer, and myself (and also to that of Chase on the same subject), I should say that the experiments were widely ignored. I think, for several reasons, that this was to be expected. Spectroscopy was still the dominant interest in experimental physics. We were all, at the time, young and unknown. Our reported results neither confirmed nor disproved any theory which was a subject of acute interest at the time.[24]

There was no specific theoretical context, at the time, into which to place these early experiments, as there was in 1957 when the explicit theoretical predictions of T. D. Lee and C. N. Yang were available. We can understand why the early experiments were not regarded as "crucial," because of the lack of theoretical predictions. What is still puzzling is why the perceived anomaly in the results did not act as a stimulus for further work, both experimental and theoretical, in the same way as the $\theta-\tau$ puzzle did in the 1950s, and why these results were ultimately ignored.

Cox also indicates that part of the reason for the lack of interest in his work was the fact that spectroscopy was considered a more important or interesting part of experimental physics. Although a detailed study of fashions in physics research is beyond the scope of this chapter, I note briefly that Cox's view seems correct. It was primarily in spectroscopy that the predictions of the new quantum theory were being tested. Even with regard to electron spin, the major evidence was taken from spectroscopy.

I wish to suggest, however, that the major reason for the neglect of these results stems from the fact that they became lost in the struggle of scientists to corroborate the predictions of Mott that there should be a forward–backward $(0°-180°)$ asymmetry in the double scattering of an electron with spin, as well as in the general problem of electron scattering from nuclei (Mott scattering). As we shall see later, this problem was of concern until the 1940s and possibly into the 1950s, with difficulties in the consistency of experimental results and with subtle and unforeseen effects in the scattering of the electrons. These experimental difficulties also led to various theoretical attempts to modify Mott's theory to explain the absence of the predicted effects. It is worth examining this history in detail, not only because it provides an explanation for the neglect of the experiments of Cox and his collaborators and of Chase but also because it gives an insight into how the physics community deals with an apparent discrepancy between experimental observations and a theory that is believed to be true for other experimental and theoretical reasons. This history also indicates that certain technological advances, namely, the development of electron accelerators with energies of hundreds of kilovolts, hindered the understanding of these early results, as contrasted with the 1950s, when the advancing technology made the understanding of the $\theta-\tau$ puzzle more important.

As indicated earlier, the experiments on the double scattering of electrons to demonstrate electron polarization, which were done before Mott's theoretical calculation in 1929, were based on analogies to optical scattering. In this case the first scattering acts as a polarizer, and the second as analyzer. With the exception of the experiment of Cox and associates, which we have discussed at length, none of those early experiments gave any evidence of electron polarization.

F. Wolf attempted to demonstrate electron polarization by aligning the electron spins by passing them through a magnetic field and observing the polarization by scattering from a metal surface, but observed no effect.[26] In fact, in a later argument, Niels Bohr showed theoretically that this experiment could not give any positive results.[27] E. Rupp did an experiment that was quite similar to that of Cox (i.e., two 90° scatters), although at low energy (up to 380 V), and observed no effect.[28] When he changed the first scattering angle to 12°–14°, he observed a difference between the 90° results and the results at 0° and 180°, which he attributed to an instrumental effect, although he was not certain of his explanation. A similar small-angle experiment (10–30°) was done by Abram F. Joffe and A. N. Arsenieva,[29] who reported that their null result was consistent with preliminary results of Davisson and Germer[30] and also with a theoretical calculation of J. Frenkel, who claimed to have shown the impossibility of producing an electron polarization by reflection from a metal surface.[31]

In 1929, Davisson and Germer gave a detailed report of their work.[32] Their experiment consisted of two 90° scatters from nickel crystals at energies of 10–200 V. They concluded that "our observation is that electron waves are not polarized by reflection." They were, however, the first authors to discuss the work of Cox and his collaborators. They did not question the correctness of Cox's results, but rather changed the emphasis to underline the differences between the counts at 0° and those at 90° or 270°, which are more consistent with their expectations from the optical analogy.

The situation changed in 1929, with the publication of Mott's theoretical calculation of the double scattering of electrons.[33] Mott's calculation was based directly on P. A. M. Dirac's relativistic electron theory and made specific theoretical predictions about the asymmetry to be observed in the double-scattering experiment.[34] He predicted that the number of electrons detected when the source (Figure 29.1) is at 180° will be greater than when the source is at 0°. Mott also specified the conditions under which this asymmetry should be observed, namely, single, large-angle scattering of high-velocity electrons from a high-Z nucleus. He also gave a formula for the magnitude of the effect expected, although he subsequently modified this and provided accurate numerical results in subsequent papers published in 1931 and 1932.[35] In this paper, Mott particularly noted that his theory did not predict any asymmetry between the 90° and 270° directions. He remarked that his calculation did not apply to Cox's experiment, but did not question the results.

Subsequent experimental work during the 1930s took on a different character following Mott's researches because there were now explicit theoretical predictions, based on an accepted theory (namely, Dirac's electron theory), with which to compare the experimental results. The experimental situation was confused, at best. We find some experimenters finding the predicted results, others doing similar experiments and obtaining null results, and some experimenters finding positive results at one time but not at others. In general, however, the trend of experimental results was in disagreement with Mott's calculation. This discrepancy between theory and experiment led not only to further experimental work but also to many attempts by theoretical physicists to give reasons for the apparent absence of the polarization effects predicted by Mott. As we shall see later, these attempts were quite unsuccessful.

The earliest paper that discussed Mott's work seems to have been Chase's experiment published in 1929.[36] He found a 4 percent difference, which he attributed to a difference in path, rather than to electron polarization. In his 1930 paper, however, Chase did claim to have observed the effect.[37]

Rupp also continued his work on electron scattering in the light of Mott's theory. His experiment consisted of the double scattering of electrons at grazing incidence (approximately $\frac{1}{3}°$).[38] He observed an increase in the polarization effect with increasing electron energy, as predicted by Mott, and obtained a difference of 11.8 percent from gold at an energy of 80 keV. Because he also observed no effect for scattering from beryllium, again in agreement with Mott, Rupp claimed to have corroborated Mott's theory. As Chase correctly pointed out, however, Mott's theory applied only to large-angle scattering.[39] Rupp subsequently corrected his numerical results, but without modifying his conclusion.[40] His data also indicated equal scattering at 90° and 270°. Rupp's results were regarded as a corroboration of Mott's theory in a theoretical paper on electron polarization by E. Fues and H. Hellmann.[41]

They noted Chase's 1929 results, mentioned earlier, in which the interpretation was uncertain, and were also aware of his later results.

F. Kirchner performed an experiment similar to that of Rupp, and although he claimed to have observed some asymmetries, he gave no quantitative results.[42] He remarked that these asymmetries were no larger than 10 percent and stated that the experimental data were insufficient to rule out Mott's theory. G. P. Thomson also attempted to reproduce Rupp's results by small-angle scattering through two gold foils. He obtained a negative result and wrote that "the experiment is in agreement with the view that the detection of polarization by such means is only possible with large angles of scattering."[43] This view was supported by Otto Halpern in his theoretical interpretation of Rupp's results.[44] He stated, correctly, that Mott's theory did not apply to Rupp's experiment and offered his own theory for scattering from a two-dimensional grating as a substitute, although he did not give any quantitative results. Rupp continued experimenting, this time with an apparatus that was closer to that required by Mott's theory.[45] He first scattered the electrons at 90° off a gold reflector and then passed the beam through a gold foil and photographed the resulting pattern. He again observed an asymmetry. In a subsequent experiment, Rupp and Leo Szilard further corroborated these results and obtained some additional results using magnetic fields to precess the electron spins, in agreement with theory.[46]

In 1931, Mott published a short note that modified his previous calculations slightly and also gave numerical results.[47] He noted briefly that E. G. Dymond had performed the experiment and found an effect that was five times too small:[48]

> The asymmetry at 70 kV is thus about five times as much as that found by Dymond. It is difficult to explain why this should be so. Multiple scattering would reduce the polarization observed, but there should not be much multiple scattering with the foils used. It is improbable also that the Dirac theory of the electron should give a wrong result when applied to the scattering by a Coulomb field, since the results for the energy levels of an electron in the same field are known to be correct.[49]

The experimental situation became further confused by Rupp's continuing experiments. He repeated his experiment on small-angle scattering from foils of different materials and again observed asymmetries, although this time he noted that Mott's theory did not apply.[50] His next experiment, however, was a reasonable approximation to the conditions required by Mott's theory.[51] The electrons were first scattered at 90° by a gold foil and scattered again at 90° by a gold wire. He obtained a clear 0°–180° asymmetry of 3–4 percent at an electron energy of 130 keV and 9–10 percent at 250 keV, with an error of 1–2 percent. These results were, however, in quantitative disagreement with Mott's numerical predictions of 15.5 percent at 127 keV and 14 percent at 204 keV.[52] This discrepancy was corroborated by more accurate measurements done somewhat later that gave 3.8 ± 0.5 percent at 130 keV and 9.6 ±

0.5 percent at 250 keV.[53] Rupp proposed that perhaps the polarization of the nucleus itself might be important and that this might explain the discrepancy, although he clearly regarded the lack of quantitative agreement as a serious problem.

Mott further modified and corrected his calculations and presented numerical results in a paper published in 1932.[54] Although he did not mention any specific experimental results, he seemed aware of the discrepancy between theory and experiment. He emphasized the conditions necessary for his theory to hold (i.e., single, large-angle nuclear scattering of high-speed electrons from a high-Z nucleus). After presenting some other possible methods for producing polarized electrons, he noted that

> These methods of producing polarized electrons are naturally quite beyond the range of present experimental techniques. Nevertheless, the theoretical existence of these models provides some evidence that electron beams can be polarized; for if they were to fail to produce a polarized beam a far more drastic revision of present-day quantum mechanics would be required than would be necessary if the double scattering experiment should give a negative result.[55]

Although Mott clearly believed that the failure to observe these new predicted effects would be more serious, he was indicating that the failure of the double-scattering experiment would also require a modification of quantum mechanics.

By this time, the experimental situation was so confused that all authors felt obliged to review the past history in some detail. J. Thibaud and associates divided previous experiments into two groups.[56] The first were those that gave either a negative result or a positive result that was within experimental error of a null result. Included in this group were the experiments of Cox and associates and Chase's 1929 experiment, which gave a null result. They made no mention of Chase's later positive result, which seems surprising, because they did refer to other later papers. The second group consisted of Rupp's experiments and were regarded by Thibaud and his collaborators as giving a positive result. Their own experiment, however, did not satisfy the conditions for Mott scattering. They performed a single small-angle scattering, followed by diffraction through a thin foil, and observed no effect.

G. O. Langstroth performed a double 90° scatter from thick tungsten targets at 10 keV and observed no effect.[57] Thus, he was in agreement with Mott's calculation, which predicted no polarization at such a low energy. His review of other work shows clearly the confusing experimental situation:

> With very much higher energies, however, an asymmetry in the intensity distribution of the secondary scattering appears to exist, although the evidence is somewhat contradictory and incomplete....
>
> Experimentally asymmetric distributions have been observed with fast electrons for both small and large scattering angles. The reported effects are of a quite different kind; varying apparently with the method of production.

> With small angle scattering from solid targets an effect has been reported which is not found when the scattering occurs through foils. Moreover, a still different effect is observed when a combination of solid target and foil is used. An additional type of asymmetry (in the 90–270° plane) was first observed by Cox et al., but was later attributed to instrumental causes. More recently, however, Chase reported that this type of asymmetry does exist. Finally, a small asymmetry (about 2%) of the type predicted by the specialized theory [Mott] has been observed by Dymond in electron scattering from thin foils.[58]

The results of Chase were regarded as valid, but they were only one set among a collection of confusing results.

After a somewhat more detailed account of these earlier experiments, Langstroth tried to offer some reasons why Mott's theory did not apply.

> In view of the fact that practical conditions may be immensely more complicated than those of Mott's theory, it is not surprising that it does not furnish a guide, even in a qualitative way, to all of the above experiments. This may be due to (a) the fact that a large proportion of the beam scattered from a thick target consists of electrons which have undergone more than one collision, (b) the insufficiency of the theoretical model, (c) the inclusion of extraneous effects in the experimental results.[59]

Langstroth discussed these possible effects in detail and mentioned other theoretical efforts, namely, those of Frenkel and Halpern, without reaching any conclusion.

The difficulty of offering a consistent interpretation of the experimental results was increased with the publication of Dymond's detailed account of the experiment mentioned earlier.[60] Dymond's apparatus came quite close to the conditions required by Mott. He used double 90° scattering by thin gold foils at an energy of 70 keV. In order to guard against instrumental or geometrical asymmetries, he also took data at 20 keV, where the predicted asymmetry is very small, and subtracted the results at 20 keV from those at 70 keV to determine the polarization. As a further check, he substituted one aluminum foil for a gold foil. Mott's theory predicted that there should be almost no polarization from aluminum, and Dymond obtained a result of 0.11 ± 0.10 percent polarization, consistent with zero. Dymond's result for the scattering off two gold foils was 1.70 ± 0.33 percent. His data showed a fair degree of consistency and were in the direction predicted by the theory (i.e., more scattering at 180°). He was greatly concerned, however, that Mott's theory predicted a polarization of approximately 10 percent at this energy and that his result was six times too small. He offered a possible explanation for his low results by considering plural scattering, where several reasonably large angle scatters add up to a 90° scattering. He noted, however, that "although it is not possible to estimate the influence of plural scattering, it does not seem likely that it can be called upon to bridge the gulf between the experimental and theoretical values."[61] This is a point we shall return to later. Dymond also believed that the theoretical calculations had been done correctly. After a brief discussion of Rupp's results and the difference between them and the results of Thomson and Kirchner, Dymond decided that "the final conclusion on the present evidence is that there is no concordance with theoretical expectations."[62]

Dymond, interestingly, performed yet another check on his apparatus by measuring the scattering at 90° and 270°, where he expected no asymmetry. To his evident surprise he found a polarization of -1.75 ± 0.98 percent, which indicated more scattering at 270°. He noted, however, that the data were quite inconsistent, with the sign of the effect changing from run to run. He again used a single aluminum foil and obtained a result of $+1.97 \pm 0.84$ percent. If one regards the aluminum data as giving the geometrical or instrumental asymmetry, then one should subtract the aluminum data from the gold data to obtain the correct asymmetry. This gives a result of -3.7 ± 1.3 percent. Although Dymond did not perform this calculation, he was obviously concerned about these results.

What is most surprising here is that Dymond made no mention at all of the similar results of Cox and associates or of Chase's 1930 result. He was obviously aware of Chase's paper because he noted early in his own work, in reference to the 0°–180° effect, that "Chase has carried out experiments with electrons of somewhat higher velocities (β-rays), but finds no significant polarization."[63]

Our previous analysis of double scattering shows that there should be no 90°–270° asymmetry for thermionic electrons, which are initially unpolarized. The effect should be observed only by using the longitudinally polarized electrons from β decay. Thus, in retrospect, Dymond was correct in doubting the correctness of his results.

Both Thomson and Rupp continued their experiments, with Thomson again failing to find any asymmetry in a 90° scattering off gold, followed by diffraction.[64] Rupp, on the other hand, repeated his double 90° scattering experiment, this time using thallium vapor rather than gold foils.[65] He again reported clear evidence of the asymmetry predicted by Mott. Although his results were still below the theoretical predictions, they were in closer agreement than the previous results with gold. It seems likely that Rupp used the thallium vapor to test his earlier hypothesis that the polarization of the nucleus was the reason for the discrepancy. He noted, correctly, that multiple scattering would be larger in the gold foil and would depress the gold values.

At approximately the same time, F. Sauter had redone Mott's calculation by another method and obtained identical results.[66] He also considered the question of whether or not screening of the nucleus by atomic electrons might cancel the polarization effects and found that, except for slow electrons, where no polarization was expected, this was not the case.

The situation became even more complicated in early 1934, when Dymond published a full repudiation of his earlier results.[67] In continuing his experiments, he found that he obtained results that were of opposite sign and increasing with increasing electron energy. Upon examining his apparatus, he discovered that there was a considerable instrumental asymmetry that cast doubt on his earlier results.

Dymond then considered possible reasons for the discrepancy between theory and experiment, namely, inelastic, stray, and plural scattering, and nuclear screening. He rejected all of these, citing Sauter's work, and he concluded that

> We are driven to the conclusion that the theoretical results are wrong. There is no reason to believe that the work of Mott is incorrect;... It seems not improbable, therefore, that the divergence of theory from experiment has a more deep-seated cause, and that the Dirac wave equation needs modification in order to account successfully for the absence of polarization.[68]

Dymond also noted the conflicting results of Rupp and Thomson, and though he obviously had doubts about Rupp's work, he drew no conclusions.

Thomson also published a comprehensive review of the field, along with his own detailed results, both on a single 90° scattering from thin foils, followed by diffraction through a foil, and on double 90° scattering.[69] In the former experiment he found no effect of the type previously found by Rupp. In the latter he found a ratio of counts at 0° to counts at 180° of 0.996 ± 0.01 at an energy of 100 keV, whereas Mott's calculations required a result of 1.15. Thomson remarked that his diffraction results were in agreement with theoretical expectations, where no polarization is predicted at small angles, but noted that Rupp's effects all appeared at higher energies. He suggested that Rupp may have had a nonuniform background of scattered electrons, but he was concerned about Rupp's results using magnetic fields. In reviewing the double-scattering experiments, Thomson pointed out that all of them, those of Langstroth, Dymond, and Rupp, previously mentioned, made use of a comparative method either by comparing their data to those obtained either at low energies or by scattering from low-Z nuclei, where no polarization effects were expected. He stated that because the results of most experiments were in disagreement with Mott's predictions, the use of the theory to analyze the data was itself open to question: "...if Mott's theory is wrong in some respects it may be wrong in others, and the assumption that aluminum gives much less polarization than gold becomes doubtful."[70] Thomson's conclusion about his own double-scattering experiment is quite revealing with regard to the problem of the discrepancy between theory and experiment:

> We have seen that there is no polarization of the kind required by Mott's theory within the limits of the probable error, which we may take as $\pm 1\%$. Certainly the 14% effect required by the theory does not exist. Most of Mott's work has been, I understand, checked independently; but unless there is some error in the calculations, we are driven to the conclusion that Dirac's theory cannot be applied to the problem of scattering by heavy atoms. It seems very unlikely that the presence of the outer electrons can make much

difference to the scattering, which must be almost entirely nuclear. . . . All the conditions of Mott's calculations were fulfilled in these experiments, but it seems possible that Dirac's theory doesn't hold in the intense field close to a gold nucleus where, of course, most of the scattering takes place. It is probable that an electron with sufficient energy can cause a breakdown in the medium in such a region and produce a pair of positive and negative electrons, as quanta are known to do. While the electrons I used would certainly not be able to do this, it is not unreasonable to suppose that the Dirac equation ceases to hold exactly for energies much below those required to cause a breakdown. It is, however, surprising that the effect should be just to cancel the polarization, and one is inclined to surmise that the polarization of a free electron may be among the unobservables.[71]

Thomson did make a brief reference to the work of Cox and to that of Chase, although again only to the 1929 publication: "The first experiments with fast electrons were made with γ-rays by Cox, McIlwraith, and Kurrelmeyer, and continued by Chase. The effect found, about 4%, in the opposite direction to that predicted by Mott, was considered to be instrumental."[72] Thomson seems to have been the first to question the validity of these results, but it is again surprising that no mention was made of Chase's later work, which was much more reliable. The lack of mention of the $90°-270°$ asymmetry is less surprising, because that casts no light on the problem he was considering.

Rupp, at this same time, continued to refine his double-scattering experiment, with particular emphasis on the effects of both longitudinal and transverse magnetic fields.[73] He regarded this experiment as a test of the relativistic formulas used to calculate the magnetic-field effects. He did not mention Mott's theoretical work at all, perhaps because of the confused situation. F. E. Myers and associates attempted to corroborate the earlier results of Rupp, and Rupp and Szilard, in a double-diffraction experiment.[74] They found no evidence of asymmetry and discussed a possible cause for the positive results of Rupp's experiments. They summarized the experimental situation as follows:

> Decided evidence of polarization is reported in all the experiments of Rupp, and Dymond[75] obtained a positive result, which, though small was significantly larger than the estimated error of observation. In the other experiments cited, the results were negative, or, as in those of Cox, McIlwraith, and Kurrelmeyer, and Chase, rather inconclusive, because of experimental difficulties.[76]

At this time, Cox and his collaborators were expressing doubts about their earlier results.

The general trend of the experimental results, which was in disagreement with Mott's theory, led to several theoretical attempts either to modify that theory or to present a new one that would be in agreement with the data. Hellmann, in 1935, offered a theory of electron scattering from a layered homogeneous potential and concluded that no polarization was to be expected from such scattering.[77] He noted that this was in contrast to Mott's work, but regarded his own calculation as more realistic, which is not the case. Halpern and Julian Schwinger modified the Coulomb potential by the addition of a repulsive term of the form $V = b/r^5$ and found that it "annihilates the polarization effect completely."[78] J. Winter also obtained a negative result for the double scattering of monochromatic Dirac plane waves from a sphere of constant potential and remarked that this explained the negative results of the then recent experiments.[79] We note here the ad hoc nature of these theoretical discussions. They do not seem to have been regarded as solving the problem of Mott scattering.

In 1937, H. Richter published what he regarded as the definitive experiment on the double scattering of electrons.[80] He claimed to have satisfied the conditions of Mott's theory exactly and found no effect. He concluded that

> Despite all the favorable conditions of the experiment, however, no sign of the Mott effect could be observed. *With this experimental finding, Mott's theory of the double scattering of electrons from the atomic nucleus can no longer be maintained.* It cannot be decided here how much Dirac's theory of electron spin, which is at the basis of Mott's theory, and its other applications are implicated through the denial of Mott's theory.[81]

Richter included the experiment of Cox and associates in his survey of previous experiments and remarked that they did observe an asymmetry. He also noted that Chase's 1929 paper had an asymmetry, which was zero within experimental error. No mention was made of Chase's 1930 results or the 90°–270° asymmetry, showing that this effect had been lost in the consideration of the problem of Mott's theory.

In 1938, F. C. Champion reviewed the entire subject of electron scattering and noted that there were discrepancies between Mott's theory of single scattering and the experimental results:[82]

> This theoretical work is, of course, based on Dirac's relativistic wave-equation for an electron, and it appears to be quite impossible that any slight modification of existing theory can be made to account for a divergence of the magnitude found experimentally...we should mention here that Richter has examined the double scattering of electrons of energies up to 0.12 mV at 90° by gold foils. The results are in complete disagreement with the theory, no polarization can be observed, whereas 16% of polarization is to be expected from the theory of Mott, based on Dirac's relativistic wave-equation.[83]

His view that there was a definite discrepancy between theory and experiment was confirmed in a 1939 paper by M. E. Rose and Hans A. Bethe.[84] They tried various ways of solving the problem and concluded that the discrepancy remained, "perhaps more glaring than before."[85]

J. H. Bartlett and R. E. Watson refined Mott's numerical calculations and observed that "our polarization values are about 10% less than those of Mott at the maximum, so that the theory still predicts an effect which has not been observed."[86]

In 1939, however, K. Kikuchi reported preliminary results that seemed to be in agreement with Mott's theory:

> Of course, it is quite beyond the range of experimental technique to realize perfectly all conditions from which Mott's theory is originated. In fact, the thickness of the gold foil may be too large to be applicable for the theory.[87]

Kikuchi's view was accepted by Rose in another paper that attempted to modify the theory.[88] Rose noted that Kikuchi had used such thick targets that his results, which were in apparent agreement with the theory, were in fact much larger than the theory predicted, because multiple scattering should result in an 80 percent depolarization. He suggested that Kikuchi's results might be attributable to an instrumental asymmetry similar to that of Dymond. Rose's attempt to modify the theory to eliminate the discrepancy resulted in failure, once again.

It is fair to say that in early 1940, the situation with respect to Mott's prediction of a polarization effect to be observed in the double scattering of electrons was as follows. Although Rupp and, later, Kikuchi had observed such effects, the preponderance of evidence was that such effects had not been observed. Various theoretical attempts had been made to resolve the apparent discrepancy between theory and experiment, with no success. We have argued earlier that it was in fact this problem, which seemed so crucially important to the Dirac theory of the electron, that led physicists to neglect the anomalous results of Cox and associates and those of Chase. This certainly is borne out by our examination of the literature of the 1930s. It is therefore the ultimate irony that the solution to the problem was begun in the work of Chase, Cox, and their collaborators.

In August 1940, Chase and Cox published a paper on the single scattering of 50-keV electrons from aluminum.[89] They noted that previous experiments on this subject "have indicated exceptions to this equation and to the closely related prediction of an asymmetry in double scattering, and it has been suggested that they show an actual invalidity of the Dirac equations in a range of electron speeds where it is hardly to be expected on other evidence."[90] In performing this experiment, they found a totally unexpected asymmetry for which they could offer no explanation:

> It appeared in a comparison of the intensities of the beams scattered at 90° on the two sides of the foil, the side on which the beam was incident and the opposite side. Single nuclear scattering should be equally intense on both sides. But the observed scattering was consistently more intense on the side on which the beam was incident.[91]

They rejected other possible scattering effects and instrumental asymmetries and ultimately found results in agreement with Mott's theory after a suitable averaging process.

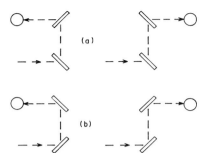

Figure 29.4. (a) An ideal transmission experiment and (b) an ideal reflection experiment.

This reflection–transmission asymmetry was then used in analyzing the results of a double-scattering experiment by Clifford G. Shull[92] in a preliminary report, and in more detail by Shull and associates.[93] In these papers they noted that the polarization observed depended on whether it was a "transmission" or a "reflection" polarization experiment (Figure 29.4). They pointed out that a much larger effect was observed in the transmission experiment and speculated that only these results should be used for comparison with theory. They analyzed their transmission data and obtained a polarization asymmetry of 1.12 ± 0.02, in good agreement with the 10.4 percent calculated by Bartlett and Watson.[94]

These authors also examined the previous experiments that gave no polarization effects, most notably those of Dymond and Richter, and noted that both experiments were reflection experiments, where the polarization effect was expected to be considerably smaller. We may understand this using the theory developed by Gerald Goertzel and Cox.[95] The polarization effects are large only for scattering angles near 90°, and thus electrons scattered at 45° will not be polarized very much. We also note that the probability of Coulomb scattering is proportional to $\sin^{-4}\theta/2$ and thus decreases rapidly with increasing angle. In a reflection experiment (Figure 29.4), a 90° scatter can be made up of either two 45° scatters or two 135° scatters. This is called plural scattering. We neglect the latter because such scattering is very small. The first 45° scattering results in the electron going along the foil, passing through a large amount of scattering material, and making the second 45° scattering quite likely. Because the electrons thus scattered are not polarized, they can mask the polarization expected by a single 90° scattering. In a transmission experiment, however, a 90° scattering can only be made up of a combination of a 45° scattering and a 135° scattering. This plural scattering will be considerably smaller than the two 45° scatters, because of the $\sin^{-4}\theta/2$ dependence of the scattering, and thus the polarization will not be masked.

It is sad to note that after this work, which initiated the resolution of the Mott scattering problem, and in which Cox and Chase played such an important role, they themselves had forgotten about their earlier scattering anomaly. Although there was a reference to the earlier attempt of Cox and associates to find evidence for the polarization of free electrons, no mention was made of the 90°–270° asymmetry found by them and subsequently by Chase.[96] Although much work remained to be done, primarily in improving the quantitative agreement between theory and experiment, it is clear that the major breakthrough in the problem of Mott scattering was provided by the work of Chase, Cox, and their collaborators.

We mentioned previously that a major factor in the failure of the physics community to understand the early results of Cox and associates and those of Chase was the development during the decade of the 1930s of electron accelerators that had energies of the order of hundreds of kilovolts. As we have seen, there were numerous difficulties in performing experiments on Mott scattering, and the development of accelerators in which the beam size, direction, and energy could be controlled precisely was seen, justifiably, as an important technological advance. With the exception of the early work of Cox and associates and Chase, who used high-energy electrons from β decay, all experiments used artificially accelerated electrons. This technological advance, however, precluded any confirmation of the anomalous results of Cox and Chase. As we have seen earlier, the 90°–270° asymmetry observed by them can occur only for electrons that are longitudinally polarized initially. Thermionic electrons, which are initially unpolarized, can give rise only to the 0°–180° asymmetry predicted by Mott. It is clear that no physicist of the time thought that the difference between thermionic electrons and those from nuclear β decay was of any significance. Cox himself confirms this view:

> For some years a small group of us at N.Y.U. continued experiments in the scattering and diffraction of electrons. But, as well as I can remember, most of our experiments were not with β-rays but with artificially accelerated electrons. Although the title of our first paper was "Apparent Evidence of Polarization in a Beam of β-rays," I did not suppose, and I do not think the others did, that β-rays were polarized on emission. I thought of the targets as having the same effect on any beam of electrons at a given speed, polarizing at the first target, analyzing at the second. Consequently I did not think of the change from a radioactive source to an accelerating tube as a radical change in my field of research.[97]

Discussion

It seems clear that the experiments of Cox, McIlwraith, and Kurrelmeyer and those of Chase show, at least in retrospect, the nonconservation of parity. It is also true that the significance of these experiments was not recognized by anyone in the physics community at the time. At least part of the reason for this was the lack of a theoretical context, such as existed in 1956 with the work of Lee and Yang, in which to place the work. I have also argued that the reason why these experimental results, which were originally considered valid and which were not predicted by any existing theory, did not lead to any further theoretical or experimental work was they were lost in the struggle to solve the discrepancy between Mott's calculation based on Dirac's electron theory and the experimental results on double scattering so important to physicists in the 1930s. The advance of technology in which electron beams of high intensity, good resolution, and controlled high energy became available through the development of electron accelerators also precluded the possibility of reproducing the results of Cox and Chase, which depended on the longitudinal polarization of electrons from β decay. At the time, no one realized that there was any difference between thermionic and decay electrons.

This episode also illustrates one possible reaction of the physics community to a seemingly clear discrepancy between experimental results and a well-corroborated theory. Dirac's theory was not rejected or regarded as refuted even after many repetitions had seemed to establish the discrepancy beyond any doubt. Repetitions of the experiment continued, under similar and slightly different conditions, and various ad hoc theoretical suggestions were made to try to solve the problem, all of which were unsuccessful. As we have seen, the discrepancy was finally resolved by an experimental demonstration, followed by a theoretical explanation, of why the earlier experimental results were wrong.

Notes

1 R. T. Cox, C. G. McIlwraith, and B. Kurrelmeyer, "Apparent Evidence of Polarization in a Beam of β-Rays," *Proc. Nat. Acad. Sci. U.S.A. 14* (1928), 544–9.

2 Carl T. Chase, "The Scattering of Fast Electrons by Metals. II. Polarization by Double Scattering at Right Angles," *Phys. Rev. 36* (1930), 1060–5.

3 L. Grodzins, "The History of Double Scattering of Electrons and Evidence for the Polarization of Beta Rays," *Proc. Nat. Acad. Sci. U.S.A. 45* (1959), 399–405, contains an excellent discussion.

4 L. de Broglie, "Ondes et Quanta," *C. R. Acad. Sci. 177* (1923), 507–10; "Quanta de Lumiere, Diffraction et Interferences," *C. R. Acad. Sci. 177* (1923), 548–50, "Sur la Definition Generale de la Correspondance entre Onde et Mouvement," *C. R. Acad Sci. 179* (1924), 39–40; "A Tentative Theory of Light Quanta," *Philos. Mag. 47* (1924), 446–8.

5 C. Davisson and L. H. Germer. "Diffraction of Electrons by a Crystal of Nickel," *Phys. Rev. 30* (1927), 705–40.

6 G. E. Uhlenbeck and S. Goudsmit, "Spinning Electrons and the Structure of Spectra," *Nature (London) 117* (1926), 264–5.

7 C. G. Darwin, "The Electron as a Vector Wave," *Proc. R. Soc. London, Ser. A 116* (1927) 227–53.

8 N. F. Mott, "Scattering of Fast Electrons by Atomic Nuclei," *Proc. R. Soc. London, Ser. A 124* (1929) 425–42; "The Polarisation of Electrons by Double Scattering," *Proc. R. Soc. London Ser. A 135* (1930), 429–58.

9 Note 1 (p. 545).

10 Note 1 (p. 548).

11 Present-day theory and experiment show that the polarization is proportional to v/c, where v is the electron velocity and c is the speed of light.

12 C. Chase, "A Test for Polarization in a Beam of Electrons by Scattering," *Phys. Rev. 34* (1929), 1069–74.

13 C. Chase, "The Scattering of Fast Electrons by Metals. I. The Sensitivity of the Geiger Point-Discharge Counter," *Phys. Rev. 26* (1930), 984–7.

14 Note 2 (p. 1064).

15 Note 1 (p. 547).
16 Note 1 (p. 548).
17 Note 3 (p. 401).
18 L. Grodzins, "A Comment on the History of Double Scattering of Beta Rays," in *Adventures in Experimental Physics, Gamma Volume*, edited by B. Maglich (Princeton: World Science Education, 1973), pp. 154–60.
19 Note 18 (p. 160).
20 R. T. Cox, private communication (1977), reprinted by permission. In a subsequent letter, Cox informed me that he was now convinced by my analysis that he had done a correct experiment, but had made a slip in the coordinate system.
21 Richard Cox, "Story of the Experiment on Double Scattering of Electrons," in *Adventures in Experimental Physics*, edited by B. Maglich (Princeton: World Science Education, 1973), p. 149.
22 B. Kurrelmeyer, private communication.
23 Wigner's paper on parity conservation was not published until 1927. The term "parity" was not in use at the time, and I do not know when the term first appeared.
24 R. T. Cox, private communication.
25 Note 18 (p. 149). Cox pointed out that this statement was an editorial emendation by Maglich and seems to convey an unreasonable reproach to the physics community, which was not his intent (private communication). Including the next sentence gives a better statement of his view: "Only by viewing it from the new theoretical framework and experimental observations of the late 50's could our results be comprehended."
26 F. Wolf, "Versuch über die Polarisations fahigkeit eines Elektronenstrahls," *Z. Phys. 52* (1928), 314–17.
27 Bohr's argument appears in Mott's 1929 paper, note 8.
28 E. Rupp, "Versuche zur Frage nach einer Polarisation der Elektronenwelle," *Z. Phys. 53* (1929), 548–52.
29 A. F. Joffe and A. N. Arsenieva, "Experiences sur la Polarisation des Ondes Electroniques," *C. R. Acad. Sci. 188* (1929), 152–3.
30. C. J. Davisson and L. H. Germer, "An Attempt to Polarise Electron Waves by Reflection," *Nature (London) 122* (1928), 809.
31 J. Frenkel, "Sur l'impossibilite de Polariser des Ondes Cathodiques par Reflexion," *C. R. Acad. Sci. 188* (1929), 153–5.
32 C. J. Davisson and L. H. Germer, "A Test for Polarization of Electron Waves by Reflection," *Phys. Rev. 33* (1929), 760–72.
33 N. F. Mott, note 8 (1929).
34 P. A. M. Dirac, "The Quantum Theory of Electrons," *Proc. R. Soc. London, Ser. A 117* (1928), 610–24.
35 N. F. Mott, "Polarization of a Beam of Electrons by Scattering," *Nature (London) 128* (1931), 454; also note 8.
36 C. Chase, note 12.
37 C. Chase, note 2.
38 E. Rupp. "Ueber eine unsymmetrisch Winkelverteilung zweifach reflektierter Elektronen," *Z. Phys. 61* (1930), 158–69.
39 C. Chase, note. 2.
40 E. Rupp, "Ueber eine unsymmetrische Winkelverteilung zweifach reflektierter Electronen," *Nature (London) 18* (1930), 207.
41 E. Fues and H. Hellmann, "Ueber polarisierte Elektronenwellen," *Z. Phys. 31* (1930), 465–78.
42 F. Kirchner, "Ein Kathodenstrahl-Interferenzapparat fur Demonstration und Strukturuntersuchungen," *Z. Phys. 31* (1930), 772–3.
43 G. P. Thomson, "Polarisation of Electrons," *Nature (London) 126* (1930), 842.
44 O. Halpern, "Zur Reflexionspolarisation der Elektronenwellen," *Z. Phys., 67* (1930), 320–32.

45 E. Rupp, "Direkte Photographie der Ionisierung in Isolierstoffen," *Nature (London) 19* (1931), 109.

46 E. Rupp and L. Szilard, "Beeinflusung 'polarisierter' Elektronenstrahlen durch Magnetfelder," *Nature (London) 19* (1931), 422–3.

47 N. F. Mott, note 35.

48 E. G. Dymond, "Polarisation of a Beam of Electrons by Scattering," *Nature (London) 128* (1931), 149.

49 N. F. Mott, note 35.

50 E. Rupp, "Versuche zum Nachweis einer Polarisation der Elektronen," *Phys. Z. 33* (1932), 158–64.

51 E. Rupp, "Neure Versuche zur Polarisation der Elektronen," *Phys. Z. 33* (1932), 937–40.

52 N. F. Mott, note 35.

53 E. Rupp, "Ueber die Polarisation der Elektronen bei zweimaliger 90°-Streuung," *Z. Phys. 79* (1932), 642–54.

54 N. F. Mott, "The Polarisation of Electrons by Double Scattering," *Proc. R. Soc. London, Ser. A 135* (1932), 429–58.

55 Ibid. (p. 432).

56 J Thibaud, J. J. Trillat, and T. von Hirsch, "Recherches sur la Polarisation d'un Faisceau d'electrons par Reflexion Cristalline," *J. Phys. Radium 3* (1932), 314–19.

57 G. O. Langstroth, "Electron Polarisation," *Proc. R. Soc. London Ser. A 136* (1932), 558–68.

58 Ibid. (pp. 559–60).

59 Ibid. (pp. 566–7).

60 E. G. Dymond, "On the Polarisation of Electrons by Scattering," *Proc. R. Soc. London, Ser. A 136* (1932), 638–51.

61 Ibid. (p. 649).

62 Ibid. (p. 650).

63 Ibid. (p. 639).

64 C P. Thomson, "Polarisation of Electrons," *Nature (London) 132* (1933), 1006.

65 E. Rupp, "Polarisation der Elektronen an freien Atomen," *Z. Phys. 88* (1934), 242–6. H. R. Post has pointed out to me an interesting aspect of Rupp's work. In 1935, Rupp published a note [*Z. Phys. 95* (1935), 801] in which he withdrew five papers because of a mental illness that allowed aspects of a dream world to intrude into his work. Rupp added that he saw no necessity to withdraw any earlier work. This was not, however, the view taken by Ramsauer [*Z. Phys. 96* (1936), 278], who stated that the group at Berlin had undertaken a critical review of Rupp's earlier work and was repeating some of his earlier experiments on electron polarization. He also urged a note of caution in using Rupp's results without independent corroboration and suggested that Rupp's collaborators on some of the work might wish to reexamine their position. No mention of this seems to appear in the subsequent literature. This revelation has no effect on the thesis that the problem of Mott scattering was the most important reason for the failure of scientists to reinvestigate the experiments of Cox and Chase. Rupp's positive results only added to the confusion concerning the problem, and by the late 1930s, as we shall see later, the consensus of the physics community was that there was a discrepancy between Mott's theory and the experimental results. Cox informed me that during the 1930s he received a letter from an official in the Forschungs Institut der Allgemeinen Elektrischen Gesellschaft that indicated Rupp's difficulties. He supposes that it was sent to other workers in the field, and although he does not recall the author or its exact content, it seems probable that it was similar to Ramsauer's published letter.

66 F. Sauter, "Ueber den Mottschen Polarisationseffekt bei der Streuung von Elektronen an Atomen," *Ann. Phys. (Leipzig) 18* (1933), 61–80.

67 E. G. Dymond, "On the Polarisation of Electrons by Scattering. II," *Proc. R. Soc. London, Ser. A 145* (1934), 657–68.

68 E. G. Dymond, note 67 (p. 666).

69 G. P. Thomson, "Experiments on the Polarization of Electrons," *Philos. Mag. 17* (1934), 1058–71.

70 Ibid. (p. 1062).

71 Ibid. (pp. 1070–1).

72 Ibid. (p. 1060).

73 E. Rupp, "Polarisation der Elektronen in magnetischen Feldern," *Z. Phys.* *90* (1934), 166–76.

74 F. E. Myers, J. F. Byrne, and R. T. Cox, "Diffraction of Electrons as a Search for Polarization," *Phys. Rev.* *46* (1934), 777–85.

75 This refers only to Dymond's 1932 report (note 60). A note added in proof indicated his 1934 retraction (note 67).

76 F. E. Myers et al., note 74 (p. 778).

77 H. Hellmann, "Bemerkung zur Polarisierung von Elektonenwellen durch Streuung," *Z. Phys.* *96* (1935), 247–50.

78 O. Halpern and J. Schwinger, "On the Polarization of Electrons by Double Scattering," *Phys. Rev.* *48* (1935), 109–10.

79 J. Winter, "Sur la polarisation des ondes de Dirac," *C. R. Acad. Sci.* *202* (1936), 1265–6.

80 H. Richter, "Zweimalige Streuung schneller Elektronen," *Ann. Phys. (Leipzig)* *28* (1937), 533–54.

81 Ibid. (p. 554).

82 F. C. Champion, "The Single Scattering of Elementary Particles by Matter," *Rep. Prog. Phys.* *5* (1938), 348–60.

83 Ibid. (p. 358).

84 M. E. Rose and H. A. Bethe, "On the Absence of Polarization in Electron Scattering," *Phys. Rev.* *55* (1939), 277–89.

85 Ibid. (p. 278).

86 J. H. Bartlett and R. E. Watson, "The Elastic Scattering of Fast Electrons," *Phys. Rev.* *56* (1939), 612–13.

87 K. Kikuchi, "A Preliminary Report on the Polarization of Electrons," *Proc. Phys.-Math. Soc. Japan 21* (1939), 524–7. Kikuchi gave a more detailed report in "On the Polarization of Electrons," *Proc. Phys.-Math. Soc. Japan 22* (1940), 805–24.

88 M. E. Rose, "Scattering and Polarization of Electrons," *Phys. Rev.* *57* (1940), 280–8.

89 C. T. Chase and R. T. Cox, "The Scattering of 50-Kilovolt Electrons by Aluminum," *Phys. Rev.* *58* (1940), 243–51.

90 Ibid. (p. 243).

91 Ibid. (p. 248).

92 C. G. Shull, "Electron Polarization," *Phys. Rev.* *61* (1942), 198.

93 C. G. Shull, C. T. Chase, and F. E. Meyers, "Electron Polarization," *Phys. Rev.* *63* (1942), 29–37.

94 J. H. Bartlett and R. E. Watson, note 86.

95 G. Goertzel and R. T. Cox, "The Effect of Oblique Incidence on the Conditions for Single Scattering of Electrons by Thin Foils," *Phys. Rev.* *63* (1943), 37–40.

96 C. G. Shull et al., note 93 (p. 29).

97 R. T. Cox, private communication.

30 K-meson decays and parity violation

RICHARD H. DALITZ

Born 1925, Dimboola, Australia; Ph.D., Cambridge University, 1950;
theoretical physics; Oxford University

The period of discovery

The first K-meson decay process to become established was the τ^+
decay mode,

$$\tau^+ \to \pi^+ + \pi^+ + \pi^- \tag{30.1}$$

observed in 1948 by Cecil F. Powell's group at Bristol in emulsion exposed to
the cosmic radiation at mountain altitudes.[1] That was only a year after they
had discovered and identified the π^+ meson, Yukawa's nuclear-force meson,
through its decay mode $\pi^+ \to \mu^+ + \nu$.[2,3] The τ^+ decay mode was soon confirmed by J. Brian Harding's report in 1950 of a second event of this type,
with a compatible release of kinetic energy.[4] In the next few years, further
events were found, both by these and by other groups, until there were eleven
events of mean total mass 495 ± 2 MeV available for comparison at the Royal
Society Discussion Meeting held at London, 29 January 1953.

The second K-meson decay process to become established was the $\varkappa^+$
decay mode,

$$\varkappa^+ \to \mu^+ + \text{neutrals} \tag{30.2}$$

reported by Cormac O'Ceallaigh in 1951, also from observations in emulsion.[5]
Because the charged secondary came to rest and decayed, it was identified
uniquely; its measured kinetic energy was only 5.9 MeV. No other identified events of this type were reported until 1954, when three further events
were described, their μ^+ kinetic energies being 33.0 MeV, 28.0 MeV, and
13.6 MeV.[6-8] This spread of μ^+ energies implied that there were at least two
neutral particles emitted in $\varkappa^+$ decay.

Many other K^+ decay modes of the general type $[(\text{particle})^+ + \text{neutrals}]$

were seen in the period up to 1954. In most cases, the secondary charged particle was fast, and escaped from the emulsion, so that its mass could be estimated only roughly from the characteristics of its track. Even so, M. G. K. Menon and O'Ceallaigh were able to conclude in 1953 that both π^+ and μ^+ secondaries occurred.[9] Separating these on a statistical basis, they found that the decay events yielding μ^+ gave a wide range of μ^+ momenta and were therefore of the general type (30.2), most probably with three particles in the final state,

$$\varkappa^+ \rightarrow \pi^+ + ? + ? \tag{30.3}$$

whereas the decay events giving π^+ were of a two-body type,

$$\chi^+ \rightarrow \pi^+ + ? \tag{30.4}$$

consistent with a unique secondary π^+ momentum. In 1954, Jean Crussard and associates reported two K^+ events giving a slow π^+ meson that came to rest in the emulsion and identified itself by its characteristic $\pi^+ \rightarrow \mu^+$ decay, the π^+ kinetic energy being about 15 MeV in both events.[7,10] Three subsequent events reported at the Padua conference (1954) gave π^+ kinetic energies of 6.0, 41.5, and 22.7 MeV, respectively.[11] It was natural to suppose that these latter events might be due to a decay mode closely related with the τ^+ mode (30.1), namely,

$$\tau'^+ \rightarrow \pi^+ + \pi^0 + \pi^0 \tag{30.5}$$

because the process $\pi^+\pi^- \rightarrow \pi^0\pi^0$ can occur through strong interactions, although other final states, such as $(\pi^+\pi^0\gamma)$ or $(\pi^+\nu\nu)$, would also have been compatible with these early data.

For this period, given that the initial K^+ mass value was poorly determined from the scattering along its track, 500 ± 150 MeV being a typical measurement, and that the mass and momentum of the secondary charged particle often were rather difficult to determine, it is not surprising that new particles were being reported almost every month. What relationships existed among these many K^+ particles, with their wide variety of decay modes and possibly with a range of mass values, were quite unclear at that time. It appeared reasonable to hope that the τ^+ and τ'^+ modes might be alternative decay modes of one K meson, the τ^+ particle, but even this was far from being proved. The senior physicists took the general attitude of Enrico Fermi: Collect evidence, but make no assumptions about the identity of any of these K mesons until the evidence allows no other possibility. They took a severe attitude toward hypothetical speculation, especially simplifying assumptions, about the K mesons, and we may recall from the experience recounted by George D. Rochester in Chapter 4 concerning the failure of the Manchester cloud-chamber group to recognize two groups of V^0 particles in their early V^0 data that this severe attitude was indeed sound at this stage of the research.

The culmination of this period was marked by the conference held at

Bagnères de Bigorre, 6–12 July 1953, the *Congrès International sur le Rayonnement Cosmique*, an IUPAP and UNESCO conference organized by the University of Toulouse. The many cosmic-ray groups engaged in this work came together at this conference, compared their techniques, their procedures of analysis, and their results, and concluded that they were all observing the same phenomena, sometimes in quite different ways, and that their results were all compatible. Possibly the most important single item reported at this conference was the fine, painstaking cloud-chamber work of Robert W. Thompson, which established the existence of a neutral K meson, with the decay mode

$$\theta^0 \to \pi^+ + \pi^- \tag{30.6}$$

and mass 496 ± 5 MeV, among the V^0 events first observed by Rochester and Clifford C. Butler in 1947, in their early cloud-chamber studies of the cosmic radiation at Manchester.[12,13]

During this conference, there was set up an ad hoc "Committee on Nomenclature for New Particles," charged with the task of rationalizing the names given to the new particles that appeared to be increasing so rapidly in variety. For the K mesons, the notation recommended was $K^Q_{xn}(M)$ for a particle with mass M and charge Q, decaying to an n-particle final state that included one charged particle x. This left open the possibility, as some evidence suggested, that there were many K mesons, with differing mass values, each of which might have several decay modes (e.g., the case of τ and τ' modes), but which also could each have different decay modes. This conservative approach was certainly appropriate, but it rather institutionalized the attitude that it would be quite unsound to attempt to identify different decay modes as due to the same K meson, unless there were powerful empirical facts compelling this identification. Such evidence, on their production and interaction characteristics and on their mass and lifetime values, for example, would have to be rather accurate in order to require such a conclusion and therefore would lie far in the future, as things appeared at that time.

In fact, although this nomenclature was adopted rather generally for the existing K-meson modes, it came into play rather little for new decay modes, because rather few were discovered. By far the most important K^+ decay mode established after the Bagnères de Bigorre Conference was the mode

$$K^+_{\mu 2} \to \mu^+ + (\nu) \tag{30.7}$$

established from the study of S particles (i.e., of those charged particles in the cosmic radiation that stopped and decayed in the plates of a multiplate cloud chamber).* The Ecole Polytechnique group, working at the Pic-du-Midi, and the MIT group, working at Echo Lake (Colorado), agreed early in 1955 that their measurements on S-particle decay were indicative of two types of

---

*   *Ed. note:* The notation (ν) means that the neutral particle is inferred to be a neutrino.

secondaries: a π^+ component with momentum roughly compatible with that observed in emulsion for the χ^+ mode (30.4) and a μ^+ component of greater penetration, with momentum corresponding to initial mass value 481 ± 6 MeV in the former decay mode, assumed to be $\chi^+ \rightarrow \pi^+ \pi^0$ and 486 ± 8 MeV in the latter, assuming that this decay mode was (30.7).[14,15]

According to the new nomenclature, the $\varkappa^+$ decay mode (30.3) was re-named $K_{\mu3}^+$. There was also established a corresponding decay mode K_{e3}^+, where an electron replaced the μ^+ in the mode (30.3), as was expected to be possible on theoretical grounds. The first K_{e3}^+ events were reported in 1954.[16,17] Accurate determination of the energy spectra for the μ^+ in $K_{\mu3}^+$ decay and for the e^+ in K_{e3}^+ decay proved to be very difficult and was still going on long after the period of development we are considering here. The emulsion technique allowed the most accurate energy measurements in the early days, but only if the particle concerned came to rest in the emulsion, an unlikely chance for the upper part of these spectra, where the momentum can run almost as high as 250 MeV/c.

The nomenclature was also used subsequently for the mode $K_{e2}^+ \rightarrow e^+ + (\nu)$, which proved to be an exceedingly rare decay mode, and for the modes K_{l4}^+,

$$K_{l4}^+ \rightarrow \begin{cases} l^+ + (\nu) + \pi^+ + \pi^- \\ l^+ + (\nu) + \pi^0 + \pi^0 \end{cases} \tag{30.8}$$

where $l^+ = $ lepton denotes either μ^+ or e^+, which are also quite rare modes, although now well known.

Following the Bagnères de Bigorre Conference, there was a long period of consolidation. Evidence was gradually collected on the nature of the neutral particles in the decay modes listed earlier. The multiplate cloud chambers allowed the possibility of sometimes detecting the γ-ray showers that result from secondary γ rays produced either directly in the decay or as a result of $\gamma\gamma$ decay for a final π^0 produced in the decay. For example, the absence of any γ rays associated with the $K_{\mu2}^+$ decay modes was powerful evidence for the identification of the neutral particle in this decay as a neutrino; this $K_{\mu2}^+$ decay appeared to be a counterpart to the well-known $\pi_{\mu2}^+$ decay mode by which the π^+ meson had first become identified and established.

The π^+ spectrum for τ'^+ decay was amenable to study by the emulsion technique, because the mean π^+ kinetic energy is only about 28 MeV, and geometrical correction factors could be applied to the spectrum observed, to take account of those τ'^+ events where the pion escaped from the emulsion. However, the rate of collecting τ'^+ events was rather slow. On the other hand, the τ^+ decay spectrum was more readily determined, and in more detail, be-cause all of the final particles were charged and therefore observable. The requirement that the three pion tracks were coplanar was a clear marker for these decay events, and the pion momenta were determined in ratio by their directions. If even one pion stopped in the emulsion, so that its energy could be accurately determined, the K^+ mass could be determined from the measurement of its energy. Each τ^+ event adds information on the way the final kinetic energy is shared between the three pions; such information is not available for τ'^+ events, where two of the pions are neutral. Finally, as we know today, τ^+ decay events occur about four times as often as τ'^+ events. It is therefore natural that study of the τ^+ decay mode was the path by which we first gained knowledge of the quantum numbers of the τ^+ particle and of its relationship with the θ^+ decay mode.

The analysis of τ^+ decay

Edoardo Amaldi made the first compilation of τ^+ decay events in his report to the Third Rochester Conference, held in December 1952.[18] These events had widespread origins: two were from the Bristol work, three from Imperial College, London, and one each from the Padua and Rome groups. From these seven events, Amaldi estimated the τ^+ mass to be 500 ± 2 MeV.

At the same Royal Society Discussion Meeting mentioned earlier, held at London in January 1953, where the eleven events were presented by various speakers, it became quite clear that $\tau^+ \rightarrow \pi^+\pi^+\pi^-$ was the first well-identified and well-established decay mode for a K^+ meson. It was an important stimulus for me to see these events all presented at the same occasion and discussed as a class of events having a common parent mass value. This had an impact far greater than separate and scattered publications could ever have had, and it gave me an early and deep understanding of the importance of personal contact between scientists and of the benefits of their direct communication, with questions and immediate answers to settle apparent differences between them. In any event, it became clear to me at that meeting that the time had come for an attempt to determine the spin parity of the τ^+ meson, because the number of events available for analysis was increasing quite rapidly.

We needed, first of all, a means to plot the events, showing at a glance how the kinetic energy $Q = (M_\tau - 3m_\pi)$ released in the decay was shared between the three final pions, so that we could see clearly the relative importance of different decay configurations. This was provided by the equilateral-triangle plot shown in Figure 30.1. In this plot, each event is represented by a point P lying within the equilateral triangle ZXY, such that

$$PA/T(\pi_1^+) = PB/T(\pi_2^+) = PC/T(\pi^-) \tag{30.9}$$

where PA, PB, and PC are perpendiculars from P to the sides of the triangle, and $T(\pi)$ denotes the relevant pion kinetic energy. The subscripts 1 and 2 are assigned such that $T(\pi_1^+) \leq T(\pi_2^+)$, so that only the right-hand half of the triangle is used. Because the sum of these three perpendiculars is the same for any point inside the equilateral triangle, being equal to $3a$, where a is the radius of the circle inscribed within the triangle, the value of the ratio (30.9) is equal to

$$(PA + PB + PC)/(T(\pi_1^+) + T(\pi_2^+) - T(\pi^-)) = 3a/Q \tag{30.10}$$

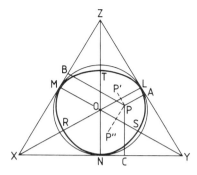

Figure 30.1. Phase space plot for τ^+ → $\pi^+\pi^+\pi^-$ events. Only the right half of the plot is used, since the two π^+ mesons are indistinguishable. Points P that correspond to physically possible states necessarily lie within the dark boundary.

Not all points P within the triangle are possible for physical events, because these must also satisfy the constraints of momentum balance. With non-relativistic kinematics, physical events are limited to points P lying within the circle inscribed to ZYX. With relativistic kinematics, physical events are limited to points P within the dark boundary line lying within this circle in Figure 30.1; relativistic effects do not have a large effect on τ^+ decay because the energy release is only 75 MeV. Points P lying on the dark boundary correspond to collinear configurations, such as that illustrated in Figure 30.2, where the three final pions have parallel momenta. The point N corresponds to an event where the final π^- meson is at rest, the two π^+ mesons then emerging with opposite momenta; the point T corresponds to an event where the two π^+ mesons have the same momentum, say q, so that they are both stationary in the $\pi^+\pi^+$ rest frame, while the π^- meson moves in the opposite direction with momentum $2q$.

This plot has one advantage that should be emphasized here. Area on the allowed region of the plot is proportional to relativistic phase space for the configurations whose representative points lie within this area; that is, for a constant matrix element M, the points are distributed uniformly on this plot. In other words, the density of events on this plot gives a measure of $|M|^2$ for the corresponding configurations.

The earliest data, up to the time of the Bagnères de Bigorre Conference, were obtained in single emulsion layers, so that it was rare for more than one pion to stop in the emulsion and so to reveal its charge. If the π^- meson stopped, the event could be determined completely and assigned to one definite point P. If only one π^+ meson stopped, then it was not known which was the π^- meson; the event might correspond to the point P', the reflection of P in OL, just as well as to P, since this reflection corresponds to interchanging the interpretation of the two outgoing tracks as π^- and π^+ mesons. If no pion stopped, the decay event could still be identified as τ^+ decay, if the momentum of one track could be well measured, but the event might correspond to P, P', or P'', where P' is the reflection of P in OL, and P'' that in OY, according to which track was that of the π^- meson. The simplest way to

Figure 30.2. A typical three-pion configuration that is collinear in the overall rest frame.

deal with this ambiguity was to fold the distribution within the right-hand semicircle across the axes OR and OM, which places both P' and P'' on top of P. Each event is then represented by a unique point P, and the region allowed for these points is then the sector LOS. The data available at the Bagnères de Bigorre Conference are plotted in this way in Figure 30.3.[19]

In order to determine the spin parity of the three-pion system from τ^+ decay, a reliable estimate of the matrix element for this decay was necessary. It was my opinion that this matrix element should be largely calculable in form, although not in magnitude, by using angular-momentum barrier-penetration arguments, since the pion momenta in this decay mode are generally sufficiently low relative to the momentum $\hbar/(lf) \approx 200$ MeV/c characteristic of hadron interactions for this to be the determining factor. The momenta and angular momenta involved are specified on Figure 30.4. The π^+ mesons have momenta $\pm k_{12}/2$ and orbital angular momentum l_{12} in the $\pi^+\pi^+$ rest frame, while the π^- meson has momentum k_3, and the $\pi^-(\pi^+\pi^+)$ system orbital angular momentum l_3 in three-pion rest frame. The total angular momentum J and the parity P of the $\pi^+\pi^+\pi^-$ system are then given by

$$\text{(a)} \quad J = l_{12} + l_3, \qquad \text{(b)} \quad P = -(-1)^{l_{12}+l_3} \tag{30.11}$$

Because the barrier-penetration amplitude at momentum k for angular momentum l is proportional to $(kr)^l$ when $(kr) \leq l$, where r measures the radial dimension of the region of strong interactions, the other factors being l-dependent but only weakly dependent on k^2, the form of the matrix element can be estimated by the form

$$M_m \sim C(l_{12}, l_3)(k_3)^{l_3}(k_{12})^{l_{12}} \, \Phi^{J,m}_{l_{12},l_3}(\theta) \tag{30.12}$$

where the angular factor Φ is uniquely determined, m being the angular-momentum component J_3, and $C(l_{12}, l_3)$ is essentially a constant to be determined empirically. In general, the complete matrix element would consist of an infinite sum of such terms for all the pairs (l_{12}, l_3) that satisfy the constraints (30.11), but it is reasonable to expect that the terms with the lowest numerical sum $(l_3 + l_{12})$ will give the dominant contributions. For example, for $(JP) = (1-)$, only the pairs $(l_{12}, l_3) = (2, 2)$, $(4, 4)$, and so forth, are possible, and the matrix element should be well estimated by the first term alone.

At this point, a few remarks about kinematics are in order. For a definite energy $T(\pi^-)$, related to the variables defined in Figure 30.4 by the equation

$$(m_\pi^2 + k_3^2)^{1/2} = m_\pi + T(\pi^-)$$

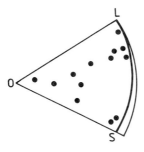

Figure 30.3. The plot for the thirteen τ decay events available in July 1953, disregarding what little knowledge there was of final pion charge signs.

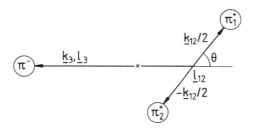

Figure 30.4. The internal momenta k_{12} and k_3 used to characterize the three-pion configuration from τ^+ decay, k_{12} being specified in the $\pi^+\pi^+$ rest frame, and k_3 in the $\pi^+\pi^+\pi^-$ rest frame; θ denotes the angle between k_3 and k_{12}, as seen in the $\pi^+\pi^+$ rest frame; l_{12} denotes the $\pi^+\pi^+$ angular momentum in their rest frame, and l_3 denotes the orbital angular momentum of the $\pi^- - (\pi^+\pi^+)$ relative motion in the $\pi^+\pi^+\pi^-$ rest frame.

the center-of-mass energy in the $\pi^+\pi^+$ system is given by

$$(4m_\pi^2 + k_{12}^2)^{1/2} = \{(M_\tau - m_\pi)^2 - 2M_\tau T(\pi^-)\}^{1/2} \qquad (30.13)$$

so that k_{12} is also definite. The only other variable needed for specifying a configuration is the angle θ, defined in Figure 30.4. When $\theta = \pi/2$, we have a configuration with an axis of symmetry along k_3, because $T(\pi_1^+) = T(\pi_2^+)$ then; its point P lies on the axis TON. When $\theta = 0$, the three momenta are parallel, and the point P lies on the dark boundary, the value of PC remaining the same. For intermediate θ, we have

$$T(\pi_1^+) - T(\pi_2^+) = \Lambda(T(\pi^-))\cos\theta \qquad (30.14)$$

where $\Lambda = k_3 k_{12}/(4m_\pi^2 + k_{12}^2)^{1/2}$. On the phase space plot, we always have $LO = NO = 2Q/3$, where $Q = (M_\tau - 3m_\pi)$. Owing to relativistic effects, the dark boundary curve is distorted from circular shape, as shown in Figure 30.1. We note here that

$$T(\pi^-)_{\text{max}} = Q(M_\tau + m_\pi)/2M_\tau \qquad (30.15)$$

On the plots in Figures 30.7 and 30.9, we use the variable

$$y = T(\pi^-)/(2Q/3) \tag{30.16}$$

which runs from 0 to $3(M_\tau + m_\pi)/(4M_\tau)$.

At the time of the Bagnères de Bigorre Conference, there were only two K-meson decay modes well established:

$$\text{(a)} \ \tau^+ \to \pi^+ + \pi^+ + \pi^-, \qquad \text{(b)} \ \theta^0 \to \pi^+ + \pi^- \tag{30.17}$$

but these states differed in charge and therefore did not necessarily have any close relationship. It was natural to suppose that the observed χ^+ decay mode (30.4) might at least sometimes be $\pi^+\pi^0$ and that the class of $K_{\pi 2}^+$ mesons might include a θ^+ meson, a charged counterpart to the θ^0 meson, but this was not necessarily so and had not been demonstrated to be the case. Showers of γ rays were sometimes seen to accompany the S^+ particles whose decay was in the $K_{\pi 2}^+$ class, and these γ rays might well have been the result of the $\gamma\gamma$ decay of an accompanying π^0, but other interpretations were quite possible. We shall not here go into the detailed history of how this question was settled, but abbreviate discussion by accepting the interpretation of the $K_{\pi 2}^+$ mesons as being dominantly θ^+ particles with decay mode $\pi^+\pi^0$. At the time, this was made psychologically easy by a remarkable event observed in the Princeton cloud chamber at Echo Lake (Colorado) in 1954 by A. Leslie Hodson and associates, as shown in one view in Figure 30.5.[20] This shows a θ^+ decay in flight in which the π^0 decayed into two direct electron–positron pairs, so that all final particles were charged and therefore observed and measurable, the complete event being

$$\theta^+ \to \pi^+ + \pi^0$$
$$\to \ (e^+e^-) + (e^+e^-) \tag{30.18}$$

This was the first observed example of this π^0 decay mode, which has branching fraction only 3×10^{-5}. This rare event demonstrated that at least some $K_{\pi 2}^+$ mesons did have the θ^+ decay mode.

The important question that could then be asked was about the possible relationship between the τ^+ and θ^+ decay modes. Was it possible that they could result from the same K^+-meson parent? Could the 3π state from the τ mode have the same spin parity (JP) as the 2π state from the θ mode? Concerning this question, the key points are as follows:[21]

(i) For total angular momentum J, the 2π system from the θ mode necessarily has parity $P = (-1)^J$.

(ii) If parity $P = (-1)^J$ holds for the 3π system from the τ mode, its decay matrix element M necessarily vanishes for all collinear configurations for the pions. In other words, the density $|M|^2$ of events per unit area on the triangular plot of Figure 30.1 must then vanish on the dark boundary of that plot.

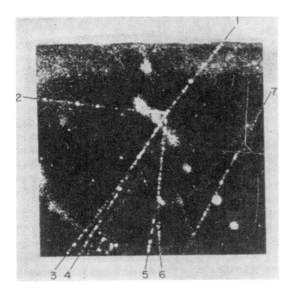

Figure 30.5. One view of the θ^+ event reported by Hodson and associates, observed in the Princeton cloud chamber at Echo Lake. It shows the two electron–positron pairs resulting from the π^0 double-internal-conversion decay mode $\pi^0 \to (e^+e^-)(e^+e^-)$ and the accompanying π^+ meson due to the decay process $\theta^+ \to \pi^+\pi^0$ [*Source*: Hodson et al. (1954)[20]].

We consider three particular configurations on this plot to illustrate this general conclusion (ii):

Point N. This corresponds to the configuration of Figure 30.6(a), where the π^- meson is at rest. The $\pi^+\pi^+$ system carries the total J and necessarily has parity $(-1)^J$. Bose statistics requires parity $(+)$ here; hence, J must be even. Including the π^-, the net parity of this configuration is necessarily $(-)$. Hence, this configuration N is not possible; that is, the matrix element for it is zero, if $J = $ odd or $P = (+)$, which includes all cases with $P = (-1)^J$, as well as others.

Point L. This corresponds to the configuration of Figure 30.6(b), where one π^+ meson is at rest. The other π^+ and the π^- carry the whole angular momentum J and have the same parity as a pion pair from a θ^0 decay from rest, with the same J, namely, $(-1)^J$. Including the π^+ at rest makes the net parity P of this configuration necessarily $P = -(-1)^J$. Hence, if $P = (-1)^J$, this 3π configuration L is not possible, and the matrix element must vanish at L.

Point T. This corresponds to the configuration of Figure 30.6(c), where the two π^+ mesons have the same momentum. In the $(\pi^+\pi^+)$ rest frame, there is no relative motion, so that the $(\pi^+\pi^+)$ system

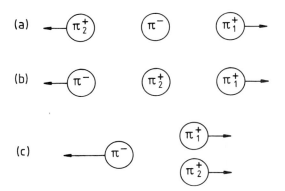

Figure 30.6. Three special collinear configurations for the final $\pi^+\pi^+\pi^-$ system from τ^+ decay at rest: (a) configuration N, where the π^- meson is at rest, (b) configuration L, where one π^+ meson is at rest, (c) configuration T, where the π^- meson has its maximum possible energy, the $\pi^+\pi^+$ kinetic energy being zero in the $\pi^+\pi^+$ rest frame.

necessarily have spin parity $(0+)$. The 3π system then consists of two objects of spin 0, but one having parity $(+)$, the other parity $(-)$. The net parity of this system is necessarily opposite that for a θ mode for the same J, so that the matrix element for configuration T must necessarily vanish.

Hence, on the question whether or not τ and θ modes can result from a common parent, the key point is whether or not the density of events is zero on the boundary of the triangular plot.

It is of interest to note here that this feature of the distribution is not affected by folding the full distribution onto one sector LOS, as depicted in Figure 30.1, where the points P' and P'' then fall onto P. In other words, to decide on this point, it is not necessary to know the charges of the pions emitted.

At the Bagnères de Bigorre Conference, the available data were plotted on the sector plot, as shown in Figure 30.3, and gave no indication that the distribution might vanish on the boundary LS; indeed, quite the contrary. It suggested already that the τ and θ modes could not stem from a common parent, but the data were rather too limited for this conclusion to be really firm at that time.

The τ^+ decay data

The Bagnères de Bigorre Conference also marked the time when a significant improvement in technique in the use of nuclear emulsion was achieved, the use of "stripped emulsions," a development worked out rather independently by the Bristol group, the Bombay group at the Tata Institute for Fundamental Research, and the group at the Naval Research Laboratory (Washington, D. C.). This involved the use of large emulsion blocks made by stacking a number of emulsion layers, each of which had been stripped off the glass on which it had been made. Each emulsion layer was separated from its neighbors by a thin film, and fiducial marks were made through the emulsion stack, so that the position of each layer relative to the next was precisely known. After exposure, the emulsion layers were separated one by one, each being mounted on glass for development, processing, and study in the usual way. This new technique with stripped emulsions allowed the possibility of following a secondary-particle track through many layers to its endpoint, where the particle came to rest. If it was a pion, its charge sign could then be determined to be positive or negative, according as it decayed or underwent nuclear interaction at the end of its range. Also, its energy could now be determined rather accurately, from measurement of its total range in the emulsion, and its mass was much better known from the multiple scattering along its track when the full track was available.

With the stripped-emulsion technique, all three pions from $\tau \to 3\pi$ decay often came to rest in the emulsion block. In all such cases, the parent τ particle was found to have positive charge, in accord with the general expectation that negatively charged hadrons coming to rest in emulsion would be attracted into atomic orbits around nuclei and would undergo nuclear interaction rather than decay. Consequently, while the source available was only the cosmic radiation, K-meson decay events were always attributed to K^+ mesons. Accepting this assumption, all three pion charge values were now generally known for τ^+ decay events, and the events could therefore be plotted unambiguously on the full plot of Figure 30.1.

My 1953 paper on τ decay[22] had accepted the limitation on experiment at that time – that the pion charges were not generally determined in single emulsion-layer experiments – and was worked out using the sector plot appropriate to this situation, with predictions of the event distribution being made for a number of low spin values. Because I had always found experiments dismissive of predictions made beyond the capabilities of their current technique, I confined attention to this limited case, because I wanted the point of these calculations to be fully appreciated. With the new stripped-emulsion technique, predictions were now needed for the case where the pion charges are known, and this work was carried out at the end of 1953, my paper[23] being held back until the Fourth Rochester Conference in the hope that it would be possible to include additional data presented there. Ettore Fabri, who was working at Rome under the supervision of Bruno Touschek, with strong encouragement from Edoardo Amaldi, published a rather similar paper at about the same time.[24] The similarity between these two papers was not at all surprising, since they both represented the logical extension of the 1953 paper to the new situation.

The use of stripped emulsions spread rapidly and gave a tremendous impetus to research on K mesons and hyperons using the cosmic radiation. A large amount of nuclear emulsion, in stacks of forty stripped-emulsion layers, was flown in balloons launched from Sardinia in June and July 1953, each balloon spending a period of typically eight hours at an altitude of about 80,000 feet. More than twenty laboratories supported the project and shared the exposed emulsion. Some of the results obtained from these emulsions were presented at a conference held at Padua in April 1954 and reported in a supplement to *Il Nuovo Cimento*. Amaldi and associates were able to report on sixteen fully identified τ^+-meson decay events, using a statistical analysis proposed by Fabri.[25,26] By the Fifth Rochester Conference,[27] held at the end of January 1955, there were forty-two of these events available from cosmic-ray work, together with the first fully identified τ^+-meson decay events from the high-energy proton accelerators, one event from Brookhaven National Laboratory, using the Cosmotron, and ten events from the emulsion work at Berkeley, using the Bevatron, which had only just come into regular operation. Even at that early stage it had become very clear how great was the advantage of the higher proton energy of the Bevatron for strange-particle studies.

In October 1954, there had been a further large balloon flight organized by the groups at the universities of Bristol, Milan, and Padua. This was a single flight carrying a large quantity of emulsion known as the G-stack, consisting of 250 emulsion layers with a total volume of fifteen liters, whose exposure to the cosmic radiation was for about six hours at about 80,000 feet. No data from this stack reached the Fifth Rochester Conference, it appears, but it was the central topic of the conference held at Pisa in June 1955 and reported in a supplement to *Il Nuovo Cimento*.[28] Amaldi's report to this conference on the τ decay data included fifty-eight fully identified events obtained from emulsions exposed to cosmic radiation; these events have been plotted in Figure 30.7.[29] There were also seventeen events, mostly with positive charge, representing τ-decay in flight, observed in cosmic-ray cloud-chamber experiments, including six events from the Ecole Polytechnique chamber at the Pic-du-Midi and five events from the Manchester chamber on the Jungfraujoch. Because the biases in cloud-chamber work were quite different from those in emulsion work, their distribution on the phase space plot was of some interest, and this is shown in Figure 30.8; it was quite compatible with that observed with emulsion. A further thirty-five τ^+ decay events were reported by several groups at Berkeley, using emulsion exposed to stopping K^+ beams of well-defined momentum from the Bevatron. The advantage of using these beams was very great because of the high yield of K^+ decay events, both absolutely and relative to background, and because of the fact that these events occurred in a quite well-defined part of the emulsion block, so reducing the search time. Indeed, it was so overwhelming that this conference marked essentially the end of significant cosmic-ray contributions to τ^+ decay studies.

At this point, the emulsion groups interested in studying the systematics of

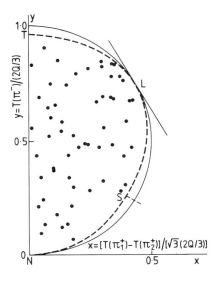

Figure 30.7. A plot of fifty-eight fully identified τ^+ decays at rest, found in nuclear emulsions exposed to the cosmic radiation at balloon altitudes.

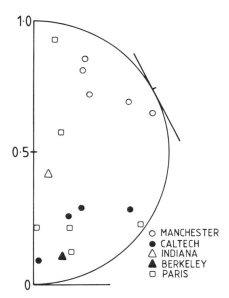

○ MANCHESTER
● CALTECH
△ INDIANA
▲ BERKELEY
□ PARIS

Figure 30.8. A plot of seventeen τ decays in flight (mostly with positive charge) produced in various cloud chambers (as specified) by the cosmic radiation at mountain altitudes.

K-meson and hyperon decays and interactions turned to work with emulsions exposed to the $K^\pm$ beams available at the Bevatron. To illustrate the rapidity of the change, we show in Figure 30.9 the plot of 219 τ^+ decay events published a year later, obtained from observations and measurements using emulsion stacks exposed to K^+ beams at the Bevatron and scanned by groups at Columbia, Berkeley, and MIT.[30] The distributions of Figures 30.7 and 30.9

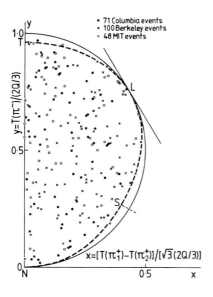

Figure 30.9. A plot of 219 τ^+ decay events found by a number of groups, in nuclear emulsion exposed to stopping K^+ beams from the Bevatron [*Source*: Orear et al. (1956)[30]; courtesy of the Institute of Physics, London].

are quite compatible. They are both rather uniform, increasing in density a little as we go from bottom to top of the figure [i.e., as $T(\pi^-)$ increases]. The density of events does not go to zero as we approach the boundary of the plot, and we could conclude quite firmly by the end of 1956 that $P = (-1)^J$ does *not* hold for the $\pi^+\pi^+\pi^-$ state reached in τ^+ decay. Thus, *the 2π state resulting from θ^0 or θ^+ decay differed in either spin J or parity (or both) from the 3π state reached in τ^+ decay.* This conclusion left us with two alternatives, as we discuss in the next section:

1. The τ^+ and θ^+ modes observed resulted from the decay of two types of K^+ mesons that differed in spin or parity (or both). Presumably these two K^+ mesons would then differ in other respects, for example, in mass and lifetime, and we could search for this difference by accurate measurements on these and other properties of the K^+ mesons, in order to confirm this conclusion and to separate the two kinds of K^+ mesons involved.

2. The τ^+ and θ^+ modes were competing decay modes of the same K^+ meson. However, the 3π and 2π final states have opposite parity for the same J; hence, this interpretation would imply that parity conservation was violated in one or other of these two K^+ decay processes.

These τ^+ decay data also allowed us to reach some conclusions on the spin parity of its final $\pi^+\pi^+\pi^-$ state, irrespective of the question of the $\tau-\theta$ relationship. For this purpose, we can confine ourselves to the data available[31] at the end of 1956, since it is our state of mind at that time that concerns us in this story. We discuss the spin-parity possibilities briefly and in four groups:

$JP = 0+$ is completely excluded. $J = 0$ required $l_{12} = l_3$, and the parity is necessarily the intrinsic parity of the three pions, namely $(-1)^3 = -1$.

$JP = 2+$ is conceivable, however. Bose statistics imply $l_{12} =$ even always, and the net parity requires $l_3 =$ odd. The simplest configuration is $(l_{12}, l_3) = (2, 1)$, the next being $(2, 3)$. The value $l_3 = 1$ requires the density of events on the plot to approach zero linearly in $T(\pi^-)$ as $T(\pi^-) \to 0$, and $l_{12} = 2$ requires the density to approach zero quadratically, like $(T_{max} - T(\pi^-))^2$, as $T(\pi^-)$ approaches its maximum allowed value T_{max}. Further, the density must vanish at all points on the boundary, the angular distribution in θ being $\sin^2\theta$. None of these expectations corresponds to the data shown in Figures 30.10–30.12. The inclusion of the $(2, 3)$ configuration can reduce this massive discrepancy a little,[32] but it was clear that the possibility $(2+)$ was quite excluded. The same remarks apply to the higher spin possibilities in this series, $(4+)$ and so forth, but such high spin values were unattractive for many reasons and were not seriously considered.

$JP = 1+$ requires $l_3 =$ odd. The simplest configuration is unique, namely $(l_{12}, l_3) = (0, 1)$. As just mentioned, the density must approach zero as $T(\pi^-) \to 0$, which is quite inconsistent with Figure 30.10. That the density does not approach zero at the very lowest $T(\pi^-)$ values is emphasized by Orear's plot[33] of the mean density in 2-MeV bins for $T(\pi^-)$ below 10 MeV, shown in Figure 30.12.

$JP = 3+$ allows $(2, 1)$ and $(0, 3)$ as the simplest configurations, giving a little more freedom for a fit to the data, but the essential argument excluding this case was that of Figures 30.10 and 30.12.

$JP = 1-$ requires both l_3 and l_{12} to be even, so that neither of them can be zero. The simplest configuration is $(2, 2)$ for which the matrix element is proportional to $(k_{12} \cdot k_3) (k_{12} \times k_3)$. The density must vanish quadratically at both $T(\pi^-) = 0$ and T_{max}, as well as on the whole of the dark boundary of Figure 30.1 (i.e., for $\cos \theta = 1$). Further, it must also vanish quadratically along TN, the central axis of symmetry of Figure 30.1, where $\cos \theta = 0$ holds. These strong requirements are all contrary to the data, and this possibility was completely excluded at quite an early stage.

$JP = 3-$ also has $(2, 2)$ as the simplest configuration, but its matrix element does not vanish for $\cos \theta = 0$, which alleviates one of the discrepancies just mentioned. However, the other three areas of discrepancy still stand, and this possibility is completely excluded. Its predictions are included among the theoretical estimates plotted in Figures 30.10 and 30.11.

$JP = 0-$ allows as simplest configuration $(l_{12}, l_3) = (0, 0)$. The density is not required to vanish anywhere, but to be more or less uniform, with some mild energy dependence and possibly some angular dependence arising from higher configurations. The data are certainly consistent with this case.

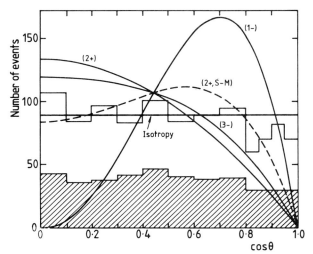

Figure 30.10. The mean density of τ^+ decay events per unit phase space, averaged over cos θ for fixed $T(\pi^-)$, is plotted versus $T(\pi^-)/T(\pi^-)_{\text{max}}$, for 892 τ^+ events, reported by a number of groups from exposures to the K^+ stopping beam at the Bevatron. The histogram is compared with curves predicted theoretically for various (JP) spin-parity values for the final 3π system. The hatched histogram denotes the events contributed by Baldo-Ceolin et al.[31] [*Source*: Dalitz (1957)[21]. Courtesy Institute of Physics].

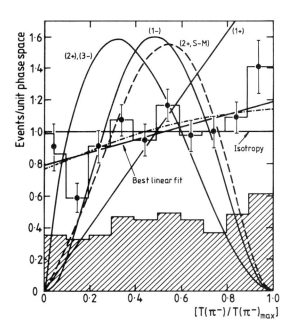

Figure 30.11. The distribution of cos θ, averaged over all values for $T(\pi^-)$ for fixed θ, for the same set of τ^+ decay events that contributed to Figure 30.10 [*Source*: Dalitz (1957)[21]. Courtesy Institute of Physics].

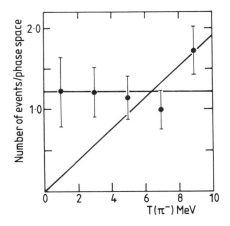

Figure 30.12. Following Orear,[23] the mean density of τ^+ decay events per unit phase space, averaged over cos θ for fixed $T(\pi^-)$, is plotted in small $T(\pi^-)$ bins for $T(\pi^-) \leqslant 10$ MeV and compared with expectation for a decay mode with $l_3 = 1$ [e.g., for the final state with $(JP) = (1+)$].

$JP = 2-$ would allow, as simplest configurations, both (0, 2) and (2, 0). Although there are points at which these matrix elements vanish, namely, at $T(\pi^-) = 0$ for the first and $T(\pi^-) = T_{\max}$ for the second, there is no point at which both must vanish. The data could be quite well fitted using these configurations, since their amplitudes could (as always) each have some modest form factors and phase factors that could be adjusted as needed to fit.

The outcome of all this discussion was clear. Only the last two spin-parity values were acceptable for the 3π system, observed, and the simplest and most natural of them was $JP = (0-)$. With $J = 0$, the 2π system necessarily has parity $(+)$. The same conclusion followed, as before: Either the 2π and 3π systems observed stem from different K mesons, having opposite parity values, or else parity conservation is violated in one or the other of these two decay modes.

The $\tau-\theta$ relationship and its interpretation

Our knowledge of the properties of the τ^+, θ^+, and $K_{\mu2}^+$ particles grew only slowly during 1955, until some appreciable experience had been gained of experiments with the Bevatron. The mass of the $K_{\mu2}^+$ was soon measured as accurately as that of τ^+ and θ^+ with the use of the stripped emulsion; the three masses agreed within the errors, which were still of the order of several MeV. During 1956, the lifetimes associated with each of the three decay modes became measured to an accuracy of the order of ± 30 percent and were compatible within that accuracy. The branching ratios for these K^+ decay modes were measured for K^+ beams with differing parameters (e.g., for beams produced at different Bevatron proton energies or at different production angles, or for K^+ beams with different initial energy or different flight times from production). In all cases, the branching ratios for the major modes were the same, within uncertainties of order ± 20 percent. The same branching ratios held for K^+ mesons that had undergone nuclear scattering out of these beams, a result indicating that the nuclear interactions of the various K^+ mesons in these beams were also rather similar, at least for τ^+, θ^+, and $K_{\mu2}^+$. All of this evidence was consistent with τ^+ and θ^+ being the same particle, but its accuracy was not high, so that the argument was not completely compelling.

Theoretical ideas were put forward to accommodate these empirical facts. For example:

(i) The hypothesis of parity doublets for strange particles, such that the θ particle coupled with a $\Lambda(-)$ hyperon [i.e., $p \leftrightarrow \theta\Lambda(-)$] had exactly the same strong interaction as the τ particle coupled with a $\Lambda(+)$ hyperon [i.e., $p \leftrightarrow \tau\Lambda(+)$], the $\Lambda(\pm)$ hyperons having the same mass and the spin-parity values $(\frac{1}{2}\pm)$, just as the θ and τ mesons had the same mass and spin parity $(0\pm)$.[34] However, this did not account for the agreement between the θ^+ and τ^+ lifetimes, although this might have been fixed up by appropriate isospin selection rules. Worse, the $\Lambda \to p\pi^-$ decays should have been seen to have two lifetimes, since the $\Lambda(+)$ hyperon would decay by emitting a p-wave π^-, whereas the $\Lambda(-)$ hyperon would decay by emitting an s-wave π^-; with a kinetic-energy release of only 37 MeV, these lifetimes should differ widely, but there was no evidence to suggest two lifetimes.

(ii) The hypothesis that there were two K^+ mesons, associated with the τ and θ modes, respectively, which differed in mass, but between which there were low-energy γ transitions. The early form of this hypothesis required that the θ particle had nonzero spin, so that $\theta^+ \to \tau^+ + \gamma$ was a possible electromagnetic transition; the observed $K_{\mu2}^+$ mode came from the parent θ^+ meson in competition with a γ transition to the lighter τ^+ particle, which gave rise to the 3π decay mode. If the γ transition were magnetic dipole [i.e., spin $(1-)$ for θ^+, $(0-)$ for τ^+], the lifetime needed implied a mass difference of ≈ 100 keV, thus requiring a rather remarkable coincidence in mass for these two states. Spin $(2+)$ for the θ^+ meson would have required a magnetic-quadrupole transition and allowed a much larger mass difference, a more reasonable proposition. The second form of this hypothesis assigned spin $(0+)$ to the θ^+ meson and gave it a mass lower than the τ^+ meson, by amount Δ (say).[35] The $\tau^+ \to \theta^+$ transition then required 2γ emission, because $(0-) \to (0+)$ is not allowed by single γ emission, whether transverse or longitudinal. This order was chosen for the mass values because the θ^+ decay lifetime was required to be short relative to that for τ^+, in order that the apparent lifetime associated with the θ^+ mode should be that of the τ^+ parent, and this fitted naturally with the idea that the decay transition $\theta \to 2\pi$ should be much faster than $\tau \to 3\pi$. This scheme required Δ to be much larger, perhaps even as large as 10 MeV. Such γ rays were sought by the Luis Alvarez group at Berkeley in their bubble chamber, exposed to a stopping K^+ beam, but they were able to demonstrate that no γ ray in the energy range 1 to 10 MeV accompanied the observed $K_{\mu 2}^+$ events with the expected rate.[36]

Many arguments were made to me against this τ^+ decay analysis and its implications, along the following lines:

(i) Most often, it was implied that some more clever approach was needed. My analysis would turn out to be incomplete, not covering all the theoretical possibilities for the mechanism of τ^+ decay, or perhaps there were some hidden assumptions I had made that would turn out not to be valid. Well, it was certainly possible to improve agreement with the data by including a large number of terms (30.12), as might be appropriate if the radius of the τ^+ particle were large relative to 1 fermi. In such a case, the density might well vanish right on the boundary, while having a large value until very close to the boundary. Spin (2+) was the most natural candidate to consider, because a competing decay mode $\theta^+ \to 2\pi$ was possible for this spin parity. As mentioned earlier, E. C. G. Sudarshan and Robert Marshak considered adding one higher configuration for this case, but the result was still very, very far from the trend of the data.[32] In any case, such a resolution of the $\tau - \theta$ dilemma would have been ad hoc, very contrived, and unsatisfactory.

(ii) Others considered parity violation to be just inconceivable and balked at it. These people assumed that parity conservation/violation must represent a property of space–time. If parity were violated in τ decay, it should therefore fail for all interactions, whereas the evidence was strong that parity was conserved in nuclear and electromagnetic interactions. This was a very deep and common confusion, and most of the discussions I was involved in during 1955–6 ended on this note: Why did we not see evidence of parity violation in nuclear and electromagnetic processes?

I stood firm against these objections. The question was how to demonstrate this parity violation in an explicit way. Well, these data on τ^+ and θ^+ modes and on the approximate equality of all the properties of the K^+ mesons that gave rise to them, all taken together, constituted a demonstration of parity violation for one or the other of these decay modes. The difficulties were that the data were not infinitely accurate and that one had to accept this theoretical analysis of the $\tau \to 3\pi$ decay. This was just not good enough to persuade the doubters!

Discussion and consequences

We were all very slow on the uptake, concerning what to do next, to settle the matter. As we recognize now, the key words were "measure a pseudoscalar."[37] On hearing these words, the matter was immediately obvious, as if scales had fallen from our eyes. However, even T. D. Lee and C. N. Yang, who had the genius to look for parity violation in quite different (semileptonic) weak interactions, did not realize that this was the key to the matter until mid-May 1956, according to their own accounts.[38,39] The rest of the story – the observation of backward/forward symmetry in the β decay of polarized ^{60}Co nuclei and in the β decay sequence $\pi^+ \to \mu^+ \nu_\mu$, followed by $\mu^+ \to e^+ \nu_e \bar{\nu}_\mu$ – is well known, and I need go no further.

As for the τ–θ puzzle, this was not settled directly by the β decay experiments, because these 2π and 3π decay modes of the K meson are nonleptonic, and parity nonconservation was demonstrated in these experiments only for weak interactions in which leptons were involved. It was not inconceivable that the parity-violation effects observed at that time might have been only a consequence of the nature of the neutrino. An explicit demonstration of parity nonconservation was still necessary for the nonleptonic decay interactions of strange particles, but there was little real doubt about how this would turn out. The mental barriers against parity nonconservation had gone, and the simplest interpretation was acceptable: that there was just one kind of K^+ meson, which was spinless and which decayed to both 2π and 3π final states through a weak interaction that violated parity conservation.

If we had accepted this conclusion earlier, on the basis of the arguments presented in the last chapter, parity nonconservation might well have been demonstrated for the nonleptonic decay interactions of strange particles before it had been demonstrated for the semileptonic and leptonic weak interactions of nonstrange particles. To demonstrate parity nonconservation, some pseudoscalar quantity must be measured to be nonzero, and this quantity will necessarily arise in the theory from the interference between two states of opposite parity. The 2π and 3π states from K^+ decay have opposite parities, but they are final states involving different numbers of particles, so that they cannot be made to interfere directly. How could we have found a pseudoscalar to measure, in these circumstances? The answer is that the two processes $K \to 2\pi$ and $K \to 3\pi$ can lead to interference if they both contribute virtually to processes that lead to the same final state. For example, we might have considered their contributions to Λ decay:[40]

$$\Lambda \to p + K^- \to \begin{Bmatrix} p + \pi^- + \pi^+ + \pi^- \\ p + \pi^0 + \pi^- \end{Bmatrix} \to p + \pi^- \tag{30.19}$$

These two amplitudes lead to a matrix element of the form

$$M = s + p\boldsymbol{\sigma} \cdot \mathbf{q}\pi \tag{30.20}$$

where σ denotes the baryon spin, and $\mathbf{q}_\pi$ is the pion momentum. Assuming pseudoscalar $(p\Lambda K)$ coupling, the amplitudes s and p refer to the emission of an s-wave or p-wave pion in the decay process (30.19), the former arising from the parity-reversing weak-interaction process $\bar{K} \to 2\pi$, and the latter from the parity-conserving weak-interaction process $\bar{K} \to 3\pi$. The two $p\pi^-$ states reached are different, of course, having spin parities $(\frac{1}{2}-)$ and $(\frac{1}{2}+)$, respectively, but they do involve the same set of particles. Their interference leads to a forward/backward asymmetry in the π^- angular distribution relative to the Λ polarization vector, in the decay of polarized Λ particles. We know today that the production process

$$\pi^- + p \rightarrow \Lambda + K^0 \tag{30.21}$$

for pion momenta of order 1,000 MeV/c does produce Λ particles with a strong polarization perpendicular to their production plane. There is therefore an up–down asymmetry for the pions resulting from Λ decay, following this production reaction, relative to the production plane. This possibility was first investigated in 1956 by Reinhard Budde and associates,[41,42] not so much for this purpose as for testing for possible biases in their experimental observations on this reaction (30.21). In an early report[41] at the Sixth Rochester Conference (April 1956), the up–down asymmetry {(up − down)/(up + down)} on nineteen ΛK^0 production events was −0.26 ± 0.2, which was considered compatible with zero, the value required by parity conservation in $\Lambda \rightarrow p\pi^-$ decay; by the final report,[42] this asymmetry had moved to −0.17 ± 0.2, closer to zero. This null result occasioned no surprise at the time, for there was no a priori expectation that the Λ polarization in the production reaction $\pi p \rightarrow \Lambda K^0$ and the up–down asymmetry in polarized Λ decay should either of them be other than very small, and the nonzero value obtained was well understood as a statistical fluctuation.

After parity nonconservation was established for polarized ^{60}Co decay and for the $\pi \rightarrow \mu \rightarrow e$ sequence, and it was realized that parity-violation effects could lead to large asymmetries, these observations on Λ decay following the Λ production reaction (30.21) were carried further, with much larger statistics. In October 1957, three groups[43–45] submitted for publication their final reports on measurements of the up–down asymmetry in Λ decay following reaction (30.21), with consistent results, altogether differing from zero by about five standard deviations, and parity nonconservation in $\Lambda \rightarrow p\pi^-$ decay thereby became well established. It is interesting, in retrospect, to realize that this line of thought and this demonstration of parity nonconservation in the nonleptonic weak interactions for strange particles could well have preceded the experiments on leptonic and semileptonic weak interactions.

Notes

1 R. Brown, U. Camerini, P. H. Fowler, H. Muirhead, C. F. Powell, and D. M. Ritson, "Observations with Electron-Sensitive Plates Exposed to Cosmic Radiation. II. Further Evidence for the Existence of Unstable Charged Particles of Mass ~1000 m_e and Observations on their Mode of Decay," *Nature (London) 163* (1949), 82–7.

2 C. M. G. Lattes, H. Muirhead, G. P. S. Occhialini, and C. F. Powell, "Processes Involving Charged Mesons," *Nature (London) 159* (1947), 694–7.

3 H. Yukawa, "Interaction of Elementary Particles. Part I," *Proc. Phys. -Math Soc. Japan 17* (1935), 48–57.

4 J. B. Harding, "Further Evidence for the Existence of τ-Mesons," *Philos. Mag. 41* (1950), 405–9.

5 C. O'Ceallaigh, "Masses and Modes of Decay of Heavy Mesons. I. $\varkappa$-Particles," *Philos. Mag.* *42* (1951), 1032–9.

6 T. F. Hoang, L. Jauneau, G. Kayas, L. Leprince-Ringuet, D. Morellet, A. Orkin-Lecourtois, and J. Trembley, "Evidence Supporting the Existence of Two Types of Heavy Mesons Giving a Secondary μ Meson" (In French), *C. R. Acad. Sci 238* (1954), 1633–5.

7 J. Crussard, M. F. Kaplon, J. Klarmann, and J. H. Noon, "Heavy Unstable Particles in Stripped Emulsions," *Phys. Rev. 95* (1954), 584.

8 E. Amaldi et al., note 11 (p. 210).

9 M. G. K. Menon and C. O'Ceallaigh, in *Proceedings Congrès International sur le Rayonnement Cosmique*, Bagnères de Bigorre (University of Toulouse, 1933), p. 124.

10 J. Crussard, M. F. Kaplon, J. Klarmann, and J. H. Noon, "Observation of a New Decay Mode of a Heavy Meson," *Phys. Rev. 93* (1954), 253.

11 G. Belliboni, B. Sechi, and B. Vitale, "Analysis of Some τ-Mesons," *Nuovo Cimento (Suppl. 2) (Ser. 9) 12* (1954), 195–201; M. DiCorato, C. C. Dilworth, and L. Scarsi, "An Example of a τ^+ and of a $^{(\tau)}K_\pi^+$," ibid., 204–4; E. Amaldi, G. Cortini, and A. Manfredini," Contribution to the *K*-Meson Investigation," ibid., 210–19.

12 R. W. Thompson, A. V. Buskirk, H. O. Cohn, C. J. Karzmark, and R. H. Rediker, in *Proceedings Congrès International sur le Rayonnement Cosmique*, Bagnères de Bigorre (University of Toulouse, 1953), p. 30.

13 G. D. Rochester and C. C. Butler, "Evidence for the Existence of New Unstable Elementary Particles," *Nature (London) 160* (1947), 855–7; G. D. Rochester and C. C. Butler, "The New Unstable Cosmic-Ray Particles," *Rep. Prog. Phys. 16* (1953), 364–407.

14 L. Leprince-Ringuet, "*S* Particles," in *High Energy Nuclear Physics, Proceedings of the Fifth Annual Rochester Conference*, edited by H. P. Noyes, E. M. Hafner, G. Yekutieli, and B. J. Raz (New York: Interscience, 1955), pp. 74–7.

15 B. Rossi, "*S* Events," in Noyes et al., note 14 (pp. 87–9).

16 M. Friedlander, D. Keefe, M. G. K. Meson, and L. Van Rossum, "Evidence for the β-Decay of a *K*-Meson," *Philos. Mag. 45* (1954), 1043–9.

17 C. Dahanayake, P. E. Francois, Y. Fujimoto, P. Iredale, C. J. Waddington, and M. Yasin, "Observations on the Associated Production of Heavy Mesons and Hyperons," *Philos. Mag. 45* (1954), 855–62.

18 E. Amaldi, "τ Mesons," in *High Energy Nuclear Physics, Proceedings of the Third Annual Rochester Conference*, edited by H. P. Noyes, M. Camac, and W. D. Walker (New York: Interscience, 1952), p. 50–1.

19 R. H. Dalitz, in *Proceedings Congrès International sur le Rayonnement Cosmique*, Bagnères de Bigorre (University of Toulouse, 1953), p. 236.

20 A. L. Hodson, J. Ballam, W. H. Arnold, D. R. Harris, R. R. Rau, G. T. Reynolds, and S. B. Treiman, "Cloud-Chamber Evidence for a Charged Counterpart of the θ^0 Particle," *Phys. Rev. 96* (1954), 1089–95.

21 R. H. Dalitz, "*K*-Mesons and Hyperons. Their Strong and Weak Interactions," *Rep. Prog. Phys. 20* (1957), 163–303.

22 R. H. Dalitz, "On the Analysis of τ-Meson Data and the Nature of the τ-Meson," *Philos. Mag. 44* (1953), 1068–80.

23 R. H. Dalitz, "Decay of τ Mesons of Known Charge," *Phys. Rev. 94* (1954), 1046–51.

24 E. Fabri, "A Study of τ-Meson Decay," *Nuovo Cimento 11* (1954), 479–91.

25 E. Amaldi, G. Baroni, G. Cortini, C. Franzinetti, and A. Manfredini, "Contribution to the τ-Meson Investigation," *Nuovo Cimento (Suppl. 2) (Ser. 9) 12* (1954), 181–94.

26 E. Fabri, "The Phenomenological Treatment of τ-Meson Decay," *Nuovo Cimento (Suppl. 2) (Ser. 9) 12* (1954), 205–6.

27 R. Dalitz, "Analysis of τ Meson Decays," in *High Energy Nuclear Physics, Proceedings of the Fifth Annual Rochester Conference*, edited by H. P. Noyes, E. M. Hafner, G. Yekutieli, and B. J. Raz (New York: Interscience, 1955), pp. 140–2.

28 Proceedings of the international conference on elementary particles held in Pisa, Italy, 12–18 June, 1955, *Nuovo Cimento (Suppl. 2) (Ser. 10) 4* (1956), 135–1078.

29 E. Amaldi, "Report on the τ-Mesons," *Nuovo Cimento (Suppl. 2) (Ser. 10) 4*, 179–214.

30 J. Orear, G. Harris, and S. Taylor, "Spin and Parity Analysis of Bevatron τ Mesons," *Phys. Rev. 102* (1956), 1676–84.

31 M. Baldo-Ceolin, A. Bonetti, W. D. B. Greening, S. Limentani, M. Merlin, and G. Vanderhaeghe, "An Analysis of 419 τ-Meson Decays," *Nuovo Cimento 6* (1957), 84–97.

32 R. E. Marshak, "Identity of θ and τ on the Assumption of Complex Structure," in *High Energy Nuclear Physics, Proceedings of the Sixth Annual Rochester Conference*, edited by J. Ballam, V. L. Fitch, T. Fulton, K. Huang, R. R. Rau, and S. B. Treiman (New York: Interscience, 1956), pp. VIII-18-19.

33 J. Orear, "Evidence against Spin 1 for the τ-Meson," *Phys. Rev. 106* (1957), 834–5.

34 T. D. Lee and C. N. Yang, "Mass Degeneracy of the Heavy Mesons," *Phys. Rev. 102* (1956), 290–1; T. D. Lee and C. N. Yang, "Possible Interference Phenomena between Parity Doublets," *Phys. Rev. 104* (1956), 822–7.

35 T. D. Lee and J. Orear, "Speculations on Heavy Mesons," *Phys. Rev. 100* (1955), 932–3.

36 L. Alvarez, "Search for the Lee–Orear γ-Ray," in *High Energy Nuclear Physics, Proceedings of the Sixth Annual Rochester Conference*, edited by J. Ballam, V. L. Fitch, T. Fulton, K. Huang, R. R. Rau, and S. B. Treiman (New York: Interscience, 1956), pp. V-28–31.

37 T. D. Lee and C. N. Yang, "Question of Parity Conservation in Weak Interactions," *Phys. Rev. 104* (1956), 254–8.

38 Chen Ning Yang, *Selected Papers 1945–1980 With Commentary* (San Francisco: Freeman, 1983), p. 28.

39 T. D. Lee, "History of Weak Interactions," in *Elementary Processes at High Energy*, edited by A. Zichichi (New York: Academic, 1971), pp. 828–40.

40 R. H. Dalitz, *Atti del Convegno Mendeleeviano 'Periodicita e Simmetrie nella Strutture Elementare della Materia,'* edited by M. Verde (Torino: Accad. Sci. Torino & Accad. Naz. Lincei, 1971), p. 341.

41 J. Steinberger, "Curious Particle Production in a Bubble Chamber: Σ^- Lifetime and Q Value; Angular Correlations," in *High Energy Nuclear Physics, Proceedings of the Sixth Annual Rochester Conference*, edited by J. Ballam, V. L. Fitch, T. Fulton, K. Huang, R. R. Rau, and S. B. Treiman (New York: Interscience, 1956), pp. VI-20–6.

42 R. Budde, M. Chretien, J. Leitner, N. P. Samios, M. Schwarz, and J. Steinberger, "Properties of Heavy Unstable Particles Produced by 1.3 BeV π^- Mesons," *Phys. Rev. 103* (1956), 1827–36.

43 F. S. Crawford, M. Cresti, M. L. Good, K. Gottstein, E. M. Lyman, F. T. Solmitz, M. L. Stevenson, and H. K. Ticho, "Detection of Parity Nonconservation in Λ Decay," *Phys. Rev. 108* (1957), 1102–3.

44 F. Eisler et al., "Demonstration of Parity Nonconservation in Hyperon Decay," *Phys. Rev. 108* (1957), 1353–5.

45 L. B. Leipuner and R. K. Adair, "Production of Strange Particles by $\pi^- - p$ Interactions Near Threshold," *Phys. Rev. 109* (1958), 1358–63.

31   The $\tau-\theta$ puzzle: an experimentalist's perspective

VAL L. FITCH

Born 1923, Merriman, Nebraska; Ph.D., Columbia University, 1954;
experimental physics; Nobel Prize, 1980, for discovery of *CP*
violation; Princeton University

It is a great pleasure to contribute to this volume on the history
of particle physics in the 1950s as one of the experimental physicists who
participated in the establishment of *the* puzzle, one that eventually was
so constrained by experiment that it had to have a far-from-trivial resolution.

The period I shall discuss starts in the summer and fall of 1953, when the
properties of what we now call K mesons were being established, and concludes in the summer of 1956 with the publication of the Lee–Yang paper on
parity violation.[1] What always paces the rate of discovery is the development
of new instruments – new devices for probing phenomena, either with a new
level of precision or by extending the range of observation into realms never
before explored. The growth of our knowledge of K mesons is a superb
example of this. The new knowledge was acquired at the rate at which new
devices were invented, developed, and applied to the problems at hand.
Figure 31.1 gives my impression of some of the important instrument developments and the time when they became useful laboratory tools. Chronological lists like this one have appeared elsewhere. Perhaps the only thing new
on the present list is the development of the end-window phototube. While in
itself it seems like a minor thing, it had a tremendous impact in the utility
of both scintillation and Cerenkov counters. All of us who tried to design a
Cerenkov counter using the old 931-A or 1P21 phototubes will understand
what I mean. The other item of special note is the diffusion cloud chamber, in
which, in its brief period of glory, associated production was first seen unambiguously (as discussed in Chapter 22 by William Fowler).

Undoubtedly, the best-read issues of the *Physical Review* for a particular
person are those in which that person has a paper. For that reason, I was very

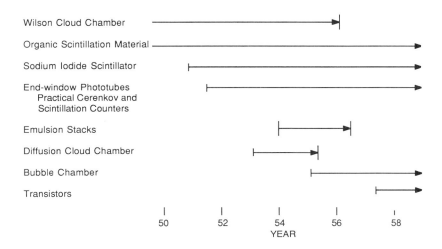

Figure 31.1. Significant instruments and instrumental components used and invented during the period of interest. These important developments, coupled with the availability of new high-energy accelerators, were responsible for the explosive increase in our knowledge of the properties of elementary particles in less than one decade.

well acquainted with the issue of 15 November 1953. It was a vintage issue, and not necessarily for the personal reason just mentioned. Contained in that issue is, for example, Murray Gell-Mann's paper with the concept of a new quantum number that came to be called strangeness.[2] It has in it the first announcement by Frederick Reines and C. L. Cowan of the detection of the neutrino by a direct reaction.[3] It has many other interesting articles, but for the present purpose I point to a paper by Louis Leprince-Ringuet and Bruno Rossi entitled "A Summary of *K*-Meson Data,"[4] containing an excellent account of just how confusing the situation was for the particles with masses about half the proton mass – a multiplicity of decay fragments, some apparently two-body, some three-body, with missing neutrals totally unidentified. It was confusing. However, these authors strove to reduce it as much as possible by finally stating:

> Without minimizing the significance of the experimental results listed above, we feel that they need a further check before we can be sure that several different kinds of *K* mesons actually exist; and the main purpose of this note is to emphasize the importance of these checks. For the moment we only wish to point out that if all the available experimental data are taken at face value, even the assumption of three different kinds of *K* mesons (such as the τ, the χ, and the ϰ particle, according to the Bristol nomenclature) does not remove all difficulties of interpretation.

It is appropriate to note that the use of the generic term "*K* meson" had apparently been decided at the Bagnères de Bigorre cosmic-ray conference the preceding summer.

This issue of the *Physical Review* has a paper by the Ralph P. Shutt group, at Brookhaven National Laboratory (BNL) devoted to π^+ scattering near 260 MeV.[5] The paper described work that was prefatory to an important paper that was submitted to *Physical Review* on 10 November 1953, "Production of Heavy Unstable Particles by Negative Pions of 1.5 BeV."[6] Utilizing production by an accelerator for the first time, the authors had observed the associated production of $\Lambda^0 + \theta^0$, and what they called $\Lambda^- + K^+$, both with a cross section of ~1 mb. This observation by Shutt's group, the associated production of strange particles in the high-pressure diffusion cloud chamber, marked the beginning of the end of what I like to recall as the romantic era of particle physics, the period when cosmic rays were the only source of the new particles, the time when experimenters took their cloud chambers and arrays of Geiger counters and made their way up the tall mountains, the Pic-du-Midi, the Aiguille-du-Midi, the Jungfrau, and Mount Evans, and other experimenters flew emulsion stacks at the edge of the planet on giant balloons. Those were the romantic years.

The observation of associated production spurred all kinds of activity. A year and a half later, by the time of the Pisa conference, 12–18 June 1955, progress had been spectacular. I have mentioned the beginning of the end of the romantic era of particle physics with the observations of Shutt's group at Brookhaven. The Pisa conference saw further romance removed with the presentations of several papers dealing with *K* mesons produced at the Berkeley Bevatron and detected in emulsions.

Edoardo Amaldi summarized the τ data, obtained from stacks of emulsions stripped from their glass-plate backing, giving a mass of 493.7 ± 0.3 MeV and lifetime limits ranging over $10^{-9} < \tau_\tau < 10^{-8}$ sec.[7] Cecil F. Powell discussed what was known of the masses of the other particles in this range of mass, namely,

τ	966 ± 3
$K^+ \to \mu^+ + 2\,\text{neutrals}$	—
$\chi \to \mu^+ + \pi^0$	$\begin{cases} 952 \pm 11 \\ 968 \pm 15 \end{cases}$
$\theta^0 \to \pi^+ + \pi^-$	966 ± 10
$K^+ \to \mu^+ + \nu$	$\begin{cases} 935 \pm 15 \\ 950 \pm 15 \\ 971 \pm 12 \end{cases}$

All the masses are quoted in units of the electron mass.[8] As seen in the tabulation, the τ and the θ masses were in good agreement, but the $K_{\mu2}$, the last entry, gave hope to those who still thought that there were several distinct parents giving rise to all these various progeny. In addition to the mass measurements, the τ data were submitted to Dalitz analysis to determine spin and parity.

Very early, the essentials of the τ–θ puzzle became evident. The decay of a $\tau^+ \to \pi^+ + \pi^+ + \pi^-$, which shows a very low energy odd-sign pion (it is highly improbable for a low-energy pion to take away any angular momentum), cannot be from the same parent as the one that decays to two π's, the θ, because of the odd intrinsic parity of the pion. Sure enough, very low energy odd-sign pions were soon observed.

Two sociological observations are of interest. First, the theoretical contributions to the conference all dealt with the real world. Second, there were eighty-five experimental papers with 340 authors, or, if one excludes double counting, 250 authors. Now, as we know, a single paper can have 250 authors.

By the summer of 1955, both at BNL and at Berkeley, counter measurements of the lifetimes as functions of the decay modes were going on. They yielded

$$\tau(K_{\mu 2}) = 12.0 \pm 0.8 \times 10^{-8} \text{ sec}$$
$$\tau(K_{\pi 2}) = 12.4 \pm 1.0$$
$$\tau(K_{\pi 3}) = 11.8 \pm 0.8$$

By early 1956, we knew that the masses of the τ and the θ were the same to within 1 percent, and the lifetimes were the same to within 15 percent. Of course, there were several theoretical suggestions as to how the difficulties could be resolved.

To get the flavor of the times, of special interest are those reports written during 1956. For example, there is "Introduction to the Physics of the New Particles" by Carlo Franzinetti and Giacomo Morpurgo, which was written in the summer of 1956.[9] Chapter 12 is entitled, "The Problem of the Bosons." In it is discussed the Lee–Orear cascade assumption, the parity doublet model of Tsung Dao Lee and Chen Ning Yang, and finally the speculation of Lee and Yang that, indeed, parity may not be conserved, with a number of specific tests suggested. As stated by Franzinetti and Morpurgo, "the fact that no one of the solutions presented up to now is entirely satisfactory has recently led [Lee and Yang] to consider another possibility: It is the assumption that parity may not be conserved in the weak interactions. Of course, in principle, this may solve all the difficulties."

I began with a discussion of the contents of a particular issue of the *Physical Review*. It is no accident that this corresponded to the time when I became sufficiently interested in the strange particles to make a new commitment. I had been working with James Rainwater and S. Koslov, measuring for the first time ever the vacuum polarization term in μ-mesic atoms. It was exciting business, and there was much to be done. Abraham Pais and then Gell-Mann came to Columbia and talked of all the curious new-particle results. Shutt came and talked about the observation of associated production. My head was turned to the new area.

I departed Columbia for Princeton, which had cosmic-ray work going on at Mount Evans in Colorado. I also had the promise of support for any activity I might initiate at Brookhaven. I spent the summer of 1954 on Mount Evans looking about for something relevant and interesting. I said that I found the cosmic-ray work in the mountains romantic. I soon recognized, as with nearly all romantic notions, that with continuous exposure romanticism is replaced with harsh realism. By the end of the summer, the enormous difficulties attendant upon obtaining useful, quantitative data in cosmic-ray work had become very apparent. And by then the Cosmotron at Brookhaven was operating steadily at 3 GeV, and counter experiments looked feasible. When I got back to Princeton, an interested graduate student, Robert Motley, and I went to BNL to see George Collins, who then directed the Cosmotron, about doing an experiment on the lifetime of the K mesons as a function of decay mode. He gave us such a favorable hearing that we went back to Princeton and started designing and building equipment. Now, I am not sure how Collins operated. I presume he had some sort of "kitchen cabinet" for advice. Certainly there was no scheduling committee to contend with, at least none that was visible to me.

When we moved our gear to BNL the following summer (1955), we did find that we were not alone; we were one of three groups interested in this same experiment. One was an emulsion group, G. Harris and Jay Orear, and the other was a newly formed counter group from BNL. The initial running was scheduled into one block of twenty-four hours that was to be divided into three pieces, eight hours for each group. Motley and I were, frankly, rather put off by this arrangement because we thought we were there first. Nevertheless, that was the way it was, and cooperatively, much to our advantage, the three groups set up the focusing magnets (which we had borrowed from Rainwater) for a very short beam line at 60° to an internal target in the machine.

The twenty-four-hour period began. Motley and I had the last eight hours, midnight to 8:00 A.M. During this period we had to roll in our detector (after the other group took theirs out), hook up the cables, stack some shielding, and do whatever we could in the remaining time.

Motley and I were very lucky. At the end of the first eight hours we had a rough differential range curve of K mesons, the first detection of K mesons using counter techniques. That was largely due to the fact that we had used a new Cerenkov counter of novel design, simple but effective. Shortly after this initial success, Motley and I got a whole twenty-four hours of machine time. When we finally published our results as a letter in *Physical Review*, there in the same issue was another letter from Luis Alvarez, Frank Crawford, Myron L. Good, and Lynn Stevenson on the same subject with similar results – also using counter techniques.[10,11] We hadn't known of their activity, just vague rumors. It must have been Alvarez's last counter experiment, because he was then already heavily into his bubble-chamber program.

That summer was an exciting one around the Cosmotron. William Chinowsky, Kenneth Lande, and Leon Lederman were discovering the θ_2 in the corn crib out in the parking lot (see Chapter 21 by Chinowsky); Melvin Schwartz and Jack Steinberger were firing up their propane bubble chamber (see Chapter 20 by Steinberger); Rodney Cool and Oreste Piccioni in one beam line, and Sam Lindenbaum in another, were doing total cross sections.

The following spring we were given still additional time for studying τ decay. In the concluding phase of that experiment, Wolfgang K. H. Panofsky walked into our lab and got us interested in trying a counter experiment on the recently discovered θ_2's to confirm some of their expected properties. Panofsky, Motley, Walter Chesnut, a BNL staff member, and I teamed up and did the experiment. It turned out that Chesnut and his wife lived next door to James Cronin and his family, and that summer I first met Cronin.* One thing does lead to another.

Notes

1 T. D. Lee and C. N. Yang, "Question of Parity Conservation in Weak Interactions," *Phys. Rev. 104* (1956), 254–8.

2 M. Gell-Mann, "Isotopic Spin and New Unstable Particles," *Phys. Rev. 92* (1953), 833–4.

3 F. Reines and C. L. Cowan, Jr., "Detection of the Free Neutrino," *Phys. Rev. 92* (1953), 830–1.

4 L. Leprince–Ringuet and B. Rossi, "A Summary of *K*-Meson Data," *Phys. Rev. 92* (1953), 722–3 (actually in the issue of 1 November 1953 – *Ed.*).

5 W. B. Fowler, R. M. Lea, W. D. Shephard, R. P. Shutt, A. M. Thorndike, and W. L. Whittemore, "$\pi^+ - p$ Scattering Near 260 MeV," *Phys. Rev. 92* (1953), 832.

6 W. B. Fowler, R. P. Shutt, A. M. Thorndike, and W. L. Whittemore, "Production of Heavy Unstable Particles by Negative Pions," *Phys. Rev. 93* (1954), 861–7.

7 E. Amaldi, "Report on the τ-Mesons," *Nuovo Cimento (Suppl.) 4* (1956), 179–214.

8 C. F. Powell, "Recent Advances in our Knowledge of Heavy Mesons and Hyperons," *Nuovo Cimento (Suppl.) 4* (1956), 337–50.

9 C. Franzinetti and G. Morpurgo, "An Introduction to the Physics of the New Particles," *Nuovo Cimento (Suppl.) 6 : 2* (1957), 469–802.

10 V. Fitch and R. Motley, "Mean Life of K^+ Mesons," *Phys. Rev. 101* (1956), 496–8.

11 Luis W. Alvarez, F. S. Crawford, Myron L. Good, and Lynn Stevenson, "Lifetime of K Mesons," *Phys. Rev. 101* (1956), 503–5.

---

*   *Ed. note:* In 1980, Cronin and Fitch received the Nobel Prize for discovery of violations of fundamental symmetry principles in the decay of neutral K mesons.

32   The early experiments leading to the $V-A$ interaction

VALENTINE L. TELEGDI

Born 1922, Budapest; Ph.D., Swiss Federal Institute of Technology, 1950; experimental physics; Swiss Federal Institute of Technology, Zurich

Never in my fondest dreams did I expect in January 1957 to be invited, twenty-eight years later, to reminisce about decisive experiments on symmetry violation in a great national laboratory directed by Leon M. Lederman. Such reminiscences not only are unnecessary (some of the key protagonists having already published their recollections in Maglich's now-defunct journal) but also can be misleading. Nobody has total recall, and personal souvenirs can get warped by things heard but not experienced. Just recall Einstein's contradictory answers to the question whether he was or was not aware of the Michelson–Morley experiment in 1905.

Theoretical prologue

My friend Laurie Brown has asked me to include in my talk some remarks about the history of the theoretical developments in the field of weak interactions, as several of the main contributors to this field are present here but have chosen not to talk. I shall oblige Laurie Brown in some sense, but still avoid giving a chronological presentation devoid of personal remini-scences. Such presentations are available in standard textbooks.

Let me not praise famous men: It might be embarrassing (or insufficient) to them and to me. Let me instead mention a few early contributions that, for one reason or another, have rarely been cited in past reviews.

Before the famous Lee–Yang paper on parity nonconservation, theorists generally stated that there existed five different four-fermion couplings. However V. B. Berestetsky and I. Y. Pomeranchuk, in a 1949 paper[1] on the β decay of the neutron, mentioned a remark by Lev D. Landau that there actually exist ten couplings "if one allows, in addition to spinors for pseudo-spinors." They added, however, "that since the latter lead to no observable results in the heavy particles (i.e., baryons), we shall not take them into account."

Similarly, before the Lee–Yang paper, it had become almost a credo that mirror invariance (parity conservation) is an a priori property of the laws of nature rather than a hypothesis to be tested by experiment. Almost, but not quite: In a short paper[2] published in 1950, E. M. Purcell and Norman F. Ramsey explicitly pushed this latter point of view and proposed a measurement of the electric dipole moment of the neutron as a check (which was then first carried out by Ramsey and James H. Smith). The only flaw in this remarkable suggestion was that the authors did not realize (as Landau did subsequently) that time-reversal invariance had to be violated as well, in order for the dipole moment not to vanish.

In an unpublished report in May 1956, the Soviet physicist I. S. Shapiro (a) assumed parity violation in weak interactions and (b) suggested that such a violation could manifest itself through an up–down asymmetry in the β decay of polarized nuclei. He did not publish this report because it also contained some erroneous ideas (mass differences connected with mirroring, Möbius strips in vacuum, etc.), which Landau immediately objected to. Shapiro's report was seen by L. B. Okun, who confirmed its existence to Western inquirers. Moral: To become famous, it is not enough to have an idea (or an experimental result) – one also has to believe in it.[3]

Next we turn to the *CPT* theorem. One standard reference is an article published by Wolfgang Pauli in 1955 in the *Festschrift* for Niels Bohr's seventieth birthday.[4] An independent paper by J. S. Bell on the same topic, published at about the same time in a highly reputable journal, is rarely, if ever, quoted.[5] It is also interesting that the introduction of the first edition of *PCT, Spin and Statistics, and All That*, by R. F. Streater and A. S. Wightman, traces the paternity of that theorem to Bruno Zumino, whereas the second edition omits that reference.

The late Bruno Touschek is well recognized for his great contributions to the theory and construction of e^+e^- storage rings. He is, however, rarely remembered for proposing in January 1957 that a suitable gauge transformation of the neutrino field, imposed to keep $m_\nu = 0$, leads to two-component neutrinos (i.e., to maximal parity violation).[6] In two subsequent papers he elaborated on lepton conservation and the equivalence, with that law, of two-component and Majorana neutrinos.[7] Finally, I would like to mention a short paper by W. Theis submitted on 20 December 1957 to the *Zeitschrift für Physik*, in which he proposed the parity-violating $V–A$ interaction by a formal argument quite analogous to that of Richard Feynman and Murray Gell-Mann (i.e., by insisting on a derivative-free coupling in terms of two-component spinors).[8]

The crucial experiments

Experiments originally proposed by T. D. Lee and C. N. Yang

Up–down asymmetry in β decay. It was proposed to measure the angular distribution

$$W(\overleftarrow{P} \cdot \vec{v}_e) = 1 + \mathscr{A}(\overleftarrow{P} \cdot \vec{v}_e)$$

where $\overleftarrow{P} = \langle \overleftarrow{J} \rangle / J$ = nuclear polarization (I use left-pointing arrows to denote axial vectors in order to emphasize pseudoscalar products), and $\vec{v}_e = \beta^\pm$ velocity. Pure Fermi ($J = 0$) transitions are of no interest, because the $e\bar{\nu}$ pair is emitted as a singlet and hence is isotropically disturbed. For the Gamow–Teller (spin-flip) transitions, it suffices to consider the $1^+ \rightarrow 0^+$ case. We represent the transition amplitude graphically as in Figure 32.1.[9]

The top part of Figure 32.1 corresponds to "tensor" coupling T (the one that was believed to hold in 1956!), the bottom part to "axial vector" coupling A (the one that actually holds in nature). A good mnemonic for these archaic designations is through the associated $e-\bar{\nu}$ angular correlations: With the T coupling, e and $\bar{\nu}$ run together; in the A case, they run antiparallel. The heavy arrows indicate the j's of the particles, the light arrows their velocities, and angular momenta are balanced. For each configuration of lepton velocities (say T, i.e., parallel), there is still a choice of helicities. The amplitudes of the two mirror-image configurations are distinguished by an index R or L, where we choose the following convention: R = left-handed antineutrinos, L = right-handed antineutrinos. (Obviously, we are referring to the helicity of the neutrino that "gets destroyed.") Angular-momentum balance fixes the handedness of the emitted electron.

Thus, $T_{R(L)}$ and $A_{R(L)}$ represent, respectively, the tensor and axial vector coupling constants for R (L) neutrinos. Clearly, with parity conservation, one has to have $T_R = T_L$ and/or $A_R = A_L$. Conversely, $A_L = 0$, and so forth, corresponds to maximal parity violation. The lepton "arrow state amplitudes" depicted in Figure 32.1 and in the rest of my discussion are defined to have squares such that

$$W(\theta) = 1 \pm v \cos(\theta) \qquad (c = 1) \tag{32.1}$$

where θ is the angle between the velocity and the angular-momentum vector (say j_z) of the lepton represented.

Thus, equation (32.1) can be read as "given that $j_z = +\frac{1}{2}$, the probability of finding the lepton at an angle with respect to the z axis is (proportional to) $1 \pm v \cos \theta$." From this, the angular-correlation coefficient $\mathscr{A}$ follows immediately:

$$\mathscr{A} = \begin{cases} -(T_R^2 - T_L^2)/(T_R^2 + T_L^2) \\ \text{or} \\ -(A_L^2 - A_R^2)/(A_R^2 + A_L^2) \end{cases} \tag{32.2}$$

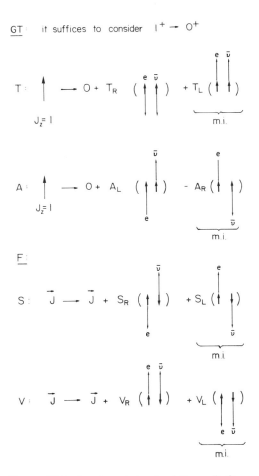

Figure 32.1. Lepton states in (allowed) Gamow–Teller and Fermi transitions. Mirror-image (m.i.) amplitudes are added if parity is conserved.

Incidentally, one cannot distinguish A from T without information about the neutrinos. Thus, for example, $T_R = 1$ gives the same $\mathscr{A}$ coefficient as A_L; the same ambiguity holds, as is evident from Figure 32.1, for the electron helicity, This distinction is made possible, for instance, by the (parity-conserving) $e-\nu$ angular correlation

$$W(\vec{v}_e \cdot \vec{v}_{\bar{\nu}}) = [1 \pm (\tfrac{1}{3})\vec{v}_e \cdot \vec{v}_{\bar{\nu}}] \begin{cases} T \\ A \end{cases}$$

The (incorrect) experimental evidence in 1956 indicated T to hold!

The first result on $\mathscr{A}$ was obtained at the National Bureau of Standards (NBS) by an NBS–Columbia team in December 1956, using polarized ^{60}Co nuclei.[10] This is a pure Gamow–Teller decay, but with the spin sequence

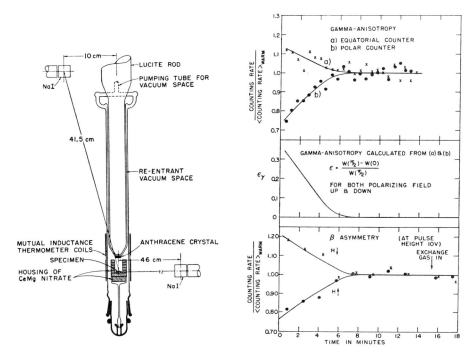

Figure 32.2. (left) Schematic drawing of the lower part of the cryostat used in the NBS–Columbia experiment; (right) results of that experiment [*Source*: Wu et al. (1957)[10]].

$5^+ \rightarrow 4^+$ instead of $1^+ \rightarrow 0^+$ as in the example discussed. This fact was actually an advantage: The β angular correlation is the same, but the final state, 4^+ in ^{60}Ni, decays with a $4^+ \rightarrow 2^+ \rightarrow 0^+$ γ cascade.

The β decay (even if parity-conserving) transfers the orientation of the initial Co nucleus to the ^{60}Ni, and the γ decay of the latter is therefore anisotropic. Its angular distribution could be, and was, used to monitor the polarization of the ^{60}Co sample, in particular its evolution with time [Figure 32.2(b)]. The NBS–Columbia result, submitted for publication on 15 January 1957, not only proved parity violation but also showed it to be a large effect, consistent even with maximal violation ($\mathscr{A} = -1$, i.e., $T_L = 0$ or $A_R = 0$).

It is interesting to ponder about the time and the place of this very important experiment. Could it have been done elsewhere? Could it have been done earlier? The answer to the first question is "yes." The answer to the second is "hardly, except by accident." The time (the fall of 1956) was ripe, and the place (NBS) was right. The art of orienting radioactive nuclei had just recently been perfected at Oxford and Leyden, and two members of the NBS team (E. Ambler and R. P. Hudson) had been "imported" from Oxford to NBS (prior to the idea of any parity-violation experiment) specifically because of their expertise in this field. In terms of experimental difficulty, the low-temperature aspect was undoubtedly much more exacting than the β decay part; in this regard, it is interesting to read a comment by N. Kürti, a veteran of Oxford low-temperature physics.[11]

Now, why was the experiment not done at Oxford or at Leyden? At Oxford it was not done at all, and in Leyden only with considerable delay and using a much less favorable decay.[12] I have never been able to discover the real reasons for this, but I suspect that in those two laboratories the theoretical urgency of the problem was far less clearly realized than at Columbia University, where C. S. Wu was in constant contact with Lee. In many laboratories, both in the United States and in Europe, parity violation was considered too wild a hypothesis to warrant an all-out experimental effort. The case of Leyden is particularly ironic: There, H. A. Tolhoek presented in 1956 a thesis on the angular asymmetries in the decay of oriented nuclei (confining himself, of course, to mirror-invariant effects), and he wrote a review paper on polarization that turned out to be most useful for the design of experiments on longitudinal electron polarization.[13]

Returning to the detailed physics of the ^{60}Co experiment, the main open questions were as follows:

(i)   What would the decay asymmetry of *positron* emitter be?
(ii) How is one to choose between the alternatives T_R (then favored) and A_L (now accepted)?

This second question could in principle be elucidated by a measurement of the analogous ν up–down asymmetry, described by

$$W(\overleftarrow{J} \cdot \vec{v}_\nu) = (1 + B\overleftarrow{P} \cdot \vec{v}_\nu)$$

here with

$$B = \begin{cases} -1 & \text{for } T_R \\ +1 & \text{for } A_L \end{cases}$$

as can easily be seen from Figure 32.1. This road was soon chosen in the case of the free-neutron decay. Another ideal experiment, much discussed in those days, was to study the decay of anti-^{60}Co. Its substitute was the much more practical comparison of μ^+ and μ^- decays.

β–γ circular-polarization correlation. Denote the circular polarization of a γ ray by P_γ, with $|P_\gamma| = +1 (-1)$ for R (L) 100 percent polarization. Then there arises in a β–γ cascade, even in the case of *unoriented* nuclei, a correlation of the form

$$W(\overleftarrow{P}_\gamma \cdot \vec{v}_e) = 1 + \mathscr{A}'\overleftarrow{P}_\gamma \cdot \vec{v}_e$$

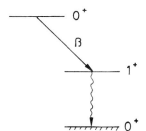

Figure 32.3. Decay scheme for $\beta-\gamma$ circular-polarization correlation.

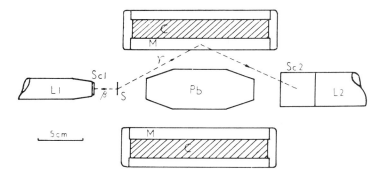

Figure 32.4. Apparatus used in Schopper's $\beta-\gamma$ experiments [*Source*: Schopper (1957)[14]].

Lee and Yang considered this experiment "too difficult" in their fundamental paper; actually, it could be realized with much simpler means than the up–down asymmetry experiment just discussed. In fact, the two experiments are conceptually quite equivalent, and $\mathscr{A}$ and $\mathscr{A}'$ are directly related. We can see this by considering the decay chain $0^+ \xrightarrow{\beta} 1^+ \xrightarrow{\gamma} 0^+$ (Figure 32.3) and "running it backward." The absorption of a circularly polarized photon fully polarizes the 1^+ state, and the latter decays with an asymmetry parameter $\mathscr{A} = -1$, as already discussed; thus, $\mathscr{A}' = -1$. When the spin sequence is less simple, there is a reduction of $\mathscr{A}'$.

The difficulty of the experiment was in the availability of a circular-polarization analyzer ($\lambda/4$ plate) for γ rays. Again, the time was just ripe: Such an analyzer had been developed, in 1952–3, by S. B. Gunst and L. A. Page and by F. L. Hereford and associates. It exploits the spin dependence of Compton scattering, using magnetized iron as a "polarized electron target."

The first to publish results by this technique was H. Schopper, the present director general of CERN, then working at Cambridge under the auspices of O. R. Frisch.[14] With the apparatus schematically shown in Figure 32.4, he obtained the following results:

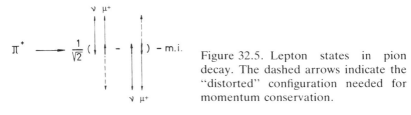

Figure 32.5. Lepton states in pion decay. The dashed arrows indicate the "distorted" configuration needed for momentum conservation.

	$\mathscr{A}'_{exp}$	$\mathscr{A}'_{th}(max)$
β^- ^{60}Co $(5 \rightarrow 4 \rightarrow 2)$	$-0.41(7)$	$-\frac{1}{3}$
β^+ ^{22}Na $(3 \rightarrow 2 \rightarrow 0)$	$+0.39(8)$	$+\frac{1}{3}$

They were, for *both* β^- and β^+, consistent with maximal parity violation! Many other groups carried out such measurements, and their results appeared in print in close succession.[15]

π–μ–e decay chain. In π–μ decay, momentum conservation and angular-momentum conservation suggest the configuration indicated in Figure 32.5 (spins and momenta antiparallel). However, in 1956, the prevailing mood favored the A configuration (contrary to the prevailing choice of T in β decay) because in that case the pion decay is greatly suppressed, as was observed. In fact, the lepton velocity v_l can be made parallel to its angular momentum with a probability

$$W(0) \sim (1 - v_l \cos \theta)|_{\theta=0} = (1 - v_l)$$

which vanishes as $v_l \rightarrow 1$ (π–e decay). This decay seemed at first not to occur even at the small predicted rate. It was observed in 1958 at the "correct" theoretical level.[16,17]

In any case, the muon would come out polarized if parity were not conserved (and indeed with the "wrong" helicity). It would, in turn, decay with a characteristic up–down asymmetry

$$W(\overleftarrow{P}_\mu \cdot \vec{v}_e) \cong [1 + a(E_e)\overleftarrow{P}_\mu \cdot \vec{v}_e]$$
$$= [1 + a\overleftarrow{P}_\mu \cdot \vec{v}_e] \quad \text{(averaged over } E_e\text{)}$$

with

$$\overleftarrow{P}_\mu = \langle \bar{\sigma}_\mu \rangle$$

The asymmetry coefficient a depends on the electron energy, because, as opposed to β decay, one has here a true three-body final state. Thus, establishing an angular correlation between the muon momentum (to which $\overleftarrow{P}_\mu$ was to be parallel or antiparallel) and the electron velocity $\vec{v}_e$, one would exhibit parity violation in both steps of the decay chain. Note that universality between β decay and π decay would call for A_L coupling in *both* cases.

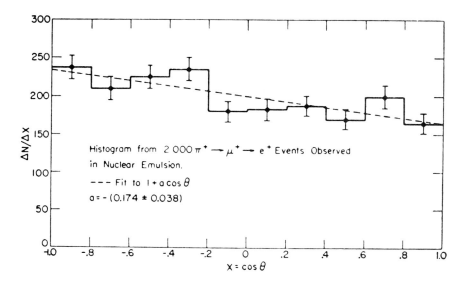

Figure 32.6. Final result of the Chicago experiment. Histogram from 2,000 $\pi^+ \to \mu^+ \to e^+$ events observed in nuclear emulsion. Dashed curve is least-squares fit to $1 + a \cos \theta$; $a = 0.174 \pm 0.038$. Indicated errors are statistical standard deviations [*Source*: Friedman and Telegdi (1957)[18]].

The Chicago experiment. The simplest way to determine the angular correlation in question is to follow the method by which the $\pi-\mu-e$ chain was first exhibited experimentally, that is, by stopping π^+'s (in our case, from a cyclotron) in a sufficiently thick nuclear emulsion. This is the approach that Jerry Friedman and I followed; we both had had lots of experience with the emulsion technique.[18] Let me put on record that Friedman was not some obscure sidekick in this venture, but a full partner – he is the same Friedman who later (at MIT/SLAC) played a major role in the deep inelastic scattering experiment that led to the parton (quark) picture of matter.

This approach had two shortcomings: It was hard to accumulate high statistics, and there was the (justified) danger of at least partial depolarization of the μ^+ by atomic magnetic fields, in particular "muonium" ($\mu^+ e^-$) formation. Thus, it could not provide a measurement of the asymmetry parameter *a* directly comparable with theory. The final result of the Chicago experiment is shown in Figure 32.6.

The simplicity of this experiment makes one wonder why it was not undertaken by many groups elsewhere. The sole explanation I can give is again that few people were prepared to take the idea of parity nonconservation seriously enough to invest their time in a tedious experiment. We, too, had been told by our senior colleagues not to waste our effort!

The Columbia experiment. In this rightly famous experiment, a μ^+ beam (actually a pion beam with muon contamination) from the Nevis (Columbia) cyclotron was stopped in a target, and the decay positrons were detected electronically with a simple scintillator telescope through a "gate" that could be delayed in time with respect to the muon stop signal.[19] The experiment involved several brilliant ideas:

(i)   the realization that the muons would emerge (in virtue of the fact that $g \cong 2$) essentially polarized from the cyclotron's magnetic field;
(ii) mapping the angular distribution by rotating the muon spin (i.e., $\vec{P}_\mu$) instead of moving the positron telescope (this involved the realization that a polarization vector moves essentialy classically);
(iii) using graphite – which turned out to be practically nondepolarizing – as a stopping target.

Figure 32.7(b) shows the statistically overwhelming result, obtained in the course of a single night. The time for this experiment was particularly ripe: The equipment existed (from the thesis work of Leon Lederman's student M. Weinrich), and the outcome of the NBS–Columbia experiment was known at Nevis before it was known to the world at large.

Up–down asymmetry in Λ decay. The angular distribution of the decay

$$\Lambda \rightarrow p + \pi^-$$

was investigated in bubble chambers by a Berkeley group[20] and a Columbia–Bologna–Pisa–Michigan collaboration.[21] The pion up–down asymmetry parameter α was determined with respect to the normal of the production plane, because the strong (parity-conserving) production process would polarize the Λ along that direction (if at all). The results were ($P = \Lambda$ polarization):

	αP
Berkeley	0.51 ± 0.15
Bologna–Columbia	0.44 ± 0.11

A null result would not have distinguished $P_\Lambda = 0$ from $\alpha = 0$, so it was a fortunate accident that P_Λ was nonzero at the experimental conditions chosen.

These results aroused particular interest at that time, because no neutrinos were involved in the decay. Thus, parity violation was indeed to be a universal property of weak interactions, as had been hypothesized from its starting point, the "$\theta-\tau$ puzzle," and not simply a manifestation of two-component neutrinos.

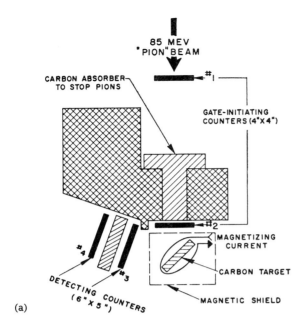

(a)

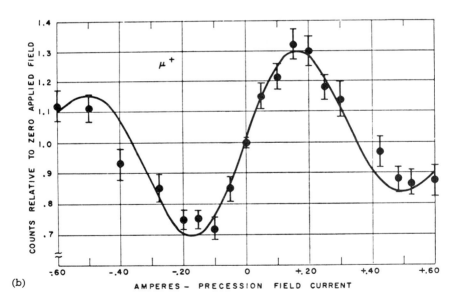

(b)

Figure 32.7. (a) Experimental arrangement of the Columbia muon asymmetry experiment. The magnetizing coil was close-wound directly on the carbon to provide a uniform vertical field of 79 G per ampere [*Source*: Garwin et al. (1957)[19]]. (b) Observed variation of gated 3–4 counting rate with magnetizing current. The solid curve is computed from an assumed electron angular distribution $1 - \frac{1}{3} \cos \theta$, with counter and gate-width resolution folded in [*Source*: Garwin et al. (1957)[19]].

Experiments not originally proposed by Lee and Yang

Electron polarization in β decay. Although Lee and Yang realized that the muons in π decay would be polarized if parity were not conserved in that process, they seem to have overlooked the analogous effect in ordinary β decay, namely, the longitudinal polarization of the decay electrons (or positrons). With our arrow states (i.e., for maximal parity violation), one expects

$$(\bar{\vec{\sigma}}_e \cdot \vec{v}) = \pm v$$

This fact, which leads to much easier experiments than some of those originally proposed, was first realized by Louis Michel in France and Lincoln Wolfenstein in the United States.

The first experiment was performed by Hans Frauenfelder and his collaborators in Urbana,[22] using the same source as in the earlier original up–down asymmetry experiment (^{60}Co). They used the well-known Mott scattering as a polarization analyzer. Because that parity-conserving, electromagnetic process is sensitive only to transverse polarization, along the normal to the scattering plane, Frauenfelder and associates used an electrostatic "spin rotator" to transform the (presumed) longitudinal polarization into transverse polarization. The time was again ripe: S. R. de Groot and Tolhoek had proposed such a "spin rotator" just a few years earlier.[13]

This experiment was rapidly followed by many others, for both e^- and e^+ emitters. The latter required special methods, because the Mott asymmetry is very small for positrons. The main conclusion was

$$(\bar{\vec{\sigma}} \cdot \vec{v}) \cong \begin{cases} -v & \text{for } e^- \ (L) \\ +v & \text{for } e^+ \ (R) \end{cases}$$

consistent with maximal parity and charge conjugation violation. The road to $V–A$ was paved, but these experiments (not involving neutrinos) could unfortunately not tell T_R from A_L, as our figures show.

$e^\pm$ polarization in μ decay. As already mentioned, a question often raised in early 1957 concerned the comparison of particle/antiparticle decays. Experiments on μ^+ decays could make this dream come true, although experimentally the task was not easy. Because of the high electron energies (30 MeV) involved, special methods of polarization analysis had to be invented. The best of these was based on the timely realization that the photons produced in a shower initiated by a longitudinally polarized $e^\pm$ would carry a corresponding circular polarization.[23] These photons could then be analyzed by the methods we mentioned in connection with $\beta–\gamma$ circular-polarization experiments.

The first trustworthy, although qualitative, measurements were done at Liverpool, with results yielding a clear indication of maximal C violation.[24]

Similar, but uncalibrated, results were obtained at Berkeley.[25] Note that around 1963 there was a brief excitement based on a report from CERN that the e^+ polarization was actually not 1, but very small. This led to a whole series of more quantitative experiments that disproved that notion (that would have ruined universality!) and showed consistency with the expected value 1.

Muon polarization in $\pi-\mu$ decay. The experiments on the $\pi-\mu$ chain did not give the *sign* of the muon polarization. This was established by an ingenious cosmic-ray experiment by A. Alikhanov and associates, who studied the Möller scattering of muons by the polarized electrons of magnetized iron.[26] The same principle was used by P. Marin and associates at the then new CERN PS, using man-made muons. A later experiment at the Columbia cyclotron used Mott scattering of muons on Pb; the necessary transverse polarization was established by choosing the appropriate kinematics of decays in flight.

Maximal parity violation: two-component neutrinos

By the summer of 1957 it had become pretty clear that parity (and charge conjugation) violation were *maximal* (i.e., that two-component neutrinos were involved). The open question was whether the (presumably universal) coupling was S and T (as the preceding twenty years had made us believe) or V and A. Several key experiments served to clear up this point quickly, at least in β decay.

Electron–antineutrino correlation in ^{35}Ar. The Illinois group measured the $e-\bar{\nu}$ correlation in the pure Fermi decay of argon 35:

$$W(\vec{v}_e \cdot \vec{v}_{\bar{\nu}}) = (1 + a\vec{v}_e \cdot \vec{v}_{\bar{\nu}})$$

and found $a \sim 1$, clearly indicating a V coupling (S gives $a = -1$). Assuming that the same-helicity neutrinos are involved in Gamow–Teller transitions, this important result suggested the solution V_L, A_L. It also suggested that the previous main evidence ($e-\bar{\nu}$ correlation in ^{6}He decay) was in error![27]

The correlations in free-neutron decay. All β^- decay is essentially that of the neutron. Thus, by studying only correlations in the decay of the *polarized* free neutron, the entire structure of β decay must become evident. This was indeed the case, as we shall try to show.

Figure 32.8 shows this β decay in the "arrow state" language. It has two channels (amplitudes): Gamow–Teller ($e\bar{\nu}$ triplet combination) and Fermi ($e\bar{\nu}$ singlet combination). The former can involve a nucleon spin flip. By adding the amplitudes properly (noting that the F and GT channels are coherent for the $m = 0$ amplitude) and using the appropriate Clebsch–Gordan coefficients, one readily finds (assuming two-component neutrinos) the following correlation coefficient for the electron up–down asymmetry:

Neutron decay

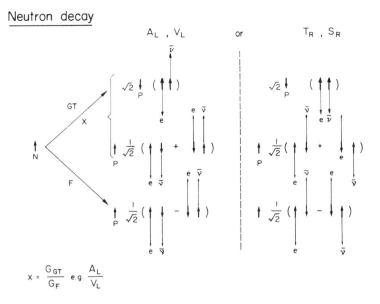

Figure 32.8. Lepton configuration in (polarized) neutron decay.

$$\mathscr{A} = -\frac{2 \times (x + 1)}{1 + 3x^2}$$

where

$$x = \text{ratio of GT to F amplitudes}$$
$$= G_{GT}/G_F \quad \text{(with sign)}$$

No neutrinos being involved, one can, of course, again not tell A from T. Note that for $x = -1$, $\mathscr{A}$ vanishes, mimicking parity conservation! For this, one has to compute the corresponding coefficient B for the antineutrino up−down asymmetry:

$$B = \pm\frac{2x(1 - x)}{1 + 3x^2}$$

where now -1 corresponds to the pair (V, A), and $+1$ to the pair (S, T). It is easy to see that with vanishing $e-\bar{\nu}$ correlation, the measurement of B is essentially a determination of the $\bar{\nu}$ helicity. A right-handed $\bar{\nu}$ will tend to run along the neutron spin, giving $B = +1$, and so forth.

The Argonne−Chicago experiment[28] was started in January 1957 and produced a value of $\mathscr{A}$ by July of that year (0.37 ± 0.11). This value was, because of a very slight misalignment of the apparatus, slightly in error (the correct final value being -0.11 ± 0.02).[28] This error, although obviously regrettable, brought me the signal distinction of being mentioned in Feynman's recent "autobiography."[29] Our group realized soon that slight misalignment could falsify $\mathscr{A}$ only if B was *large* and positive. Thus, we had a strong hint that the interaction was (essentially) $V-A$, with the relative sign being supplied by us for the first time. By November 1957, Vic Krohn of our group announced this publicly at an American Physical Society meeting. (At the time of the Venice meeting, in September of that year, where Robert Marshak courageously announced his and E. C. G. Sudarshan's theory, things were not yet so clear.)

It is interesting to compare the definitive experimental values of these coefficients with the theoretical values corresponding to $|x| = 1.25$.

	$S + T$	$S - T$	$V + A$	$V - A$	Experiment
$\mathscr{A}$	-1	-0.10	-1	-0.10	-0.11 ± 0.02
B	-0.10	-1	0.10	$+1$	$+0.88 \pm 0.15$
	$\underbrace{}_{\nu_R}$		$\underbrace{}_{\nu_L}$		

When I first went out to Argonne to persuade some people to undertake this experiment, I learned that Roy Ringo, a specialist on polarized neutrons, had considered it years before, but that he had been dissuaded by their senior theorist on the grounds that because of mirror invariance, no observable asymmetries could arise! In fact, nature is so wicked that even with maximal parity violation, but for exact $V-A$ (i.e., $x = -1$), the up–down asymmetry parameter vanishes. On the other hand, Ringo and his colleagues proposed initially to use Felix Bloch's old method for polarizing the slow neutrons, whereas I insisted on the cobalt-mirror method that had been developed a few years before at Argonne itself, by M. T. Burgy and Donald J. Hughes. The Co mirrors existed at that time only at Argonne and contributed enormously to the rapid success of the experiment.

In this context it should be mentioned that John M. Robson – *the* great pioneer of free-neutron decay – and his collaborators obtained a relatively inaccurate value of D, the time-reversal asymmetry parameter, presumably because of the less advantageous polarization method that they had adopted.[30] They later switched to Co mirrors, and fully confirmed the Argonne results.

Direct measurement of the neutrino helicity. This seemingly impossible measurement was performed in an experiment of unsurpassed elegance at Brookhaven by Maurice Goldhaber, Lee Grodzins, and Andrew W. Sunyar.[31] Its key idea is to transfer, by angular-momentum conservation, the neutrino helicity to the helicity of a γ-ray following neutron emission. The essential steps are the following:

(i) Choose a K-capture A $(J = 0) + e \rightarrow B^*$ $(J = 1) + \nu$, followed by a deexcitation B^* $(J = 1) \rightarrow B$ $(J = 0) + \gamma$.

(ii)   The recoiling nucleus B^* is polarized opposite to the neutrino spin (i.e., along its direction of flight for ν_R, and opposite to it for ν_L).

(iii) The deexcitation γ ray emitted "along the recoil direction" is R (L) circular for ν_R (ν_L).

(iv)   The "forward"-emitted γ rays are identified through their Doppler shift toward higher energy.

In practice, the Doppler shift is established by scattering the γ rays off a target containing the nucleus B. This scattering occurs with appreciable probability (resonance fluorescence) only if the effective photon energy (i.e., allowing for recoil losses) matches within the level width, the excitation energy E^* of B^*. It is easy to see from momentum balance that the resonance condition is met if the neutrino energy E is essentially equal to E^*.

So to carry out the experiment, one has to find not only a decay chain with the right spin sequence but also one with the proper energetics. That is not all! The spin-parity sequence must be $0^- \rightarrow 1^- \rightarrow 0^+$ for the γ emission to be $E1$ (sufficient level width) and the K capture to be an allowed one. Furthermore, the nuclide B must be a stable and abundant isotope to provide a practical scatterer. Even today, some thirty years later, only one K capture meeting all these conditions is known. It has $A = {}^{152}$Eu, B $= {}^{152}$Sm; at the time of the helicity experiment, it was known only to the Goldhaber group, who had just established its decay scheme. Once again, the time was ripe. Figure 32.9(a) shows the apparatus used; Figure 32.9(b) shows the γ-ray yield. The result showed that the ν is left-handed, with a helicity (0.88 ± 0.20) well consistent with 1.

In concluding the somewhat detailed discussion of this remarkable experiment, two points should be mentioned. First, an identical experiment had been independently proposed by Page,[32] who presumably did not know of a suitable K capture. Second, Goldhaber and associates quoted in their letter the (then unpublished) Argonne–Chicago evidence for $V–A$; rarely does one find such a combination of ingenuity and integrity!

Furthermore, Burgov proposed a resonance scattering method involving a $\beta–\gamma$ cascade,[33] with which he and Terekhov later obtained reasonable evidence for A coupling.[34] A somewhat analogous experiment involving the *muonic* neutrino, using the sequence ^{12}C $+ \mu^- \rightarrow {}^{12}$B $+ \nu_\mu$, ^{12}B $\rightarrow$ ^{12}C $+ e + \nu_e$, was proposed by Jackson, Treiman, and Wyld for determining the sign of the muon polarization.[35] This experiment, and its extension to measure the neutrino helicity, could be carried out only rather recently.[36]

The final steps to V–A

The relative sign of V and A in muon capture. The universal faith in electron–muon universality and the general euphoria about the $V–A$

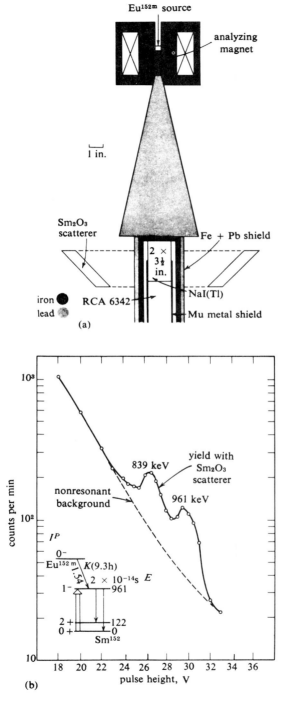

Figure 32.9. The Brookhaven ν helicity experiment: (a) apparatus used; (b) γ-ray yield versus pulse height [*Source*: Goldhaber et al. (1958)[31]].

structure of β decay and muon decay left little doubt that basically the same coupling would hold in case of muon capture, which is much more involved, especially in the case of complex nuclei. It was still of considerable interest to establish as much as possible experimentally.

The key was the so-called hyperfine effect, first pointed out by Y. B. Zel'dovitch and S. S. Gershtein, for the basic reaction $p + \mu^- \rightarrow n + \nu$. For *exact V–A* (i.e., $G_A = -G_V$), the capture rate Λ_t for an initial triplet ($F = 1$) state *vanishes*, whereas the singlet rate Λ_s is finite.[37] In hydrogen, there exists a rapid triplet–singlet conversion before capture, and the average capture rate, $\bar{\Lambda}_{\text{cap}} = \frac{1}{4}(3\Lambda_t + \Lambda_s)$ becomes Λ_s. Roger H. Hildebrand's pioneering bubble-chamber experiment yielded a capture rate in liquid H_2 in agreement with the $V–A$ prediction and the Zel'dovitch hypothesis, but could, of course, not exhibit the conversion process.[38]

The case of complex nuclei was also considered,[39] and subsequently it was pointed out that in most mesic atoms, a fast electron conversion process connects the hyperfine doublet members of the ground state.[40] The case of ^{19}F is particularly attractive: With one proton outside a closed shell, it constitutes the "poor man's hydrogen." Furthermore, the conversion rate R is comparable to the capture rates, thus making the observation of the hyperfine effect relatively easy. Figure 32.10, taken from the work of the Chicago group,[41] clearly shows the growth of the muon capture rate with time (i.e., $\Lambda_s > \Lambda_t$). Detailed analysis provided quantitative agreement.

Not only $V–A$, but also current–current interaction. Whereas a $V–A$ Fermi coupling had been proposed also by several other theorists, Feynman and Gell-Mann were unique in advocating "current-times-current" dynamics. This predicted so-called self-terms, for example, $(e\bar{\nu})(e\bar{\nu})^+$, $(\bar{p}n)(\bar{p}n)^+$ (i.e., new effects). In particular, the $(\bar{p}n)(\bar{p}n)^+$ term leads to weak, parity-violating "nuclear forces." The first convincing evidence (four standard deviations) for the latter was provided in 1964 by Y. G. Abov and associates at ITEP, who measured the forward–backward asymmetry of the γ rays emitted in thermal neutron capture by ^{144}Cd.[42] An earlier, statistically weaker, experiment at BNL had failed to detect the effect. Many other delicate experiments, in particular on the spontaneous circular polarization of nuclear γ rays, followed. Their detailed interpretation still is not settled to date.

Conclusion

The few years following the discovery of parity violation were most exciting. Theory and experiment marched forward hand in hand. Crucial experiments could be performed by small teams, using modest apparatus, but often entirely novel techniques. "It was a time when," to quote a statement made by P. A. M. Dirac in another context, "average people could make outstanding contributions."

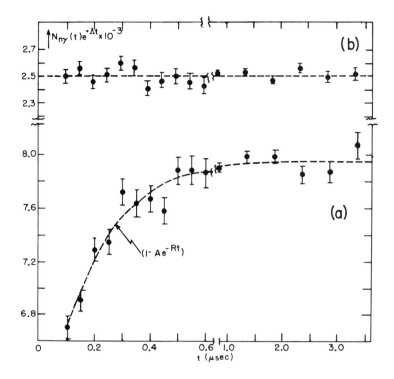

Figure 32.10. (a) Time dependence (corrected for the asymptotic $e^{-\Lambda t}$ behavior) of muon capture in ^{19}F. (b) Analogous data with ^{16}O as the capturing nucleus [*Source*: Culligan et al. (1961)[41] (for a,b)].

Notes

1 V. B. Berestetsky and I. Y. Pomeranchuk, "On β-decay of the Neutron" (in Russian), *Zh. Eksp. Teor. Fiz. [Sov. Phys.–JETP] 19* (1949), 756–7.

2 E. M. Purcell and N. F. Ramsey, "On the Possibility of Electric Dipole Moments for Elementary Particles and Nuclei," *Phys. Rev. 78* (1950), 807.

3 Murray Gell-Mann emphasized to me, after I wrote this paragraph, that I. S. Shapiro most strenuously objected to the parity-violation idea when Gell-Mann presented the latter in 1956 in the Landau seminar as one of the possible solutions to the $\tau-\theta$ puzzle.

4 Wolfgang Pauli, *Niels Bohr and the Development of Physics* (London; Pergamon, 1955), p. 30.

5 J. S. Bell, "Time Reversal in Field Theory," *Proc. R. Soc. London, Ser. A 231* (1955), 479–95.

6 B. F. Touschek, "Parity Conservation and the Mass of the Neutrino," *Nuovo Cimento 5* (1957), 754–5.

7 B. F. Touschek, "The Mass of the Neutrino and the Nonconservation of Parity," *Nuovo Cimento 5* (1957), 1281–91; L. A. Radicati and B. F. Touschek, "On the Equivalence Theorem for the Massless Neutrino," *Nuovo Cimento 5* (1957), 1693–9.

8 W. R. Theis, "A Classification of the Various Terms in Weak Interactions," *Z. Phys. 150* (1958), 590–2.

9 The "arrowmanship" was developed by the author in 1957 and so far never published by himself. It was inspired by B. T. Feld, "Kinematics of β Decay and Parity Nonconservation in Weak Interactions," *Phys. Rev. 107* (1957), 797–804.

10 C. S. Wu, E. Ambler, R. W. Hayward, D. D. Hoppes, and R. P. Hudson, "Experimental Test of Parity Conservation in Beta Decay," *Phys. Rev. 105* (1957), 1413–15.

11 N. Kürti, "Nuclear Orientation and Nuclear Cooling," *Phys. Today 11* (1958), 19–25.

12 H. Postma, W. J. Huiskamp, A. R. Miedema, M. J. Steenland, H. A. Tolhoek, and C. J. Gorter, "Asymmetry of the Positron Emission by Polarized ^{58}Co Nuclei," *Physica (Utrecht) 23* (1957), 259–60; H. Postma, W. J. Huiskamp, A. R. Miedema, M. J. Steenland, H. A. Tolhoek, and C. J. Gorter, "Asymmetry of the Positron Emission from Polarized ^{58}Co and ^{52}Mn Nuclei," *Physica (Utrecht) 24* (1958), 157–68.

13 H. A. Tolhoek, "Electron Polarization, Theory and Experiment," *Rev. Mod. Phys. 28* (1956), 277–98; H. A. Tolhoek and S. R. de Groot, "On the Theory of Beta-Radioactivity. III. The Influence of Electric and Magnetic Fields on Polarized Electron Beams," *Physica (Utrecht) 17* (1951), 17–32.

14 H. Schopper, "Circular Polarization of γ-Rays: Further Proof for Parity Failure in β-Decay," *Philos. Mag. 2* (1957), 710–13.

15 Hans Frauenfelder and R. M. Steffen, "The Helicity of β-Particles," α, β, γ *Ray Spectroscopy*, edited by K. Siegbahn (Amsterdam: North Holland, 1966), pp. 1431–52.

16 T. Fazzini, G. Fidecaro, A. W. Merrison, H. Paul, and A. V. Tollestrup, "Electron Decay of the Pion," *Phys. Rev. Lett. 1* (1958), 247–8.

17 G. Impeduglia, R. Plano, A. Prodell, N. Samios, M. Schwartz, and J. Steinberger, "β Decay of the Pion," *Phys. Rev. Lett. 1* (1958) 249–50.

18 J. I. Friedman and V. L. Telegdi, "Nuclear Emulsion Evidence for Parity Nonconservation in the Decay Chain $\pi^+ - \mu^+ - e^+$," *Phys. Rev. 105* (1957), 1681–2; *106* (1957), 1290.

19 R. L. Garwin, L. M. Lederman, and M. Weinrich, "Observation of the Failure of Conservation of Parity and Charge Conjugation in Meson Decays: The Magnetic Moment of the Free Muon," *Phys. Rev. 105* (1957), 1415–17.

20 F. S. Crawford, Jr., M. Cresti, M. L. Good, K. Gottstein, E. M. Lyman, F. T. Solmitz, M. L. Stevenson, and H. K. Ticho, "Detection of Parity Nonconservation in Λ Decay," *Phys. Rev. 108* (1957), 1102–3.

21 F. Eisler, R. Plano, A. Prodell, N. Samios, M. Schwartz, J. Steinberger, P. Bassi, V. Borelli, G. Puppi, G. Tanaka, P. Woloschek, V. Zoboli, M. Conversi, P. Franzini, I. Mannelli, R. Santangelo, V. Silvestrini, D. A. Glaser, C. Graves, and M. L. Perl, "Demonstration of Parity Nonconservation in Hyperon Decay," *Phys. Rev. 108* (1957), 1353–5.

22 H. Frauenfelder, R. Bobone, E. Von Goeler, N. Levine, H. R. Lewis, R. N. Peacock, A. Rossi, and G. de Pasquali, "Parity and the Polarization of Electrons from Co^{60}," *Phys. Rev. 106* (1957), 386–7.

23 K. W. McVoy and F. J. Dyson, "Longitudinal Polarization of Bremsstrahlung and Pair Production at Relativistic Energies," *Phys. Rev. 106* (1957), 1360–1.

24 G. Culligan, S. G. F. Frank, and J. R. Holt, "Longitudinal Polarization of the Electrons from the Decay of Unpolarized Positive and Negative Muons," *Proc. Phys. Soc. London, 73* (1959), 169–77.

25 P. C. Macq, K. M. Crowe, and R. P. Haddock, "Helicity of the Electron and Positron in Muon Decay," *Phys. Rev. 112* (1958), 2061–71.

26 A. Alikhanov, Y. V. Galaktionov, Y. V. Gorodkov, G. P. Eliseev, and V. A. Lyubimov, "Measuring of the Chirality of the Mu Meson," *Zh. Eksp. Teor. Fiz. [Sov. Phys.–JETP] 38* (1960), 1918–20.

27 W. B. Herrmansfeldt, D. R. Maxson, P. Stähelin, and J. S. Allen, "Electron–Neutrino Angular Correlation in the Positron Decay of Argon 35," *Phys. Rev. 107* (1957), 641–3.

28 M. T. Burgy, R. J. Epstein, V. E. Krohn, T. B. Novey, S. Raboy, G. R. Ringo, and V. L. Telegdi, "Measurement of Beta Asymmetry in the Decay of Polarized Neutrons," *Phys. Rev. 107* (1957), 1731–3; "Measurements of Asymmetries in the Decay of Polarized Neutrons," *Phys. Rev. 110* (1958), 1214–16; "Measurements of Spatial Asymmetries in the Decay of Polarized Neutrons," *Phys. Rev. 120* (1960), 1829–38.

29 Richard P. Feynman, *"Surely You're Joking Mr. Feynman!"* *Adventures of a Curious Character* (New York: Norton, 1985).

30 M. A. Clark, J. M. Robson, and R. Nathans, "Investigation of the Time-Reversal Invariance in the Beta Decay of the Neutron," *Phys. Rev. Lett. 1* (1958), 100–1.

31 M. Goldhaber, L. Grodzins, and A. W. Sunyar, "Helicity of Neutrinos," *Phys. Rev. 109* (1958), 1015–17.

32 L. A. Page, "Proposed Experiment Bearing Directly on Helicity of Neutrinos," *Nuovo Cimento 7* (1958), 727–8.

33 N. A. Burgov, "Resonance Scattering of γ-Rays," *Zh. Eksp. Teor. Fiz. 33* (1957), 655–9 [*Sov. Phys.–JETP 33* (1958), 502–5].

34 N. A. Burgov and I. V. Terekhov, "Types of Interaction in Beta Decay. The Decay of Na^{24}," *Zh. Eksp. Teor. Fiz. 35* (1958), 932–9 [*Sov. Phys.–JETP 35* (1959), 651–5].

35 J. D. Jackson, S. D. Treiman, and H. W. Wyld, Jr., "Proposed Experiment to Determine the Direction of μ-Meson Polarization in Pion Decay," *Phys. Rev. 107* (1957), 327–8.

36 L. P. Roesch, V. L. Telegdi, P. Truttman, A. Zehnder, L. Grenacs, and L. Palffy, "Direct Measurement of the Helicity of the Muonic Neutrino," *Am. J. Phys. 50* (1982), 931–5.

37 Y. B. Zel'dovitch and S. S. Gershtein, "Nuclear Reactions in Cold Hydrogen. I. Mesonic Catalysis," *Usp. Fiz. Nauk* [*Sov. Phys.–Usp.*] *3* (1961), 593–623, 953.

38 R. H. Hildebrand, "Observation of μ^- Capture in Liquid Hydrogen," *Phys. Rev. Lett. 8* (1961), 34–7.

39 J. Bernstein, T. D. Lee, H. Primakoff, and C. N. Yang, "Effect of the Hyperfine Splitting of a μ-Mesonic Atom on Its Lifetime," *Phys. Rev. 111* (1958), 313–15.

40 R. Winston and V. L. Telegdi, "Fast Atomic Transitions Within μ-Mesonic Hyperfine Doublets, and Observable Effects of the Spin Dependence of Muon Absorption," *Phys. Rev. Lett. 7* (1961), 104–7.

41 G. Culligan, J. F. Lathrop, V. L. Telegdi, R. Winston, and R. A. Lundy, "Experimental Proof of the Spin Dependence of the Muon Capture Interaction and Evidence for Its (F–GT) Character," *Phys. Rev. Lett. 7* (1961), 458–60.

42 Y. G. Abov, P. A. Krupchitsky, and Y. A. Oratovsky, "On the Existence of an Internucleon Potential not Conserving Spatial Parity," *Phys. Lett. 12* (1964), 25–6.

33   Midcentury adventures in particle physics

E. C. G. SUDARSHAN

Born 1931, Kottayam, India; Ph.D., University of Rochester, 1958;
theoretical physics; University of Texas, Austin

It was in the spring of 1952 that I joined the Tata Institute of Fundamental Research (TIFR) as a student. The shining light in theoretical research was Homi Bhabha, who had then completed his work on cosmic-ray
cascades and relativistic wave equations. Bhabha was deeply involved in
organizing a nuclear research team in Bombay, but he was still very much the
academic leader and presided over the weekly colloquia. He also brought
to Bombay many distinguished physicists, like Paul A. M. Dirac, Maria
Goeppert-Mayer, Maurice Levy, Robert Marshak, Joseph E. Mayer,
Abraham Pais, Wolfgang Pauli, and Sin-itiro Tomonaga, and he convinced
Bernard Peters to join the TIFR faculty. I worked with Peters, and I quickly
graduated from developing and scanning nuclear emulsions to developing
theoretical aspects of experimental particle physics. Several τ mesons were
discovered in our laboratory; for myself, I began to study in detail Richard
Dalitz's paper on τ decay analysis.

Enrico Fermi's thermodynamic model and Werner Heisenberg's shock-
wave model of multiple meson production were of great interest at TIFR. A
large "star" was discovered in an emulsion stack; the study of the distribution
and interaction of secondaries was among my first lessons in particle physics.[1]
But all this time the glamour topics were relativistic wave equations and
quantum theory of higher-spin fields.

During my second year in Bombay, Marshak came to lecture at TIFR and
told us about the Chicago experiments on pion–nucleon scattering and the
3-3 resonance. This was the beginning of an association that has had a decisive
effect on my scientific career. Marshak suggested that I come to Rochester as

Work supported by DOE grant DE-F605-85ER40200.

a graduate student. I could go one year later, which gave me the chance to be apprenticed to Kundan Singwi in statistical mechanics. The year was very usefully spent studying Dirac's many dazzling ideas, including his work on constrained dynamic systems.[2] Along with the papers of Dirac and Bhabha, books by Edward Michael Corson, *Tensors, Spinors and Relativistic Wave Equations*, and by Elie Joseph Cartan, *Leçons sur la Theorie des Spineurs*, enabled my immersion into the mysteries of spinors.[3] The remarkable facts that what we now call chiral decomposition is respected by the purely kinematic terms in the Lagrangian, and hence by the anticommutation relations, and that the mass is very much a zero-frequency, zero-wave-number scalar field, were already noticed at this time.

Graduate study at Rochester

I joined the University of Rochester in the fall of 1955 with my bride Lalita, who also became a graduate student at Rochester. Murray Gell-Mann's classification of the hadrons based on the Gell-Mann–Nishijima scheme was the first news that I got. It was a relief to switch from the cosmic-ray physics notation to the new notation. Abdus Salam came to Rochester that winter to lecture about dispersion relations and to attend the Sixth Rochester Conference. He discussed with Marshak and me the calculation of magnetic moments and mass differences of hyperons, with a view to explaining the mass difference of the charged Σ hyperons. With this "new" calculation, I became involved with computations in particle physics (as distinguished from quantum electrodynamics)[4] and also became interested in the first application of symmetry breaking and the Wigner–Eckart theorem for magnetic moments.

This involvement came about through a curious event. Marshak had asked me to calculate the magnetic moments and mass differences of hyperons with a variety of cutoff techniques. While in an airplane going to Brookhaven, Marshak noticed that the neutral Σ magnetic moment came out to be the average of the charged Σ moments. He returned to Rochester and mentioned this curious fact. While I was busy checking this simple relation and verifying it to various orders of perturbation theory, Susumu Okubo pointed out that this could be the Wigner–Eckart theorem in operation. We verified this, and the paper that Marshak, Okubo, and I published on that was to be the forerunner of many more papers.[5] Okubo went on to generalize the result to the SU(3) group, giving the relation now known as the Gell-Mann–Okubo formula.[6] Alan Macfarlane and I later used this relation on electromagnetic properties of SU(3) multiplets, generalizing the Coleman–Glashow relations.[7]

Another dramatic result was G parity, which formed the subject of a short paper by T. D. Lee and C. N. Yang. Michel had already stated the result in a different form, but in any case I should have known the result from Cartan's book on spinors.[8]

In search of the Universal Fermi interaction

Nevertheless, the big excitement was in weak interactions. Soon after I joined the University of Rochester, I had to take an exhaustive (and exhausting) qualifying examination, along with Robert Knox, Fred Seward, and Giorgio Giacomelli. This exam, which demanded a thorough knowledge of nuclear physics, and in particular β decay, was good for me; through it I became acquainted with the crucial experimental results. So when parity violation was discovered not long afterward, and β decay became a hot topic in particle physics, I had an advantage. Marshak, who had earlier done pioneering work on unique forbidden β decays,[9] recognized the importance of the calculation by Malvin Ruderman and Robert J. Finkelstein of the electron–muon branching ratio for pseudoscalar and axial vector interactions.[10] Because I showed little interest in the study of nucleon–nucleon forces or in devising dispersion calculations, Marshak suggested that we study weak interactions. It became clear to us from muon decay that the $V-A$ interaction was the natural choice; muon capture had to have an A (or P) interaction, for otherwise the pion could not decay. We set about examining β decay data in detail to see if we could challenge the β decay assignment of S, T that was generally accepted.

It was early recognized that the ^{19}Ne and neutron–electron–neutrino angular correlation could be due to either S, T or V, A.[11] If only ^{6}He angular correlation, which was consistent with T, could change![12] The new experiment on ^{35}Ar angular correlation,[13] a dominantly Fermi transition, was consistent with V rather than S. It suggested that the β decay angular-correlation results were not mutually consistent, and one result at least had to be wrong. It also suggested that the V, A combination was the correct combination. Some were so sure that the Gamow–Teller interaction was T that they were willing to consider V, T for β decay, disregard ^{19}Ne and neutron angular correlations, and abandon all hope of a universal Fermi interaction. We were, however, quite convinced that it was V, A.

Then came the Seventh Rochester Conference, at which I was to present this theory in outline. But a few days before the conference, Marshak realized that as a graduate student I was not a delegate to the conference, and he felt that it was therefore not possible for me to make a presentation. Marshak, the chairman of the conference, was giving the principal talk on the nucleon–nucleon interaction and could not present another contribution in which he was a coauthor. So Marshak and I jointly requested Paul Mathews, then a visiting professor at Rochester, to make the presentation. Mathews never did, and I could not get to speak! I would have to wait approximately twenty-eight years before getting a chance to talk about it.

Soon after the Seventh Rochester Conference, 1 had to complete the formalities for my Ph.D., which included a German language test. I got hold of flash cards for vocabulary and some books on syntax and somehow managed to get through the examination. These months helped to rule out the *V*, *T* situation. That satisfied Marshak, who was cautious in accepting the necessity of an *A* interaction from the evidence of pion decay. In fairness to Marshak, I should say that the β decay of the pion was yet to be conclusively demonstrated.

The Padua–Venice *V–A* paper

In the summer of 1957, Marshak made an offer to Ronald Bryan and me: If we could get to Los Angeles on our own, he would support us as research assistants for two months. We did, and he did. At UCLA we were assigned desks. Steve Moszkowski was there, but others were away. My task was to write up the first draft of the paper on universal weak interactions. A large number of experiments were being carried out, and we kept track of the results and their implications. A systematic analysis of all data showed that *not all accepted experimental results were compatible*. We had to make a considered choice. In the paper written up for the forthcoming Padua–Venice conference on mesons and newly discovered particles in September, we identified four experiments as the best candidates to be in error and in need of being redone:[14]

1. the electron–neutrino correlation in the Gamow–Teller ^{6}He decay
2. the sign of the positron polarization in muon decay
3. the branching ratio of the electron and muon modes in pion decay
4. the asymmetry from polarized neutron decay

Of these, the ^{6}He experiment was old. Others were new. Some were by quiet people, and some were by very unquiet people, but all people of proven ability. So it was quite hazardous to point to such experiments as being wrong. Fortunately for us all, the experiments were repeated within a period of eighteen months, and the new results endorsed the *V–A* theory.

During the first week of July, Marshak invited Bryan and me to join him for lunch with Gell-Mann. I was told that Gell-Mann had graciously set up the luncheon with Berthold Stech, Felix Boehm, Marshak, and myself. Leona Marshall was also present at this lunch. It was a pleasure for me to be able to discuss our conclusions with Gell-Mann, who was most cordial and appreciative. Boehm gave us much reassurance that the $\beta-\gamma$ correlation for ^{46}Sc gave a much larger value than ^{58}Co, so that *V*, *A* was more acceptable than *V*, *T* or *S*, *A*. This was the final experimental confirmation to our theory.

The weekend found me busy completing the brief four-page paper for the Padua–Venice conference; I was not invited to the conference, but Marshak was going. The very next day, Lalita and I went on a much needed month-long vacation. The manuscript remained with Marshak until he returned to Rochester several weeks later.

The inscrutable Occidentals

After completing my Ph.D. thesis defense, I went to Harvard University as a Corporation Fellow, apprenticed to Julian Schwinger. It was amazing that at that place no one would be bothered that there was a new theory of weak interactions and that one of the authors of the theory was at Harvard. The few who knew about the theory conveniently forgot that Marshak and I were its authors. Yang was Loeb Lecturer at Harvard, and toward the end of his characteristically beautiful lecture, he mentioned the ideas he had heard from Richard Feynman. They were a subset of the conclusions of our Padua–Venice paper presented in a session chaired by Lee. I looked up, down, and sideways at the audience and finally decided that this humble foreigner should speak up. I did. Not that it made any difference.

During my first weeks at Harvard, I heard from Sheldon Glashow that Feynman and Gell-Mann had sent in a paper to the *Physical Review* suggesting a $V-A$ interaction. I telephoned Marshak to ask about the status of our paper. He said that it was already presented at Padua–Venice and was also issued as a Rochester preprint. He assured me that it was a sufficient way to inform all concerned about our work. Years later, when it became usual for people to make no reference at all to our work, or to refer to a later second paper,[15] rather than the Padua–Venice paper, it puzzled me a lot how anyone who had read our paper and seen our analysis of experimental data and the subsequent outline of the chiral $V-A$ theory could ignore it, or quote the paper of Feynman and Gell-Mann[16] for their theory (rather than for their excellent proof of nonrenormalization of the vector coupling constant). Robert Oppenheimer, more forthright than most, told Marshak that he had never read our paper. Even more curious is the mischief of quoting the work on mass reversal along with ours. At least one of the authors of mass reversal had a preprint of our paper in his hands before he wrote his paper.

Back to the $V-A$ interaction. There were three anomalies regarding universality. One was the apparent suppression of strangeness-changing leptonic decays; this was parameterized by Nicola Cabibbo[17] within the context of SU(3) and is now understood on the basis of the Kobayashi–Maskawa mass matrix.[18] The second was the absence of the electron–photon mode of the muon; that was resolved by the recognition that the muon neutrino is distinct from the electron neutrino.[19] The third concerns the magnitude of the vector and axial vector coupling constants in β decay. Feynman and Gell-Mann explained the vector nonrenormalization in terms of a conserved vector current of isospin, an idea earlier mentioned and discarded by S. S. Gershtein and Y. B. Zel'dovich.[20] The conserved vector current and its relation to weak magnetism was developed by Gell-Mann subsequently and verified experimentally by L. W. Mo, Y. K. Lee, and C. S. Wu.[21] The axial vector renormalization constant was related to pion nuclear scattering via the PCAC hypothesis and current algebra by Stephen Adler, by William Weisberger, and by Yukio Tomozawa.[22]

"Nonlocal" weak interactions

My involvement with weak-interaction physics continued for the next few years. During the two years that I spent at Harvard, I used to take the night bus to Rochester, where we had a small working group consisting of Marshak, Okubo, and myself. We looked at several problems in weak interactions. In his travels, Marshak met a young theorist named Steven Weinberg, who joined for some sessions. The late Werner Teutsch was the fifth in our five-man collaboration, which resulted in two papers.[23] Weinberg wrote the draft of one paper, and I did the other. But the major effort was contained in an unpublished manuscript written with a carbon copy. (Xerox was not easily available to us in its own home town.) The starting point was a derivative coupled realization of partially conserved axial vector current (PCAC), and the paper identified mass as a zero-frequency, constant external scalar field that violated chirality. I am ashamed that today I do not even have a copy of this working paper. I could not get my colleagues at that time to be too enthused about mass being only an external scalar field. And none of us at that time had any ideas about the virtues of spontaneously broken Yang–Mills theories. Nor could I impress on my colleagues the notion that the chiral components were independent kinematic entities that happened to get coupled by the interaction with the "mass."

I made some effort to interest Schwinger in our work; Schwinger and I used to have lunch, along with Paul Martin, Stanley Deser, and Walter Gilbert, thrice a week. To my surprise, Schwinger seemed to believe that in his paper "Fundamental Interaction" he had predicted $V-A$![24] The nonrenormalization of the vector coupling he took in his stride, but he did not like at all the use of the second-order Dirac equation by Feynman and Gell-Mann.[16] He pointed out to me that if such an equation was derived from the action, this would imply that the field would have vanishing equal time anticommutators and thus an indefinite metric. Meanwhile, Schwinger suggested that I investigate integral representations of Geen's functions in quantum field theory. This involved me in a program for the next year or so, into which I invited Deser and Gilbert.[25]

Gatlinberg and after

Meanwhile, Okubo, Marshak, and I continued our research collaboration on various aspects of weak interactions, in particular the three-body decays of kaons. It was in this second postdoctoral year for me that the Gatlinberg conference on weak interactions took place. I was invited to chair the session where Marvin Goldberger and Henry Primakoff spoke. I was flattered, but failed to note that I was not invited to talk at this conference, even though our work was the reason the conference was held at all! Marshak was not even invited! One was like a witness in a court not allowed to speak even when what one had to say was important. But worse was to come.

Several years later, Argonne National Laboratory held a conference on the weak interactions in which neither Marshak nor I had even a ceremonial role, not to mention an invited talk at the meeting. When I asked my friend Kameshwar Wali, a kingpin of this conference, "How come?" his response was, "George, you make it difficult for me to be your friend."

My ceremonial role was restored at the Kiev conference in 1967 and the Paris conference on the history of particle physics in 1982. Marshak got the speaking role at the Racine meeting in 1984. I am puzzled.

At Gatlinberg, Feynman gave a leisurely evening lecture on β decay, and while it contained some account of a second-order Dirac equation, he concentrated on the numerical calculation of decay rates. The next evening, Feynman and I sat next to each other; he very graciously said that he had been told that I had had the essential $V-A$ ideas much earlier than anyone else. I was very pleased to meet him, and I told him that even as he spoke at the seventh Rochester conference I could have presented the full solution. He could not understand how, if I was in Rochester, I could not speak up. I recall with satisfaction that consistent with this conversation at Gatlinberg, Feynman spoke at the 1974 neutrino conference at the University of Pennsylvania. A more patient and systematic review is given in my paper at the 1984 Wingspread conference, "Origin of the Universal $V-A$ Theory."[26]

Hadron resonances

Already at this time, strong interactions had shifted from the one-dimensional dispersion relation to the Mandelstam representation, Geoffrey Chew's analytic function models, and the subsequent development of Regge-pole formalism. The eightfold-way reincarnation of SU(3) attracted everyone's attention after the brilliant presentation by Gell-Mann at the La Jolla conference, where Okubo presented the general theory of broken SU(3). Charles Goebel, at Rochester, was friend, guide, and philosopher to the nonperturbative S-matrix models; my expertise in deciphering Okubo was extended to cover Goebel, also to my great personal advantage. During this time I also did some work on higher symmetry groups applied to particle physics in collaboration with Macfarlane and Narasimha Mukunda.[27]

But my more interesting involvement was with pion resonances. With the deep impression created by the study of the Dalitz papers on τ decay[28] and my background in multiple meson production,[29] it became clear to me that if pion–pion and three-pion resonances existed, the best evidence for them was to be obtained from two-pion and three-pion effective-mass plots. With two graduate students, Gabriel Pinsky and Kalyana Mahanthappa, I calculated such plots and presented them at the Tenth Rochester Conference.[30] William Chinowsky and Frank Solmitz were at the session, and they categorically denied any evidence for resonance in effective-mass plots. My efforts to obtain Lawrence Radiation Laboratory films for analysis at Rochester were not successful. But some months later, Bogdan Maglich telephoned me from Berkeley in great enthusiasm, saying that they had read my paper in the proceedings of the Tenth Rochester Conference, carried out the mass plots, and found the two-pion resonances, but not the three-pion resonance. I told him that according to my calculations the three-pion resonance would have a mass lower than the California estimates, and I urged him to look at the data again. He called me back the next day in great excitement and told me that he found it within a few million electron volts of the value I had quoted! He sent me, two days later, the draft of a paper in which my contribution was given due prominence. But two weeks later I got a "revised" version, with more authors, and with all reference to me eliminated in relation to this discovery.[31] Maglich was very apologetic and said that the matter was beyond his control, but that if he himself were ever to write about it, he would surely bring out the whole story.

In the next few years I got interested in Heisenberg's nonlinear spinor field theories, in axiomatic field theory, and of course in classical mechanics[32] and quantum optics.[33] I returned·to symmetries on my sabbatical at Bern and Madras and worked on the origin of symmetries inverting the Smushkevich theorem[34] and on combining the Lorentz group with broken symmetries.[35] My second paper on the latter topic got tangled up with a moronic referee for *Journal of Mathematical Physics*, and the result[36] was reincarnated with mathematical sophistications as O'Raifeartaigh's theorem. I had met Lochlain O'Raifeartaigh in Europe and taken him to Madras and subsequently Syracuse.

Concluding comments

It has been a sad but enlightening experience to recognize that the universality of science does not imply unbiased acclaim for scientific truth and a true history of science, and that if one has neither powerful alliances nor influential sponsors, he should learn to do science for its own sake and not be depressed by lack of appreciation. Over the years I have developed this skill and have contributed to many areas of physics even outside particle physics.

Notes

1 S. Biswas, E. C. George, and B. Peters, "An Improved Method for Determining the Mass of Particles from Scattering Versus Range and Its Application to the Mass of *K*-Mesons," *Proc. Ind. Acad. Sci. 38* (1953), 418–39; S. Biswas, E. C. George, B. Peters, and M. S. Swamy, "Mass Determination on Steeply Dipping Tracks in Emulsion Block Detectors," *Nuovo Cimento (Suppl.) 12* (1954), 369–73; R. R. Daniel, E. C. George, and B. Peters, "The Range–Energy Relation in Nuclear Emulsions," *Proc. Ind. Acad. Sci. 41* (1955), 45–8.
2 P. A. M. Dirac, "Generalized Hamiltonian Dynamics," *Canad. J. Math. 2* (1950), 129–48; P. A. M. Dirac, "The Hamiltonian Form of Field Dynamics," *Canad. J. Math. 3* (1951), 1–23.

3 E. M. Corson, *Tensors, Spinors and Relativistic Wave Equations* (New York: Hafner, 1953); E. Cartan, *Leçons sur la Theorie des Spineurs* (Paris: Hermann, 1937; English translation Dover, 1981).

4 E. C. G. Sudarshan and R. E. Marshak, "Mass Difference of $\Sigma^{\pm}$ and Their Anomalous Magnetic Moments," *Phys. Rev. 104* (1956), 267–9.

5 R. Marshak, S. Okubo, and E. C. G. Sudarshan, "Consequences of Charge Independence for the Magnetic Moments and Masses of Σ Hyperons," *Phys. Rev. 106* (1957), 599–601.

6 M. Gell-Mann, California Institute of Technology Synchrotron Laboratory report CTSL-20 (1961); Susumu Okubo, "Note on Unitary Symmetry in Strong Interactions," *Prog. Theor. Phys. 27* (1962), 949–66.

7 A. J. Macfarlane and E. C. G. Sudarshan, "Electromagnetic Properties of Stable Particles and Resonances According to the Unitary Symmetry Theory," *Nuovo Cimento 31* (1964), 1176–96; A. J. Macfarlane, N. Mukunda, and E. C. G. Sudarshan, "Electromagnetic and Decay Properties of G_2 Multiplets," *Phys. Rev. 133* (1964), B475–7.

8 T. D. Lee and C. N. Yang, "Charge Conjugation, a New Quantum Number G, and Selection Rules Concerning a Nucleon–Antinucleon System," *Nuovo Cimento 3* (1956), 749–53; L. Michel, "Selection Rules Imposed by Charge Conjugation," *Nuovo Cimento 10* (1953), 319–39.

9 R. E. Marshak, "β-Ray Spectrum of K^{40} and Theory of β-Decay," *Phys. Rev. 70* (1946), 980.

10 M. Ruderman and R. Finkelstein, "Note on the Decay of the π-Meson," *Phys. Rev. 76* (1949), 1458–60.

11 D. R. Maxson, J. S. Allen, and W. K. Jentschke, "Electron–Neutrino Angular Correlation in the Beta Decay of Neon[19]," *Phys. Rev. 97* (1955), 109–16; Myron L. Good and Eugene J. Lauer, "Electron–Neutrino Angular Correlation in the Beta Decay of Neon-19," *Phys. Rev. 105* (1957), 213–16; W. Parker Alford and Donald R. Hamilton, "Recoil Spectrum in the Beta Decay of Ne[19]," *Phys. Rev. 95* (1954), 1351–3; J. M. Robson, "Angular Correlation in the Beta Decay of the Neutron," *Phys. Rev. 100* (1955), 933–4.

12 Brice M. Rustad and Stanley L. Ruby, "Correlation between Electron and Recoil Nucleus in He[6] Decay," *Phys. Rev. 89* (1953), 880–1.

13 W. B. Herrmannsfeldt, D. R. Maxson, P. Stähelin, and J. S. Allen, "Electron–Neutrino Angular Correlation in the Positron Decay of Argon 35," *Phys. Rev. 107* (1957), 641–3.

14 E. C. G. Sudarshan and R. E. Marshak, "The Nature of the Four-Fermion Interaction," in *Proceedings of the Padua–Venice Conference on Mesons and Newly Discovered Particles*, edited by N. Zanichelli (Bologna, 1958); reprinted in *The Development of Weak Interaction Theory*, edited by P. K. Kabir (New York: Gordon & Breach, 1963), pp. 118–28.

15 E. C. G. Sudarshan and R. E. Marshak, "Chirality Invariance and the Universal Fermi Interaction," *Phys. Rev. 109* (1958), 1860–2.

16 R. P. Feynman and M. Gell-Mann, "Theory of the Fermi Interaction," *Phys. Rev. 109* (1958), 193–8.

17 Nicola Cabibbo, "Unitary Symmetry and Leptonic Decays," *Phys. Rev. Lett. 10* (1963), 531–3; see also A. Garcia and P. Kielanowski, "The Beta Decay of Hyperons," *Sprinter Lecture Notes 222* (1985).

18 Makoto Kobayashi and Toshihide Maskawa, "*CP*-Violation in the Renormalizable Theory of Weak Interaction," *Prog. Theor. Phys. 49* (1973), 652–7.

19 G. Danby, J. M. Gaillard, K. Goulianos, L. M. Lederman, N. Mistry, M. Schwartz, and J. Steinberger, "Observation of High-Energy Neutrino Reactions and the Existence of Two Kinds of Neutrinos," *Phys. Rev. Lett. 9* (1962), 36–44.

20 S. S. Gershtein and Y. B. Zel'dovich, "On Corrections from Mesons to the Theory of β-Decay," *Zh. Eksp. Teor. Fiz. 29* (1955), 698–9 [translation, *Sov. Phys.–JETP 2* (1956), 576].

21 Y. K. Lee, L. W. Mo, and C. S. Wu, "Experimental Test of the Conserved Vector Current Theory on the Beta Spectra of B^{12} and N^{12}," *Phys. Rev. Lett. 10* (1963), 253–8.

22 Stephen L. Adler, "Sum Rules for the Axial-Vector Coupling-Constant Renormalization in β Decay," *Phys. Rev. 140* (1965), B736–47. William I. Weisberger, "Unsubtracted Dispersion Relations and the Renormalization of the Weak Axial-Vector Coupling Constants," *Phys.*

Rev. 143 (1966), 1302–9; Y. Tomozawa, "Axial-Vector Coupling Constant Renormalization and the Meson–Baryon Scattering Lengths," *Nuovo Cimento 46A* (1966), 707–17.

23 S. Weinberg, R. E. Marshak, S. Okubo, E. C. G. Sudarshan, and W. B. Teutsch, "Divergenceless Currents and *K*-Meson Decay," *Phys. Rev. Lett. 1* (1958), 25–7; S. Okubo, R. E. Marshak, E. C. G. Sudarshan, W. B. Teutsch, and S. Weinberg, "Interaction Current in Strangeness-Violating Decays," *Phys. Rev. 112* (1958), 665–8.

24 Julian Schwinger, "A Theory of the Fundamental Interactions," *Ann. Phys. (N.Y.) 2* (1957), 407–34.

25 S. Deser, W. Gilbert, and E. C. G. Sudarshan, "Structure of the Forward Scattering Amplitude," *Phys. Rev. 117* (1960), 266–72; S. Deser, W. Gilbert, and E. C. G. Sudarshan, "Integral Representations of Two-Point Functions," *Phys. Rev. 117* (1960), 273–9.

26 E. C. G. Sudarshan and R. E. Marshak, "Origin of the Universal $V-A$ Theory," in *50 Years of Weak Interactions, Wingspread Conference 1984*, edited by David B. Cline and Gail M. Riedasch (Madison: University of Wisconsin, 1986), pp. 1–14.

27 C. Dullemond, A. J. Macfarlane, and E. C. G. Sudarshan, "Relative Weights of the Decays of Certain Resonances in Theories with Broken Symmetry," *Phys. Rev. Lett. 10* (1963), 423–5; A. J. Macfarlane, E. C. G. Sudarshan, and C. Dullemond, "Weyl Reflections in the Unitary Symmetry Theory of Strong Interactions," *Nuovo Cimento 30* (1963), 845–58; A. J. Macfarlane and E. C. G. Sudarshan, "Electromagnetic Properties of Stable Particles and Resonances According to the Unitary Symmetry Theory," *Nuovo Cimento 31* (1964), 1176–96; A. J. Macfarlane, N. Mukunda, and E. C. G. Sudarshan, "Electromagnetic and Decay Properties of G_2 Multiplets," *Phys. Rev. 133* (1964), B475–7; A. J. Macfarlane, N. Mukunda, and E. C. G. Sudarshan, "Generalized Shmushkevich Method: Proof of the Basic Results," *J. Math. Phys. (N.Y.) 5* (1964), 576–80; N. Mukunda, A. J. Macfarlane, and E. C. G. Sudarshan, "Representation Mixing for SU_3 and G_2," *Phys. Rev. 138* (1965), B665–6.

28 R. H. Dalitz, "The Decay of the τ-Meson," *Proc. Phys. Soc. London, Sect. A 66* (1953), 710–13; R. H. Dalitz, "On the Analysis of τ-Meson Data and the Nature of the τ- Meson," *Philos. Mag. 44* (1953), 1068–80; R. H. Dalitz, "Decay of τ Mesons of Known Charge," *Phys. Rev. 94* (1954), 1046–51.

29 George Sudarshan, "Annihilation of Antinucleons," *Phys. Rev. 103* (1956), 777–9; Prem Prakash Srivastava and George Sudarshan, "Multiple Production of Pions in Nuclear Collisions," *Phys. Rev. 110* (1958), 765–6.

30 G. Pinski, E. C. G. Sudarshan, and K. T. Manhantappa, "Pion–Pion Interaction and Antinucleon Annihilation," in *Proceedings of the Tenth International Conference on High Energy Physics*, Rochester, edited by E. C. G. Sudarshan et al. (New York: Interscience, 1960), pp. 173–7.

31 B. C. Maglich, L. W. Alvarez, A. H. Rosenfeld, and M. L. Stevenson, "Evidence for $AT = 0$ Three-Pion Resonance," *Phys. Rev. Lett. 7* (1961), 178–82.

32 E. C. G. Sudarshan, "Principles of Classical Dynamics," University of Rochester Report NYO-10250; E. C. G. Sudarshan and N. Mukunda, *Classical Dynamics: A Modern Perspective* (New York: Wiley, 1974).

33 John R. Klauder and E. C. G. Sudarshan, *Fundamentals of Quantum Optics* (New York: W. A. Benjamin, 1968).

34 E. C. G. Sudarshan, L. O'Raifeartaigh, and T. S. Santhanam, "Origin of Unitary Symmetry and Charge Conservation in Strong Interactions," *Phys. Rev. 136* (1964), B1092–6.

35 Meinhard E. Mayer, Howard J. Schnitzer, E. C. G. Sudarshan, R. Acharya, and M. Y. Han, "Concerning Space-Time and Symmetry Groups," *Phys. Rev. 136* (1964), B888–92.

36 E. C. G. Sudarshan, "Concerning Space-Time, Symmetry Groups, and Charge Conservation," *J. Math. Phys. (N.Y.) 6* (1965), 1329–31.

PART VIII

THE PARTICLE PHYSICS COMMUNITY

Take careful heed, fresh Ph.D.'s, aflame for fortune's prizes:
The laurel crown is not for sale in well-assorted sizes.
Hard work alone is not enough, no matter how severe,
The lightning may descend on him who idly quaffs a beer . . .
Now all who love the wild surmise, the mad conjecture free,
Who do not like research in groups of more than twenty-three,
Remember still the theorem that Glaser proved again –
To hell with organization man, there's nothing like a brain!
   – © Arthur Roberts, "Birth of the Bubble Chamber" (1956)

34 The postwar political economy of high-energy physics

ROBERT SEIDEL

Born 1945, Kansas City, Missouri; Ph.D., University of California,
Berkeley, 1978; history of science; Los Alamos National Laboratory,
University of California

> Six billion dollars will buy two nuclear powered aircraft carriers, 19 months
> of research on the Star Wars antimissile system – or the ultimate dream
> machine for the 2,000 or so high-energy physicists in the United States. It is a
> particle accelerator – an "atom smasher" in the old vernacular – and if
> funded would be the biggest machine ever built....Physicists say the
> machine would help them fathom such basics as the nature of matter and the
> origin of the universe. But the government may well balk at spending such a
> sum for a device that produces mere scientific findings.

This statement, from *Newsweek* (22 April 1985), epitomizes the political
economy of postwar high-energy physics. Its practitioners have had to allege
that more than "mere scientific findings" will result from the construction of
their atom smashers – the ultimate dream machine has changed with time – in
order to satisfy their federal patrons. The increasingly energetic and costly
machines built since the war have testified to the physicists' success in linking
their own with the national purpose. There has been, inevitably, a trade-off
between scholarly and state purposes that can be understood by an historical
analysis of the political economy of high-energy physics in the late 1940s and
early 1950s, when the modern patterns of patronage of particle physics first
manifested themselves.

We can discuss three elements that entered into the trade-off. The first was
manpower: Young men, attracted to physics by the glamour of probing the
limits of the universe, have been trained with the intention of putting their
talents to use in more mundane and sometimes more military pursuits, like
nuclear reactors or nuclear weapons. Second, those who trained scientific
manpower for these purposes, but pursued the higher calling of particle
physics, were called upon in times of crisis by their patrons to consecrate their

talents to these purposes, and served, therefore, as a sort of "national guard" to reinforce the weapons laboratories. Third, the instruments of discovery, particle accelerators, were turned into instruments of war, when defense seemed to require turning plowshares into swords. Three episodes from the history of particle physics in the decade following World War II illustrate the bargain.

Negotiations for the Bevatron

At the end of World War II, accelerators, which had been used to design atomic weapons and produce nuclear explosives, were high on the agenda of science. Ernest Lawrence, the inventor of the cyclotron, an instrument turned to electromagnetic isotope separation during the war, convinced the Manhattan Engineer District (MED) commander, General Leslie Groves, in the fall of 1945 that it was in the national interest to support research to refill the reservoir of knowledge exhausted by wartime applications. He won the general's support for the completion of the 184-inch cyclotron and the construction of two new machines utilizing war-surplus equipment and ideas, the linear accelerator and the synchrotron. All three devices were generously supported by the MED and its successor, the Atomic Energy Commission (AEC) at a level thirty times that of the prewar laboratory budget, or 31 percent of the wartime budget.[1]

The AEC's first director of research, James Fisk, balked at building additional machines, however, until Lawrence, in a meeting at Bohemian Grove, California, in August 1947, convinced the commission that $15 million ought to be set aside for new accelerators. Although this sum was intended to support the Office of Naval Research (ONR) accelerator program as well as those in the AEC laboratories, Lawrence had in mind a 10-GeV proton-synchrotron, designed by William Brobeck, his laboratory's chief engineer, that had been estimated to cost more than the total appropriated. To avoid the appearance of greed, Lawrence halved its energy to 5 GeV. Edwin McMillan and Wolfgang Panofsky suggested that 6 GeV, the energy thought necessary for nucleon creation, might be a more scientific goal.[2] To build the 6-GeV proton-synchrotron, Lawrence proposed a $9 million construction program to the AEC's General Advisory Committee (GAC) in November of that year.

However, Enrico Fermi questioned the need for more energetic machines before those under construction at the Berkeley Radiation Laboratory were completed and their energy ranges explored, and he worried that "it would harm science to have [the GAC] endorse what appear[ed] to be an unthoughtful project."[3] Isidor I. Rabi wanted support for a similar machine at Brookhaven. He pressed for a delay until a proposal was forthcoming from that laboratory. When the GAC agreed to entertain more developed proposals for high-energy accelerators from both Brookhaven and Berkeley, Lawrence, Brobeck, and McMillan revised the Berkeley Radiation Laboratory proposal. The second bid called for a 1.8-GeV machine that could later be taken apart, its magnet doubled in radius, and its generators quadrupled in capacity to provide the higher energy.[4]

At the same time, Lawrence intervened to prevent the University of California regents from abandoning the Los Alamos Laboratory, which was distant and difficult to manage and could expose them to possible liabilities if nuclear accidents should occur. From 1945 onward, the university had sought to terminate the contract that bound it to the birthplace of the bomb. The competition for the proposed giant proton-synchrotron introduced a new consideration. Robert Underhill, the secretary of the regents of the University of California, was warned in December 1947 that there was a feeling in the East that it was "time to break the University of California atomic trust" by awarding the large accelerator to Brookhaven. As a result, he was prepared to give up his previously unrelenting opposition to continuing to run the Los Alamos Laboratory in peacetime. At Lawrence's instance, university president Robert Gordon Sproul undertook to persuade the regents "to co-operate [with the AEC], particularly if it would assist in the University's obtaining its desire in the matter of the proper financing of the [Bevatron]."[5]

Although AEC General Manager Carroll Wilson assured Sproul, Lawrence, and Underhill "that the University was under no compulsion or pressure . . . to renew the contract; and that the continuance of the University in this project had no connection whatever with contracts for the cyclotron [Bevatron]," Lawrence personally went to Los Alamos. He endeavored to straighten out some of the contract difficulties and to determine if a new contract was needed, rather than an extension, as Los Alamos director Norris Bradbury and his associate director, John Manley, wanted "for morale purposes."[6] At the end of January, Underhill rejected their proposed draft.[7] This left the Los Alamos contract still in limbo at the beginning of February 1948, when the GAC met in Washington to decide the fate of Lawrence's Bevatron proposal. It is unlikely that its chairman, Robert Oppenheimer, did not know about these negotiations. He had helped Lawrence in previous disputes with the AEC.[8]

At Brookhaven National Laboratory (BNL), Rabi and AEC commissioner Robert Bacher had overruled M. Stanley Livingston's scheme for a 750-MeV synchrocyclotron and now urged the laboratory to seize the competitive advantage over Berkeley afforded by the laboratory's proximity to a larger portion of the high-energy physics community and its relations to nine leading universities. In two months, a sophisticated, lengthy proposal for a 2–3-BeV machine, with enough energy to manufacture pion pairs, was readied.[9] Although based on Brobeck's design, BNL's design was more innovative, and the GAC preferred it to Berkeley's expanding machine. Rabi argued that the Brookhaven accelerator would serve a sizable fraction of the American scientific community, unlike the University of California machine. McMillan and Brobeck warned that if the machine were not built at Berkeley, "the [Radiation Laboratory] shop and engineering staff would have to be decreased markedly,"[10] thereby dismantling an important technical resource for the electromagnetic separation project during the war and central to the AEC accelerator effort in peacetime. Rabi responded that the California engineers could be transferred to Los Alamos or Argonne, where they would be useful in other programs.[11]

However, the morale of the Berkeley Radiation Laboratory had to be considered. Morale at Los Alamos, Argonne, and Oak Ridge was low. Fifty-five hundred of the 7,100 wartime scientists and technical personnel of the MED had left, and it had been necessary to recruit on a large scale to rebuild laboratory staff. The GAC's decision at the end of 1947 to centralize reactor design at Argonne had sapped Oak Ridge élan and imposed Washington's direct supervision on the Chicago laboratory. Los Alamos was struggling through water and personnel shortages, and the University of California had not yet signed a new contract to run it. Oppenheimer and Glenn Seaborg convinced their GAC colleagues that "discouragement of the Berkeley group would result in the loss of something valuable to the national scientific health." Whether that was high morale or the Los Alamos contract, the GAC took the course that would safeguard both and recommended that "two machines, aimed at substantially different maximum energies, should be built," one at each laboratory.[12] One month later, when the energy ranges were decided, Lawrence, whose 184-inch synchrocyclotron had manufactured a pion, claimed the larger prize. A month later, AEC chairman David Lilienthal traveled to San Francisco to finalize the revised arrangement to operate Los Alamos under the "contractual supervision" of the university.[13]

The need to sustain morale, personnel, and legitimacy in the AEC laboratories complemented the ONR interest in nuclear propulsion, which led them to create their own ambitious program of accelerator support. ONR and AEC programs in high-energy physics at universities and "academic" laboratories like Berkeley and Brookhaven attracted young physicists who were trained in fields vital to the development of nuclear energy for peaceful and military applications. The University of California management of Los Alamos gave it the simulacrum of "academic atmosphere" its scientists wished. This permitted the remote mesa laboratory to retain and to recruit better talent than any civil service laboratory.[14]

By virtue of its support of AEC laboratories like Berkeley and Brookhaven, the AEC got "secrecy, and big groups of scientists who will take orders, and big equipment," as commissioner-scientist Henry Smyth told AEC laboratory representatives in 1949.[15] The laboratories used that big equipment for the research in particle physics that made possible many of the accomplishments discussed in this volume.

Materials-testing accelerator

In 1949, when the explosion of the Soviet atomic bomb sent shock waves through the AEC laboratory system, the effect of the quid pro quo between high-energy physics and its military applications was manifested by laboratory reactions. While the Radiation Laboratory remobilized for national defense, Brookhaven, which had lost sight of the ultimate rationale of its existence, or rather, substituted a scientific one for it, sought to evade enlisting in the national defense.

In the wake of Joe I, the first Soviet atomic bomb, Lawrence and Edward Teller launched their well-known campaign for a super bomb. Lawrence initially committed his laboratory to the construction of heavy-water reactors to make tritium for the super, and then, after the famous GAC decision not to pursue a crash program, to production accelerators. He persuaded the AEC's director of research, University of California chemist Kenneth Pitzer, that an accelerator might produce critical materials. At Livermore, forty-five miles from Berkeley, he created the materials-testing accelerator (MTA). Although this attempt to extrapolate accelerator technology was not successful – 25-MeV 1-A proton linear accelerators tend to spark when hundreds of thousands of volts are placed on their drift tubes, and it was found cheaper to subsidize uranium exploration in the West than to power such huge machines – Livermore was converted into a second weapons laboratory when Teller and Lawrence joined forces again in 1952. The transfer of personnel from Berkeley to Livermore validated the Radiation Laboratory as a "national guard" of the AEC weapons program.

Lawrence's enthusiasm, his influence with the regents of the University of California, and his prestige among the faculty made the expansion of the university's weapons laboratories possible. When the issue of academic freedom posed by contractual work was reviewed by an all-university conference in 1949, Lawrence persuaded the assembled dons that "gold is where you find it!"[16] When Livermore was converted to a weapons design laboratory, the regents embraced the new and lucrative assignment.[17]

At Brookhaven, the mix of duty and pleasure was slightly richer. Although Leland Haworth sought to dispel "a widespread impression that [his] laboratory desired to hold itself entirely aloof from the practical problems" faced by the AEC in the emergency, he argued that "a vigorous program of fundamental research" was "appropriate" at the northeastern laboratory. "We can be most useful by attacking problems involving investigation, measurement, and so on."[18] AEC commissioner Smyth agreed that there was need for "a certain amount of basic work" in the laboratory system during the crisis, and although "it would be better to tell the people at Brookhaven that they should drop their basic work than to pull people out of the universities and put them into applied work, . . . it would be more effective to pull some of the Argonne staff [out of basic research] and put them on other work than to do this at Brookhaven."[19]

The commission persuaded BNL to investigate the civil defense of New York City, to measure inelastic neutron cross sections for Los Alamos, and to study chemical separation of fissionable elements, heavy-water production, and weapons metallurgy. To protect itself from further incursions on its program, the laboratory drafted a "Guide for the Operations of Brookhaven National Laboratory" that emphasized its fundamental research responsibilities and restricted programmatic work to "applied or engineering research of the kind appropriate to a university environment [and] such programmatic researches as are mutually agreed upon [with the AEC] which utilize the laboratory's unique facilities or special talents of its staff."[20]

Of course, the two laboratories were not unique sources of AEC personnel. To supplement that resource, the AEC expanded its research programs in universities in the 1950s. Joseph Platt, the chief of the Physics and Mathematics Branch of the AEC, argued in 1951 that the commission should support off-site research in nuclear physics "to the limit set by the availability of able research men in the field." Assuming that one in ten of the 10,000 members of the American Physical Society (APS) were available, he estimated this would cost $10 million per year.[21]

Assuming that half of these, or 500 nuclear physicists, were "willing, able, and eager to use particle accelerators, and that on the average five such men per accelerator is an effective team," Platt saw a need for 100 accelerators, 36 more than the 64 built since the war. He modestly proposed that the AEC "should contribute to the support of one or two more accelerator programs per year for the next several years, if the international situation remain[s] roughly as at present."[22] He expected that the ONR, the National Science Foundation (NSF), and universities would support the construction of other machines. The "national guard" would be further strengthened by these peacetime exercises.

Strong focusing and CERN

As John Blewett and Ernest Courant point out in Chapters 10 and 11, Brookhaven's completion of the Cosmotron in 1952 was shortly followed by the reinvention of strong focusing. The Radiation Laboratory had overlooked the first suggestion of this concept by Nicholas Christofilos in 1950, perhaps because the Berkeley physicists were still busy with the MTA, which also delayed completion of the Bevatron. This oversight assured both short-term and long-term losses of leadership in high-energy physics to the northeastern laboratory.

The discovery of strong focusing also caught the AEC off guard. When the "secret" was published, "some of the Washington staff [of the AEC had] the feeling that any idea developed in an unclassified area at Brookhaven National Laboratory [would] be lost before the Commission had time to evaluate its significance," and Brookhaven was warned that any new idea that might be classified should be reported to the AEC before publication.[23] Even where information was not "born classified," it appeared, it could be baptized in the faith.

The commission's report on its high-energy accelerator program at the end of 1952 revealed the Januslike nature of the discovery. On the one hand, interest in strong-focusing machines had been stimulated by cosmic-ray investigations that had shown the existence of new elementary particles that their more energetic beams might manufacture. On the other hand, the strong-focusing principle might "have applications...to an accelerator for protection against A-bombs." This concept was under study by an ONR–AEC committee.[24]

In February 1953, Lee Dubridge, Oppenheimer, J. R. Zacharias, Haworth, R. P. Johnson, and General James McCormack, head of the AEC Military Liaison Committee, met at Caltech to discuss the possibility suggested by the study that "the illumination of a falling atomic bomb by an intense beam of gamma radiation coming from a suitable electron accelerator aimed at the bomb by a radar-computer-servo-mechanism" would do the job.[25] Although some at the meeting were skeptical, follow-up investigations were pursued at several laboratories. Cornell's Robert Wilson and Hans Bethe reported to the commission that it was "not possible to rule out, on technical considerations alone, the possibility that a reasonably effective countermeasure [of this sort] might be developed."[26] The AEC's director of research drew the obvious lessons:

> It is clear that a major share of the responsibility for military and developmental aspects of this program is to be carried by the Department of Defense. However, it is also clear that the problem involves the development of considerable basic information in the fields of ultra-high energy nuclear physics and particle accelerator development...and it is appropriate that the Division [of Research] cooperate with the DOD in the basic processes of accelerator development relevant to the development of such devices for air defense purposes.

He suggested that the "military usefulness of particle accelerators should be included as an additional justification in further defense of ultra-high energy particle accelerator construction items before the Bureau of Budget and the Congress."[27]

Although the Bureau of the Budget had removed the $5 million AEC budget line item for the Brookhaven alternating-gradient synchrotron for the FY 1954 budget, a rider later added by Congress provided plant and engineering funds that the AEC could use at its discretion "for the construction of particle accelerators without regard to any other provision of this act," making $2 million available to begin construction of the $20 million accelerator.[28]

While the interest in particle-beam weapons seemed to lend urgency to the construction of the alternating-gradient synchrotron (AGS), the AEC's collateral interest in production accelerators was a potential barrier to scientific communication of the details of accelerator design. At the 24 July 1953 meeting of the AEC, the chairman, Lewis Strauss, questioned the wisdom of transmitting "unclassified accelerator drawings to CERN." Smyth argued that CERN's leadership was "friendly to the United States," that the Communists were hostile to CERN, and that "since the [drawings] were unclassified there was essentially no reason why they should not be published." Smyth pointed out that the idea of a production accelerator "would not be suggested by the blueprints of the BNL machine but by the studies which CERN was certain to conduct whether they had the BNL drawings or not." He concluded that "the proposed transmittal would gratify our friends abroad and would help to create a spirit of close cooperation, while a refusal would gratify our enemies." Strauss, who found this argument "unsubstantial," deferred action until he could consult the father of the MTA, Lawrence. Presumably, Lawrence set him straight.[29]

The creation of CERN added another national purpose for bigger machines: international competition. By October 1953, the AEC's director of research was arguing that although the Cosmotron had only recently come on line and the Bevatron was yet to operate, the commission should start immediately on the AGS because

> American scientists have held the lead in nuclear science since the invention of the cyclotron and they do not wish to fall behind. The European scientists have already had authorized for construction a synchrotron, which will probably be for about 30 BEV, and its completion is planned for about 1960. An immediate start in this country would result in the completion of a machine in 1958 or 1959.[30]

Personnel and morale were still crucial: A delayed decision "might result in degradation of interest" and the loss of "scientists who have close familiarity with the problems as well as engineers and technicians with special talents and skills."[31]

While the development of strong-focusing machines went on at Brookhaven, CERN, and elsewhere to serve the purposes of particle physics, international competition, personnel training, and laboratory morale, the high-intensity accelerator technology useful for production purposes and weapons work disappeared, along with Christofilos, into the secret world of the Defense Department and weapons laboratories. By 1960, for example, the Lawrence Livermore laboratory claimed "unusual competence in the development of high current accelerators," which would allow it to "carry out high current beam research"[32] for weapons applications faster and cheaper than any other laboratory. Although military interest in such weapons has seesawed, the Chair Heritage and White Horse programs are linked to this ancient tradition.

Unpeeling the onion

How important were such considerations to the scientists who built and used particle accelerators? To most working high-energy physicists, particle production rather than weapons production was of immediate importance. Those who voted with their feet in joining the weapons laboratories substituted patriotism or pragmatism for particle physics. Many scientists would, to be sure, have agreed with the goal of training more nuclear physicists for weapons and reactor development, at least until the 1960s.

In looking beyond the physics community, however, these reasons for supporting the arcane enterprise of high-energy physics loom larger. To a laboratory director like Lawrence, the fight for support was a consuming task, requiring him to seek those resonances between scientific and national needs that could redound to the benefit of his laboratory. To the AEC, whose staff often came from and returned to its laboratories, convincing the Congress and the Bureau of the Budget of the need for research funding was taking out insurance for the future of American science.

To Congress, which often poorly understood the relations between accelerators and national needs, hints such as those offered by Lawrence in defending the MTA project in 1951 were often the primary basis for decision. As North Carolina congressman Carl Durham, presiding over a July 1951 session of the Joint Committee on Atomic Energy, told Lawrence: "We have to base our opinions and our actions on the information we receive from people like you." Lawrence responded:

> We are all trying to do the best we can. I endeavored to give you sound opinion. There is always a danger of overstating things. However, if these processes work [in the MTA], they can be used in peacetime for the production of fissionable material, but when war comes, they could overnight be producing radioactive materials in large quantities; and it could be done on a scale so that we could do all our fighting with radioactive materials and not use atomic bombs. That would be a great thing. If we had a hundred grams of neutrons a day, we could kill all the people we want to without causing physical destruction; we can disrupt great populations of Europe; we can avoid destroying Paris or London if they were occupied.

Ohio congressman John Bricker enthused: "We could move in and take it over." Congressman Carl Hinshaw of California, mindful of the persistence of radioactivity, warned, "Later."[33]

Although his interest in radiological warfare was generally in radiological agents as barriers to troop movements, as Lawrence himself remarked, in the relations between scientists and their patrons, "there is always a danger of overstating things." However, the understanding of the purposes of high-energy physics, as mediated by public spokesmen like Lawrence, seems to have been crucial to the success of the scientific enterprise. It is neither possible nor desirable to discuss the history of particle physics in the 1950s without taking these elements of its political economy into account.

Notes

1 The first postwar budget authorized by Groves for the Radiation Laboratory at Berkeley was $630,000 for operation of the laboratory for six months. This did not include the work of John

Lawrence's Donner Laboratory or the Department of Chemistry, which ran at about $20,000 per month.

2 Interview with McMillan by the author, June 1981; interview of W. M. Brobeck by the author, 26 March 1985.

3 AEC General Advisory Committee (GAC) 7, 21–23 November 1947, DOE Archives.

4 E. O. Lawrence, E. M. McMillan, and W. M. Brobeck, "Proposal for Construction of a Proton Synchrotron (Bevatron)" 5 February 1948, to Chicago Directed Operations, AEC, DC Files, LBL Archives: "This project contemplates a 1.8 Bev accelerator, with provisions of space to allow eventual expansion to 2.8 and then to 6.5 Bev. This would ensure the production of 1.8 Bev particles in the shortest possible time, estimated at about two and a half years. . . . The output of this installation could later be increased to 2.8 Bev simply by doubling the generator capacity at an additional cost [to the original $4,650,000] of $600,000. The increase from 2.8 to 6.5 Bev would require doubling the size of the magnet [from a 25-foot radius to a 50-foot radius] and again doubling the generator capacity."

5 Underhill to Sproul, 29 December 1947, and Sproul's memorandum of a meeting with Carroll Wilson, Robert Bacher, Carroll Tyler, Lawrence, and Underhill, 9 January 1948. For earlier attempts to divest the university of the laboratory, see Underhill to Oppenheimer, 22 March 1944, K. D. Nichols to Underhill, 20 August 1945, Underhill to Norris Bradbury, 26 September 1945, Underhill to Col. H. C. Gee, acting manager, Santa Fe Area, USAEC, 7 February 1947, Carroll Wilson to Underhill, 15 February 1947, Underhill Files, LANL.

6 Underhill to Sproul, 19 January 1948, and Sproul's memorandum of a meeting with Carroll Wilson, Robert Bacher, Carroll Tyler, Lawrence, and Underhill, 9 January 1948, Underhill Files, LANL.

7 Sproul to Wilson, 24 January 1948, Underhill Files, LANL.

8 Oppenheimer to Lawrence, 24 September 1947, Oppenheimer Papers, Library of Congress.

9 "Program Report on High Energy Accelerator Design," 15 January 1948, BNL. I am grateful to Allan Needell for providing me with a copy of this document.

10 GAC 8, 6–8 February 1948, DOE Archives.

11 Ibid.

12 For more on this episode, see Robert W. Seidel, "Accelerating Science; The Postwar Transformation of the Lawrence Radiation Laboratory," *Historical Studies in the Physical Sciences 13:2* (1983), 375–400. For the ONR's motives in funding particle accelerator research, see Harvey Sapolsky, "Academic Science and the Military; The Years Since the Second World War," in *The Sciences in the American Context: New Perspectives*, edited by Nathan Reingold (Washington, D.C.: Smithsonian Institution Press, 1979), p. 383.

13 David E. Lilienthal, *The Atomic Energy Years 1945–1950*, Vol. 2 of *The Journals of David E. Lilienthal* (New York: Harper & Row, 1964), p. 332n.

14 A. W. Kelley, Los Alamos's employment director, reported on 17 July 1946 that he had had to hire 1,192 out of 1,557 project employees since 1 January, that top-grade scientists were the hardest to find, and that "our best recruiting efforts have not offset the general exodus of staff members who are returning to academic careers or who have sought other research employment." A. W. Kelley to E. J. Dennison, 17 July 1946, Robert Underhill Papers, LANL. See also Darroll Froman and L. E. Seeman to Cooksey, 22 July 1946, Underhill Papers, LANL: "There is a scarcity of high grade scientific talent accentuated at this laboratory by well meaning efforts of other laboratories, both academic and industrial, to employ personnel affiliated with this laboratory in its pioneer scientific work." One may contrast with this the rosy situation pictured by J. H. Manley in "The Los Alamos Scientific Laboratory," *Bulletin of the Atomic Scientists 5* (April 1949), 101–5, and the testimony of Bradbury before the Joint Committee on Atomic Energy, Investigation into the United States Atomic Energy Project (Washington, 1949), 816–17.

15 Henry D. Smyth, "The Role of the National Laboratories in Atomic Energy Development," *Bulletin of the Atomic Scientists 6* (1950), 5–8. This comment was made in response to a letter from Alvin Weinberg dated 27 September 1949, Smyth Papers 1:3, American Philosophical Society, Philadelphia. I am grateful to Bruce Wheaton for a copy of this letter.

16 "Contract Research," report of study committee no. 4, in *The University of California in the Next Ten Years* (proceedings of the fourth all-university faculty conference held at Davis, California, 28–30 April 1949) (Berkeley: UC, June 1949), p. 24.

17 W. B. Reynolds to R. M. Underhill, 16 April 1954, enclosure "Background Notes on Livermore Site," DC 1954, Donald Cooksey Files, LBL Archives.

18 Leland J. Haworth to Shields Warren, 12 May 1950, BNL-DO.

19 "Minutes of the [Argonne National] Laboratory Executive Committee," 15 August 1950, p. 2. I am grateful to Bruce Wheaton for copies of these documents.

20 John D. Jameson to Frank D. Fackenthal, 2 October 1950, enclosure "Statement of Objectives, Policies and Procedural Principles of Associated Universities, Inc.," 10 October 1950, Director's Office, Brookhaven National Laboratory (BNL-DO).

21 Joseph B. Platt to Paul W. McDaniel, re "Extent of Off-Site Research Program in Physics," 27 July 1951, Oppenheimer Papers, box 175, folder "AEC Research Division," Library of Congress.

22 Ibid., pp. 2, 5, 6.

23 E. L. van Horn to L. J. Haworth, 19 September 1952, BNL-DO.

24 AEC 603: "High Energy Accelerator Program," 1 December 1952, p. 2, DOE Archives.

25 AEC 603/2: "High Energy Accelerator Program," 26 February 1953, p. 1, DOE Archives.

26 Ibid., p. 3. See also "Resume and Minutes of October 3 [1952?] AEC ONR Conference on the Military Utilization of Particle Accelerators," 30 December 1952. DOE Archives.

27 AEC 603/2: "Military Utilization of Particle Accelerators," and "Status of Decisions and Their Implementation: AEC 603/2," 22 June 1953, DOE Archives.

28 AEC 603/9: "High Energy Accelerator Program," 8 October 1953, DOE Archives.

29 Minutes of Atomic Energy Commission meeting no. 894, 24 July 1953, pp. 454–5, DOE Archives. Cf. AEC minutes of 28 April 1948, p. 109–10. On the earlier occasion, Smyth appealed to Lawrence; see Smyth to Lawrence (not dated), Smyth Papers, American Philosophical Society, Philadelphia.

30 AEC 603/9: "High Energy Accelerator Program," 8 October 1953, p. 11, DOE Archives.

31 AEC 603/9: "High Energy Accelerator Program," 8 October 1953, p. 11, DOE Archives.

32 C. M. Van Atta to E. C. Shute, 29 March 1960, LBL Archives.

33 "Production Particle Accelerators," transcript of hearing of the Joint Committee on Atomic Energy, United States Congress, 11 April 1951, Ernest Lawrence Papers, Bancroft Library, University of California, Berkeley, 33:14.

35   The history of CERN during the early 1950s

EDOARDO AMALDI

Born 1908, Piacenza, Italy; *laurea*, University of Rome, 1929; nuclear and
cosmic-ray physics; University of Rome

Introduction

The history of CERN is of considerable interest, not only for high-
energy physics but also more generally, because CERN is the first example of
an intergovernmental research laboratory created in Europe that has been
operated successfully for more than thirty years – a remarkable model for the
creation of international organizations.

Armin Hermann, with the help of a few younger historians, is preparing a
complete history of CERN in two volumes.* From this extensive presentation, John Krige of Hermann's group, with the help of his colleagues, will
extract a more concise CERN history, contained in a single volume of 300
pages and addressed to a wider public.

My account here is of a completely different nature. I did not consult the
archives of the foreign ministries or of the research councils of the member
states of CERN or of other intergovernmental organizations. The material
here is based on a few well-known documents, on my personal diary of that
period, and on a few reports and lectures prepared years ago by Lew Kowarski or by me.[1]

Everyone agrees that the early history of CERN can be divided into three
periods. The first period encompassed the first initiatives and extended from
the middle of the 1940s to 15 February 1952. On that date, the representatives
of eleven European governments signed in Geneva the agreement establish-

---

*   *Ed. note:* The first volume of A. Hermann, J. Krige, U. Mersits, and D. Pestre,
     History of CERN, Launching the European Organization for Nuclear Research, was
     published in 1987 by North Holland. A second volume covering the years 1954–65 is to
     appear in 1989.

ing a provisional organization with the aim of planning an international laboratory and organizing other forms of cooperation in nuclear research.[2] The second period encompassed the planning stage, extending from February 1952 to 1 July 1953, when the convention establishing the final organization was signed by the representatives of twelve European governments.[3] During the third period, usually called the interim stage, from July 1953 to 29 September 1954, a prescribed point of the ratification procedure by the parliaments of the member states was reached, and the convention went into effect.

I limit my discussion to these three early phases, because of their peculiar nature, and because I was directly involved in them.

The first initiatives

It is not easy to fix a date for the beginning of the first stage. It is certainly true that even the founding of the United Nations immediately after the end of World War II and the foundation of its specialized agency UNESCO were essential for arriving, a few years later, at the creation of CERN. Already in October 1946, the French delegation to the "UN Economic and Social Council" submitted a proposal for creating "United Nations Research Laboratories." The "UN Atomic Energy Commission," also created in 1946, with the aim of bringing atomic weapons under some form of international control, was a typical circle where the problem of international collaboration, especially in nuclear physics, was raised and discussed. These initiatives, as well as another that appeared in 1949 and involved the Netherlands, Belgium, and the Scandinavian countries, refer mainly to low-energy nuclear physics and its applications and should therefore be considered, in my opinion, more as precursors of the idea of Euratom than of CERN.

In those years in many European countries, especially in France, Italy, West Germany, and Belgium, the idea of moving toward some form of economic and/or political unification of at least a considerable part of the old Continent was considered of primary importance by many authoritative politicians, who adopted it as a guiding principle of their immediate and long-range program of action. Another favorable element was the scientific, technical, and administrative experience gained in a few countries, during and immediately after the war, of wide and complex organizations operating in the field of the nuclear sciences and their applications. That experience had brought about the creation in the United States of a few large research laboratories such as Argonne National Laboratory and Brookhaven National Laboratory. In particular, Brookhaven National Laboratory had been created, and run very successfully, by the Associated Universities Incorporated. Furthermore, the dimensions of the geographic region involved and of the laboratory were very similar to those of a possible future European research establishment. For Europe, however, the chosen program of research had to be completely free of any limitation or restriction originating from military, political, or even industrial secrecy.

At the same time, in many European and overseas places, scientists were becoming aware of the continuously increasing gap between the means available in Europe for research in the field of nuclear physics and elementary particles and those available in the United States, where a few high-energy accelerators had started to produce results, while others had already reached advanced stages of construction or design (Table 35.1). It was becoming more and more evident that such a situation could be changed only by a considerable effort made jointly by many European nations.

A further element was important. Immediately after World War II, cosmic-ray research was on a very high level in Western Europe and was in part being carried out through successful international collaborations. The discovery in 1946 by Marcello Conversi, Ettore Pancini, and Oreste Piccioni that the cosmic-ray mesotron is a weak-interacting particle, and not a Yukawa particle, the discovery in 1947 by C. M. G. Lattes, G. P. S. Occhialini, and Cecil F. Powell of the π meson and of its decay into the muon, and the discovery of the first two strange particles by George Rochester and Clifford Butler (a Λ^0 and a θ^0) are among the most brilliant results of an extensive and rich experimentation in Western Europe.

As I pointed out in a lecture at the Varenna school in the summer of 1972, the mountain laboratories in Switzerland, France, and Italy, and, even more, the nuclear-emulsion laboratories of the universities of Bristol and Brussels, led, respectively, by Powell and by Occhialini, had become in those years points of encounter for young physicists from many countries.[4] The collaborations in mountain huts and the coordination of the experiments planned by different groups paved the way to the idea of wider and more ambitious collaborations, which were indeed arising in various places in Europe.

Cosmic-ray physicists, of course, were interested in a European laboratory only if it were devoted to high-energy physics. I remember that in the years 1948–50 the various aspects of the problem, including energy and cost of machines, were examined in Rome, in frequent discussions between Bruno Ferretti and myself, and in letters exchanged with Gilberto Bernardini, who in those years was at Columbia University at the invitation of Isidor I. Rabi.

I remember becoming aware that similar problems were discussed in other European countries, in particular in France, at the time I heard of the European Cultural Conference, held in Lausanne in December 1949. At this meeting, Raoul Dautry, administrator of the French Commissariat à l'Energie Atomique, read a message from Louis de Broglie in which the proposal was made to create in Europe an international research institution, without mentioning, however, nuclear physics or fundamental particles. To de Broglie's message, Dautry added the proposal that the conference study ways of strengthening collaboration in astronomy and astrophysics, by building powerful telescopes and all necessary auxiliary material, and in the field of atomic energy, by setting up a center with all the required modern apparatus. As we see from these examples, at that time opinions were still rather

Table 35.1 *Accelerators in operation before 1956 capable of producing at least π mesons*

Institution	Type of machine	Energy (MeV)	Date first operated
In the United States			
Univ. Calif., Berkeley	synchrocyclotron	350	1946
Univ. Rochester	synchrocyclotron	240	1948
Columbia Univ.	synchrocyclotron	400	1950
Univ. Chicago	synchrocyclotron	460	1951
Carnegie Inst. Tech.	synchrocyclotron	450	1952
Brookhaven Nat'l. Lab. (Cosmotron)	proton-synchrotron	3,200	1952
Univ. Calif., Berkeley (Bevatron)	proton-synchrotron	5,700	1954
		6,200	1955
Univ. Calif., Berkeley	electron-synchrotron	320	1949
Cornell Univ.	electron-synchrotron	300	1951
MIT	electron-synchrotron	300	1952
Univ. Michigan	electron-synchrotron	300	1953
Purdue Univ.	electron-synchrotron	300	1954
Univ. Illinois, Urbana	betatron	300	1950
In Europe			
Univ. Liverpool	synchrocyclotron	400	1954
Univ. Glasgow	electron-synchrotron	350	1954

Note: For more information, see, for example, M. S. Livingston and J. P. Blewett: *Particle Accelerators* (New York: McGraw-Hill, 1962).

confused about the type of research to be tackled by the new organization and even more confused about the nature of the collaboration to be established.

In June 1950, the General Assembly of UNESCO was held in Florence. Rabi, a member of the delegation from the United States, made an important speech about "the urgency of creating regional centers and laboratories in order to increase and make more fruitful the international collaboration of scientists in fields where the effort of any one country in the region was insufficient for the task." In the official statement, approved unanimously by the General Assembly along the same lines, neither Europe nor high-energy physics was mentioned, although this specific case was clearly intended by many, in particular by Rabi himself and by Pierre Auger, who was director of the Department of Natural Sciences of UNESCO.

I had known Rabi very well since 1936, when I visited Columbia University for the first time. I had since then visited him in his laboratory a few times, both before and after World War II. In June 1950, on the occasion of his trip to Italy, I had a thorough conversation with him about the future European laboratory.

A further endorsement of this idea came from the International Union of Pure and Applied Physics (IUPAP), whose president at that time was H. A. Kramers. I was one of the vice-presidents, and at the beginning of the summer of 1950 I had asked Kramers to include the discussion of Rabi's proposal, with specific reference to Europe and high-energy physics, in the agenda of the meeting of the Executive Committee of IUPAP that was to take place at the beginning of September of the same year in Cambridge, Massachusetts. Although Kramers could not preside at the meeting because of poor health, the problem was discussed at length under the chairmanship of C. G. Darwin assisted by Secretary General Fleury.[5] At the conclusion of this meeting, I was asked to contact Rabi and other physicists in various European countries in order to clarify the aims and structure of the new organization and to help in the coordination of the different efforts.

My first step was to write to Auger, who in the meantime had presented the problem to the conference on nuclear physics held in Oxford during the month of September.[6] Auger now had the authority to act, but no money was appropriated on the scale required for a detailed study of such a project. In December 1950, Denis de Rougemont, director of the European Cultural Centre (ECC) (which was founded at the Lausanne meeting of 1949 mentioned earlier), agreeing with Auger, called at Geneva a commission for scientific cooperation for discussing Rabi's proposal. The participants, in addition to a few people from the ECC, were P. Auger (France), P. Capron and Verhaege (Belgium), B. Ferretti and M. Rollier (Italy), H. A. Kramers (Netherlands), P. Preiswerk (Switzerland), and G. Randers (Norway). I was invited but was unable to attend.

The commission concluded its work with a series of recommendations, the most important of which was "the creation of an international laboratory centred on the construction of an accelerator capable of producing particles of an energy superior to that foreseen for any other accelerator already under construction," namely, the Brookhaven Cosmotron (3 GeV) and the Berkeley Bevatron (6 GeV). The commission also discussed and endorsed the estimates brought by Ferretti of the cost of such a machine obtained by comparison with the cost of the two American machines mentioned earlier.

Immediately after this meeting, G. Colonnetti, president of the Consiglio Nazionale delle Ricerche of Italy, R. Dautry, administrator of the French Commissariat à l'Energie Atomique, and J. Willelms, director of the Fond National de la Recherche Scientifique in Belgium, made available to Auger some funds that together amounted to about $10,000. This sum, although very modest, was sufficient for Auger to initiate the first steps for arriving at the planning and construction of a large particle accelerator.

At the beginning of 1951, Auger established a small office at UNESCO and invited me to Paris at the end of April to discuss the constitution and composition of a working group of European physicists interested in the problem. The first meeting of this "Board of Consultants" was held at UNESCO, in Paris, at the end of May 1951. The board members were E. Amaldi (Italy), Capron (Belgium), O. Dahl (Norway), Frank Goward (U.K.), F. A. Heyn (Netherlands), L. Kowarski and Francis Perrin (France), Preiswerk (Switzerland), and S. Alfven (Sweden) in place of I. Waller.

Two goals were immediately established: a long-range, very ambitious goal for an accelerator second to none in the world; in addition, the construction of a less powerful and more conventional machine that could allow experimentation in high-energy physics to begin at an earlier date by European teams.

In this and a few successive meetings of the Board of Consultants, a few other features of the new organization were also examined as preparatory work for a Conference of Delegates of Governments convened by UNESCO in Paris, in December 1951. This conference took place under the chairmanship of François de Rose, a French diplomat who had been involved in discussions with French scientists (Auger, Perrin, Kowarski) about the future European laboratory from the time of the UN commissions mentioned earlier. He was elected, a few years later, president of CERN. The meeting led to the signing of the CERN agreement, which took place, as I said, in Geneva in February 1952.

The first problem tackled by the council of the provisional organization that had been created by the CERN agreement was the nomination of the officers responsible for the appointment of the remainder of the staff and for planning the laboratory. C. J. Bakker, from the Netherlands, was nominated director of the Synchro-Cyclotron Group; Dahl, from Norway (with Goward from the United Kingdom as deputy), was nominated director of the Proton-Synchrotron Group; Kowarski, from France (with Preiswerk, from Switzerland, as deputy), was nominated director of the Laboratory Group, which had to take care of site, buildings, workshops, administrative forms, financial rules, and so forth; Niels Bohr was nominated director of the Theoretical Group; E. Amaldi was nominated secretary general, with the task of maintaining cohesion between the four groups of which the provisional organization was composed. Almost all these people had worked on the Board of Consultants, nominated by UNESCO during the first stage, and contributed later to the creation of the new organization and to the development of its activities.

In July of the same year, 1952, an international nuclear physics conference was held in Copenhagen. On that occasion, Werner Heisenberg presented to the council, which held its second meeting in Copenhagen immediately after the conference, a proposal on the type of accelerator to be built as the main goal. Thus, the decision was made that the Proton-Synchrotron (PS) Group should explore the possibility of constructing a 10-GeV proton-synchrotron, which, at that time, would represent the biggest machine in the world.

During the month of August, Dahl and Goward went to Brookhaven in order to study the almost completed Cosmotron (with a maximum energy of about 3 GeV) in detail. During their two-week visit (and in some way connected with the discussions going on in relation to the European project, see Chapter 11 by Ernest Courant), Courant, M. S. Livingston, and Hartland Snyder invented the "strong-focusing principle."[7] This important discovery came soon enough to allow a change in the plans of the provisional organization. With the approval of the council, the PS Group embarked on the study of a strong-focusing PS of 20–30 GeV, instead of the weak-focusing 10-GeV machine considered earlier.

During the summer of 1952, four sites were offered for the construction of the new laboratory: one near Copenhagen, one near Paris, one in Arnhem in the Netherlands, and one in Geneva. After long and lively discussions, the site in Meyrin, near Geneva, was unanimously selected.[8]

Another point I would like to touch on concerns the participation of the European nations. All the European members of UNESCO had been invited to the conference, which opened in Paris in December 1951 and closed in Geneva on 15 February 1952, but no response came from the countries of Eastern Europe. Furthermore, the reader probably has noticed that whereas the CERN agreement was signed at that date by the representatives of eleven countries, the CERN convention establishing the permanent organization was signed (1 July 1953) by twelve countries.

The difference involved the United Kingdom, which at the beginning was rather cautious in committing itself to take part in the new organization (see Chapter 36 by Hermann). The U.K. government preferred to remain in the formal position of an observer during the first two stages, whereas by signing the CERN convention it became a full-rights member of the permanent organization.

I remember that in the autumn of 1952 it was decided that Bakker, Dahl, and I should go to Brookhaven to take part in the dedication of the Cosmotron that was foreseen for 15 December.[9] Our trip was already arranged when I was called on the telephone from London by John Cockcroft, who had from the start been very much in favor of the participation of the United Kingdom in the new venture. As a consequence of our telephone conversation, I decided to leave earlier and pass through London on my way to Brookhaven.

In London, I went to the Department of Scientific and Industrial Research (DSIR), where I met its chairman, Ben Lockspeiser, who a few years later was elected president of the permanent CERN. After rather long discussions about various organizational and financial aspects of the project as a whole, Cockcroft and Lockspeiser brought me to Lord Cherwell, who, at that time, was a member of the Churchill government.[10] Lord Cherwell appeared to be very clearly against the participation of the United Kingdom in the new organization. As soon as I was introduced to him in his office, he said that the European laboratory was to be one more of the many international bodies consuming money and producing a lot of papers of no practical use. I was annoyed, and I answered rather sharply that it was a great pity that the United Kingdom was not ready to join such a venture, which, without doubt, was destined for full success, and I went on by explaining the reasons for my convictions. Lord Cherwell concluded the meeting by saying that the problem had to be reconsidered by His Majesty's government. When we left the Ministry of Defence, where the meeting had taken place, I was rather unhappy about my lack of self-control, but Cockcroft and Lockspeiser were rather satisfied and tried to cheer me up.

A few weeks later, Lockspeiser wrote an official letter asking that the status of observer be given to the United Kingdom in the provisional organization, and the DSIR started to regularly pay "gifts," as they were called, corresponding exactly to the U.K. share calculated according to the scale adopted by the other eleven countries. Thus, in spite of its particular legal position, in practice the United Kingdom gave fundamental support to the provisional organization and was the first among the European countries to ratify the CERN convention.

The interim stage

At the beginning, the four groups had worked at the institutions of the corresponding directors. In October 1953, the PS Group staff was assembled in Geneva, partly at the Institute of Physics of the university and partly in temporary huts built in its vicinity. At that time, the transition was made from mainly theoretical work to experimentation and technical designing. John and Hildred Blewett of the Brookhaven National Laboratory arrived in Geneva in September 1953 and joined the PS Group, which benefited from their experience in design and construction of accelerators for about one year.

In October, an international conference on proton-synchrotrons was organized at the Institute of Physics by the PS Group staff. In March 1954, Goward died after a short and tragic illness and was succeeded by John Adams (1920–84), who became director of the PS Division in 1955, when Dahl went back to Norway to direct the design and construction of the first Norwegian nuclear research reactor.

During October 1953, an administrative nucleus began to function in a temporary Geneva office and, from January 1954 on, in the Villa Cointrin at the Geneva airport. First, instrumentation workshops, and then a library and a few laboratories were gradually set up also at the airport in the summer of 1954. On 12 August 1954, France completed its ratification procedure of the CERN convention, and the situation appeared sufficiently promising to dare starting, on 13 August 1954, major excavation work on the Meyrin site for the foundations of the first permanent buildings, in particular, the one to house the synchrocyclotron. I had taken this decision, with the agreement of the council, under the pressure of the Geneva weather months before all ratification formalities were completed. Lockspeiser was clearly pleased and commented: "Now we have another task – keeping Amaldi out of jail."

To appoint the director general of the permanent organization, the council of the provisional organization set up, in June 1953, a Nomination Committee composed of Bohr, Cockcroft, Colonnetti, Heisenberg, Perrin, and P. Scherrer, with Perrin as convener. The attention of the Nomination Committee soon focused on Felix Bloch (1905–83), who had been proposed by Bohr and supported with enthusiasm by all the other members.

On the occasion of a short visit to Europe by Bloch in March 1954, on the invitation of the Nomination Committee, he and I examined together the steps to be taken for creating in Geneva, as soon as possible, the proper atmosphere of a research institution. We agreed that a few of the steps had to be taken before the beginning of Bloch's tenure.

The first point regarded the gradual displacement from Copenhagen to Geneva of the Theoretical Group (or Division, as it was called by then). As the first step, B. d'Espagnat was engaged from 1 September 1954 and J. Prentki from 3 January 1955. Both were on leave of absence from the Centre National de la Recherche Scientifique in Paris and worked together on meson production and scattering.

The other step Bloch and I agreed on was to create a two-pronged cosmic-ray experimental program at CERN.[11] One project was to set up, in collaboration with the Scientific and Technical Services Division directed by Kowarski, the nucleus for an experimental physics group. The first scientists involved in this group, starting in the summer of 1954, were Y. Goldschmidt-Clermont from Belgium and G. von Dardel from Sweden, with Charles Peyrou, then at the University of Bern, as consultant. Until the accelerators in CERN could be brought into operation, this group could conduct scientific research and gain experience useful for future experiments.

The second project consisted of taking over, in August 1955, the U.K. cloud-chamber group working at Jungfraujoch Station, under J. A. Newth from Manchester. The Geneva group collaborated with this team in a new experiment for measuring the lifetime of K mesons, using a Cerenkov detector for triggering the cloud chamber on K-meson decays.

The provisional organization suddenly came to an end while the permanent organization was not yet in existence. All the assets of the provisional CERN were suddenly masterless. For eight days I had the honor, as secretary general, of having sole responsibility for all the properties and liabilities on behalf of a newborn organization.

Then, at the first meeting of the permanent council, assembled in Geneva on 7 October 1954, the secretary general presented his final report. Bloch was nominated director general, and CERN entered its final permanent form.[12] The synchrocyclotron operated for the first time in 1958, and a proton beam at full energy circulated in the proton-synchrotron on 24 November 1959, almost one year ahead of schedule.

In conclusion, I would like to add that from my personal point of view, it was a great thing to work for CERN in those early times. My duties as secretary general, although not always easy, brought me in contact with many people of high class, like de Rose, R. Valeur (another French diplomat), Lockspeiser, Willelms, J. H. Bannier from the Netherlands, and many others. All of these have contributed enormously to the creation of CERN and, in some way, to enrich my life. In addition, I related constantly with eminent scientists, like Rabi, Auger, P. M. S. Blackett, Bohr, Cockcroft, Heisenberg, Perrin, Powell, and many others, whose primary role or external support was essential for the success of the enterprise.

Highly rewarding and stimulating also was the eagerness with which many young physicists and engineers accepted my invitation to join the new provisional organization, to contribute to the construction of a laboratory still in a rather indefinite form. I could give a list of names of people recruited in that period who later contributed remarkably to the development of CERN. I shall, however, recall a single case. In December 1952, I was in London for conversations with the British authorities and for recruiting qualified staff. Cockcroft invited me to lunch at the Seville Club and introduced to me a young man who had distinguished himself as an engineer at Harwell. The young man was John Adams, who later joined CERN and devoted his whole life to its development.

Notes

1 E. Amaldi, "CERN, the European Council for Nuclear Research," lecture given 2 August 1954, pp. 339–54 of the proceedings of the International School of Physics, held in Varenna, 18 July to 7 August 1954, *Nuovo Cimento (Suppl.) 2* (1955), 2–469. Similar lectures had been given by Amaldi on 28 April 1954, at the National Academy of Sciences in Washington, D.C., and on 9 July 1954, at the Royal Institution in London; L. Kowarski, "An Account of the Origin and Beginning of CERN," CERN 61–10, 10 April 1961; E. Amaldi, "CERN, Past and Future," in *Topical Conference on High Energy Collisions of Hadrons*, edited by L. Van Hove (Geneva: CERN, 1968), p. 415; E. Amaldi, "First International Collaborations between Western European Countries after World War II," in *History of Twentieth Century Physics*, edited by C. Weiner (New York: Academic Press, 1977), pp. 326–51; L. Kowarski, "New Forms of Organization in Physical Research after 1945," in Weiner, ibid., pp. 370–401.

2 Belgium, Denmark, Federal Republic of Germany, France, Greece, Italy, the Netherlands, Norway, Sweden, Switzerland, and Yugoslavia.

3 The countries listed in note 2, plus the United Kingdom.

4 Amaldi (1977), note 1.

5 C. G. Darwin, professor of theoretical physics at the University of Cambridge, is well known for many important contributions, including his treatment of the anomalous Zeeman effect by means of the Dirac equation, and in general the study of solutions of the Dirac equation in the case of low velocities, so that the four-component Dirac spinors can be replaced by two-component spinors of which one is large, the other small. For a biography of Darwin, see Sir G. P. Thomson, "C. G. Darwin (1887–1962)," *Biographical Memoirs of the Royal Society 9* (1963), 69–85.

6 My letter to Auger of 3 October 1950 is given in Note 10 of Amaldi (1977), note 1. In the *Proceedings of the Harwell Nucleon Physics Conference, September 1950*, edited by E. W. Titterton, AERE 9/M/68, P. Auger, B. Ferretti, and L. Kowarski appear as members of the

conference, but there is no mention of Auger's speech, which, according to Ferretti, found enthusiastic support from many, particularly from the young physicists; see Amaldi (1977), note 1. However, Kowarski had quite a different recollection.

7 Ernest D. Courant, M. Stanley Livingston, and Hartland S. Snyder, "The Strong-Focusing Synchrotron – A New High Energy Accelerator," *Phys. Rev. 88* (1952), 1190–6. The strong-focusing principle had been discovered by the Greek engineer Christofilos in 1950. In the same year, he had applied for a U.S. patent, but his application was not noticed by the experts in the field and therefore remained ignored.

8 Minutes of the Third Session, Amsterdam, 4–7 October 1952, CERN/GEN/4, Rome, 15 February 1953. The meeting took place under the chairmanship of P. Scherrer.

9 The dedication ceremony for the Cosmotron took place on 15 December 1952. On that evening, I was asked to give a dinner speech on what was going on in Europe about the creation of CERN. An international conference on high-energy accelerators took place at Brookhaven 16–17 December and was followed by a high-energy physics conference that was held in Rochester (actually the third annual Rochester conference) 18–20 December 1952. At the banquet (the evening of the 19th) I gave another dinner speech on CERN.

10 Before becoming an eminent personality in public life, Lord Cherwell was well known as a physicist under the name of F. A. Lindemann. Among other contributions, his work on the theory of melting is now classical. For a biography of Lord Cherwell, see George P. Thomson, "Frederich Alexander Lindemann, Viscount Cherwell (1886–1957)," *Biographic Memoirs of the Royal Society 4* (1958), 45–71.

11 Two international expeditions to Sardinia for launching balloons carrying packages of nuclear emulsions for studying high-energy events produced by cosmic rays at high altitude had been made already, in 1951 and 1952, under the sponsorship of CERN. For more detail, see Amaldi (1977), note 1.

12 Final report to member states on the progress of activities of the European Council for Nuclear Research up to 29 September 1954 (end of the interim period), CERN/GEN/15, Rome, 30 June 1955.

36   Arguments pro and contra the European laboratory in the participating countries

ARMIN HERMANN

Born 1933, Vernon, B.C., Canada; Ph.D., Munich University, 1963; history
of science and technology; Stuttgart University, Historisches Institut,
Lehrstuhl für Geschichte der Naturwissenschaften und Technik

Although some of you already know about the CERN history project, please allow me a few words of description. Like the object of our study,
the history team is supranational: John Krige from the United Kingdom,
Ulrike Mersits from Austria, Dominique Pestre from France, and myself
from Germany. The funding comes from science foundations in several
European countries. CERN is not financing the project, but provides offices
and other help.

Our results are being issued in a series of reports called *Studies in CERN
History*, to stimulate criticism and discussion before the final manuscript
is produced. The prehistory, up to the official foundation of CERN on 29
September 1954, is nearly finished and will be published as the first of two
volumes.* One part is an analysis of the emergence of the idea to create a
European laboratory from scientific needs and the desire for European unity.
The two principal champions who struggled for the laboratory were Pierre
Auger and Edoardo Amaldi. Amaldi's account at this symposium is included
here as Chapter 35.

The present chapter is based on an analysis of the decision-making process
in four of the participating countries – the studies by Krige for the United
Kingdom, Pestre for France, Lanfranco Belloni for Italy, and Hermann for
Germany.[1] The work is based (as are all the studies of the team) on source
material in many governmental and private archives, as well as in the CERN

---

*   *Ed. note*: The first volume of A. Hermann, J. Krige, U. Mersits, and D. Pestre,
    History of CERN, *Launching the European Organization for Nuclear Research*, was
    published in 1987 by North Holland. A second volume covering the years 1954–65 is to
    appear in 1989. A concise history addressed to a wider audience is also planned.

archives, and will form part of the first volume of CERN's history. I shall review the main arguments used in the four countries for and against joining the European laboratory, and we shall see that these arguments depended on the political conditions and the state of science in those countries.

Before doing so, a look back. Vis-à-vis their governments, the project's pioneers were in a situation that had been typical for the scientist for centuries. In the time of Sir Isaac Newton, scholars constantly appealed to their sovereigns for money. What is granted to honest and reasonable people who attain perfection in science, Gottfried Leibniz said, is to please God. The scholars promised to gain insight into the secrets of God's creation and, by applying nature's laws to machines and manufacture, to improve the living conditions of humankind. However, demonstrating the utility of physics turned out not to be so easy. Frederick the Great joked as follows:

> The English have built ships with the best profile according to Newton's opinion, but their admirals assured me that they did not sail as well as those built according to experience. I wished to have a fountain for my court-garden: Leonhard Euler calculated the wheels of the device, but it could not pump one drop of water.[2]

Only in the nineteenth century did the practical value of science become clear to everyone. The chemist Justus von Liebig argued in 1840 that in granting support, the government must favor the disciplines from which an impact on industry can be expected.[3] That is exactly what happened. Nevertheless, the demands always exceeded the funding. At the end of the eighteenth century, physicists wanted support for apparatus and a single room in the university to store it; around 1850, they explained to the government the need for·their own building, still modest in size; and around the turn of the twentieth century they had big research institutes in mind. As Liebig had suggested, physics and chemistry came to be favored over philosophy and classics. The economic benefit was decisive. In addition, the national prestige and defense value of science began to play major roles.

By the time we stepped into the "atomic age" at the end of World War II, the important role of science had become completely obvious. In his inaugural address, President Harry S. Truman said that "greater production is the key to prosperity and peace. And the key to greater production is a wider and more vigorous application of modern scientific and technological knowledge."[4] In a document of the Quai d'Orsay of 1950, we read that "scientific research is a prime necessity of all modern countries."[5]

The arguments used for the European laboratory fell into two categories: scientific and political. The scientific ones, which focused on an explanation of the benefits of the new installation for science, had a political context as well, because the national value of science, especially nuclear physics, was taken for granted. I shall first discuss the scientific arguments in the different countries.

In the United Kingdom, whose weight in nuclear physics was at least equal to that of all the continental countries combined, the consensus on a big new machine was not favorable.[6] "We have already devoted more effort than was wise on these big machines," George P. Thomson said; "I feel it is time to stop engineering and do some physics."[7] In addition to this main argument against British participation, a certain skepticism of the technological abilities of the continental people played a role. "The idea," wrote P. M. S. Blackett, "that in three years' time, if we voted the money, we could have a better machine than the Americans is too crazy to be taken seriously."[8] Nevertheless, the members of the Nuclear Physics Committee believed the United Kingdom should cooperate, if not as a member, at least in some way, with the European laboratory. International cooperation was one of the best traditions in science.

In that connection, the British thought that the 400-MeV Liverpool synchrocyclotron could be made available. In discussions about that offer, H. W. B. Skinner was surprised to meet not dubious UNESCO personnel but "competent people" like Niels Bohr and Paul Scherrer, from whom he was willing "to accept advice."[9] Although British scientists still opposed construction of the "big machine," at the end of 1951 they proposed to their government that the country join. The British government, however, decided otherwise on the basis of a political argument, to which we shall come.

The conditions in France, Italy, and Germany for scientific work in nuclear physics were poor compared with those in the United Kingdom. The scientific argument was essentially the same in these three countries: Without accelerators, no research. A French document explains that serious research would be possible in the near future only at the installations of the Commissariat à l'Energie Atomique, not at the Collège de France and the universities.[10] Facilities in Italy and Germany were even worse. So, on the Continent, the conclusion was as follows: If working conditions remain unchanged, every nuclear physicist is condemned to inefficiency, unable to stimulate younger people, and thus inclined to emigrate. Because no single country could afford the expensive research on its own, the only solution was cooperation. The main push to European cooperation was the financial pressure, as W. Heisenberg summarized.[11]

Scientific collaboration of European physicists in France, Italy, and Germany was also highly desirable from the political standpoint. The European laboratory was just in line with French foreign policy, as François de Rose explained.[12] I now move on to the political arguments.

The main political argument

In France, Italy, and Germany, the idea of a united Europe was given high priority, and it was thought that the European laboratory project could serve as a model for European cooperation. France was the principal champion of European unity, the Quai d'Orsay states; it followed that she must play a preeminent role in the realization of this laboratory.[13] The first aim of German foreign policy was to again become a member of the European family of nations. It was stated explicitly at the Deutsche Forschungsgemeinschaft that "Herr [Otto] Haxel does not regard physics as the main value for a German participation," and, quoting further, "Heisenberg believes that one has to judge about 80% of the cost from the point of view of European collaboration."[14] Such physicists were convinced that science was the only remaining credit that the country had. "Today, research is the only possible way to make foreign policy," the Research Council explained in an aide-mémoire for Konrad Adenauer. For Heisenberg, the project offered "a realistic and inspiring starting point for European collaboration."[15]

In the United Kingdom, however, things were different: Hardly anyone felt inspired by a United Europe, the great ideal and hope for young people in France, Italy, and Germany. The Labour Party was hostile to all European initiatives. Even Sir Winston Churchill, the founder of the European Movement, regarded Europe as a matter only of the continental countries. Thus, the planned laboratory project did not fit into the lines of British policy; its European character was but one aspect. Everyone in Britain was convinced that it was much better to remain independent and to collaborate with the Commonwealth and the United States. In February 1952, the Cabinet Steering Committee announced that the standing policy of Her Majesty's government is "against joining or promoting international organizations unless they were shown to be absolutely necessary."[16] The United Kingdom remained in the position of an observer until the organization was officially founded in September 1954.

Other political arguments

The European character of the planned laboratory was not the only political argument. Three other arguments, already used for generations, were made: that technical applications might one day arise from the developing new field; that the project would promote national influence and prestige; and that a developed science would guarantee a high level in war technology.

With respect to technical applications, the argument was simple, and nearly the same in every country. In those years, the field was regarded as a part of nuclear physics, and there were no doubts about the economic value of nuclear physics.

The national prestige argument was more complex. That a flourishing science yields intellectual influence had long been realized. Adolf von Harnack complained in 1909, in his famous aide-mémoire for William II, that the French "Instituts Pasteur" had spread French national culture and influence in the world, and to continue this tendency was a reason to create, under the name of the Kaiser Wilhelm Society, research institutes for all branches of science.[17] In the early fifties it was again explicitly argued in France that the

CERN project would prevent French science from suffering further damage, which would be disastrous for the country's intellectual independence. If the laboratory were established on French soil, use of the French language would be promoted, and the intellectual output of the country stimulated. The project had been launched primarily by Frenchmen, and so the Quai d'Orsay wished to preserve a certain degree of French influence. (Communists later lamented that France could decide only on the length and width of doors and windows.[18])

The French connection was not liked much across the Channel. "If the French want to have a nuclear physics research laboratory," Skinner said, "why don't they go ahead with the cooperation of any other nation interested....We in this country should have nothing to do with the scheme."[19] Italians and Germans, aware of the weak political position of their countries, were more inclined to make concessions. But for Heisenberg it also seemed important "to gain influence," and he thought that at least one of the eight top positions should be occupied by a German.

Finally, there are abundant examples in the documents of the old argument that a developed science guarantees a high level in war technology. This argument was used by Gustavo Colonnetti, the president of the Consiglio Nazionale delle Ricerche (CNR). Italy had suffered terrible damage during the last two wars because it had failed to mobilize scientists and engineers, Colonnetti wrote to Alcide de Gaspari: "This mobilization is now a demand for national defense." Such an argument had been used in France after the Franco-Prussian War, and in Germany after 1918.

Of course, Colonnetti did not have in mind that the project itself should serve military purposes, but he was convinced that in a modern industrialized state, a high level of science and technology is a prerequisite for military power.[20] As is well known, of paramount importance for the champions of the project, especially for Amaldi, was that the laboratory be devoted only to pure fundamental research, without any military concern; that was stated explicitly in the CERN convention.[21] Nevertheless, it is obvious that in our "atomic age" it is impossible to separate from physics what Wolfgang Pauli called "die böse Hinterseite der Physik," the unfortunate backside of physics.

My concluding remarks are on an old subject: "Knowledge is power." The ambition to rule nature had been, right from the beginning of Galileo's "Nuovo Scienza," an integral part of physics. Even adherents of pure science cannot totally deny this motivation, as Pauli suggested.[22] In May 1950, the U.S. Department of State published a document on "Science and Foreign Relations" in which it was stated that "the security of the United States, and the security of free peoples everywhere...is dependent on a national policy which adequately reflects the potentialities of science. Developed intellectual capacity arising anywhere has the potential power of producing fundamental creative ideas that bear on our national welfare and security."[23] After thirty-five years, such "ideas that bear on national security" have indeed emerged from particle physics. We all sincerely hope that in another thirty-five years, a complete and definitive history of high-energy physics can be written, describing all the wonderful discoveries that the famous physicists among us have made, and that nothing else need be described – that will be the complete story.

Notes

1 A. Hermann, J. Krige, U. Mersits, and D. Pestre, *History of CERN, Launching the European Organization for Nuclear Research*, Vol. 1 (Amsterdam: 1987). We quote, in this paper, from the preliminary reports *Studies in CERN History*, especially John Krige, "Launching the European Laboratory Project," in *Studies in CERN History 4* (hereafter denoted as *CHS-4*, etc.) (Geneva: CERN); "Britain's Physicists Respond," *CHS-6*; "The Change in Policy of British Physicists," *CHS-11*; Dominique Pestre, "Les attitudes françaises," *CHS-15*; Lanfranco Belloni, "The Italian Scenario," *CHS-8*; Armin Hermann, "The Role of Germany in the Foundation of CERN," *CHS-5*.

2 *Briefwechsel Friedrichs des Grossen mit Voltaire*, edited by R. and H. Droysen, Part 3 (Leipzig: 1911), p. 427.

3 Justus von Liebig, *Reden und Abhandlungen* (reprint Wiesbaden: 1965), p. 9.

4 Quoted from "Science and Foreign Relations," Department of State publication 3860, released May 1950, p. 21.

5 "La recherche scientifique apparait dans tous les etats modernes comme une imperieuse necessité." Cf. *CHS-15*, p. 19.

6 Note du Quai d'Orsay pour l'Ambassadeur de France à Londres, 4 April 1953.

7 *CHS-4*, p. 7.

8 *CHS-6*, p. 5.

9 *CHS-11*, p. 43.

10 Le president du Conseil des Ministres au Ministre des Affaires Etrangères, 20 November 1950. Cf. *CHS-15*, p. 19.

11 Aide-mémoire of the Deutsche Forschungsgemeinschaft, 11 March 1953.

12 Project de lettre au Ministre du Budget, 8.12.1951. Cf. *CHS-15*, pp. 26f.

13 The author is François de Rose. *CHS-15*, p. 27.

14 *CHS-5*, p. 4.

15 Aide-mémoire of the Deutsche Forschungsgemeinschaft, 11 March 1953.

16 *CHS-11*, p. 50.

17 Denkschrift von Harnack an den Kaiser, 21 November 1909, in *50 Jahre Kaiser-Wilhelm-Gesellschaft* (Göttingen: 1961), p. 85.

18 *CHS-15*, p. 46.

19 *CHS-6*, p. 8.

20 *CHS-8*, p. 24.

21 Convention, article 2, paragraph 1. Cf. *CERN 61–10*, Annex IV.

22 Wolfgang Pauli, "Die Wissenschaft und das abendlandische Denken," *Aufsätze und Vortrage über Physik und Erkenntnistheorie* (Braunschweig: 1961), p. 108.

23 Department of State, Science and Foreign Relations, Department of State publication 3860, released May 1950, p. 21. It is of interest to note how the Communists reacted to the ideas developed here, "Un plan U.S.A. de mainmise sur la science" (Paris: 1953).

37 Physics and the excellences of the life it brings

ABDUS SALAM

Born 1926, Jhang, Pakistan; Ph.D., Cambridge University, 1952; theoretical physics; Nobel Prize, 1979, for the electroweak theory and the weak neutral current; International Centre for Theoretical Physics, Trieste, and Imperial College, London

The title of this essay is taken from Robert Oppenheimer. He had three types of excellence in mind: first, for the theoretician, the excellence of new ideas while he attempts to read Allah's design; for the experimenter, the excellence of new discoveries, the sheer pleasure of the search of carrying an experimental technique to its limits and beyond; and, finally, of the desire to spite the theorist. Oppenheimer had these forms of excellence in mind, but also much more: He emphasized the opportunity physics affords to come to know internationally a class of great human beings whom one respects not only for their intellectual eminence but also for their personal human qualities – a reflection of their greatness in physics. In addition, he had in mind the opportunities that physics uniquely affords for involvement with humankind – in the parlance of today, in engaging in problems of development and of enhancing the human ideal.

I wish to speak here on some aspects of Oppenheimer's thoughts from a personal point of view. I shall illustrate these by recalling my induction into research on the renormalization of meson theories and the excellent men I was privileged to meet while pursuing this research. I also wish to speak on excellence through world development. In particular, I want to speak about the International Centre for Theoretical Physics, whose creation, under the auspices of the United Nations, I was privileged to suggest in September 1960. The Centre came into being only in October 1964 – beyond the cutoff date of this symposium's coverage. However, the ideas that went into the Centre's creation, and the political battles that had to be won, as well as the physics milieu of the early sixties, with its desire to keep alive the internationalism of the subject, with its emphasis on science rather than on technology, and with its perception of brain drain of high-level talent, particularly to the United States, do fall within the period covered.

The notion of a center that would cater particularly to the needs of physicists from developing countries had lived with me from 1954, when I was forced to leave my own country because I realized that if I stayed there much longer I would have to leave physics, because of sheer isolation. At the September 1960 Rochester conference, in his banquet speech, John McCone, then chairman of the U.S. Atomic Energy Commission, made a reference to the desirability of creating international centers in physics. He had principally in mind accelerator establishments, which might be created under joint U.S.–USSR–European auspices. After the banquet, over coffee, I remember a conversation with Hans Bethe, Robert Sachs, and Nicholas Kemmer in the beautiful hall of the women's residence at Rochester University. We discussed the practical possibility of creating such centers and came to the conclusion that the simplest approach would be to think in terms of an international theoretical center.

That same month, I had the privilege of being able to voice, on behalf of the government of Pakistan, this visionary ideal in the form of a resolution at the annual conference of the International Atomic Energy Agency (IAEA) at Vienna. We were fortunate to receive cosponsorship of the resolution from the governments of Afghanistan, the Federal Republic of Germany, Iran, Iraq, Japan, the Philippines, Portugal, Thailand, and Turkey. As the list of sponsors indicates, the setting up of such a center was of interest not only to the less privileged countries but also to some of the developed countries. The hope was that a center of this type, besides providing an avenue for collaborative international research for the East and the West, might also help in resolving one of the most frustrating problems that active scientists in poorer countries face – the problem of isolation. Such scientists, supported by international funds, would come fairly frequently to the center to renew their contacts and engage in active research in their fields.

Right from the beginning, we received enthusiastic support from the world physics community. Niels Bohr, before his death, expressed his wholehearted support. Scientific panels, convened in 1961 and again in 1963 by the IAEA's physicist director general, Dr. Sigvard Eklund, forcefully recommended the creation of the International Centre. The members of the 1961 panel were Aage Bohr, Paolo Budini, Bernard Feld, Leopold Infeld, Maurice Levy, and Walter Thirring; in 1963, the "three wise men panel" included Robert Marshak, Leon Van Hove, and Jayme Tiomno.

Unfortunately, there was not the same unanimous response from the atomic energy commissions around the world. At the 1962 annual conference of the IAEA (where these commissions represent their governments), even though the creation of a center was accepted in principle after a divided vote (by and large the industrialized countries voted against, and the developing voted for, the Centre) the IAEA's board of governors voted the princely sum of $55,000 to set up an International Centre for Theoretical Physics. The United Nations Educational, Scientific and Cultural Organization

(UNESCO) voted $27,000. Thus, additional offers of financial assistance from interested member states had to be solicited. Of the five offers received (from the government of Italy for a center to be located in Trieste, from Austria for Vienna, from Denmark for Copenhagen, from Pakistan for Lahore, and from Turkey for Ankara), the most generous was the Italian government's offer of around $300,000 plus a prestigious building, with Budini, professor of physics at the University of Trieste, as the moving spirit behind it. This last offer was accepted in June 1963, and the Centre started functioning on 1 October 1964, with a four-year charter. Oppenheimer served on the Centre's first scientific council. He came to Trieste in spite of his terminal illness and there helped to draft the Centre's charter. One admired him and his felicity of phrase – even in such legal drafting. Other members of the first Scientific Council were Aage Bohr, A. Matveyev, V. G. Soloviev, Sandoval Vallarta, and Victor Weisskopf. Dr. Alexander Sanielevici, from Romania, was one of the scientific secretaries.

I shall speak little about the Centre and its functioning, for surely this will be one of the subjects covered by the future conferences in this series. In 1964, when we assembled in Trieste in a rented building, the whole enterprise seemed like a dream. Once again the world's theoretical community rallied around us – plasma physicists as well as particle physicists. We cared not for frills, only for physics. Our goal was to acquire scientific visibility. In this we succeeded. Thus, one year after the Centre's inception, Oppenheimer could comment to the Centre's council:

> It seems to me that the Centre has been successful in these eight or nine months of operation in three important ways. It has cultivated and produced admirable theoretical physics, making it one of the great foci for the development of fundamental understanding of the nature of matter. The Centre has obviously encouraged, stimulated and helped talented visitors from developing countries who, after rather long periods of silence, have begun to write and publish during their visit to the Centre in Trieste. This is true of physicists whom I know from Latin America, from the Middle East, from Eastern Europe and from Asia. It is doubtless true of others. The Centre has become a focus for the most fruitful and serious collaboration between experts from the United States and those from the Soviet Union on the fundamental problems of the instability of plasmas, and of means for controlling it. Without the Centre in Trieste, it seems to me doubtful that this collaboration would have been initiated or continued. In all the work at the Centre of which I know, very high standards prevail. In less than a year it has become one of the leading institutions in an important, difficult and fundamental field.[1]

To continue the story briefly, in the twenty years of its existence, the Centre has flourished, with physicists from 100 countries, East and West, North and South, ranging over all disciplines of physics – from fundamental physics to physics on the interface of technology, environment, energy, the living state, and applicable mathematics. The Centre welcomes around 1,000 physicists from industrialized countries and 1,000 physicists from developing countries every year for research courses, workshops, and meetings and for conducting research for periods ranging from a few months to a few years. In addition, from a generous grant from the Italian government, we provide 100 fellowships in experimental physics, tenable at Italian laboratories. We are federated with approximately 100 institutes, mostly in developing countries. In addition, our scientific council selects 200 physicists (whom we call associates of the Centre) and awards them the privilege of coming to the Centre three times in six years, for periods up to three months per visit, at times of their own choosing, provided they are living and working in developing countries. The Centre's current budget is of the order of $5 million: $3 million comes from the government of Italy, $1 million from IAEA, $0.5 million from UNESCO, and the rest from other government agencies. The U.S. Department of Energy gives us a special grant of $50,000 for U.S. physicists.

Although in the founding and running of the Centre we have depended on the volunteer help of the world's leading physicists, it remains a sad fact that the physics communities of the developed countries have, by and large, rendered little assistance in an organized form to the cause of physics in developing countries, including the Centre. I wish to stress the word "organized," lest I fail to pay a heartfelt tribute to the continued work of great individuals – men like Marshak – who have made real sacrifices in this cause. There is no question but that the real amelioration of the worsening situation for physics research in developing countries lies within the countries themselves, and the role of the Centre and any other outside agency can only be to help generate self-reliant communities. But outside help, particularly if it is organized help, can make a crucial difference. This can take various forms. For example, the physical societies could help by donating 200–300 copies of their journals to the deserving institutions and individuals and by waiving publication charges. The American Physical Society, in fact, does provide its publications at half cost to thirty-one physicists from thirteen least developed countries. The International Union of Pure and Applied Physics (IUPAP) has been helping the Centre defray postage costs for distribution of old runs of journals donated by generous individuals. These schemes should, however, be extended by other societies and laboratories to also cover equipment, and in fact CERN has recently signified its willingness to donate some of its used equipment to laboratories in developing countries. Most important, the research laboratories and the university departments in developed countries could finance visits of their staffs to the institutions in developing countries in an organized manner and, reciprocally, by creating schemes like the three-month associateship scheme we run at the Centre, at the least for their alumni now working at laboratories in the developing countries. Leon Lederman has initiated a scheme at Fermilab whereby a number of Latin American experimental physicists are regularly brought over to be trained in techniques of particle physics and ancillary disciplines. And then there are the excellent cooperative schemes of training, like the one T. D. Lee runs for China. These could perhaps be extended to other developing countries.

May I be forgiven for thinking in the following terms: that the physics institutions in developed countries may consider contributing in their own ways, according to the norms of the well-known United Nations formula, whereby most developed countries have pledged to spend 1 percent of their gross national product (GNP) resources for world development. In the end, it is a moral issue whether the better-off segments of the physics community are willing to look after their own deserving but deprived colleagues, not only helping them materially to remain good physicists but also joining them in their battle to obtain recognition within their own communities, as valid professionals who are important to the development both of their countries and of the world. So much for the excellence of a life of physics for realizing the ideals of development.

I would now like to turn to the second aspect of Oppenheimer's thought, to some of the excellent and humanly great physicists I came to know internationally in the early part of my research. It occurs to me that during the period covered by this symposium (1950–64) there have been five major developments in theory: first, the rise and the fall of Yukawa's standard model of pions and nucleons. Connected with this was the rise (and the later fall) of the S-matrix theory. The second major development was the understanding of the role of flavor symmetries, in particular, of flavor SU(3). The third development concerned the emergence of chirality; the fourth, the Nambu–Goldstone spontaneous symmetry-breaking phenomena; and the fifth, the Yang–Mills–Shaw gauge theory and its application to electroweak unification.

I have told the story (at least of my humble part in it) with respect to the last three developments, the rise of chiral symmetry, of spontaneous symmetry breaking, and of the electroweak unification, in the Stockholm lecture of 1979, including the story of interactions with Wolfgang Pauli, Rudolph E. Peierls, John C. Ward, Steven Weinberg, Sheldon Glashow, and others.[2] I shall not repeat this, except to say that I take legitimate pride that the Yang–Mills theory and the flavor eightfold way were independently invented by two of my illustrious pupils within the ethos of my research groups at Cambridge and London.

Here, I shall concentrate mainly on the story of the short-lived rise of the pion–nucleon theory as the standard model of 1950–1, in consequence of the proof that this was the only theory that could be renormalized then. The people concerned with my story were P. A. M. Dirac, Kemmer, and Paul Matthews at Cambridge, besides Freeman Dyson, who was visiting Birmingham, and Ward at Oxford.

The immediate postwar generation – our generation – was brought up to believe implicitly in the Yukawa model of the nuclear forces. The only open question at that time concerned the spin of the meson and the precise form of the nucleon–meson interaction. After Hideki Yukawa, Kemmer had made the most crucial contributions toward defining this problem, at least outside Japan. In a classic paper written at Imperial College, London, in 1938, he had classified the Yukawa interactions according to meson spins and parities, and whether they were direct or derivative couplings.[3]

When I started research in October 1949, Kemmer was at Cambridge. Surprising though it may seem, I had started out as an experimental research student in the Cavendish, with orders to scatter tritium against deuterium for Samuel Devons, now a professor at Columbia. Finding myself as an experimenter was in accordance with the revered Cambridge tradition, handed down from Ernest Rutherford's days: Those who fared well in the physics tripos became experimentalists; those who got third classes were consigned to theoretical research. Soon after starting, I knew the craft of experimental physics was beyond me; I sadly lacked the sublime quality of patience, particularly with the recalcitrant equipment of the Cavendish. Reluctantly, I turned my papers in, and started instead on quantum field theory, with Kemmer, in Dirac's exciting department.

But theory research was not that easy. Those were the great days of renormalization theory, with the papers of Sin-itiro Tomonaga, Julian Schwinger, Richard Feynman, and Dyson providing feverish excitement. At Cambridge, Kemmer was the only senior person interested in these developments. He had behind him not only the kudos of having tabulated all possible meson interactions but also the reputation of being a prince among men, of generosity to a fault to his students. So I went to Kemmer and asked that he accept me for research. He said he had eight research students already and could not take any more, and he suggested that I go to Birmingham to work with Peierls. But I could not bear to leave Cambridge – principally because of the beauty of the rose gardens behind my college, St. John's. (Incidentally, Dirac was also at St. John's College.) I asked Kemmer: "Would you mind if I worked with you peripherally for the time being?" He graciously assented. In my first interview with Kemmer, he said that "all theoretical problems in quantum electrodynamics have already been solved by Schwinger, Feynman and Dyson. Matthews has applied their methods to renormalize meson theories. He is finishing his Ph. D. this year. Ask him if he has any problem left."

This was early 1950. So I went to Matthews and asked him what he was working on and if he had any crumbs left. The first piece of advice Matthews gave me was to forget the papers of Schwinger and Feynman and to concentrate on Dyson's two classic papers, particularly his most recent, in 1949, where he had shown that quantum electrodynamics was renormalizable to all orders in α.[4] He told me he had spent one and a half years, already, trying to renormalize meson theories. He had found that only spin zero might work. He was writing up his one-loop calculations for his Ph. D. thesis and had shown that the theory of spin-zero mesons was indeed renormalizable up to the second order.

Matthews had at that time already tabulated which theories might possibly be renormalizable with the techniques then known. He had come to the conclusion that no derivative coupling meson theory could be renormalized at all, and that among the direct coupling theories with nucleons, the only hopefuls were either the spin-zero or the neutral-vector meson theories with conserved currents for nucleons. No charged-vector theory (with massive mesons) could be renormalizable. He had also shown that the neutral-vector meson theory with mass was a replica of electrodynamics, and one could take over the work of Dyson more or less intact and show its renormalizability. Regarding the spin-zero theories, he had shown that one would, at the least, need an additional $\lambda\phi^4$ term, where ϕ is the meson field. The corresponding term for electrodynamics $(e^4 A^4)$ was gauge-variant, as had been remarked on by Dyson, with Ward actually proving that the corresponding infinity did not exist.

The ϕ^4 term for spin-zero mesons would, however, be a new fundamental interaction term, with a new fundamental constant λ. A new fundamental constant simply appeared too radical those days, and we agonized over this. But the real question was this: Could one be sure that even with this new interaction term, all the infinities could be assimilated to a renormalization of the meson-nucleon coupling constant, the new constant λ, the masses of the mesons and the nucleons, plus a renormalization of their wave functions? Matthews had worked with one-loop diagrams and shown that renormalizability appeared possible. He could not go beyond one loop because overlapping infinities started to come in for higher loops, and one had to solve this basic problem before progress could be made. This was the situation around March 1950.

Matthews had his Ph. D. exam shortly afterward. His external examiner was Dyson, who was visiting Birmingham at that time. Dyson used to spend a few months at Birmingham, and the rest of the year in the United States. In the exam, Dyson had asked Matthews about overlapping loop infinities: "Have you come across these infinities? And if so, how do you resolve the problems posed by these?" Matthews had replied: "You have claimed in your paper on quantum electrodynamics that these infinities, which occur in the self-energy graphs, can be properly taken care of. I am simply following you." No further question on these infinities was asked; both Dyson and Matthews kept silent after this brief exchange.

Now, overlapping infinities had indeed appeared in quantum electrodynamics (QED), where a general self-energy graph can be viewed as an insertion of a modified vertex at either end of the lowest-order self-energy graph. Insertions of modified vertices at both ends would be tantamount to double counting. But Dyson, in his paper, while discussing these, had recommended precisely this – that one should subtract the vertex-part sub-infinities twice before subtracting the final overall self-energy infinity. Dyson must be right, but why? What made life awkward was that whereas this troublesome overlap occurred only for self-energies in QED, for meson theories the overlaps of the infinities were everywhere.

With characteristic generosity, of which I became a lifelong recipient, Matthews said to me, "My exam is over. After my degree, I'm going off, to take a few months holiday. And then I'll go to Princeton. You can have this problem of renormalizing meson theories till I get back to work in the fall. And if you don't solve it by then, I'll take it back."

That was the sort of gentleman's agreement we parted on. So I had to get to the bottom of the overlapping infinity problem. I thought that the best thing for me would be to ask Dyson's direct help. So I rang him. I said: "I am a beginning research student; I would like to talk with you. I am trying to renormalize meson theories, and there is this problem of overlapping divergences which you have solved. Could you give me some time?" He said, "I am afraid I am leaving tomorrow for the United States. If you wish to talk, you must come tonight to Birmingham." So I traveled from Cambridge to Birmingham that evening. Richard Dalitz and his gracious wife put me up for the night.

Next morning, Dyson came to the department. This was the first time I met him. I said, "What is your solution to the overlapping infinity problem?" Dyson said, "But, I have no solution. I only made a conjecture." For a young student who had just started on research, this was a terrible shock. Dyson was our hero. His papers were classics. For him to say that he had only made a conjecture made me feel that my support of certainty in the subject was slipping away. But he was being characteristically modest about his own work. He explained to me what the basis of his conjecture was. What he told me was enough to build on and show that he was absolutely right. I traveled with him to London that afternoon. He was due to catch his boat from Southampton later that day. I think it was during that train journey, in conversation with Dyson, that I appreciated for the first time how weak the weak forces really are.

At Cambridge, amid the summer roses behind the colleges, I went back to the overlapping infinity problem, to keep the tryst with Matthews's dateline. Using a generalization of Dyson's remarks, I was able to show that the spin-zero meson theories were indeed renormalizable to all orders. At that time, transatlantic phone calls for physics research had not yet been invented. So I had a vigorous correspondence with Dyson, with the fullest participation of Kemmer, my supervisor. Exciting days indeed!

The subtraction procedure that I designed worked in momentum space. A crucial element of the proof was to associate with a given graph a set of integration variables in momentum space such that for the entire graph or for any of the subgraphs contained in it, every possible infinity could be associated on a one-to-one basis with a single subintegration. Assuming that this was possible, the subtraction procedure left behind an absolutely convergent remainder, absolutely convergent in the mathematical sense. To prove this one-to-one relationship, one had to consider the topology of the graphs. I could show, with Res Jost's help, that this result certainly holds for the so-called renormalizable theories. I have always felt very proud of this particular part of the proof,[5] but to my knowledge, the paper embodying this has never been referred to by anybody. I can only assume that the result has been taken on trust and that no one has ever rechecked it.

Contemporaneous with my work was the work at Oxford of Ward, who devised a most ingenious scheme of regularization. This depended on differentiation with respect to external momenta, the technique used later by Murray Gell-Mann and Francis Low in their beautiful work on the renormalization group.[6] Later still, other regularization schemes were devised in x space, notably by Klaus Hepp, E. R. Speer, N. N. Bogoliubov, and O. S. Parasiuk.[7] My procedure, however, was a straightforward subtraction in momentum space; the use of the technique I had devised would make the final integrals absolutely convergent, and it would also permit a count of the wave-function renormalization (Z) factors correctly in all conceivable situations.[8] Matthews and I wrote a brief review of these developments for *Reviews of Modern Physics*, October 1951,[9] in which we stated the following criterion for acceptability of a proof in this subject: "The difficulty...is to find a notation which is both concise and intelligible to at least two people, of whom one may be the author." We left it unsaid that the other person may be the coauthor.

I can here tell a story about this work being considered deep and being believed in, but seldom read. I was invited to the Institute for Advanced Study at Princeton in January 1951. I had, by then, applied my technique to renormalize spin-zero electrodynamics. I took a manuscript copy of the new paper to Oppenheimer to read and, if he approved, to send to the *Physical Review*. I then realized that I had given him a copy with no diagrams in it. So I went to his office to retrieve the manuscript. I had to wait for some while, because he had visitors, but then he came out of his inner office, saw me and said, "I enjoyed reading your paper. It is a fine paper." I should have kept quiet, but like a fool I said, "I am sorry, I gave you a copy in which there were no diagrams. I don't think you could have understood it." Oppenheimer visibly changed color. But he only said, "The results are surely true and also intelligible even without diagrams."

This proof of the renormalizability of spin-zero meson–nucleon direct-coupling theory had come at an opportune time. With the discovery by Cecil Powell of the pion and the subsequent determination of its spin as zero, theory and experiment seemed to converge to a definitive standard model of nucleons and pseudoscalar pions, with a direct Yukawa plus the Matthews interaction. Our elation, however, was short-lived. The Yukawa coupling, which nature seemed to favor, was not the direct renormalizable pseudoscalar coupling, but the unrenormalizable pseudovector coupling. The two couplings were, of course, equivalent in the lowest order, but with the large coupling parameter $g^2/4\pi \approx 14$, was order-by-order perturbation of any practical significance?

Then came the discovery of the $\Delta(3/2, 3/2)$ resonance, plus the discovery of the form factor for the nucleon by Robert Hofstadter. These coups de grace finally killed the model. Influential in our thinking was also the paper of Enrico Fermi and C. N. Yang that questioned heretically whether the pion was a fundamental entity or merely a nucleon–antinucleon composite.[10]

For me personally, the disenchantment with the pion–nucleon theory had started much earlier. One of the postwar texts on nuclear physics was Leon Rosenfeld's, which I believe he wrote in a war cellar during 1944–5 in Belgium.[11] This was a 600-page book, which then cost £6 – the equivalent of something like eighty dollars today. As a research student, I had invested in the book with great reluctance; it had burnt a hole in my meager pocket. The book consisted of the theory of the deuteron, a complete analysis of meson-theoretic nuclear forces, with Møller–Rosenfeld mixtures and the like, and a description of pion–nucleon-scattering phase-shift analyses below 1 MeV. Then Bethe came to lecture at the Cavendish. During that lecture, he made the categorical statement that all known deuteron parameters, as well as any phase-shift analyses below 1 MeV, could determine no more than two parameters of the nuclear potential: the scattering length and the effective range. While listening to the lecture, I kept thinking that surely this result Bethe has announced makes a book like that of Rosenfeld irrelevant. The thought crossed my mind that just after the lecture finishes, everyone who has acquired a copy of the Rosenfeld book will be trying to dispose of it. So immediately after Bethe finished, I rushed to my lodgings at St. John's College, retrieved my copy, and sprinted to Heffer's Bookshop, from which I had purchased the book. The sharks at Heffer's offered me £3 to buy the book back, even though it was in mint condition. I accepted, but of course now I feel very sorry that I sold it, because the book contained marvelous tables on harmonic functions.

I started my remarks with Dirac, who did not believe in the renormalization ideas that we were pursuing in 1950–1. He listened to us, but always maintained the hope for a finite theory. He has now recently been proved right by the rise of supersymmetry theories, some of which are completely finite, among them the $N = 2$ and $N = 4$ supersymmetry theories. In three decisive years, 1925, 1926, and 1927, with three papers, Dirac laid, first, the foundations of quantum physics as we know it, second, the foundation of the quantum theory of fields, and, third, that for the theory of elementary particles, with his famous equation of the electron.[12] No person except Albert Einstein has had such a decisive influence in such a short time on the course of physics in this century. But additionally, for me, Dirac, whom I later came to know better at the Trieste Centre, represented the highest reaches of personal integrity of any human being I have ever met. Knowing him has been one of the excellences in my life in physics.

I shall conclude with a story of Dirac and Feynman that perhaps will convey, in Feynman's words, what we all thought of Dirac. I was a witness to it at the 1961 Solvay conference. Those who attended the old Solvay conferences will know that, at least then, we sat at long tables that were arranged as if we were sitting to pray. Like a Quaker gathering, there was no fixed agenda; the expectation – seldom belied – was that someone would be moved to start off the discussion spontaneously.

At the 1961 conference, I was sitting at one of these long tables next to Dirac, waiting for the session to start, when Feynman came and sat down opposite. Feynman extended his hand toward Dirac and said, "I am Feynman." It was clear from his tone that it was the first time they were meeting. Dirac extended his hand and said, "I am Dirac." There was silence, which from Feynman was rather remarkable. Then Feynman, like a schoolboy in the presence of a master, said to Dirac, "It must have felt good to have invented that equation." And Dirac said, "But that was a long time ago." Silence again. To break this, Dirac asked Feynman, "What are you yourself working on?" Feynman said, "Meson theories," and Dirac said, "Are you trying to invent a similar equation?" Feynman said, "That would be very difficult." And Dirac, in an anxious voice, said, "But one must try." At that point the conversation finished, because the meeting had started.

Notes

1 Statement made by J. R. Oppenheimer to the Scientific Council of the International Centre for Theoretical Physics, which met in 1965. The members of the Council were the late Professor M. Sandoval Vallarta (Mexico), Prof. A. Abragam (France), Prof. Oppenheimer (USA), Professor Abdus Salam, Prof. Soloviev (USSR), and Prof. V. Weisskopf (USA).

2 A. Salam, "Gauge Unification of Fundamental Forces," *Rev. Mod. Phys. 51* (1980), 525–38.

3 N. Kemmer, "Charge-Dependence of Nuclear Forces," *Proc. Cambridge Philos. Soc. 34* (1938), 354–64.

4 F. J. Dyson, "The Interaction of Nucleons with Meson Fields," *Phys. Rev. 73* (1948), 929–30; "The Radiation Theories of Tomonaga, Schwinger and Feynman," *Phys. Rev. 75* (1949), 486–502; "The S Matrix in Quantum Electrodynamics," *Phys. Rev. 75* (1949), 1736–55.

5 A. Salam, "Divergent Integrals in Renormalizable Field Theories," *Phys. Rev. 84* (1951), 426–31.

6 J. C. Ward, "The Scattering of Light by Light," *Phys. Rev. 77* (1950), 293; "On the Renormalization of Quantum Electrodynamics," *Proc. Phys. Soc. London, Sect. A 64* (1951), 54–6.

7 Klaus Hepp, *Theorie de la Renormalisation* (Berlin: Springer-Verlag, 1969).

8 P. T. Matthews and A. Salam, "Renormalization," *Phys. Rev. 84* (1954), 185–91.

9 P. T. Matthews and A. Salam, "The Renormalization of Meson Theories," *Rev. Mod. Phys. 23* (1951), 311–14.

10 E. Fermi and C. N. Yang, "Are Mesons Elementary Particles?" *Phys. Rev. 76* (1949), 1739–43.

11 Leon Rosenfeld, *Nuclear Physics* (Amsterdam: North Holland, 1948).

38 Social aspects of Japanese particle physics in the 1950s

MICHIJI KONUMA

Born 1931, Tokyo, Japan; Ph.D., University of Tokyo, 1958; theoretical physics; Keio University, Japan

Introduction

Japan changed considerably during the decade of the 1950s, and research in science and technology was no exception in this regard. World War II came to an end in August 1945 with the surrender of Japan, which in 1950 was still under occupation by the allied powers. The economic situation was miserable: Inflation was rampant, black markets were thriving, and people were literally starving. However, the morale of theoretical physicists was high, for during the 1940s they had produced the two-meson theory, the super-many-time theory, and the covariant renormalization theory. In 1949, Hideki Yukawa was honored with the Nobel Prize in physics. In contrast to this flourishing theoretical culture, the level of experimental particle and nuclear physics was very depressed. For example, on 23 November 1945, all the cyclotrons in Japan – two at the Institute for Physical and Chemical Research (Riken) in Tokyo, one at Osaka University, and one under construction at Kyoto University – were dismantled by order of the occupation forces, who threw the dismembered parts into the sea (possibly also into a lake). On 30 January 1947, the Far Eastern Committee, which was the organization for policy making of the allied powers in Japan, decided that "all research in Japan of either a fundamental or applied nature in the field of atomic energy should be prohibited."[1] This prohibition, which obviously put severe restrictions on the resumption of research in nuclear physics, was in force until the peace treaty became effective in 1952.

Over the ten years after the signing of the peace treaty in 1952, by which Japan recovered its independence, economic conditions improved significantly. By 1960, the most important economic problem was the adjustment to rapid large-scale expansion, and the hottest political issue was revision of the U.S.-Japan security treaty. Experimental physics improved also. Some cyclotrons had been rebuilt, and the first Japanese electron-synchrotron was nearly completed. The physics community had started to discuss future plans for nuclear and particle physics. In 1950, I became an undergraduate physics student at the University of Tokyo. Ten years later, when I was a research associate at that university, I was involved in planning discussions as one of the secretaries of the National Committee for Nuclear Physics of the Science Council of Japan.[2]

Military occupation

The earliest Japanese scientific activity after World War II was the investigation of the effects of the atomic bombing of Hiroshima and Naga-saki. Japan's leading nuclear physicist, Yoshio Nishina, flew with his group to Hiroshima on 8 August 1945, two days after it had been bombed, and identified the explosion to be that of a nuclear-fission bomb. At first, the Japanese military authorities, and later the general headquarters (GHQ) of the supreme commander for the allied powers (SCAP), forbade the pub-licizing of Nishina's mission, which was not disclosed until the recovery of independence.

A general effect of the occupation on the scientific community resulted from directive no. 3, issued by GHQ on 22 September 1945.[3] In this directive, all laboratories, research institutes, and similar scientific and technical or-ganizations were ordered to submit reports to GHQ on the first day of each month, detailing the projects in which their personnel and facilities had been engaged during the preceding month and giving the results of such work. In the same directive, research and development work on mass separation of uranium 235 or any other radioactive elements was prohibited. University professors of physics were required to submit very detailed reports of their activities. There remain fragments of such reports prepared by Yukawa, who wrote about his scientific activities during the war, on the aims of his current research, and on the progress made by his research group at Kyoto University.[4]

Scientific journals

GHQ opened the first Civil Information and Education (CIE) library in Tokyo on 26 February 1946.[5] This was the only place in Japan that gave access to current issues of the *Physical Review* and other scientific journals. Users enthusiastically read these journals, as well as the library's books, and often made copies by hand. Soon other CIE libraries were opened in other large cities, and in 1947 scientific journals, both current issues and back numbers, began to be donated to the main universities. By the end of 1949, subscriptions could be obtained, with special allotment of foreign currency, through bookstores having import licenses. Russian periodicals and books became available at the office of the Soviet representative after February

1947, and French publications soon after that. The CIE libraries were actively used through the early 1950s.

Toward the end of the war, Japanese industry was almost entirely at a standstill because of serious shortages of personnel and resources, and also as a result of destruction by bombing raids. The printing of scientific journals was no exception. The Physico-Mathematical Society of Japan, with head-quarters in Tokyo, found it was unable to continue to publish its proceedings and asked Yukawa in Kyoto if he could arrange for their publication in western Japan. Yukawa received a positive response from a publisher in Kyoto, but before negotiations were completed, the war ended. As the immediate situation for publishing was not any better, Yukawa then decided to publish a physics journal privately, and he founded *Progress of Theoretical Physics*. The first issue of twenty-six pages appeared in July 1946. In that year, four issues were produced. Thereafter it was issued bimonthly, until in June 1952 it became a monthly.

Manuscripts on theoretical physics written during the war, some of which had already been published in Japanese, were translated into English and rushed to Kyoto for publication. These included Sin-itiro Tomonaga's paper on super-many-time theory and the two-meson theories that had been proposed by Yasutaka Tanikawa and by Shoichi Sakata and Takesi Inoue. Because of the demand on the limited publishing resources, long articles like Tomonaga's were separated into parts and were published serially. In 1948, Tomonaga described this state of affairs with feeling in an essay entitled "Sad Realities":

> Shortage of paper and of printing facilities is a serious problem for us. . . . At this moment there are only two journals in physics. . . . It takes a terribly long time before even very short articles are published. . . . I submitted several articles to these journals on my research during and also after the wartime, and of those submitted in 1946, two years ago, only one was published two or three months ago. Articles I sent to the other journal in the same year have not yet appeared. We are not sure whether we can continue to publish under the pressure of the inflation of the costs of paper and printing. . . . Sad realities![6]

(The high rate of inflation during that time is shown in Figure 38.1.)

Tomonaga and others sent early issues of the *Progress of Theoretical Physics* to foreign physicists. As Freeman Dyson later recalled:

> In that spring of 1948 there was another memorable event. Hans [Bethe] received a small package from Japan containing the first two issues of a new physics journal, *Progress of Theoretical Physics* . . . printed in English on brownish paper of poor quality. They contained a total of six short articles. The first article in issue No. 2 was called "On a Relativistically Invariant Formulation of the Quantum Theory of Wave Fields," by S. Tomonaga of Tokyo University [of Education]. Underneath it was a footnote saying, "Translated from the paper . . . (1943) appeared originally in Japanese."

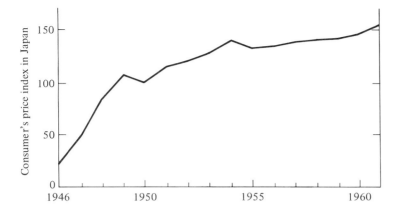

Figure 38.1. The rate of inflation in Japan.

Hans gave me the article to read. It contained, set out simply and lucidly without any mathematical elaboration, the central idea of Julian Schwinger's theory. The implications of this were astonishing. Somehow or other, amid the ruin and turmoil of the war, totally isolated from the rest of the world, Tomonaga had maintained in Japan a school of research in theoretical physics that was in some respects ahead of anything existing anywhere else at that time. He had pushed on alone and laid the foundations of the new quantum electrodynamics, five years before Schwinger and without any help from the Columbia experiments. He had not, in 1943, completed the theory and developed it as a practical tool. To Schwinger rightly belongs the credit for making the theory into a coherent mathematical structure. But Tomonaga had taken the first essential step. There he was, in the spring of 1948, sitting among the ashes and rubble of Tokyo and sending us that pathetic little package. It came to us as a voice out of the deep.[7]*

A few weeks later, Robert Oppenheimer at Princeton received a letter from Tomonaga describing their recent work. Oppenheimer replied immediately by a telegram, sent on 4 April 1948, with the following text:

> Grateful for your letters and papers. Found most interesting and valuable mostly paralleling much work done here. Strongly suggest you write a summary account for prompt publication *Physical Review*. Glad to arrange. Most constructive development here application by Schwinger of your relativistic formalism to self consistent subtraction to obtain several definite quantitative results.[8]

The work of the Tomonaga group on the renormalization theory helped greatly to increase the prestige and the circulation of the *Progress of Theo-*

---

* Reprinted from p. 57 of F. Dyson, *Disturbing the Universe*. Copyright © 1979 by Freeman J. Dyson. Reprinted by permission of Harper & Row.

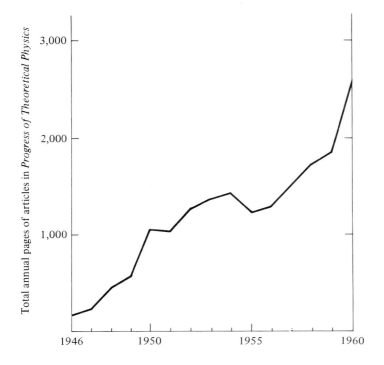

Figure 38.2. The rapid expansion of the *Progress of Theoretical Physics.*

retical Physics. The issue of July–August 1950 commemorated the fifteenth anniversary of Yukawa's meson theory, which happened to occur just after Yukawa won the Nobel Prize in physics for 1949. There were many contributors from abroad to this special issue, including Enrico Fermi, Wolfgang Pauli, Werner Heisenberg, Gregor Wentzel, Leon Rosenfeld, Walter Heitler, and Robert Marshak. This increase in scientific exchange with foreign physicists was very encouraging to the Japanese physicists, who had for so long been isolated. Figure 38.2 shows the rapid growth of the *Progress of Theoretical Physics.*

Overseas travel

During World War II, travel abroad was impossible for Japanese scientists. Afterward, GHQ approved only a few overseas visits for scientists who were invited, beginning in 1948. One of these was Yukawa, with an invitation from the Institute for Advanced Study at Princeton. In September 1949, Nishina became the first scientist after the war to represent Japan at an international meeting when he attended the General Assembly of the International Council of Scientific Unions at Copenhagen on behalf of the Science Council of Japan. In December of the same year, the resumption of foreign trade was approved by GHQ, and funds for foreign travel were included in the government's budget. With this support, the numbers of scientists and engineers who traveled abroad increased: in fiscal year 1949 (January–March 1950), 65; in fiscal year 1950 (April 1950–March 1951), 233; in fiscal year 1951 (April 1951–March 1952), 300.[5]

The United States began to help students and young researchers to study abroad in 1949, using funds designated as "government account for relief in occupied areas," and this type of assistance was continued under the Fulbright Act after the end of the occupation. The first Fulbright Scholars went to the United States in September 1952. Study abroad at public expense was extended to France in 1950, the United Kingdom in 1951, West Germany and Italy in 1952, and so on.

The Japanese government also started to send researchers abroad for longer stays in 1951; in that year they sent fifteen to the United States, two to France, and one each to Great Britain, India, and Denmark. Tomonaga went on invitation to the Institute for Advanced Study at Princeton in 1949–50, Seishi Kikuchi to Cornell in 1950–1, Satio Hayakawa to MIT in 1950–2, Tatsuoki Miyajima to Birmingham in 1952–3, and Hiroomi Umezawa to Manchester in 1953–5. Yoichiro Nambu and Toichiro Kinoshita went to the Institute at Princeton in 1952, and later took permanent positions in the United States.

An important role for graduate students was played by Marshak, who asked Tomonaga to recommend graduate students for the University of Rochester. Tomonaga screened both theoretical and experimental applicants, testing them in physics, English, and a second foreign language. The first three graduate students were sent in 1953, and Tomonaga's recommendations continued until 1959. Among the outstanding students (with their present affiliations) were Masatoshi Koshiba (Tokyo), Susumu Okubo (Rochester), Taiji Yamanouchi (Fermilab), and Bunji Sakita (CCNY).

During the 1950s, the numbers of Japanese graduate students and visiting scientists working abroad gradually increased, as did the number of Japanese attending international conferences. However, certain difficulties should not be minimized. Besides the occupation in the early 1950s, there was the Korean War in 1950–3. During the cold war, passport and visa applications from physicists thought to have left-wing sympathies were frequently denied. Visits to the Soviet Union and the People's Republic of China were under close restriction.

Research group on elementary particle theory

The Japanese word *soryushiron* means the theory of elementary particles (and is taken to include nuclear theory). To understand the history of particle physics in Japan, one must know about the informal Soryushiron Group, which was especially active and important in the 1950s, and which still exists today.[9] To be a voting member of the group (one of three grades of membership), one must have published at least one research article on *soryushiron* and also be on a registration list. The voting members elect an Advisory Committee and also the Research Planning Committee of the interuniversity institutes, which will be discussed later. The second grade of membership requires a research paper on *soryushiron* in a refereed journal; that allows the member to apply for research funds for projects supported by the Soryushiron Group, the funds being obtained as grants-in-aid from the Ministry of Science, Education, and Culture and other sources. Finally, and most loosely, anyone who regards himself or herself as a *soryushiron* researcher is a member.

The origin of the group can be traced to the so-called Meson Club, which was active in 1941–3, when Tomonaga and Yukawa brought together some twenty younger physicists in informal gatherings after the semiannual meetings of Riken, in order to discuss meson theory and quantum field theory in general.[10] After the resumption of the regular meetings of the Physical Society of Japan (the successor to the Physico-Mathematical Society) in 1946, Seitaro Nakamura, Ziro Koba, and others decided to continue the informal spirit of the Meson Club by publishing a mimeographed circular, *Soryushiron Kenkyu* (*kenkyu* means research). The first issue, containing eighty-three pages of summaries of talks to be presented at a meeting on elementary particles of the Physical Society, was published in October 1948. The second issue (near the beginning of 1949) contained twenty articles. It also printed letters from abroad addressed to various Japanese physicists, including Tomonaga, from Oppenheimer, Abraham Pais, Heitler, Heisenberg, and especially from Yukawa at Princeton.

Soryushiron Kenkyu became an indispensable medium for the exchange of new information and new ideas. Having no refereeing system, it played a role similar to the distribution of preprints. This journal was considered to be the bulletin of the Soryushiron Group. It published announcements of positions available, and it was important to the users of the interuniversity institutes, to be described later. In 1950, it was a bimonthly, and in 1952 a monthly, and during 1953–7 the number of papers required the publication of two or three issues per month. The section of letters from abroad expanded rapidly. Figure 38.3 shows the growth of this key informal journal, and Figure 38.4 shows the increased membership of the group from 130 in 1950 to 360 in 1960.

A great impetus was given to particle physics research in Japan by the international conference on theoretical physics held in Kyoto and Tokyo in 1953. The first purely scientific international conference ever held in Japan, it had parallel sessions for five days in Kyoto in three sections: field theory and nuclear physics, statistical mechanics, and solid-state physics. It was sponsored by the International Union of Pure and Applied Physics and was jointly organized by the Science Council of Japan, the Physical Society of Japan, and Kyoto University. Supported by UNESCO and the Rockefeller Foundation, it was attended by fifty-five foreign participants from thirteen

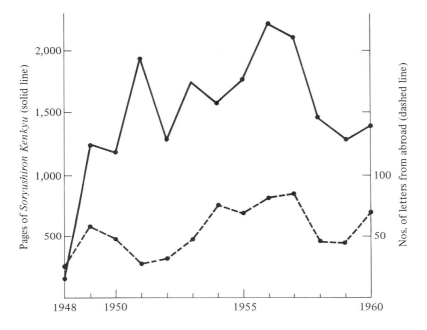

Figure 38.3. Mimeographed circular *Soryushiron Kenkyu:* number of published pages and number of letters from abroad.

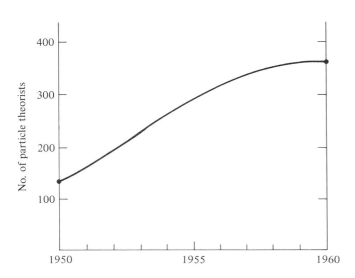

Figure 38.4. The increase in numbers of particle physicists in Japan.

countries. About 600 Japanese physicists participated, "who in preparation ...formed study groups...and in the work of completing their respective papers to be read, did not forget to...brush up their English."[11]

Establishment of interuniversity institutes

While Yukawa was spending the year 1948–9 at the Institute for Advanced Study, he wrote in a letter to Tanikawa as follows:

> When many new universities are established, there is a tendency toward a lack of communication among researchers. I think that institutes similar to the Institute at Princeton should be created at several places in eastern and western Japan, where temporary members could concentrate on research for a certain term.[12]

After Yukawa became Japan's first Nobel laureate in December 1949, Kyoto University, the Science Council, and the Japanese Diet all had ideas for a commemorative project. Gradually, these took the form of a new interuniversity institute such as Yukawa had suggested.

This new research center, named Yukawa Hall, started operating at Kyoto University in July 1952. It was inaugurated with a ten-day seminar with 250 participants; two research associates were appointed, and visitors started coming.[13] A year later it became the Research Institute for Fundamental Physics (RIFP), when Yukawa returned from Columbia University to become its director. (The name of the building it occupies remains Yukawa Hall.) The international conference held in 1953 was a major event for the new institute in Kyoto.

Simultaneously with the opening of RIFP, another interuniversity institute, the Cosmic Ray Laboratory of the University of Tokyo (CRL), was established near the top of Mount Norikura, at an altitude of 3,026 m. Actually, experiments at that location had been conducted as early as 1949 by Osaka City University physicists. They were joined in 1950 by Kobe University, Nagoya University, and the Institute of Scientific Research (the Tokyo group formerly known as Riken) to establish there the Asahi Laboratory, taking its name from the Asahi newspaper company, which provided financial support. The shift to a governmental interuniversity institute meant that the government provided increased support.[14]

The third interuniversity institute, the Institute for Nuclear Study, University of Tokyo (INS), was established in 1955. As noted earlier, experimental nuclear research had been suspended after World War II by GHQ and the Far Eastern Commission, and its resumption was delayed by the severe economic distress. In 1951, Ernest O. Lawrence visited Japan and suggested that the cyclotron of the Institute for Scientific Research in Tokyo be reconstructed. Discussions began, and in 1952 the result was a 3.7-MeV cyclotron, financed by the Ministry of International Trade and Industry.[15] At the same time, the Ministry of Education supported the construction of a 13-MeV

Table 38.1 *Scientists visiting RIFP, 1953–60*

Year	No.	Partial list of visiting scientists
1953	50	E. Amaldi, R. P. Feynman, R. E. Marshak, A. Pais, E. P. Wigner, C. N. Yang
1954	3	A. H. Compton
1955	8	R. P. Feynman, E. O. Lawrence, A. I. Oparin, L. Pauling
1956	11	C. F. Powell, R. R. Wilson
1957	3	
1958	4	
1959	5	V. F. Weisskopf
1960	9	M. A. Markov, J. R. Oppenheimer

cyclotron at Osaka University, while private industry paid for a 14-MeV cyclotron at Kyoto University.[16]

At the time of the start of RIFP, nuclear physicists began thinking about building an accelerator, perhaps an electron-synchrotron or proton-synchrotron, that would be too large for any single university to manage. This idea crystallized in the establishment of INS, whose first director, Kikuchi, returned to Japan after a two-year stay at Cornell University. In 1958, a synchrocyclotron with proton energy from 51 to 57 MeV was completed. An electron-synchrotron begun in 1956 was completed in 1962 and reached a maximum energy of 750 MeV, permitting the first experiments in Japan to obtain artificial pion tracks in nuclear emulsion. The energy was increased to 1.3 GeV in 1966. The INS was especially important in providing a gathering place for theorists and experimentalists, and for both high- and low-energy nuclear and cosmic-ray physicists.[16]

Finally, the interuniversity institutes invited visiting foreign physicists for longer or shorter periods, both senior and younger researchers. The numbers of scientists visiting RIFP in the earlier years, and some of their names, are shown in Table 38.1.[17]

Planning the next stage

The second half of the 1950s showed great improvements in science and technology, improvements realized because of the increased demand for science and technology and made possible by the rapid improvement of the Japanese economy. The research-and-development budget started to increase markedly in 1956, as shown in Figure 38.5 (which gives only data on governmental spending). Most of the increase went to the budget for nuclear energy.[18]

The next stage of planning for particle, nuclear, and cosmic-ray physics

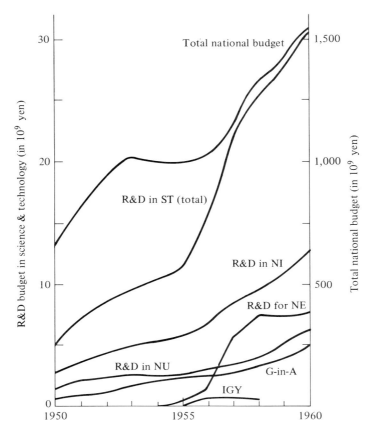

Figure 38.5. Research-and-development (R&D) budgets in science and technology (ST) and the total national budget. NE, nuclear energy; NI, national institutes; NU, national universities; G-in-A, grants-in-aid; IGY, International Geophysical Year.

began in May 1958, when Kikuchi suggested that one should consider the next step to be taken after the completion of the electron-synchrotron.[19] In October of the same year, Yukawa, who had just returned from a European trip, made a strong appeal for the development of fundamental science, pointing out the unfavorable comparison of Japan's largest accelerator, the 60-MeV synchrocyclotron, with the much larger machines in Europe and America. In April 1962, the Science Council of Japan presented to the prime minister a recommendation for a "Future Plan of Nuclear Physics" (which term included particle and cosmic-ray physics). It recommended the creation of the Research Institute for Particle Physics, for conducting research in high-energy and cosmic-ray physics. Eventually this plan resulted in the establishment in 1971 of KEK, the national laboratory for high-energy physics.

We conclude with some insights to be gained from a study of the social aspects of particle physics. In the years 1945–50, many younger physicists were attracted to theoretical particle physics, but very few posts were available. This unemployment problem was solved at a stroke when the present-day university system was initiated in 1949, expanding greatly the number of university chairs. In 1955–60, however, a second stage of unemployment resulted, when all the new positions had been filled. The earlier pattern reasserted itself in 1960–5, when there was renewed emphasis on science and technology in universities, and the physics departments in all the major universities doubled in size.

On the other hand, the expansion of industry had no impact on particle physics, and although the establishment of KEK produced many new positions for accelerator physicists and experimental particle physicists for a short time in the 1970s, unemployment soon reappeared. Each time, the too rapid development of the field created a shortage of positions available for the succeeding generation – and so it has continued through the 1980s. Thus we are faced with a dilemma in Japan in our effort to maintain a high level of activity in particle physics.

Notes

1 *Documents Concerning the Allied Occupation and Control of Japan, Vol. II, Political, Military and Cultural*, compiled by the Division of Special Records, Foreign Office, Japanese government (Tokyo: Toyo Keizai Shinposha, 1949), p. 231.

2 *Science Council of Japan: The First 25 Years (1949–1974)* (Tokyo: Science Council of Japan, 1974), p. 434.

3 *Documents Concerning the Allied Occupation and Control of Japan, Vol. I, Basic Documents*, compiled by the Division of Special Records, Foreign Office, Japanese government (Tokyo: Toyo Keizai Shinposha, 1949), p. 79.

4 Hideki Yukawa, unpublished documents deposited at Yukawa Hall Archival Library, Research Institute for Fundamental Physics, Kyoto University.

5 Tetu Hirosige, *Scientific Movements in the Post-War Japan* (Tokyo: Chuo Koronsha, 1960).

6 Sin-itiro Tomonaga, "Sad Realities," *Shizen 4: 1* (1949), 1.

7 Freeman Dyson, *Disturbing the Universe* (New York: Harper & Row, 1979), p. 57.

8 Robert Oppenheimer, telegram to S. Tomonaga, 28 May 1948, *Soryushiron Kenkyu 2* (1948), 144.

9 Yoshinori Kaneseki, "The Elementary Particle Theory Group" (in Japanese), *Shizen 4* (1950), 48; translated and compiled in *Science and Society in Modern Japan*, edited by Shigeru Nakayama, David L. Swain, and Eri Yagi (University of Tokyo Press, 1974), p. 221.

10 Rokuo Kawabe and Michiji Konuma, "Birth of the Meson Theory," *Butsuri 37* (1982), 265.

11 *Proceedings of the International Conference of Theoretical Physics, 1953, Kyoto and Tokyo* (Tokyo: Science Council of Japan, 1954), p. xiii.

12 Hideki Yukawa, letter to Yasutaka Tanikawa, 18 January 1949, *Soryushiron Kenkyu 3* (1949), 236.

13 "Yukawa Hall in the Year of Its Opening," *Soryushiron Kenkyu 61* (1980), 965.

14 Koichi Kamata, "On the Institute for Cosmic Rays, University of Tokyo," *Gakujutsu Geppo 37* (1984), 638.

15 Hidehiko Tamaki, "Reconstruction of Small Cyclotron" (in Riken), *Shizen 33:13* (1978), 101.

16 *History of the First 20 Years of Institute for Nuclear Study, University of Tokyo* (Tokyo: INS, 1978).

17 *Research Institute for Fundamental Physics, Kyoto University, 1953–1978* (Kyoto: RIFP, 1978), p. 93.

18 "Governmental Budget for the Promotion of Science and Technology Seen from Figures and Tables," *Gakujutsu Geppo 12* (1959), 3; *13* (1960), 2; *14* (1961), 2. See also note 5 (p. 122).

19 Michiji Konuma, "Steps of Planning on Nuclear Research," *Butsuri 27* (1972), 251.

PART IX

THEORIES OF HADRONS

We had weak coupling, strong coupling,
Wrong as we knew all along coupling,
Each month our troubles are quadrupling –
   Some people don't know where to begin.
        – © Arthur Roberts, "Some People
        Don't Know Where to Stop" (1952)

39   The early S-matrix theory and its propagation (1942–1952)

HELMUT RECHENBERG

Born 1937, Berlin, Germany; Ph.D., University of Munich, 1968; theoretical physics, history of physics; Werner-Heisenberg-Institut für Physik, Max-Planck-Institut für Physik und Astrophysik, Munich

Introduction

Study of the formation and propagation of S-matrix theory in the 1940s helps one to understand the rapid development, under radically different scientific and sociological conditions, of the theory of elementary particles in the fifties. The S matrix is a two-dimensional array whose elements are the transition amplitudes between states of atomic or elementary particle systems. According to a standard textbook of the midfifties, "Such a scattering matrix was first introduced by Wheeler (1937) in connection with the problems of nuclear structure and scattering. It has been reinvestigated in great detail by Heisenberg (1943) in connection with the theory of elementary particles."[1] During World War II and the first postwar years, few people were interacting on problems of pure physics; only on rare occasions, often accidentally, did they meet for discussions, or even exchange letters. Despite these obstacles, they did interesting and original work that led to the resumption of particle physics by a larger group after World War II. The S matrix played a comparatively large role in the discussions around 1945, but was abandoned in the early fifties. Toward the end of the fifties, however, several developments led to renewed interest in it.

The origin of the S-matrix theory of elementary particles[2]

In a paper submitted in the summer of 1937, John A. Wheeler proposed to derive the properties of light nuclei, such as energy levels and transition probabilities, with the help of a unitary "scattering matrix." The elements of this matrix were connected with the transition from incoming to outgoing groups, consisting of a few protons and neutrons within the nuclei in question.[3] Five years later, in a paper submitted to *Zeitschrift für Physik* in September 1942, Werner Heisenberg introduced a similar, also unitary, "characteristic matrix S," describing the scattering and emission problem in the theory of elementary particles.[4] Despite the similarity of the two S-matrix conceptions – which led to nearly identical mathematical notations – the relativistic elementary particle theory of Heisenberg did not emerge directly from Wheeler's nonrelativistic nuclear theory. (Indeed, when I asked Heisenberg in 1961 if he had recalled in 1942 Wheeler's earlier work in nuclear theory, he answered, "No, I learned about the connection only later. Wheeler's is a fine work.")

Part of the background of Heisenberg's introduction of the S matrix was the unsuccessful effort during the 1930s to arrive at a satisfactory quantum field theoretical description of elementary particle properties and reactions. All the schemes considered then – whether quantum electrodynamics, Fermi's β decay theory, or Yukawa's meson theory of nuclear forces – exhibited fundamental difficulties. Heisenberg, in thinking about radical means to remove these difficulties, proposed two ideas, which he assumed to be somehow connected: first, the existence of a universal length l_0, such that the known quantum and relativity theory broke down at distances smaller than l_0; second, the necessity of involving nonlinear features in the mathematical description of interacting elementary particles.[5] In the later work on the scattering matrix, however, he did not want to propose a program incorporating those ideas; rather, as he expressed it in the introduction to his first publication, "the present paper attempts to isolate from the conceptual scheme of quantum theory of wave fields those concepts that probably will not be hit by the future alteration (of the theory of elementary particles) and which may therefore represent a constituent also of the future theory."[6]

Because one did not yet know how to formulate a divergence-free theory describing elementary particles, he took the approach of asking "the question as to which concepts of the present theory might be kept in future, and this question is roughly identical with the other question as to what quantities in the present theory are observable."[7] That is, Heisenberg tried to revive a philosophical program that had earlier led him from the old quantum theory to quantum mechanics, hoping now to replay the earlier success in elementary particle theory.[8]

Among the unobservable quantities to be kept out of the theory, Heisenberg mentioned distances smaller than 10^{-13} cm and time intervals smaller than 3×10^{-24} sec. On the other hand, he took as safely observable (i) the discrete energy values of stationary states of closed systems and (ii) the asymptotic behavior of wave functions in scattering, emission, and absorption processes. The properties (i) and (ii) could be related to each other, because the phase differences between the incoming and outgoing wave functions had to yield the discrete energy states as well, if the scattering system were closed by a sphere of large radius around the scatterer.

The mathematical formulation of this idea, when applied to relativistic systems of scattering elementary particles, led Heisenberg to introduce his unitary, "characteristic" S matrix. By writing this matrix as

$$S = e^{i\eta} \tag{39.1}$$

he defined a Hermitian matrix η, which allowed a particularly simple description of the physical situation. The matrix η should then replace the defective Hamiltonian formulation of scattering problems in the quantum theory of wave fields; it contained only observable quantities.[9] The obvious problem now was how to calculate the relativistically invariant η matrix.

Several weeks after the submission of his first paper, Heisenberg sent a second with the same title to the *Zeitschrift für Physik*.[10] In it, he discussed examples of relativistic η matrices for

- (i) a δ-function-like interaction leading to pure scattering of the particles;
- (ii) a certain distance-dependent interaction resulting in finite cross sections for the scattering of particles with arbitrarily high energies; and
- (iii) an interaction implying the creation of new particles.

The last two examples could not be adequately represented in the known quantum field theories, nor did they possess an analogue in the classical theory. The S- or η-matrix theory thus provided a suitable tool to treat what Heisenberg called "explosive showers" in cosmic radiation.[11]

Heisenberg dedicated his first paper on "the observable quantities" to Hans Geiger on the occasion of the latter's sixtieth birthday. A little later he wrote to Geiger that "I had nothing else available that was connected more closely with your research work, e.g., on cosmic radiation, for your *Festschrift*. But I have for some time been concerned with these purely theoretical considerations, which I have now carried further."[12] These remarks reveal that he had already been thinking about applying the practical philosophy of observables to elementary particle theory for several years. Indeed, in a paper that he submitted in January 1938, he speculated about possible new concepts emerging from the assumption of a universal length, and he concluded:

> Perhaps one may remember to advantage, in attempting to find new concepts, that in mathematical formulae, we are now confronted with the task of finding computational rules, by which we can connect the cross sections of cosmic radiation processes partly with each other, and partly with other simple observational data.[13]

In 1942, those ideas received a mathematical formulation in the S-matrix theory.

Since the beginning of the war, Heisenberg had been heavily involved in the German uranium research program.[14] He conducted this work mainly at the University of Leipzig and at the Kaiser-Wilhelm-Institut für Physik in

Berlin–Dahlem. But he did not give up thinking about elementary particle problems; for example, in 1941 and 1942 he conducted a seminar in Berlin on cosmic-ray physics, whose material was published in a book, *Vorträge über kosmiche Strahlung* (Berlin: Springer, 1943).

In June 1942, he received a letter from Gian Carlo Wick, a former associate of Heisenberg, then in Rome.[15] Wick wrote that Arnold Sommerfeld had invited him to come to Munich later that month and that he would like to continue from Munich to meet Heisenberg in Leipzig or Berlin.[16] Heisenberg replied that he would be happy to see Wick and asked him to present a seminar in Leipzig; the audience would consist "only of Mr. Hund, two young gentlemen and myself, but in compensation for that you will have the more attentive auditors."[17] He added, "Perhaps I can still organize a colloquium lecture for you before a larger circle; however, for that I have to obtain the permission of higher authorities." If Wick could not make this, Heisenberg suggested that he come to Berlin, where Heisenberg spent half of each week working on the uranium project: "My invitation remains also in this case, and we have to see to it that you tell us a little on cosmic rays at the Dahlem Institute." "The most important thing for me is that we can have a thorough conversation once again," he concluded. Wick came both to Leipzig and to Berlin and provided his host a welcome chance for scientific and personal discussions, a rare event in the war times; in particular, they debated Heisenberg's new ideas on the *S*-matrix theory.[18]

The conversations with Wick had important consequences. For example, Wick told Heisenberg of earlier work (e.g., a paper by Gregory Breit, supporting some of the consequences of the new *S*-matrix theory). Most crucial, however, was that he encouraged Heisenberg to write down his ideas and publish them. Heisenberg therefore was "deeply obliged to Wick for discussions" and continued to correspond with his younger colleague,[19] to whom he sent in February 1943 the proofs of his first two papers on the *S*-matrix theory[4,10] and a letter discussing problems of cosmic-ray physics.[20] In their following correspondence, Wick and Heisenberg exchanged ideas on the *S*-matrix theory and other physical topics.[21] Later on, the war times became very hard, and the exchange of mail between Berlin and Rome was finally interrupted. Wick's last letter from Rome, on 16 April 1944, closes with the words "May God protect you and your family."

Difficult scientific exchange during World War II

In his first communication on the *S*-matrix theory, Heisenberg wrote that "Wick has called my attention to a paper of Breit (*Phys. Rev.* 1941), in which similar considerations are made about the scattering matrix; the paper of Breit has so far not been accessible to me, unfortunately."[22] Breit's paper had already been published in the *Physical Review* on 15 December 1940, but Heisenberg did not see it until two years later,[23] because soon after the outbreak of World War II, the exchange of scientific literature between Germany and the United States was interrupted, even though the two countries were not at war with each other until December 1941. The cutoff was not perfect, as some exchange took place through intermediate neutral countries, such as Switzerland, Sweden, or Ireland, but that depended on accidental circumstances.[24] To illustrate a particularly difficult connection, let us examine the mail situation between Germany and Japan, two allies in World War II. Before Germany invaded Russia (i.e., before 22 June 1941), mail could be sent via rail to East Asia. Until Japan attacked the United States, mail could pass by ship. But after 6 December 1941, essentially every possible direct path was closed between Germany and Japan. Still, we know that the Japanese physicists received, for example, copies of Heisenberg's two S-matrix papers by late 1943.[25] How did these publications arrive in the Far East?

The most likely possibility is that submarines carried them. Between April 1942 and July 1944, the Japanese and the German navies tried to establish a systematic contact by submarines, mainly for the purpose of exchanging important war material, such as information on rare metals or technical know-how. However, in five navigations, only one Japanese boat succeeded in getting from the Japanese port Ku-Re to the French port Brest and back (1 June–21 December 1943), while others made it halfway, either from Germany to Japan[26] or from Japan to France.[27] One of the latter boats probably carried a letter from Heisenberg addressed to Yoshio Nishina, his former colleague from earlier Copenhagen days. Heisenberg wrote: "Several times in the past years I have tried to write you a letter, but found no possibility of connecting with you. Now, by a lucky accident, namely through the kind visit of Korvettenkapitän Nomaguti,[28] I have been offered the opportunity to write you." Heisenberg said he was "happy to seize this opportunity for thanking you [i.e., Nishina] and the other Japanese friends heartily for the papers that you have prepared for the December 1941 issue of the *Zeitschrift für Physik* and which have appeared a little later." In closing the letter, Heisenberg wished "you all to get safely through this war, and that in the not too distant future a peaceful collaboration in the field of physics will again be possible"[29] Unlike the two papers of Heisenberg, the letter to Nishina never reached Japan. The papers, however, were carefully studied by the Japanese physicists.[30]

A less exotic method of exchanging scientific news operated in central Europe. The easiest access to German physics was in Switzerland, which received all the German journals and had regular visitors from Germany to give lectures and seminars on scientific topics. Thus, Heisenberg spent the week of 17–25 November 1942 on a lecture tour in Zurich, Basel, and Berne. In particular, he spoke on the afternoon of 18 November at the Zurich colloquium on "The Observables in Particle Physics Theory."

While many Swiss colleagues, among them Gregor Wentzel in Zurich and André Mercier in Berne, heard Heisenberg present his recent approach to particle theory, Ernst Carl Gerlach Stueckelberg, from Geneva, perhaps the one most interested of all, missed it.[31] "I regret very much that my doctor did not allow me to leave Geneva and that I therefore missed the interesting talk you have given in Zurich," he wrote a little later, adding that "Mr. Wentzel has promised me to present your papers in our physics seminar during the spring semester."[32]

In any case, Stueckelberg studied Heisenberg's work carefully and found a relation between his new quantum theoretical formalism of a point electron with finite self-energy and the theory of the scattering matrix, which he published in a paper that appeared in February 1944.[33] Soon afterward, he wrote a letter to Heisenberg, thanking him for having received a copy of the latter's book, *Vorträge über kosmiche Strahlung*, and saying that "with the same mail I am sending you the reprint on my subtraction theory. I have advanced a good step forward and I believe to be able to demonstrate how one can describe the static field and understand bound levels. This is connected with the eigenvalue problem of my α-matrix (your eta)."[34]

In his paper, Stueckelberg discussed in detail a model of a point electron that interacted with the Maxwell field and had no divergence problems. No Hamiltonian function could be given for this model, but a description was possible through a function $\alpha(\beta)$ of a Hermitian operator β, such that the unitary matrix

$$S(\beta) = e^{-\alpha(\beta)} \tag{39.2}$$

(with α a real polynomial function of β) represented the "characteristic matrix" of the problem. In later work, Stueckelberg studied general properties of the scattering matrix, partly stimulated by Heisenberg's further papers on that subject.[35]

"Political" visits to Holland and Denmark

Heisenberg's excursion of November 1942 to Switzerland turned out to be fruitful in many respects. It enabled him to renew contact with old friends and colleagues, informed him of what had happened in science abroad, and provided him with the opportunity to expose his own ideas. Politically, the travel to a neutral country did not involve great problems, as did the visits to Holland and Denmark in late 1943 and early 1944. Both of these latter countries had been under German occupation since the spring of 1940, so that Heisenberg's mission was burdened with many delicate problems.[36]

The invitation already contained touchy issues. In the spring of 1943, Heisenberg received a letter from the general secretary of the Dutch Ministry of Education, Science, and Cultural Affairs. This letter began as follows:

> Even under the present circumstances, which make the international contact between scholars in the various fields of science so difficult, I am driven by the desire to do everything possible to enable reputable scientists working at Dutch universities to exchange ideas with their eminent colleagues abroad. I have tried especially to establish this in the field of physics.[37]

After this ponderous introduction, Heisenberg was invited for a visit of at least a week to Holland in the summer of 1943; he was to be a guest of the Dutch ministry and to have the opportunity to meet Dutch colleagues. Heisenberg, after four weeks, accepted the invitation in principle, but asked to shift the visit to the fall of 1943[38] and asked for the names of the Dutch colleagues who "wish to see me in Holland."

Heisenberg soon learned who stood behind the invitation: a dear friend from the 1920s at Bohr's institute, Hendrik Kramers, of Leiden, with whom he had not had any contact since the outbreak of the war. Kramers also wrote a letter, which opened with these words:

> For years I have wanted to write to you, but I hesitated again and again. Now, in connection with your invitation to Holland, I seize the opportunity to get in touch with you, especially today, because yesterday I had a conversation on the topic with Herr L. P. D. Op ten Noort. . . . It is this gentleman who runs the whole thing on the Dutch side.[39]

Kramers reported that the difficulties in the working conditions for the Dutch physicists had caused Op ten Noort to suggest the reestablishment of personal contact between Dutch and foreign scientists. Kramers had immediately discussed the idea with his colleagues Johannes Wander de Haas, Hendrik Casimir, and Eliza Cornelis Wiersma and reported to Heisenberg that "all have sympathized with a visit from you in this context." He then outlined the details of the planned visit, including the question of travel and living expenses, food ration coupons, visits of Dutch and German authorities, and the places and scientific institutions to be covered (Leiden, Delft, Utrecht, Eindhoven, Amsterdam, Groningen), and finally the appropriate time (i.e., the first half of September 1943): "Now I believe I have described the thing accurately enough, so that you understand that you are really welcome, which is – as I found out from your letter to the General Secretary – also important for you."[40]

With this letter in hand, Heisenberg accepted "the invitation of the Dutch Ministry with pleasure."[41] This made Kramers very happy, and he worked out a "splendid program" for the tour, suggesting that Heisenberg come on 18 October 1943 via Appeldoorn, paying the first visit to Op ten Noort.[42] The plan worked perfectly: On 19 October, Heisenberg was in Utrecht, on the 20th and 21st in Leiden, on the 22nd and 23rd in Delft, on the 24th again in Utrecht, and on the 25th in Amsterdam. After having met, besides Kramers, his old friends and colleagues Ralph Kronig (Delft) and Léon Rosenfeld (Utrecht) and the cosmic-ray specialist Jacob Clay (Amsterdam), he returned to Berlin on 26 October 1943.[43]

The visit to Holland was a success in every respect. Not only did Heisenberg have the chance to speak with his various hosts, both experimentalists and theoreticians, but also he participated in seminars and colloquia (e.g., in the Utrecht colloquium of 24 October where Kramers spoke on ferromagnetism). He saw the new equipment in the Dutch laboratories, such as the high-voltage generator for nuclear reactions in Utrecht (not yet operating) and the cosmic-ray experiments in Amsterdam. His lecture on the S-matrix theory, presented on 21 October in Leiden, attracted guests from other universities (e.g., Adriaan Fokker from Haarlem). He made his deepest impression, however, on Kramers, who wrote him soon afterward

> First of all, I would like to tell you once more, how happy your visit has made me, again stimulating old ideals. Also the music has given me pleasure. The only regrettable thing is that we do not meet and see each other more often. . . . I would very much like to discuss with you ideas on the problem and whole formalism of the S-matrix.[44]

He continued, mentioning an old calculation that he had made with his student S. A. Wouthuysen on the quantum mechanical scattering from a central potential, which could be used to determine the η matrix in that case.

The exchange of ideas with Kramers in October 1943 also excited new activity on Heisenberg's part: In January 1944, Heisenberg wrote to Kramers: "Your nice letter has unfortunately reached me only after some detours. . . in following your discussion remark, I calculated for several weeks on the problem of the η-matrix and wrote up a smaller paper, which I enclose in this letter."[45] He claimed that the paper was still "rather incomplete" and might not come out in print because of war damage to the Leipzig printing companies. "Therefore I would really prefer, if we could approach the difficulties in the paper more closely and perhaps write a joint paper."[46]

Kramers answered four months later. He had studied Heisenberg's paper in detail only very recently, forced by the fact "that I have to talk the day after tomorrow on difficulties in the theory of elementary particles at a symposium of our Physical Society."[47] He commented on various parts of Heisenberg's manuscript and criticized, for example, the proof given there for the commutation relations of a generalized quantum field theory as being based, perhaps, on too specific assumptions. He suggested incorporating in the theory a kind of time-reversal symmetry. He also discussed Heisenberg's three-body example, arguing in an appendix to his letter that a nonrelativistic calculation (involving an assumption that Heisenberg had made) seemed to yield wrong results. Finally, he raised the question of describing scattering problems in external electric and magnetic fields with the help of the η matrix. On Heisenberg's proposal of a joint publication, he said that "I can only rejoice that my remark on the analytic character of S has borne fruits for your considerations, and if you write this in a publication, my role in the thing is also described." Overall, he concluded that "the moment is not yet ripe for a joint publication."[48] Having received this answer, Heisenberg submitted his manuscript, with a few alterations, to the *Zeitschrift für Physik*.

The main progress in the third communication on observables in elementary particle theory – which was received 12 May 1944 and published in issue 1–2 of Volume 123 toward the end of the year – consisted in answering the question whether or not the η matrix of a given system, besides yielding all scattering cross sections, also described its bound states appropriately.[49] "This gap existing in the present considerations is closed by a remark of Kramers, according to which one can treat the matrix $(k_i'|S|k_i'')$ as an *analytic* function of the state variables k_i' and k_i'' and derive the stationary states from its behavior in the complex plane," Heisenberg declared. He explained further: "The zeros of the matrix S for imaginary k_i' provide the position of the stationary states," and "for the eigenvalues of the matrix η this result means that the poles of η lying on the imaginary k-axis determine the position of the stationary states."[50]

The postulate of an analytic S matrix restricted the possible forms of η matrices describing given relativistic problems. Thus, Heisenberg found, in particular, that his examples treated in the second communication did not involve smooth enough functions. Hence, he had to construct more suitable models, and he presented in his new paper an η-matrix model theory for two particles having momenta k_1 and k_2, showing that the analytic continuation in the variable $(k_1 + k_2)$ indeed provided the bound states.[51] A similar model for the three-body case also seemed to give reasonable results;[52] however the author commented in a footnote added in proof: "The treatment of the three-body problem presented below is, as I meanwhile found out, not consistent. Since, for external reasons, I cannot make larger corrections in the text, I shall discuss the situation more accurately in a later paper."[53] In spite of such technical difficulties, Heisenberg was confident that more general examples, involving the production and annihilation of particles with spin and electric charge, could be treated as well. Finally, he emphasized that "the difficulty of the theory consists in the correct incorporation of the manifold of phenomena, but not in avoiding divergencies."[54]

In addition to its scientific value for both sides, Heisenberg's visit to Holland produced other gains for the Dutch physicists. Rosenfeld reported in a letter that "today I received a visa to go to Liège....I thank you again for your effective help."[55] A few months later, Rosenfeld was again permitted, "as a beneficial after-effect of your visit,"[56] to visit his home country Belgium, from Holland. Still more important than those personal favors were the effects on the working conditions of Dutch physicists. "You have rendered us big services, and I am happy to transfer to you our heartiest thanks," wrote Kramers on 5 July 1944 to Heisenberg. The latter had met in Berlin a German official, General-Kommissar Fischböck, and asked him to help his Dutch friends; Fischböck had then immediately visited the Dutch laboratories and had removed the ban on experimental work, as Kramers reported.

If Heisenberg's October visit to Holland led to a considerable improvement in the working conditions of Dutch physicists, his two later visits to Copenhagen proved to be essential for the survival of Bohr's institute, which had been closed in fall 1943 after Bohr escaped from Denmark. As Heisenberg noted before he went first in January 1944, three possible outcomes could be imagined:

(i)   Bohr's institute would be taken over by the Germans.
(ii)   The institute would be returned to the Danish scientists under certain conditions and restrictions.
(iii)   The institute would be returned without conditions.

On 1 February 1944, Heisenberg wrote to Hans Jensen in Hanover: "In the past week I was in Copenhagen for three days. The final result of the negotiations was very satisfactory. Bohr's institute has been returned to the Danes without official conditions. [J. K.] Böggild, who was taken into custody for seven weeks, has been released." Happy with the success, Heisenberg arranged another, less political, visit to Copenhagen, for the purpose of delivering several talks.[57]

From a document in the Heisenberg archives, dated 17 April 1944 and signed by Heisenberg in his capacity as Abwehrbeauftragter (protection official) for the Kaiser-Wilhelm-Institut für Physik, we learn that he took with him to Copenhagen: "1. Proofsheets of 'Die Atomtheorie vom Altertum bis zum Ende des 19. Jahrhunderts,' 32 pages;[58] 2. Letter of Professor Kramers to Professor Heisenberg, 5 sheets; 3. Letter of Professor Christian Møller to Professor Heisenberg, 14 sheets; 4. Five white sheets." Of this material, the Kramers letter was the one of 12 April 1944, mentioned earlier.[47] The third item, Møller's fourteen-page letter, was written on 28 December 1943 by the Danish physicist, whom Heisenberg knew well from previous visits to Copenhagen.

Møller's letter discussed Heisenberg's first two communications on the *S*-matrix theory and suggested a detailed and general formulation of Heisenberg's ideas through which many of the assertions were clarified or proved. Evidently, Møller was eager to learn Heisenberg's opinion about this work. He received this during the latter's second visit to Copenhagen, together with an introduction to the new concept of an analytic *S* matrix. Møller then accepted the analyticity property; on 10 July 1944, he wrote another letter to Heisenberg containing several results that he had achieved meanwhile, and announcing that he wanted to write a paper on it during the summer vacations. Møller promised to send a copy to Heisenberg as soon as it was available.

End of the war and new beginning

In a footnote to the third communication on *S*-matrix theory, mentioned earlier, Heisenberg announced a fourth paper investigating further the three-body problem. No such publication can be traced in the literature. However, the Heisenberg archive does have a manuscript of twenty-one handwritten pages, entitled "Die Behandlung von Mehrkörperproblemen mit Hilfe der *S*-matrix," containing this material.

The aim of the paper is described in the introduction:

> The stationary states of closed systems and the cross sections in collisions are determined by a unitary scattering matrix S or by a Hermitian matrix η, which is related with S by the equation $S = \exp(i\eta)$. If one is dealing with systems containing more than two elementary particles, then it does not suffice, of course, to define say the η-matrix only for a pair of elementary particles; it must rather be added how the function η is composed in the presence of several particles from the interaction of each pair of them. Possibly, also, many-body forces play a role. In the following we wish to discuss various difficulties, which occur in this context in many-body systems.[59]

Evidently, the program outlined here agrees exactly with the one that Heisenberg had announced in the footnote of the third communication. In a later paper, Heisenberg referred to a part IV of his series "Die beobachtbaren Grössen in der Theorie der Elementarteilchen";[60] a further reference is in a paper by Stueckelberg.[61] We must assume, therefore, that the manuscript is indeed identical with the fourth communication on S-matrix theory.[62]

As we have mentioned, the crucial problem of the entire scattering-matrix approach to elementary particles was the actual construction of the S or η matrices for a given system. Although the η matrix for particles *without* interaction might be composed from the sum of the matrices for individual particles, an interaction between the particles could not be expressed – as in the existing Hamiltonian quantum field theory – by adding a perturbative potential.[63] Upon investigating the particular case of a weak interaction in the usual nonrelativistic quantum theory (where a Schrödinger equation exists) within his formalism, Heisenberg found that in general, the η matrix of the interacting particles is a somewhat complicated function of the η matrix of the free particles and of the interaction potential; he noticed especially that the calculation of the probability for processes involving bound states "in general presumes the accurate knowledge of the S-matrix" and that "one cannot, as a rule, be satisfied with approximations in deriving the S-matrix from the η-matrix, since an approximation solution, which for real momenta represents the S-matrix well, may still have different singularities (in the complex plane) than the exact solution."[64]

The latter remark applied a fortiori to systems with interactions that could not be described by the usual quantum field theory. The construction of the scattering matrix for a system of n interacting particles from the corresponding Hermitian η matrix must be supplemented by assuming additional matrix elements associated with transitions to discrete states, that is, of the η matrix describing the transition to a system with $n - 1$ particles. As a consequence, he derived, for example, the result "that one does not obtain in principle any interaction of the two-particle system with a third particle, if one composes the η-matrix additively from terms each referring only to a pair of particles." Hence, the existence of many-body forces seemed to be inevitable, "and the problem of the η-matrix for many particles...thus remains for the moment, as before, covered by an impenetrable darkness."[65]

The investigation of the η matrix for many-body systems was completed and submitted to the *Zeitschrift für Physik* after Heisenberg's visit to Copenhagen in April 1944, and before another visit to Switzerland in December of the same year.[66] We have information about this visit from a rather unusual source, namely, the American Secret Service agent Moe Berg. Weakly connected with the ALSOS mission – the intelligence project to discover and destroy the alleged German uranium bomb efforts – Berg came in late 1944 to Switzerland and made contact with Paul Scherrer, the experimental physicist at the Eidgenössische Technische Hochschule (ETH).[67] He managed to have Otto Hahn and Heisenberg invited to Switzerland, hoping to learn from them the status of the German project. Heisenberg indeed came and delivered a talk at the Zurich colloquium on 18 December 1944. Berg met him and talked privately with him. Because he did not obtain any statement from Heisenberg indicating an involvement by Heisenberg in bomb construction, he gave up the plan to kill the German scientist.

Whether the details of the Berg story are correct or partly fictional, Heisenberg failed to notice any personal danger. He rather enjoyed the opportunity to escape, for at least a few days, war-beaten Germany and the difficulties of work and life there and to relax with his Swiss friends by discussing with them his latest results on the S-matrix theory.[68] He left copies of two manuscripts in Switzerland on "Die beobachtbaren Grössen in der Theorie der Elementarteilchen," namely, parts III and IV, perhaps thinking that they were safer in Zurich than in Germany.

Heisenberg returned from Switzerland before Christmas 1944. During the following months he had a hectic life, oscillating between Berlin and Hechingen (or Haigerloch, where the last experiments for establishing a chain reaction in a uranium–heavy-water pile were set up) and Urfeld, where the family had been staying for a couple of years. At the end of the war, on 3 May 1945, he was taken prisoner at Urfeld by a unit of the ALSOS mission. Together with Kurt Diebner, Walther Gerlach, and Paul Harteck, who were also captured in Bavaria, he joined other prominent members of the German uranium project (Erich Bagge, Otto Hahn, Horst Korsching, Max von Laue, Carl Friedrich von Weizsäcker, and Karl Wirtz) in France. After some interrogation, the German scientists were brought to England on 23 July 1945 and interned at the country estate Farm Hall near Cambridge, where they were well treated and well watched. Their detainment lasted until 3 January 1946, when the scientists were released to the British Zone in Germany. Heisenberg, together with von Laue, von Weizsäcker, and Wirtz, went to Göttingen to reestablish the Kaiser-Wilhelm-Institut für Physik there.

During their enforced stay in England, the physicists began to discuss purely scientific topics again; in particular, Heisenberg talked on superconductivity with von Laue and on turbulence with von Weizsäcker. They did not have access to a physics library or to new issues of journals, some of which they had not seen for years, but they were visited by a few British colleagues, such as Patrick M. S. Blackett and Geoffrey Ingram Taylor, who tried to supply some literature. In this closed and remote situation, Heisenberg advanced his *S*-matrix program, and he wrote up a review of the subject in a paper entitled "On the Theory of Elementary Particles." In a letter to Bohr in October 1945, he wrote that "this manuscript is a summary of four papers, which... reviews the results of a fine work of Møller and considerations of Kramers." He added: "By the way, I have written the paper in (bad) English, partly in order to practice and partly since I hoped to be able to show it occasionally to English or American physicists. The references have been...left out, because I did not have available the necessary literature."

The letter to Bohr is undated, but Heisenberg wrote it, as he himself mentioned, a few days before Bohr's sixtieth birthday (7 October 1945); he intended to transmit his own congratulations and those of his interned friends. Heisenberg also mentioned that he had sent, through a friend, manuscripts of papers by von Laue, von Weizsäcker, and himself (e.g., the one mentioned earlier) as a tribute to Bohr: "We hope that these papers some time or other may appear in German journals; for the moment it would be very good, if you could receive the papers and take the opportunity to show them to other physicists, in order that they do not totally disappear in the drawer."[69]

In the same letter to Bohr, Heisenberg inquired whether or not Møller's paper on the *S* matrix had already appeared in print. "My correspondence with Møller unfortunately came to a standstill last fall because of the external catastrophes," he wrote, adding, "I ask you, however, to greet Møller and to tell him that I hope soon to reassume the correspondence." More than half a year later, he wrote Møller personally from Göttingen:

> In our Göttingen circle we have debated...much about the problems of the η-matrix and of the elementary particles, and we have also discussed in detail the problems of your previous letter....[Richard] Becker here has found a nice method, of making transparent the mathematical treatment of the problem posed by a given η-matrix, by considerations in Hilbert space.[70]

He further inquired if Møller had meanwhile published the paper announced in his earlier letter and whether or not he had made further progress since that time.

Møller replied soon, stating that "I have written two papers on the subject – as soon as possible I shall send you reprints (at the moment one is not allowed to send reprints)."[71] He also reported that in February 1946 he had given lectures in England on Heisenberg's theory of the scattering matrix and his "own modest contributions" to it. He concluded that "they were very much interested in this matter in England, and at the conference in Cambridge at the end of July the problem of how to determine the *S*-matrix will be one of the points of discussion."

This was quite a bit of news for Heisenberg, and although he did not receive the reprints of Møller's two papers until another year had passed,[72] the scientific world soon learned about the *S*-matrix theory, mainly through those Danish publications.[73,74]

The main content of Møller's first communication was to establish the ideas of Heisenberg's papers I and II in a systematic way. The author proved in particular the Lorentz invariance of the scattering matrix and studied in detail the constants of collision and the incorporation of symmetry properties – those usually assumed in the theory of elementary particles – in the *S*-matrix scheme. In his second communication, Møller focused on the analyticity properties; he showed how complex variables could be introduced in relativistic quantum mechanics; he established the calculation of bound states in *S*-matrix theory, for both real and complex eigenvalues – the latter to describe radioactively decaying states. In short, the two lengthy papers presented nothing less than the complete and detailed formalism of the *S*-matrix theory of elementary particles, based on Heisenberg's pioneering ideas, fully applicable in practical cases.[75] Heisenberg himself was very pleased: "It has made a great impression on me, how detailed and carefully you have discussed mathematically the various aspects of the *S*-matrix problem. Especially your considerations on the transformation properties of the *S*-matrix and the cross sections were, as far as details are concerned, new to me."[76]

Møller was the appropriate spokesman for the *S*-matrix theory at the first international conference after World War II, "Fundamental Particles and Low Temperature Physics," held at the Cavendish Laboratory in Cambridge, 22–27 July 1946.[77] It was too soon for Heisenberg to attend the conference, as his travels were still being restricted. As Møller had noticed earlier, people in England were extremely interested in the ideas of the scattering matrix; besides Møller, Wolfgang Pauli, Walter Heitler, and Stueckelberg also spoke at the conference on related topics;[78–80] 1946 was perhaps *the* year of *S*-matrix theory, when a large number of people became involved. In Germany, for example, Pascual Jordan looked for a connection between the *S*-matrix theory and an operator theory in a von Neumann space.[81] Heisenberg's former teacher, Max Born, wrote in his first postwar letter from England that "I have heard from you indirectly and seen from your papers that you have fine ideas as before. I have read a couple of your papers on the *S*-matrix and have learned further things from Møller's publications. I tried at the Cambridge Congress in July to connect these things with my own, rather nebulous ideas."[82] In Norway, Harald Wergeland, a former associate of Heisenberg, made a few model calculations with the *S*-matrix formalism.[83] The scattering matrix was finally discussed in some detail by Wentzel in his review on the meson theory.[84]

Pauli's crusade against *S*-matrix theory

Although Pauli devoted the main part of his report at the Cambridge conference in the summer of 1946 to the *S*-matrix method,[85] he did not advocate the theory as Møller did. On the contrary, he declared at the beginning of the presentation that "Heisenberg did not give any law or rule which determines mathematically the S-matrix in the region where the usual theory fails because of well-known divergencies. Hence his proposal is at present still an empty scheme."[86] Indeed, Pauli had been skeptical since the time in 1943 when he first heard about the new scheme.[87] Thus, he had stressed in a letter to Paul Dirac the absence of a method for obtaining the scattering matrix for elementary particles, and he even claimed that it might be dangerous to include the energy shell among the assumptions of a fundamental theory.[88]

When in September 1945 Pauli heard about the progress of the S-matrix work in Germany and Copenhagen through a letter from Møller, he replied:

> The idea of an analytic continuation of S in order to determine the discrete states (and their energies) by the condition $S(K_n', l')$ (in your notation $[K_n', l'$ denote mass and angular momentum of the bound state]), was quite new to me and interested me very much. I am, however, not sure whether this analytic continuation in the present quantum theory leads to branching points or other singularities before one reaches the zeros of S.[89]

Pauli even went further, criticizing the concept of the analytic continuation to be "a bit alien to physics," since one never knows whether it works or not.[90] Later, after he had returned from Princeton to Zurich in the spring of 1946, he wrote again to Møller, stating that he found the manuscripts of parts III and IV of Heisenberg's S-matrix papers – which the latter had "left here on his visit at Xmas 1944" – and commented: "I am still not convinced, not only the whole frame of concepts is empty (no theory is given which determines S), but the rather complicated formalism does not contain the classical mechanics as a limiting case."[91]

To find the reason for Pauli's rejection of the scattering-matrix approach to elementary particle physics, one must go back more than fifteen years. In the late twenties, he had tried to develop with Heisenberg a consistent relativistic quantum field theory of interacting particles, but they had soon noticed that incurable defects, especially infinities, showed up. During the thirties, Pauli and Heisenberg had embarked on the problem of removing the infinities by rather drastic assumptions – such as a lattice structure of space – again without success. They had, however, speculated that a consistent, finite theory should exist only for given values of the coupling constants (i.e., the theory must eventually yield the empirical values of the electromagnetic fine-structure constant, the mass of the electron, etc.). Toward the end of the thirties, Heisenberg emphasized the role of a universal length limiting the application of the relativistic quantum theory. Because this idea had not resulted in giving the desired theory, Pauli had taken a lukewarm attitude toward the concept of the universal length. Now, in the midforties, he considered the new S-matrix theory to be a formulation of that idea, and the fact that it did not provide the tools to determine the matrix elements confirmed his doubts about the whole approach. Pauli, therefore, was happy to discover faults in the Heisenberg–Møller scheme, in order to prevent Heisenberg and others from investing too much labor in the pursuit.

One weak point was the failure to prove analyticity of the S matrix, and a second point soon showed up, as he wrote to Møller: "Before I left Princeton, S. T. Ma found an interesting example (potential $\exp(-\mu r)$, S-term), where the analytic continuation of the S-matrix gives too many zeros: Some redundant zeros don't correspond to eigenvalues of wave mechanics."[92]

The redundant zeros discovered at Princeton played a crucial role in Pauli's subsequent correspondence with Heisenberg.[93] He first informed Heisenberg in some detail about Ma's result in a letter dated 9 September 1946. Besides obtaining unwanted zeros, Ma had also discovered some errors in Heisenberg's fourth, unpublished communication. Pauli then mentioned that Ma's unpleasant zeros disappeared in the case that the potential had just a finite range [i.e., became strictly zero for $r > R$ (finite number)]. He added: "It appears, however, to me to be very unsatisfactory that in S-matrix theory there is an essential difference whether one uses *sharply* cut-off potentials or those that decrease gradually."[94]

Ma's discovery disturbed the Copenhagen partisans of S-matrix theory. The young Dutch visitor Dirk ter Haar proved, however, that the redundant zeros did not satisfy a condition that Møller had established to hold for the value of the scattering matrix at the resonance points; he also detected other difficulties in the S-matrix approach (e.g., in the definition of bound states in the presence of a Coulomb potential).[95] Res Jost, from Zurich, investigated the situation further[96] and summarized the results in a letter as follows: "In general, it seems to me that there does not exist a simple relation between $S(k)$ and the stationary states."[97]

Pauli, in his September letter to Heisenberg quoted earlier, also commented on the radiation-damping theory of Heitler, which seemed to be related to the theory of the scattering matrix.[79] He noted that this theory led to unsatisfactory results, such as the wrong magnetic moment for the proton. Altogether, he emphasized the following negative conclusions:

> In general I have arrived at the opinion that the S-matrix is not a concept, of which we may expect that it occurs in a future theory as a primary fundamental concept. It indeed has the character of something complicated and derived and therefore might hardly be suitable to lead us beyond the present wave mechanics.[98]

He added: "Should you be able to teach me something better, I would be delighted." Heisenberg could not. Four months later, Pauli, after returning from Stockholm (where he had received the Nobel Prize for 1945), declared further:

> I belive that no real progress is possible any more without having a theory that determines the quantity $e^2/\hbar c$. In your point of view of the universal length, I miss, however, the connection with the $e^2/\hbar c$ problem; in addition I

do not know whether perhaps different lengths are associated with different particles.[99]

On his way from Stockholm to Zurich, Pauli had stopped in Copenhagen and met Møller, who had meanwhile abandoned work on the *S*-matrix theory, "since he does not see any possibility of getting on with it."[100]

Heisenberg, on the other hand, did not give up the approach completely. He continued to study specified problems connected with it; and with respect to the redundant zeros, which bothered Pauli so much, he concluded simply "that the existence of a singularity of *S*, though being a necessary, is not a sufficient condition for the existence of a stationary state."[101] He also thought unsuccessfully about conditions to determine the scattering matrix in the case of many-body systems. He further defended the idea of a universal length, arguing "that it must be put into the foundations of the future theory at a central place, and that from this length all others (like the mass of the individual elementary particles) must follow in the same sense as the energy levels of the hydrogen atom follow from the Hamilton function of this atom."[102] In pursuing this goal, he soon turned back to a quantum field theory that was a generalization of the conventional ones in the sense indicated in the third communication on the *S*-matrix theory.[103]

Application to quantum electrodynamics and new prospects

In a letter in the spring of 1948, Pauli reported to Heisenberg that "Weisskopf will come to Zurich in the middle of May...and will tell us about the work by himself, Schwinger, and others. I have the feeling that now one might really be able to guess something about the new formalism."[104] Pauli was speaking about the rapid progress in quantum electrodynamics that occurred after the summer of 1947, especially after the measurement of the splitting of the $2S$ levels and $2P$ levels of hydrogen by Willis E. Lamb and Robert Retherford became known.[105] The work of many physicists, notably Julian Schwinger and Richard P. Feynman, combined with earlier investigations of Sin-itiro Tomonaga, led to the establishment of a renormalized theory of electrons and photons. Freeman J. Dyson, in uniting the three approaches in 1949, made effective use of the concept of the *S* matrix.[106,107]

In his paper "The *S*-Matrix in Quantum Electrodynamics," Dyson carried out the calculations of the matrix elements, including their renormalization.[107] In the introduction, he remarked that "the Feynman method is essentially a set of rules for the calculation of the elements of the Heisenberg *S*-matrix and can be applied with directness to all kinds of scattering problems," adding in a footnote that "the idea of using standard electrodynamics as the starting point for an explicit calculation of the *S*-matrix has been previously developed by Stueckelberg," and referring especially to the papers of the Swiss author published between 1943 and 1946.[108] Thus, he established contact with the earliest application of the *S*-matrix scheme in particle physics.

The success of renormalization in quantum electrodynamics initiated attempts to obtain a satisfactory renormalized meson theory of nuclear forces. However, it yielded few reasonable results. As a consequence, copying Dyson's approach in meson theory led only to qualitative conclusions (e.g., that a bound nucleon–meson state should exist).[109] All these investigations may be summarized as follows: Given a certain quantum field theory – or its (often nonrelativistic) approximation – then the S-matrix elements can be derived. Or, as Pauli noted more negatively, "the S-matrix, although it might exist in a future theory, seems to be completely unfit to constitute the point of departure for a [new] theory. It is not *the* quantity which will occur in the general laws of nature, but a late consequence of them."[110]

In contrast to the interest it aroused at the Cambridge conference of 1946, the S-matrix theory did not play an important role at later international meetings on elementary particles until the late fifties. Rather, quantum electrodynamics, and especially the newly discovered particles such as the π meson, occupied the interest of the physicists at, say, the eighth Solvay conference in the fall of 1948. "I was at Brussels, where I also met [Edward] Teller and where Bohr and [Robert] Oppenheimer talked much about the elementary particles in the general discussion," Pauli informed Heisenberg afterward. Pauli concluded that he did not expect any radical change to emerge from new discoveries, but "what future experiments may teach us will be empirical rules," admitting at the same time that "it would be easier, of course, to find a [new] theory, if one possessed quantitative empirical laws."[111]

Heisenberg was not asked to attend the Brussels conference, but in the following year (1949) he participated in the International Congress on the Physics of Cosmic Rays in Como, Italy. Around that time, his official scientific isolation ended, and that of the German physicists in general.[112] Already in December 1947, however, he had been invited to deliver lectures in Cambridge, Bristol, and Edinburgh. In Cambridge, in particular, he presented a general talk on "The Present Situation in the Theory of Elementary Particles," dealing mainly with the S-matrix approach.[113] A young student at Peterhouse College, Richard John Eden, who had become interested in the subject, was introduced to him and proved to be helpful during his stay. Eden continued his contact with Heisenberg between the spring of 1948 and the spring of 1950, exchanging numerous letters on the status and progress of work on the S-matrix theory.

Eden began the correspondence by sending Heisenberg a manuscript dealing with a formula that had been used by Møller in his proof of the unitary condition, and Eden asked Heisenberg's opinion about it.[114] A month later he presented to Heisenberg the manuscripts of two papers on the analytic behavior of the S matrix, with the same request.[115] Heisenberg replied by letters and discussed the work in detail with Eden during the latter's visit to Göttingen, 23–30 June 1948.[116] "I had not had such an enjoyable and interesting week for a long time," Eden characterized his visit in a letter to Heisenberg a few months later. He further reported that "Professor Dirac has recently returned here from America, and has kindly consented to supervise my research during the coming year, but I have not yet had the opportunity to discover his opinion on *S*-matrix theory."[117]

Less than three months later, Eden completed a thesis that he wanted to submit for a fellowship at his college. He asked Heisenberg, whom he thanked "for the many discussions which helped me either directly or indirectly in writing my thesis," to comment on the work.[118] Heisenberg looked through the work and found no objections.[119] Eden submitted and published his thesis, which was communicated by Dirac to the Royal Society of London and published later in its proceedings.[120]

In the first paper, Eden investigated the structure of the scattering matrix of a fixed number of particles; he found two types of outgoing waves, one of them involving an interference effect between the incident wave of one particle and the outgoing wave of another. Both types led to different bound states. While this study was rather straightforward and unproblematic in principle, the consideration of systems with a varying number of particles gave rise to greater difficulties. Ning Hu, a former collaborator of Heitler, had recently obtained some results, but Eden doubted their reliability, because a nonunitary *S* matrix was involved.[121] He therefore started a detailed calculation of a particular model, in which he considered the discontinuous process occurring "when an incident particle is scattered on a fixed centre of force which itself has internal excited states of different energies" or "when the centre can then absorb some of the kinetic energy of the particle and cause inelastic scattering" – in his second paper.[122] As a consequence, new aspects appeared in the scattering-matrix theory; for example, "certain matrix elements of *S*, treated as functions of energy, have natural bounds at the threshold values for inelastic scattering," and one can "define those functions to be zero outside their natural bounds." Hence, "these matrix elements are then non-analytic functions of energy, but it does not follow that every matrix element of *S* becomes non-analytic at each threshold value, or that the eigenvalues of *S* are non-analytic."[123]

Eden's work extended the scattering-matrix formalism of elementary particles as established by Heisenberg and Møller. During the next several years, he abandoned work on *S*-matrix problems, but in the summer of 1951 he submitted a paper analyzing the threshold behavior of the *S* matrix for systems in which massive particles can be created.[124] "At the thresholds of such creation processes the *S*-matrix will have a complicated behaviour," he concluded from an evaluation of the scattering-matrix elements in renormalized quantum field theory, explaining further:

> For a typical matrix element there are thresholds of two main types. The first is a creation threshold below which the element is zero on account of energy-momentum conservation. . . . The second is an interference threshold above which a competing process has non-zero probability. . . . Between the

threshold values it is shown that the *S*-matrix elements are analytic functions of the energies and momenta of the incident particles.[125]

Heisenberg had followed Eden's work with great interest. It even stimulated him – as he occasionally mentioned in the correspondence – to work on some specific questions. After 1952, the original scattering-matrix program was abandoned (even by Eden). It was left in that dormant state until, in the late fifties, important new aspects, such as the inclusion of crossed channels and the idea of an analytic behavior in the complex angular-momentum plane, arose and caused the rise of a new *S*-matrix program.[126]

During most of the fifties, another aspect of the early *S*-matrix theory, discussed first by Kronig – whom Heisenberg had visited in October 1943 at Delft – played an important role. In a short note submitted in the fall of 1946 to *Physica*, Kronig expounded a connection between the analytic properties of the scattering matrix and causality.[127] The author derived a relation between the real and the imaginary parts of the scattering amplitudes, similar to one observed twenty years earlier in the diffraction of light by atoms. Those so-called dispersion relations, developed from 1954, allowed the description of the scattering of strongly interacting elementary particles that could not be accounted for at that time by any renormalizable quantum field theory.

In spite of its genesis in a most difficult time, when communication between physicists was extremely poor, the *S*-matrix theory rendered useful services, not only for that moment, when most physicists were concerned with more practical matters than thinking about high-energy physics, but also for the future growth of elementary particle physics in the early fifties and beyond.

Notes

1 S. Schweber, *Mesons and Fields, Volume I: Fields* (Evanston: Row, Peterson & Co., 1955), p. 163.

2 There exist some other discussions on the *early* development of *S*-matrix theory. See Reinhard Oehme's "Theory of the Scattering Matrix (1942–1946)," in Werner Heisenberg, *Collected Works, Volume A II* (Berlin: Springer-Verlag, 1988), for an analysis of the physical contents of Heisenberg's pioneering work. For more "external" historical accounts, see Inge Grythe, "Some Remarks on the Early *S*-Matrix," *Centaurus 26* (1982), 198–203, and James T. Cushing, "The *S*-Matrix Program – Anatomy of a Scientific Theory," unpublished: University of Notre Dame preprint, April 1985, and "The Importance of Heisenberg's *S*-Matrix Program for the Theoretical High Energy Physics in the 1950's," *Centaurus 29:2* (1986), 110–49.

3 J. A. Wheeler, "On the Mathematical Description of Light Nuclei by the Method of Resonating Group Structure," *Phys. Rev. 52* (1937), 1107–22 (received 17 August 1937, published in the issue of 1 December 1937).

4 W. Heisenberg, "Die 'beobachtbaren Grössen' in der Theorie der Elementarteilchenphysik," *Z. Phys. 120* (1943), 513–38 (received 8 September 1942, published in the issue of 25 March 1943).

5 Heisenberg analyzed these points very clearly in his "Bericht über die allgemeinen Eigenschaften der Elementarteilchen" (Report on the General Properties of Elementary Particles), which he prepared for the eighth Solvay conference, planned for the fall of 1939. See Werner Heisenberg, *Collected Works, Volume B*, edited by W. Blum, H.-P. Dürr, and H. Rechenberg (Berlin: Springer-Verlag, 1984), pp. 346–58.

6 See note 4 (p. 513).

7 See note 4 (p. 514).

8 The similarities to the story of the twenties goes even further. Albert Einstein, in discussing the foundations of quantum mechanics with Heisenberg in April 1926, had refuted the principle of using only observables in a physical theory: "In reality the very opposite happens. It is the theory which decides what we can observe." See Werner Heisenberg, *Physics and Beyond – Encounters and Conversations* (New York: Harper & Row, 1971), p. 63. In 1942, Heisenberg recalled this argument when he wrote in his first paper on the *S* matrix that "admittedly it is always the final theory that decides which quantities are really observable." Yet he argued: "But already before we know the future theory, the study of the difficulties of the previous theory may provide hints that certain concepts will not be touched. Thus one may arrive at relations between observable quantities, which are not only part of the old but also probably of the future theory." See note 4 (p. 514).

9 The *S* matrix for elementary particle scattering exhibits also trivial singular parts, corresponding to transitions in which some incoming and outgoing states are identical (in momentum, energy, etc.). The nontrivial (and nonsinglular) parts or elements represent essentially the phase differences between incoming and outgoing waves describing the scattering objects. By the way, Wheeler defined the scattering matrix of his nuclear problem exactly this way.

10 W. Heisenberg, "Die beobachtbaren Grössen in der Theorie der Elementarteilchen. II," *Z. Phys. 120* (1943), 673–702 (received 30 October 1942, published in the issue of April 1943).

11 Since 1936, Heisenberg had claimed that so-called explosive showers (i.e., direct multiparticle production) existed in cosmic radiation. See his paper "Zur Theorie der 'Schauer' in der Höhenstrahlung," *Z. Phys. 101* (1936), 533–40. Although the empirical evidence was, by 1942, not yet convincing for most physicists, Heisenberg insisted on their reality. Explosive showers seemed to be a natural consequence of the alleged nonlinearity of elementary particle theory.

12 Heisenberg to Geiger, 13 October 1942: "Leider habe ich für diese Festschrift nichts gehabt, was in engerem Zusammenhang mit Ihrer Forschungsarbeit, z.B. über kosmische Strahlung, gestanden hätte. Aber ich habe mich schon längere Zeit mit diesen rein theoretischen Ueberlegungen beschäftigt, die sich jetzt weiter fortgesetzt haben."

13 W. Heisenberg "Ueber die in die Theorie der Elementarteilchen auftretende universelle Länge," *Ann. Phys. 5:32* (1938), 20–33, especially 33.

14 In September 1939, Heisenberg was drafted by the *Heereswaffenamt* (Army Weapons Bureau), which until the spring of 1942 directed a secret uranium project (*Uranverein*). After preliminary theoretical and experimental studies, it was decided (in the summer of 1942) that the members of the *Uranverein* should attempt the construction of a uranium reactor. Heisenberg was appointed leader of this project in July 1942 and was called to the Kaiser-Wilhelm-Institut für Physik in Berlin–Dahlem, the center of this still-secret work.

15 Wick had spent some time as a visitor at the University of Leipzig, after obtaining his doctorate in 1930. He was later professor in Palermo (1937), Padova (1938), and finally Rome (1940). After the war, he emigrated to the United States.

16 G. C. Wick to W. Heisenberg, undated, Werner-Heisenberg-Archiv, Werner-Heisenberg-Institut für Physik, Munich.

17 W. Heisenberg to G. C. Wick, 19 June 1942, Werner-Heisenberg-Archiv.

18 G. C. Wick to W. Heisenberg, 22 February 1943. That Wick came to Berlin and saw the Kaiser-Wilhelm-Institut can be derived from his stay at the Harnack-Haus, the guest house of the Kaiser-Wilhelm-Gesellschaft (see the cable of Victor Regener to Wick, dated 6 July 1942, Werner-Heisenberg-Archiv).

19 See note 4 (p. 533, footnote 1).

20 W. Heisenberg to G. C. Wick, 10 February 1943, Werner-Heisenberg-Archiv.

21 G. C. Wick to W. Heisenberg, 22 February and 15 March 1943; W. Heisenberg to G. C. Wick, 8 March 1943, Werner-Heisenberg-Archiv.

22 W. Heisenberg, note 4 (p. 533, footnote 1).

23 G. Breit, "Scattering Matrix of Radioactive States," *Phys. Rev. 58* (1940), 1068–74 (received 28 October 1940, published in the issue of 15 December 1940). Breit used Wheeler's scattering-matrix theory (to which he refers explicitly in footnote 2 on p. 1068) to describe the decay modes of radioactive nuclei. He found in a special example that the scattering matrix can be composed of two terms, "resonance" and "background" scattering, the latter referring to elastic scattering.

24 Thus, Heisenberg wrote in the summer of 1943 from Berlin a letter to Gregor Wentzel in Zurich, with the following request: "Since we are here with respect to foreign literature completely sitting on the dry, I want to ask you whether you have received new interesting papers on cosmic radiation, elementary particles and the like. It would be very kind of you, if you keep us a little bit informed in this respect" (W. Heisenberg to Wentzel, 12 August 1943). However, the Swiss situation at that time was not much better; e.g., Fritz Coester remarked in a paper submitted in December 1943 that "American publications since May 1942 are presently in Switzerland not available." Coester, "Ueber die Stabilität schwerer Kerne in der Mesonentheorie," *Helv. Phys. Acta 17* (1944), 35–58.

25 According to the material kept in the Yukawa Hall Archival Library, Yukawa had those copies in hand by the end of 1943. Especially, box no. 3 of his papers contained a photoprint of Heisenberg's first paper (note 4) and a mimeographed copy of the second article (note 10), dated 13 December 1943 and distributed by the Science Bureau, Investigation Section, Japanese Ministry of Education. The latter was stamped "secret for outside the department." Yukawa received copy no. 20 of that paper; other recipients were S. Sakata (no. 3), K. Ariyama (no. 5), M. Kobayashi (no. 15), and S. Nakamura (no. 18). Yukawa was, in those days, a member of the Liaison Council of Physical Research, connected with the Ministry of Education. For detailed information on this matter and on the following items concerning the German–Japanese exchange, I wish to thank Prof. Rokuo Kawabe, Yukawa Hall Archival Library, Kyoto, Japan 606.

26 For example, in August 1943, the German submarine U-511, which was a gift from Germany to Japan.

27 The Japanese class I submarine no. 29, which arrived 11 March 1944 at the French port of Lorient; it left 16 April, and was sunk 26 July 1944 in the Balitang Channel close to Hong Kong by an American torpedo attack.

28 M. Nomaguti, a captain of the Imperial Japanese Navy and then at the Japanese Embassy in Berlin, was a brother of the physicist Kaneteka Ariyama, who had been a visitor in Heisenberg's Leipzig Institute during the thirties. Nomaguti promised to take care of the letter. The Japanese papers mentioned in Heisenberg's letter appeared in issue no. 3–4 of *Zeitschrift für Physik*, Volume 119, dated 9 July 1942: K. Ariyama, "Zur Theorie der Supraleitung," pp. 174–81; Y. Fujioka, "Molekülspektren des D_2-Wasserstoffs im ultravioletten Gebiet und Isotopeffekt des D_2-Normalzustandes," pp. 182–4; S. Kikuchi, Y. Watase, and J. Itoh, "Ueber die Winkelabhängigkeit der zwei γ-Quanten, die von einem Atomkern kaskadenweise ausgestrahlt werden," pp. 185–7; S. Misushima and Y. Morino, "Ueber die innermolekulare Drehbarkeit bei einfacher Kohlenstoffbindung," pp. 188–94; Y. Nishina, K. Kimura, and M. Ikawa, "Einige Spaltprodukte aus der Bestrahlung des Urans mit schnellen Neutronen," pp. 195–200; H. Yukawa, "Bemerkungen über die Natur des Mesotrons," pp. 201–5. They were intended for a *Festschrift* issue celebrating Heisenberg's fortieth birthday, but this publication was delayed – the date of celebration being 5 December 1941 – and the intention was not mentioned in the July 1942 issue.

29 W. Heisenberg to Y. Nishina, 18 January 1944, Werner-Heisenberg-Archiv.

30 For example, Sin-itiro Tomonaga used the concept of a "characteristic matrix" in his work, on "A General Theory of Ultra-Short Wave Circuits" during the war, submitted in August 1946 and published in *J. Phys. Soc. Jpn. 2* (1947), 158–71; *J. Phys. Soc. Jpn. 3* (1948), 93–105. He referred to Breit's paper of 1940 (note 23), but not to Heisenberg's publications of 1943, although his investigations were carried out in 1944, i.e., when copies of those papers were available in Japan. We assume that Tomonaga also had either heard of Heisenberg's *S*-matrix papers or seen them, because he used the phrase "characteristic matrix" (which occurs only in Heisenberg's publications).

31 After hearing about Heisenberg's prospective visit to Zurich in the fall of 1942, Stueckelberg wrote to Heisenberg and invited him to come to Geneva as well (E. C. G. Stueckelberg to W. Heisenberg, 8 August 1942, Werner-Heisenberg-Archiv). The latter agreed first to do so, announcing his visit to Geneva for the end of November and offering either a talk on the present status of cosmic-ray physics (the same lecture he gave in Zurich on 19 November) or – just for a more specific audience of theoreticians – a talk on his recent work on *S*-matrix theory (W. Heisenberg to E. C. G. Stueckelberg, 14 October 1942, Werner-Heisenberg-Archiv). Although a date for Heisenberg's Geneva lecture was fixed, 21 November 1942 (see W. Heisenberg to E. C. G. Stueckelberg, 16 November 1942, Werner-Heisenberg-Archiv), Heisenberg did not come to Geneva.

32 E. C. G. Stueckelberg to W. Heisenberg, 13 February 1943, Werner-Heisenberg-Archiv.

33 E. C. G. Stueckelberg, "Une modèle d'electron ponctuel II," *Helv. Phys. Acta 17* (1944), 3–26 (received 29 October 1943, published in the issue of 29 February 1944). Stueckelberg pointed out the connection between his electron theory and *S*-matrix theory first in a paper read at the Schaffhausen meeting of the Swiss Physical Society, 28–29 August 1943: "Le freinage du rayonnement en théorie des quanta," *Helv. Phys. Acta 16* (1943), 427–8.

34 E. C. G. Stueckelberg to W. Heisenberg, 23 March 1944, Werner-Heisenberg-Archiv.

35 E. C. G. Stueckelberg, "Mécanique fonctionelle," *Helv. Phys. Acta 18* (1945), 195–220 (received 27 January 1945, published in the issue of 28 May 1945); "Une propriété de l'operateur *S* en mécanique asymptotique," *Helv. Phys. Acta 19* (1946), 242–3 (presented on 4 May 1946 at the Aarau meeting of the Swiss Physical Society, published in the issue of 31 July 1946); E. C. G. Stueckelberg and D. Rivier, "Opérateurs non linéaires en théorie des quanta," *Helv. Phys. Acta 19* (1946), 240–2 (presented on 4 May 1946 at the Aarau meeting of the Swiss Physical Society, published in the issue of 31 July 1946). In the paper presented at the Aarau meeting, Stueckelberg reformulated Eq. (39.2) as

$$S = \frac{1 - i/2\alpha}{1 + i/2\alpha}$$

in order to avoid negative energy state in his theory. The use of such states for curing the divergence evils of quantum field theory was proposed by Paul Dirac in his Bakerian Lecture of 1941. Stueckelberg received two copies of the published paper, "The Physical Interpretation of Quantum Mechanics," *Proc. R. Soc. London, Ser. A 180* (1942), 1–40, one from Dirac himself and one from Erwin Schrödinger in Ireland. He then sent one copy to Max von Laue in Berlin; he did the same with a subsequent paper by Wolfgang Pauli, "On Dirac's New Method of Field Quantization," *Rev. Mod. Phys. 15* (1943), 175–207. He studied the papers of Dirac and Pauli carefully and concluded that "I do not think highly of it, since I am now convinced that one is able to carry out covariant subtractions without complicated hypotheses on light quanta of negative energy" (E. C. G. Stueckelberg to W. Heisenberg, 23 March 1944, Werner-Heisenberg-Archiv). Stueckelberg's work on the covariant subtraction procedure paved the way to renormalized quantum electrodynamics, in which, indeed, a procedure without using negative energy states (more accurately, without states of negative probability) seemed to be possible.

36 At this point, we are not referring to the ordinary problems of a German scientist who wanted to travel abroad in those days. One had first to ask the authorities (at the university or the Ministry of Education) for permission to make a particular visit, then apply for an appropriate amount of foreign currency, which often was not granted because of lack of funds in Germany or because of other, less rational, reasons. If one really did go abroad, one had to make official visits to the German representatives (at the embassies or consulates or at offices of the National Socialist Party). Finally, after returning to Germany, one had to write a more or less detailed report on the visit to the Reichsminister für Erziehung und Volksbildung (minister of education). For a man like Heisenberg, who was involved in the secret German reactor project, the whole situation required particular caution and care in taking his steps and actions before, during, and after each visit abroad.

37 General secretary of the Dutch Ministry of Education, Science, and Cultural Affairs to W. Heisenberg, 28 May 1943, Werner-Heisenberg-Archiv.

38 W. Heisenberg to the general secretary of the Dutch Ministry of Education, Science and Cultural Affairs, 21 June 1943, Werner-Heisenberg-Archiv.

39 H. Kramers to W. Heisenberg, 29 July 1943, Werner-Heisenberg-Archiv.

40 Kramers further requested in his letter copies of two papers (one of which had been published in the *Physical Review* of 1941, the other in the Italian journal *Ricerche Scientifiche* of the same year) and expressed the expectation that Heisenberg might bring with him much scientific news. Clearly, in Holland the situation with foreign journals during the war was even worse than in Germany. (All the Heisenberg–Kramers correspondence is in German.)

41 W. Heisenberg to H. Kramers, 20 August 1943, Werner-Heisenberg-Archiv.

42 H. Kramers to W. Heisenberg, 5 October 1943, Werner-Heisenberg-Archiv.

43 The foregoing itinerary is from Heisenberg's "Bericht über eine Reise nach Holland," dated 10 November 1943 and addressed to the Ministry of Education (see note 36).

44 H. Kramers to W. Heisenberg, 1 December 1943, Werner-Heisenberg-Archiv.

45 W. Heisenberg to H. Kramers, 10 January 1944, Werner-Heisenberg-Archiv.

46 Ibid.

47 H. Kramers to W. Heisenberg, 12 April 1944, Werner-Heisenberg-Archiv, Kramers gave his talk, "Fundamental Difficulties of a Theory of Particles," in Dutch on 14 April 1944 at the symposium on elementary particles in Utrecht, published in the July–August issued of *Ned. Tid. Natuurk. 11* (1944), 134–40. In it, he spoke on (1) difficulties in the classical electron theory, (2) difficulties in the modern theory of particles, and (3) attempts to solve the difficulties. Among the latter attempts, he mentioned (1) new theories of Fritz Bopp and Stueckelberg, (2) ideas of Max Born and Leopold Infeld, of Paul Dirac, and of W. Opechowski and himself to introduce suitable changes in the classical electron theory before quantization (in order to get a finite quantum electrodynamics), and (3) Heisenberg's theory of the scattering matrix.

48 Ibid.

49 W. Heisenberg, "Die beobachtbaren Grössen in der Theorie der Elementarteilchen. III," *Z. Phys. 123* (1944), 93–112 (received 12 May 1944, published in the issue of 10 October 1944).

50 Note 49 (pp. 94–5). We find the first published statement about the analyticity property of the *S* matrix and its relation to the bound-state problem in Kramers's Utrecht report, where he wrote that "it is interesting that the scattering matrix is also able in principle to answer the question, in which stationary states the particles considered can be bound together. These are related to the existence and the position of zeros and poles of the eigenvalues of the scattering matrix, considered as a complex function of its arguments." See note 47 (pp. 139–40). After that statement, Kramers immediately pointed out that, meanwhile, Heisenberg had treated a particular example of a two-particle system "in which a perfectly sharply relativistically determined stationary state occurs, while there are no divergence difficulties whatsoever."

51 Note 49 (pp. 96–8).

52 Note 49 (pp. 98–103).

53 Note 49 (p. 98, footnote 1).

54 Heisenberg's paper (note 49, p. 104) also contains a part II, which is not directly connected with the *S*- or η-matrix approach (pp. 104–11). In this part, the author discussed generalized field theories, e.g., those leading to linear integral equations or differential equations of higher than second order (or than had been considered in the conventional field theory). Although it was not sure that the equations suggested actually possessed finite or nontrivial solutions at all, Heisenberg tried to probe this possibility of avoiding the known divergent results of quantum field theory. He argued that "on the whole the two parts of the paper therefore should answer the question how far *maximally* and *minimally* the present theory has to be altered; that is, they should include the future theory in its general, formal scheme between as narrow bounds as possible" (p. 94).

55 L. Rosenfeld to W. Heisenberg, 10 December 1943, Werner-Heisenberg-Archiv.

56 L. Rosenfeld to W. Heisenberg, 14 April 1944, Werner-Heisenberg-Archiv.

57 One of Heisenberg's duties in Copenhagen was to deliver a talk at the Deutsches Wissen-

schaftliches Institut. There he spoke on 19 April 1944 about a general topic, probably "Die Atomtheorie vom Altertum bis zum Ende des 19. Jahrhunderts" (Atomic Theory from Antiquity to the End of the Nineteenth Century), a subject that he had treated in his recent book *Die Physik der Atomkerne* (Braunschweig: Fr. Vieweg & Sohn, 1943). See also note 58.

58 The proofsheets (item 1) were those of Chapter I of *Die Physik der Atomkerne*.

59 "Die Behandlung von Mehrkörperproblemen mit Hilfe der *S*-matrix," Werner-Heisenberg-Archiv.

60 W. Heisenberg, "Der mathematische Rahmen der Quantentheorie der Wellenfelder," *Z. Naturforsch. 1* (1946), 608–22 (received 5 August 1946, published in issue no. 11–12 of 1946) (see especially footnote 3 on p. 608).

61 E. C. G. Stueckelberg, "Mécanique fonctionelle," note 35 (p. 220). The author knew about the work from Heisenberg's previous Zurich visit, as described later.

62 I suspected for a while that the manuscript in the Werner-Heisenberg-Archiv would give the content of the announced, but unpublished, part IV. This suspicion was confirmed by the existence of a typescript copy of a manuscript by Heisenberg to which James T. Cushing of the University of Notre Dame has kindly drawn my attention; he also supplied me with a copy. On checking the contents of both items, the identity of the texts has been proved. Also, the author's address on the typescript (Berlin–Dahlem, Max-Planck-Institut der KWG) confirms that it was composed at the right time. Professor Cushing received the typescript from Fritz Coester, now at Argonne National Laboratory. Coester was a student in Zurich and heard Heisenberg's lecture on the *S*-matrix theory in November 1942. In 1944, having completed his doctorate, he worked in Winterthur, but visited Zurich occasionally. He recalled: "My best guess is that Heisenberg left the manuscript with Gregor Wentzel during his second visit in Zurich and that I acquired it from him [Wentzel] some time in 1945" (F. Coester to H. Rechenberg, 18 July 1985). I would like to thank him and Professor Cushing for the information clearing up this point. The Heisenberg manuscript on the many-body problem in the η-matrix theory was published in Werner Heisenberg, *Collected Works, Volume A II* (Berlin: Springer-Verlag, 1988).

63 For systems of free particles, Heisenberg observed that the η matrix is indeed the sum of those for single particles. In determining the asymptotic behavior of the system, however, one needs not only the η matrix of the full system but also the η matrices of the parts separately. (See section I of the manuscript.)

64 Note 59 (p. 15).

65 Note 59 (manuscript, pp. 20–1). Kramers, in his letter to Heisenberg of 12 April 1944, had expressed doubt concerning a statement opposite to this one in Heisenberg's third communication. See note 49 (p. 98).

66 This visit cannot be traced in the documents kept in the Werner-Heisenberg-Archiv. However, it is confirmed by several independent sources. For example, Stueckelberg referred to it in a postscript to a paper that he finished in December 1944: "In the manuscripts of two recent papers III and IV (which should appear in *Zeitschrift für Physik*, 1945), and in a talk presented to the theoretical seminar in Zurich, Heisenberg has completed his theory proposed in (1943) I and II." See "Mécanique fonctionelle," note 35 (p. 220) Also, Gregor Wentzel mentioned it in a footnote to a later review, "Recent Research in Meson Theory," *Rev. Mod. Phys. 19* (1947), 1–18, especially footnote 69 on p. 17. The reason for the absence of documents in the Heisenberg files may be that the second Zurich visit happened toward the end of the war; the official correspondence may have been displaced or lost, since Heisenberg was oscillating at that time between Berlin and the region of Hechingen in southwest Germany, where the Kaiser-Wilhelm-Institut für Physik had been moved in the middle of 1943.

67 For the report of Berg's activities in Switzerland, see Louis Kaufman, Barbara Fitzgerald, and Tom Sewell, *Moe Berg – Athlete, Scholar, Spy* (Boston: Little, Brown, 1974), Chapter 7.

68 The recollection of Heisenberg, reported in the book on Moe Berg (note 67, p. 196), is not correct; the incident happened during his second wartime visit to Switzerland, which he did not remember later.

69 W. Heisenberg to N. Bohr, undated, Werner-Heisenberg-Archiv; Heisenberg actually sent

two manuscripts of his own to Bohr, one on superconductivity and the other (mentioned earlier) on the *S*-matrix theory of elementary particles.

70 W. Heisenberg to C. Møller, 1 June 1946, Werner-Heisenberg-Archiv.

71 C. Møller to W. Heisenberg, 7 July 1946, Werner-Heisenberg-Archiv.

72 See Heisenberg's letter to Møller, dated 23 June 1947: "For sending us the reprints, which were brought by Mr. Fraser [a British liaison officer for the German scientists] from Copenhagen, I thank you and the whole Institute.... But your theoretical papers I have seen [now] for the first time."

73 C. Møller, "General Properties of the Characteristic Matrix in the Theory of Elementary Particles, I," *K. Dan. Vidensk. Selsk. Mat.-Fys. Medd. 23* (1945), 1–48 (received 1 December 1944, published in July 1945).

74 C. Møller, "General Properties of the Characteristic Matrix in the Theory of Elementary Particles. II," *K. Dan. Vidensk. Selsk. Mat.-Fys. Medd. 22:19* (1946), 1–46 (received 17 April 1946, published in October 1946).

75 Møller's lectures on "Recent Developments in Relativistic Quantum Theory," delivered in the spring of 1946 at the H. G. Wells Physical Laboratory of Bristol University – which he mentioned in the letter to Heisenberg – give a simplified presentation of the formalism.

76 W. Heisenberg to C. Møller, 23 June 1947, Werner-Heisenberg-Archiv.

77 C. Møller, "On the Theory of the Characteristic Matrix," in *Report of an International Conference on Fundamental Particles and Low Temperatures, held at the Cavendish Laboratory, Cambridge, on 22–27 July 1946. Volume I: Fundamental Particles* (London: The Physical Society, 1947), pp. 194–198.

78 W. Pauli, "Difficulties of Field Theories and of Field Quantization," in *Report*, note 77 (pp. 5–9).

79 W. Heitler, "The Quantum Theory of Damping as a Proposal for Heisenberg's *S*-Matrix," in *Report*, note 77 (pp. 189–93).

80 E. C. G. Stueckelberg, "The Present State of the *S*-Operator Theory," in *Report*, note 77 (p. 199).

81 P. Jordan to W. Heisenberg, 2 April 1946, Werner-Heisenberg-Archiv.

82 M. Born to W. Heisenberg, 2 October 1946, Werner-Heisenberg-Archiv.

83 H. Wergeland to W. Heisenberg, 5 September 1946, Werner-Heisenberg-Archiv.

84 See G. Wentzel, note 66.

85 See W. Pauli, note 78.

86 Ibid., p. 6.

87 W. Pauli to P. Dirac, 21 December 1943, Werner-Heisenberg-Archiv.

88 Pauli knew, as he stressed in the letter to Møller, dated 24 September 1945, and written from Princeton, Heisenberg's two 1943 papers (notes 4 and 10).

89 See note 88.

90 Pauli also said that he was not able to prove the absence of essential singularities of the *S*-matrix in a quantum mechanical model of two particles with mass $\varkappa$ outside the original physical region $K' = [(2\varkappa)^2 + |\vec{K'}|^2]^{1/2}$ for a natural choice of the scattering potential, except for a box potential.

91 W. Pauli to C. Møller, 18 August 1946, copy in the Werner-Heisenberg-Archiv.

92 W. Pauli to C. Møller, 18 April 1946, Werner-Heisenberg-Archiv.

93 Heisenberg opened the correspondence by congratulating Pauli on his Nobel Prize for the year 1945; Pauli answered from Zurich on 26 May 1946 in a letter. The paper of S. T. Ma, "Redundant Zeros in the Discrete Energy Spectra in Heisenberg's Theory of Characteristic Matrix," was a letter, dated 29 May 1946, to *Phys. Rev. 69* (1946), 668.

94 W. Pauli to W. Heisenberg, 9 September 1946, Werner-Heisenberg-Archiv.

95 D. ter Haar, "On the Redundant Zeros in the Theory of the Heisenberg Matrix," *Physica 22* (1946), 501–8 (received 24 August 1946, published in the issue of November 1946).

96 R. Jost, "Bemerkungen zur vorstehenden Arbeit," *Physica 12* (1946), 509–10 (received 5 October 1946, published in the issue of November 1946); "Ueber die falschen Nullstellen der Eigenwerte der *S*-Matrix," *Helv. Phys. Acta 20* (1947), 256–66 (received 20 January 1947, published in the issue of 4 March 1947).

97 R. Jost to W. Heisenberg, 27 January 1947, Werner-Heisenberg-Archiv.
98 W. Pauli to W. Heisenberg, 9 September 1946, Werner-Heisenberg-Archiv.
99 W. Pauli to W. Heisenberg, 25 December 1946, Werner-Heisenberg-Archiv.
100 Ibid.
101 W. Heisenberg to W. Pauli, 20 June 1947, Werner-Heisenberg-Archiv.
102 W. Heisenberg to W. Pauli, 29 April 1948, Werner-Heisenberg-Archiv.
103 Heisenberg's program led him, via nonlocal quantum field theories, to nonlinear ones with indefinite metric (i.e., those that possessed a generalized Hilbert space including negative norm states, as had been first suggested by Dirac, see note 35). In a period in 1957, Pauli would join Heisenberg again, as in the thirties, in the hope that both might ultimately work out the future theory of elementary particles, a theory in which also the fine-structure constant could be computed. However, in early 1958, Pauli withdrew from the particular nonlinear spinor theory that they had found and criticized it as an "empty frame." Heisenberg continued to work on the new theory with his younger collaborators.
104 W. Pauli to W. Heisenberg, 4 May 1948, Werner-Heisenberg-Archiv.
105 One may find a review of events, e.g., in Julian Schwinger's report for the last Fermilab symposium, "Renormalization Theory of Quantum Electrodynamics," in *The Birth of Particle Physics*, edited by L. M. Brown and L. Hoddeson (Cambridge University Press, 1983), pp. 329–53.
106 Freeman Dyson, "The Radiation Theory of Tomonaga, Schwinger and Feynman," *Phys. Rev.* 75 (1949), 476–502 (received 6 October 1948, published in the issue of 1 February 1949). Dyson introduced the operator that "is identical with Heisenberg's *S*-matrix" at the end of section III of this paper (p. 489).
107 Freeman Dyson, "The *S*-Matrix in Quantum Electrodynamics," *Phys. Rev.* 75 (1949), 1736–55 (received 24 February 1949, published in the issue of 1 June 1949).
108 Dyson, note 107 (p. 1736). It might be mentioned at this point that Dyson learned of the scattering-matrix theory while still a student of Nicolas Kemmer at Cambridge (1946–7). For more details of the Dyson story, we refer to the second article by James T. Cushing cited in note 2 (especially footnote 28).
109 The first calculation of this sort was performed by Walter Heitler and Ning Hu in Dublin using Heitler's radiation-damping theory, which had some relation to the *S*-matrix theory, as we mentioned earlier: "Proton Isobars in the Theory of Radiation Damping," *Proc. Irish Acad. A41* (1947), 123–40 (read 27 May 1946, published 2 May 1947). Later on, a calculation on the basis of a scalar meson theory yielded a similar result; see S. Kamefuchi, H. Nakai, and R. Kawabe, "*S*-Matrix and Nuclear Isobar," *Prog. Theor. Phys.* 6 (1951), 891.
110 W. Pauli to W. Heisenberg, 20 October, 1948, Werner-Heisenberg-Archiv. Similarly, Wentzel had earlier called Heisenberg's *S*-matrix program "very incomplete – it is like an empty frame for a picture yet to be painted" (see note 66, p. 15). Wentzel had further called the attempts of Heitler (on meson theory) and Stueckelberg (on quantum electrodynamics) steps to partly fill this frame. In the same sense, Freeman Dyson wrote in a letter that "I believe it to be probable that the Feynman theory will provide a complete fulfillment of Heisenberg's *S*-matrix program" (F. Dyson to J. R. Oppenheimer, 17 October 1948).
111 W. Pauli to W. Heisenberg, 20 October 1948, Werner-Heisenberg-Archiv.
112 The first physics conference abroad that Heisenberg attended after the war was a meeting of the Zurich Physical Society, held in June 1948.
113 W. Heisenberg, "The Present Situation in the Theory of Elementary Particles," in *Two Lectures* (Cambridge University Press, 1949), pp. 9–25.
114 R. J. Eden to W. Heisenberg, 28 April 1948, Werner-Heisenberg-Archiv. Heisenberg had used this formula in a wrong way in an earlier publication (see note 60).
115 R. J. Eden to W. Heisenberg, 29 May 1948, Werner-Heisenberg-Archiv.
116 For the planning of Eden's visit, we refer to the letter exchange. Thus, Eden wrote to Heisenberg on 20 April 1948: "I have just written to Dr. Frazer asking his assistance and advice for my visit to Göttingen and have suggested the dates 23rd and 30th June." Heisenberg replied: "Your visit to Göttingen between 23rd and 30th of June would be very convenient to me" (W. Heisenberg to R. J. Eden, 3 May 1948). Eden finally fixed the time of

arrival in Göttingen as "shortly after 10 a.m. on 23rd June" (R. J. Eden to W. Heisenberg, 9 June 1948).

117 R. J. Eden to W. Heisenberg, 4 October 1948, Werner-Heisenberg-Archiv.

118 R. J. Eden to W. Heisenberg, 30 December 1948, Werner-Heisenberg-Archiv.

119 W. Heisenberg to R. J. Eden, 19 January 1949, Werner-Heisenberg-Archiv.

120 R. J. Eden, "Heisenberg's *S*-Matrix for a System of Many Particles," *Proc. R. Soc. London, Ser. A 198* (1949), 540–59 (received 12 March 1949, published in the issue of 7 September 1949); "The Analytic Behaviour of Heisenberg's *S*-Matrix," *Proc. R. Soc. London, Ser. A 199* (1949), 256–71 (received 25 May 1949, published in the issue of 25 October 1949).

121 N. Hu, "On the Application of Heisenberg's Theory of the *S*-Matrix to the Problems of Resonance Scattering in Nuclear Physics," *Phys. Rev. 74* (1948), 131–40 (received 15 April 1948, published in the issue of 15 July 1948).

122 See note 120, second paper (p. 257).

123 In his second paper, Eden also discussed the scattering matrices involving photons as incident particles and derived results on the line width and the relative intensities of spectral lines.

124 R. J. Eden, "Threshold Behaviour in Quantum Field Theory," *Proc. R. Soc. London, Ser. A 210* (1952), 388–404 (received 27 July 1951, published in the issue of 7 January 1952).

125 Eden, note 124 (p. 388).

126 The story of the early stages of this theory is discussed by Geoffrey Chew in Chapter 41.

127 R. Kronig, "A Supplementary Condition in Heisenberg's Theory of Elementary Particles," *Physica 12* (1946), 543–4 (letter dated 15 September 1946, published in the issue of November 1946).

40 From field theory to phenomenology: the history of dispersion relations

Born 1948, Coventry, England; Ph.D., University of Edinburgh (science studies), 1984, and University of London (theoretical physics), 1973; history of modern physics/sociology of scientific knowledge; University of Illinois

> Men make their own history, but they do not make it just as they please; they do not make it under circumstances chosen by themselves, but under circumstances directly encountered, given and transmitted from the past. The tradition of all the dead generations weighs like a nightmare on the brain of the living.[1]

The history of strong-interaction theory in the 1950s is one of a continuous struggle to bring the tradition handed down from the dead generations into conformity with contemporary needs. The tradition was quantum field theory (QFT), and the need was to confront the masses of data on strong-interaction processes emerging from the growing number of high-energy particle accelerators. The struggle originated in the observation that the strong interactions were indeed strong: The pion–nucleon coupling constant, for example, could be estimated to be appreciably greater than unity.[2] The perturbative methods developed in the late 1940s for applications of renormalized quantum electrodynamics therefore failed when applied to the strong interaction. As Steven Weinberg put it in his 1977 *Notes for a History of Quantum Field Theory*, "it was not that there was any difficulty in thinking of renormalizable quantum field theories that *might* account for the strong interactions – it was just that

The research reported here was conducted while I was a member of the Program in Science, Technology and Society, Massachusetts Institute of Technology, from which I gratefully acknowledge the award of an Exxon Fellowship. For interviews, discussions, and the provision of biographical material, I thank Professors Geoffrey Chew, Bernard Feld, Murray Gell-Mann, Marvin Goldberger, Francis Low, and Yoichiro Nambu. I am especially indebted to James Cushing and Silvan S. Schweber for access to their work prior to publication and for much historical advice.

having thought of such a theory, there was no way to use it to derive reliable quantitative predictions, and to test if it were true."[3]

Weinberg proceeded from this observation to remark that "the uselessness of the field theory of strong interactions led in the early 1950s to a widespread disenchantment with quantum field theory" that lasted for almost twenty years.[4] But this is a misleading oversimplification. The strong-interaction theory literature of the early 1950s is saturated with QFT: the development of increasingly sophisticated QFT formalisms appropriate to strong-interaction processes, and calculations in simplified renormalizable QFT models. What one sees, throughout the 1950s and into the 1960s, is a fascinating process wherein field-theory investigations repeatedly gave birth to quasi-autonomous traditions of phenomenology and further theorizing. As Murray Gell-Mann described it in an often-repeated metaphor, "we may compare this process to a method sometimes employed in French cuisine: a piece of pheasant meat is cooked between two pieces of veal, which are then discarded."[5] The veal stands for QFT, while the pheasant meat is a free-standing phenomenological or theoretical tradition.

Emerging from the veal-and-pheasant approach to strong-interaction theory, one can distinguish two major lines of relatively independent work. In an aphorism attributed to Richard Feynman, "there are two types of particle theorists: those who form groups and those who disperse." "Those who formed groups" passed through global symmetry and the "eightfold way" to arrive in the mid-1960s at quarks.[6] I focus here on "those who disperse," and on the history of a complex of developments clustering around investigations of the analyticity properties of the S matrix. Figure 40.1 schematizes the main developments in this area between the early 1950s and the early 1960s. The horizontal line represents the mainstream development of QFT. The lines coming from above represent decisive individual contributions – from Walter Thirring (1954), Stanley Mandelstam (1958), and Tullio E. Regge (1959) – each of which led to the formation of a distinctive field of practice: single-variable dispersion relations, double dispersion relations, and Regge poles, respectively. Note that after each intervention, a split took place between the phenomenological application of conjectured results and attempted proofs and extensions of those conjectures.

The focus of this account is on the first split indicated in Figure 40.1. Consistent with the dictum that "men make their own history," I shall seek to avoid representing the development of dispersion theory as that of the unfolding of disembodied ideas through their own internal logic. Instead, I shall attempt to sketch an explanation of why particular theorists adapted and extended their cultural heritage in particular ways in terms of their backgrounds and contexts.[7] This means paying attention to who did what. We shall see that the history of dispersion relations was marked by successive social translocations of knowledge, as common themes were taken up by different groups of physicists, and that these translocations were accompanied by transforma-

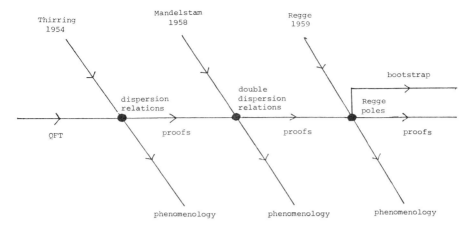

Figure 40.1. Schematic history of dispersion relations (time flows from left to right).

tions of the knowledge in question, as the different groups elaborated each theme in characteristically different ways.[8]

The genesis of dispersion relations
In 1960, John Hamilton began an introductory review of "Dispersion Relations for Elementary Particles" with the remark that "it may indeed be true that the first words spoken by the best theoretical physicists are not 'Ma' or 'Da' or 'more,' but rather the magic words 'Kramers–Kronig.'"[9] Hendrik A. Kramers and Ralph de Laer Kronig had, in 1926 and 1927, written down dispersion relations for the scattering of light.[10] Kramers, for example, arrived at the following prototypical dispersion relation:

$$R_e[f(\omega) - f(0)] = \frac{2\omega^2}{\pi} P \int_0^\infty \frac{\mathrm{Im}\, f(\omega')\, d\omega'}{\omega'(\omega'^2 - \omega^2)}$$

Here, $f(\omega)$ is the forward-scattering amplitude at frequency ω, and the characteristic feature of this result is that the real part of $f(\omega)$ is given by a weighted integral over the imaginary part. Later derivations of such relations concentrated on the role of causality and analyticity.

The early impact of the work of Kramers and Kronig was limited.[11] But in the mid-1940s, interest began to increase, in part through the efforts of Kramers and Kronig themselves. This effort by Kronig was made in the context of Werner Heisenberg's S-matrix program.[12] Attempting to escape from the divergence problems of contemporary QFT, Heisenberg considered the possibility that some kind of fundamental length might appear in nature, and he set up a formalism for the scattering or S matrix that might accommodate this. Foreshadowing Gell-Mann's veal-and-pheasant approach, Heisenberg aimed to use QFT as a guide to understanding the properties of the S matrix, preparatory to discarding QFT and with it its divergence problems. In 1944, Kramers suggested to Heisenberg that the S matrix should be an analytic function of the relevant variables, on the basis of a causality argument analogous to that lying behind the original Kramers–Kronig relation.[13] In a 1946 publication, Kronig made explicit the analogy between the earlier work on optics and contemporary particle theory: "It...would seem reasonable," he wrote, "to postulate for the scattering of elementary particles a connection of the real and imaginary parts of the scattering amplitudes of the same type as in optics."[14]

In the work of Kramers, Kronig, Heisenberg, and others in the latter half of the 1940s one can see the basis being laid for an elementary particle theory independent of quantum fields. But a major distraction lay ahead – the renormalization in the late 1940s of QFT.[15] This work opened up vistas in the technical development of QFT that had been blocked in prewar calculations by the divergences that appeared beyond the leading order in perturbation theory. Encouraged by the quantitative success of higher-order calculations in renormalized quantum electrodynamics, many theorists set out to explore those vistas. In this new context, Heisenberg's S matrix came to be seen as a quantity to be calculated in QFT, and his ideal of an S-matrix theory autonomous from QFT remained undeveloped: "very incomplete...like an empty frame for a picture yet to be painted," as Gregor Wentzel described it in 1947.[16] Not until the late 1950s did Geoffrey Chew begin to resurrect the autonomous S matrix in his bootstrap program, and then only after the unbootstraplike and field-theoretic developments indicated in Figure 40.1.

What were those developments? Three lines of theoretical work converged on dispersion relations in the early 1950s:

(i)   explorations of the properties of the S matrix itself,
(ii) attempts to extract some understanding of strong-interaction physics from QFT, and
(iii) formulation of simplified QFT models of strong-interaction processes and associated phenomenological investigations.

The first of these lines of work was associated primarily with Princeton; we shall return to it later. From our perspective, the most important contributions to the second and third came from the midwestern United States. At the University of Chicago, Gell-Mann and Marvin L. Goldberger made important contributions to the development of a QFT formalism appropriate to the analysis of the strong interactions, as did Francis Low at the University of Illinois at Urbana.[17] Also at Urbana was Chew, who from 1954 onward developed to great phenomenological effect a field-theory-inspired model of pion–nucleon interactions.[18] As Gian Carlo Wick put it in 1955, "thanks to Chew's efforts, the Yukawa theory has at last achieved some contact with quantitative aspects of meson physics."[19] After some sophisticated field-

Table 40.1. *Copublications between Gell-Mann (MGM), Goldberger (MLG), Chew (GFC), and Low (FEL), 1948–60*

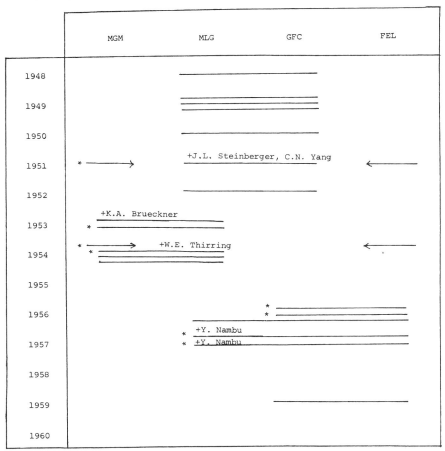

Note: Each horizontal bar represents a publication jointly authored by the theorists under whose initials it appears (Gell-Mann and Low copublications are indicated by arrowed broken bars). Names of additional coauthors are added explicitly. Bars marked with an asterisk correspond to publications cited in the text.

theory input from Low, Chew's fixed-extended-source model became, in 1956, the Chew–Low model.[20]

These four theorists – Chew, Gell-Mann, Goldberger, and Low – were at the heart of the developments summarized in Figure 40.1, with Gell-Mann and Goldberger especially prominent in the development of single-variable dispersion relations. It is therefore of interest to note the extent to which they formed an interconnected social group. Table 40.1 summarizes the extent of collaborative publication among the four between 1948 and 1960.[21] The

Table 40.2 *Institutional locations of Gell-Mann, Goldberger, Chew, and Low, 1950–60*

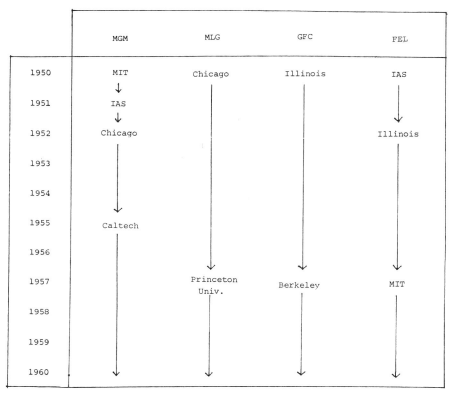

	MGM	MLG	GFC	FEL
1950	MIT ↓	Chicago	Illinois	IAS
1951	IAS ↓			↓
1952	Chicago			Illinois
1953				
1954				
1955	Caltech			
1956				
1957		Princeton Univ.	Berkeley	MIT
1958				
1959				
1960				

Note: IAS = Institute for Advanced Study, Princeton.

appearance of Yoichiro Nambu among the authors of two 1957 papers points to the extended contribution of Chicago theorists to the development of dispersion relations in the latter half of the 1950s. Another important member of the Chicago group was Reinhard Oehme, who had studied at Göttingen with Heisenberg in the late 1940s.[22]

Table 40.2 traces the institutional locations of the Big Four through the 1950s. Note the overlaps in the Midwest in the early 1950s, the critical period in the formulation of the dispersion-relation program. In the latter half of the 1950s, the dispersion theorists dispersed – Gell-Mann to Caltech, Goldberger to Princeton, Chew to Berkeley, and Low to MIT – taking their distinctive theoretical orientations to the four corners of the particle physics world, but leaving behind a viable group of dispersion theorists at Chicago.

We can pick up the history of dispersion relations in the collaboration of Gell-Mann and Goldberger at the University of Chicago in the early 1950s.

As Gell-Mann has described it, they were engaged "in a program of extracting as many general results as possible from local field theory, by proof if possible but otherwise by finding rules that held to every order of perturbation theory."[23] One such result was the crossing relation among scattering amplitudes, announced by Goldberger at the 1954 Rochester conference.[24]

In the course of their work, Gell-Mann and Goldberger found in the literature a 1952 publication by theorists Fritz Rohrlich and Robert L. Gluckstern, at Cornell, in which dispersion methods were used to effect an economical calculation of forward photon scattering in the Coulomb field of the nucleus.[25] Gell-Mann and Goldberger began an investigation of the status of dispersion relations in QFT that brought to their attention a whole set of publications from theorists at Princeton on causality and the analyticity properties of Heisenberg's S matrix.

The shift in the geographical location of the S-matrix program from Europe (where Heisenberg had moved on to his nonlinear spinor theory) to Princeton requires more analysis than I can offer. It appears to have been partly, as least, a consequence of the formulation and use in the late 1930s and 1940s of S-matrix-type formalisms by John A. Wheeler and Eugene Wigner, both at Princeton University.[26] In any event, a considerable body of work emerged from Princeton in the early 1950s on the analytic properties of the S matrix. In 1951, W. Schutzer and Jayme Tiomno discussed the relation between causality and analyticity for scattering from a potential of finite radius. In two 1953 publications, N. G. van Kampen at the Institute for Advanced Study carried through a similar analysis, using what he considered to be a more adequate definition of causality, again for a finite-range potential.[27] Goldberger later recalled that he considered this work "very dull" because "the radius of the interaction region entered the dispersion relations obtained."[28] At around the same time (1952), John S. Toll, a student of Wheeler, completed an unpublished but often cited Ph.D. thesis on the connection between causality and dispersion relations for the scattering of light.[29]

Goldberger spent the academic year 1953–4 on leave from Chicago as a visitor at Princeton. He arrived there with an active interest in dispersion-relation problems and stepped into an atmosphere of active work on the relation between causality and the analyticity properties of the S matrix. He began a collaboration with another visitor, Thirring, and it was Thirring who supplied the key element in the QFT development of dispersion relations.[30] Authors like Schutzer, Tiomno, and van Kampen had attempted to implement classical intuitions of causality, straightforwardly insisting that effect could not precede cause at the macroscopic level. Thirring, in contrast, suggested that one might attempt to specify causality conditions at the microlevel. In particular, he suggested that one enforce the condition that the commutator of two field operators should vanish for spacelike separations of their arguments.

The Thirring commutator condition, unlike earlier definitions of macro-causality, was directly tailored for implementation in contemporary formulations of QFT, which were indeed constructed around products of field operators. A series of three publications followed, one jointly authored by Gell-Mann, Goldberger, and Thirring (GGT) in 1954, and two by Goldberger alone, in 1955.[31] The appearance of these papers quickly transformed dispersion relations from the minority pursuit of a handful of theorists into a major growth industry within particle physics. As Goldberger noted in 1961, "there followed [from the work of GGT] an explosion of activity as is evident to the most casual reader of journals over the past seven years."[32] I shall briefly review the contents of the three papers and comment on their distinctive features, before discussing the "explosion of activity" that followed from them.

The 1954 publication of GGT deployed the apparatus of contemporary field theory in order to explore the consequences of Thirring's microcausality condition. GGT were able to elucidate the analytic properties of the photon forward-scattering amplitude, and hence to derive dispersion relations à la Kramers–Kronig to lowest order in perturbation theory for the interactions of the electromagnetic field. In his first publication of 1955, Goldberger showed that the restriction to lowest order was unnecessary: The existing apparatus for QFT calculations was sufficient to demonstrate that the GGT results could be obtained without resort to perturbative techniques.[33] This paper sufficed to integrate further the dispersion-relation approach into the mainline of field-theory development, but accomplished little beyond that. In Goldberger's concluding words:[34]

> It is, of course, not surprising that the Kramers–Kronig relations which follow from the above results are correct, independent of perturbation theory, and under very general assumptions on the form of the coupling between the electromagnetic field and the matter field. It is gratifying, however, that such a simple derivation may be given.

Goldberger's second paper of 1955, in contrast, marked a decisive step away from the rederivation of already established results, and into a new area. Here, he extended his previous derivation of dispersion relations for zero-mass particles to non-zero-mass case, writing down equations such as that reproduced in the Appendix to this chapter.[35] These equations asserted relations between the real and imaginary parts of, say, pion–nucleon scattering amplitudes, relations that could, in principle, be tested against experimental data.[36] Goldberger's work thus pointed the way forward to a QFT-based analysis of strong-interaction physics.

A distinctive feature of this sequence of publications is its pragmatic orientation. The central concern is to obtain phenomenologically useful results, and the derivations offered are heuristic rather than rigorous proofs from QFT, as is evident from subsequent attempts to remove their deficiencies, as described later. The adoption of Thirring's commutator condition is one example of this pragmatism: It fitted very nicely into the framework of contemporary QFT manipulations – it facilitated the derivation of useful results – but beyond that, its relation to physical intuitions on the nature of causality was, in Goldberger's words, "far from clear."[37] Likewise, at the level of technical manipulations, the early paper on dispersion relations was more than suspect. As Goldberger recalled in 1969, speaking of his extension of the dispersion-relation approach to the scattering of finite-mass particles, "of course, my derivation was not really correct; [but] since the result was correct, I was sure that someone would prove it eventually."[38]

A pragmatic attitude toward theory continued to mark many of the important developments that followed from the first articulations of single-variable dispersion relations, and it is interesting to speculate on its origins. S. S. Schweber has already noted the pragmatism of the postwar generation of U.S. theorists.[39] But beyond this overall characteristic, it is relevant to note the institutional location of much of the early work on dispersion relations: the Institute for Nuclear Studies and the Department of Physics at the University of Chicago. Enrico Fermi was at Chicago from 1946 until his death in 1954, and he was known for his emphasis on bringing theory into contact with data. There was also at Chicago the group of experimenters working at the cyclotron, primarily concerned with exploring the features of pion–nucleon scattering in the vicinity of the 3-3 (now the Δ) resonance. This 3-3 resonance was a central concern in strong-interaction phenomenology; it was the main success of the Chew–Low theory, and the Chicago theorists were in close, sometimes collaborative, contact with their experimental colleagues. The proximity and social ties between experimenters and theorists left their conceptual mark in the pragmatic thrust of the dispersion-relation program.[40]

Theory and phenomenology: the first split

So much for the genesis of the publications that ushered in the modern era of dispersion relations. What of the response? How should we understand the "explosion of activity" that followed? The point to note is that, as articulated at Chicago by Goldberger and others, dispersion relations hung suspended between the twin poles of contemporary QFT and experimental data on the strong interactions. In consequence, Goldberger's 1955 dispersion relations represented different objects to different groups of physicists. To mathematical physicists, dispersion relations represented the endpoints of putative proofs from field-theory axioms. To those more phenomenologically inclined, the relations represented a set of tools to be used, in one way or another, in the analysis of data. Thus, from the pioneering work of Goldberger and associates flowed two relatively distinct streams of work: one theoretical, aiming at proofs of various conjectured dispersion relations; the other phenomenological, aiming at the application of those relations to data analysis. This division between theoretical and phenomenological practice is indicated by the first split in Figure 40.1.

The points of contact between the theoretical and phenomenological traditions were the dispersion equations themselves, and through these contacts the two traditions generated mutually reinforcing contexts for one another. But it is worth emphasizing that, by and large, the connection in practice between the two was weak. The dispersion equations marked a cutout as well as a point of contact between the two traditions: applications of the equations were almost entirely independent of their theoretical derivations, and vice versa. The abyss between the two realms of practice was expressed clearly in Chew's 1959 review of "The Pion–Nucleon Interaction and Dispersion Relations." Chew began by remarking that "the systematic derivation of the new dispersion relations is complicated and not at all suitable to a review of this kind," – namely, the *Annual Review of Nuclear Science* – and continued:[41]

> For the reader who wishes to see all the essential steps in a complete and yet economical derivation of the pion–nucleon dispersion relation, the following use of the published literature is recommended: (a) Read the first and about half of the second section of [Bremermann, Oehme, and Taylor, 1958], up to the point where dispersion relations have been obtained for imaginary mass. (b) Switch here to a recent paper by Lehman [1958] which uses the Dyson representation not only to carry out the necessary extension in the mass variable but also to justify the use of Legendre polynomials in implementing dispersion relations. (c) If any strength remains, read the Dyson paper [1958].... At present it remains true that the methods of implementation of dispersion relations are elementary and quite unrelated to the sophisticated mathematical techniques required for their derivation. Such a situation may not persist indefinitely, but it motivates the decision to avoid in this review the mathematics of derivation.*

Because of the extent of this split between theory and application, I shall discuss the two traditions separately, starting with the latter: phenomenology.

Dispersion phenomenology
How were dispersion relations phenomenologically implemented? The 1954 and 1955 publications of GGT and Goldberger aimed to demonstrate the existence of dispersion relations for elementary particle scattering. The first attempt to put such relations to use – to make significant contact with empirical data – came not from Chicago but from the University of California at Berkeley. In a paper received for publication on 31 January 1955, predating Goldberger's second 1955 publication by nine weeks, Robert Karplus and Malvin Ruderman proposed the existence of dispersion relations for the forward scattering of finite-mass particles and applied their formalism to the existing data on pion–nucleon scattering.[42] Because the imaginary part of

---

*   Reproduced, with permission, from the *Annual Review of Nuclear Science*, Vol. 9. © 1959 by Annual Reviews Inc.

the forward-scattering amplitude could be directly related to the total cross section via the optical theorem, Karplus and Ruderman were able to express the real part of the forward-scattering amplitude as an integral over the corresponding total cross sections – which were known from experiment, at least at low energies. They thus deduced that the sign of the real part of the forward-scattering amplitude was the same as that already inferred from experimental observations on the interference between nuclear and Coulomb scattering. Unfortunately, as Goldberger pointed out to them, Karplus and Ruderman had failed to take into consideration a subtlety in the application of Gell-Mann and Goldberger's crossing theorem. This failure implied that their formalism was appropriate only to neutral-pion scattering, whereas they had taken the charged-pion cross sections as input to their computations.[43]

The next step toward application of dispersion relations to forward pion–nucleon scattering came from the Chicago collaboration of Goldberger, H. Miyazawa, and Oehme.[44] Received for publication thirteen days after Goldberger's paper extending dispersion relations to finite-mass particles, this paper spelled out the spin and isospin dependence of pion–nucleon dispersion relations, taking the requirements of crossing more carefully into account than had Karplus and Ruderman. One of Goldberger, Miyazawa, and Oehme's dispersion relations is reproduced here in the Appendix.[45] The Chicago theorists noted the connection between their findings and the phenomenologically successful analyses of Chew and Low (discussed later), and they referred ahead to a forthcoming application of their formalism to the analysis of pion–nucleon phase shifts.

The promised analysis was received for publication three weeks later, on 11 May 1955.[46] Three Chicago physicists, Herbert L. Anderson, William C. Davidon, and Ulrich E. Kruse, had used the measured pion–proton cross sections as input to the dispersion relations just written down by Goldberger and associates to calculate the real part of the forward-scattering amplitude. Taking their value of the pion–nucleon coupling constant ($f^2 = 0.08$) from Chew's work, they obtained agreement with the resonant phase shifts of F. de Hoffmann and associates.[47] Anderson and associates further remarked that if one stood the logic of their argument on its head, and regarded dispersion relations as an analytic tool rather than as a theoretical result to be tested, then "the dispersion relations can serve as a means to evaluate the strength of the pion–nucleon coupling."[48]

In subsequent applications of forward-dispersion relations to pion–nucleon scattering, different physicists ran the argument both ways. Either way, the upshot was an intimate entanglement of dispersion phenomenology with experimental practice – an entanglement symbolized in the fact that the first detailed application of dispersion relations to data analysis came from the three experimenters, Anderson, Davidon, and Kruse. In their work, one can begin to see the shift of dispersion relations from the theoretical to the experimental particle physics community, and their final transformation from objects to be calculated in field theory to resources in determining experimental strategies.[49]

Forward pion–nucleon scattering did not, of course, remain the sole focus for dispersion investigations. Considerable efforts were devoted to extending the range of the dispersion-relation approach to a class of phenomena as encompassing as possible. In 1955 and 1956, many theorists began to develop the apparatus required for the analysis of nonforward pion–nucleon scattering: Goldberger, Nambu, and Oehme at Chicago; Gell-Mann and John C. Polkinghorne at Caltech; Abdus Salam and W. Gilbert in Cambridge, England; and R. H. Capps and G. Takeda at Berkeley and the Brookhaven National Laboratory.[50] At the same time, others began to develop a formalism appropriate to nonforward pion photoproduction: E. Corinaldesi at Glasgow; and A. A. Logunov and B. M. Stepanov in the Soviet Union.[51] Rather than discuss these steps along the way, let me conclude this section of the discussion with some remarks on the two "classic" publications of 1957 from the Chicago–Illinois collaboration of Chew, Goldberger, Low, and Nambu (CGLN, for short).

Published back-to-back, these papers offered a dispersion analysis of nonforward pion–nucleon scattering and pion photoproduction, respectively. One of the CGLN dispersion relations for pion–nucleon scattering is reproduced here in the Appendix.[52]

In the first instance, the CGLN publications represented an investigation of the relation between the Illinois Chew–Low model and the Chicago dispersion-relation approach to strong-interaction physics.[53] The Chew–Low model was successful in describing the main features of pion–nucleon scattering and pion photoproduction.[54] But it remained a nonrelativistic model, based on simplified QFT calculations, in which it was not clear how to implement gauge invariance in the analysis of photoproduction. The two CGLN publications discussed how well the Chew–Low results could be reproduced in a relativistic dispersion analysis. The answer was "well enough," if one assumed that the dispersion integrals were dominated by the contribution from the 3-3 resonance.

The CGLN publications were, then, a source of immediate satisfaction to the authors: The phenomenological success of the Chew–Low model was understood within the Chicago dispersion-relation approach. But the long-term significance of CGLN's work lay elsewhere. Already CGLN had noted that in their analysis of photoproduction, "with luck, if we have made no serious mistakes, the final amplitude written down above may have an accuracy of $\sim 5–10\%$ in the subresonance region. It will certainly deteriorate rapidly above resonance."[55] Events were to confirm those remarks in an unexpected way. The 1960s population explosion of hadronic resonances revealed that the 3-3 was only the first of an apparently indefinite sequence of resonant contributions to scattering and photoproduction cross sections.[56] And the CGLN publications were "classics" in the sense that they offered a set of phenomenological tools ready-made to cope with this circumstance. Instead of assuming that the dispersion integrals were dominated by the 3-3 alone, one had simply to assume that they were dominated by whatever set of resonances were currently believed to exist. Such resonance-saturated dispersion relations played an important role in the phenomenological analysis of resonance-region data, a region to which many experimenters devoted their resources in the 1960s and early 1970s.[57] Nonforward, like forward, dispersion relations served to endow experimental pion–nucleon and photoproduction physics with theoretical significance and, within that context, to mark out targets for further experiment. In return, the data thus generated served to sustain a growing tradition of phenomenological analysis.

The work discussed thus far typifies the dispersion-relation phenomenology of the latter half of the 1950s, but by no means exhausts it. As J. D. Jackson put it in his lectures at the 1960 Scottish Universities' Summer School, "once the obvious derivations and applications to the pion–nucleon problem and photon interactions had been made, the theoretical physicists were forced to look for other areas of application." Jackson then listed work published in 1957 and 1958 by the Chicago group and others on the dispersion analysis of pion–deuteron scattering, nucleon–nucleon scattering, kaon–nucleon scattering, and nucleon electromagnetic form factors, as well as work on the weak interactions of hadrons by N. N. Bogoliubov, Logunov, and S. M. Bilenky in the Soviet Union and by Goldberger and S. B. Treiman in the United States – including the first derivation of the famous Goldberger – Treiman relation.[58]

Rather than go into the details of these applications, I want now to turn from the phenomenological to the theoretical dispersion-relation tradition as it developed in the latter half of the 1950s and briefly sketch the subsequent history of theoretical innovations, followed by theory/phenomenology splits that characterized this era.

Dispersion theory: more splits and the decline of field theory

The prototypical dispersion relations of Goldberger and associates were heuristically derived rather than rigorously proved from QFT. In parallel with the phenomenological application of the relations, several authors sought to improve their theoretical justification. I shall only outline the consequent theoretical developments here. One point to stress in advance is that once more we shall encounter social translocations and transformations of knowledge. The burden of justifying dispersion relations quickly passed from their creators to a separate group of mathematicians and mathematical physicists, whose work was often almost incomprehensible and totally irrelevant to the users of dispersion relations.

The first rigorous proof of forward-scattering dispersion relations was offered in 1957 by the German theorist Kurt Symanzik, a student of Heisenberg educated in the tradition of Schwinger, who was then a visitor at Chicago.[59]

In 1957 and 1958, various proofs were offered showing the validity, under certain conditions, of already conjectured dispersion relations for nonforward scattering. These proofs came from Bogoliubov and associates in the Soviet Union, from Bremermann, Oehme, and Taylor at Princeton, and from the German theorist Lehmann (then visiting the Institute for Advanced Study at Princeton).[60]

The next major innovation in dispersion theory came in 1958, with Mandelstam's conjectured "double dispersion relations."[61] The Mandelstam representation mapped out the analytic properties of scattering amplitudes as functions of two variables rather than just one. As indicated in Figure 40.1, Mandelstam's work marked a second branch point in the development of strong-interaction theory, from which two distinct traditions of practice emerged: a phenomenological tradition, which used double dispersion relations as tools in the extraction of information from data; and a theoretical tradition, which aimed to prove the validity of the Mandelstam representation in field theory (never with full success).[62] From this theoretical tradition emerged the work of Regge, extending the analysis of the S matrix into the complex angular-momentum plane (Figure 40.1).[63] And from Regge's work there eventually came into being three distinct theoretical traditions: the Regge pole (and later cut) phenomenology, which dominated the analysis of high-energy soft-scattering data through the 1960s and into the 1970s;[64] a program of theoretical investigations in QFT of the singularity structure of the S matrix in the complex angular-momentum plane; and, most strikingly, the anti-QFT bootstrap program of Chew. At the La Jolla conference of 1961, Chew advocated the view that the "empty frame" of Heisenberg's original S-matrix program had been sufficiently filled in by fifteen years of work on the analyticity properties in QFT. Field theory could, and should, now be forgotten. The autonomous S matrix should be the central object of theorists' concern, its properties to be elucidated according to the bootstrap requirement of self-consistency.[65]

This is as far as I shall take the history of dispersion relations here. To close, some comment is appropriate on the social fortunes of field theory during the period we have been discussing. In the early 1950s, the dominant theoretical orientation within the particle physics community was QFT; by the early 1960s, QFT was in deep decline, at least as far as strong-interaction physics was concerned.[66] One can easily understand the reasons for this. The three phenomenological traditions I have discussed – single-variable dispersion relations, double dispersion relations, and Regge – existed in an intimate state of mutual reinforcement with the dominant programs of strong-interaction experiment – resonance physics at low energies, soft scattering at high energies. The growth of these phenomenological traditions was thus driven by the rapid expansion of experimental facilities and data-production rates in particle physics and was sustained by a continuing increase in the size of the particle physics community. QFT, in contrast, in strong-interaction physics, received no such direct stimulus from experiment. Field theorists struggled to explore the general analytic features of the S matrix, but never arrived at the detailed analysis required to make contact with the specifics of data.

At the same time, QFT itself became increasingly specialized, technical, and mathematical. It became increasingly difficult for outsiders, such as graduate students, to make a significant contribution to QFT discourse. Even insiders began to look for new topics, leaving behind a remnant of axiomatic field theorists so mathematically inclined as to be hardly recognizable as physicists – certainly not people like Goldberger, who had, in his own words, learned complex-variable theory "in the gutter."[67] Low expressed a typical sentiment in speaking of attempts to prove the Mandelstam representation: QFT "wasn't any fun any more."[68] And beyond the lack of empirical stimulus and technical difficulty, QFT became overshadowed by the bootstrap program. In the early 1960s, the bootstrap program appeared to offer an equally fundamental but more promising approach to the physics of elementary particles.

Of course, all of the phenomenological traditions discussed earlier, as well as the bootstrap, had their origins in field theory, but in different ways, each had achieved its independence. The pheasants were kept, and the veal of QFT discarded. This has been the story of field theory's contribution to its own demise in strong-interaction physics.

Appendix: Evolution of pion–nucleon dispersion relations

1. Goldberger (1955): prototype dispersion relation for forward meson scattering

$$D_{\alpha\beta}^{(1)}(k;\lambda) - D_{\alpha\beta}^{(1)}(0;\lambda) = \frac{k^2}{2\pi} \int_0^\infty dk' \frac{\sigma_{\alpha\beta}^{(1)}(k';\lambda)}{k'^2 - k^2}$$
$$+ 2\sum_n (E_n - E_p) f_{\alpha\beta}^{(+)}(E_n - E_p, n; \lambda)$$
$$\times \left[\frac{1}{(E_n - E_p)^2 - k^2 - \mu^2} - \frac{1}{(E_n - E_p)^2 - \mu^2} \right]$$

Here the forward scattering of mesons is described in terms of two independent amplitudes, $D^{(1,2)}$; a similar relation is given for $D^{(2)}$; α and β are meson isospin indices, λ is a target isospin index, and μ is the mass of the meson; $\sigma_{\alpha\beta}^{(1,2)}$ are appropriately defined total cross sections. The sum is over all possible poles of the S matrix.

2. Goldberger, Miyazawa, and Oehme (1955): dispersion relation for forward pion–nucleon scattering

$$D_+(k) - \frac{1}{2}\left(1 + \frac{\omega}{\mu}\right)D_+(0) - \frac{1}{2}\left(1 - \frac{\omega}{\mu}\right)D_-(0)$$
$$= \frac{k^2}{4\pi^2}\int_\mu^\infty \frac{d\omega'}{k'}\left[\frac{\sigma_+(\omega')}{\omega' - \omega} + \frac{\sigma_-(\omega')}{\omega' + \omega}\right] + \frac{2f^2}{\mu^2}\frac{k^2}{\omega - \mu^2/2M}$$

Here the two independent amplitudes have been redefined as $D_{\pm}$; a similar relation is given for a second linear combination of amplitudes; ω is the total pion energy in the laboratory system; $k^2 = \omega^2 - \mu^2$. M is the nucleon mass. The single-nucleon pole term has been written out explicitly; f is the pion–nucleon coupling constant.

3. *CGLN (1957): dispersion relation for nonforward pion–nucleon scattering*

$$\operatorname{Re} B^{(\pm)}(\nu, k^2) = \frac{g_r^2}{2M}\left(\frac{1}{\nu_B - \nu} \mp \frac{1}{\nu_B + \nu}\right)$$

$$+ \frac{P}{\pi}\oint_{1-k^2/2M}^{\infty} d\nu' \operatorname{Im} B^{(\pm)}(\nu', k^2)\left(\frac{1}{\nu' - \nu} \pm \frac{1}{\nu' + \nu}\right)$$

Here the two independent amplitudes are denoted as A and B. A similar relation is given for A. The superscript is an isospin index. Note that for nonforward scattering, the amplitudes are functions of two kinematic variables – energy (ν) and momentum transfer (k^2) – rather than one.

Notes

1 K. Marx, *The Eighteenth Brumaire of Louis Bonaparte*, in K. Marx and F. Engels, *Selected Works in One Volume* (London: Lawrence & Wishart, 1970), p. 94.
2 See, for example, Hans A. Bethe and Frederic de Hoffmann, *Mesons and Fields*, Vol. 2 (Evanston: Row, Peterson & Co., 1955), p. 20.
3 S. Weinberg, "The Search for Unity: Notes for a History of Quantum Field Theory," *Daedalus* (Fall 1977), 17–35.
4 Weinberg (note 3) cites the nonrenormalizability of weak-interaction theory as the second contributing cause of the general decline of QFT. The twenty-year gap extends up to 't Hooft's work on the renormalization of gauge theory in 1971.
5 M. Gell-Mann, "The Symmetry Group of Vector and Axial Currents," *Physics, 1* (1964), 63–75. Gell-Mann was writing here of his strategy in formulating the current-algebra approach to the weak interactions of hadrons, but the metaphor applies equally well to his own work on dispersion relations, as described later, and the eightfold way [see A. Pickering, *Constructing Quarks: A Sociological History of Particle Physics* (University of Chicago Press, 1984), pp. 56–60] and to the overall development of strong-interaction theory in general. Weinberg (note 3, p. 31) notes that "in both lines of work [weak and strong interactions], quantum field theory was used heuristically, as a guide to general principles, but not as a basis for quantitative calculations."
6 On the early development of the quark model, see M. Gell-Mann, "Particle Theory from *S*-Matrix to Quarks," talk presented at the First International Congress on the History of Scientific Ideas, held at Sant Feliu de Guixols, Catalunya, Spain, September 1983, edited by M. G. Doncel, A. Hermann, L. Michel, and A. Pais (Barcelona: Bellaterra, 1987), pp. 474–97; Y. Ne'eman, "Hadron Symmetry, Classification and Compositeness," ibid.; and A. Pickering (note 5, Chapter 4).
7 For an explicit discussion of the sociological model of the dynamics of research practice that underlies this account, see A. Pickering (note 5, Chapter 1).
8 There is a growing secondary literature on the history of dispersion relations. I have consulted M. Cini, "The History and Ideology of Dispersion Relations: The Pattern of Internal and External Factors in a Paradigm Shift," *Fundamenta Scientiae 1* (1980), 157–72; J. T. Cushing, "Models and Methodologies in Current High-Energy Physics," *Synthese 50* (1982), 5–101; J. T. Cushing, "The *S*-Matrix Program – Anatomy of a Scientific Theory," working draft, University of Notre Dame, April 1985; Gell-Mann, "Particle Theory from *S*-Matrix to Quarks" (note 6); M. L. Goldberger, "Fifteen Years in the Life of Dispersion Relations," in

Subnuclear Phenomena, International School of Physics "Ettore Majorana," Vol. 2, Erice, Sicily, 3–19 July 1969, edited by A. Zichichi (New York: Academic Press, 1970), pp. 684–93; J. D. Jackson, "Introduction to Dispersion Relation Techniques," in *Dispersion Relations: Scottish Universities' Summer School, 1960,* edited by G. R. Screaton (New York: Interscience, 1961), pp. 1–63; M. L. G. Redhead, "Some Philosophical Aspects of Particle Physics," *Studies in the History and Philosophy of Science 11* (1980), 279–304; and S. S. Schweber, Chapter 46 of this volume. James Cushing is currently engaged in a major study of the history of *S*-matrix theory, with particular emphasis on the evolution of the bootstrap conjecture.

9 J. Hamilton, "Dispersion Relations for Elementary Particles," *Prog. Nucl. Phys. 8* (1960), 143–95.

10 R. Kronig, "On the Theory of Dispersion of X-Rays," *J. Opt. Soc. Am. 12* (1926), 547–57; H. A. Kramers, "La Diffusion de la Lumiere par les Atomes," in *Congresso Internazionale de Fisici,* Vol. 2, Como (Bologna: Nicolo, 1927), pp. 545–57. See also H. A. Kramers, "The Law of Dispersion and Bohr's Theory of Spectra," *Nature (London) 113* (1924), 673–4, reprinted in *Sources of Quantum Mechanics,* edited by B. L. van der Waerden (Amsterdam: North Holland, 1967), pp. 177–80; and H. A. Kramers and W. Heisenberg, "Ueber die Streuung von Strahlen durch Atome," *Z. Phys. 31* (1925), 681–708, reprinted as "On the Dispersion of Radiation by Atoms," in *Sources of Quantum Mechanics* (cited in this note).

11 See J. D. Jackson (note 8, p. 162) and M. Cini (note 8).

12 W. Heisenberg, "Die 'beobachtbaren Grössen' in der Theorie der Elementarteilchen," *Z. Phys. 120* (1943), 513–38; part II, ibid., 673–7; part III; ibid., *123* (1944), 93–111. On the early history of *S*-matrix theory, see J. T. Cushing (note 8, p. 48); I. Grythe, "Some Remarks on the Early *S*-Matrix," *Centaurus 26* (1982–3), 198–203; R. Oehme, "Theory of the Scattering Matrix: An Introduction to Heisenberg's Papers," in *Collected Works, Volume AII* (Berlin: Springer-Verlag, 1988); and H. Rechenberg, Chapter 39 of this volume.

13 J. T. Cushing (note 8, p. 47).

14 R. Kronig, "A Supplementary Condition in Heisenberg's Theory of Elementary Particles," *Physica 12* (1946), 543–4.

15 On the history of renormalization, see J. T. Cushing (note 8); S. Hayakawa, "The Development of Meson Physics in Japan," in *The Birth of Particle Physics,* edited by L. M. Brown and L. Hoddeson (Cambridge University Press, 1983), pp. 82–107; M. L. G. Redhead (note 8); S. S. Schweber, "Some Chapters for a History of Quantum Field Theory: 1938–1952," in *Relativity, Groups and Topology, II,* edited by B. S. DeWitt and R. Stora (Amsterdam: North Holland, 1984), pp. 37–220; S. S. Schweber (note 8); J. Schwinger, "Renormalization of Quantum Electrodynamics: An Individual View," in *The Birth* (this note), pp. 329–53; J. Schwinger, "Two Shakers of Physics: Memorial Lecture for Sin-itiro Tomonaga," *The Birth* (this note, pp. 354–75); and S. Weinberg (note 3).

16 G. Wentzel, "Recent Research in Meson Theory," *Rev. Mod. Phys. 19* (1947) 1–18; quoted in J. D. Cushing (note 8, p. 6).

17 See, for example, M. Gell-Mann and F. E. Low, "Bound States in Quantum Field Theory," *Phys. Rev. 84* (1951), 350–4; M. Gell-Mann and F. E. Low, "Quantum Electrodynamics at Short Distances," *Phys. Rev. 95* (1954), 1300–12. M. Gell-Mann and M. L. Goldberger, "The Formal Theory of Scattering," *Phys. Rev. 91* (1953), 398–408; and F. E. Low, "Boson–Fermion Scattering in the Heisenberg Representation," *Phys. Rev. 97* (1955), 1392–8.

18 G. F. Chew, "Renormalization of Meson Theory with a Fixed Extended Source," *Phys. Rev. 94* (1954), 1748–54; G. F. Chew, "Method of Approximation for the Meson–Nucleon Problem when the Interaction Is Fixed and Extended," *Phys. Rev. 94* (1954), 1755–9; G. F. Chew, "Improved Calculation of the *p*-Wave Pion–Nucleon Scattering Phase-Shifts in the Cut-Off Theory," *Phys. Rev. 95* (1954), 285–6; G. F. Chew, "Comparison of the Cut-Off Meson Theory with Experiment," *Phys. Rev. 95* (1954), 1669–75.

19 G. C. Wick, "Introduction to Some Recent Work in Meson Theory," *Rev. Mod. Phys. 27* (1955), 339–62.

20 G. F. Chew and F. E. Low, "Effective-Range Approach to the Low-Energy *p*-Wave Pion–Nucleon Interaction," *Phys. Rev. 101* (1956), 1570–9; G. F. Chew and F. E. Low, "Theory of

Photomeson Production at Low Energies," *Phys. Rev. 101* (1956), 1579–87. The Chew–Low papers of 1956 provide a nice illustration of the interplay between technical developments in QFT and the elaboration of simplified models of the strong interaction. The first of the 1956 papers reexamines Chew's fixed-extended-source model using Low's QFT formalism.

21 Tables 40.1 and 40.2 are based on biographical and publication data provided by Professors Chew, Gell-Mann, Goldberger, and Low. As is evident from the acknowledgments within the publications of the four theorists, the copublication data of Table 40.1 underestimate the density of social and intellectual interconnections among the group. For a brief account of some of the personal ties existing between individuals, see M. L. Goldberger, "Francis E. Low – A Sixtieth Birthday Tribute," in *Asymptotic Realms of Physics: Essays in Honor of Francis E. Low*, edited by A. H. Guth, K. Huang, and R. L. Jaffe (Cambridge, Mass.: MIT Press, 1983), pp. xi–xv.

22 R. Oehme (note 12, p. 5).

23 M. Gell-Mann (note 6, p. 9).

24 M. L. Goldberger, "Validity of Pseudoscalar Meson Theory with Pseudoscalar Coupling," in *Proceedings of the Fourth Annual Rochester Conference on High Energy Nuclear Physics*, University of Rochester, 25–27 January 1954, edited by H. P. Noyes et al. (University of Rochester, 1954), pp. 26–32.

25 F. Rohrlich and R. L. Gluckstern, "Forward Scattering of Light by a Coulomb Field," *Phys. Rev. 86* (1952), 1–9; H. A. Bethe and F. Rohrlich, "Small Angle Scattering of Light by a Coulomb Field," *Phys. Rev. 86* (1952), 10–16; M. Gell-Mann, "Particle Theory from *S*-Matrix to Quarks" (note 6, p. 9); M. L. Goldberger, interview.

26 J. A. Wheeler, "On the Mathematical Description of Light Nuclei by the Method of Resonating Group Structure," *Phys. Rev. 52* (1937), 1107–27; E. P. Wigner, "Resonance Reactions and Anomalous Scattering," *Phys. Rev. 70* (1946), 15–33. See J. T. Cushing (note 8, pp. 3–10) for a discussion of this work and references to the original literature. Note that several of our principals were members of the Princeton Institute for Advanced Study in the early 1950s, including Gell-Mann (1951–2), Low (1950–2), and Nambu (1952–4).

I. Grythe (note 12, pp. 200–1) argues that Pauli encouraged work on the *S* matrix at Princeton during and immediately after World War II. Goldberger notes that Pauli was a visitor at Princeton during the academic year 1953–4 and encouraged the extension of dispersion relations to the scattering of massive particles, as described later. In an unpublished manuscript, Schweber mentions an important talk given at Princeton in 1941 by Richard Feynman. Feynman reported on a paper by Bode published that year in the *Bell System Technical Journal* that discussed the relation between causality and analyticity in the context of amplifier design. Wheeler recalled this talk in 1947.

Schweber also notes (private communication) that causality and analyticity were important considerations in wartime work on microwave theory and that the postwar interest in dispersion relations may have been in part a reflection of wartime experience. (I thank also Martin Krieger for a discussion on this point.) On the relation of microwave research and the renormalization of quantum electrodynamics, see J. Schwinger, "Two Shakers of Physics" (note 15, pp. 364–6). A different connection between applied research and elementary particle theory is evident in Cini's autobiographical account (note 8, p. 162): In 1949, Cini derived dispersion relations in an industrial setting, studying deformations of suspensions of rubber; three years later, he had changed fields and begun to investigate the possibility of deriving dispersion relations for elementary particle scattering.

27 W. Schutzer and J. Tiomno, "On the Connection of the Scattering and Derivative Matrices with Causality," *Phys. Rev. 83* (1951), 249–51; G. van Kampen, "*S*-Matrix and Causality Condition. I. Maxwell Field," *Phys. Rev. 89* (1953), 1072–9; G. van Kampen, "*S*-Matrix and Causality Condition. II. Nonrelativistic Particles," *Phys. Rev. 91* (1953), 1267–76.

28 M. Goldberger (note 8, p. 686).

29 J. S. Toll, "The Dispersion Relation for Light and Its Application to Problems Involving Electron Pairs," unpublished Ph.D. thesis, Princeton University, 1952. See also J. S. Toll, "Causality and the Dispersion Relation: Logical Foundations," *Phys. Rev. 104* (1956), 1760–70. In "Distribution of Recoil Nucleus in Pair Production by Photons," *Phys. Rev. 80* (1950),

189–96, R. Jost, J. M. Luttinger, and M. Slotnick of Princeton University acknowledge the assistance of Toll and Wheeler in developing the dispersion techniques applied to the calculation of forward photon scattering by themselves and, later, by Rohrlich and Gluckstern. See also F. Rohrlich and R. L. Gluckstern (note 25).

30 M. L. Goldberger (note 8, p. 687).

31 M. Gell-Mann, M. L. Goldberger, and W. E. Thirring, "Use of Causality Conditions in Quantum Theory," *Phys. Rev. 95* (1954), 1612–27; M. L. Goldberger, "The Use of Causality Conditions in Quantum Theory," *Phys. Rev. 97* (1955), 508–10; M. L. Goldberger, "Causality Conditions and Dispersion Relations. I. Boson Fields," *Phys. Rev. 99* (1955), 979–85.

32 M. L. Goldberger, "Theory and Applications of Single Variable Dispersion Relations," in *The Quantum Theory of Fields, Proceedings of the Twelfth Solvay Conference on Physics*, University of Brussels, October 1961 (New York: Interscience, 1961), pp. 179–95.

33 M. L. Goldberger, "The Use of Causality Conditions" (note 31). Goldberger (p. 510) notes that his result is a generalization of similar findings by Low and Nambu.

34 Ibid. (p. 510).

35 M. L. Goldberger, "Causality Conditions and Dispersion Relations. I. Boson Fields" (note 31). In "Use of Causality Conditions in Quantum Theory" (Note 31), Gell-Mann, Goldberger, and Thirring claimed that their results were valid for the scattering of massive particles. However, they included a "note added in proof" (p. 1613) acknowledging that "it has been pointed out to us by Professor N. G. van Kampen that in this section the arguments pertaining to particles with mass are lacking in rigor. In the following sections, however, we use only results pertaining to massless fields." In "Causality Conditions and Dispersion Relations. I. Boson Fields" (note 31, p. 979), Goldberger went further: "Another problem treated in GGT was the dispersion relation for Bose particles with mass. Only a very idealized situation was considered and the discussion was very involved and not very satisfactory. In addition, some erroneous and misleading statements were made in connection with the results. One of the purposes of the present work is to give a simplified and more satisfactory derivation of the generalized dispersion relations for particles with mass based on a general formalism analogous to that used in ["The Use of Causality Conditions in Quantum Theory" (note 31)]. Although written by a single author, Goldberger's second 1955 publication makes evident the interconnectedness of the Chicago theory group's work at the time. Thus, Goldberger refers forward to a future copublication with Oehme on dispersion relations for fermion-scattering processes (p. 979), thanks Nambu for elucidation of a technical point in QFT (n. 4), acknowledges the genesis of his work in his collaboration with Gell-Mann, and expresses his gratitude to Nambu, Oehme, and Hironari Miyazawa (a Japanese theorist then visiting Chicago) for "very many helpful suggestions" (p. 985).

36 Ibid. (p. 983, equation 2.40).

37 M. L. Goldberger (note 32, p. 183).

38 M. L. Goldberger (note 8, p. 688).

39 S. S. Schweber (note 8).

40 This argument is elaborated and extended by Cini (note 8, p. 157), who concludes that "the dominant ideology in the U.S. lent itself particularly, through the mechanism of unbridled competitivity and the rat race, to the acceptance of a utilitarian and pragmatic, but fragmentary, concept of science with the consequent abandoning of its traditional aim of unification of knowledge."

41 G. F. Chew, "The Pion–Nucleon Interaction and Dispersion Relations," *Annu. Rev. Nucl. Sci. 9* (1959), 29–60.

42 R. Karplus and M. A. Ruderman, "Application of Causality to Scattering," *Phys. Rev. 98* (1955), 771–4. Goldberger's second 1955 paper was received for publication 7 April 1955.

43 See the "note added in proof" in Karplus and Ruderman (note 42, p. 772).

44 M. L. Goldberger, H. Miyazawa, and R. Oehme, "Application of Dispersion Relations to Pion–Nucleon Scattering," *Phys. Rev. 99* (1955), 986–8.

45 Ibid. (p. 987, equation 2.15).

46 H. L. Anderson, W. C. Davidon, and U. E. Kruse, "Causality in the Pion–Proton Scattering," *Phys. Rev. 100* (1955), 339–43.

47 F. de Hoffmann, N. Metropolis, E. Alei, and H. A. Bethe, "Pion–Hydrogen Phase Shift Analysis between 120 and 217 MeV," *Phys. Rev. 95* (1954), 1586–605.

48 Anderson et al. (note 46, p. 343).

49 To take the story of forward pion–nucleon dispersion relations a little further: At Chicago, Davidon and Goldberger, in "Comparison of Spin-Flip Dispersion Relations with Pion–Nucleon Scattering Data," *Phys. Rev. 104* (1956), 1119–21, used dispersion relations as a consistency check, favoring the Fermi set of phase shifts over the alternative proposed by Yang; at Illinois, U. Haber-Schaim, in "Pion–Nucleon Coupling Constant and Scattering Phase Shifts," *Phys. Rev. 104* (1956), 1113–15, used dispersion relations to extract from the data a value for the pion–nucleon coupling constant, $f^2 = 0.082 \pm 0.015$ (as usual, in agreement with the Chew–Low figure); and W. Gilbert, at Harvard, in "New Dispersion Relations for Pion–Nucleon Scattering," *Phys. Rev. 108* (1957), 1078–83, used a novel formulation of dispersion relations both to extract the pion–nucleon coupling constant ($f^2 = 0.084$) and to compute s-wave scattering lengths from the p-wave data. (Gilbert had become involved in dispersion-relation work through his collaboration with Salam; see note 50.)

The alternative perspective on the relation between dispersion relations and scattering data was to regard the latter as testing the former. For a review of the analysis of pion–nucleon data from this perspective, see G. Puppi, "The Nucleon and Its Interaction with Pions, Photons, Nucleons and Antinucleons; through 3/2, 3/2 Resonance – Experimental I," in *Proceedings of the 1958 Annual International Conference on High Energy Physics* (Geneva: CERN, 1958), pp. 39–64. The importance of dispersion relations as a resource in determining experimental strategies is evident from Puppi's conclusion (pp. 45–6): "As many people suggest, the solution [to the problem of testing dispersion relations or of extracting a unique value of f^2 from the data] cannot come from further manipulation of the algebra, but only from more precise experimental information. In particular it seems important to check the total cross-section data for $\pi^{\pm}$ around the resonance and at least make a good experiment on π^{-} elastic scattering around 135 MeV. The accuracy required. . .would be of a few percent."

50 A. Salam, "On Generalized Dispersion Relations," *Nuovo Cimento 3* (1956), 424–9; A. Salam and W. Gilbert, "On Generalized Dispersion Relations II," *Nuovo Cimento 3* (1956), 607–11; R. H. Capps and G. Takeda, "Dispersion Relations for Finite Momentum-Transfer Pion–Nucleon Scattering," *Phys. Rev. 103* (1956), 1877–96. The Chicago and Caltech work went unpublished, but is referred to in J. D. Jackson (note 8, p. 3), M. L. Goldberger (note 8, p. 688), and M. Gell-Mann (note 6, p. 9). Goldberger has suggested (interview with author) that Salam became interested in dispersion relations through conversations between the two men during a visit to Chicago by Salam. Polkinghorne's collaboration with Gell-Mann at Caltech perhaps marks a first step toward the extensive perturbative investigations of the analytic structure of the S matrix carried out by Cambridge theorists in the 1960s and expounded in R. J. Eden, P. V. Landshoff, D. I. Olive, and J. C. Polkinghorne, *The Analytic S-Matrix* (Cambridge University Press, 1966). Cushing (note 8, p. 25) suggests that Olive became interested in S-matrix theory while attempting to construct a renormalizable Yang–Mills theory of the weak interaction.

51 E. Corinaldesi, "Dispersion Relations for Photoproduction of Mesons," *Nuovo Cimento 4* (1956), 1384–98; A. A. Logunov and B. M. Stepanov, "Dispersionnie Sootnoshenia dlya Reaktsii Photoroshdeniya – Mesonov," *Dok. Akad. Nauk SSSR 110* (1956), 368–70. It appears from Corinaldesi's paper that he had no direct contact with U.S. dispersion theorists; instead, his interest derived from that of his experimental colleagues working with Glasgow's newly operational (1954) 350-MeV electron-synchrotron.

52 G. F. Chew, M. L. Goldberger, F. E. Low, and Y. Nambu, "Application of Dispersion Relations to Low-Energy Meson–Nucleon Scattering," *Phys. Rev. 106* (1957), 1337–44; G. F. Chew, M. L. Goldberger, F. E. Low, and Y. Nambu, "Relativistic Dispersion Relation Approach to Photomeson Production," *Phys. Rev. 106* (1957), 1345–55. The dispersion relation reproduced in the Appendix is equation 3.2 of the first of these papers (p. 1339). The adjective "classic" is a self-description from Goldberger (note 8, p. 688): "The absence of real derivations had not deterred [Chew, Goldberger, Low, and Nambu] from completing what are usually referred to as our classic papers on pion–nucleon scattering and photopion production."

53 Goldberger and Low interviews with author.

54 G. F. Chew and F. E. Low (note 20).

55 Chew et al. (note 52, p. 1355).

56 On the population explosion, see A. Pickering (note 5, pp. 47–50, 102–7).

57 Quantitative estimates of data-production rates from the 1950s to the 1970s are given in A. H. Rosenfeld, "The Particle Data Group: Growth of Operations – Eighteen Years of Particle Physics," *Annu. Rev. Nucl. Sci. 25* (1975), 555–98, and summarized in A. Pickering (note 5, p. 121, n. 54).

58 J. D. Jackson (note 8, p. 4) provides references to the original literature. See M. L. Goldberger (note 8, pp. 689–90) for his interest in dispersion calculations of electromagnetic form factors and hadronic weak interactions. See also S. B. Treiman, Chapter 27 of this volume.

59 K. Symanzik, "Derivation of Dispersion Relations for Forward Scattering," *Phys. Rev. 105* (1957), 743–9.

60 N. N. Bogoliubov, B. V. Medvedev, and M. K. Polivanov, unpublished lecture notes issued in translation by the Institute for Advanced Study, Princeton (1957); N. N. Bogoliubov, B. V. Medvedev, and M. K. Polivanov, "Probleme der Theorie der Dispersionsbeziehungen," *Fortschr. Phys. 6* (1958), 169–245; see also N. N. Bogoliubov and D. V. Shirkov, *Introduction to the Theory of Quantized Fields* (New York: Interscience, 1959); H. J. Bremermann, R. Oehme, and J. G. Taylor, "Proof of Dispersion Relations in Quantized Fields," *Phys. Rev. 109* (1958), 2178–90; H. Lehmann, "Analytic Properties of Scattering Amplitudes as Functions of Momentum Transfer," *Nuovo Cimento 10* (1958), 579–89. Note that both the Soviet Union and Germany were countries with strong theoretical traditions, but with few resources for elementary particle experiment.

61 S. Mandelstam, "Determination of the Pion–Nucleon Scattering Amplitude from Dispersion Relations and Unitarity. General Theory," *Phys. Rev. 112* (1958), 1344–60. For an earlier attempt along similar lines, see Y. Nambu, "Structure of Green's Functions in Quantum Field Theory," *Phys. Rev. 100* (1955), 394–411. The relation between the work of Nambu and Mandelstam is discussed by J. T. Cushing (note 8, p. 20).

62 For extended reviews of theoretical and phenomenological approaches deriving from Mandelstam's work, see the lectures collected in *Dispersion Relations: Scottish Universities' Summer School*, edited by G. R. Screaton (note 8).

63 T. Regge, "Introduction to Complex Orbital Angular Momentum," *Nuovo Cimento 14* (1959), 951–76.

64 See A. Pickering (note 5), pp. 46–7, 73–8, 121(n. 59), 122(n. 63).

65 On the history of the bootstrap, see F. Capra, "Bootstrap Physics: A Conversation with Geoffrey Chew," in *A Passion for Physics: Essays in Honor of Geoffrey Chew*, edited by C. DeTar, J. Finkelstein, and C. I. Tan (Philadelphia: World Scientific, 1985), pp. 247–86; G. F. Chew, Chapter 41 of this volume; J. T. Cushing (note 8); M. Gell-Mann (note 6); M. L. G. Redhead (note 8).

   In "The *S*-Matrix Program – Anatomy of a Scientific Theory" (note 8, p. 20), Cushing notes that at the 1959 Kiev conference the Soviet theorist Landau supported the trend of Chew's thought away from QFT. Gell-Mann (note 6, p. 8) also remarks that Landau and his collaborators regarded Bogoliubov as "the enemy." It would be interesting to explore whether or not such an antagonism in Soviet theory was instrumental in maintaining Landau's anti-QFT stance. For a short biography of Landau, see F. Janouch, "Lev D. Landau: His Life and Work," CERN Yellow Report 79-03 (1979).

66 I emphasize that this section is concerned with the social decline of field theory – with the declining number of physicists actively working in QFT, and with their declining impact on the practice of the particle physics community. For more on this topic, see A. Pickering (note 5, pp. 107–8). On the major conceptual results obtained in rigorous QFT during the 1950s, see A. S. Wightman, Chapter 42 of this volume.

67 Interview with Goldberger by author.

68 Interview with Low by author.

41   Particles as *S*-matrix poles: hadron democracy

GEOFFREY F. CHEW

Born 1924, Washington, D. C.; Ph. D., University of Chicago, 1948;
theoretical physics; University of California, Berkeley

The idea of hadron democracy – that all hadrons are "composites" and nonfundamental – is not the same as the idea that all physically observable particles correspond to singularities of an analytic scattering matrix. Nevertheless, the histories of those two ideas, both of which belong to the decade under study at this symposium, are intertwined. I give here a personal recollection of the connection. I am indebted to James Cushing for allowing me a look at a preliminary version of his case study of the *S*-matrix program. I am forbidden from quoting Cushing and take full responsibility for what I shall say here, but his efforts have been helpful in compensating my poor memory.

Let me begin by recalling that when Marvin L. Goldberger and I worked closely together from 1946 to 1948 as students of Enrico Fermi at the University of Chicago, we learned, as did other students of that epoch, that there were a few elementary particles out of which everything was built. Among them were neutrons and protons – the building blocks of nuclei. No one then doubted the elementarity of nucleons; Fermi never expressed such a doubt to me. By the end of the fifties, there was a growing belief that no hadron deserved to be called elementary. The distinction between protons and deuterons had become blurred. I propose here to recall how the evolving understanding of the *S* matrix contributed to that blurring.

All the ingredients for the new understanding of the *S* matrix had existed during the 1940s. As Helmut Rechenberg has explained in Chapter 39 in this

This work was supported by the Office of High Energy Research, Office of High Energy and Nuclear Physics, Division of High Energy Physics, of the U.S. Department of Energy under contract DE-AC03-76SF00098.

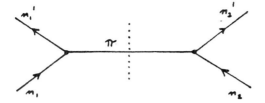

Figure 41.1. Landau graph for pion pole in nucleon–nucleon scattering amplitude.

volume, Werner Heisenberg had in the early forties defined the S matrix and recognized unitarity and Poincaré invariance as key general properties. Hendrick A. Kramers had suggested the importance of analyticity, and Ralph de Laer Kronig had connected analyticity with causality. The idea that the S matrix might be a framework for a complete theory – replacing field theory and circumventing its divergences – had been stated by Heisenberg. But where was the S-matrix counterpart of *force* between particles? Hideki Yukawa in the midthirties had proposed a meson-field basis for the force between nucleons. How could the S matrix, which deals with asymptotic states where particles are outside regions of interaction, incorporate the equivalent of a Yukawa force? A similar question could be asked about electromagnetic forces.

A decisive S-matrix step of the fifties, for the most part occurring after Fermi's death, connected force with the singularities of an analytic S matrix through recognition of the so-called *crossing principle* – that when a particle's energy, on which an analytic S-matrix element depends, is continued from positive to negative values, an outgoing particle changes into an ingoing antiparticle. That idea was one aspect of the more general principle, recognized by the end of the fifties, but not at the beginning, that graphs of the type invented by Richard Feynman for perturbative evaluation of a Lagrangian field theory are relevant to the analytic S matrix, independent of any approximation based on a small coupling constant. These graphs describe S-matrix singularities, in a manner compatible with crossing (tree graphs correspond to poles, and loops to branch points). Lev D. Landau seems to have been the first, in his 1959 paper,[1] to formalize the connection between Feynman-like graphs and S-matrix singularities, but Landau did not claim to discover the idea, which emerged from the area of theoretical activity that has been called dispersion relations, as reviewed in this volume by Sam Treiman (Chapter 27) and Andy Pickering (Chapter 40).

Before attempting to identify historical ingredients in the discovery of graph rules for S-matrix singularities, I show in Figure 41.1 the Landau graph representation of the Yukawa force. Lines in the graph correspond to physical hadrons; there is no renormalization to be considered. This graph

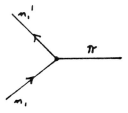

Figure 41.2. Graph for pion–nucleon coupling constant.

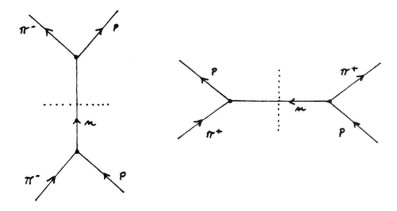

Figure 41.3. Graphs for nucleon poles in pion–nucleon scattering amplitudes.

denotes the position and residue of a pion pole in a nucleon–nucleon scattering amplitude. The subgraph of Figure 41.2 depicts one factor building the pole residue; this factor was called the pion-nucleon coupling constant.

During the fifties, it gradually dawned on the collective consciousness of a subset of particle theorists that physical consequences from a meson-exchange force, such as had been proposed by Yukawa, follow if the pole of Figure 41.1 is present in the nucleon–nucleon scattering amplitude. It furthermore became understood that such a pole *must* be present if the S matrix is to be simultaneously analytic, unitary, and Poincaré-invariant.

An experimentally persuasive part of the story was that the pion–nucleon coupling constant, defined as a factor in a pole residue, could be measured in a variety of different reactions. There was not only nucleon–nucleon scattering (Figure 41.1) but pion–nucleon scattering (Figure 41.3) and pion photoproduction (Figure 41.4). The latter two processes are extensively referred to in this symposium. The poles shown here lie close enough to experimentally accessible regions that careful measurements allowed their residues to be determined. (Sufficiently close to an isolated pole of an analytic function, the pole residue determines the value of the function.) As measure-

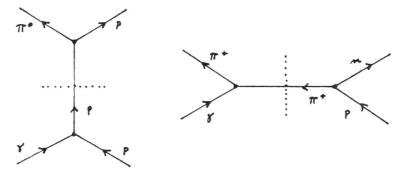

Figure 41.4. Graphs for poles in amplitudes for pion photoproduction from protons.

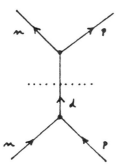

Figure 41.5. Deuteron pole in neutron–proton elastic scattering.

ments gradually became more and more accurate, the pion–nucleon coupling constant determined by very different experiments converged to a single value. The correctness of the graphical pole–particle correspondence, never "proved" during the fifties from any accepted set of general principles, slowly became compelling. Little by little the idea took hold in some fraction of the particle physics community that analyticity, together with unitarity and Poincaré invariance, determines the forces acting between particles, once particle quantum numbers have been specified. Incompleteness of proofs based on field theory became uninteresting. (During the sixties, an independent axiomatic analytic S-matrix framework was developed, taking off from Landau's 1959 paper.[1])

Heisenberg and other S-matrix enthusiasts of the forties failed to appreciate the generality of the pole–particle correspondence. Although they recognized the deuteron as a pole within the neutron–proton scattering amplitude (Figure 41.5), they did not appreciate the pion pole of Figure 41.1 or the notion that there would be amplitudes in which neutrons and protons themselves appear as poles (e.g., Figures 41.3 and 41.4). And, curiously, although the 1953 dispersion relations formulated by Murray Gell-Mann,

Goldberger, and Walter Thirring explicitly manifested particle poles, no emphasis at first was given to this feature. It would take several years before "pole consciousness" would develop.

Before the awareness came of general graphical rules for S-matrix singularities, Gell-Mann was stressing the dynamical content of dispersion relations. Gell-Mann, Goldberger, Thirring, and their followers in the fifties did not speak of an analytic S matrix, but of dispersion relations.[2] Not until the end of the decade was the connection between these two notions appreciated: Dispersion relations are Cauchy-Riemann formulas expressing an analytic S-matrix element in terms of its singularities. Amazingly, the S-matrix thinking of the forties had no impact on the dispersion-relations developments of the fifties. Crossing was overlooked in the forties; when appreciated in the early fifties, the term "S matrix" was not in vogue. Until 1960, no connection was made between dispersion relations and Heisenberg's work.

The term "crossing" was used first by Gell-Mann and Goldberger[3] in 1954, in connection with dispersion relations. At the 1956 Rochester conference, Gell-Mann[4] stressed the power of this and other general principles, reviving (without awareness of his predecessor) the Heisenberg idea that the S matrix might replace field theory. Francis Low and I at that point had been working at the University of Illinois on a semirelativistic static model of the pion–nucleon interaction, where the notion of force was explicit in the traditional sense of a Lagrangian theory.[5] We had found a formulation of the model that, by employing analytic functions, allowed more direct contact with experimental data than was usual for field theories of strong interactions. The poles of our analytic functions were the key to such contact; we had associated our pole residue with *force strength*, and pole position with particle mass (in this case, pion mass). We did not know how to make our model fully relativistic, but were struck by the fact that Goldberger's completely relativistic dispersion relation for pion–nucleon scattering involved an analytic function whose properties looked similar to those of the function in our model.[6]

I had earlier worked closely with Goldberger, both at Chicago and at Berkeley, before meeting Low, and Goldberger was again close by at the University of Chicago. It was natural for the three of us to join forces in connecting dispersion relations with the static model, and Goldberger involved Yoichiro Nambu, also at Chicago, in this project. That 1956 collaboration by Chew, Goldberger, Low, and Nambu (CGLN), which led to two papers, in my recollection yielded the first clear statement that force – in the sense of Yukawa – resides in the singularities of an analytic S matrix.[7] From that point on, I never believed that the description of interhadronic forces demanded a Lagrangian. Stanley Mandelstam's paper of 1958 gave powerful reinforcement to this belief.[8] Although I failed to recognize until 1960 that the CGLN papers and that of Mandelstam were dealing with the concept identified by Heisenberg in the early forties, my thinking for two decades, starting in 1956, became based on the analytic S matrix.

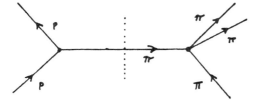

Figure 41.6. Pion pole in amplitude for pion production from pion–nucleon collision.

Curiously, even people who believed in dispersion relations during the midfifties did not with heart and soul always accept the amazing connection of Feynman-like graphs with *S*-matrix singularities. In 1958, I wrote a paper *conjecturing* that the nucleon–nucleon scattering pole of Figure 41.1 could be verified by extrapolation of scattering data, but I remember finding it difficult to believe that such would actually work.[9] (It did.) Slightly later, Low and I made a corresponding conjecture about the pole of Figure 41.6.[10] The fact that Low expected this latter conjecture to be verified was comforting to me; I had enormous respect for Francis's judgment.

In connection with *S*-matrix poles, I recall a remark by Landau, made privately to me during the 1959 Kiev conference. Landau had been scolding me for wasting time on approximate dynamical models and stated that recognition of the pole-particle correspondence was a momentous achievement that should not be blurred by unreliable model calculations. Landau seemed to be giving somebody in the United States credit for discovering the general pole-particle correspondence, but it has never been clear to me that any individual deserves credit. Maybe somebody at this meeting will stand up and assign priority. Certainly Gregory Breit and Eugene Wigner have some claim, although they did not know about graphs or crossing when they proposed their celebrated 1936 formula.[11]

The *S*-matrix models that I and others spent so much time on in the fifties never (as Landau foretold) achieved a reliable status, but they contributed to a changing attitude about the nature of neutrons, protons, and pions. It was found that when other neighboring meson singularities were added to the pion pole of a nucleon–nucleon scattering amplitude, the combined neutron–proton force was approximately that needed to bind the deuteron. Earlier, Low and I had found that the force of Figure 41.3(b) could generate the Δ resonance as a pion–nucleon bound state.[5] Then, in 1959, just before the Kiev conference, Mandelstam and I encountered a mind-boggling phenomenon.[12] We found that a spin-1 $\pi\pi$ resonance could be generated by a force due to Yukawa-like exchange of this same resonance. Later, such a resonance was named the ϱ meson, and although we did not use the name ϱ in 1959, Figure 41.7 sketches how the dispersion-relation summation over an infinite

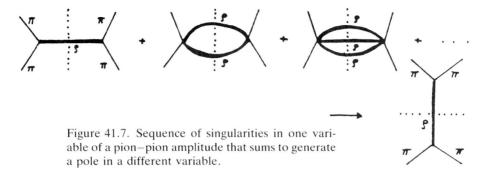

Figure 41.7. Sequence of singularities in one variable of a pion–pion amplitude that sums to generate a pole in a different variable.

sequence of ϱ discontinuities in a $\pi\pi$ elastic amplitude can generate a ϱ pole. We here were using a boundary condition that later was associated with the name of Tullio Regge. The phenomenon represented in Figure 41.7 is analogous to that of generating a bound state through a Schrödinger equation with an attractive potential. (The idea that summing over an infinite sequence of discontinuities in one variable can generate a pole in another variable came in the sixties to be called *duality*.) Any particle corresponding to such a pole could be regarded as a *bound state* of other particles.

The mechanism typified by Figure 41.7 was called bootstrap, because ϱ as a force generates ϱ as a particle.[13] It did not take long to ask, Cannot *any* hadron be so regarded as a bound state of other hadrons – due to (Yukawa-like) hadron-exchange forces? The pole-particle correspondence, following from general S-matrix principles, makes no distinction between elementary and composite particles. The question was, Are observed hadron masses, spins, and coupling constants compatible with bound-state status? Model-based estimates yielded an affirmative answer for all the known hadrons, including neutrons and protons. No methods were ever developed for summing all important S-matrix discontinuities, but S-matrix theorists of the early sixties saw the neutron and proton as bound states, in a sense qualitatively similar to that of the deuteron. The puzzle to be resolved was no longer one of elementary hadrons, but of the internal quantum numbers carried by hadrons.

Three related but different statements about hadrons were heard at the end of the fifties:

1. There is hadron democracy – all hadrons having an essentially equivalent status.
2. Hadrons are bound states of other hadrons sustained by hadron-exchange forces.
3. Hadrons are self-generated by an S-matrix bootstrap mechanism that determines all their properties.

To the present time, none of these statements has achieved precise meaning, but none has been shown false. The hope kindled in the fifties, that general principles such as *S*-matrix unitarity allow no arbitrariness in particle properties, remains very much alive today.

Notes

1 L. D. Landau, "On Analytic Properties of Vertex Parts in Quantum Field Theory," *Nucl. Phys. 13* (1959), 181–92.
2 M. Gell-Mann, M. L. Goldberger, and W. Thirring, "Use of Causality Conditions in Quantum Theory," *Phys. Rev. 95* (1954), 1612–27.
3 M. Gell-Mann and M. L. Goldberger, "Scattering of Low-Energy Photons by Particles of Spin ½," *Phys. Rev. 96* (1954), 1433–8.
4 M. Gell-Mann, in *High Energy Nuclear Physics, Proceedings of the Sixth Annual Rochester Conference, April 3–7, 1956*, edited by J. Ballam, V. L. Fitch, T. Fulton, K. Huang, R. R. Rau, and S. B. Treiman (New York: Interscience, 1956), pp. III-30–6.
5 G. F. Chew and F. E. Low, "Effective-Range Approach to the Low Energy *p*-Wave Pion-Nucleon Interaction," *Phys. Rev. 101* (1956), 1570–9; G. F. Chew, "Pion-Nucleon Scattering When the Coupling Is Weak and Extended," *Phys. Rev. 89* (1953), 591–3, 904.
6 M. L. Goldberger, "Use of Causality Conditions in Quantum Theory," *Phys. Rev. 97* (1955), 508–10; M. L. Goldberger, "Causality Conditions and Dispersion Relations. I. Boson Fields," *Phys. Rev. 99* (1955), 979–85.
7 G. F. Chew, M. L. Goldberger, F. E. Low, and Y. Nambu, "Application of Dispersion Relations to Low-Energy Meson-Nucleon Scattering," *Phys. Rev. 106* (1957), 1337–44, and "Relativistic Dispersion Relation Approach to Photomeson Production," ibid., 1345–55.
8 S. Mandelstam, "Determination of the Pion-Nucleon Scattering Amplitude from Dispersion Relations and Unitarity. General Theory," *Phys. Rev. 112* (1958), 1344–60.
9 G. F. Chew, "Proposal for Determining the Pion-Nucleon Coupling Constant from the Angular Distribution for Nucleon-Nucleon Scattering," *Phys. Rev. 112* (1958), 1380–3.
10 G. F. Chew and F. E. Low, "Unstable Particles as Targets in Scattering Experiments," *Phys. Rev. 113* (1959), 1640.
11 G. Breit and E. P. Wigner, "Capture of Slow Neutrons," *Phys. Rev. 49* (1936), 519–31.
12 G. F. Chew and S. Mandelstam, "Theory of the Low-Energy Pion-Pion Interaction," *Phys. Rev. 119* (1960), 467–77.
13 G. F. Chew, "Theory of Strong Coupling of Ordinary Particles," in *Proceedings of the Ninth International Conference on High Energy Physics, Kiev, July 15–25, 1959* (Moscow: Academy of Science, USSR, 1960), pp. 313–52.

42   The general theory of quantized fields in the 1950s

ARTHUR S. WIGHTMAN

Born 1930, Rochester, N. Y.; Ph.D., Princeton University, 1949; theoretical physics; Princeton University

Introduction

At the beginning of the 1950s, the outlook for elementary particle theory was strongly influenced by the success of renormalization theory for quantum electrodynamics. As a result of the efforts of Sin-itiro Tomonaga, Julian Schwinger, Richard P. Feynman, Freeman Dyson, and others, the quantum field theory of Dirac spin-$\frac{1}{2}$ particles interacting with radiation had been solved, in the sense that well-defined formal power series in the fine-structure constant could be fairly unambiguously attributed to various quantities of interest and gave spectacular agreement with experiment, for the gyromagnetic ratio of the electron and the Lamb shift in hydrogen, in particular.[1] Attention turned to the problem of giving a nonperturbative treatment. This was urgent, because the burgeoning demands of strong-interaction physics made it unlikely that perturbative methods would be of practical significance in that area. The responses were varied; as examples, consider dispersion theory, axiomatic field theory, the Schwinger action principle, and the Feynman path integral. What many of these efforts had in common was then somewhat novel: They attempted to encapsulate the currently existing understanding of the relativistic quantum theory of fields in a general theory of quantized fields. From a logical point of view, they can be viewed as attempts to systematize and bring to logical completion the process that had occurred in renormalization theory, in which the initially ambiguous equations of quantum electrodynamics were given a precise meaning and greater predictive power.

The purpose of this review is to describe some of the developments of the 1950s that were of significance for such a putative general theory of quantized fields, with particular emphasis on those developments that made possible a mathematically rigorous theory. Of course, the general theory was heavily influenced by experimental and theoretical discoveries in the concrete problems of particle physics, and so it is somewhat artificial to describe the general theory in isolation. Nevertheless, I shall mainly stick to the creation of the language in which particle physics is described; it is the task of others to review the particle physics itself.

The state of the art as of 1950 and some questions to which it gives rise

What tools were then available for those who wished to construct a general theory of quantized fields? I shall mention only the following:

(a) The most basic tool was certainly the Hamiltonian formulation of quantum dynamics. Expressed in terms of the Hamiltonian H, it can be described as follows: If H is a self-adjoint operator in a Hilbert space $\mathcal{H}$, then $U(t) = e^{iHt}$ for t real is unitary,

$$U(t)^{-1} = U(t)^{+} \tag{42.1}$$

depends continuously on t, and forms a one-parameter group

$$U(t)U(t') = U(t + t') \tag{42.2}$$

Stone's theorem says that, conversely, every continuous unitary one-parameter group arises from a unique self-adjoint Hamiltonian.[2] In rough physical terms, this says that the requirement that a quantum dynamics conserve probability is equivalent to the existence of a self-adjoint Hamiltonian. Nowadays, one finds assertions about the self-adjointness of Hamiltonians in the *Physical Review*, but in the late forties and fifties such matters were regarded as too mathematical to appear there (and so the fundamental paper of T. Kato in which self-adjointness was first proved for the Schrödinger n-body problem ended up in the *Transactions of the American Mathematical Society*[3]). The practical importance of this insistence on conservation of probability will appear later in connection with a discussion of indefinite metric.

(b) Another main tool was canonical quantization. Here there is a story of some complexity involved, which I digress to relate.

For boson fields, the basic ideas were introduced by Max Born, Werner Heisenberg, and Pascual Jordan, by Paul A. M. Dirac, and by Wolfgang Pauli and Victor F. Weisskopf.[4-6] The method was to decompose the field into contributions from oscillators and to quantize the oscillators independently according to the usual method of quantum mechanics. This leads to the canonical commutation relations (CCR). An analogous procedure leading to the canonical anticommutation relations (CAR) was adopted for fermion fields following the ideas of Jordan and Eugene Wigner.[7] When the method is properly formulated and applied to a system of a finite number of degrees of freedom, it leads to an essentially unique result, as was proved by John von Neumann for the CCR and by Jordan and Wigner for the CAR.[8] More explicitly, von Neumann's result may be stated as follows: Let $U(\alpha)$ and $V(\beta)$ be unitary operators in a separable Hilbert space $\mathcal{H}$, defined for all α in a real vector space $\mathcal{S}$ of finite dimensions and for β in the dual vector space $\mathcal{S}'$. Suppose $U(\alpha)$ and $V(\beta)$ are continuous in α and β and satisfy

$$
\begin{aligned}
U(\alpha)U(\alpha') &= U(\alpha + \alpha') \\
V(\beta)V(\beta') &= V(\beta + \beta') \\
U(\alpha)V(\beta) &= [\exp i\beta(\alpha)]V(\beta)U(\alpha)
\end{aligned}
\tag{42.3}
$$

where $\beta(\alpha)$ is the linear functional β evaluated at the vector α. Then $\mathcal{H}$ is a direct sum of subspaces invariant under U and V in which U and V act irreducibly. Every such irreducible representation of the relations, (42.3) is unitary-equivalent to the Schrödinger representation in which $\mathcal{H}$ is $L^2(R^n)$, n = dimension of $\mathcal{S}$, and

$$
\begin{aligned}
(U(\alpha)\Phi)(x) &= \Phi(x + \alpha) \\
(V(\beta)\Phi)(x) &= \left(\exp i \sum_{j=1}^{n} \beta_j x_j \right)\Phi(x)
\end{aligned}
$$

if $\alpha = (\alpha_1, \ldots, \alpha_n)$ and $\beta = (\beta_1, \ldots, \beta_n)$.[*]

The realization that the situation is completely different for systems of an infinite number of degrees of freedom came only gradually and over a long time. In retrospect, one can find inequivalent irreducible representations of the CAR implicit in a remarkable lecture of von Neumann in 1934 on systems of an infinite number of degrees of freedom.[9] Examples were explicitly constructed by Kurt Friedrichs in his pioneering studies of the mathematical foundations of quantum field theory, although he did not discuss their inequivalence.[10] Various inequivalent representations were independently rediscovered by I. E. Segal, R. Haag, and myself in the period 1952–4. Haag studied, in particular, the properties of the so-called Fock representations, in which there is a no-particle state, and showed that for a large class of Euclidean invariant-field-theory models, the Hamiltonian cannot be well defined unless a representation inequivalent to a Fock representation is used (Haag's theorem[11]). D. Hall and I generalized this result and showed that two relativistic quantum field theories that use unitary-equivalent representations of the CCR necessarily have equal two-point, three-point, and four-point vacuum expectation values.[12] The idea of the proof is to show that the unitary operator that makes the representations of the CCR in the two theories equivalent necessarily makes the representations of the Euclidean group equivalent and therefore maps the vacuum state of one theory on that of the other. From this follows the equality of the n-point vacuum expectation values of the two theories when all times in them are equal. An analytic

---

[*]   *Ed. note:* The result quoted here is the precise formulation of the operator realization of the Bose commutation rules. It is phrased in terms of unitary operators instead of the customary creation and annihilation operators, because the latter are unbounded.

continuation argument then yields, for $n = 2, 3, 4$, equality for all values of the arguments. Thus, one could expect to have to use an inequivalent representation for each distinct value of coupling constant or mass. The moral of this story is that in relativistic quantum field theory, the kinematics (representation of the CCR and/or CAR) is all mixed up with the dynamics. Although L. Gårding and I, for the CCR and CAR, and Segal for the CCR, gave classification theories for the representations, little progress was made in using the new representations to solve models.[13,14] By the end of the 1950s, the idea that one should first pick the appropriate representation of the CCR (or CAR), and then show, using it, that the Hamiltonian of a model makes sense, had been pretty generally abandoned. There remained the method that practical-minded people had been using all along: To solve a model, introduce cutoffs in it until it becomes essentially a system of a finite number of degrees of freedom; quantize with a Fock representation (there is no other choice possible!); study the limit of the resulting theory as the cutoffs are removed.

(c) The third basic tool was perturbative renormalization theory. It provided guidance and suggested conjectures for the nonperturbative theory. For example, the use of Green's functions and retarded functions to express the content of quantum field theories turned out to be very natural, and their renormalization in perturbation theory was a natural extension of the renormalization of S-matrix elements. As a second example, consider the question whether or not quantized fields make sense when restricted to a fixed time. Perturbative renormalization theory indicated that in many field theories, the fields make sense as operators when smeared with smooth test functions on space–time, but *not* when smeared with smooth test functions on space at a fixed time. This suggested that Green's functions might provide a more natural description than the transition amplitudes that were common in many papers of the period, because the initial and final states involved in the transition amplitudes usually were labeled by field values at a fixed time. This behavior also casts doubt on the reliability of the CCR and CAR in the theories in question, because the right-hand side of the relations would become $(1/Z)\delta(\mathbf{x} - \mathbf{y})$, with $Z = 0$, instead of $\delta(\mathbf{x} - \mathbf{y})$.

(d) A fourth basic tool was mathematical: L. S. Schwartz's theory of distributions.[15] Although the recognition that vacuum expectation values of products of fields are generalized functions containing Dirac δ functions or worse goes back at least to Jordan and Pauli in the 1920s and its physical significance was explored by Niels Bohr and L. Rosenfeld in the 1930s in a celebrated paper, there was, at that time, no general mathematical theory of such generalized functions.[16,17] The mathematicians themselves used restricted classes of generalized functions for problems in partial differential equations (Friedrichs, Sobolev, Leray) without having a general theory. That was all changed by the work of Schwartz. The appearance of the grand treatise in 1951 appeared to me to be an act of Providence. (Gårding was well aware of the Schwartz theory, because he had attended seminars by J. W. Tukey at Princeton in 1947–8 based on the original Schwartz publications.[18]) From the spring of 1952 on, Gårding and I felt we had a clear idea of what we meant by a quantized field: It was an operator-valued distribution satisfying certain conditions of relativistic invariance, locality, and completeness.* We delayed a long time publishing our inductive treatment of the axioms because we felt we ought to have nontrivial examples, a desideratum not so easily obtained. Eventually, our account appeared.[19] We were not the only ones who saw in distribution theory a key to mathematical precision in quantum field theory.[20,21]

So much for some of the tools that students of field theory inherited for use in the 1950s. More will be said later in connection with specific developments. Here are some of the questions that naturally occurred to people at that time:

1. What are the indispensable attributes of a quantized field and of a quantum field theory?
2. Can quantum field theory be developed without reference to the canonical formalism? Without reference to an action principle?
3. If a quantum field theory is formulated in terms of Green's functions (or retarded functions or vacuum expectation values of products of fields) alone, how can one be sure that it describes a quantum theory in which probability is positive and conserved?
4. Are composite particles describable by local fields? If so, how can their S-matrix elements be calculated?

Here are a few preliminary comments on these questions. First of all, although they were unusual in their generality, they were very conservative compared with the analogous questions of Heisenberg's program for an S-matrix theory of 1943, which proposed to develop a theory of elementary particles without any interacting fields at all.[22] Second, they reflect an attitude toward the formulation of theories that was more common in mathematics than in physics at that time. To make this remark more specific, let us accept a partial answer to question 1 by asserting that a quantized field is an operator-valued distribution on space–time; that is, for each test function f on space–time belonging to a certain test-function space $\mathscr{S}$, it gives an operator $\phi(f)$ in the Hilbert space of states of the theory such that $\phi(f)$ depends linearly on f, and matrix elements of $\phi(f)$ vary continuously when f varies. If one proceeds in this way, one has to specify in advance which space $\mathscr{S}$ one is talking about. If $\mathscr{S}$ is Schwartz's space of infinitely smooth functions of rapid decrease, then ϕ is what is called a tempered field. To specify precisely in advance the space in which one was looking for solutions was

---

* *Ed. note:* Careful, complete definitions of all these concepts can be found in an earlier work.[76]

normal in mathematics. It was much more typical of physics to calculate formally and see what came out.

Disengaging the Poincaré invariance; understanding Wigner

In 1940, Wigner published a famous analysis of the irreducible unitary-ray representations of the Poincaré (inhomogeneous Lorentz) group. He found that omitting tachyons and representations in which translation is trivially represented, they are characterized by mass, m, and spin, s, for $m > 0$, and by helicity for $m = 0$.[23] He interpreted this result as a classification of the possible transformation laws of elementary particles under Poincaré transformations. One of the first problems that had to be solved in setting up a general relativistic field theory was how to generalize this interpretation from a representation appropriate to the states of a single elementary particle to those of a whole quantum field theory.

Nowadays, we solve this problem for a theory of massive particles as follows. Suppose the particles have masses and spins $m_1 s_1, \ldots, m_n s_n$. Then the representation $\{a, \Lambda\} \to U(a, \Lambda)$ for the full interacting theory is unitary-equivalent to the representation for a free-field theory of those masses and spins, because scattering theory tells us that the infields corresponding to these stable particles are free fields and transform according to their field character with the very same U as the interacting fields. For example, for a free scalar field ϕ^{in},

$$U(a, \Lambda)\phi^{\text{in}}(x)U(a, \Lambda)^{-1} = \phi^{\text{in}}(\Lambda x + a)$$

for all $\{a, \Lambda\}$ in the restricted Poincaré group. Now the asymptotic completeness of the scattering states of the theory makes the infields an irreducible set that uniquely determines the representation U up to phase factors as that appropriate to free particles. For the sake of completeness, let me give the construction explicitly. Let the single-particle Hilbert space of the particle of mass m_i and spin s_i be $\mathscr{H}_i^{(1)}$, and let the single-particle unitary-ray representation in it be $\{a, \Lambda\} \to U_i^{(1)}(a, \Lambda)$. It is uniquely determined up to equivalence by the foregoing theorem of Wigner. Then the theory of arbitrarily many particles of kind i has a state space $\mathscr{H}_i$ that is a Fock space

$$\mathscr{H}_i = \mathscr{F}_s^{}(\mathscr{H}_i^{(1)}) = \bigoplus_{n=0}^{\infty} [(\mathscr{H}_i^{(1)})^{\otimes n}]_s^{}$$

where s stands for symmetric, and a for antisymmetric, chosen to be consistent with the spin-statistics theorem. $U_i^{(1)}$ induces a representation in $\mathscr{H}_i$:

$$U_i = \bigoplus_{n=0}^{\infty} [(U_i^{(1)})^{\otimes n}]_s^{}$$

The full state space is then $\otimes_i \mathscr{H}_i$, and the full ray representation of the Poincaré $U = \otimes_i U_i$.*

In 1952, it was already possible to make this argument formally complete because of the work of Gunnar Källen and Chen Ning Yang and David Feldman on the infield and the outfield equations.[24,25] However, it took me quite a while during 1951–2 to overcome my doubts. Haag, whom I met at Copenhagen in June 1952, had already become convinced that the argument should be taken seriously. As will be recounted later, by the time another ten years had passed, it became possible to justify it rigorously using the Haag–Ruelle scattering theory.

Wigner's analysis showed that for the restricted Poincaré group $P_+^\uparrow$ (no inversions of space or time), one can replace ray representations of the group by uniquely determined unitary representations of its covering group, whose elements are $\{a, \Lambda\}$, with $a \in |R^4$ and $A \in SL(2, C)$. Thus, Gårding and I wrote, for the transformation law of the field ϕ,

$$U(a, A)\phi(f)U(a, A)^{-1} = \phi(\{a, A\}f)$$

Here the test functions have been generalized to n-tuples to cope with general spinor and tensor fields, and the action $f \to \{a, A\}f$ of the group element $\{a, A\}$ on f is assumed to be

$$(\{a, A\}f)_j(x) = \sum_k S(A^{-1})_{kj} f_k(\Lambda(A^{-1})(x - a))$$

This version of the transformation law expresses directly a connection between the transformation law of states $\Phi \to U(a, A)\Phi$ and the transformation law of the field $\phi \to S(A)\phi$ (or, equivalently, $f \to \{a, A\}f$). At the time, most accounts of relativistic field theory interpreted relativistic invariance in terms of the form invariance of equations of motion and for the action principle from which they were usually derived. In principle, the two formulations should be equivalent, and for noninteracting fields that is easy to prove, but for interacting fields it involves deep technical problems (pointwise multiplication of distributions). I want to emphasize that modulo these technical difficulties, the new formulation brought nothing that was not contained in the work of Schwinger and G. Lüders, except the recognition that the relativistic transformation law of fields can be expressed very simply in terms of the relativistic transformation law of states.[26,27]

Another of the advantages of this form of the tranformation law is that it also works for inversions and charge conjugation, with appropriate changes

---

* *Ed. note:* The remarkable theorem referred to here can be (almost correctly) understood in this way: n identical noninteracting particles possess a product wave function $\phi(q_1)...\phi(q_n)$. Strictly, the product is a tensor product. The products are symmetrical or antisymmetrical, indicated here by s and a subscripts. The space of all such states is the direct sum $\oplus$, over n. For many types of particles, the state space for the full set of particles is a tensor product over particle types.

for the definition of $f \rightarrow \{a, A\}f$. For example, for a charged scalar field ϕ, the PCT transformation acts according to

$$\Theta\phi(f)\Theta^{-1} = \phi^+(jf)$$

where $(jf)(x) = \overline{f(-x)}$, and the bar denotes complex conjugation. Schweber and I wrote out the transformation laws under inversions and charge conjugation,[28] but we regarded it as an exercise in the principles Wigner had stated and written out for the nonrelativistic case in his paper of 1932.[29] It should be noted that in his 1940 paper, Wigner did not discuss the antiunitary operations.

There is a simple direct argument based on the group-representation property that the positivity of the energy forces time inversion to be antiunitary and space inversion to be unitary. If $U(I)$ is the representative of the inversion I,

$$U(I)U(a, 1)U(I)^{-1} = U(Ia, 1)$$

so

$$U(I)P^\mu U(I)^{-1} = I_\nu^\mu P^\nu \quad \text{if } U \text{ is unitary}$$
$$-U(I)P^\mu U(I)^{-1} = I_\nu^\mu P^\nu \quad \text{if } U \text{ is antiunitary}$$

Thus, in any state Φ,

$$\pm I_0^0(U(I)\Phi, P^0 U(I)\Phi) = (\Phi, P^0\Phi)$$

from which the statement follows.[30]

Green's functions, retarded functions, and vacuum expectation values; asymptotic conditions and reduction formulas; reconstruction theorems; Borchers algebra

The first objective of this section is to describe the process by which the Green's functions, retarded functions, and vacuum expectation values emerged as objects to be studied in their own right. It was very complicated, and an account that would give appropriate weight to every brick carried to the wall would require more space than I can claim here. Let me illustrate the point with the example of the two-point functions or propagators. Nowadays, one usually defines them as the expectation values in the physical vacuum, Ψ_0, of a product, time-ordered, or retarded or not; that is, for two Bose fields A and B,

$$(\Psi_0, (A(x)B(y))_+\Psi_0) \quad \text{or} \quad (\Psi_0, \theta(x^0 - y^0)[A(x), B(y)]\Psi_0)$$
$$\text{or} \quad (\Psi_0, A(x)B(y)\Psi_0) \tag{42.4}$$

For a Dirac electron in an external electromagnetic field, the study of the analogous object goes back to Dirac in 1934.[31] It was reexamined at great length in the context of quantum electrodynamics by Tomonaga, Schwinger, Feynman, Dyson, and Källen, with an accompaniment of perturbative-renormalization-theory complications that were important but distracting. Eventually, Murray Gell-Mann and Francis Low emphasized how a ratio of vacuum expectation values in the bare vacuum, which when expanded yields all the contributions from the standard Feynman diagrams, is equal to an expectation value in the physical vacuum, Ψ_0, of a product of Heisenberg picture operators.[32] That probably was not much of a surprise to the preceding five authors, but I can remember how informative and satisfying I found it in 1951. Thus, when H. Umezawa and S. Kamefuchi wrote a spectral representation whose analogue for the functions in equation (42.4) would be, say,

$$(\Psi_0, A(x)B(y)\Psi_0) = -i \int da^2 [\varrho(a^2)/\Delta^+(a^2, x - y)]$$

it should have been fairly obvious that it is a consequence of relativistic invariance and the spectral conditions, even though they described it in the context of a detailed dynamical theory of vacuum polarization.[33] That was also implicit in the extensive use Källen made of its analogues for spinor and vector fields in his efforts to put quantum electrodynamics on a nonperturbative footing.[34] However, it was H. Lehmann who freed the derivation of irrelevant dynamical trappings and showed how simple general conclusions could be drawn from it about the singularities of the propagator.[35] So it is the Umezawa–Kamefuchi–Källen–Lehmann representation, as Källen used to say with evident relish, and we could add Gell-Mann and Low.[36] Recognizing the contributions of all these authors, and possibly others, I would propose to call it the spectral representation. In my opinion, it deserves a name not because it is so deep but because it is so simple, being a consequence of relativistic kinematics.

A similar process took place with the three-point functions. They played a key role as vertex functions in the perturbative analysis of quantum electrodynamics, and Källen used them heavily in his attempts at a nonperturbative analysis.[37]

However, so far as I know, Lehmann, K. Symanzik, W. Zimmermann, and V. Glaser were the first who undertook a systematic study of Green's function and retarded functions of field theory and, in particular, undertook to express a theory's S-matrix elements directly in terms of them. This led Lehmann, Symanzik, and Zimmermann (LSZ)[38] to two fundamental results: the asymptotic condition and the reduction formula for Green's functions.

The LSZ asymptotic condition for a neutral scalar field ϕ with a single stable particle of mass m expresses an S-matrix element for the collision of particles of energy–momentum $k_1, \ldots, k_r$ to produce particles of momentum $k'_1, \ldots, k'_s$ as

$$(a^{\text{out}}(k'_1)^* \ldots a^{\text{out}}(k'_s)^* \Psi_0, a^{\text{in}}(k_1)^+ \ldots a^{\text{in}}(k_r)^+ \Psi_0) \tag{42.5}$$

LSZ then use the following asymptotic condition to express the out and in creation operators directly in terms of the field ϕ:

$$\lim_{t \to \pm \infty} (\Phi, \phi_f(t)\Psi) = (\Phi, \phi_f^{\substack{\text{out} \\ \text{in}}} \Psi)$$

where

$$\phi_f(t) = -i \int_{x^0=t} \phi(x) \, \overleftrightarrow{\partial}/\partial x^0 \overline{f}(x) \, d^3x$$

and similarly for

$$\phi_f^{\substack{\text{out} \\ \text{in}}}$$

Here, f is a normalized positive-energy solution of the Klein–Gordon equation

$$(\Box + m^2) f(x) = 0$$

A by-now classical series of manipulations using Green's theorem on space–time shows equation (42.5) to be proportional to the Fourier transform of the $(r + s)$-point Green's function

$$\int \exp\left\{i\left[\sum_{j=1}^{s} k_j' \cdot y_j - \sum_{j=1}^{r} k_j \cdot x_j\right]\right\} \prod_{j=1}^{s} (\Box_{y_j} + m^2) \prod_{k=1}^{r} (\Box_{x_k} + m^2)$$
$$\times (\Psi_0, (\phi(y_1)\ldots\phi(y_s)\phi(x_1)\ldots\phi(x_r))_+ \Psi_0)$$
$$\times (d^4y_1 \ldots d^4y_s d^4x_1 \ldots d^4x_r)$$

evaluated on the mass shell, that is, for $k_j^2 = m^2$, $k_j^0 > 0$, $j = 1, \ldots, n$, $k_j'^2 = m^2$, $k_j'^0 > 0$. The reduction formula was met with considerable enthusiasm because it gave such a simple expression for a quantity of physical importance, which previously had been rather inaccessible. It was the starting point of innumerable dispersion-theory calculations.

The paper also contained an interesting system of nonlinear equations for the Green's functions and showed how they could be solved in a formal power series in a coupling parameter. The discussion represented a considerable advance because it contained none of the ambiguities and infinities that abound in the usual Hamiltonian or Lagrangian field theory.

The second LSZ paper undertook to develop a relation between the S matrix and local interpolating field by making systematic use of retarded products.[39] It also deduced reduction formulas for the $2 \to n$ particle reactions in terms of retarded functions. These formulas were for $n = 2$, the starting point for proofs of dispersion relations, as will be discussed in the next section.

There is a third paper in this series, the Glaser, Lehmann, and Zimmermann paper.[40] It has rather different objectives. It proposes a set of nonlinear conditions on retarded functions that permit one to recover local relativistic field operators in a Hilbert space. The conditions in question are both necessary and sufficient. That is what has been called a reconstruction theorem. It was later generalized by K. Nishijima to include composite particles.[41] Such reconstruction theorems are important in principle because they permit one to reduce problems from the realm of operator-valued distributions to problems on sets of ordinary distributions.[43,67]

Now I come to the alternative theory, which uses unordered products, objects that naturally arise if one tries to compute expectation values in the physical vacuum of products regarded as arising from operator-valued distributions, for example,

$$(\Psi_0, \phi(f_1)\phi(f_2)\ldots\phi(f_n)\Psi_0) \tag{42.6}$$

For this to make sense, one needs some information on the domain of vectors to which $\phi(f)$ may be applied. All the evidence indicates that, in general, the operator $\phi(f)$ is unbounded and cannot be defined on every vector. One possibility, the one I chose, is to assume the existence of an appropriate common dense domain, D. Another possibility is to abandon the traditional quantized fields in favor of bounded operators – Haag did just that, as I shall relate later. There remains the possibility of ignoring the problem, as most of the field-theory literature does. What other properties is it reasonable to ask of D? One certainly wants it to be a dense linear set in the Hilbert space of states. To make it possible to calculate $(\Psi_0, \phi(f_1)\ldots\phi(f_n)\Psi_0)$, one wants $\phi(f)$ defined on D (at least) and to map vectors in D into vectors in D; that is, $\phi(f)D \subseteq D$, and $\Psi_0 \in D$. In addition, we ask that $\phi(f)*DCD$ and $U(a, \Lambda)D \subseteq D$. These assumptions might have seemed a little technical and ad hoc at the time, but they turned out better than one could have hoped: First, they sufficed for the construction of a mathematically rigorous general theory, including a reconstruction theorem for the vacuum expectation values (42.6); second, two decades later, they were proved valid in the quantum field theory models in space–time dimensions 2 and 3, which were studied with the methods of constructive field theory.[42]

Here is the reconstruction theorem in question for a Hermitian scalar field.[43]

Theorem*. Let $\{\mathscr{W}^{(n)}\}$, $n = 1, 2, 3$, be a sequence of tempered distributions, where $\mathscr{W}^{(n)}$ depends on n four-vector variables $x_1, x_2, \ldots, x_n$. Suppose the $\mathscr{W}^{(n)}$ satisfy the following conditions:

(a) *Relativistic invariance*

$$\mathscr{W}^{(n)}(x_1, \ldots, x_n) = \mathscr{W}^{(n)}(\Lambda x_1 + a, \ldots, \Lambda x_n + a) \quad \text{for } \{a, \Lambda\} \in P_+^\uparrow.$$

$$\tag{42.7}$$

---

*   *Ed. note:* For an elaboration of this discussion, see Streater and Wightman.[76]

(b) Spectral conditions

$$\hat{\mathscr{W}}^{(n)}(p_1, \ldots, p_n) = (2\pi)^4 \, \delta\left(\sum_{j=1}^{n} p_n\right)$$
$$\times \hat{W}(p_1, p_1 + p_2, \ldots, p_1 + p_2 + \ldots + p_{n-1})$$

and

$$\hat{W}(q_1, \ldots, q_{n-1}) = 0$$

if any q_i is outside the future cone.

(c) Hermiticity conditions

$$\mathscr{W}^{(n)}(x_1, \ldots, x_n) = \overline{\mathscr{W}^{(n)}(x_n, \ldots, x_1)}$$

(d) Local commutativity

$$\mathscr{W}^{(n)}(x_1, \ldots, x_j, x_{j+1}, \ldots, x_n) = \mathscr{W}^{(n)}(x_1, \ldots, x_{j+1} x_j, \ldots, x_n)$$
$$\text{if } (x_j - x_{j+1})^2 < 0 \quad \text{for } j = 1, 2, \ldots, n - 1$$

(e) Positive definiteness conditions

$$\sum \int \ldots \int dx_1 \ldots dx_j \, dy_1 \ldots dy_k \, \overline{f_j(x_1, \ldots, x_j)}$$
$$\times \mathscr{W}^{(j+k)}(x_j, \ldots, x_1, y_1, \ldots, y_k) f_k(y_1, \ldots, y_k) \geq 0$$

(f) Cluster decomposition property

$$\lim_{\lambda \to \infty} [\mathscr{W}^{(n)}(x_1, \ldots, x_j, x_j + \lambda a, \ldots, x_n + \lambda a)$$
$$- \mathscr{W}^{(j)}(x_1, \ldots, x_j) \mathscr{W}^{(n-j)}(x_{j+1}, \ldots, x_n)] = 0$$

where a is spacelike. Then there exists a separable Hilbert space $\mathscr{H}$, a continuous unitary representation of the restricted Poincaré group $\mathscr{P}_+^{\uparrow}$, $\{a, \Lambda\} \to U(a, \Lambda)$, a unique state $\Psi_0 \in \mathscr{H}$ invariant under $U(a, \Lambda)$, and a local Hermitian scalar field ϕ such that

$$U(a, \Lambda)\phi(x)U(a, \Lambda)^{-1} = \phi(\Lambda x + a)$$

and

$$(\Psi_0, \phi(x_1), \ldots, \phi(x_n)\Psi_0) = \mathscr{W}^{(n)}(x_1, \ldots, x_n)$$

Furthermore, any other field theory with these vacuum expectation values is unitary-equivalent to this one.

The general theory of quantized fields, as exemplified in the work of Lehmann, Symanzik, Zimmermann, Glaser, and Wightman and that of Haag, H. Araki, and D. Kastler, yet to be discussed, came to be referred to as axiomatic field theory, as opposed to the standard Lagrangian field theory that started with a classical Lagrangian and tried to quantize it. At the end of the 1950s, when there had been very satisfactory progress in the axiomatic theory, there was still some hope that it might be possible to obtain a general structure theory for local fields. There was a further somewhat fainter hope that, in the end, one might arrive at the solutions of Lagrangian field theories by finding them as special representations in an appropriate axiomatic framework. One can see this in Symanzik's program of structure analysis.[44] Neither the decade of the fifties nor that of the sixties was to see anything like a completed program of such a character. (There has been slow, but important, progress since then along these lines, as reviewed by Bros.[45])

The operation of passing from a quantum field theory to its vacuum expectation values is analogous to passing from observables to correlation functions in statistical mechanics or analogous to passing from an operator algebra to a state (positive linear form) on it. If this analogy is developed, one can interpret the reconstruction theorem as the analogue of the construction that leads from a state on an operator algebra to a cyclic representation of that algebra. This idea was brought to fruition by H. J. Borchers, who made an algebra of sequences of test functions in such a way that a state is a set of vacuum expectation values satisfying (a)–(e) of the reconstruction theorem.[46] This will probably be the way people look at the reconstruction theorem in the future: It is the GNS construction of a cyclic representation of the Borchers algebra.[47]

Analytic functions and field theory: dispersion relations; Euclidean field theory

There are two families of analytic functions in field theory: those in x-space and those in p-space. For both kinds, the fundamental mathematical fact is that the Laplace transform of a distribution that vanishes outside a convex cone is analytic in a tube. Explicitly,

$$F(x + iy) = \int e^{ip \cdot (x+iy)} F(p) \, dp$$

If $F(p) = 0$ for p outside the convex cone V, then the exponential damps for every y in the dual cone V'. (Here, $y \in V'$ if $p \cdot y \geq 0$ for every $p \in V$; x is arbitrary.) For the future-light cone in Minkowski space, $V' = V$.

For the x-space analytic functions, the $\hat{W}(q_1, \ldots, q_{n-1})$ described in the reconstruction theorem vanishes unless each q lies in the closure of the future light, $\bar{V}_+$. That is a consequence of the "spectral condition," which says that every physical-energy momentum lies in $\bar{V}_+$. Thus, the Laplace transform of W is analytic in the variables $\xi_1' = \xi_1 + i\eta_1, \ldots, \xi_{n-1}' = \xi_{n-1} + i\eta_{n-1}$ for $\eta_1, \ldots, \eta_{n-1} \in \bar{V}_+$. The x-space analytic functions have a short history; they were first described in 1956.[43]

The p-space analytic functions have been around longer. They go back at least to the work of Schwinger in the 1940s and are implicit in the field-theory calculations of the 1930s. They are defined as Laplace transforms of retarded functions. The retarded functions $r(x; x_1, \ldots, x_n)$ are defined with a distinguished point x, so that they vanish except when $x - x_1, \ldots, x - x_n$ lie in the future-light cone V_+. This vanishing is partly a consequence of the assumed local commutativity of fields and partly is forced by multiplying by a suitable product of θ functions, which chops off contributions from the past.

If one is given n different fields, one might expect $n!$ different x-space analytic functions and n different p-space analytic functions. However, all $n!$ of the former can be obtained from each other by permutations of argument in the regions where all fields commute, and this implies that the $n!$ analytic functions are also obtainable from each other by permutation of argument.[43] The reasoning is not quite so simple for the p-space analytic functions, but the result is analogous: All p-space analytic functions are obtained from each other by permutation of argument.[48,49] The $n!$ different vacuum expectation values and the Fourier transforms of the n different retarded functions are just different boundary values of the x- and p-space analytic functions, respectively. Although the analyticity domains of the x- and p-space three-point analytic functions are the same in general, beginning with the four-point functions, there are considerable differences, first explored by O. Steinmann and D. Ruelle in their theses in Zurich.[50–52] This was the beginning of a major industry of the 1960s, the systematic study of the p-space analytic functions and their boundary values, work that provided the foundation for the study of the analyticity properties of S-matrix elements.

The most important physical application of the p-space analytic function is, of course, to dispersion relations. It was first used in this sense to obtain the forward-dispersion relations by Gell-Mann, Marvin Goldberger, and Walter Thirring.[53] Rigorous proofs of dispersion relations came later, with Symanzik,[54] N. N. Bogoliubov, B. V. Medvedev, and M. K. Polivanov,[55] H. Bremermann, R. Oehme, and J. G. Taylor,[56] and Lehmann.[57] The extensive application of dispersion relations to the analysis of particle reactions was certainly one of the most characteristic features of the physics of the 1950s.

The most important applications of the analyticity properties of the x-space analytic functions probably are to the proofs of the *PCT* theorem, the spin-statistics theorem, and the Reeh–Schlieder theorem.* For these purposes, a theorem of Bargmann, Hall, and Wightman played a significant role.[58] It showed that a function f of n four-vector variables analytic in the forward tube T_+,

$$\zeta_1 = \xi_1 + i\eta_1, \ldots, \zeta_n = \xi_n + i\eta_n, \qquad \eta_1, \ldots, \eta_n \in V_+$$

and invariant under the restricted Lorentz group $L_+^\uparrow$,

$$f(\Lambda\zeta_1, \ldots, \Lambda\zeta_n) = f(\zeta_1, \ldots, \zeta_m)$$

where Λ is a real Lorentz transformation, with det $\Lambda = +1$, $\Lambda_0^0 \geqslant 1$, is also analytic and single-valued in the extended tube whose points are $\Lambda\zeta_1, \ldots, \Lambda\zeta_n$, with $\zeta_1, \ldots, \zeta_n \in T_+$, but Λ in the proper complex Lorentz group $L_+(\mathbb{C})$, which consists of all complex linear transformations Λ satisfying $\Lambda^T G\Lambda = G$, det $\Lambda = +1$, $G = \mathrm{diag}(+1, -1, -1, -1)$. Furthermore, the invariance (42.7) holds for such complex Λ. Because -1 lies in $L_+(\mathbb{C})$, this implies, in

---

*   *Ed. note:* For a more detailed discussion, see Streater and Wightman.[76]

particular, that the analytic functions whose boundary values are the n-point vacuum expectation values satisfy

$$W^{(n)}(\zeta_1, \ldots, \zeta_{n-1}) = W^{(n)}(-\zeta_1, \ldots, -\zeta_{n-1}) \tag{42.8}$$

throughout the extended tube. The usefulness of this relation will be seen in the next section.

An application of these results on the analyticity domain of the $W^{(n)}$ is that all points of the form

$$\begin{Bmatrix} it_1 \\ \mathbf{x}_1 \end{Bmatrix} \begin{Bmatrix} it_2 \\ \mathbf{x}_2 \end{Bmatrix} \cdots \begin{Bmatrix} it_n \\ \mathbf{x}_n \end{Bmatrix}$$

with t_j, $\mathbf{x}_j$ ($j = 1, \ldots, n$) real and distinct lie in the analyticity domain of $W^{(n)}$. These are the so-called Euclidean points, because regarded as a function of $\mathbf{x}_j$ and t_j, the Minkowski scalar product becomes a Euclidean scalar product. The $W^{(n)}$ restricted to the Euclidean points are called Euclidean Green's functions or Schwinger functions. They are analytic functions of their arguments, except where two or more points coincide. It was Schwinger and T. Nakano who had the idea that perhaps a Euclidean field theory might exist in which the Euclidean Green's functions would be vacuum expectation values.[59,60] Such a theory would have the advantage of being invariant under $O(4)$ instead of the Lorentz group. Viewed in retrospect, that was a splendid idea. Fortunately, Symanzik took it seriously and carried the torch for it through the decade of the sixties when few others paid much attention.

Applications of the general formalism: *PCT* theorem, theorem on spin and statistics

The *PCT* theorem was discovered by Lüders and Pauli (a version was found by Schwinger[61], but the modern and more general form of it was found by Res Jost.[61–63] Jost's argument is so simple that I can repeat it here for the case of a neutral scalar field ϕ. The conditions

$$\Theta\Psi_0 = \Psi_0 \quad \text{and} \quad \Theta\phi(x)\Theta^{-1} = \phi(-x)$$

with Θ antiunitary, imply

$$(\Psi_0, \phi(x_1)\ldots\phi(x_n)\Psi_0) = (\Psi_0, \phi(-x_n)\ldots\phi(-x_1)\Psi_0)$$

This set of relations is the expression of *PCT* symmetry in terms of vacuum expectation values. It implies, for the analytic functions,

$$W^{(n)}(\zeta_1, \ldots, \zeta_{n-1}) = W^{(n)}(\zeta_{n-1}, \zeta_{n-2}, \ldots, \zeta_1) \tag{42.9}$$

throughout the analyticity domain, and conversely this relation on the analytic functions implies *PCT* because one can pass to the boundary values. On the other hand, the invariance (42.8) implies

$$W^{(n)}(\zeta_1, \ldots, \zeta_{n-1}) = W^{(n)}(-\zeta_{n-1}, \ldots, -\zeta_1) \tag{42.10}$$

which evaluated for $x_1, \ldots, x_n$ relatively spacelike is

$$(\Psi_0, \phi(x_1)\ldots\phi(x_n)\Psi_0) = (\Psi_0, \phi(x_n)\phi(x_{n-1})\ldots\phi(x_1)\Psi_0)$$

a relation Jost calls weak local commutativity. Thus, the locality of the field ϕ implies the weak local commutativity identity at spacelike separated points, which implies (42.10), which implies (42.9), which is equivalent to *PCT* symmetry. The contrast between this simple, almost kinematical, argument and the detailed examination of admissible terms in the interaction Lagrangian that is essential in preceding proofs is striking.

The proof of the spin-statistics theorem was simplified in a similar way.[64,65] The original Fierz–Pauli proof was only for free fields. Using the x-space analytic function, N. Burgoyne and Lüders and Bruno Zumino could again put the argument into a form that was essentially purely kinematical.

There were other general facts about field theory that were easily proved in similar fashion, for example, the Reeh–Schlieder theorem.[66] It says that if vectors of the form $P(\phi(f)\ldots)\Psi_0$ with arbitrary test functions are dense in the Hilbert space of states, then the same is true when the supports of the test functions are restricted to lie in a fixed arbitrarily small region of space–time.

The x-space analytic function did not turn out to be of much use when it came to the detailed dynamics of particular quantum field theories, but for general principles it was very useful.

Collision theory; Haag–Ruelle theory; Borchers classes

The LSZ asymptotic condition was immediately accepted by most of those whose primary concern was pursuing the particles, but there were others who had grave doubts. I have in my files a long letter from Källen summarizing several years of discussion in which he expressed suspicions that the behavior of the p-space analytic function near the mass shell was decent enough to be controlled by so simple an *Ansatz* as the LSZ asymptotic condition. For those of us who started from the x-space vacuum expectation values, the problem was to find what had to be added to the axioms[67] to arrive at the LSZ asymptotic condition. In the unpublished thesis of O. W. Greenberg, a definite proposal of this kind was made, but in my opinion, we bet on the wrong horse.[68] Haag had a deeper insight into the physical situation and proposed versions of scattering theory[69-71] characterized by increasing sharpness of mathematical focus. This was the starting point for Ruelle, who, in a tour de force of analysis showed that within the framework of the axioms, the existence of collision states could be established for arbitrary numbers of whatever stable particles the theory contained.[72] (If the theory has no discrete-mass states in its representation of the Poincaré group, Ruelle's theory is empty.) Thus, for massive stable particles, no additional assumptions have to be added to the axioms to assure the existence of collision states. Of course, this Haag–Ruelle theory does not guarantee that the collision states are complete; that desideratum has to be assumed (or proved in specific cases). The theory does not cover zero-mass particles nor massive particles whose mass is smudged by clouds of massless particles. Those problems were still under attack in the seventies (and eighties). Furthermore, as it stands, the Haag–Ruelle theory does not prove the validity of the LSZ asymptotic condition. That was finally done by K. Hepp in 1965.[73]

Another important insight into collision theory was achieved by Borchers.[74] He showed that local fields could be divided into equivalence classes based on the relation of relative locality: A and B are relatively local if

$$[A(x),\ B(y)]_- = 0$$

for all x and y such that $(x - y)^2 < 0$ (i.e., $x - y$ spacelike). Two sets of fields are relatively local if each field of one set is relatively local with respect to every field of the other set. A second part of the Borchers assertion is that two sets of fields that are relatively local have the same S matrix provided that both have the vacuum as cyclic vector. The theory of Borchers classes can be regarded as the first step toward the determination of all theories that have a given S matrix.

There are two books that summarize the general theory of quantized fields based on the axioms up to the early sixties, one by Jost and one by R. F. Streater and Wightman.[67,75,76] Jost's book contains an account of the Haag–Ruelle theory, which ours does not. Later on, there appeared the treatise of Bogoliubov, A. Logunov, and I. Todorov, which covers also the Haag–Araki and Haag–Kastler theory of local algebras of bounded operators.[77] The reader interested in more details will find them there, as well as in a useful review by Streater.[78]

Local algebras

During the course of the events recounted earlier, Haag was pursuing his own investigation on the foundations of fields. One can find a summary of his thinking up to the midfifties in reference 71. In my opinion, he was never very comfortable with technicalities about the domains of unbounded operators that I kept insisting on, but at that stage he did not quite have a natural language with which to replace it. In the end, he came to the idea of using as the basic objects of his theory the algebras of bounded operators generated by observables associated with regions in space–time local algebras. For these, no domain questions arise, all the operators being everywhere defined. (An intermediate stage is visible in one source,[79] where he considered local algebras of unbounded operators.) In collaboration with Araki and then with Kastler, he showed that the theory naturally lent itself to a treatment of superselection rules and displayed interesting analogies with statistical mechanics.[80]

Because their explanation of superselection rules is a matter of great importance and often does not receive the attention it deserves, let me explain in a little detail.

Suppose that one were given a complete solution of the quantum electro-dynamics of electrons in the vacuum vector (i.e., for the subspace of states of charge zero). Because all observables of the theory carry states of charge zero into states of charge zero, there is no way, using observable operators, to make states of nonzero charge from those of the vacuum vector. How could one know that to get a complete theory, one should adjoin all the states of charge different from zero? How could one know that the observable of electric charge is capable of values other than zero? This is a special case of a general question asked by Haag: Given a theory with superselection rules, what is the relation between the observables in the different coherent subspaces? (In the example, the charge defines the superselection rule, and the different coherent subspaces are labeled by their charges.) The answer that Haag and Kastler gave was that there is a basic C^* algebra of quasi-local observables and that representations of it in different coherent sub-spaces are unitary-inequivalent. To decide, given the vacuum sector, whether or not other (charged) sectors exist, one looks to see if the quasi-local algebra of observables in the vacuum sector has other unitarily inequivalent representations.[80]

A decade after Haag and Kastler published their theory, it was discovered that the sine–Gordon model in space–time of two dimensions is the vacuum sector of the massive Thirring model, a beautiful example of Haag and Kastler's theory.

Quantum field theory in spaces of indefinite metric; the Gupta–Bleuler formalism

To close this review, let me mention an achievement that was not highly regarded in the 1950s, but that turned out to be absolutely fundamental for the quantum gauge field theory of the 1970s and 1980s: the indefinite metric formalism. Although the use of this formalism was proposed by Dirac in the early 1940s and studied systematically by Pauli, they did not connect it specifically with quantum electrodynamics; that was done by S. N. Gupta and K. M. Bleuler.[81–83] The essential feature of this formalism is a triple of spaces $\mathcal{H}''$, $\mathcal{H}'$, and $\mathcal{H}$, with $\mathcal{H}'' \subseteq \mathcal{H}' \subseteq \mathcal{H}$, and an indefinite inner product, $\langle \cdot, \cdot \rangle$, on $\mathcal{H}$ such that

$$\langle \Phi, \Phi \rangle \geqslant 0 \quad \text{for } \Phi \in \mathcal{H}'$$

and $\mathcal{H}''$ the subspace of $\mathcal{H}$ consisting of Φ for which $\langle \Phi, \Phi \rangle = 0$. The physical Hilbert space of the theory is identified with the quotient space $\mathcal{H}_{\text{phys}} = \overline{\mathcal{H}'/\mathcal{H}''}$, where the bar signifies closure in the metric $\|\Phi\| = \sqrt{\langle \Phi, \Phi \rangle}$. Observables are operators, A, which are self-adjoint in the sense that

$$\langle \Phi, A\Phi \rangle = \langle A\Phi, \Phi \rangle$$

for every vector Φ on which A is defined. If $\mathcal{H}' = \mathcal{H}$ and $\mathcal{H}'' = \{0\}$, one has the usual Hilbert-space formalism, but the added flexibility of the general theory turns out to be indispensable for the definition of gauge fields. In verifying the conservation of probability in such theories, the Stone theorem with which we dealt at the beginning of this review plays an essential role. An axiomatic version of this formalism was given by Gårding and Wightman.[19]

Concluding remarks

During the 1950s, the general theory of quantized fields developed into a systematic mathematical theory with well-defined fundamental concepts. It went from being murky, arcane, and mysterious to being clear and hard. It lacked worked-out nontrivial examples; they appeared in the form of constructive quantum field theory only after another decade and a half of hard work.

As far as the concrete applications to physics are concerned, the most important contributions of axiomatic field theory probably are the following:

(a) the LSZ asymptotic condition and the associated reduction formulas,
(b) the Haag–Ruelle theory of scattering, including the treatment of composite particles,
(c) the Borchers theory of equivalence classes of local fields,
(d) the general theory of p-space analytic functions and the consequent restrictions on S-matrix elements, in particular, the derivation of dispersion relations,
(e) the modern forms of the *PCT* theorem and the theorem on spin and statistics, which display these results as essentially kinematical in nature, and
(f) the Reeh–Schlieder theorem and Haag's theorem, which reveal the significance of the vacuum state and of the representations of the CCR and CAR.

Notes

1 The problem of overlapping divergences had not been completely solved by Dyson.
2 M. H. Stone, "On One-Parameter Groups in Hilbert Space," *Ann. Math. 2:33* (1932), 643–8; for a modern exposition, see Michael Reed and Barry Simon, *Methods of Modern Mathematical Physics*, Vol. I (New York: Academic Press, 1972), p. 266.
3 T. Kato, "Fundamental Properties of Hamiltonian Operators of Schrödinger Type," *Trans. Am. Math. Soc. 70* (1951), 195–211.
4 M. Born, W. Heisenberg, and P. Jordan, "Zur Quantenmechanik II," *Z. Phys. 35* (1926), 557–615.
5 P. A. M. Dirac, "The Fundamental Equations of Quantum Mechanics," *Proc. R. Soc. London, Ser. A 109* (1926), 642–53.
6 W. Pauli and V. Weisskopf, "Ueber die Quantisierung der skalaren relativistischen Wellengleichung," *Helv. Phys. Acta 7* (1934), 109.
7 P. Jordan and E. Wigner, "Ueber das Paulische Aequivalenzverbot," *Z. Phys. 47* (1928), 631–51.
8 J. von Neumann, "Die Eindeutigkeit der Schrödingerschen Operatoren," *Math. Ann. 104* (1931), 570–8.
9 J. von Neumann, "The Quantum Mechanics of Infinite Systems," in *Theory of the Positron and Related Topics* (Princeton: W. Pauli Institute for Advanced Study, 1936), pp. 147–72.

10 Kurt Otto Friedrichs, *Mathematical Aspects of the Quantum Theory of Fields* (New York: Interscience, 1953).

11 R. Haag, "On Quantum Field Theories," *K. Dan. Vidensk. Selsk. Mat.-Fys. Medd. 29* (1955), 1–37.

12 D. Hall and A. Wightman, "A Theorem on Invariant Analytic Functions with Applications to Relativistic Quantum Field Theory," *K. Dan. Vidensk. Selsk. Mat.-Fys. Medd. 31* (1957), 1–41.

13 L. Gårding and A. S. Wightman, "Representations of the Anticommutation Relations," *Proc. Nat. Acad. Sci. U.S.A. 40* (1954), 617–21; "Representations of the Commutation Relations," ibid., 622–6.

14 I. E. Segal, "Tensor Algebras over Hilbert Spaces. I," *Trans. Am. Math. Soc. 81* (1956), 106–34.

15 L. Schwartz, *Theorie des Distributions*, Vols. I and II (Paris: Hermann, 1957, 1959).

16 P. Jordan and W. Pauli, "Zur Quantenelektrodynamik ladungsfreier Felder," *Z. Phys. 47* (1928), 151–73.

17 N. Bohr and L. Rosenfeld, "Zur Frage der Messbarkeit der elektromagnetischen Feldgrossen," *K. Dan. Vidensk. Selsk. Mat.-Fys. Medd. 12* (1933), 1–65.

18 L. S. Schwartz, "Generalisation de la Notion de Fonction de Derivation, de Transformation de Fourier, et Applications Mathematiques et Physiques," *Annales Univ. Grenoble 21* (1945), 57–74.

19 A. S. Wightman and L. Gårding, "Fields as Operator-Valued Distributions in Relativistic Quantum Field Theory," *Ark. Fys. 28* (1964), 129–84.

20 W. Güttinger, "Quantum Field Theory in the Light of Distribution Analysis," *Phys. Rev. 89* (1953), 1004–19.

21 K. Baumann, "Quantentheorie der Felder als Distributionstheorie," *Nuovo Cimento 4* (1956), 860–86.

22 W. Heisenberg, "Die 'beobachtbaren Grössen' in der Theorie der Elementarteilchen. I," *Z. Phys. 120* (1943), 513–38; part II, ibid., 673–702.

23 E. Wigner, "Unitary Representations of the Lorentz Group," *Ann. Math. 40* (1939), 149.

24 G. Källen, "Formal Integration of the Equations of Quantum Theory in the Heisenberg Representation," *Ark. Fys. 2* (1950), 371–410.

25 C. N. Yang and D. Feldman, "The *S*-Matrix in the Heisenberg Representation," *Phys. Rev. 79* (1950), 972–8.

26 J. Schwinger, "Quantum Electrodynamics. I. A Covariant Formulation," *Phys. Rev. 74* (1948), 1439–61.

27 G. Lüders, "Zur Bewegungsumkehr in quantisierten Feldtheorie," *Z. Phys. 133* (1952), 325–9.

28 S. Schweber and A. S. Wightman, "Configuration Space Methods in Relativistic Quantum Field Theory. I," *Phys. Rev. 98* (1955), 812–37, especially 831–7.

29 E. Wigner, "Ueber die Operation der Zeitumkehr in der Quantenmechanik," *Göttingen Nach.* (1932), 546–59.

30 A. S. Wightman, "Relativistic Invariance and Quantum Mechanics," *Nuovo Cimento (Suppl. 14)* (1959), 81–94, especially 92–4.

31 P. A. M. Dirac, "Discussion of the Infinite Distribution of Electrons in the Theory of the Positron," *Proc. Cambridge Philos. Soc. 30* (1934), 50–163.

32 M. Gell-Mann and F. Low, "Bound States in Quantum Field Theory," *Phys. Rev. 84* (1951), 350–4.

33 H. Umezawa and S. Kamefuchi, "The Vacuum in Quantum Electrodynamics," *Prog. Theor. Phys. 6* (1951), 543–58.

34 G. Källen, "On the Definition of the Renormalization Constants in Quantum Electrodynamics," *Helv. Phys. Acta 25* (1952), 417–34.

35 H. Lehmann, "Ueber Eigenschaften von Ausbreitungsfunktionen und Renormierungskonstanten quantisierter Felder," *Nuovo Cimento 9:10* (1954), 342–57.

36 M. Gell-Mann and F. Low, "Quantum Electrodynamics at Small Distances," *Phys. Rev 95* (1954), 1300–12.

37 G. Källen, "On the Magnitude of the Renormalization Constants in Quantum Electro-dynamics," *K. Dan. Vidensk. Selsk. Mat.-Fys. Medd. 27* (1953), 1–18.

38 H. Lehmann, K. Symanzik, and W. Zimmermann, "Zur Formulierung quantisierter Feldtheorie," *Nuovo Cimento 10:1* (1955), 205–25.

39 H. Lehmann, K. Symanzik, and W. Zimmermann, "On the Formulation of Quantized Field Theories. II," *Nuovo Cimento 10:6* (1957), 319–33.

40 V. Glaser, H. Lehmann, and W. Zimmermann, "Field Operators and Retarded Functions," *Nuovo Cimento 10:6* (1957), 1122–8.

41 K. Nishijima, "Formulation of Field Theories of Composite Particles," *Phys. Rev. 111* (1958), 995–1011.

42 J. Glimm, A. Jaffe, and T. Spencer, "The Wightman Axioms and Particle Structure in the $P(\phi)_2$ Quantum Field Model," *Ann. Math. 100* (1974), 582–632; J. Feldman and K. Osterwalder, "The Wightman Axioms and the Mass Gap for Weakly Coupled $(\phi^4)_3$ Quantum Field Theories," *Ann. Phys. 97* (1976), 80–135.

43 A. S. Wightman, "Quantum Field Theory in Terms of Vacuum Expectation Values," *Phys. Rev. 101* (1956), 860–6.

44 K. Symanzik, "On the Many-Particle Structure of Green's Functions in Quantum Field Theory," *J. Math. Phys. (N.Y.) 1* (1960), 249–73.

45 J. Bros, "Analytic Structure of Green's Functions in Quantum Field Theory," in *Mathematical Problems in Theoretical Physics, Lecture Notes in Physics*, edited by K. Osterwalder (Berlin: Springer-Verlag, 1980), pp. 166–99.

46 H. J. Borchers, "On the Structure of the Algebra of Field Observables," *Nuovo Cimento 24* (1962), 214–36.

47 The Gelfand–Naimark–Segal (GNS) construction is the standard method of obtaining a cyclic representation of a C^* algebra from a state on the algebra. See A. Wightman, "Constructive Field Theory: Introduction to the Problems," in *Fundamental Interactions in Physics and Astrophysics*, edited by G. Iverson, A. Perlmutter, and S. Mintz (New York: Plenum Press, 1973), pp. 46–53.

48 For the three-point functions: R. Jost, "Ein Beispiel zum Nukleon-Vertex," *Helv. Phys. Acta* (1958), 263–72; or note 50.

49 For the four-point function see note 51 (I); for the *n*-point function, see notes 51 (II), 52.

50 G. Källen and A. S. Wightman, "The Analytic Properties of the Vacuum Expectation Value of a Product of Three Scalar Local Fields," *Mat. Fys. Skrift. Dan. Vid. Selsk. 1:6* (1958), 1–57.

51 O. Steinmann, "Ueber den Zusammenhang zwischen den Wightman Funktionen und den retardierten Kommutatoren. I," *Helv. Phys. Acta 33* (1960), 257–98; part II, ibid., 347–62.

52 D. Ruelle, "Connection between Wightman Functions and Green Functions in Momentum Space," *Nuovo Cimento 10:19* (1961), 356–76. See also H. Araki and N. Burgoyne, "Properties of the Momentum Space Analytic Functions," *Nuovo Cimento 10:18* (1960), 342–6; H. Araki, "Generalized Retarded Functions and Analytical Function in Momentum Space in Quantum Field Theory," *J. Math. Phys. (N.Y.) 2* (1961), 163–77; A. S. Wightman, "Wightman Functions, Retarded Functions and Their Analytic Continuations," *Prog. Theor. Phys. (Suppl.) 18* (1961), 83–125.

53 M. Gell-Mann, M. Goldberger, and W. Thirring, "Use of Causality Conditions in Quantum Theory," *Phys. Rev. 95* (1954), 1612–27.

54 K. Symanzik, "Derivation of Dispersion Relations for Forward Scattering," *Phys. Rev. 105* (1957), 743–9.

55 N. N. Bogoliubov, B. V. Medvedev, and M. K. Polivanov, "Probleme der Theorie der Dispersionsbeziehungen," *Fortschr. Phys. 6* (1959), 159–245.

56 H. Bremermann, R. Oehme, J. G. Taylor, "Proof of Dispersion Relations in Quantized Field Theories," *Phys. Rev. 109* (1958), 2178–90.

57 H. Lehmann, "Analytic Properties of Scattering Amplitudes as Functions of Momentum Transfer," *Nuovo Cimento 10* (1958), 579–89.

58 D. Hall and A. S. Wightman, "A Theorem on Invariant Analytic Functions with Applications to Relativistic Quantum Field Theory," *K. Dan. Vidensk. Selsk. Mat.-Fys. Medd. 31* (1957), 1–41.

59 J. Schwinger, "On the Euclidean Structure of Relativistic Field Theory," *Proc. Nat. Acad. Sci. U.S.A. 44* (1958), 956–65.

60 T. Nakano, "Quantum Field Theory in Terms of Euclidean Parameters," *Prog. Theor. Phys. 21* (1959), 241–59.

61 W. Pauli, "Exclusion Principle, Lorentz Group, and Reflection in Space–Time and Charge," in *Niels Bohr and the Development of Physics*, edited by W. Pauli (London: Pergamon Press, 1955), pp. 30–51.

62 G. Lüders, "On the Equivalence of Invariance under Time Reversal and under Particle–Antiparticle Conjugation for Relativistic Field Theories," *K. Dan. Vidensk. Selsk. Mat.-Fys. Medd. 28* (1954), 1–17.

63 R. Jost, "Eine Bemerkung zum *CTP* Theorem," *Helv. Phys. Acta 30* (1957), 409–16.

64 N. Burgoyne, "On the Connection between Spin and Statistics," *Nuovo Cimento 8* (1958), 607–9.

65 G. Lüders and B. Zumino, "Connection between Spin and Statistics," *Phys. Rev. 110* (1958), 1450–3.

66 H. Reeh and S. Schlieder, "Bemerkungen zur unitäräquivalenz von Lorentz invarianten Feldern," *Nuovo Cimento 22* (1961), 1051–68.

67 A. S. Wightman, "Quelques problèmes mathematiques de la Theorie Quantique des Champs," in *Colloques Internationaux du Centre National de la Recherche Scientifique* (Paris: CNRS, 1959), pp. 1–38.

68 O. W. Greenberg, "The Asymptotic Condition in Quantum Field Theory" (unpublished Ph.D. thesis, Princeton University, 1956).

69 W. Brenig and R. Haag, "Allgemeine Quantentheorie des Stossprozesse," *Fortschr. Phys. 7* (1959), 183–242.

70 R. Haag, "Discussion des 'Axiomes' et des Propriétés Asymptotiques d'une Théorie des Champs locales avec Particules Composées," in *Les Problèmes Mathématiques de la Théorie Quantique des Champs* (Paris: CNRS, 1959), pp. 151–62.

71 R. Haag, "Quantum Field Theories with Composite Particles and Asymptotic Conditions," *Phys. Rev. 112* (1958), 669–73.

72 D. Ruelle, "On the Asymptotic Condition in Quantum Field Theory," *Helv. Phys. Acta 35* (1962), 147–63.

73 K. Hepp, "On the Connection between LSZ and Wightman Quantum Field Theory in Axiomatic Field Theory," in *1965 Brandeis University Summer Institute in Theoretical Physics, Vol. I* (New York: Gordon & Breach, 1966), pp. 137–246.

74 H. J. Borchers, "Ueber die Mannigfaltigkeit der interpolierenden Felder zu einer kausalen *S*-Matrix," *Nuovo Cimento 15* (1960), 784–94.

75 R. Jost, *The General Theory of Quantized Fields* (Providence: American Mathematical Society, 1965).

76 R. F. Streater and A. S. Wightman, *PCT, Spin and Statistics and All That* (New York: Benjamin, 1964; enlarged edition, Benjamin/Cummings, 1978).

77 N. Bogoliubov, A. Logunov, I. Todorov, *Introduction to Axiomatic Quantum Field Theory* (Reading, Mass.: Benjamin, 1975).

78 R. Streater, "Outline of Axiomatic Relativistic Quantum Field Theory," *Rep. Prog. Phys. 38* (1975), 771–846.

79 R. Haag, note 70.

80 H. Araki, "Local Quantum Theory. I," in *Local Quantum Theory*, edited by R. Jost (New York: Academic, 1969), pp. 65–96.

81 P. A. M. Dirac, "The Physical Interpretation of Quantum Mechanics," *Proc. R. Soc. London, Ser. A 180* (1942), 1–40.

82 W. Pauli, "On Dirac's New Method of Field Quantization," *Rev. Mod. Phys. 15* (1943), 175–207.

83 S. N. Gupta, "Theory of Longitudinal Photons in Quantum Electrodynamics," *Proc. Phys. Soc. London 63* (1950), 681–91; K. M. Bleuler, "Eine neue Methode zur Behandlung der longitudinalen und skalaren Photonen," *Helv. Phys. Acta 23* (1950), 567–86.

43   The classification and structure of hadrons

YUVAL NE'EMAN

Born 1925, Tel Aviv; Ph.D., University of London, 1962; theoretical physics; Tel Aviv University

I give a sketch here of the discovery of SU(3) symmetry and quarks. I first review three failed attempts to use dynamical models directly; these were replaced by the quark model, once the SU(3) pattern was identified in the phenomenological hadron spectrum. I then analyze the factors that led to SU(3) and discuss the related roles of classification in mathematics and in physics. I describe the emergence of the quark model and conclude with a discussion of the present form of the "flavor" and generations problem.

The structural or dynamical attempts

There are no real prescriptions in the construction of science, since it is truly an exploration of the unknown, rather than an engineering feat. Some years ago, I realized that the entire endeavor of scientific research precisely fulfills the role of the randomized mechanism necessary in any evolutionary machinery.[1] Since the advent of "man the toolmaker," evolution has become the evolution of human society, rather than the Darwinian evolution of genetic features of man the primate (although the two may soon merge again, with the development of genetic engineering). Society "mutates" through technological innovation, and the random process characteristic of any mutationary mechanism is, in this case, scientific research. "Good" mutations, such as the invention of flint tools or of computers, survive and turn the evolutionary wheel.

It is thus presumptuous of me to try to systematize the scientific method, and yet some broad lines may be drawn, with the qualification that things

Supported in part by the U.S.-Israel Binational Science Foundation and by U.S. DOE grant DE-FG05-85ER40200.

often work out differently. The point I would like to stress – and I am certainly influenced by my personal experience – is that we often require a three-stage sequence: general explorations, the detection of regularities, and the understanding of structure.[2] First there were the astronomical data, with Tycho Brahe's measurements as their culmination; then came Johannes Kepler's discovery of regular patterns in planetary motion (his three laws) and Galileo's description of inertial behavior; and third came Isaac Newton with a dynamical theory explaining it all. Note that I have not mentioned Nicolaus Copernicus: He represented the third and last stage of a previous sequence, started by the Greek observations and the detection of retrograde motions, and so forth (patterns), by Ptolemy and others. Copernicus finally identified the structural model.

Sometimes the patterns stage represents classification. This was the case for botany, zoology, and chemistry. It happened in geology, where we have only recently entered the third stage, based on the dynamics of continental drift. Yet there are exceptions. Take the Yukawa case. The discovery of the neutron in 1932 implied that the nuclear glue was not electromagnetic. Within three years, Hideki Yukawa (and, independently, E. C. G. Stueckelberg) had offered a structural model, namely pion exchange, inspired by the exchange of photons in atoms, or of electrons in molecules. That inspired guess at dynamics succeeded. It became an encouragement for the less patient physicist, a hint that structure might be reached directly. This was indeed the route followed by many in the fifties and sixties.

First, there were those who assumed that the new internal quantum numbers – isospin, strangeness – represented simply internal mechanical motion. Louis de Broglie, Takehiko Takabayasi, Herbert C. Corben, and Yukawa himself – these were some of the leaders following this approach.[3] Herman Feshbach tried a more sophisticated solution in that line: interpreting strangeness as parastatistics.[4] Another school chose the fundamental constituents among the more familiar particles and fields and tried to construct everything else from these basic bricks. Enrico Fermi and C. N. Yang made the first such theoretical experiment, assuming that pions were nucleon–antinucleon compounds.[5] With the advent of strangeness, a "strange" brick had to be added, and after tries by Maurice Goldhaber and Robert F. Christy, the Sakata model appeared to be best suited for that role.

The most popular alternative in the sixties was the S-matrix bootstrap approach.[6] Here the assumption was that only on-mass shell dynamics could work in a strong interaction. The difficulties of relativistic quantum field theory (RQFT) in the midfifties, mostly relating to the difficulty in ensuring off-mass shell unitarity, were taken to imply its early demise, after what would then have been the "accidental" success in quantum electrodynamics (QED). It had to be a return to quantum mechanics, with the addition of some axioms abstracted out of RQFT amplitudes: crossing symmetry, analyticity (in various dynamical variables), and so forth. The internal quantum numbers were conjectured to result from imposing such conditions on amplitudes.[7]

It is intersting to note that each of these structural approaches turned out to be appropriate, but at a different level and for other purposes. Quarks rather than baryons make up the basic bricks in the hadron system. The quark–antiquark or three-quark systems do display a rich spectrum of angular-momentum excitations. The $SU(3)_{color}$ forces producing confinement are equivalent to parastatistics. The bootstrap approach produced Regge trajectories and Veneziano amplitudes, which then led to "strings" as a candidate theory of quantum gravity.

Unitary symmetry

I have related elsewhere the personal sequence of events that landed me, in May 1960, at Imperial College in London, with a one-year fellowship (at the very late age of thirty-five) to work on a thesis on the physics of particles and fields under Abdus Salam.[8,9] Considering the time constraints, Salam wanted me to work on a specific calculation relating to the mass in Yang–Mills fields, an issue resolved at Imperial College in 1964–7 by P. Higgs and T. Kibble, and by others working elsewhere. I was more interested in the general picture.

The strong-interaction Lagrangian was then believed to consist of the pseudoscalar meson couplings (π, K, $\bar{K}$), with eight independent couplings $g_1, \ldots, g_8$ for the eight baryons (four isospin-hypercharge multiplets N, Λ, Σ, Ξ) and seven mesons (three multiplets). "Global symmetry," an attempt by Murray Gell-Mann and Julian Schwinger to circumscribe these couplings by the imposition of a certain higher symmetry, had just failed.[10] Salam had pointed out that the experimental s-wave phase shifts were very different in π-hyperon scattering, and this contradicted the symmetry's basic assumption of symmetric π couplings as against symmetry-breaking K couplings (also responsible for the pattern of the mass spectrum).

I read the 1954 Yang–Mills paper a short time after I had settled at Imperial College. I liked the aesthetic purity of this approach; it evoked in me feelings similar to what I had experienced when first exposed to Albert Einstein's general relativity. I used to think it was the geometrization that had captivated me over there, but I could now see that it was really covariance, a local gauge symmetry, that gave the theory its beauty. Jun John Sakurai's preprint, "A Vector Theory of the Strong Interactions," arrived about that time and appeared to indicate that the true strong interaction might well be such a Yang–Mills theory.

The recently determined "$V-A$" nature of the weak interactions also appeared to point to a Yang–Mills mechanism for that interaction.[11,12] Moreover, the "conserved-vector-current" idea implied that these weak-interaction currents were the currents generated by the strong-interaction symmetry through the Noether theorem.[12,13] Therefore, both led to a Yang–

Mills symmetry of the strong interactions. Somehow, the weak interaction then had to repeat the Yang–Mills mechanism a second time in order to produce its own intermediate bosons, a problem that seemed to imply duplication, with which I later dealt.[14]

It is interesting that Lie groups and algebras had attracted very little attention. The various suggestions for global symmetry and the like had been formulated as assumptions constraining the $g_1, \ldots, g_8$ without identifying a formal group structure. Jayme Tiomno had, however, shown that the Gell-Mann–Schwinger *Ansatz* was equivalent to a global SO(7) symmetry.[15] Salam and John C. Ward had then further generalized it to local SO(8) and SO(9), invoking the Yang–Mills mechanism.[16] Ryoyu Utiyama had reformulated the local gauge for any abstract Lie group and shown that even general relativity could be regarded as a local gauge for the Poincaré group (a much more complicated problem, with which I dealt in 1977–9).[17,18]

I had tried to work out some better scheme, but I was just reproducing attempts whose existence I had not been aware of in 1957–60. At Salam's suggestion, once he realized I was interested in this general issue, I took a different path and launched into a study of Lie algebras and groups per se. I used the works of E. B. Dynkin, who had redone Elie Cartan's nineteenth-century classification of the semisimple finite-dimensional Lie algebras and introduced an extremely useful graphic method to describe the algebras and their representations.[19] Dynkin had also classified the maximal subalgebras, and so forth.

It is interesting to note how useful classification can be. Here was a classification in mathematics that was immediately applicable to physics, producing yet another classification there. I realized that my task consisted in selecting a candidate symmetry G out of the Cartan catalogue, as presented by Dynkin. The various irreducible unitary representations of G would then characterize the particle multiplets, a classification for physics. A local gauge based on that same group G, now $G(x)$, would yield the dynamical Yang–Mills strong-interaction theory, plus a multiplet of vector mesons to mediate that interaction. It would also provide an assignment for the weak and electromagnetic charges and currents, as parts of a multiplet of the same symmetry. In fact, both charge currents and mediating Yang–Mills vector bosons had to be assigned to the "regular" (or adjoint) representation ad(G).

Realizing that all reactions allowed by the conservation of isospin and strangeness appeared to occur in nature, I decided I wanted a rank-two Lie algebra with an appropriate inclusion of the isospin hypercharge U(2). There were five rank-two Lie algebras: A_2 [generating SU(3)], B_2 [generating SO(5)], C_2 [generating Sp(4)], D_2 [generating SO(4) or SU(2) × SU(2), a semisimple algebra], and G_2, one of the five exceptional Lie algebras and groups. I was interested by the "exceptional" nature of G_2 (the same excitement occurred in the seventies with E_6 as a candidate gauge-unifying theory, and in 1985 with $E_8 \times E_8$ for quantized strings) and by the fact that its root diagram turned out to be a star of David! However, it included unobserved transitions in the weak current, with $\Delta S/\Delta Q = -1$ (S is strangeness, Q is electric charge), and so forth. A_2 appeared to fit perfectly; that is, the symmetry group would then be SU(3). It reproduced an experimental factor of $\frac{1}{3}$ that Sakurai had pointed out as the ratio observed between the isospin and the hypercharge gauge couplings.

The good fit required the baryons to be assigned to an octet, that is, not to the group's fundamental representation. This appeared puzzling. It would mean one of two things: Either the group would turn out to be SU(3)/Z(3), something like having only integer-spin representations in SU(2), that is, representations of SO(3) $\sim$ SU(2)/Z(2), or, alternatively, the baryons would be composites.

I was prepared to follow either path. The first seemed nice mathematically, but the phase independence of quantum mechanical probabilities appeared to rule it out. The second possibility, composite protons and neutrons, seemed revolutionary, but I could as yet see no reason for this to be impossible. I decided to postpone this investigation and complete the verification of the octet assignment itself.

By the beginning of the 1960–1 academic year, I had done it, and I was confident in my identification of SU(3) and the baryon octet. The mesons fitted another octet, thus predicting an eighth 0^- meson [the η^0 (550 MeV), found in mid-1961]. I was also predicting an octet (and singlet) for an as yet unseen 1^- vector meson.

I have described elsewhere the events leading to the publication of my model in the journal *Nuclear Physics* (received 13 February 1961) and the simultaneous and independent suggestion of the same model by Gell-Mann in a paper that was to remain unpublished.[8,9,20,21] I had first presented the model to Salam in November and had then written it up in December. The delay between December 1960 and early February 1961 was caused by Salam's first intending to coauthor the paper, adjoining an idea of his about an SU(3) gauge for the Sakata model, but then deciding to publish that idea in another paper (with Ward, with whom it apparently had been initiated) and releasing my draft for separate publication. Gell-Mann had circulated a first draft at Caltech on 20 January. He submitted a second version to the *Physical Review* on 27 March, then withdrew it in June when Σ–Λ relative-parity measurements appeared to invalidate the octet assignments. Gell-Mann ultimately resubmitted a very different version (20 September) in which he presented the idea of using the commutation relations between the unitary symmetry charges, between charges and currents, and between local currents as a sort of matrix mechanics.[22] The Sakata model and the octet are introduced as two candidate realizations of SU(3).

Gell-Mann's method of "current algebra" was an important step in providing a justification for strong-interaction symmetries, even if one had to abandon RQFT, following the criticism and advice from the Berkeley *S*-

matrix school. Field theory being in bad odor, we were soon dropping the Yang–Mills dynamics, preserving the SU(3) classification solely. What was sad was that the predicted octet and singlet of 1^- particles had been found in the meantime, and their couplings were indeed precisely "universal," as required by a local gauge. Of course, the mesons were massive, and such a mass term would break the symmetry. It all had to wait until M. J. G. Veltman and Gerard 't Hooft had completed the renormalized quantum version of the Yang–Mills interaction, including "soft" mass terms.[23] Meanwhile, current-algebra methods could explain the vector-meson couplings as an "effective" approximate result. Several years later, the vector mesons themselves came to be regarded as quark-antiquark systems, "accidentally" fitting a Yang–Mills picture.

It should be noted that a search based on the classification of Lie algebras was also initiated somewhat later by D. R. Speiser and J. Tarski, who noted the A_2 possibility, but were more attracted by G_2.[24] R. E. Behrends and A. Sirlin used the weak transitions as a guide and also somehow settled on G_2.[25] Gell-Mann and Sheldon Glashow at Caltech and Penelope Ionides (partly with my advice) at Imperial College showed that the Yang–Mills local gauge required a compact Lie group.[26]

Since 1979, one may perhaps again think of unitary symmetry as a Yang–Mills gauge. It has been suggested by 't Hooft that "fundamental" and "composite" are definitions holding within some energy range.[27] In the low-energy world up to several billion electron volts, the vector mesons can be considered as "fundamental," together with the pseudoscalar,[28] and 't Hooft has suggested equations relating levels in which a set of particles appears as fundamental in one and composite in the other.

In the years 1961–5, unitary symmetry gave hundreds of predictions for multiplet assignments, strong-interaction branching ratios, masses, electromagnetic mass differences, magnetic moments, weak transitions, and so forth.[29,30] It all fitted beautifully.

Triplets (the future quarks)

In the late spring of 1961, I returned to the question of the nucleon's compositeness. Looking at the systematics suggested by J. Wess for the Sakata–Thirring model, I noticed that the baryon octet (including the nucleon) could be "made of" three leptonlike fundamental triplets.[31] I was puzzled, as I was thinking of the (v^0, e^-, μ^-) set as candidates (with a B^{++} boson to set the charges right). After my return to Israel in August 1961, I tried a hypothetical set with atomic mass (baryon) number $B = \frac{1}{3}$. With H. Goldberg, we worked out the group representation structure:

(2, 1, 0) for the baryon octet
(1, 0, −1) for the mesons
(0, −1, −2) for antibaryons

Here, the sum of (h_1, h_2, h_3) corresponds to the "quark" number; $\lambda = h_1 - h_2$ and $\mu = h_2 - h_3$ yield the usual SU(3) quantum numbers. The triplet field is introduced as a sakatonlike fundamental field, with particles appearing "as if" they are made of three pseudosakatons, a pseudosakaton-antipseudosakaton pair, or three antipseudosakatons. The dynamical picture, however, was still rather confused, as I had difficulty thinking of the "production" of mesons in baryon–antibaryon scattering and their "composition" in terms of a triplet–antitriplet pair. Several years later, it was indeed noticed that baryon–antibaryon pairs do give rise, in the main, to three mesons, the scattering then consisting in a rearrangement of the triplets. Harry J. Lipkin[31] has written of that period that

> Goldberg and Ne'eman then pointed out that the octet model was consistent with a composite model constructed from a basic triplet with the same isospin and strangeness quantum numbers as the sakaton, but with baryon number $B = \frac{1}{3}$. The baryon octet was constructed from three triplets. However, equations...show that particles having third-integral baryon number must also have third-integral electric charge and hypercharge. At that time the eightfold way was generally considered to be rather far-fetched and probably wrong. Any suggestion that unitary symmetry was based on the existence of particles with third-integral quantum numbers would not have been considered seriously. Thus the Goldberg–Ne'eman paper presented this triplet as a mathematical device for construction of the representations in which the particles were classified.
>
> Several years later, new experimental data forced everyone to take SU(3) more seriously. The second baryon multiplet was found including the Ω^-, and with spin and parity $\frac{3}{2}^+$. Gell-Mann and Zweig then proposed the possible existence of the fundamental triplet as a serious possibility and gave it the name of quark.

This gives a good description of the background that made me hesitate and wonder whether or not to take my own "fundamental" model seriously. I have described elsewhere the event leading to the prediction of the Ω^- with its quantum numbers and mass, by Gell-Mann in the 1962 CERN conference plenary, and by me in a document presented at that same conference to Gerson Goldhaber the day before. Goldhaber has now related these events in a publication I received as a gift on my sixtieth birthday.[32] I refer the reader to previous histories of the SU(3) story for a report on the multiple experimental verifications.[8,9]

The mystery of flavors and generations

It is much too early to draw conclusions, even after twenty-five years. With the discovery of asymptotic freedom in Yang–Mills interactions, another SU(3) gauge, that of the superimposed color-SU(3), was assumed to hold. This quantum chromodynamics (QCD) theory appears to be a good description of interquark (unitary-symmetric) forces, even though it cannot yet be considered completely proven, as long as there is no mathematical proof of its confining nature, except on a lattice. Since 1974, our original "flavor" SU(3) has been relegated to an "accidental" symmetry, resulting from the small mass differences between the $u^{2/3}$, $d^{-1/3}$, and $s^{-1/3}$ quarks. However, where do these masses originate? Assuming that the "initial" mass of $u^{2/3}$, $d^{-1/3}$ is zero, this is the problem of the mass of the strange quark, the SU(3)-breaking mass term. I had suggested that this is due to another, the "fifth" interaction, which could also generate the muon mass.[33] Indeed, in the "generations" picture, the strange quark and the muon again come together, as in the old Gamba–Marshak–Okubo picture. The conventional wisdom is that all this is due to the Higgs couplings. We face here what appears to be an extremely arbitrary selection for a large set of parameters: three generations, with six quark masses, three (or six) lepton masses, two quark mixing angles (Cabibbo's and the Kobayashi–Maskawa mechanism for *CP* violation), two allowed lepton mixing angles, and at least one angle and one vacuum expectation value for the *W* and Z^0 mesons, although this may well relate to a different issue, together with Θ_{QCD}, α_{QCD}, α_{EM}.

The origin of unitary symmetry is thus a part of the general unanswered question of sequential generations, perhaps related to the issue of more basic constituents – haplons, rishons, alphons, or whatever. In most presentations, all of this is hidden under the cover of the "Higgs sector." It is an issue that has been with us for almost fifty years, since the discovery of the muon in 1937. I guess its resolution will provide some of the excitement in the next fifty years.

Notes

1 Y. Ne'eman, "Science as Evolution and Transcendance," presented at the Fairchild Symposium on the Relevance of Science, Pasadena, 1977, *Acta Scientifica Venezolana 31: 1–3* (1980).

2 Y. Ne'eman, "Patterns and Symmetry in the Structure of Matter," *Proc. Academia dei Lincei (Rome) 8* (1975), 1–10.

3 See, for example, L. de Broglie, D. Bohm, P. Hillion, F. Halbwachs, T. Takabayasi, and J. P. Vigier, "Rotator Model of Elementary Particles Considered as Relativistic Extended Structures in Minkowski Space," *Phys. Rev. 129* (1963), 438–50; "Space–Time Model of Relativistic Extended Particles in Minkowski Space. II. Free Particle and Interaction Theory," ibid., 451–66.

4 H. Feshbach, CERN report Th. 321 (1962).

5 E. Fermi and C. N. Yang, "Are Mesons Elementary Particles?" *Phys. Rev. 76* (1949), 1739–43; M. Goldhaber, "Compound Hypothesis for the Heavy Unstable Particles. II," *Phys. Rev. 101* (1956), 433–8. Similar suggestions were made by G. Gyorgyi, R. Christy, G. Derdi, M. A. Markov, and Y. B. Zel'dovich. S. Sakata, "On a Composite Model for the New Particles," *Prog. Theor. Phys. 16* (1956), 686–8.

6 See, for example, Geoffrey F. Chew, *S-Matrix Theory of Strong Interactions* (New York: Benjamin, 1961).

7 See, for example, R. E. Cutkosky, "A Mechanism for the Induction of Symmetries Among the Strong Interactions," *Phys. Rev. 131* (1963), 1888–90.

8 Y. Ne'eman, *Proc. Israel Acad. Sci. Hum. 21* (1983).

9 Y. Ne'eman, in *Symmetries in Physics (1600–1980)*, edited by M. Garcia-Doncel et al. (Singapore: World Scientific, 1987), pp. 499–540.

10 M. Gell-Mann, "Model of the Strong Couplings," *Phys. Rev. 106* (1957), 1296–300; J.

Schwinger, "A Theory of the Fundamental Interactions," *Ann. Phys. (N.Y.) 2* (1957), 407–34.

11 R. E. Marshak and E. C. G. Sudarshan, *Proc. Padua-Venice Conf. (1957)*; (Padua-Venice: Soc. Ital. de Fisica, 1958).

12 R. P. Feynman and M. Gell-Mann, "Theory of the Fermi Interaction," *Phys. Rev. 109* (1958), 193–8.

13 S. S. Gershtein and Y. B. Zel'dovich, "On Corrections from Mesons to the Theory of β-Decay," *Zh. Eksp. Teor. Fiz. 29* (1955), 698–9 [English translation: "Meson Corrections in the Theory of Beta Decay," *Sov. Phys.–JETP 2* (1956), 576].

14 Y. Ne'eman, "Unified Interactions in the Unitary Gauge Theory," *Nucl. Phys. 30* (1962), 347–9.

15 J. Tiomno, "On the Theory of Hyperons and *K*-Mesons," *Nuovo Cimento 6* (1957), 69–83.

16 A. Salam and J. C. Ward, "On a Gauge Theory of Elementary Interactions," *Nuovo Cimento 19* (1961), 165–70.

17 R. Utiyama, "Invariant Theoretical Interpretation of Interaction," *Phys. Rev. 101* (1956), 1597–667.

18 Y. Ne'eman and T. Regge, "Gravity and Supergravity as Gauge Theories on a Group Manifold," *Phys. Lett. 74B* (1978), 54–6; "Gauge Theory of Gravity and Supergravity on a Group Manifold," *Riv. Nuovo Cimento (Ser. 3) I* (1978), 1–43; Y. Ne'eman "Gravity, Groups, and Gauges," in *General Relativity and Gravitation, Einstein Centennial*, edited by A. Held (New York: Plenum Press, 1980), pp. 309–28. See also F. W. Hehl, P. von der Heyde, G. D. Kerlick, and J. M. Nester, "General Relativity with Spin and Torsion: Foundations and Prospects," *Rev. Mod. Phys. 48* (1976), 393–416.

19 E. B. Dynkin, "Semisimple Subalgebras of Semisimple Lie Algebras," *Ann. Math. Soc. Transl. (Ser. 2) 6* (1957), 111–244; E. B. Dynkin, "Maximal Subgroups of the Classical Groups," *Ann. Math. Soc. Transl. (Ser. 2) 6* (1957), 245–379; E. B. Dynkin, *Usp. Mat. Nauk 2 : 4* (1947), 59–127.

20 Y. Ne'eman, "Derivation of Strong Interactions from a Gauge Invariance," *Nucl. Phys. 26* (1961), 222–9; "Yukawa Terms in the Unitary Gauge Theory," *Nucl. Phys. 26* (1961), 230–2; Israel AEC report IA-698 (August 1961).

21 M. Gell-Mann, Caltech report CTSL-20 (1961) (unpublished).

22 M. Gell-Mann, "Symmetries of Baryons and Mesons," *Phys. Rev. 125* (1962), 1067–84.

23 G. 't Hooft, "Renormalization of Massless Yang–Mills Fields," *Nucl. Phys. B33* (1971), 173–99; "Renormalizable Lagrangians for Massive Yang-Mills Fields," *Nucl. Phys. B35* (1971), 167–88; G. 't Hooft and M. Veltman, "Regularization and Renormalization of Gauge Fields," *Nucl. Phys. B44* (1972), 189–213.

24 D. R. Speiser and J. Tarski, IAS (Princeton) report (1961).

25 R. E. Behrends and A. Sirlin, "Weak-Coupling Currents and Symmetries of Strong Interactions," *Phys. Rev. 121* (1961), 324–36; V. M. Shekhter, "Symmetry Properties of Strong Interactions," *Sov. Phys.–JETP 14* (1962), 582–8.

26 Sheldon L. Glashow and Murray Gell-Mann, "Gauge Theories of Vector Particles," *Ann. Phys. 15* (1961), 437–60; P. Ionides, Imperial College, Ph. D. thesis.

27 G. 't Hooft, "Naturalness, Chiral Symmetry, and Spontaneous Chiral Symmetry Breaking," in *Recent Developments in Gauge Theories*, edited by G. 't Hooft et al. (New York: Plenum, 1980), pp. 135–57.

28 S. Weinberg, "Dynamical Approach to Current Algebra," *Phys. Rev. Lett. 18* (1967), 188–91.

29 M. Gell-Mann and Y. Ne'eman, *The Eightfold Way* (New York: Benjamin, 1964).

30 Y. Ne'eman, *Algebraic Theory of Particle Physics* (New York: Benjamin, 1967).

31 H. J. Lipkin, "Quarks for Pedestrians," *Physics Reports 8C* (1973), 173–268, especially 180.

32 G. Goldhaber, "The Encounter on the Bus," in *From SU(3) to Gravity*, edited by E. Gotsman and G. Tauber (Cambridge University Press, 1986), pp. 103–6.

33 Yuval Ne'eman, "The Fifth Interaction: Origins of the Mass Breaking Asymmetry," *Phys. Rev. 134* (1964), B1355–7.

44   Gauge principle, vector-meson dominance, and spontaneous symmetry breaking

YOICHIRO NAMBU

Born 1921, Tokyo; Sc. D., University of Tokyo, 1952; elementary particle physics; University of Chicago

With the benefit of hindsight, I would like to speak on certain theoretical developments that occurred during the late 1950s. The subject matter I shall discuss centers around the views regarding the meaning and role of symmetries, or the lack thereof. I shall talk in particular about the happenings in Chicago, not only because they are what I experienced at first hand but also because one of the participants, Jun John Sakurai, unfortunately cannot be heard any more. Let me begin by stating that, at the risk of oversimplification, I regard Ernest Lawrence and Hideki Yukawa as the two founding fathers of particle physics, in that they respectively established the basic experimental and theoretical methodologies in this field.[1] That these are the basic methodologies still holds true, with some qualifications that I shall come to in a moment.

Limiting myself to the theoretical side only, Yukawa's way was to freely invent (or postulate) new particles in order to explain phenomena that are new or not yet understood. Although Yukawa stopped pursuing this direction after his success with the meson theory, the philosophy behind it was articulated and practiced by his collaborator Shoichi Sakata, yielding further successes. The two-meson theory was one such example.[2] At any rate, Yukawa's approach was phenomenological and ad hoc, in that it lacked a theoretical guiding principle of its own, which was perhaps the reason why he stopped pursuing it. This contrasts with the current situation in which gauge theory has established itself as the supreme principle.

The origin of the new trend goes back to the work of Chen Ning Yang and Robert Lawrence Mills in 1954, and it became a dominant paradigm in the

Work supported in part by the National Science Foundation, PHY-83–01221, and the Department of Energy, DE F602–84 ER-45144.

1970s.[3] But in the late 1950s, there were already signs of a painful clash between Yukawa's phenomenologically driven approach and the theoretical and deductive approach based on the gauge principle. Almost simultaneously with Yang and Mills, Ryoyu Utiyama developed his own ideas about non-Abelian gauge theory as the general dynamical principle, which in his view also encompassed Einstein's gravity.[4] Recently, in a personal conversation with me, he confided how poorly he had been received by the Yukawa establishment in Japan at that time, and consequently how hesitant he had been to publish his work.

This brings me to the work of Sakurai. He came to Chicago from Cornell as a fresh Ph. D. and assistant professor in the late 1950s. He was already known for his involvement with the $V-A$ theory, but in his pursuit of symmetries in strong interactions he was gradually led to the gauge principle of Yang and Mills as the basis of strong-interaction dynamics. In present-day language, his theory may be called an $SU(2) \times U(1) \times U(1)$ gauge theory of strong interaction, representing isospin, hypercharge, and baryonic charge as its strict symmetries. [Flavor $SU(3)$ was not around yet.] He considered the vector mesons corresponding to these symmetries to be the basic gauge bosons of strong interactions, and he applied the resulting vector-meson-dominance model to phenomenology with considerable success. His own historical account is found in his lecture notes, but I would like to quote a few paragraphs from his 1960 paper, which was written before the discoveries of the ω, ϱ, and ϕ mesons, and reveals his bold vision better than his later papers.[5,6] With the tone of a brash young physicist, he writes:

> After having spent considerable time (and energy) on various symmetry models, the present author is convinced that there are no simple patterns in the Yukawa-type Lagrangians in which pions and K particles are coupled linearly to baryons, and that all those symmetry models proposed up to now are mere mental exercises devoid of any physical significance whatsoever.
>
> Why is it that the Creator was so supremely imaginative when He declared, "Let there be light," while He did not use any imagination when He switched on the γ_5 couplings of the pion field? It appears that only by forsaking the idea that the Yukawa interaction is fundamental can we restore the depth, simplicity, beauty and elegance that are so characteristic of true physical theories.
>
> Now we must start from the very beginning as though we did not know any meson theory of the Yukawa form. What should be the guiding principles in constructing a new theory of strong interactions? First of all, the theory should be deeply rooted in symmetry laws that hold exactly in the absence of the electromagnetic and weak couplings. Instead of looking for artificial higher symmetries, we should take the existing symmetries more seriously than ever before and exploit them to the utmost limit.

The problem with his theory, however, was how to account for the mass of those gauge bosons. He knew the problem, but did not know how to solve it. On this point he remarked:

> Since our ideas are rather novel, it is not too surprising that there are difficulties associated with our theory. It would be a pity to give up our theory on account of the *B* [vector-meson] mass problem, just as it would be a pity to give up Bohr's atomic model on account of the difficulties associated with the notion of "quantum jumps."

I would not know how to counter such an argument. It is also worth pointing out that this paper contains speculations concerning the problem of baryon excess of the universe as being due to baryon number (and energy) non-conservation.

Turning now to my involvement in symmetries, it is more accidental.[7] It happened one day when Robert Schrieffer, a finishing graduate student at the University of Illinois, was invited to give a seminar at Chicago on the still unpublished BCS theory.[8] I had always been interested in superconductivity, as had Gregor Wentzel, the senior theorist at Chicago in those days. My recollection about this seminar on the BCS theory is that I was very disturbed by the authors' disrespect of gauge invariance in a theory that was supposed to explain the Meissner effect.

Eventually I came to understand the symmetry-breaking aspect of the BCS theory and how the emergence of the Goldstone mode restores the symmetry in a way, a fact that was discovered also by other people like Philip W. Anderson and G. Rickayzen.[9,10] The analogy between the Bogoliubov–Valatin quasi-particle equation and the Dirac equation, on the one hand, and the analogy between the induced Goldstone current and the pion contribution to the nucleon axial vector current (which leads to the Goldberger–Treiman relation for the pion decay constant), on the other hand, naturally led me to the model in which chiral symmetry breaking, massive nucleon, and massless pion are logically linked together.[11] Perhaps the part that required the most courage on my part was to challenge the dogma of axiomatic theories concerning the properties of the vacuum.

It is not surprising that V. G. Vaks and A. I. Larkin in the Bogoliubov school also found a rather similar model.[12] I was very much aware of the symmetry breaking and the associated Goldstone mode as a general phenomenon, and I was trying to collect examples in condensed-matter physics before writing a paper devoted to this general aspect. But Jeffrey Goldstone's paper appeared in the meantime.[13] Robert Marshak has graciously pointed out that my ideas are already on record in the proceedings of the 1959 Kiev conference. I had completely forgotten about it, probably because its proceedings took two years to come out, only after the proceedings of the 1960 conference.

One point that may perhaps be worth clarifying is the following. It has to do with the motivation for the four-fermion interaction model that Giovanni Jona-Lasinio and I adopted. It is true that the model is directly analogous to that of BCS as well as to Werner Heisenberg's nonlinear unified theory. But Heisenberg's theory never appealed to me, nor did I take it seriously. I chose the model only because the mathematical aspect of symmetry breaking could be most clearly demonstrated there. Actually I had considered quantum electrodynamics as an example, but could not convince myself of the existence of a solution to the gap equation because of vacuum polarization (running charge) effects.

Besides symmetry breaking and fermion mass generation, there is another aspect in the BCS theory, that is, the mass generation for the gauge fields, resulting from the mixing of two originally massless fields, the gauge field and the Goldstone mode. This fact was more or less obvious to me, because it is essentially the familiar plasmon phenomenon. But in the context of relativistic field theory, the mechanism was recognized as a general one by Anderson,[14] François Englert and Roger Brout,[15] and Peter W. Higgs.[16] In particular, Higgs's rediscovery of the Ginzburg–Landau effective field theory has since become the standard formalism to be used for building unified field theories. As it turned out, the true relevance of the gauge principle with symmetry breaking was to be found in the weak interactions, rather than in the strong interactions as Sakurai and I had envisioned.

Notes

1 Y. Nambu, "Concluding Remarks," *Phys. Reports 104* (1984), 237–58.
2 S. Sakata, "On the Relation between the Meson and the Yukawa-Particle," *Bull. Phys.-Math. Soc. Japan 16* (1943), 232 (in Japanese).
3 C. N. Yang and R. L. Mills, "Conservation of Isotopic Spin and Isotopic Gauge Invariance," *Phys. Rev. 96* (1954), 191–5.
4 R. Utiyama, "Invariant Theoretical Interpretation of Interaction," *Phys. Rev. 101* (1956), 1597–667.
5 J. J. Sakurai, "Vector Mesons 1960–1968," in *Lectures in Theoretical Physics XI-A*, edited by K. T. Mahanthappa et al. (New York: Gordon & Breach, 1969), pp. 1–21.
6 J. J. Sakurai, "Theory of Strong Interactions," *Ann. Phys. (N.Y.) 11* (1960), 1–48.
7 Yoichiro Nambu, "Symmetry Breakdown and Small Mass Bosons," *Fields and Quanta 1* (1970), 33–51; Yoichiro Nambu, "Superconductivity and Particle Physics," *Physica 126B* (1984),328–34.
8 J. Bardeen, L. N. Cooper, and J. R. Schrieffer, "Microscopic Theory of Superconductivity," *Phys. Rev. 106* (1957), 162–4.
9 P. W. Anderson, "Coherent Excited States in the Theory of Superconductivity: Gauge Invariance and the Meissner Effect," *Phys. Rev. 110* (1958), 827–35.
10 G. Rickayzen, "Meissner Effect and Gauge Invariance," *Phys. Rev. 111* (1958), 817–21.
11 Yoichiro Nambu, "Axial Vector Current Conservation in Weak Interactions,"*Phys. Rev. Lett. 4* (1960), 380–2; Y. Nambu and G. Jona-Lasinio, "Dynamical Model of Elementary Particles Based on an Analogy with Superconductivity. I," *Phys. Rev. 122* (1961), 345–58; Y. Nambu and G. Jona-Lasinio, "Dynamical Model of Elementary Particles Based on an Analogy with Superconductivity. II," *Phys. Rev. 124* (1961), 246–54.
12 V. G. Vaks and A. I. Larkin, "On the Application of the Methods of Superconductivity Theory to the Problem of the Masses of Elementary Particles," *Sov. Phys.–JETP 13* (1961), 192–3.
13 J. Goldstone, "Field Theories with Superconductor Solutions," *Nuovo Cimento 19* (1961), 154–64.
14 P. W. Anderson, "Plasmons, Gauge Invariance, and Mass," *Phys. Rev. 130* (1963), 439–42.
15 F. Englert and R. Brout, "Broken Symmetry and the Mass of Gauge Vector Mesons," *Phys. Rev. Lett. 13* (1964), 321–3.
16 Peter W. Higgs, "Broken Symmetries and the Masses of Gauge Bosons," *Phys. Rev. Lett. 13* (1964), 508–9.

PART X

PERSONAL OVERVIEW

The beginning is easy to recite for us;
The ending is nowhere in sight for us.
   And though the answers may some day be nearer,
   Things will get worse, before they get clearer.
                    – © Arthur Roberts, "Some People Don't
                              Know Where to Stop" (1952)

45   Scientific impact of the first decade of the Rochester conferences (1950–1960)

ROBERT E. MARSHAK

Born 1916, New York City; Ph.D., Cornell, 1939; theoretical physics;
Virginia Polytechnic Institute and State University

Introduction

The first Rochester conference on high-energy nuclear physics was
held 16 December 1950, twenty months after the last of the three Shelter
Island conferences on the foundations of quantum mechanics, organized
by Robert Oppenheimer (1947–9). Five high-energy accelerators ("high
energy" was defined as >100 MeV thirty-five years ago!) had been completed
since the end of World War II and were producing research results: two at
Berkeley (340-MeV synchrocyclotron and 300-MeV electron-synchrotron),
the Rochester 240-MeV and Harvard 150-MeV synchrocyclotrons, and the
Cornell 300-MeV electron-synchrotron. Many more accelerators with even
higher energies were under construction. Cosmic-ray experimentalists
throughout the world were making extremely important contributions to
the field.

It seemed clear to me that the Shelter Island conferences – which had been
limited to a small number of theorists, with a couple of "token" experimentalists – should be replaced by a new series of conferences that would give
equal weight to the participation of accelerator experimentalists, cosmic-
ray experimentalists, and theorists. As the newly installed chairman of the
Rochester Physics Department – with one of the few meson-producing
machines in operation and a very active cosmic-ray group – I decided to
organize an invitational conference that would thoroughly discuss the latest
developments in high-energy physics.

The first two Rochester conferences were small (fifty to seventy-five participants), were of short duration (two days), were supported by local industry,
and still were not truly international (Europe and Japan were recovering from
the war). Their success and the increasing worldwide interest in the field led

to the raising of sights in all respects. The third Rochester conference acquired a more pretentious title: "Third Annual Conference on High Energy Nuclear Physics." It had 150 participants, lasted three days, and received support for the first time from a government agency, the National Science Foundation. It also went international, with the attendance of representatives from Great Britain, France, Italy, Australia, Holland, Japan, and several other countries.

As accelerators of higher and higher energy came into operation in the United States (Cosmotron, Bevatron, etc.) and high-energy accelerators were constructed in other countries, the annual Rochester conference expanded in scope, duration, and international attendance. By April 1957, Rochester VII, the seventh Rochester conference,* grew to 300 invited guests from twenty-four countries and lasted five days. Although the number of participants had increased fivefold from 1950 to 1957, I would guess that the Rochester conference was equally élitist, in view of the rapid growth of high-energy physics during those years.[1]

Some striking features of the Rochester conferences during those years were the genuine collegiality that prevailed, the informal discipline accepted by the participants to focus on major developments, and the subtle ways in which the entire roster of conferees served as a global planning group, a sort of high-energy physics advisory panel (HEPAP) for the world research activities during the ensuing year. Correlative with this last function, it was deemed essential that the conference proceedings should be made available to all nonparticipants within two or three months after the conference, and it was remarkable how readily the various American laboratories accepted responsibility for these editorial (and other organizational) chores.

The seventh conference was the last of the Rochester-generated annual conferences in high-energy physics. The rapid growth of interest in the field, particularly in Western Europe (as represented by CERN) and in the Soviet Union, naturally led to the desire to host an international conference of the Rochester type. Indeed, in 1956, the sixth Rochester conference (held in April) was followed by the Moscow conference "High Energy Particles" (held in May), which was in turn followed by a CERN high-energy physics conference (held in June). The scientific programs of all three conferences were essentially identical, although, of course, the distribution of participants was correlated with the location of the conference. This duplication was time-consuming for many participants, was costly, and seemed to be wholly unnecessary.

As a result, at the International Union of Pure and Applied Physics (IUPAP) General Assembly held in Rome in September 1957, as a member of the American delegation, and with the support of the head of the Soviet

---

* Rochester $N = N$th Rochester conference held in 1950 + N, except for $N = 1$ (held in December 1950) and $N = 3$ (held in December 1952).

delegation, A. L. Ioffe, I proposed the creation of a high-energy physics commission. In addition to placing the Rochester conference under its aegis, IUPAP gave two further charges to the newly established commission:[2] to encourage rapid exchange of the latest scientific results and to encourage international collaboration among the various high-energy laboratories to insure the best use of the facilities of these large and expensive installations. These two charges were taken rather seriously and led to the most extensive preprint system in operation in any scientific field and to the creation later of the International Committee on Future Accelerators (ICFA).

To return to the Rochester conference, the six members of the initial IUPAP high-energy commission were apportioned equally among the United States, the Soviet Union, and Western Europe. (Some years later, several members were added to represent the "rest of the world" in high-energy physics.) The membership of this first IUPAP commission consisted of C. J. Bakker as chairman, Rudolph E. Peierls, Robert E. Marshak as secretary, W. K. H. Panofsky, I. E. Tamm, and V. I. Veksler. At our first meeting, it was decided to have an automatic three-way rotation of the conference, starting with CERN in 1958, Kiev in 1959, and back to Rochester in 1960. The 1958 conference in Geneva became the "Eighth Annual International Conference on High Energy Physics," taking on the sequential ordering of the first seven Rochester conferences. Except for the change from annual to biennial after 1960, the present system of Rochester conferences was in place in 1957.

The number of participants for Rochester VIII, IX, and X was limited to 300, despite the explosive growth of the field and the involvement of scientists from thirty countries. The corollary of this exclusiveness was to arrange an elaborate system of rapporteurs and scientific secretaries in order to provide a full account of the scientific discussions as rapidly as possible. The tenth conference was the last to be held in Rochester. With its eight days of meetings, its thirty-six scientific secretaries, and its comprehensive coverage of topics in high-energy physics, the Rochester conference had become a major undertaking. The scientific secretaries at Rochester X were an impressive group, consisting of the brightest young physicists at that time, two of whom – J. Cronin and S. Weinberg – were awarded the Nobel Prize in later years. It is not surprising that the scientific impact of the Rochester conferences during the decade of the 1950s was so far-reaching and felicitous.

I now turn to a summary account of the scientific highlights of the Rochester conferences during their first decade of operation, precisely the period in the history of particle physics under review at this symposium. Because the early Rochester conferences made a conscious effort each year to discuss the full panoply of themes in the field of high-energy physics, trying to give an account of the scientific impact of the conferences is tantamount to trying to sketch a history of the field itself during the 1950s. In delineating the special role of the Rochester conferences in contributing to that early history,

I make no pretense at completeness. I must confess that I have been guided in my choice of topics by personal taste and by anticipating the degree of coverage of a given topic by a scheduled speaker at this symposium.

With these caveats, I describe the scientific highlights of the Rochester conferences during the 1950s under the following five headings: nucleon; pion; strange particles; parity violation and universal $V-A$ weak interaction; SU(3) flavor, baryon–lepton symmetry, and compositeness of hadrons.

Nucleon

A goodly portion of the discussion at the first two Rochester conferences was devoted to the latest experimental results (from the Berkeley, Rochester, and Harvard synchrocyclotrons) on pp and np scattering (p here refers to proton, n to neutron, and N to nucleon). Those performing the experiments had all been brought up as nuclear physicists, and the opportunity to probe more deeply into the two-nucleon interaction (validity of charge independence at higher energies, exchange character, spin dependence, etc.) was pursued vigorously. Some of the initial findings – such as the peaking of the np differential cross section and its approximate symmetry around 90° (always in the center-of-mass system) – could be explained qualitatively through the simple exchange of a Yukawa meson. However, the near isotropy and approximate constancy of the pp differential cross section were more difficult to understand, and it soon became apparent that quantum electrodynamics (QED) could not serve as a model for calculations with meson theory. If the pion–nucleon interaction was pseudoscalar – so that the theory was renormalizable – the QED model was rendered meaningless by the huge value of the coupling constant. On the other hand, pseudovector coupling of the pion–nucleon system yielded a nonrenormalizable theory. (I shall return to the theoretical treatment of the pion–nucleon interaction in the next section.)

Under these circumstances, a more phenomenological approach had to be taken to the NN interaction problem. On the experimental side, it was realized that the spin of N opened up the possibility of doing a double-scattering experiment, and by the spring of 1953, Charles Oxley at Rochester had successfully carried out the first pp double-scattering experiment (at 210 MeV), observing a surprisingly large polarization.[2] This set in motion a worldwide NN scattering program extending to triple scattering and correlation experiments.[3] On the theoretical side, L. D. Wolfenstein showed that a minimum of five such experiments would determine the S matrix for NN scattering (i.e., all the phase shifts) at a given energy.[4]

The subsequent success of the experimental NN scattering program encouraged the theorists to try to find the semiphenomenological NN potential that would match all the known phase shifts, say up to 300 MeV. By "semiphenomenological" is meant that one would start with the most general velocity-dependent potential consisting of five terms (ordinary, spin-dependent, tensor, spin-orbit, quadratic spin-orbit) and then be guided in the choice of the spatial dependence of each term by meson theory.[5] Thus, by 1957, the Signell–Marshak potential, building on the Case–Pais spin-orbit potential (which could explain the near isotropy of the *pp* differential cross section), as well as the Gartenhaus potential (derived on the basis of the Chew–Low *p*-wave coupling of the pion to an extended nucleon source), were able to give a reasonably good fit of all *pp* and *np* scattering experiments in the region of several hundred million electron volts.[6,7] The semiphenomenological approach was continued for many years after 1957 and finally led to a fairly simple formulation in terms of the one-boson exchange (OBE) potential, in which the vector mesons ϱ and ω (in addition to π and a "scalar" meson) were employed to motivate the choice of the various terms in the two-nucleon potential.[8] The OBE potential has been particularly useful in nuclear physics applications.

While the *NN* elastic-scattering program did not yield any deep theoretical insights (at least from the vantage point of particle physics), the nucleon itself became an important object of investigation. I am referring here in the first instance to the work of Robert Hofstadter and collaborators on the electromagnetic structure of N.[9] As is well known, knowledge of the electromagnetic form factors of N is indispensable for understanding the high-energy interaction cross sections of leptons with nucleons and has provided an important test of new theoretical ideas. It is amusing that Hofstadter felt that the initial reception of his work was cool. Thus, the following sentence will be found in the proceedings of Rochester VI:[9]

> Hofstadter closed by saying that he felt their group had been "sort of working in a vacuum," because there had been no reflection from the outside world on their experimental results.

The trouble was that Hofstadter's unexpected results had to be digested by the theoretical community, and a year later, Yoichiro Nambu argued for the existence of the ω meson (which he called π_0') as follows:[10]

> I would like to present here another attempt at explanation of the nucleon charge distribution found in the Stanford electron scattering experiments. I will assume that there is another neutral meson, which I shall call π_0. It is a neutral vector meson of isotopic spin zero, and a mass two to three times that of the ordinary pion.

The complexity of the early situation was underlined by Peierls's questioning of Nambu:[10]

> You seem to know so much about this particle; about its decay modes and lifetime. But one has the impression that the discrepancies that this was invented to cure are not yet very firmly established.

Within several years, the ϱ meson also entered the parametrization of the proton and neutron form factors.[11] (Hofstadter discusses his pioneering work on the electromagnetic structure of the nucleon in Chapter 7 of this volume.)

With its complex structure, it was important to check that the antiparticle $\bar{N}$ fulfilled all the predictions of the Dirac equation for a spin-$\frac{1}{2}$ particle (equal mass, opposite sign of the charge and magnetic moment, etc.). Even though there was sufficient energy in the cosmic radiation to produce antiprotons, no one had succeeded in detecting them. Emilio Segrè was able to announce the detection of $\bar{p}$ on the Bevatron at Rochester VI:[12]

> Segrè opened the session by giving a discussion of the experiments performed at the Bevatron leading to the discovery of the anti-proton and to some of its properties. Theory had predicted the existence of anti-protons long ago and there had been serious attempts to find them in the cosmic radiation. Only three events which could have been anti-protons had been reported by cosmic ray workers and all of these events suffered from lack of some information that would have pinned down the determination of the identity of the particle concerned.

(Owen Chamberlain tells his story of the discovery of the antiproton with Segrè in Chapter 17 of this volume.)

Pion

By 1950, the mass of the charged pion had been well measured, the cross sections for its production in NN and γN collisions were roughly those expected for a Yukawa meson, and the lifetime for the dominant decay process $\pi \rightarrow \mu + \nu$ had also been measured. The neutral pion had not yet been seen (although it was expected, from the charge independence of the NN interaction). With the Berkeley and Rochester synchrocyclotrons (soon to be joined by the Columbia machine) in operation, it became a matter of the highest priority to completely fix the properties of the charged and neutral pions, such as their masses, lifetimes, decay products, spins, and parities. With the help of these accelerators, this program was essentially accomplished by the time of Rochester III.

The full story is told, and the references given, in two books on meson physics that appeared in 1952, and I shall here only mention a few highlights.[13] The copious production of charged pions by the accelerators permitted precise measurement of the mass, and nature's kindness in enforcing the condition $m(\pi^+) - m(\pi^0) > m(n) - m(p)$ meant that an accurate determination could also be made of the π^0 mass by studying the energy spectrum of the two decay photons from the π^0 produced in the reaction $\pi^- + p \rightarrow n + \pi^0$ (Panofsky effect).[14] The lifetime for the dominant π^+ decay process, $\pi^+ \rightarrow \mu^+ + \nu$, was easily measured, but only an upper limit could at first be placed on the lifetime for $\pi^0 \rightarrow 2\gamma$, namely, 10^{-14} sec. The spin of π^+ was measured directly by applying detailed balancing to the reactions $\pi^+ + d \rightarrow p + p$.[15] (In Chapter 20 of this volume, Jack Steinberger discusses the pion-spin experiment as one of several important experiments in pion physics carried out during the 1950s.) Parity determinations were made by measuring the branching ratios for the reactions $\pi^- + p \rightarrow n + \gamma$ and

$\pi^- + d \to n + \pi^0$ (in hydrogen) and $\pi^- + d \to 2n$, $\pi^- + d \to 2n + \gamma$, and $\pi^- + d \to 2n + \pi^0$ (in deuterium).[14] That settled the question of the pseudo-scalar character of the charged and neutral pions.

While the program to measure the properties of the charged and neutral pions was being completed, Enrico Fermi and his collaborators were measuring the relative cross sections for the three πN scattering processes: $\pi^+ + p \to \pi^+ + p$, $\pi^- + p \to n + \pi^0$ and $\pi^- + p \to \pi^- + p$ in the energy region 100–200 MeV for the pion.[16] They came up with the striking ratio $9 : 2 : 1$ (within experimental error). Fermi, speaking for the Chicago group, said at Rochester II:[17]

> If one assumes charge independence, i.e., that the isospin is a good quantum number, the two possible isospins, namely $I = \frac{3}{2}$ and $I = \frac{1}{2}$ scatter independently. If moreover, one assumes that the isospin $I = \frac{1}{2}$ does not scatter at all, one gets just the ratio $9 : 2 : 1$; on the other hand, to assume that the $I = \frac{3}{2}$ does not scatter at all, would lead to the ratio $0 : 1 : 2$. This conclusion is independent of angular momentum, spin correlation, or anything else. One can therefore interpret the experimental results by postulating the existence of a broad resonance level $I = \frac{3}{2}$ in the band of energy 100–200 MeV, with the consequence that practically all the scattering comes through $I = \frac{3}{2}$ in this energy region.

Some evidence for this resonance had been found by the Cornell group in measuring the energy dependence of the photoproduction cross section for charged and neutral pions in hydrogen, although the Chicago results appeared to be more convincing at the time. Soon after Rochester II, C. N. Yang pointed out an ambiguity in the phase-shift analysis of the πN scattering experiments, and it took several years before the $I = \frac{3}{2}$, $J = \frac{3}{2}$ πN resonance (i.e., the Δ resonance) was placed on a completely sure footing.[18] (Robert Walker recounts in Chapter 6 of this volume how the Caltech photoproduction experiments, together with those at Cornell, contributed to this clarification.) But it is fair to say that after Rochester II, the concept of isospin invariance of the πN interaction, and consequently the search for other symmetry principles, moved into the forefront of theoretical thinking in particle physics. Moreover, the methods developed to confirm the Δ resonance were used to establish the existence of scores of hadronic resonances in succeeding years.

The determination that the pion was a pseudoscalar particle, that the πN interaction was isospin-invariant, and that the first excited state of N possessed the quantum numbers $I = \frac{3}{2}$, $J = \frac{3}{2}$ seemed to provide a reasonable starting point for a dynamical theory of the strong πN interaction. Hope was high that the successes of QED could somehow be emulated, even though the mediating boson was pseudoscalar rather than vector and the global symmetry was SU(2) rather than U(1). Theorists tried to imitate the renormalization strategy of QED, but, unfortunately, underestimated the important role of gauge invariance in QED. There were some temporary successes (e.g., the Levy potential for the NN interaction), but it soon became apparent that the large value of the πN coupling constant was catastrophic for the perturbation-theoretic approach characteristic of QED. By Rochester IV, Fermi was joking that:[19]

> He had perhaps broken precedent with himself by attending a theoretical session because there was a point that, shall I say, gave me hope. Mr. Goldberger was making a gallant attempt at killing the pseudoscalar theory with pseudoscalar coupling and if such an attempt should succeed it would be a great boon to physics, second only to a definite proof that this theory had something to do with experiment.

Several months later, the Yang–Mills paper was published, but it seemed irrelevant to the pseudoscalar π. It is true that the proceedings of Rochester V record that:[20]

> Yang reported on some considerations he made with [T. D.] Lee. They raised the question of whether it is true that every conservation law is related to a gauge transformation. This has already been discussed in a published paper in connection with isotopic spin conservation. The essential point is that the conservation law is related to invariance under certain transformations, which implies that there is some indeterminacy in the phase. What has to be asked is whether this indeterminateness of phase should have a local character.

But the connection of Yang–Mills with strong interactions was many years off and only entered the picture after the electroweak interaction had put it to good use.

The effort to develop a theory of strong interactions along the lines of QED was abandoned after much travail, and a new tack was taken. Beginning in the mid-1950s, and taking a cue from Werner Heisenberg's S-matrix theory, a theoretical push was made (by Murray Gell-Mann, Marvin L. Goldberger, Lev D. Landau, Geoffrey Chew, Stanley Mandelstam, and others) to identify certain general properties of local field theory – such as crossing, analyticity, and unitarity – that would, hopefully, have a validity beyond the first few orders of perturbation theory. This dispersion–theoretic approach could yield fairly rigorous predictions in strong-interaction physics, provided one was willing to accept a considerable amount of phenomenological input. Much time was spent at the Rochester conferences during the late 1950s discussing various types of dispersion relations and their experimental confirmations. At Rochester VIII, Chew unveiled Mandelstam's generalization of the dispersion relations embodied in a representation of the scattering amplitude as a function of the energy and momentum transfer in the whole two-dimensional manifold. It seemed that Mandelstam's representation, when supplemented by unitarity, could yield a "bootstrap" system of equations for scattering.

Indeed, Chew was so taken with the Mandelstam representation that he argued that S-matrix theory could provide a complete theory of the strong interactions. As he puts it in his book published in 1961:[21]

So that there can be no misunderstanding let me say at once that I believe the conventional association of fields with strongly interacting particles to be empty. I do not have firm convictions about leptons or photons, but it seems to me that no aspect of strong interactions has been clarified by the field concept. Whatever success theory has achieved in this area is based on the unitarity of the analytically continued S-matrix plus symmetry principles.... The general goal then is, given the strong-interaction symmetry principles to make a maximum number of predictions about physical singularities in terms of a minimum amount of information about unphysical singularities....We have absolutely no ideas as to the origin of the strong-interaction symmetries, but we expect that promising developments here can be incorporated directly into the S matrix without reference to the field concept....Which of the strongly interacting particles are elementary?...I am convinced that there can be only one sensible answer, and that is that none of them is elementary.

(The last statement motivates Geoffrey Chew's concept of "nuclear democracy," on which he expands in Chapter 41 of this volume.) Apart from Chew's claim for the independence of S-matrix theory from field theory – which was vigorously debated at the time – S-matrix theory could not explain the origin of the symmetry principles that govern the strong interactions. It was precisely the understanding of symmetry principles that became increasingly important as progress was made in strange-particle physics in the 1950s.

Strange particles

It is no exaggeration to say that the two greatest surprises in particle physics during the decade of the 1950s were the discovery of strange particles and the confirmation of parity violation in weak interactions. It should also be noted that these two surprise developments were not unrelated, as the $\theta-\tau$ dilemma was to demonstrate. The blossoming of strange-particle physics during the early 1950s was a consequence of a stream of initial observations by cosmic-ray physicists, followed by the rapid exploitation of the new results in accelerator experiments. The first indication of the existence of particles in the cosmic radiation heavier than the pion was the observation of two V particles in the cloud chamber by George Rochester and Clifford C. Butler.[22] The hunt began in earnest in 1949, when dozens of V particles were observed in cloud chambers, under special triggering conditions, by the Caltech group.[23] The dramatic debut of the nuclear-emulsion technique as a cosmic-ray detector in 1947, with the discovery of $\pi \to \mu$ decays and the development of electron-sensitive plates a couple of years later, made it possible to search for V particles in nuclear emulsions.[24]

By 1952, a combination of experiments with cloud chambers and nuclear emulsions uncovered a menagerie of these V particles – with masses greater than that of the pion or even the nucleon, having their origin directly in nuclear collisions, and decaying into a great variety of end products with lifetimes not much shorter than the charged-pion lifetime. Representatives of all the groups engaged in this worldwide search were brought together at Rochester II for a thorough discussion of every clue, every shred of evidence that would throw light on the differences and similarities among various incarnations of the *V* particle. (Some facets of this exciting period in cosmic-ray physics are covered by Rochester and Donald Perkins in Chapters 4 and 5 of this volume.)

Throughout Rochester II, the new particles were referred to as "megalomorphs" because they had so much structure and because Fermi "had become bored with the name elementary particles." Oppenheimer introduced the theoretical session on *V* particles at Rochester II in his characteristically succinct and evocative fashion:[25]

> Three alternative explanations have been proposed which may be schematized as (1) the live parent; (2) the heavy brother; and (3) selection rules. The first turned out to be the correct explanation of the π meson. This was the conjecture of Marshak and Bethe that there was a step after production in which the strongly interacting particle could turn into something else that did not interact strongly with nuclear matter. This explanation does not, however, work out well for the *V* particles.

Oppenheimer then called on Abraham Pais to present an account of his "ordering principle for megalomorphian zoology" to explain the long lifetimes of the *V* particles, despite their large production cross section. Pais basically proposed to combine explanations (2) and (3) on Oppenheimer's list, namely, associated production ensured through the conservation of a multiplicative quantum number. As we now know, Pais's multiplicative quantum number had to be replaced by the additive strangeness quantum number of Gell-Mann and Nishijima.

With the completion of the Cosmotron and the Bevatron by the mid-1950s, the center of gravity in strange-particle physics rapidly shifted from cosmic rays to man-made laboratories. Indeed, the ascendance of accelerator over cosmic-ray experiments in strange-particle physics developed so rapidly that at Rochester VI, Robert Leighton remarked that

> next year those people still studying strange particles using cosmic rays had better hold a rump session of the Rochester conference somewhere else.

(Luis Alvarez explains how he helped to bring this about in Chapter 19 of this volume.)

Clarification of the strange-particle situation came rapidly as the accelerator experiments swung into action. As Oppenheimer put it at Rochester V,[26]

> It is only three years ago that we could raise, somewhat foggily, the question of were there selection rules which accounted for the stability of these particles. It is only two years ago that we noticed the rather peculiar structuring of the mean lives of the particles between 10^{-8} and 10^{-10} second. It is only last year, when it was clear enough that the selection rules had some connection with isotopic spin, that we talked about charge doublets and charge

triplets. This year I think there is a new point which will come out, and that is, in addition to the charge degeneracy, there appear to be other degeneracies, or quasi-degeneracies which do have a connection with the theory of the stability of these particles.

In a word, the supporting evidence for the Gell-Mann–Nishijima relation $Q = I_3 + Y/2$ (Q is the charge, I_3 is the third component of the isospin, and Y is the hypercharge) was speedily in hand.

With the Cosmotron and Bevatron in steady operation, it became possible to perform experiments on hyperon–nucleon and kaon–nucleon interactions, and the whole field of strange-particle resonances began to flourish. (Gerson Goldhaber and William Chinowsky give fuller accounts of these experiments in Chapters 16 and 21 of this volume.) All results were consistent with the conservation of isospin and hypercharge [i.e., the group $SU(2)_I \times U(1)_Y$]. Furthermore, the strange-particle resonances provided the underpinning for the larger unifying group $SU(3)_{\text{flavor}}$ [containing $SU(2)_I \times U(1)_Y$ as a subgroup] presented by Y. Ohnuki at Rochester X. I shall discuss later this important consequence of the discovery of strange particles.

Two other major developments in particle physics during the 1950s owe their origin to the discovery of strange particles: (1) the phenomenon of the $K^0-\overline{K^0}$ particle mixture (or $\theta^0-\overline{\theta^0}$ mixture, as it was known in the 1950s) and (2) the $\theta-\tau$ dilemma. I shall only mention the first development in passing. (Val Fitch covers the second subject in Chapter 31 of this volume.) Suffice to say, the Gell-Mann–Pais idea was picked up by Oreste Piccioni, who suggested how the effect could be measured. At Rochester VII, Alvarez reassured his audience that the phenomenon was real:[27]

> Last year there was only one bit of information which was not very strong. But, in the past year the work has been done in several places, both in bubble chambers and in emulsions...the Pais–Piccioni phenomena seem to be in good shape. The Gell-Mann, Pais particle mixture seems to be well confirmed. There was a while when there was a big flap in the business a few months ago when parity and charge conjugation seemed to be on their way out. It looked as though the particle mixture theory was not going to be true but it does seem to be almost true in the way it was originally stated.

It is amazing how much rich physics has been revealed by the neutral-kaon system. I shall be more expansive about the $\theta-\tau$ dilemma, which led to the discovery of parity violation in all weak interactions.

Parity violation and the universal ($V-A$) interaction

As was noted in the last section, the Rochester conferences during the early 1950s devoted a great deal of attention to the similarities and differences among the various types of strange particles – with regard to their masses, lifetimes, and modes of decay. Two strange particles were of particular interest, the θ meson (defined by the 2π decay mode) and the τ

meson (defined by the 3π decay mode). As the θ and τ measurements on the Cosmotron and Bevatron began to accumulate, it became increasingly clear that the masses and lifetimes of θ and τ were identical within experimental error.

When Rochester VI convened, the excitement was high. It was obvious to everyone that if the spins of both θ and τ were zero, θ had to be scalar and τ pseudoscalar. Alvarez reported that T. D. Lee and Jay Orear's suggestion to explain the θ–τ puzzle through the photonic decay of τ into θ was not supported by experiment. Richard Dalitz reported on his analysis of τ decay and the great difficulty of solving the θ–τ puzzle within the framework of parity conservation, even with a high spin (e.g., $J = 2^+$, proposed by E. C. G. Sudarshan and myself).

Some sense of the intensity of the θ–τ debate is conveyed by this account of the session "Theoretical Interpretation of New Particles" at Rochester VI:[28]

> *Yang* felt that so long as we understand as little as we do about the θ–τ degeneracy, it may perhaps be best to keep an open mind on the subject. Pursuing the open mind approach, *Feynman* brought up a question of Block's: Could it be that the θ and τ are different parity states of the same particle which has no definite parity, *i.e.*, that parity is not conserved. That is, does nature have a way of defining right or left-handedness uniquely? *Yang* stated that he and Lee looked into this matter without arriving at any definite conclusions. Wigner. . . has been aware of the possible existence of two states of opposite parity, degenerate with respect to each other because of space–time transformation properties. So perhaps a particle having both parities could exist. But how could it decay, if one continues to believe that there is absolute invariance with respect to space–time transformations? Perhaps one could say that parity conservation, or else time inversion invariance, could be violated. Perhaps the weak interactions could all come from this same source, a violation of space–time symmetries. . . . *Gell-Mann* felt that one should also keep an open mind about possibilities like the suggestion by Marshak. . . that the θ and τ may, without requiring radical assumptions, turn out to be the same particle. . . . The chairman [Oppenheimer] felt that the moment had come to close our minds. . . he suggested: Perhaps some oscillation between learning from the past and being surprised by the future of this [θ–τ dilemma] is the only way to mediate the battle.

The rest is history! After Rochester VI, Lee and Yang proceeded to carry out their comprehensive analysis of the consequences of parity violation in β decay.[29] Within months, the confirmation of parity violation in ^{60}Co by C. S. Wu and associates had been made, and preliminary results supporting parity violation were known for other β emitters and for μ decay.[30] (A more complete account of the decisive parity-violating experiments in weak interactions is given by Valentine L. Telegdi in Chapter 32 of this volume.) These parity-violating experiments – apart from their revolutionary consequences for the future development of particle physics – placed a new set of constraints on the form of the interaction in β decay and in μ decay. In β decay, the parity-violating experiments required a combination of S and T with a right-handed neutrino or a combination of V and A with a left-handed neutrino. For μ decay, a combination of V and A was called for. So far, so good. However, an earlier experiment in β decay that did not measure parity violation, the $e-\nu$ angular correlation experiment in ^{6}He, required the Gamow–Teller interaction to be T.

Clearly, the outlook for a simple universal interaction that would govern all weak processes was not bright, as emphasized by Lee in his introductory talk at the session "Weak Interactions" at Rochester VII:[31]

> We...turn to the universal Fermi interaction, which is an attempt to gain a more unified understanding of certain of the weak interactions. We draw the famous triangle representing the interactions of interest:

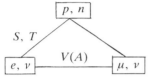

> Beta decay information tells us that the interaction between (p, n) and (e, ν) is scalar and tensor, while the two-component neutrino theory plus the law of conservation of leptons implies that the coupling between (e, ν) and (μ, ν) is vector. This means that the universal Fermi interaction cannot be realized in the way we have expressed it...at this moment it is very desirable to recheck even the old beta interactions to see whether the coupling is really scalar, a point to which we shall return later.

It is interesting that Lee did not question the T interaction, but was open-minded on S or V for the Fermi-type interaction.

This point was reiterated by Wu in her summary talk at Rochester VII, in which she remarked that[32]

> The evidence on the relative strengths of scalar and vector components in the Fermi interaction is no longer so convincing as we previously had thought....The decay of [^{35}Ar] would furnish a much more sensitive test.

Actually, a rumor was circulating at Rochester VII that an Illinois group had measured the electron–neutrino angular correlation coefficient λ from ^{35}Ar (a dominantly Fermi transition) and had found $\lambda = -1$ (as required by the V interaction) instead of $\lambda = +1$ (as required by the S interaction). However, if the β interaction was predominantly V and T, one would be forced to assign opposite helicities to the neutrinos emitted in Fermi- and Gamow–Teller-type β transitions, so that the form of the β interaction would not be simple, and one would still have trouble with μ decay. The answer, of course, was that the ^{6}He experiment was mistaken and that the universal weak interaction is $V-A$.

The full story of how Sudarshan and I arrived at this conclusion by the time of Rochester VII, but held up its announcement for several months because of the V, T confusion in β decay, was told last year at the Wingspread Conference, and I shall only repeat several relevant points here.[33] (Sudarshan gives his personal reminiscences about $V-A$ in Chapter 33 of this volume.) Somewhat delayed, the paper entitled "The Nature of the Four-Fermion Interaction," by Sudarshan and myself, was submitted in July 1957 to the Padua–Venice conference ("Mesons and Newly Discovered Particles"). In our paper, we stated that the only possible candidate for a universal weak interaction was $V-A$ with a left-handed neutrino. When I gave the talk about the $V-A$ theory in Padua in September 1957, I brashly proclaimed that the universal $V-A$ interaction could survive only if four experiments were "murdered": $e-\nu$ angular correlation from ^{6}He decay, the $\pi \to e\nu/\pi \to \mu\nu$ ratio, the sign of polarization of e from μ decay, and electron asymmetry from oriented-neutron decay. I recall being castigated by several experimental colleagues for theoretical arrogance, but the fact remains that all four experiments turned out to be wrong.

By the time of Rochester VIII (held at CERN – the first outside Rochester), three of the four experiments (all except the $\pi \to e\nu/\pi \to \mu\nu$ ratio) had been redone, and the measurements were in full accord with the $V-A$ theory. In addition, Maurice Goldhaber and collaborators had made a direct determination of the neutrino helicity – and found it to be left-handed – thereby giving strong support to the $V-A$ theory.[34] When Goldhaber gave the summary talk on weak-interaction experiments involving leptons at Rochester VIII, he devoted most of his time to the new measurements on ^{6}He, electron polarization from μ decay and electron asymmetry from oriented-neutron decay, and his own ingenious experiment on the neutrino helicity. This led Oppenheimer, in his concluding remarks at Rochester VIII, to state that[35]

> I think that the most productive of the experimental developments, between last year's conference and this, has been the revelation of the extent to which, within the framework of the strangeness-conserving weak interactions, a reasonable universal interaction describes what is found. The beautiful report of Goldhaber made that clear. Now this has really changed the situation, because this very success has enabled one to ask a great many questions which a year ago we would have been unable to formulate.

By the time of Rochester IX (held in Kiev – the first in the Soviet Union – under the rotation scheme adopted by the IUPAP commission on high-energy physics), the $\pi \to e\mu/\pi \to \mu\nu$ ratio had been remeasured, and the result was again in striking agreement with $V-A$ theory. The $V-A$ interaction with baryon and lepton conservation became the accepted theoretical framework for understanding all weak processes. One could now turn to more subtle questions in weak-interaction physics: the failure to observe $\mu \to e + \gamma$, the greater strength of A compared with V interaction, the form-factor effects in muon absorption by protons, the origin of the $\Delta Q/\Delta S = 1$ rule and the reduced strength of the strangeness-changing (compared with the strangeness-conserving) hadronic current in semileptonic decays, and some unexpected branching ratios in nonleptonic decays.[36] I was the theoretical rapporteur for the session on weak interactions at Rochester IX and concluded my talk by pointing out that the absence of most unobserved weak processes satisfying all the conservation laws could be explained by postulating the coupling of an intermediate (charged) vector boson (IVB) with zero baryon number and zero lepton number to the hadron and lepton currents.

However, the IVB hypothesis could not explain the absence of $\mu \rightarrow e + \gamma$, unless one was willing to accept the existence of distinct μ and e neutrinos. Bruno M. Pontecorvo emphasized this point when he commented, at Rochester IX,[37] that

> The intermediate boson in weak interactions would be a nice thing to have, in order to explain that the weak interaction current is a charge exchange one... the negative experiment on the $\mu \rightarrow e + \gamma$ process is difficult to reconcile with the existence of the IVB meson but this difficulty is present only if there exists one pair of neutrinos. If there are two pairs of neutrinos...there are no arguments against the existence of [IVB] mesons.

But if the two-neutrino hypothesis was correct, I remarked at the end of my rapporteurial talk,[38]

> It would destroy a useful symmetry principle [the baryon–lepton symmetry principle in weak interactions] which Gamba, Marshak and Okubo[39] have called attention to....I believe that the symmetry principle may really be significant and it evidently can not be reconciled with the two-neutrino theory (because two types of protons would be necessary).

The absence of a second "proton" at that time biased me against a second neutrino!

In 1964, after the discovery of ν_μ, the baryon–lepton symmetry principle was revived by James D. Bjorken and Sheldon L. Glashow.[40] In 1970, after the quark was firmly established, quark–lepton symmetry (into which baryon–lepton symmetry had evolved) was invoked to predict the existence of the charmed quark and to motivate the GIM mechanism that proved so successful in explaining the absence of strangeness-changing neutral weak decays.[40]

SU(3)$_\text{flavor}$, baryon–lepton symmetry compositeness of hadrons

By the mid-1950s, enough experimental data had been accumulated concerning the strange particles to feel confident about the enlarged SU(2)$_I$ × U(1)$_Y$ strong-interaction group with its accompanying Gell-Mann–Nishijima relation $Q = I_3 + Y/2$. The Fermi–Yang model of the pion had been around for some time, and it seemed reasonable to inquire if some subset of the strongly interacting particles could explain the properties of all the other observed particles, at least as far as the quantum numbers were concerned.[41] Because the strong-interaction group was as large as SU(2)$_Y$ × U(1)$_Y$, it was clear that an elegant choice would be (p, n, Λ), which is a combination of a nonstrange baryon isodoublet and a baryon isosinglet of strangeness $= -1$. That was the Sakata model, first published in 1956.[42] It was straightforward to construct the known mesons and hyperons (as of 1956) out of p, n, Λ and their antiparticles. The Sakata model also had some predictive power in weak interactions; for example, it predicted an $I = \frac{1}{2}$ strangeness-changing weak current, so that the $\Delta Q/\Delta S = 1$ rule was automatically obeyed. It also led directly to SU(3)$_{\text{flavor}}$ as well as to the baryon–lepton symmetry principle. I shall briefly discuss the last two consequences of the Sakata model.

One of the highlights of Rochester X was the paper by Ohnuki, "Theories of Elementary Particles," which was given at the last plenary session of the conference. Pointing out that the Sakata triplet consisted of three spin-$\frac{1}{2}$ baryons with masses fairly close to each other, Ohnuki stated that[43]

> We shall introduce the following assumption: In the limit of m_N and m_Λ, the theory satisfies the invariance under exchanges of Λ and proton, and of Λ and neutron in addition to the usual charge independence and conservation of the baryon number. Then it is easily seen that this is equivalent to the following statement: the theory is invariant under the transformation
>
> $$\chi_i'(x) = \sum_{j=1}^{3} U_{ij}\chi_j(x), \qquad (i = 1, 2, 3)$$
> $$\chi_1(x) = p(x), \qquad \chi_2(x) = n(x), \qquad \chi_3(x) = \Lambda(x) \tag{3.1}$$
>
> in the limiting case of $m_N = m_\Lambda$, where $p(x)$, $n(x)$, $\Lambda(x)$ stand for the proton, neutron, Λ particle fields respectively and U is an arbitrary unitary matrix of degree three. Thus if the irreducible representation of this group U(3) is obtained, all states belonging to it should have the same energy, spin and parity in the limit of $m_N = m_\Lambda$. So by examining the properties of U(3), we shall make clear some aspects of the kinematical features of Sakata's model.

Ohnuki then proceeded to write down the octet and singlet representations of the composite meson system, as well as the 15, 3, $\bar{6}$, 3′ representations of the composite baryon system that followed from SU(3).

The seven known pseudoscalar particles (π, K^-, and K) fitted nicely into the octet of meson states, and it followed that a new pseudoscalar particle, called $\pi^{0\prime}$ (later named η) should exist. The assignment of the known eight baryons (N, Λ, Σ, and Ξ) was less successful, especially since the spin of Ξ was not known at that time.

Abdus Salam was present at Rochester X and told his graduate student, Yuval Ne'eman, about Ohnuki's talk.[44] Ne'eman was to write later in his book with Gell-Mann that[45]

> in 1959–60, Ikeda, Ogawa and Ohnuki in Japan and Thirring, Wess, and Yamaguchi in Europe found a proper mathematical phrasing of the symmetry aspect of the construction process in this model. This was the group SU(3), with an eight-parameter algebra, whose representations would correspond to the structures allowed within the Sakata model. It assumed invariance of the particle-building process with respect to unitary unimodular

transformations of the three "fundamental" complex fields, identified with Sakata's (p, n, Λ) set. The mass spectrum was assumed to derive from the propagation of the $\Lambda-N$ mass difference.*

Evidently, Ohnuki's talk at Rochester X contributed importantly to the articulation of the "eightfold way." The subsequent development of the quark model went beyond the eightfold way and received independent impetus from the baryon–lepton symmetry principle, as we shall see.

The reason that the Sakata model could serve as a stepping-stone to the very important $SU(3)_{flavor}$ group stems from the fact that Sakata made a judicious choice of the carriers of the group-theoretic quantum numbers characterizing the strong interaction. Indeed, the Sakata triplet, p, n, Λ baryons are the lowest-mass states of the three-quark composites $(ud)u$, $(ud)d$, $(ud)s$. This is, of course, the reason why the Sakata and quark models of $SU(3)_{flavor}$ yield the same predictions for the meson states but not for the baryon states. For similar reasons, the baryon–lepton symmetry principle in weak interactions,[39] based as it was on the Sakata triplet, translates immediately into quark–lepton symmetry, with wide ramifications in present-day particle physics. I alluded to baryon–lepton symmetry in the last section, but did not really explain it. Let me do so.

In the paper "On a Symmetry in Weak Interactions," published in 1959, A. Gamba, Marshak, and S. Okubo made the following observation:[39]

> The (Λ, n, p) baryon triplet in order of decreasing mass appears to be the minimum number of strongly interacting particles which are necessary to explain conservation of charge, isotopic spin, and strangeness (in strong interactions). In weak interactions, conservation of isotopic spin and strangeness no longer hold and the baryon triplet (Λ, n, p) bears a striking similarity to the lepton triplet (μ, e, ν) in several respects. We postulate the following symmetry principle: All weak interactions are invariant under the following simultaneous transformation:
>
> $$\Lambda \to \mu, \qquad n \to e, \qquad p \to \nu \ldots \tag{A}$$
>
> We have found that the symmetry principle (A) between the baryon triplet (Λ, n, p) and the lepton triplet (μ, e, ν) is completely consistent with existing experiments. Favorable interactions lead to favorable interactions and the negative is also true. The success of principle (A) makes it tempting to introduce formally the notion of isotopic spin and weak hypercharge for leptons. Assigning isotopic spin one-half for the (e, ν) pair and zero isotopic spin for the muon, the charge Q for any member of the baryon and lepton triplet can be written as:
>
> $$Q = I_3 + (T + B - L)/2 \tag{B}$$
>
> This formula holds for both baryons and leptons if we assign $T = -1$ for the muon and $T = 0$ for the electron and neutrino, and if we identify T with the strangeness S in the case of the baryon.

* Reproduced, with permission, from M. Gell-Mann and Y. Ne'eman, *The Eightfold Way.* © 1964 by Addison-Wesley Publishing Company.

Note how useful the Sakata triplet was in establishing the correspondence with the known lepton triplet (we recall that there were only one neutrino and two charged leptons in 1959 and no quarks). Because the leptons are not strongly interacting, they could immediately be regarded as the carriers of the weak (and electromagnetic) interactions, just as the Sakata triplet embodied the group-theoretic properties of the strong interaction. Note also how Sakata's choice of the "elementary" baryons facilitated the introduction of the concepts of weak isospin and weak hypercharge. The translation of the Sakata triplet into the u, d, s quark triplet, the reinterpretation of the weak Gell-Mann–Nishijima relation (A) in terms of quarks and leptons, the enlargement of these ideas to the three generations of quark and lepton doublets, and finally the application of modern gauging strategy to $B-L$ lead to interesting physical results that are summarized elsewhere.[46]

Although the full impact of so-called Kiev symmetry (the name given to baryon–lepton symmetry because it was fully aired at Rochester IX in Kiev) was not felt until later, it focused attention at the time on the minimal number of fields required to construct a viable model of the observed particles, and it served as an antidote to the "bootstrap" approach to strong interactions. Two quotes from Gell-Mann, who was so active during that period in both strong- and weak-interaction physics, are relevant in this regard. Thus, in his Caltech paper on "The Eightfold Way: A Theory of Strong Interaction Symmetry," Gell-Mann uses the lepton triplet as a paradigm for $SU(3)_{flavor}$:[45]

> If we now "turn off" the μ-e mass difference, electromagnetism, and the weak interactions, we are left with a physically vacuous theory of three exactly similar Dirac particles with no rest mass and no known couplings. This empty model is ideal for our mathematical purposes, however, and is physically motivated by the analogy with the strongly interacting particles, because it is at the corresponding stage of total unitary symmetry that we shall introduce the basic baryon mass and the strong interactions of baryons and mesons.*

Later (in note 21 of the same paper) he allows that

> Earlier attempts to draw a parallel between leptons and baryons in the weak interactions have been made by A. Gamba, R. E. Marshak, and S. Okubo, *Proc. Nat. Acad. Sci. 45*, 881 (1959), and Y. Yamaguchi, unpublished. Dr. S. L. Glashow reports that Yamaguchi's scheme has much in common with the one discussed in this paper.*

Gell-Mann's Catalunya talk is even more interesting in revealing the extent to which the point of view expressed in Kiev symmetry paralleled his own thinking concerning the *S*-matrix theory. In the Catalunya talk, Gell-Mann comments that[47]

---

* Reproduced, with permission, from M. Gell-Mann and Y. Ne'eman, *The Eightfold Way.* © 1964 by Addison-Wesley Publishing Company.

> In 1963, when I developed the quark proposal, with confined quarks, I realized that the bootstrap idea for hadrons and the quark idea with confined quarks can be compatible with each other, and that both proposals result in "nuclear democracy," with no observable hadron being any more fundamental than any other. . . . What worried me about the bootstrap was that it distinguished hadrons sharply from leptons, while the weak and electromagnetic interactions treated them nearly alike. A feature that I liked about the quarks (the current quarks), when I found them, was that they presented an analogy with the leptons, in electromagnetic and weak interactions. . . . In other words, in discovering quarks we have found hadron building blocks that are just as elementary as the leptons. At very small distances, of course, there might turn out to be constituents for both, but surely the quarks and leptons are equally elementary or equally composite.

It would be tempting to continue the story of how global $SU(3)_{\text{flavor}}$ was augmented by local $SU(3)_{\text{color}}$ (Gell-Mann does this in Chapter 47 of this volume) and how quark–lepton symmetry has been extended in a variety of directions. However, that would take us too far away from the decade in the history of particle physics that is under review.

I should like to conclude by mentioning the early efforts of Heisenberg and Nambu to come to grips with the compositeness of hadrons. In the last session, "Theories of Elementary Particles," at Rochester X [the same session at which Ohnuki spoke about the newly invented $SU(3)_{\text{flavor}}$], Heisenberg and Nambu gave major presentations of their respective composite models of hadrons.[48]

Both speakers proposed to deduce finite hadron masses from the self-interaction of a massless field (or fields). They both looked for nonperturbative solutions of the nonlinear (four-fermion) field equations with which they started. They both built additional symmetries into the Lagrangian (besides Lorentz invariance, etc.), understood the difference between the symmetry of the Lagrangian and the symmetry of the vacuum, and recognized that the degeneracy of the vacuum permits asymmetric solutions corresponding to a symmetric Lagrangian. However, Heisenberg started with one massless Dirac spinor (really two Weyl spinors) and argued that isospin could be identified as a hidden symmetry of the field equations, strangeness could be deduced from the asymmetry of the vacuum, and the discrete transformations could be associated with a new type of "scale" transformation. Furthermore, he was willing to introduce an indefinite metric into Hilbert space to avoid an arbitrary cutoff in the expression for the finite hadron mass. Some of Heisenberg's ideas have been carried over to present-day gauge theories and their spontaneous symmetry breakings, but the detailed program has been abandoned in favor of the new gauge theories.

Nambu, in his 1960 talk, was more limited in his objectives. He was willing to introduce isospin invariance and strangeness ad hoc (by working with massless Dirac fields) and accepted an arbitrary cutoff in his model "nonperturbative" solution. His emphasis was on pushing the analogy between the chirality invariance of his nonlinear field equation, the finite-mass solution, and the zero-mass pseudoscalar boson (pion), on the one hand, and (ordinary) gauge invariance, the energy gap, and collective oscillations in superconductivity, on the other. As a result, Nambu gave dynamic content to chirality invariance (Sudarshan and I had used "chirality invariance" in a "kinematical" way to derive the $V-A$ interaction[33]) and came upon the "Nambu–Goldstone" boson (before Goldstone), which plays a key role in QCD and other branches of particle theory. It is worth quoting Nambu's key statement at Rochester X:[49]

> It is an old and attractive idea that the mass of a particle is a self-energy due to interaction. According to the present analogy, it will come about because of some attractive correlation between massless bare particles, and will be determined in a self-consistent way rather than by simple perturbation. Since a free massless fermion conserves chirality, let us further assume that the interaction also preserves chirality invariance, just as the electron-phonon system preserves gauge invariance [in superconductivity]. Then if an observed fermion (quasi-particle) can have a finite mass, there should also exist collective excitations of fermion pairs. Such excitations will behave like bosons, of zero fermion number, so that they may be called mesons. They will play the role of preserving the overall conservation of chirality, and from this we will be able to infer that they are pseudoscalar mesons, like the pions found in nature.

This prescient statement by Nambu is a good way to end this personalized account of the highlights of a heroic decade in particle physics as seen through the prism of the Rochester conferences.

Notes

1 See S. S. Schweber: "The Shelter Island Conferences were the precursors of the Rochester Conferences. But they differed in important ways from the later Rochester conferences: they were small, closed and elitist in spirit, in contrast to the more professional and democratic outlook of the Rochester conferences. In a sense they mark the postponed end of an era, that of the thirties, with its characteristic style of doing physics: small groups and small budgets. The Rochester conferences mark the beginning of the new era: the large group efforts and the large budgets involved in machine physics. Also, whereas Shelter Island and Pocono looked upon QED as a self-contained discipline, the Rochester conferences witnessed 'particle physics' coming into its own, with QED as one of its subfields – albeit one in a privileged, paradigmatic position." Schweber, "Some Chapters for a History of Quantum Field Theory 1938–1952," in *Relativity, Groups and Topology, II*, edited by B. S. DeWitt and R. Slora (Amsterdam: North Holland, 1984), pp. 37–220.
2 R. E. Marshak, "The Rochester Conferences: The Rise of International Cooperation in High Energy Physics," *Bull. Atomic Scientists 26* (1970), 92–8.
3 C. L. Oxley and R. D. Schamberger, "Proton–Proton Scattering at 240 MeV," *Phys. Rev. 85* (1952), 416–23; O. Chamberlain, E. Segrè, R. D. Tripp, C. Wiegand, and T. Ypsilantis, "Experiments with 315-MeV Polarized Protons: Proton–Proton and Proton–Neutron Scattering," *Phys. Rev. 105* (1957), 288–301.
4 L. D. Wolfenstein, "Possible Triple-Scattering Experiments," *Phys. Rev, 96* (1954), 1654–8.
5 S. Okubo and R. E. Marshak, "Velocity Dependence of the Two-Nucleon Interaction," *Ann. Phys. (N.Y.) 4* (1958), 166–79.

6 P. S. Signell and R. E. Marshak, "Phenomenological Two-Nucleon Potential Up to 150 MeV," *Phys. Rev. 106* (1957), 832–4; also, J. L. Gammel and R. M. Thaler, "Spin-Orbit Coupling in the Proton–Proton Interaction," *Phys. Rev. 107* (1957), 291–8.

7 K. M. Case and A. Pais, "On Spin-Orbit Interactions and Nucleon–Nucleon Scattering," *Phys. Rev. 80* (1950), 203–11; S. Gartenhaus, "Two Nucleon Potential from the Cut-Off Yukawa Theory," *Phys. Rev. 100* (1955), 900–5.

8 See R. A. Bryan and B. L. Scott, "Nucleon–Nucleon Scattering from One-Boson Exchange Potentials," *Phys. Rev. 135* (1964), B434–50.

9 R. Hofstadter, "Scattering of High Energy Electrons from Hydrogen, Deuterium and Helium," in *High Energy Nuclear Physics. Proceedings of the Sixth Annual Rochester Conference April 3–7, 1956*, edited by J. Ballam, V. L. Fitch, T. Fulton, K. Huang, R. R. Rau, and S. B. Treiman (New York: Interscience, 1956), pp. IX-1–7, especially IX-6.

10 Y. Nambu, "Speculations on a New Neutral Meson," in *High Energy Nuclear Physics. Proceedings of the Seventh Annual Rochester Conference*, edited by G. Ascoli et al. (New York: Interscience, 1957), p. I-26; R. F. Peierls, discussion, ibid., p. I-28.

11 Murray Gell-Mann, discussion remark to S. D. Drell, "Nuclear Structure. Theoretical II," in *1958 Annual International Conference on High Energy Physics at CERN*, edited by B. Ferretti (Geneva: CERN, 1958), p. 33.

12 E. Segrè, "Introductory Survey" [Antinucleons Session], *High Energy Nuclear Physics. Proceedings of the Sixth Annual Rochester Conference, April 3–7, 1956*, edited by J. Ballam, V. L. Fitch, T. Fulton, K. Huang, R. R. Rau, and S. B. Treiman (New York: Interscience, 1956), p. VII-1.

13 A. M. Thorndike, *Mesons* (New York: McGraw-Hill, 1952); Robert E. Marshak, *Meson Physics* (New York: McGraw-Hill, 1952); Robert E. Marshak, "Particle Physics in Rapid Transition: 1947–1952," in *The Birth of Particle Physics*, edited by L. M. Brown and L. Hoddeson (Cambridge University Press, 1983), pp. 376–401.

14 W. K. H. Panofsky, R. L. Aamodt, and J. Hadley, "The Gamma-Ray Spectrum Resulting from Capture of Negative π-Mesons in Hydrogen and Deuterium," *Phys. Rev. 81* (1951), 565–74.

15 D. L. Clark, A. Roberts, and R. Wilson, "Cross Section for the Reaction $\pi^+ + d \rightarrow p + p$ and the Spin of the π^+ Mesons," *Phys. Rev. 83* (1951), 649; R. Durbin, H. Loar, and J. Steinberger, "The Spin of the Pion via the Reaction $\pi^+ + d \rightleftharpoons p + p$," ibid., 646–8. (The Rochester measurement preceded the Columbia measurement.)

16 H. Anderson, E. Fermi, R. Martin, and D. Nagle, "Angular Distribution of Pions Scattered by Hydrogen," *Phys. Rev. 91* (1953), 155–68.

17 E. Fermi, in "Proceedings of Second Rochester Conference, Jan. 1952," unpublished mimeograph, edited by A. M. L. Messiah and H. P. Noyes, p. 26.

18 S. J. Lindenbaum, *Particle-Interaction Physics at High Energies* (Oxford: Clarendon Press, 1973).

19 E. Fermi, "Impression of the Theoretical Session," in *Proceedings of the Fourth Annual Rochester Conference on High Energy Nuclear Physics, January 25–27, 1954*, edited by H. P. Noyes, E. M. Hafner, J. Klarmann, and A. E. Woodruff (Rochester: University of Rochester, 1954), p. 92.

20 C. N. Yang and R. L. Mills, "Conservation of Isotopic Gauge Invariance," *Phys. Rev. 96* (1954), 191–5; C. N. Yang, "Gauge Transformation Connected with Conservation of Heavy Particles," in *High Energy Nuclear Physics. Proceedings of the Fifth Annual Rochester Conference, January 31–February 2, 1955*, edited by H. P. Noyes, E. M. Hafner, G. Yekutieli, and B. J. Raz (New York: Interscience, 1955), p. 66.

21 G. F. Chew, *S. Matrix Theory of Strong Interactions* (New York: Benjamin, 1967).

22 G. D. Rochester and C. C. Butler, "Evidence for the Existence of New Unstable Elementary Particles," *Nature (London) 160* (1947), 855–7.

23 A. J. Seriff, R. B. Leighton, C. Hsiao, E. W. Cowan, and C. D. Anderson, "Cloud-Chamber Observations of the New Unstable Cosmic-Ray Particles," *Phys. Rev. 78* (1950), 290–1.

24 C. M. G. Lattes, H. Muirhead, G. P. S. Occhialini, and C. F. Powell, "Processes Involving Charged Mesons," *Nature (London) 159* (1947), 694–7.

25 J. R. Oppenheimer, in "Proceedings of Second Rochester Conference, Jan. 1952," unpublished mimeograph, edited by A. M. L. Messiah and H. P. Noyes, p. 85.

26 J. R. Oppenheimer, "Introductory Remarks" [Elementary Particles Session], *High Energy Nuclear Physics. Proceedings of the Fifth Annual Rochester Conference January 31–February 2, 1955*, edited by H. P. Noyes, E. M. Hafner, G. Yekutieli, and B. J. Raz (New York: Interscience, 1955), p. 124.

27 L. Alvarez, "Introductory Remarks" [Strange Particle Interactions Session], *High Energy Nuclear Physics. Proceedings of the Seventh Annual Rochester Conference*, edited by G. Ascoli et al. (New York: Interscience, 1957), p. VI-10.

28 J. Ballam, V. L. Fitch, T. Fulton, K. Huang, R. R. Rau, and S. B. Treiman (eds.), *High Energy Nuclear Physics. Proceedings of the Sixth Annual Rochester Conference, April 3–7, 1956* (New York: Interscience, 1956), pp. VIII-27–9.

29 T. D. Lee and C. N. Yang, "Question of Parity Conservation in Weak Interactions," *Phys. Rev. 104* (1956), 254–8.

30 C. S. Wu, E. Ambler, R. W. Hayward, D. D. Hoppes, and R. R. Hudson, "Experimental Test of Parity Conservation in Beta Decay," *Phys. Rev. 105* (1957), 1413–15. For μ decay, see R. Garwin, L. Lederman, and M. Weinrich, *Phys. Rev. 105* (1957), 1415.

31 T. D. Lee, "Introductory Talk," in *High Energy Nuclear Physics. Proceedings of the Seventh Annual Rochester Conference*, edited by G. Ascoli et al. (New York: Interscience, 1957), p. VII-7.

32 C. S. Wu, "Summary Talk," in *High Energy Nuclear Physics. Proceedings of the Seventh Annual Rochester Conference*, edited by G. Ascoli et al. (New York: Interscience, 1957), p. VII-22.

33 E. C. G. Sudarshan and R. E. Marshak, "Origin of the Universal $V-A$ Theory," in *50 Years of Weak Interactions, Wingspread Conference 1984*, edited by David B. Cline and Gail M. Riedasch (Madison: University of Wisconsin, 1986), pp. 1–14. Our original $V-A$ paper was published in the proceedings of the Padua–Venice conference on "Mesons and Recently Discovered Particles" (1957, p. V-14) [reprinted in P. K. Kabir, *Development of Weak Interaction Theory* (New York: Gordon & Breach, 1963), p. 118]. The paper by R. P. Feynman and M. Gell-Mann, "Theory of the Fermi Interaction," was published in *Phys. Rev. 109* (1958), 193. Ironically, the scientific impact of Rochester VII turned out to be negative in the case of the $V-A$ theory, for two reasons: The Rochester conference had become the principal forum for presentation of new results in particle physics, and its proceedings the chief communications channel. Thus, our decision to postpone the announcement of our $V-A$ theory to the Padua-Venice conference, held in September 1957 and sandwiched between two Rochester conferences, was aggravated by the delayed publication of our original paper in the proceedings of that conference. It appears that most physicists only read the brief follow-up note by Sudarshan and myself on "Chirality Invariance and the Universal Fermi Interaction" published in *Phys. Rev. 109* (1958), 1860–2, after Feyman and Gell-Mann's excellent paper and soon found themselves referring to the $V-A$ theory as the Feynman–Gell-Mann theory. In fact, Feynman has acknowledged [*Theory of Fundamental Processes* (1962), p. 158] that our $V-A$ proposal preceded the Feynman–Gell-Mann $V-A$ proposal.

34 M. Goldhaber, L. Grodzins, and A. Sunyar, "Helicity of Neutrinos," *Phys. Rev. 109* (1958), 1015–17.

35 J. R. Oppenheimer, "Concluding Remarks," in *1958 Annual International Conference on High Energy Physics at CERN*, edited by B. Ferretti (Geneva: CERN, 1958), pp. 291–4.

36 Solved by the Cabibbo angle in 1963: N. Cabibbo, "Unitary Symmetry and Leptonic Decays," *Phys. Rev. Lett. 10* (1963), 531–3.

37 B. M. Pontecorvo, Discussion remark following Plenary Session VIII-a, in *Ninth Annual International Conference on High Energy Physics*, Vol. 2 (Moscow: Academy of Science USSR, 1960), pp. 233–6.

38 R. E. Marshak, "Theoretical Status of Weak Interactions," in *Ninth Annual International Conference on High Energy Physics*, Vol. 2 (Moscow: Academy of Science USSR, 1960), pp. 269–305, especially p. 297.

39 A. Gamba, R. E. Marshak, and S. Okubo, "On a Symmetry in Weak Interactions," *Proc. Nat. Acad. Sci. U.S.A. 45* (1959), 881–5. See also Y. Yamaguchi, *Prog. Theor. Phys. (Suppl.) 11* (1951), 1.

40 J. D. Bjorken and S. L. Glashow, "Elementary Particles and SU(4)," *Phys. Letters (Netherlands) 11* (1964), 255–7; S. L. Glashow, J. Illiopoulos, and L. Maiani, "Weak Interactions with Lepton–Hadron Symmetry," *Phys. Rev. D2* (1970), 1285–92.

41 E. Fermi and C. N. Yang, "Are Mesons Elementary Particles?" *Phys. Rev. 76* (1949), 1739–43.

42 S. Sakata, "On a Composite Model for the New Particles," *Prog. Theor. Phys. 16* (1956), 686–8.

43 Y. Ohnuki, "Composite Model of Elementary Particles," in *Proceedings of the 1960 Annual International Conference on High Energy Physics at Rochester*, edited by E. C. G. Sudarshan, J. H. Tinlot, and A. C. Melissinos (New York: Interscience, 1960), 843–50; S. Ogawa, "A Possible Symmetry in Sakata's Composite Model," *Prog. Theor. Phys. 21* (1959), 209–11; M. Ikeda, S. Ogawa, and Y. Ohnuki, "A Possible Symmetry in Sakata's Model for Bosons–Baryons System," *Prog. Theor. Phys. 22* (1959), 715–24.

44 Salam was aware of the Japanese work on $SU(3)_{flavor}$ as early as 1959; see his summary talk "Strange Particle Theory," in *Ninth Annual International Conference on High Energy Physics*, Vol. 2. (Moscow: Academy of Science USSR, 1960), pp. 540–76, especially 573.

45 M. Gell-Mann and Y. Ne'eman, *The Eightfold Way* (New York: Benjamin, 1964).

46 R. E. Marshak and R. N. Mohapatra, "Whither Grand Unification: Experimental Tests of $B-L$ Gauge Groups," in *From SU(3) to Gravity: Festschrift in Honor of Yuval Ne'eman*, edited by E. Gotsman and G. Tauber (Cambridge University press, 1985), pp. 173–81.

47 M. Gell-Mann, "Particle Theory from *S*-Matrix to Quarks," in *Symmetries in Physics (1600–1980), Proceedings of the First International Meeting on the History of Scientific Ideas*, edited by M. G. Doncel, A. Hermann, L. Michel, and A. Pais (Barcelona: Bellaterra, 1987), pp. 474–97.

48 W. Heisenberg, "Recent Research on the Nonlinear Spinor Theory of Elementary Particles," in *Proceedings of the 1960 Annual International Conference on High Energy Physics at Rochester*, edited by E. C. G. Sudarshan, J. H. Tinlot, and A. C. Melissinos (New York: Interscience, 1960), pp. 851–7; Y. Nambu, "Dynamical Theory of Elementary Particles Suggested by Superconductivity," pp. 858–66.

49 Y. Nambu, note 48 (p. 859); Nambu had already suggested these ideas in a preliminary way in 1959; see discussion remark following Bruno Touschek, "Remarks on the Neutrino Gauge Group," in *Ninth Annual International Conference on High Energy Physics*, Vol. 2 (Moscow: Academy of Science USSR, 1960), pp. 117–19.

46   Some reflections on the history of particle physics in the 1950s

SILVAN S. SCHWEBER

Born 1928, Strasbourg, France; Ph. D., Princeton University, 1952;
theoretical physics and history of science; Brandeis University

Introduction

Even a superficial comparison of the twelfth Solvay conference on physics, held in Brussels in October 1961 to discuss the quantum theory of fields,[1] with the eighth Solvay conference, held in September 1948 to discuss elementary particles, highlights the dramatic changes in particle physics that had taken place in the decade of the 1950s.[2]

The rapporteurs at the later conference were Richard Feynmen, Murray Gell-Mann, Marvin L. Goldberger, Walter Heitler, Gunnar Källen, Stanley Mandelstam, Abraham Pais, and Hideki Yukawa, all of whom (except for Heitler and Yukawa) received their graduate training in the 1940s. The list of invited members included a large number of younger theoreticians: Geoffrey Chew, Marcello Cini, Freeman Dyson, Tsung Dao Lee, Francis Low, Harry Lehmann, Yoichiro Nambu, Abdus Salam, Julian Schwinger, Leon Van Hove, Arthur Wightman, and Chen Ning Yang. This was in marked contrast with the 1948 conference (organized in the fall of 1947, and scheduled to be held in April 1948).[3] There, some thirty-nine physicists were invited – the youngest being Homi J. Bhabha, who was thirty-eight at the time. There were few Americans on the invited list, and all these were émigrés[4] (besides Albert Einstein, only Hans Bethe, Felix Bloch, Enrico Fermi, Bruno Rossi, and Edward Teller), because the resources of the Solvay Institute could not cover travel expenses from the United States.[5] Everyone invited was a member of the prewar élite – among the theoreticians were Niels Bohr, Paul A. M. Dirac, Hans A. Kramers, Oscar Klein, Wolfgang Pauli, Rudolph Peierls, and Erwin Schrödinger. When it was discovered that the April date interfered

P. Galison, J. Heilbron, E. F. Keller, F. Low, F. Manuel, A. Pickering, and H. Schnitzer made helpful criticisms of a first draft of this chapter. I thank them.

with other commitments of many of the participants, Sir Lawrence Bragg, the president of the Conseil de Physique Solvay, rescheduled the meeting to late September 1948.[6] Several new participants were asked to give reports (in particular, Robert Serber and Robert Oppenheimer), and an invitation was also sent to Schwinger (who declined). The youngest persons in attendance were Bruno Ferretti and Serber.

In addition to the marked differences in age and national representation between the participants at the two meetings, there was also a difference in tone. Even though at the earlier meeting Serber gave an account of recent findings from the 184-inch Berkeley synchrocyclotron, and Bloch reported on latest measurements of nuclear magnetic moments for the proton, neutron, and deuteron at Stanford and Columbia – experiments performed on apparatus and machines made possible by the wartime developments in microwave technology and electronics – the intellectual atmosphere was very similar to that at meetings held before World War II. It was a small élite conference – very much resembling in conception and execution the one held in Warsaw in the spring of 1938,[7] assembled by the Institut International de Cooperation Intellectuelle, "Frontiers of Physics." The emphasis was on "fundamental" matters, often divorced from experiments.

Bohr opened the 1948 meeting with a lecture on epistemological problems raised in atomic physics, "On the Notions of Causality and Complementarity." Most of the theoretical presentations were formal and abstract: Heitler discussed his theory of damping, a "heuristic attempt at eliminating the infinities from present quantum theory of fields."[8] Kramers presented his reformulation of the quantum electrodynamics of extended nonrelativistic charged particles, in which only structure-independent parameters (i.e., the renormalized mass and charge) of the particles appeared in the equations. Oppenheimer reviewed the formal developments that had taken place since the Shelter Island conference of June 1947. Peierls presented formal aspects of the self-energy problem in QED. Bhabha reported on relativistic wave equations for higher-spin particles. Louis de Broglie discussed his approach to the description of higher-spin particles, the method of fusion. Except for Kramers, all were reporting, for the most part, on work done by *others*.[9]

On the other hand, the reports in 1961 were delivered by the principal contributors: Feynman, Pais, Gell-Mann, Goldberger, Mandelstam. All the reports by American contributors were dominated and propelled by the impressive experimental findings of the previous decade, and all of them stressed the applications of formalism to the quantitative interpretation of experimental data. By contrast, the reports of Källen, Heitler, and Yukawa were on more formal topics and were not concerned with experiments.

That theoretical physics was being done differently in 1961 than in 1948 must have been apparent to the organizing committee, for they invited Bohr to give an historical talk, rather than philosophical and epistemological reflections, containing reminiscences of previous Solvay meetings he had attended, with an assessment of the role of the philosophical discussions at those meetings on the development of quantum mechanics.

I would like to explore the dynamics of these changes – how they came about and how they fit into their wider historical context. There was nothing inevitable in the process. The changes should not be seen as a natural unfolding of events.

World War II altered the character of science in a fundamental and irreversible way: The importance and magnitude of the contributions of scientists and engineers, particularly physicists, to the American war effort changed the relationship between the scientists and the military, industry, and government.[10] The Department of Defense, realizing that the security of the nation depended on the strength and creativity of the scientific community and its institutions, invested heavily in both their support and control.[11]

The period from the end of World War II to the beginning of the 1960s was unique in physics – particularly in the United States, the new Atlantis. There, Bacon's "College of the Six Days' Work" was being built. Essentially unlimited resources were being allocated to realize the goal of that foundation: to apprehend "the knowledge of causes, and secret motions of things, and the enlarging of the bounds of human empire, to the effecting of all things possible."[12]

The instruments created in this American house of Solomon were indeed Bacon's fantasy world reified: "high towers," "chambers," "perspective houses, where demonstrations of all lights and radiations [were made]," "means of seeing objects as far off, as in the heaven and remote places," "mathematical house[s]" with "exquisitely" made instruments. Bacon's Merchants of Light became scientific attachés in their modern incarnation. "Depredators," "Mystery-men," "Pioneers," "Compilers," and "Dowry-men" were to be found in both houses, attesting to some constant features of human nature. His interpreters of nature are our theoreticians, "those that raise the former discoveries by experiments into greater observations, axioms and aphorisms." As befitted their Baconian ancestry, the scientists of this new house had "their consultations" to decide "which of the inventions and experiences which [they had] discovered shall be published and which not." But whereas in Bacon's utopia the Elders of Solomon's house had the right to conceal from the state those inventions they thought fit to keep secret, in the new Atlantis the scientists could not. Oppenheimer's downfall – the dominant political event in the life of the scientific community during the 1950s – made it clear that although physicists had powers, power rested with the state.

The initial funding for this great postwar American enterprise came from the Office of Naval Research (ONR), the air force, the army, and the Atomic Energy Commission (AEC).[13] One of their aims was the maintenance of an adequate scientific manpower pool to meet the defense and technological needs of the United States. The support was so lavish that all needs were met: those of the universities, those of industry, and those of the state. One of the most distinctive features of the decade, the emergence of high-energy physics, was an outgrowth of this bounteous ministration by the state. Government funds made available through the ONR or the AEC paid most of the expenses for most of the high-energy machines that the universities acquired from 1945 onward.

It had been widely perceived that the success of the great wartime laboratories, such as the MIT Radiation Laboratory, the "Met Lab" at Chicago, and Los Alamos, reflected the ability of experimenters and theorists to collaborate intimately and to cooperate closely, a tradition that had been fostered in the nuclear research laboratories that had been established during the thirties at various American universities, and particularly at Berkeley in Lawrence's Radiation Laboratory.[14] One of the motivations for the funding of high-energy research was the maintenance of centers in which this kind of relation between theory and experiment would continue to flourish in large-scale projects. Incidentally, Bell Laboratories, for similar reasons, was the principal beneficiary of the postwar military funding in the development of the transistor.[15]

By 1950, four large synchrocyclotrons were in operation in the United States (at Berkeley, Rochester, Harvard, and Columbia), and two were under construction (at Chicago and Carnegie Tech). By the middle of the decade, two large proton-synchrotrons had come into operation (the Cosmotron at Brookhaven and the Bevatron at Berkeley), and final plans were being readied for the construction at Brookhaven of a 30-GeV alternating-gradient proton-synchrotron that was expected to be completed in 1960. By the end of the decade, some twenty universities had major high-energy installations.[16]

The "high-energy" theoreticians educated during this period were trained with close links to the experimental practice; they were instilled with a pragmatic utilitarian outlook and were taught to take an instrumentalist view of theories. This was the legacy of the approach that had proved so successful at Berkeley, Columbia, Harvard, and other leading American universities during the thirties.

This pragmatic stance was something constitutive of the American theoretical physics community. It was an imprint given it by the people who were responsible for its creation, its nurturing, and its growth in the period before World War II, namely, experimentalists such as Percy Bridgman, K. T. Compton, Robert A. Millikan, and Harrison M. Randall, and theorists such as E. Kemble, Gregory Breit, J. H. Van Vleck, Oppenheimer, and John C. Slater, and, later, the like-minded "refugee" theorists who had been offered positions in the United States after Hitler's rise to power in Germany: Bethe, Bloch, Teller. This pragmatic stance was reinforced by the institutional context in which it developed: departments of physics, in which experimenters and theorists were housed together and plied their trade together under one roof, in contrast with the European situation.[17] In nuclear physics, in particular, a symbiotic relationship grew between theorist and experimenter, in part because essentially no nuclear theory existed before 1933, and partly because the explanation and understanding of the data coming out of the cyclotrons and β-ray spectrographs required fairly recondite theory.[18]

This "American" style of doing physics was characteristic of the great wartime laboratories: the Radiation Laboratory at MIT, the Met Lab in Chicago, and Los Alamos.[19] It was in these wartime laboratories that many of the outstanding theoreticians of the 1950s were molded: Feynman, Goldberger, Chew, Robert Marshak. It is a style that became institutionalized at all the leading departments during the fifties and became the national norm.

The theoretical high-energy physics community that came into being during the 1950s was large. Its very size was something new. As compared with the handful of theoretical physicists working in "high energy" during the thirties, when the Rochester conferences[20] were initiated one could speak of a community of particle theorists numbering on the order of 100 active workers in the United States.[21] It was composed of bright, talented young people, deeply concerned with experimental data and sensitive to the needs of the experimental community. At the leading high-energy centers, their fortunes and future were often determined by their skill in explaining experimental results, and more generally by their usefulness to their experimental colleagues; the latter had invested enormous energy, skills, and government resources in building their high-energy machines. As important, the experimental results and discoveries that were coming from the machines at Berkeley, Carnegie Tech, Chicago, Cornell, Nevis, and, later, Brookhaven were so novel and interesting that it was clear that the challenges lay in explaining these findings and that these pastures were the greenest.

In the years immediately following World War II, from 1946 to 1950, funding from the ONR and AEC (the principal sources of support for physics in academic institutions) was determined by considerations that were *internal* to the discipline, and no explicit demands were attached to grants or contracts. The cold war, the detonation of the first Russian atomic bomb, the Korean War, and the Joseph McCarthy era changed the political climate. During the 1950s, the character of the support changed. Although the research supported at universities by the Department of Defense (DOD) and the AEC remained for the most part unclassified and publishable, applied research began to receive greater emphasis. National priorities and defense needs, in particular, began to determine more directly the areas and directions that were to be sponsored.[22] The establishment of the National Science Foundation in 1950 reinforced the tendency of the DOD to support basic research only in those areas likely to benefit the military needs of the country.

During the 1950s, the AEC and the DOD began wooing the best and brightest of the younger theoreticians to work during the summer on various defense-related projects at Los Alamos, at the Rad Lab in Berkeley, and at other DOD- or AEC-supported laboratories. Defense projects are by their very nature applied, practical, and gadget-oriented. The tools in the theoretician's armory essential for this enterprise often are the same as those that render the theoretician valuable to his experimental colleagues in the high-energy laboratories.

The defense connection during the 1950s reinforced the pragmatic, utilitarian, instrumental style so characteristic of theoretical physics in the United States. The successes of this mode of doing theoretical physics help explain its diffusion to Europe and elsewhere. The pragmatic ideal of American physics that had been visible from early on now became not only the national norm but in fact hegemonic worldwide.

The period under consideration also witnessed the disappearance of the Newton-like "geniuses," the "off-scale" creative individuals who by themselves work out most of the details and consequences of their brilliant insights and ideas. Consequences and applications now become worked out by a collective effort. Community and communal activity become key features in the activities of high-energy theorists. A confluence of factors – the greatly increased size of the theoretical community, the stiffer competition this entails, the pressure generated by the huge and costly high-energy apparatus to account for the data that are being poured out, the pressure to suggest new experiments, the pressure to maintain funding – results in the theoretical community pouncing on new likely avenues and working out attractive suggestions as if it were a coordinated whole. It is as though most of the members of the community consider it more worthwhile to work out the approach suggested by the intellectual leader at the moment (e.g., Gell-Mann, Mandelstam, Chew) than to work on their own ideas or on longer-range programs of research. As early as 1951, Feynman called it the "pack" effect.[23] The work on dispersion relations after Gell-Mann's and Goldberger's initial papers is an example of this phenomenon;[24,25] the almost wholesale adoption of Chew's *S*-matrix program is another.[26] The community at any one time seems to be dominated by the outlook and suggestions of a single individual. Gell-Mann, Goldberger, Lee, Yang, and Chew were the dominant figures from the mid-1950s to the mid-1960s, a role Steven Weinberg assumed in the late sixties.

All these developments were, of course, not independent of one another. National work-force needs, defense needs, funding, and the relation between institutional support and work on defense-related projects by the leading scientists of that institution are clearly interlinked factors. It is the historian's task to elucidate the genesis and interrelationships of these factors. The thesis is often asserted, especially by the practitioners, that context affects physicists, but *not* the physics. Recent historical studies relating to the development of the transistor, the laser, and quantum electronics more generally[27] suggest that for these fields, where the foundations (quantum mechanics and quantum electrodynamics) were secure, this thesis is open to question. To what extent and in what way this is also the case in the fields of physics, where foundational issues are open (as with particle physics), is one of the challenging tasks confronting the historian of physics.[28]

This chapter is organized as follows: The first section deals with post–World War II funding during the 1950s. The second section attempts to relate the intellectual output of the theoretical community to some of these factors. My intention is to raise questions and stimulate further investigations. Elucidation of the connections among funding, social and ideological constraints, and intellectual output awaits future detailed studies of individuals and institutions, their development, and their interactions with the larger context.

Funding after the war

World War II, a "gigantic war in which science and scientists as never before played a dominant role,"[29] ended in August 1945, and the United States emerged as *the* world power, in possession of a weapon with which it believed it could impose peace on the world. Victory had been achieved by realizing the tremendous military potentialities that were latent in the scientific developments of the late 1930s. But, as Fermi perceptively remarked

> Nobody [in 1939] had any basis for predicting the size of the effort that would be needed and it may well be that civilization owes its survival to the fact that the development of atomic bombs requires an industrial effort of which no belligerent except the United States would have been capable in time of war.[30]

Similar, perhaps more appropriate, remarks could be made regarding the development of radar, proximity fuses, bombsights, and the other devices that actually won the war, not to mention chemicals such as high-octane gasoline or pharmaceuticals such as the sulfa drugs or atabrine.

The leaders who had organized the American scientific war effort – Vannevar Bush, James Conant, Karl Compton, Frank Jewett – were haunted by the fact "that in fundamental science, in particular, and to a large extent in applied science, also, [the United States had] lost irrevocably the better part of a generation of creative research men and the better part of a generation of creative additions to [its] stockpile of fundamental knowledge."[31] They noted that the cadre of highly trained researchers in fundamental and applied science had not increased during the war, that its average age was four or five years greater, and they believed that its creative potential had been diminished. They were also aware that there was a complete absence of fully trained men in the next lower age group – a bracket filled by those who had been partially trained when they went into war work. The war had resulted in a dearth of men who would normally have completed their undergraduate training and been candidates for advanced training.

The Moe committee, one of the task forces organized by Bush in the winter of 1944 to answer the question "Can an effective program be proposed for discovering and developing scientific talent in American youth so that the continuing future of scientific research in this country may be assured on a level comparable to what has been done during the war?" issued its findings in early May 1945 in a report entitled "Gather the Spilled Corn." Its opening statement was that

> The greatest cost of the war, in terms of high living standards, full employment, and improved health for the American people in the future, has been the interruption of scientific training of men from 17 to 25 years of age – the most critical years.... [It] is most urgent to take immediate steps for the recovery of scientific talent which has been diverted in its formative stages....

It continued:

> While our objective is peacetime prosperity, it may be valuable to recall that nearly all the fields with which we are concerned have a special value to national security. As Dr. Bush has testified to a Congressional Committee: "Today it is evident to all thinking people that the evolution of new weapons may determine not only the outcome of battles, but even the total strategy of war....Tomorrow the impact of new weapons may be even more decisive....It is imperative, therefore, that after this war, we begin at once to prepare intelligently for the type of modern war which may confront us with great suddenness sometime in the future.[32]

By the winter of 1944, the need to revitalize and strengthen American physics had become a major concern within government circles and the physics community. The American Institute of Physics, which during the war had become "the focal point for urgent questions and requests from agencies responsible for the conduct of the war,"[33] in the spring of 1945 applied for and received a $29,300 three-year grant from the Rockefeller Foundation to have its War Policy Committee (hereafter to be known as the Policy Committee) address the postwar challenges, particularly the physics personnel problem.

At the national level, in the spring of 1945, Bush submitted to President Harry S. Truman a comprehensive program for postwar scientific research and scientific education. This report, *Science: The Endless Frontier*, had been written by Bush at the request of President Franklin D. Roosevelt after lengthy deliberations with his associates. The principal message was the importance of basic research. Bush stressed:

> Basic research leads to new knowledge. It provides scientific capital. It creates the fund from which the practical applications of knowledge must be drawn.... Today it is truer than ever that basic research is the pacemaker of technological progress.... *A nation which depends upon others for its new basic scientific knowledge will be slow in its industrial progress and weak in its competition in world trade, regardless of its mechanical skill.*[34]

Basic research would provide the foundation for continued economic growth and for the development of new products and new technologies. Knowledge of the methods and techniques of basic research is an essential ingredient in the training of all research investigators, be they in industry, in government laboratories, or in universities.

World War II had made evident that modern warfare "requires the use of the most advanced scientific techniques." In fact, just before the end of the war, the secretaries of war and the navy had communicated to the National Academy of Sciences their view:

> To insure continued preparedness along farsighted technical lines, the research scientists of the country must be called upon to continue in peacetime some substantial portion of those types of contributions to national security which they have made so effectively during the stress of the present war.[35]

Bush, in *Science: The Endless Frontier*, emphasized that it was essential for the national security of the United States "that there must be more – and more adequate – military research 'in peacetime." He suggested that this could best be done through a "civilian-controlled organization with close liaison with the Army and Navy, but with funds direct from Congress, and the clear power to initiate military research which will supplement and strengthen that carried on directly under the control of the Army and Navy."[36]

Bush urged that the government accept new responsibilities for promoting the flow of new scientific knowledge and the development of scientific talent and recommended the establishment of a national research foundation to implement his recommendations for the governmental support of basic research. He concluded the summary of his report with this statement:

> On the wisdom with which we bring science to bear in the war against disease, in the creation of new industries, and in the strengthening of our Armed Forces, depends in large measure our future as a nation.

Although Bush's detailed plan for a national research foundation (which included a division of national defense) encountered opposition from many different quarters, it was nonetheless the blueprint on which the postwar legislation dealing with scientific research and scientific education was based.[37] Most of his recommendations for a national research foundation (except those for the administrative oversight, and that for a division of military research) were eventually incorporated into the legislation establishing the National Science Foundation and the National Institutes of Health. However, it would take five years of legislative debate to arrive at a bill acceptable to the Congress, the executive branch, and the scientific community.

When the war ended, Los Alamos, Argonne, and the other installations that were part of the Manhattan Engineering District continued to be funded by that branch of the Department of the Army. The Office of Scientific

Research and Development (OSRD) made provisions to extend the contracts that had supported the major wartime laboratories under its supervision, in particular the Radiation Laboratories at MIT and Columbia. It made clear, however, that such support would be continued only for a limited time.

Inside the Department of the Navy, a group of young naval officers, who during the war had recognized that Allied superiority from 1943 on was based on the introduction of new technologies into warfare, and that these technological advances had originated with scientists and engineers who had been doing basic or applied research in universities and industrial laboratories, sensitized their superiors to the navy's dependence and reliance on technological progress to carry out its mission. Realizing that it would be a long time before a national research foundation could be enacted into law, staffed, supplied with funds, and opened for business, they convinced the secretary of the navy that the ONR should be established to carry out for the navy in peacetime the functions that OSRD had performed during the war.[38]

Public law 580 establishing the ONR was passed by Congress on 1 August 1946. Even prior to that date, the Office of Research and Inventions of the Department of the Navy (the immediate predecessor of ONR) was giving

> active financial support to scientific research workers in all fields of science – including medicine – in all parts of the country. Its contracts emphasized "the fundamental nature of research, and [were] carefully designed to preserve the freedom of inquiry and action so essential to the spirit and methods of research work."[39]

Captain R. Conrad, the director of the ONR's Planning Division, could justifyingly boast in the spring of 1946:

> I believe that the services are acting with wisdom and judgment, that this support will not degenerate into military control of scientists, and that a firm base is being laid for the National Science Foundation of the future.[40]

At the hearings to establish the ONR, Admiral Harold G. Bowen, who was to become the first chief of naval research, outlined its goals. One of these was the development of a pool of skilled scientific personnel (larger and more adequately trained than the pool that existed before World War II) in case the nation needed to mobilize for war again. Bowen indicated that "most of the contracts are with universities...and the work done in many cases may be accomplished by graduate students who become acquainted with naval problems and may form a future scientific group to assist the Navy."

As early as 1944, in close contact with various government agencies, and with their active support, universities began to plan how to salvage the wartime laboratories after the cessation of hostilities and how to reconstitute themselves for peacetime activities. As an example, I briefly recount here the story of MIT.[41] Similar stories can be told for Berkeley, Cornell, and elsewhere.

By the spring of 1945, plans had been formulated to establish an inter-disciplinary "Research Laboratory of Electronics," with J. A. Stratton as its head, and with support from the army, air force, and navy, to continue in the postwar period the research activities the Radiation Laboratory had carried out during the war. A Division of Basic Research was also set up to consider other, more general activities. Intensive discussions took place within the Physics Department to plan for the expansion of the department to accom-modate the expected large influx of graduate and undergraduate students in physics. Particular consideration was given to the problem of attracting (and supporting) the somewhat "older" young men, many of them with families, who had worked at the wartime laboratories with only B. S. degrees and had gained considerable experience there. The position of research associate (paying three-fourths of their wartime salary) was redefined to meet this problem. By the fall of 1945, 25 research associates had been appointed, and 90 graduate students were enrolled; by the spring term of 1946, over 125 physics graduate students were working for advanced degrees in the Department of Physics.

In the fall term of 1945, the Laboratory of Nuclear Science and Engineering (LNS&E) was established, with Jerrold R. Zacharias as its head, to oversee research activities and opportunities in the nuclear sciences.[42] The initial stimulus for the establishment of such a laboratory came from Conrad, who had contacted George R. Harrison, the dean of science, and Slater, the former chairman of the Physics Department. Interest in the field of nuclear physics and engineering was of predictable importance to the navy. The objectives of LNS&E were (a) to construct and make available to the de-partments of chemistry, physics, metallurgy, biology, chemical engineering, electrical engineering, and mechanical engineering the necessary facilities for modern research in nuclear physics, (b) to undertake new research (not carried out in the various departments) that would contribute to the development of nuclear science, and (c) to train competent personnel in nuclear science and engineering able to design, operate, and use nuclear devices.

A number of scientists, attracted in part by the promise and scope of the new LNS&E and the prospect of ample funding, accepted Zacharias's offer to come to MIT. Among these were Victor Weisskopf, Rossi, Herman Feshback, H. S. Bridge, Matt Sands, R. W. Williams, Robert W. Thompson, George E. Valley, and Ivon Getting. The budget of the laboratory for 1946 was about $300,000. Thereafter, the annual budget climbed to $1.2 million and remained fairly steady from 1948 to 1958. In 1958, after Sputnik, the AEC became the prime financier for the laboratory, and its budget was expanded.

One of the valuable resources LNS&E offered to the Department of the Navy is indicated by listing some of the defense projects that were directed by high-ranking members of the laboratory. In 1946, under the auspices of the laboratory, Clark Goodman, then an assistant professor of physics, gave a *classified* course in nuclear reactor design. In 1947, LNS&E offered the first university-based course in reactor physics; one of the by-products of this course was the first textbook on reactor design and operation.[43] These activities also led to the establishment of the Department of Nuclear Science at MIT a few years later. In April of 1948, Goodman helped to initiate a high-priority program, led by Captain H. G. Rickover, for the design of nuclear-powered submarines and the training (at LNS&E) of personnel for such vessels. In 1950, with Zacharias as director, the laboratory organized a summer study, in which some thirty scientists and engineers participated, to find ways to meet the threat that Russian submarines posed to overseas maritime shipping. Important suggestions for combatting this danger were made at that time, resulting in the development of Lofar (low-frequency directional listening arrays), in the use of helicopters for antisubmarine warfare, and in the construction of high-speed (Mariner) freighters. A similar study, Project Lamplight, with Zacharias as technical director, was established at Lincoln Laboratory in 1954.

But the most important dividend that LNS&E yielded probably was the personnel trained there. Between 1948 and 1958, roughly 400 persons participated in research at LNS&E as part of the teaching staff or as graduate students or assistants. Of the 300 or so whose careers could be traced, it was found that in 1980 some 117 were professors in 38 United States and 10 foreign universities, 106 had key positions in 74 industrial corporations, and 57 were in government service. From 1946 to 1958, some 78 Ph.D.'s were awarded, and over a thousand undergraduates did their research for senior theses in the laboratory.[44]

ONR funded other projects at MIT; during 1946–7, some thirty were being supported. One of the largest was Project Whirlwind. The project's original mission was the development of a new high-speed training system, a "universal flight trainer-aircraft simulator," which would allow a pilot to "obtain a feel for flying a particular airplane before it was ever constructed."[45] The objectives changed during the project to the development of a general-purpose computer.[46]

Similar centers, but with somewhat different interests and orientations, were established on other campuses, for example, at Chicago and at Cornell. All these "nuclear" laboratories had one feature in common. All (and, in particular, LNS&E at MIT, the Newman Laboratory for Nuclear Studies at Cornell, and the Nuclear Lab at Chicago) attempted to re-create the atmosphere of Los Alamos – the sense of community, the exhilaration and joy in doing physics – and to structure themselves so that they would be doing physics the way it was done at Los Alamos – with experimenters and theorists working closely together and interacting closely with one another.

One further consequence of Los Alamos was apparent at these laboratories: Theoreticians and experimentalists had become peers. The ultimate democratization of physics had been effected. In the 1930s, this democratic ideal had been realized by having experimentalists and theoreticians all members of the same department (but with experimentalists almost always chairmen). At Los Alamos – where Oppenheimer, a theoretician, was the guiding spirit – the powers of the theoretical physicist became apparent. Thus, theoretical physicists emerged from the war with a new status. Robert R. Wilson and Bethe provided the leadership at Cornell, Zacharias and Weisskopf at MIT, Fermi and Teller at Chicago.

Even more lavish support by the AEC was responsible for the growth of the Radiation Laboratory at Berkeley,[47] as well as Argonne and Oak Ridge, for the establishment of Brookhaven,[48] and for construction of all the high-energy machines. The scale of the support, its sources, and the requirements for its maintenance altered the character of science in the United States. What was meant by "doing physics" was changed irreversibly. With the intensification of the cold war in the late 1940s and the early 1950s – particularly after the explosion of the first Russian atomic bomb in 1949, and the outbreak of the Korean War – the AEC intensified its research activities in fusion processes. Many of the senior theoreticians who had worked on the atomic bomb project participated in these efforts. Many of the younger theoreticians were invited to joint projects at Los Alamos, at the Radiation Laboratory at Berkeley, at Princeton, and elsewhere to work either full-time or during the summer. Thus, Chew, Goldberger, Low, and K. Watson, among others, spent several summers at Los Alamos. By the middle of the decade, the value of such contributions was clear, and the secretary of defense and the joint chiefs of staff made a request for an organized university effort "as a public service" to support and strengthen the Weapons Systems Evaluation Group of the Department of Defense.

The Institute for Defense Analysis (IDA) was created to fulfill this request. IDA is an association of universities that was formed "to promote in the field of Defense studies, a more effective relationship between the national security and scientific learning."[49] Incorporated in April of 1956 and supported by a Ford Foundation grant of $500,000, IDA was to be a link between the scientific and technological community and the Department of Defense. Its first annual report, issued in 1957, gave the rationale for its mission:

> During World War II the process of converting the discoveries of science and the developments of technologies into new elements of military power – including the necessary readjustments in military strategy and tactics – went at an undreamed of pace. Developments in weapons and military equipment found their way so quickly into the pattern of military operations that the whole shape of military power was radically changed in the span of half a dozen years. . . .
>
> Present military capabilities based on these new technologies in hostile hands present our country with a threat that is entirely unfamiliar: heavy destruction by direct attack. Moreover the era of war in peace in which vast shifts in the world powers, aggravated by ruthless Communist ambitions of world domination, have brought us military responsibilities far beyond the direct defense of our own territory. . . .

> Such are the reasons why it is of paramount importance not only to give all necessary supports to weapons research, but also the maintenance of the most effective possible bridge between military strategy and the total of technology – for converting technical advances into new elements of military power – for guiding technology in the creation of new foundations of strategy.

IDA's purpose was "to assist in maintaining this bridge."

In 1958, following Sputnik, IDA signed a two-year contract with the National Security Agency to extend its activities. In the summer of 1959, Charles H. Townes, then at Columbia, wrote to Garry Norton, the president of IDA, that he and Marvin Stern had met with a group of scientists at Los Alamos (K. Brueckner, Goldberger, Watson, Norman Kroll, and Sam Treiman)

> to explore possible arrangements whereby more first rate scientists and perhaps highly trained professional persons may contribute effectively and creatively to the National Security.... A particular need... is good contact and communication with the younger established scientists who are in their most productive period.

The names suggested were Brueckner, Chandrasekhar, Fitch, Garwin, Gell-Mann, Goldberger, Karplus, Kroll, Lewis, Lieberman, Longmire, Low, Parker, Reines, Rosenbluth, Treiman, and Watson. Thus were many of the young Jasonites born.

The issue I am concerned with is not whether it was right or wrong for physicists to have worked for IDA. There is surely an important role to be played by physicists in defense and security matters relating to their country. The issue I want to address is this: Did the involvement of many of the leading American high-energy theorists in defense matters reinforce a particular kind of theoretical orientation – pragmatic, phenomenological, with "*S* matrix theory" as its most impressive statement – to the exclusion of others? Did it affect developments in theoretical high-energy physics?

I would suggest that the fragmentation of interests by these leading theorists, stemming from their consulting and their involvement in defense matters, hindered – and to a certain extent prevented – their maintaining a sustained focus and effort on fundamental theory. Also, in their capacity as reviewers of research proposals, and by virtue of their dominance in the funding process, they tended to reinforce their dominant view. At the very least, I would argue that the tempo of developments and the rate of initiation of new viewpoints in the field were affected. To do so, let me briefly review the history of development during the fifties.

Hopes and aspirations

After its great initial success, the quantum theory of fields came into a period of decline just before World War II, in the face of stubborn and seemingly insurmountable divergence difficulties and its inability to account for nuclear phenomena. During the war, a program for a relativistic quantum-mechanical treatment of matter as particles was formulated by Werner Heisenberg.[50] His hope was to construct a theory that dealt only with *observables*, such as scattering cross sections and energy levels. From quantum field theory, he aimed to abstract those features that would be valid in a future, more correct, theory. The S matrix, a quantity previously shown by John A. Wheeler[51] to contain most of the information required for a description of scattering phenomena, became the focus of Heisenberg's researches.

Kramers and S. A. Wouthuysen made important contributions to the undertaking by pointing out how the bound-state energies could be obtained by analytically continuing the S matrix into the complex energy plane.[52] Ralph Kronig, in 1946, bolstered the approach by suggesting that the analytic properties of the S matrix would be partially determined by causality requirements.[53] Christian Møller further clarified Heisenberg's research program by explicitly relating the analytic structure of the S matrix in the complex energy plane to the usual wave-mechanical description based on a Hamiltonian.[54]

These advances, made on the eve of the great successes of renormalization theory in overcoming the divergence difficulties, were not followed up. The experiments of Willis Lamb and Robert Retherford[55] and those of John Nafe, Edward Nelson, and I. I. Rabi[56] were, of course, crucial and were responsible for the great impact of the Shelter Island conference of 1947. Incidentally, these developments should be seen as a great *American* success story. The Columbia experiments were made possible by the wartime developments in microwave technology. The contributions of Bethe, Schwinger, Feynman, and Dyson were grounded in the pragmatic, practical, conservative philosophical outlook that characterized American theoretical physics. The Shelter Island conference itself was organized to "bring out" the young American theoreticians who made such important contributions to the war effort at Los Alamos and elsewhere. The self-confidence that the American theoretical physics community had acquired as a result of its wartime successes is surely an important factor in explaining the postwar contributions.[57] The precise and reliable value for the $2S$-$2P$ level shift that Lamb obtained posed a challenge.

It is interesting to compare the concerns of field theorists in the United States and elsewhere after the initial work on renormalization theory by Schwinger, Feynman, and Dyson. The international colloquium "Elementary Particles and Nuclei," organized by CNRS and held in Paris in April 1950, gives us a vantage point for this comparison.[58] Dirac, Pauli, Bohr, Kramers, Heisenberg, and Peierls were the leading theorists in Western Europe – and for the most part they were primarily concerned with formal questions.[59]

Dirac delivered the opening address at the session "General Methods and Ideas in Relativistic Quantum Mechanics." He pointed out that for a long time, progress in the theory of elementary particles had been stopped by the divergences that appear in the solutions of the "exact equations that describe the motion of these particles." Since the early 1930s, different methods had been proposed to overcome these difficulties; recently, "the summum of perfection" in subtraction physics seems to have been achieved "in the method of Lamb, Schwinger, and others that permits the calculation of the small residual effects that subsist after the elimination of the infinities." This great success might lead one to believe that the Lamb-Schwinger method was the best possible for the elimination of infinities and that one might not expect a new, better, substantial improvement in this direction. "Nonetheless," Dirac went on, "this method does not constitute a satisfactory base for the final theory of elementary particles, because it is too artificial and unaesthetic from the mathematical point of view. It is normal that a fundamental physical theory should rest on simple and direct mathematical concepts. Thus, the Einstein theory of gravitation is based on the concept of the curvature of space-time." Dirac then outlined his new Hamiltonian theory for extended charge distributions. In his comments after Dirac's talk, Pauli expressed his agreement with Dirac's position, though he did not think that Dirac's new formulation – which relied on a parametric representation of the state vectors defined on spacelike surfaces – was likely to be very helpful.

Pauli, in his paper "The Present State of Quantum Field Theory: Renormalization," explored the limits of QED. He noted that the renormalized, or "physical," field operators $F_{\mu\nu}^{(R)}$ differed from the "mathematical" operators $F_{\mu\nu}$ by the factor e/e_0, the ratio of the renormalized to the bare charge. He reported that R. Jost had calculated the vacuum expectation value of $F_{\mu\nu}^{(R)}$ (averaged over a finite space–time value) and found it to be finite, while that of $F_{\mu\nu}$ diverged. He commented that this was a satisfactory result and was in accordance with Bohr and Léon Rosenfeld's analysis of the measurements of fields and charges in QED. He concluded:

> I have the impression that we have exhausted the range of applicability of mass and charge renormalization. New progress can only be realized by totally new ideas that will allow the theoretical determination of the masses of the particles and very likely also the value of the fine structure constant. In effect, the lower limit of the size of the space-time volume over which the averages of the field operators are taken [in Jost's work] can only be determined in the framework of a general theory of the masses of elementary particles which we do not have today.

Heisenberg, working in war-ravaged Germany, was interested in seeing whether or not the renormalization procedure could explain the value of the fine-structure constant, a concern he shared with many other European theoreticians, in particular Pauli.[60] This was a question that had been frequently discussed during the 1930s. Rosenfeld and Bohr were continuing their analysis of the measurability of field strengths;[61] Peierls and many of his students were studying the structure of the self-energy divergences in QED.[62]

By contrast, in the United States, Karplus and Kroll[63] were calculating the anomalous magnetic moment of the electron, and several groups were working on calculations of the Lamb shift,[64] of the radiative corrections to the hyperfine structure of hydrogen, and on the radiative corrections to various electrodynamic processes (e.g., Compton scattering, bremsstrahlung). The situation in Japan – judging from what is published in *Progress of Theoretical Physics* – is more variegated and merits an extended analysis.

Experimental results had been the stimulus for the postwar developments in QED. Similarly, the experimental data on π mesons that were being generated by the cyclotrons that came into operation in the early 1950s were responsible for the extensive theoretical activities of that period. Marshak, in one of the first review articles on meson physics, written in 1951, noted that

> The year 1950 saw tremendous advances in the laboratory study of mesons. It was the first year in which accelerator physicists outdistanced the cosmic ray physicists insofar as their contributions to our understanding of their production, and the nature of their interaction with nucleons are concerned. The fact that much greater intensities of artificially produced mesons were available was beginning to make itself felt in a significant and impressive way. It was true, of course, that the laboratory study of mesons was restricted to energies less than about 100 MeV and that higher energy mesons were still the monopoly of cosmic ray research. However, the possibility of performing more controlled, more accurate, and more searching experiments in the laboratory reaped a rich harvest of outstanding accomplishments in 1950.[65]*

The successes of QED had raised the hope that π-mesonic phenomena would be explained by the (renormalization) *PS–PS* charge-symmetric Yukawa theory. But that hope quickly faded, because the coupling of mesons to nucleons is much stronger than the electromagnetic interactions between charged particles, and the problem of formulating reliable calculational methods proved intractable.

In an article on pions in the *Scientific American*, written in the middle of the decade, Marshak conveyed a vivid picture of the situation:

> The glue that holds the nucleus. . .is a mystery that defies all our experience and knowledge of the physical world. It is a force so unlike any we know that we can hardly find words to describe it. We do have a clue, however, to which we can give a name. It is the pi meson, or pion. In some way, not yet understood, pions are certainly involved in the nuclear binding force.

But he concluded his article in a more pessimistic tone:

> The chief difficulty is the fact that we must deal with a swarm of pions. Our mathematical techniques cannot efficiently handle more than one pion at a time. Beyond this, things become much too complicated. The problem appears to be a basic one, and it seems that only some radically new idea will enable us to solve it. And so the pion, while providing a tantalizing glimpse into nuclear forces, serves only to deepen our ignorance.[66]

---

* Reproduced, with permission, from the *Annual Review of Nuclear Science*, Vol. 1. © 1952 by Annual Reviews Inc.

By 1955, the abandonment of field theory by particle theorists was accelerated by the fact that increasing numbers of elementary particles were being discovered. The prospect of introducing a new quantum field for each new particle did not seem very convincing. The failure of field theory (considered as a "fundamental" approach) to account for the strong interactions set the stage for the resurgence of S-matrix theory in the midfifties, in the form of dispersion relations. Self-confidence, enthusiasm, and charisma were factors in the "selling" of dispersion relations. Goldberger (a product of the Met Lab) began his lecture at the "Midwest Conference on Theoretical Physics" in March of 1956 in Iowa City with the statement that

> It is possible to write down [dispersion relations for finite-mass particles and for nonforward scattering] and to give plausibility arguments for their validity but rigorous proofs do not exist yet.[67]

After indicating how an "ideal derivation" of dispersion relations for the meson–nucleon scattering amplitude of $M_{\alpha\beta}(K_0, \Delta^2)$ would proceed, and pointing out the obstacles to a rigorous proof, Goldberger asserted the correctness of the stipulated results and gave the following five reasons:

1. There would be no point in continuing this talk if they were not.
2. Reductio ad absurdum: Assume that $M(K_0, \Delta^2)$ does not have the desired character; that's obviously absurd. Q.E.D.
3. Every term in perturbation theory does have the correct analytical behavior.
4. Because the results are correct, it must be possible to make a proof.
5. The results agree with experiment.

A further indication of the attitude of many of the particle theorists is given in "a note to the discerning reader" appended by the editor to Goldberger's assertion that the dispersion relation must be correct even though no rigorous proofs exist: "The only rigor in theoretical physics is rigor mortis."

A veritable dispersion-relation industry sprang up after 1955. All the major universities in the United States offered a course on dispersion relations during the 1956–7 academic year. I have in my possession some ten sets of lecture notes on dispersion relations (by Salam, J. Hamilton, Low, J. C. Taylor, J. G. Taylor, and others) dating from 1956 to 1959. These probably constitute but a small sample. All stress the positivistic commitment and utilitarian value of the approach.

The advances in "elementary particle" theory during the fifties were such that by 1960 Feynman could say that "we live in an heroic, a unique and wonderful age of excitement, an exciting age of very vital development in fundamental physics and the study of fundamental law."[68] To have a heroic age, "we must have a series of successes." Success breeds confidence. The enormous confidence that the theoretical high-energy community in the United States possessed during the 1950s was based on a series of successes:

the war effort, the Lamb shift, QED, the two-meson hypothesis, the phenomenology of meson physics of the early 1950s, Chew–Low theory, dispersion relations, pololology, parity nonconservation, V–A, the Gell-Mann systematization of hadrons, the eightfold way, the Goldberger–Treiman relation, and so forth. Successful theoretical explanations fed on important experimental advances, and, conversely, theory provided useful guidelines for experiments. These impressive theoretical advances were made by a community that Feynman, as early as 1951, characterized as "mov[ing] in packs from one problem to another."

But there were costs to the efficient, normalized (one might say renormalized!), utilitarian new Atlantean way of doing theoretical physics. The dominant research program attracted the large majority of the community, leaving other areas impoverished. Thus, general relativity got very little attention during most of the 1950s.[69] The same can be said of axiomatic field theory.[70] It is true that the action was elsewhere, that much quicker rewards could be gained by meeting the challenges of high-energy particle theory. The point I want to stress is that general relativity and axiomatic field theory were not valued very highly.[71]

I believe the "pack effect" is a factor in explaining why such important contributions as Gell-Mann and Low's work on the renormalization group lay fallow for well over a decade.[72] Other examples can be given: the early work on gauge theories by Yang and R. L. Mills, Schwinger, Sheldon Glashow, and others.[73] Note that all of these contributions were *field-theoretic* in their approach. There was another important cost to this way of doing physics: the loss of creativity. The contributions I assess as constituting the point of discontinuity in the periodization of the theoretical work during the time frame of our concerns were made by "outsiders" – Nambu and J. Goldstone, in their work on broken symmetry.[74]

It is the dominance of the major research program that I want to emphasize. I would like to suggest that it was a confluence of several factors that was responsible for its tenacity. Among these factors were

(a) the plethora of new, exciting, and challenging experimental findings. Under those circumstances, a phenomenological, pragmatic approach is clearly the most efficient and effective way to structure and give order to a new field driven by experimental data.[75]

(b) an educational philosophy that emphasized the empirical, the experimental practice, and the value of the theorist to the experimentalist.[76]

(c) a system of funding, and a peer-review system that tended to reinforce the dominant view.

Of course, any system of funding will impose constraints and will tend to channel interest. Much more research needs to be done before the linkages among (a), (b), and (c) are clarified and before the following questions can be answered: Did the source of the funding inadvertently constrict other approaches in high-energy physics? Did it limit the questions that could be asked, and did it help legitimate what counted as explanation and understanding? I am not in search of a Marxist explanation. I believe that the persistence of a strong positivistic strand in the British and American philosophical outlook is much closer to the mark. Why it emerged again in the midfifties with such vitality is an interesting historical question.[77]

That things were being done differently again in the early sixties is clear. Not only were quark models offering valuable dynamical insights into a host of phenomena, but "fundamental" questions in the Pauli tradition were once again beginning to be addressed, and, equally as important, students were exposed to such concerns.

Let me conclude with some excerpts from a lecture Feynman gave at the Ettore Majorana summer school in Erice in 1964 on the status of strong, electromagnetic, and weak interactions.[78] I could find no lecture similar in tone in the earlier period, by him or anyone else. Feynman began his lecture with the statement that

> The first thing to do in physics is to distinguish the fundamental problems from the less fundamental ones. There are certain fields, say superconductivity, solid state physics, or chemical physics, which can be understood on principle, as soon as the fundamental laws are known....
>
> What are the areas of fundamental physics, as of today? First of all there is gravitation, although many of you might have forgotten about it. Other areas are electromagnetism, weak interactions, and the strong interactions – and that is all. Maybe someday there will be another branch of physics which has to do with the development of the universe as a whole: cosmology, but today this is nothing but astronomical history.

He summarized the extant knowledge of the four basic interactions with the statement that

> Everything below 1 GeV is known and everything above it is unknown. That is to say: we know physics down to distances of the order of 10^{-14} cm, and all troubles come from what is inside that length.

After reviewing the consequences that followed from the assumed validity of the principle of special relativity and the superposition of quantum mechanical amplitudes, he turned to look at the fundamental problems:

> You can roughly divide these questions into two types, namely existential questions and dynamical problems. The separation is not clear cut; the problems move from one class to another. An existential problem appears in the form: "Why is such and such a thing there?" Clearly this is not the kind of question which is answered experimentally in physics. Nevertheless their first appearance is in this vague philosophical form.

Next came questions relating to the numbers and kinds of strongly interacting particles, and then one final general question relative to the violation of all the symmetries, that is, to the understanding of partly satisfied symmetries.

He mentioned in connection with possible dynamical explanations Heisenberg's and Schwinger's work in the late 1950s (Nambu and Goldstone should also have been referred to):

> ...the breaking of the symmetry may come from the dynamics. Unsymmetrical physics can result from a symmetry-preserving Hamiltonian.

The asymmetry can be attributed to the state vector of the world, not to the physical law. Goldstone and Nambu's work was indeed similar to Kepler's breaking of the circles. The world is seen differently thereafter, and only those who could see it the new way could contribute to the development of the field thereafter.

Feynman concluded his lecture at Erice with some advice to the future generation of physicists, sitting there at his feet:

> First, learn to calculate. If you are theorists I would urge you to connect yourselves to nature by calculating numbers. You should develop a feeling for the subtle interplay between the general and the specific. You should cultivate both.

> Finally, have pride in [your] work. When you publish a paper you should have at least the confidence that what you say is right and will last one year.

In that last bit of advice the message was clear: high-energy theoretical physics is more than a profession; it is a calling.

Notes

1 *The Quantum Theory of Fields*, proceedings of the twelfth conference on physics at the University of Brussels, October 1961 (New York: Interscience, 1961).

2 *Les Particules Elémentaires*, Rapports et Discussions publiés par les Secretaires du Conseil. Huitième Conseil de Physique tenu à l'Université de Bruxelles du 27 Septembre au 2 Octobre 1948. (Brussels: R. Stoops, 1950).

3 L. Bragg to H. A. Bethe, 14 November 1947. This letter is in the Bethe Papers in the archives of the Cornell University Library, Ithaca, New York, 14/22/1976.

4 Ibid.

5 L. Bragg to H. A. Bethe, 29 January 1948. Bethe Papers, Cornell, 14/22/1976. The official letter inviting the participants from outside Europe informed them that UNESCO had been contacted in order to obtain funds from this organization to defray their travel costs.

6 van den Duggen to H. A. Bethe, 26 February 1948. Bethe Papers, Cornell.

7 *New Theories in Physics* (Paris: International Institute for Intellectual Cooperation, 1939).

8 W. Heitler, "Quantum Theory of Damping and Collisions of Free Mesons," in *Les Particulars Elémentaires*, note 2 (p. 159–78).

9 Casimir, in his summary of the conference in the form of a ballad, wittily captured some of the "atmospherics" of the meeting (p. 383):

La situation théorique était claire
Quand Oppi a dit qu'il pourra se defaire
De toute divergence et de l'infinité
Une fois qu'il sera devenu grand-pére;
Et c'est la le but du Conseil Solvay.

10 The literature on World War II and the scientific community is by now vast. One of the best points of entry is Daniel Kevles, *The Physicists* (New York: Knopf, 1978). Particularly valuable is the "Essay on Sources" in that volume. See also A. Hunter Dupree, "The Great Instauration of 1940: The Organization of Scientific Research for War," in *The Twentieth Century Sciences*, edited by G. Holton (New York: Norton, 1972), pp. 443–67; Carroll Pursell, "Science Agencies in World War: The OSRD and Its Challenges," in *The Sciences in the American Context: New Perspectives*, edited by Nathan Reingold (Washington, D. C.: Smithsonian Institution Press, 1979). See also Alex Roland, "Technology and War: A Bibliographic Essay," in *Military Enterprise and Technological Change*, edited by Merritt Roe Smith (Cambridge, Mass.: MIT Press, 1985), pp. 347–79.

11 Michael S. Sherry, *Preparing for the Next War: American Plans for Postwar Defense, 1941–45* (New Haven: Yale University Press, 1977).

12 F. Bacon, *Essays and New Atlantis* (New York: Walter J. Black, 1942), p. 288.

13 Even a cursory examination of the sources for the support of the research published in the *Physical Review* during these years gives an indication of the overwhelming extent of the military and AEC support. See also A. Hunter Dupree, "The Structure of the Government–University Partnerships after World War II," *Bull. His. Med. 39* (1965), 245–51.

14 J. L. Heilbron, R. W. Seidel, and Bruce R. Wheaton, *Lawrence and His Laboratory: Nuclear Science at Berkeley, 1931–1961* (Berkeley: Office for History of Science and Technology, University of California, 1981).

15 Thomas J. Misa, "Military Needs, Commercial Realities, and the Development of the Transistor, 1948–1958," in Smith, note 10 (pp. 253–88).

16 For a history of high-energy accelerators, see M. S. Livingston, *Particle Accelerators: A Brief History* (Cambridge, Mass.: Harvard University Press, 1969). For a contemporary history of the postwar development at Berkeley, see G. F. Chew and B. J. Moyer, "High Energy Accelerators at the University of California Radiation Laboratory," *Am. J. Phys. 18* (1950), 125–31; see also the review articles on high-energy accelerators in the first two volumes of the *Annual Reviews of Nuclear Science*. In particular, see M. Stanley Livingston, "Synchrotron," and "Proton Synchrotrons," *Annu. Rev. Nucl. Sci. 1* (1952), 163–8, 169–74; E. L. Chu and L. I. Schiff, "Recent Progress in Accelerators," *Annu. Rev. Nucl. Sci. 2* (1953), 79–92. This article also contains an extensive bibliography of the literature until 1952. For the situation later during the decade, see J. P. Blewett, "Recent Developments in Proton Synchrotrons," *Annu. Rev. Nucl. Sci. 4* (1954), 1–12; G. A. Behman, "Particle Accelerators. I. Bibliography. II. List of Accelerators – Installations," U.S. AEC document UCRL-8050, 1958; *Proceedings of the CERN Symposium on High Energy Accelerators and Pion Physics*, Vol. 1, edited by E. Regenstreif (Geneva: European Organization for Nuclear Research, 1956); see also the second annual report of the European Organization for Nuclear Research (Geneva: CERN, 1956), pp. 9–11; D. L. Judd, "Conceptual Advances in Accelerators," *Annu Rev. Nucl. Sci. 8* (1958), 181–216. Behman, in his review article on accelerators written in 1957, tabulated the data on accelerator installations throughout the world. He listed information on a total of 500 machines in use or under construction. The *Proceedings of the CERN Symposium on High Energy Accelerators and Pion Physics* (Geneva: CERN, 1956) gave information on some two dozen laboratories at which high-energy accelerators (> 50 MeV) existed or were under construction.

17 S. S. Schweber, "Empiricism Regnant: The American Theoretical Physics Community, 1918–1948," *Historical Studies in the Physical Sciences 17 : 1* (1986), 55–98.

18 See, for example, R. H. Stuewer (ed.), *Nuclear Physics in Retrospect* (Minneapolis: University of Minnesota Press, 1979).

19 For the Rad Lab, see, for example, John Burchard, *Q.E.D.: MIT in World War II* (New York: Wiley, 1948); Phinney Baxter, *Scientists Against Time* (Cambridge, Mass.: MIT Press, 1968); E. Pollard, *Radiation* (Durham: Woodburn Press, 1982). For the laboratories connected with the atomic bomb, see R. G. Hewlett and O. E. Anderson, *The New World 1939/1946. A History of the United States Atomic Energy Commission, Vol. 1* (University Park: Pennsylvania State University Press, 1962). For Los Alamos, in particular, see David Hawkins, *Project Y: The Los Alamos Story. Part I, Toward Trinity* (Los Angeles: Tomash

Press, 1983); S. Groueff, *Manhattan Project: The Untold Story of the Making of the Atomic Bomb* (Boston: Little, Brown, 1967).

20 R. Marshak, "The Rochester Conferences," *Bull. Atomic Scientists 26* (June 1970), 92–8.

21 For statistics on the high-energy community, see A. Pickering, *Constructing Quarks* (University of Chicago Press, 1984).

22 The situation for quantum electronics has been analyzed convincingly by Paul Forman, "Behind Quantum Electronics: National Security as a Basis for Physical Research in the United States, 1940–1960," *Historical Studies in the Physical and Biological Sciences 18 : 1* (1987), 149–229. I thank him for showing me a copy of this paper before publication. See also Paul Forman, "Atomichron: The Atomic Clock from Concept to Commercial Product," *Proc. IEEE 73* (1985), 1181–204. For the transistor, see Misa, note 15. For the laser, see Joan Lisa Bromberg, "Engineering Knowledge in the Laser Field," *Technology and Culture 27* (1986), 798–818; Joan L. Bromberg, "Research Efforts that Led to Laser Development," *Laser Focus* (October 1984), 58–60.

23 R. P. Feynman, "The Present Situation in Fundamental Theoretical Physics," *Ann. da Acad. Brasileira de Ciencias 26* (1954), 51–9.

24 M. Gell-Mann, M. L. Goldberger, and W. E. Thirring, "Use of Causality Conditions in Quantum Theory," *Phys. Rev. 95* (1954), 1612–27. M. L. Goldberger, "Use of Causality Conditions in Quantum Theory," *Phys. Rev. 97* (1955), 508–10.

25 M. L. Goldberger, "Causality Conditions and Dispersion Relations. I. Boson Fields," *Phys. Rev. 99* (1955), 979–85.

26 G. F. Chew, "Double Dispersion Relations and Unitarity as the Basis for a Dynamical Theory of Strong Interactions," UCRL-9289 (1960). A series of lectures delivered at the Summer School of Theoretical Physics, Les Houches, July 1960.

27 See Misa, note 15, and the sources cited in note 22. See also David F. Noble, "Command Performance: A Perspective on Military Enterprise and Technological Change," in Smith, note 10 (pp. 328–16); David F. Noble, *Forces of Production* (New York: Knopf, 1984).

28 For the development of quantum mechanics, the relevance of context has been forcefully argued by Paul Forman, "Weimar Culture, Causality and Quantum Theory, 1918–1927: Adaptation by German Physicists and Mathematicians to a Hostile Intellectual Environment," *Historical Studies of the Physical Sciences 3* (1971), 1–115. Marcello Cini has written a stimulating paper on the relation between context and the development of quantum theory during the 1950s: M. Cini, "The History and Ideology of Dispersion Relations. The Pattern of Internal and External Factors in a Paradigmatic Shift," *Fundamentae Scientiae 1* (1980), 157–72.

29 Frank B. Jewett, "The Future of Scientific Research in the Post War World," in *Science in Progress*, edited by G. A. Baitsell (New Haven: Yale University Press, 1947), pp. 3–23.

30 E. Fermi, "The Development of the First Chain Reacting Pile," *Proc. Am. Philos. Soc. 90* (1946), 20–4.

31 Jewett, note 29; see also Arthur H. Compton, "Science and Our Nation's Future," *Science 101* (1945), 107–209.

32 Bush-Moe Committee. "The Transition, 1945." Barton Collection, box LXXXIX, folder 1, "Gather the Spilled Corn," American Institute of Physics (AIP), New York.

33 H. A. Barton to Warren Weaver, 6 April 1949. Attached to the letter is the report to the Rockefeller Foundation on the "Work of the Policy Committee 1945–1948." Barton Collection, box LXXXIX, folder 1, AIP.

34 V. Bush, *Science: The Endless Frontier*. A report to the president on a program for postwar scientific research, July 1945. Reprinted July 1960, National Science Foundation, Washington, D. C., p. 19.

35 Ibid., p. 4.

36 Ibid., p. 6.

37 See, for example, *The Politics of American Science: 1939 to the Present*, edited by J. L. Penick, Jr., Carroll W. Pursell, Jr., Morgan Sherwood, and Donald C. Swain (Cambridge, Mass.: MIT Press, 1972). It should be noted that the contract and grant system that had been devised by Bush at OSRD – the means by which the resources of universities were brought

to the service of the nation without impinging on the autonomy of the institutions themselves – was perhaps his greatest contribution to the management of university–government relationships.

38 Bruce S. Old, "The Evolution of the Office of Naval Research," *Physics Today 14* (August 1961), 30–5. See also Harold G. Bowen, *Ship, Machinery and Mossbacks* (Princeton University Press), pp. 30–5; Mina Rees, "Early Years of the Mathematics Program at ONR," *Naval Research Review 30* (1977), 22–9; Harry M. Sapolsky, "Academic Science and the Military: The Years Since the Second World War," in *The Sciences in the American Context: New Perspectives*, edited by Nathan Reingold (Washington, D. C.: Smithsonian Institution Press, 1979), pp. 379–99.

39 Quoted in S. A. Glantz, "How the Department of Defense Shaped Academic Research and Graduate Education," in *Physics Careers, Employment and Education* (Penn State, 1977), edited by Martin L. Perl (New York: AIP, 1978), pp. 109–122.

40 Ibid.

41 See Philip M. Morse, *In at the Beginnings: A Physicist's Life* (Cambridge, Mass.: MIT Press, 1977); James R. Killian, *The Education of a College President: A Memoir* (Cambridge, Mass.: MIT Press, 1985). See also box 3 of the Morse Papers at the MIT Archives, MC 75, which contains a history of the Physics Department at MIT.

42 *Nuclear Engineering at MIT: The First Twenty Five Years* is a history of these activities. The volume was privately printed in 1982 by MIT.

43 M. Deutsch, R. D. Evans, B. T. Feld, F. Friedman, and C. Goodman, *The Science and Engineering of Nuclear Power* (Cambridge, Mass.: MIT Press, 1948).

44 "Return on Investment in Basic Research – Exploring a Methodology," report to ONR, Department of the Navy, by Bruce S. Old, Associates, Inc., November 1981. Contract N00014–79–C–0192. See also Mina Rees, "Early Years of the Mathematics Program at ONR," note 38.

45 Rees, note 38 (p. 11).

46 K. C. Redmond and T. M. Smith, *Project Whirlwind – The History of a Pioneer Computer* (Maynard, Mass.: Digital Press, 1980).

47 Heilbron, Seidel, and Wheaton, note 14.

48 Allan A. Needell, "Nuclear Reactors and the Founding of Brookhaven National Laboratory," *Historical Studies in the Physical Sciences 14* (1983), 93–122.

49 From the second annual report. All the materials relating to IDA are in box 8 of the Morse Papers, archives, MIT. Morse was a trustee of IDA.

50 W. Heisenberg, "Die 'beobachtbaren Grössen' in der Theorie der Elementarteilchen," *Z. Phys. 120* (1943), 513–38.

51 J. A. Wheeler, "On the Mathematical Description of Light Nuclei by the Method of Resonating Group Structure," *Phys. Rev. 52* (1937), 1107–22.

52 H. A. Kramers's contributions are acknowledged in Heisenberg's S-matrix papers. See also D. Ter Haar, "On the Redundant Zeros in the Theory of the Heisenberg Matrix," *Physica 12* (1946), 501–8; R. Jost, "Bemerkund zu der Vorstehenden Arbeit," *Physica 12* (1946), 509–10.

53 R. Kronig, "A Supplementary Condition in Heisenberg's Theory of Elementary Particles," *Physica 12* (1946), 543–4.

54 C. Møller, "General Properties of the Characteristic Matrix in Theory of Elementary Particles. I," *Kl. Dan. Vidensk. Selsk. Mat.-Fys. Medd. 1* (1945); part II, ibid., *19* (1946).

55 W. E. Lamb, Jr., and R. C. Retherford, "Fine Structure of the Hydrogen Atom by a Microwave Method," *Phys. Rev. 72* (1947), 241–3.

56 J. E. Nafe, E. B. Nelson, and I. I. Rabi, "The Hyperfine Structure of Atomic Hydrogen and Deuterium," *Phys. Rev. 71* (1947), 914–15.

57 S. S. Schweber, see Chapter 4 of "Some Chapters for a History of Quantum Field Theory 1938–1952," in *Relativity, Groups and Topology II*, edited by B. S. DeWitt and R. Slora (Amsterdam: North Holland, 1984), pp. 37–220.

58 *Particules Fondamentales et Noyaux* (Paris: Editions du Centre National de la Recherche Scientifique, 1953).

59 S. S. Schweber, "Shelter Island, Pocono and Oldstone: The Emergence of American Quantum Electrodynamics after World War II," *Osiris 2* (1986), 265–302.

60 W. Heisenberg, letter to F. J. Dyson, 19 December 1949.

61 N. Bohr and L. Rosenfeld, "Field and Charge Measurements in Quantum Electrodynamics," *Phys. Rev. 78* (1950), 794–8.

62 R. E. Peierls and H. McManus, "Electrodynamics without Singularities," *Phys. Rev. 70* (1946), 795A; H. McManus, "Classical Electrodynamics without Singularities," *Proc. R. Soc. London, Sev. A 195* (1948), 323–36.

63 R. Karplus and N. Kroll, "Fourth-Order Corrections in Quantum Electrodynamics and the Magnetic Moment of the Electron," *Phys. Rev. 77* (1950), 536–49.

64 M. Baranger, "Relativistic Corrections to the Lamb Shift," *Phys. Rev. 84* (1951), 866–7; M. Baranger, H. A. Bethe, and R. P. Feynman, "Relativistic Corrections to the Lamb Shift," *Phys. Rev. 92* (1953), 482–501; R. Karplus and A. Klein, "Electrodynamic Displacement of Atomic Energy Levels. I. Hyperfine Structure," *Phys. Rev. 85* (1952), 972–84; R. Karplus, A. Klein, and J. Schwinger, "Electrodynamic Displacement of Atomic Energy Levels. II. Lamb Shift," *Phys. Rev. 86* (1952), 288–301.

65 R. E. Marshak, "Meson Physics," *Annu. Rev. Nucl. Sci. 1* (1952), 1–42.

66 R. E. Marshak, "Pions," *Sci. Am. 8* (January 1957), 83–92.

67 *Proceedings of the Midwest Conference on Theoretical Physics*, edited by J. M. Jauch and Y. Takahashi (Department of Physics, State University of Iowa, 9–10 March 1956).

68 R. P. Feynman, "Conclusions of the Conference," in *The Aix-en-Provence International Conference on Elementary Particles*, edited by E. Cremieu-Alcan, P. Falk-Vairant, and O. Lebey (Gif-Sur-Yvette: C.E.N. Saclay, 1961), pp. 205–10.

69 Bryce S. DeWitt, in his introductory remarks on the "Conference on the Role of Gravitation in Physics," which was held at the University of North Carolina, 18–23 January 1957, noted that "a noticeable increase of interest and activity in the theory of gravitation and related matters has taken place only recently, following a period of relative quiet." See *Rev. Mod. Phys. 29* (1957), 351. Dicke's experiments were influential in this revival of interest. See R. H. Dicke, "Principle of Equivalence and the Weak Interactions," *Rev. Mod. Phys. 29* (1957), 355–62; P. G. Bergman, "Summary of the Chapel Hill Conference," *Rev. Mod. Phys. 29* (1957), 352–4.

70 A. S. Wightman, "Looking Back at Quantum Field Theory," *Physica Scripta 24* (1981), 813–16. Also see Wightman, Chapter 42, this volume.

71 Stanley Deser, who was a postdoctoral fellow at the Institute for Advanced Study from 1953 to 1955 and was very interested in general relativity, reports that "Oppie was vehement against Einstein." Interview with the author, 16 April 1985. J. A. Wheeler becoming a general relativist in the mid-1950s was an important turning point in the "valuation" of general relativity. See J. A. Wheeler, "Some Men and Moments in the History of Nuclear Physics: The Interplay of Colleagues and Motivations," in *Nuclear Physics in Retrospect*, edited by Roger H. Stuewar (Minneapolis: University of Minnesota Press, 1979), p. 259. Goldberger, in 1961, characterized Bogoliubov's axiomatic theoretic derivation of non-forward-scattering dispersion relations as "ugly, involved, unrewarding and uninstructive" in *The Quantum Theory of Fields*, note 1 (p. 192).

72 M. Gell-Mann and F. E. Low, "Quantum Electrodynamics at Small Distances," *Phys. Rev. 95* (1954), 1300–12.

73 C. N. Yang and R. L. Mills, "Conservation of Isotopic Spin and Isotopic Gauge Invariance," *Phys. Rev. 96* (1956), 191–5; J. Schwinger, "A Theory of Fundamental Interactions," *Ann. Phys. (N.Y.) 2* (1957), 407–34.

74 Y. Nambu, "Dynamical Theory of Elementary Particle Suggested by Superconductivity," in *Proceedings of the 1960 Annual International Conference on High Energy Physics at Rochester*, edited by E. C. G. Sudarshan, J. H. Tinlot, and A. C. Melissinos (University of Rochester, 1960), pp. 856–66; Y. Nambu and G. Jona-Lasinio, "Dynamical Model of Elementary Particles Based on an Analogy with Superconductivity. I," *Phys. Rev. 122* (1961), 345–58; J. Goldstone, "Field Theories with 'Superconductor' Solutions," *Nuovo Cimento 19* (1961), 154–64; See also B. Zumino, "Field Theories with a Degenerate Vac-

uum," in *Werner Heisenberg und die Physik unserer Zeit* (Braunschweig: F. Vieweg, 1961).

75 Note that this mode of doing theory was also practiced during the 1930s in the development of nuclear theory. Breit and Wigner's and Kapur and Peierls's theory of nuclear reactions and later Wigner's *R*-matrix theory are forms of dispersion theory. Dispersion relations are a natural outgrowth of an epistemological outlook that is embedded in quantum mechanics. The latter is the prototype of a theory in which only questions of the form "If I do this, what may I expect to happen?" are considered valid.

76 Howard Schnitzer, who was a graduate student at Rochester from 1955 to 1959, characterized the attitude of that outstanding department as "pragmatic. . .[one] looked at data, empirical facts were fundamental. Marshak embodied that tradition." Interview with S. Schweber, 9 April 1985. Similar statements could be made about Illinois, Columbia, and so forth. Goldberger had a similar impact on Princeton when he came there in 1955. That this was the prevalent attitude can be gauged from the fact that Brookhaven was *the* summer meeting place of the theoretical high-energy community during the fifties.

77 For example, the impact of McCarthyism cannot be underestimated.

78 R. P. Feynman, "Present Status of Strong, Electromagnetic and Weak Interactions," in *Symmetries in Elementary Particle Physics*, edited by A. Zichichi (New York: Academic Press, 1965), pp. 400–18.

47   Progress in elementary particle theory, 1950–1964

MURRAY GELL-MANN

Born 1929, New York City; Ph.D., Massachusetts Institute of Technology, 1951; theoretical physics; Nobel Prize, 1969, for his contributions and discoveries concerning the classification of elementary particles and their interactions; California Institute of Technology

I should like to begin by expressing regret for not having been able to be present for the last Fermilab meeting, at which I had been asked to give the final talk about what it was like to be a student of theoretical physics in the late forties, the end of the period covered by that meeting. I still hope to present that material somewhere. Between that conference and this one there were two others, one in Paris in the summer of 1982, where I gave a talk about my experiences with strangeness,[1] and one in 1983, in Sant Feliu de Guíxols in Catalunya, where I spoke on the subject "Particle Theory from *S*-Matrix to Quarks."[2] That second conference was called a "trobada" in the Catalan language, a word that reminds us of the troubadours of the Middle Ages who flourished in Catalunya. It reminds us also that we have become much like those medieval minstrels. We spend a great deal of time now traveling from one orgy of reminiscence to another, each one held in the capital of some princely state. Here I am helping to close another "trobada."

Let me repeat something I said in Catalunya: I shall once again commit the sin of hindsight in the eyes of some of those historians of science who come from the historical tradition. Like most scientists who try to provide material for the history of science, I have, of course, no objection to situating the discoveries in the intellectual context and even in the social, economic, and political contexts of the times in which they were made, but I, along with many other physicists, think it is important to inquire also how close we came to what we now perceive are the right ways of looking at things, to list what we missed, to speculate about why we missed it, and to note things we got right and why we got them right.

Work supported in part by the U.S. Department of Energy under contract DE-AC03-81-ER40050 and by the Alfred P. Sloan Foundation.

In Catalunya, I talked about the mistakes and hesitations and confusions that some theorists went through, particularly the ones that I experienced myself during the fifties and early sixties. The mood of the presentation was rather sad; I mentioned that I was like a person who could not concentrate on the doughnut but kept looking at the hole. In this chapter, I shall deal instead with the doughnut and try to emphasize what I would call the envelope of understanding that some theorists were able to achieve during the years that we are discussing. [Actually, I was assigned to discuss a period (1950–64) different from that discussed by everybody else, and I have taken that mission seriously.] What I mean by treating the envelope of understanding is that instead of talking about the hesitations and the backtracking, I shall pretend that when one of us got a correct idea for a good reason, the idea stuck. I shall include each such idea, and not stress that we doubted it or that we forgot it for a while. Of course, I shall try to point out when we failed to understand something, or when we needed another crucial idea in order to make things work.

If we add together the ideas and the omissions that I list, then we get pretty much the standard model of 1972–3. That model, as I pointed out at Shelter Island a couple of years ago,[3] solves virtually all the problems that we considered to be such when I was a graduate student. But, of course, there are other problems, just as serious, that are left. Few of us believe today that the standard model is anything but the low-energy phenomenology of a much better theory, and we hope that much better theory can be found. Ideally, it will be a completely unified, parameter-free, finite theory of all the interactions, including gravity, a theory that, for all we know, may have been finally written down during the past year. I shall return to that point at the end of this chapter. Let me first review briefly what we learned up to 1964.

I am discussing a close partnership of experiment and theory in a period during which there were sensational experiments. Just to mention a few, there were experiments on strange particles and their decays, antinucleons, hadron scattering cross sections, nucleon form factors, explicit neutrino discovery, neutrino collisions, hadron resonances, parity violation, K^0 and $\bar{K}^0$, CP violation, and the Ω^-. Hearing about these experiments, in some cases anticipating them, interacting with the people who were carrying them out, was a fantastic adventure. However, I propose to address the subject here only from the point of view of the theoretician, in fact from the point of view of a very small number of theoreticians who were in the forefront of speculation about the fundamentals of particle physics. (It may be difficult for some young people today to realize what a small set of theoreticians it was.) In most cases I shall adopt my own point of view, even in describing advances with which I did not have a great deal to do.

At the beginning of the period we are discussing, a world picture was developing according to which all physical phenomena could be ascribed to ν and $\bar{\nu}$, $e^{\pm}$ (along with the puzzling $\mu^{\pm}$), the neutron and proton and their presumed antiparticles, the pions $\pi^\pm$ and π^0, the photon, possibly some kind of intermediate bosons for the universal weak interaction, and presumably the graviton (although it was not polite to talk about that).

The pion was strongly emphasized. It was hoped that the success of the perturbative renormalization program in QED would carry over to a field theory of nucleon and pion treated as elementary particles. As the pseudo-scalar and isovector character of the pion became clear from experiment, speculations about the nuclear force centered on the renormalizable meson theory with the coupling

$$ig\bar{N}\gamma_5\tau N \cdot \pi \tag{47.1}$$

Here, I use Enrico Fermi's notation in which the field is denoted by the particle to which it corresponds. The nonrelativistic approximation is

$$\frac{g}{2M_N} \bar{N}\sigma\tau N \cdot \nabla\pi \tag{47.2}$$

which had already been used in 1942 by Wolfgang Pauli and Sidney M. Dancoff. What a heavy responsibility that poor pion had to bear during those years, and how some theorists laughed at mixed meson theories in which one combined the pion with vector mesons, for example, in order to correct singularities in the nuclear force at short distances.

From that somewhat naïve and disconnected picture of elementary particle phenomena, we progressed by 1964 to a notion of fundamental leptons and quarks, coupled to the vector fields of a generalized Yang–Mills theory of the strong, electromagnetic, and weak interactions, as in the standard model of today, but lacking several crucial ideas, especially color (which removes the clash between the Yang–Mills theories of the strong and the electroweak interactions by making the two kinds of charge variables independent of each other, and which also can lead to quark confinement), asymptotic freedom (which involves effective charge decreasing with decreasing distance, as in confinement, and which allows quarks to be nearly free in the deep interior of the nucleon), and the role of charm in canceling the neutral strangeness-changing current.

I have listed in the following sections a number of advances in understanding during the period in question, advances that helped us to progress from the earlier picture to the later one. By using square brackets, I have indicated from time to time ingredients that were still lacking or comparisons with the present level of understanding. Following some of the items in the list of advances, I have inserted comments and reminiscences about those advances and particularly about how they appeared to me at the time. I have divided my remarks into two sections: quantum field theory, and interactions and elementary particles.

Quantum field theory

A: P, C, *and* T *are empirical symmetries,*
although local field theory requires PCT *invariance*

Around 1949, it was not clear to everyone that *P*, *C*, and *T* separately are merely possible symmetries of particular field theories and may be conserved or not depending on the form of the interaction. My experience with *P* may be illuminating. When I was a graduate student at MIT in 1949, I took a course taught by Herman Feshbach, who assigned as a problem proving parity conservation by reflection of coordinates without specifying the theory. I worked on the problem over the weekend and was unable to solve it; on Monday, I turned in a statement that conservation of parity does not follow from transformation of coordinates, but is an empirical law that depends on the transformation properties of the Hamiltonian. I do not remember what impression that made, or what kind of grade I received in the course, but I retained from then on the idea that parity is an empirical symmetry, and charge conjugation as well. Neither Fermi nor P. A. M. Dirac ever believed in parity conservation as a fundamental principle, but the impression was surprisingly widespread that such symmetries were fundamental rules that could not be violated.

B: *Program for extracting exact results from field theories*

1. *"Dispersion-theory program," just quantum field theory formulated on the mass shell (renamed "S-matrix program" and misunderstood by some as a departure from quantum field theory)*
   (a) *Crossing relations*
   (b) *Dispersion relations,*
       forward light scattering, generalized to spin-flip, nonzero mass, nonforward, and so forth
   (c) *Unitarity relations generalized*
       with amplitudes extended to unphysical values of momentum, but still on the mass shell
   (d) *Boundary conditions as various momentum variables approach infinity help specify which field theory is meant*

In Catalunya, I discussed the program that Marvin Goldberger and I worked on, starting around 1952, for extracting exact results from field theories. We abstracted the crossing relations from diagrams to all orders. The dispersion relations were known, but only for forward elastic light scattering, and we generalized them to the spin-flip case. After I left Chicago, Goldberger and his colleagues generalized them to the case of nonzero mass, and then many of us extended them to non-forward dispersion relations. Next we added unitarity, and generalized unitarity so as to carry the amplitudes out of the physical region to imaginary momenta while always keeping them on the mass shell. Finally we said that there needed to be boundary conditions as various momentum variables go to infinity. This was the dispersion-theory program that I outlined at the Rochester meeting in 1956, pointing out that it was a way of formulating field theory on the mass shell. At Rochester, I showed how one could in fact obtain the whole set of scattering amplitudes on the shell iteratively starting with the particle poles, but admitted that I did not know what the boundary conditions would have to be for a particular theory. I casually mentioned Werner Heisenberg's *S*-matrix program, almost as a joke, but later on, after we had convinced Geoffrey Chew of the value of the program, he renamed it "*S*-matrix theory." As we know, L. D. Landau (around 1959) and Chew (around 1961) insisted that it was somehow distinct from field theory.

2. *Renormalization group [but possibility of asymptotic freedom overlooked]*

In 1953, I went on with extracting exact results from field theory, working with Francis Low on the renormalization group. But here, in brackets, we find the first of the crucial gaps in our understanding. Nobody noticed until the early 1970s that the renormalization group function has the sign reversed in a Yang–Mills theory, leading to asymptotic freedom.

C: *Yang–Mills theory*

1. *Group*
   (a) *SU(2) trivially generalized to products of SU(2) and U(1) factors*
   (b) *Generalized to products of arbitrary simple Lie groups and U(1) factors*
2. *Properties*
   (a) *Renormalizable if exact*
   (b) *Perfect gauge invariance*
   (c) *Massless vector bosons*
   (d) *Other multiplets degenerate with respect to the group*
3. *Soft-mass mechanism called for*
   that would break symmetries (b), (c), and even (d)
   renormalizably
   [but not found until later]

The Yang–Mills theory was developed in 1954, using SU(2), and it was immediately clear that one could generalize gauge theory to any product of SU(2) and U(1) factors, with U(1) behaving as in quantum electrodynamics. Later on, at the end of the decade, Sheldon Glashow and I showed that the SU(2) factors could be replaced by arbitrary simple Lie groups. Around the same time, it was shown by Abdus Salam and K. Kumar that Yang–Mills theory is renormalizable if the gauge invariance is exact, but that any masses for the vector mesons, or any other known kind of violation of symmetry, ruined the renormalizability. That was a very serious matter, because we did not know how to make use of perfectly unbroken non-Abelian gauge symmetry. I advertised for a soft-mass mechanism, something that would allow the renormalizability of the pure Yang–Mills theory to persist in the presence of vector-meson masses, but it took a few years for the relevance of the so-called Higgs mechanism to the soft-mass problem to become clear.

D: *Spinless bosons and symmetry breaking*

1. *Exact continuous symmetries,*
   when not producing degeneracy,
   can have massless spinless "Nambu–Goldstone bosons" instead
2. *Approximate symmetry can correspond*
   to low-mass spinless bosons
3. *"Higgs–Kibble . . . Anderson mechanism":*
   way of breaking continuous symmetry
   without degeneracy or massless spinless bosons
   if these bosons are eaten by vector gauge bosons,
   which then acquire mass [not yet recognized as a solution to the soft-mass problem]

In 1959–60, while Yoichiro Nambu and Jeffrey Goldstone were pursuing their separate investigations of how exact continuous symmetry could fail to produce degeneracy, but lead instead to the presence of massless spin–0 bosons, I was studying the problem in Paris with Maurice Lévy, Jeremy Bernstein, Sergio Fubini, and Walter Thirring. We considered the isovector axial vector current in the weak interaction of hadrons and its relation to the pion. Inspired by the success of the Goldberger–Treiman relation and the lack of a convincing derivation of it, we concentrated on the idea of a "partially conserved" axial vector current with a divergence dominated by the low-mass pion pole. In the limit of exact conservation, the pion would become massless, and thus we found the Nambu–Goldstone boson independently, and we saw how approximate conservation would lead to a low-mass boson. The models that Lévy and I used as examples included the nonlinear σ model, later to become a favorite subject of research (even more important today) and the ordinary σ model, which was useful later on in the theoretical discovery of "Higgs bosons."

Those spinless bosons represented still another way that exact continuous symmetry could manifest itself, without either degeneracy or massless spinless bosons. The work of Philip Anderson, Peter W. Higgs, T. W. B. Kibble, and many others showed how the new spinless bosons could be eaten by massless vector bosons, which would then acquire mass. In Chapter 44, Nambu describes how he and his colleague were actually thinking of this mechanism as early as 1959–60.

In any case, the "Higgs" mechanism was available by the end of the period we are discussing, but it was not yet understood that it was the soft mass that we needed for Yang–Mills theory, to supply symmetry breaking and gauge boson masses without destroying renormalizability. Steven Weinberg, in his letter about electroweak theory in 1967, suggested that such might be the case, and Gerard 't Hooft finally proved it around 1971.

E: *"Reggeism"*

1. *Composite particles tend to lie on "Regge trajectories" in a plot of J versus M^2*

During the 1960s we learned a great deal about singularities in the complex angular-momentum plane and their variation with mass. Here I consider them in their relation to quantum field theory (especially in dispersion theory or "S-matrix" theory form), and later I mention them again in connection with hadron physics and then once more at the end of this chapter.

These singularities are just as important as ever in spite of the fact that they are a less fashionable research topic than they once were. I remember how excited many of us were when we studied, around 1961 and 1962, the relationship between hadron trajectories in the domain of positive mass squared, where stable particles and resonances lie along them, and the high-energy behavior of scattering amplitudes in the cross channel with nontrivial quantum numbers exchanged. (That work was initiated by Goldberger and R. Blankenbecler, and developed and emphasized by Chew and S. C. Frautschi.)

2. *In simple problems in nonrelativistic quantum mechanics, these correspond to "Regge" or "Sommerfeld" poles in the complex J plane*
3. *In quantum field theory, there are still trajectories, but the poles tend to become more complicated singularities (like poles accompanied by branch points) [details not clarified until much later]*

As it became clear that in field theory the poles are accompanied by cuts that are only logarithmically weaker at high energies, the situation became somewhat less striking, but still not too bad.

It was after our assigned period that the fear was expressed, at the Berkeley conference of 1966, that the cuts would actually be dominant away from $t = 0$, but that turned out to be excessively pessimistic. The situation was cleared up much later by Fredrik Zachariasen and his collaborators, in work that has not received sufficient attention.

4. *In vector field theories,*
    elementary spinor fields give Regge trajectories,
    not fixed poles as in lowest order

To return to the early 1960s: Chew wanted to identify Regge poles as peculiar to composite particles, so that the Regge behavior of all hadrons would be evidence that they were all composed of one another in accordance with the bootstrap picture. I was delighted with the bootstrap idea, but not with the notion that the elementary particles of field theory were somehow doomed to inhabit fixed poles, while only composite particles were allowed the mobility of "Regge" singularities. Goldberger and I, and later several collaborators, pointed out that spinor elementary particles in a vector field theory would become "Reggeized" by radiative corrections. We had some difficulty in getting our work published, perhaps because it contradicted a popular dogma. Later on, of course, other "Reggeizations" were demonstrated.

5. *"Regge behavior" can provide boundary conditions*
    at high momenta for the dispersion-theory program
6. *High-energy behavior of scattering amplitudes*
    depends on leading "Regge" singularities in the cross channel
    as a function of momentum transfer
    (for a pole $s^{\alpha(t)}$ or $s^{\alpha(u)}$)
    [pole dominance over cuts at very high s
    not clarified until much later
    except at $t = 0$ or $u = 0$]

Interactions and elementary particles

A: *Division into strong, electromagnetic, weak, and gravitational interactions*

At the beginning of our period, it was clear to some of us that the known interactions should be divided into strong, electromagnetic, and weak (and, of course, gravitational), each with its characteristic symmetry properties. That notion was not at all universal, however, and in my first letter on strangeness in 1953, I had to describe the classification, labeling the first three interactions (i), (ii), and (iii), before going on to suggest how each behaved with respect to isotopic spin.

1. *All conserve baryon number,*
    but not understood why conservation exact
    [now thought to be probably approximate!]
2. (a) *Lepton number conserved*
        [now thought to be probably approximate]
    (b) *Electron and muon number conserved*
        [now thought to be probably approximate]

It was believed, of course, that all the interactions conserved baryon number, with the proton lifetime being not simply very long (as was known from observation) but infinite. In the absence of any known massless gauge boson coupled to baryon number (and the Eötvös experiment sets a rather stringent limit on such a coupling), exact baryon conservation was rather mysterious. Nowadays, of course, we solve that mystery by believing that the conservation is only approximate. Lepton number and also (with the knowledge that $\nu_e \neq \nu_\mu$) electron number and muon number, separately, are in the same situation. In the standard model, of course, all these quantities are conserved, and the hypothetical violations are always connected with a more general, at least partially unified theory.

B: *Electromagnetism*

1. *QED renormalizable for spin $\frac{1}{2}$, also for spin 0 if $\lambda(\phi^+\phi)^2$ renormalized*

By about 1950 it was known that QED is renormalizable for charged spinor particles (and also for charged scalar particles if an additional parameter is renormalized). The second-order renormalizability of the charge in QED had been established in 1934 by Dirac and by Heisenberg, and that of the mass by a number of authors in 1948. Of those, the first ones to complete correct relativistic calculations of the Lamb shift (Willis Lamb and Norman Kroll and J. Bruce French and Victor Weisskopf) actually used the clumsy old non-covariant method. The place where the new covariant methods played a crucial role, particularly those of E. C. G. Stueckelberg and Richard P. Feynman, which are still used today, was in permitting calculations to be done quickly, especially calculations to fourth and higher orders (which would have been impractical with the old methods), and in making possible the proof of renormalizability to all orders. The last was accomplished by Salam and P. T. Matthews in a very long paper and by John Ward in concise form, following a crude sketch by Freeman Dyson.

2. *Electromagnetic interaction minimal (Ampère's law)*
   $p_\mu \to p_\mu - eA_\mu$ *(always right, but sometimes ambiguous)*

I described at the Paris meeting how in fixing the isotopic spin properties of the electromagnetic interaction in the presence of strange particles (in 1952–3) it was necessary to exclude arbitrary new electromagnetic interactions by insisting on a "minimal" interaction obtained by the rule $p_\mu \to p_\mu - eA_\mu$, a modern version of Ampere's law. This prescription is not always unique, but it always includes the right answer.

3. *C and P conserved*
   to the extent they are conserved elsewhere

4. *QED of vector charged particles*
    thought to be related to broken Yang–Mills theory, including the
    photon
    [but not known how to do this renormalizably:
    same as problem of soft mass]

In 1957–8, when some of us were constructing theories of the weak interaction mediated by a spin-1 charged intermediate boson (which Feynman and
I called $X^{\pm}$, and which I still call $X^{\pm}$, although some authors seem to use
$W^{\pm}$), we were faced with the question of how to resolve the ambiguity in the
electromagnetic coupling of that boson, and several of us adopted the prescription that comes from a Yang–Mills theory of $X^{\pm}$ and the photon. That
prescription, which is now part of the standard model, leads to finite radiative
corrections (even without the Higgs mechanism) in some situations where
other prescription would yield infinities.

C: *Strong interaction and "strongly interacting particles" or*
"*hadrons*"

1. (a) *Strong interaction conserves isotopic spin*

The conservation of isotopic spin by the strong interaction, leading to charge
independence of the nuclear force, and probably involving an isovector
meson, was an old idea that was confirmed at the beginning of our period
by the discovery of $\pi^{\pm}$ and π^{0} and then by the π–N scattering experiments,
especially the ones at Chicago. (I remember Fermi searching in 1952 for a
convenient way to describe charge-independent elastic scattering in terms of
real quantities and fixing on phase shifts for $J = \frac{1}{2}$ and $J = \frac{3}{2}$ separately and for
$I = \frac{1}{2}$ and $I = \frac{3}{2}$ separately.)

(b) *Isotopic spin multiplets can be displaced*
     in center of charge by S/2;
     then S conserved by strong interaction
(c) *Because electromagnetism is minimal and Q is linear in I_z,*
     electromagnetism obeys $|\Delta I| = 0, 1; \Delta I_z = 0 \rightarrow \Delta S = 0$
(d) *Strong interaction conserves C and P*
     [today, its conservation of CP to high accuracy looks
     remarkable,
     and the mechanism is much discussed – Higglet (axion), etc.]
(e) *Particle–antiparticle symmetry really works*
     for nucleons!

During the early 1950s, the failure to observe antinucleons in cosmic rays
gave rise to occasional doubts about the applicability of C to the nucleon, but
the experimental discovery of $\bar{p}$ and $\bar{n}$ at the Bevatron laid those doubts to
rest.

2. (a) *"Nuclear democracy" or hadronic egalitarianism;*
      no observable hadron is more fundamental than any other

I have described the dispersion-theory program, renamed the "*S*-matrix" program, and how it requires boundary conditions as various momentum variables tend toward infinity. The condition of having amplitudes dominated in those limits by expressions corresponding to moving singularities in the *J* plane was suggested by Chew and Frautschi as the way to implement the bootstrap idea that all the hadrons are composed of one another in a consistent way, with no (observable) hadron being more fundamental than any other. (The insertion of the word "observable" is, of course, my modification, and it makes the bootstrap notion perfectly compatible with the existence of quarks and gluons that are fundamental but confined.) The success of QCD has brought research on the dispersion theory of hadrons pretty much to an end; so we do not know what kinds of conditions on that program would reproduce QCD – but presumably there are some that would do the job. It would be fascinating to know exactly how to formulate QCD on the mass shell.

3. *Rich spectrum of baryon and meson states*
   classified by J^P, I, S, Q (or I_z), G (for mesons)

The first known excited hadron state was, of course, the $J = (\frac{3}{2})^+$, $I = (\frac{3}{2})$ resonance in $\pi-N$ scattering. It had been predicted theoretically from a strong or intermediate coupling of the foregoing form (47.2) long before it was discovered experimentally by Fermi and his collaborators. Curiously, Fermi did not accept the idea of a resonant state. Starting with the experimental cross sections, he found a solution for the phase shifts that had the $p_{3/2,3/2}$ phase shift increasing with energy to about $60°$ and then falling again. Later on, one of the graduate students, after a conversation with a visiting theorist (Hans Bethe, I believe), discovered that there was another perfectly good solution in which the $p_{3/2,3/2}$ phase shift increased through $90°$ as in a resonance.

   (c) *Baryons and mesons lie on*
       nearly straight Regge trajectories extending high in spin
       [still rather mysterious]
   (d) *Approximate exchange degeneracy*
       [still rather mysterious]
   (e) *Hadron high-energy scattering amplitudes*
       really dominated by appropriate Regge singularities in the cross
       channel
   (f) *Hadron high-energy elastic scattering*
       dominated by "Pomeranchuk" singularity
       (J = 1 for t = 0)

    (g)   *At $t = 0$, this could not be a simple pole;*
        total cross sections could not be asymptotically constant
        [situation not further clarified until much later]

3. (a)   *Strong interaction has approximate "flavor" SU(3) invariance*
   (b)   *Baryons in 8 ($J^P = \frac{1}{2}^+$), 10 ($J^P = \frac{3}{2}^+$),*
       and higher multiplets
   (c)   *Mesons in 8 and 1 ($J^P = 0^-$), 8 and 1 ($J^P = 1^-$),*
       and higher multiplets
   (d)   *Violation of SU(3) by octet eighth component*
       [and a little bit of third]
   (e)   *First order predominates, giving mass formula*
       [still a little mysterious]

4. (a)   *Strong interaction has*
       quite good "chiral SU(2) $\times$ SU(2)" invariance
       and fair "chiral SU(3) $\times$ SU(3)" invariance
   (b)   *Approximate symmetry under "vector current" charges*
       gives approximate degeneracy, but
       approximate symmetry under "axial vector current" charges
       gives nearly massless modified Nambu–Goldstone pseudoscalar
       bosons: the pseudoscalar mesons
   (c)   *So terms violating conservation of axial vector current*
       are somehow soft – "PCAC": divergence of axial vector current
       has matrix elements dominated by pseudoscalar meson pole;
       they transform like $(3, \bar{3})$ and $(\bar{3}, 3)$ under SU(3) $\times$ SU(3)
   (d)   *But there is not a very light pseudoscalar boson corresponding to*
       the total axial vector current charge
       [explained much later – in QCD – using anomalous divergence of
       the current]
   (e)   *Just as form factor*
       for divergence of axial vector current
       is dominated at low q^2 by pseudoscalar meson pole,
       so form factor of vector current
       is dominated at low q^2 by vector meson poles ("vector
       dominance")
   (f)   *These are important, of course,*
       because vector current occurs
       in electromagnetic and weak couplings,
       and axial vector current occurs in weak couplings

5. *Strong interaction thought to come from a vector theory,*
  most likely Yang–Mills,
  but how can that be arranged in "flavor" space
  without clashing with weak and electromagnetic gauge theory?
  [this difficulty was overcome when we found color]

D: *Weak interaction*

1. (a) *Charge-exchange weak interaction obeying "Puppi triangle"*

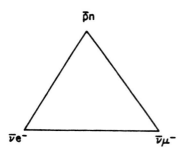

$\bar{p}n$

$\bar{\nu}e^-$ $\bar{\nu}\mu^-$

generalized to "tetrahedron"

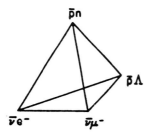

$\bar{p}n$

$\bar{p}\Lambda$

$\bar{\nu}e^-$ $\bar{\nu}\mu^-$

(b) *Leptonic interactions of "$\bar{p}n$," $|\Delta I_z| = 1$, $|\Delta I| = 1$, $\Delta S = 0$; leptonic interactions of "$\bar{p}\Lambda$," $|\Delta I_z| = \frac{1}{2}$, $|\Delta I| = \frac{1}{2}$, $(\Delta S/\Delta Q) = +1 \rightarrow$ nonleptonic interactions have no $|\Delta S| = 2$ and nonleptonic interactions with $|\Delta S| = 1$ have $|\Delta I| = \frac{1}{2}, \frac{3}{2} +$ electromagnetic corrections*

(c) *Approximate predominance of nonleptonic $|\Delta I| = \frac{1}{2}$ [still somewhat mysterious]*

The charge-exchange weak interaction was described around 1949 by the "Puppi triangle," with vertices $\bar{\nu}e^-$, $\bar{\nu}\mu^-$, and $\bar{p}n$, also discussed by Tsung Dao Lee, Marshall Rosenbluth, and Chen Ning Yang, by Jayme Tiomno and John A. Wheeler, and perhaps by others. Around 1954, N. Dallaporta and I, among others, generalized the triangle to a tetrahedron, with something like $\bar{p}\Lambda$ at the fourth vertex, to take account of the strangeness-changing weak interaction, both leptonic and nonleptonic.

Most of us assumed that some kinds of intermediate bosons were involved and that therefore the vertices would interact with themselves as well as with one another. The $\Delta S/\Delta Q = +1$ property of the weak strangeness-changing current would then prevent $|\Delta S| = 2$ nonleptonic interactions from occurring. (We were aware first of the rather mild limitation on $|\Delta S| = 2$ from the failure to observe $\Xi \rightarrow \pi \rightarrow + N$, and only later of the very stringent limitation connected with $K^0 \leftrightarrow \bar{K}^0$.) The rules $|\Delta \mathbf{I}| = 1$ and $|\Delta \mathbf{I}| = \frac{1}{2}$ at the hadronic vertices give rise to $|\Delta \mathbf{I}| = \frac{1}{2}, \frac{3}{2}$ for the nonleptonic strangeness-changing interaction (apart from electromagnetic corrections), as I understood as early as 1954, but I was unable to account in this way for the predominance of $|\Delta \mathbf{I}| = \frac{1}{2}$ over $|\Delta \mathbf{I}| = \frac{3}{2}$, which I described in an unpublished but widely distributed letter in the fall of 1953. It never occurred to me that more than thirty years later the reasons for that approximate $|\Delta \mathbf{I}| = \frac{1}{2}$ rule would still be a subject of research, while the rest of the issues would be understood in the framework of a dynamical theory like the standard model.

2.   (a) *K^0 and $\bar{K}^0$ made and absorbed as such, but decay as K_1^0, K_2^0*
            (defined by C at first)
    (b) *P and C maximally violated in lepton weak interactions with*
            hadrons
            and with other leptons
            (only two components of neutrino utilized –
            maybe only two components exist)
    (c) *PC apparently conserved (K_1^0, K_2^0 now defined by CP)*
    (d) *PC slightly violated;*
            beautiful clean-cut result, but not well understood
            at the time
            [present favorite explanation requires three fermion families
            (at least),
            but whole CP question is complicated, as mentioned earlier];
            now K_1^0, K_2^0 slightly mixed

3.   (a) *Coupling is $\gamma_\alpha(1 + \gamma_5)$ (V−A), universal*
    (b) *Most likely charged spin-1 intermediate boson*
    (c) *As its mass $\rightarrow \infty$, $(G/\sqrt{2}) J_\alpha^+ J_\alpha$, with $J_\alpha = \bar{\nu}\gamma_\alpha(1 + \gamma_5)e +$*
            $\bar{\nu}\gamma_\alpha(1 + \gamma_5)\mu$ + hadronic V−A current
            with preceding properties
    (d) *Hadron $\Delta S = 0$ vector current*
            is just component of isotopic spin current,
            conserved and not changed by renormalization
    (e) *Two neutrinos different*
            to prevent $\mu \rightarrow e + \gamma$ when intermediate boson exists;
            $J_\alpha = \bar{\nu}_e \gamma_\alpha (1 + \gamma_5)e + \bar{\nu}_\mu \gamma_\alpha (1 + \gamma_5)\mu$

4.   (a) *"$\bar{p}\Lambda$" term in hadronic current smaller than "$\bar{p}n$" term*
            ratio $(\varepsilon\sqrt{1 + \varepsilon^2})$ to $(1/\sqrt{1 + \varepsilon^2})$ or $\sin \theta$ to $\cos \theta$, $\theta \approx 15°$
    (b) *Whole hadronic current must be current of a generator of some*
            sensible algebra.
            Looks like

$$\bar{\nu}_e \gamma_\alpha (1 + \gamma_5)e + \bar{\nu}_\mu \gamma_\alpha (1 + \gamma_5)\mu$$

$$+ \text{``}\bar{p}\gamma_\alpha (1 + \gamma_5)\left(\frac{n}{\sqrt{1 + \varepsilon^2}} + \varepsilon\frac{\Lambda}{\sqrt{1 + \varepsilon^2}}\right)\text{''}$$

   (c)   *In fact, total weak and electromagnetic currents*
          probably correspond to generators of SU(2) × U(1)
          (as today)

   (d)   *Most likely neutral weak interaction*
          through intermediate boson Z^0
          (as today),
          but if SU(2) × U(1), why no appreciable
          strangeness-changing neutral current?
          [not understood until 1970]

   (e)   *Z^0 and γ result of mixture using weak angle*
          (as today)

   (f)   *Z^0 and $X^{\pm}$ masses ∼ 100 GeV*
          (as today)

   (g)   *SU(2) × U(1) thought to be broken Yang–Mills theory (as*
          today) [but soft breaking that gives fermion and boson masses
          not understood]
          [and neutral-current problem → trouble]

   (h)   *Is SU(2) × U(1) embedded in a larger*
          weak and electromagnetic gauge group
          including heavier boson?
          [still not really known]

E: *Quarks (1963–4)*

1.

"flavors"		
u	*d*	*s*
Q $+\frac{2}{3}$	$-\frac{1}{3}$	$-\frac{1}{3}$

2.   *Statistics peculiar, Bose-like for three quarks*
     like "parafermions of rank 3,"
     but where are the baryons with peculiar statistics
     that would result? [not understood until much later
     that paraquarks with parabaryons, etc., suppressed,
     equivalent to color with colored hadrons suppressed,
     and that quark confinement is related to color suppression]

3.   (a)   *Confinement ("mathematical quarks")*
          as opposed to emergence of single quarks ("real quarks")

     (b)   *"Mathematical quarks" defined first (1963) as limit as quark*
          mass → ∞, then (1965–6) as limit of infinite potential barrier
          (as today)

     (c)   *If quarks never emerge as "real" particles,*
          then compatible with "nuclear democracy"

4.   *Easily see (with peculiar statistics)*
     8 ($\frac{1}{2}^+$) and 10 ($\frac{3}{2}^+$)

with p-states above, etc.,
for baryons, and
8 and 1 (0⁻ and 1⁻)
with p-states above, etc., for mesons

5. (a) *Many properties abstracted*
        from theory of quarks
        coupled to single neutral vector gluon
        (keep pheasant, throw away veal)
        [unfortunately not Yang–Mills color octet
        of neutral vector gluons, as today]

With the proposal of quarks in 1963, it became possible to sum up in a very few hypotheses a long list of properties of hadrons and their weak and strong interactions that we had put together painfully over the preceding fifteen years or so. Even before that, I had suggested that we could abstract many of these properties from a quantum field theory of three flavors of spin-$\frac{1}{2}$ fermions coupled to a single neutral vector gluon field. With the quark idea, it became possible to abstract even more correct properties from that theory. In discussing the use of that field theory for such a purpose and then throwing it away, I used a simile in which I borrowed from my friend Valentine Telegdi the notion of a French recipe for cooking a pheasant between two slices of veal and then throwing away the veal and eating the pheasant. If I had used the eight neutral vector gluons of an SU(3) Yang–Mills theory of color, the veal would have been QCD, and there would have been no need to discard it, but the single-gluon theory evidently was not completely correct.

(b) *Chiral SU(3) × SU(3) broken by mass terms with right behavior*
     (3, $\bar{3}$) and ($\bar{3}$, 3) [but not yet understood why soft]

(c) *Vector current charges generate SU(3) of flavor,*
     masses give octet-breaking, weak current now

$$\bar{v}_e\gamma_\alpha(1 + \gamma_5)e + \bar{v}_\mu\gamma_\alpha(1 + \gamma_5)\mu + \bar{u}\gamma_\alpha(1 + \gamma_5)d'$$
$$d' = d\cos\theta + s\sin\theta$$

electromagnetic current now, of course,

$$\tfrac{3}{2}\bar{u}\gamma_\alpha u - \tfrac{1}{3}\bar{d}\gamma_\alpha d - \tfrac{1}{3}\bar{s}\gamma_\alpha s$$

(d) *1964 (D. J. Bjorken and S. Glashow): charm suggested;*
     add $+\tfrac{2}{3}\bar{c}\gamma_\alpha c$ to electromagnetic current; add $\bar{c}$
     $\gamma_\alpha(1 + \gamma_5)s'$ to weak current;
     $s' = s\cos\theta - d\sin\theta$; [fits beautifully with SU(2) × U(1)
     Yang–Mills theory
     of weak and electromagnetic currents and solves problem of
     neutral current
     having no appreciable strangeness-changing term,
     but those consequences not noticed until 1970]

In closing, let me say a word about what may be the most important result to emerge from the theoretical research done during the period with which we are concerned. It has been mentioned that we do not know what hypotheses should have been introduced into the mass-shell formulation of field theory in order to obtain QCD, but we do not know what did actually happen in the evolution of the bootstrap idea. I remember complaining during the middle 1960s that although the bootstrap idea was a good one, the implementation by the Berkeley group was mostly focused on one channel (two pions) and one intermediate state (the ϱ meson). I favored instead an infinite number of states, approximated by narrow resonances, for the intermediate states in all channels and for all the external legs. This rough idea was made much more concrete in the seminal "duality" paper by the three Caltech postdoctoral fellows R. Dolen, D. Horn, and C. Schmid. It was the search for a concrete realization of the dual bootstrap that led G. Veneziano to his famous model, with its remarkable properties.

Of course, the Veneziano model contained a particle with negative mass squared, included no fermions, and turned out to work only in twenty-six dimensions. In 1971, A. Neveu and J. H. Schwarz suggested a different model, with both bosons and fermions and a critical dimension of 10. After the proof by F. Gliozzi, J. Scherk, and D. I. Olive that a sector of that theory could be consistently omitted, it was shown that the remaining theory contained no particles of negative mass squared or negative probability. This "superstring theory" thus appeared to be a consistent bootstrap theory, but it had serious difficulties (such as predicting a massless spin-2 particle) as a description of hadrons. (Besides, QCD was then available and looked very promising.)

Scherk and Schwarz then suggested that the characteristic slope of Regge trajectories in the theory be modified by a factor of 10^{38} or so, and they showed that the theory then becomes a generalization of a quantum version of Einsteinian gravitation, in fact containing supergravity (which had just been proposed) as an approximation, provided the extra six dimensions spontaneously curl up into a tiny structure comparable in size to the Planck length. Moreover, the theory contains fields that might describe the elementary particles we know, such as quarks, leptons, and gauge bosons.

More recently, as a result of the work of Green and Schwarz, it has been found that there are at least five such theories, at least some of which seem to be finite in perturbation theory, not even requiring renormalization. At least one of them, proposed by the Princeton "string quartet" at the end of 1985, offers a serious hope of being the long-sought unified field theory of all the elementary particles and interactions of nature.

Notes

1 M. Gell-Mann, "Strangeness," in Colloque International sur l'Histoire de la Physique des Particules, *J. Phys. (Paris)* 43 (December 1982), C8-395–408.

2 M. Gell-Mann, "Particle theory from *S*-Matrix to Quarks," in *Symmetries in Physics (1600–1980), Proceedings of the First International Meeting on the History of Scientific Ideas*, edited by M. G. Doncel, A. Hermann, L. Michel, and A. Pais (Barcelona: Bellaterra, 1987), pp. 474–97.
3 Murray Gell-Mann, "From Renormalizability to Calculability?" in *Shelter Island II, Proceedings of the 1983 Shelter Island Conference on Quantum Field Theory and Fundamental Problems of Physics* (Cambridge, Mass.: MIT Press, 1985), pp. 3–23.

Name index

Abashian, Al, 286
Abov, Yu. G., 481
Adair, Robert K., 320
Adams, John B., 162, 173–4, 183, 197
Adenauer, Konrad, 52
Adler, Stephen, 490
Ageno, Mario, 191
Akeley, Edward, 209
Alfven, S., 513
Alikhanov, A., 476
Allen, James, 198
Allison, Samuel, K., 202
Allkofer, O. C., 242
Alston, Margaret, 304
Altman, Sidney, 415
Alvarez, Luis W., 12, 15, 21, 40, 94, 129, 166,
    168, 170, 176, 185–6, 191, 193, 196, 198,
    218–20, 232–5, 240–1, 243–4, 261, 265,
    271, 289, 291–2, 345, 349, 452, 462–3,
    515, 517, 654–6
Amaldi, Edoardo, 17, 40, 78, 152, 265, 267,
    275, 278
Amaldi, Ugo, 185, 191, 193, 195, 197–200,
    438, 445–6, 460, 513, 516, 519, 523
Ambler, E., 469
Amman, Fernaud, 153
Ammar, Raymond, 337
Amundsen, Roald, 305
Andersen, Jan, 173
Anderson, Bert, 151
Anderson, Carl D., 7, 43, 57–8, 63–4, 66–7,
    79, 86, 215, 273, 308, 313, 317, 350

Anderson, Herbert L., 9, 40–1, 117, 190, 198,
    204, 589
Anderson, Philip, 641–2, 699
Araki, Gentaro, 90
Araki, H., 619, 624
Armenteros, R., 69, 78, 83
Arsenieva, A. N., 418
Ashkin, J., 313
Auger, Pierre, 172, 511–13, 517, 519

Bacher, Robert, 12, 128, 150–1, 499
Bagge, Erich, 225–6, 242, 257, 562
Baker, W., 199
Bakker, C. J., 513–14, 647
Baldin, A. M., 155
Bannier, J. H., 517
Barber, C., 155
Barbour, I., 95
Bargmann, V., 373
Barkas, Walter H., 260, 267, 305
Barker, K. H., 61
Barnes, R. Bowling, 127
Bartlett, J. H., 426, 428
Bartlett, M. S., 78
Barton, Mark, 206
Bay, Zoltan Lajos, 199
Bayley, Don, 128
Beams, J. W., 156
Becker, Richard, 563
Becquerel, Henri, 89
Behrends, R. E., 635
Bell, Jocelyn, 304, 465

Subject index

accelerator, 9–15, 31, 89, 95, 105–6, 111, 115, 149–212, 218, 232, 308–14, 349, 498, 500, 502–4, 510, 645–7, 653–4, 684
acoustic chamber, 245–6
AdA, 153, 155
Aharonov–Bohm effect, 46
Aiguille du Midi, 460
almost conserved axial vector current, *see* PCAC
ALSOS mission, 562
alternating-gradient (AG) focusing, 13, 152–5, 157, 163, 169–71, 173, 175, 182, 186–9, 197, 202–4, 262, 264, 278, 287–8, 292, 502, 514
alternating-gradient synchrotron (AGS), 13, 155, 157, 162, 169, 175–8, 180–4, 196–7, 326, 503–4, 671
American Institute of Physics, 675
American Physical Society, 218, 269, 285, 289, 299, 349, 478, 502, 528
Amsterdam, University of, 557–8
analyticity, 30, 559, 565, 581–2, 585–6, 601–3, 630, 652, 682, 685
anomalous magnetic moment, 10, 19, 145, 273, 608, 683
Annello d'Accelerazione, *see* AdA
anthracene, 186, 190–1
anticommutation relations (CAR), 609–11, 626
antimatter, *see* antiparticle
antineutrino, 40, 191, 466
antineutron, 21, 40, 288, 295

antinucleon, 378, 696, 703
antiparticle, 19, 22, 23, 29, 50, 275, 498, 660, 696
antiproton, 11, 13, 21, 40, 50–2, 167, 191, 234, 265, 269, 273–95, 301, 379–80, 385–6, 650
Argonne National Laboratory (ANL), 14, 16, 17, 154, 175, 196, 205, 208, 252, 477–9, 491, 500–1, 509, 676, 680
associated production, 8, 15, 19, 20, 41, 78, 83, 106, 182, 300, 318, 320, 343–4, 351–2, 370, 458, 460, 654
asymptotic condition, 615–16, 623, 626
asymptotic freedom, 698
atomic bomb, *see* nuclear weapon
atomic energy, 11
Atomic Energy Commission (AEC), 11, 13, 31, 40, 50, 51, 150, 180–1, 206, 210, 219–20, 235, 302–3, 311, 380, 498–505, 526, 670–3, 678, 680
  GAC, *see* General Advisory Committee
Atoms for Peace conference (1955), 196
axial vector current, 30, 386–8, 641, 699, 705
axial vector interaction, 28
axiomatic field theory, 30, 375, 492, 593, 608, 619–20, 626, 686

Bagnères de Bigorre conference, 8, 75–8, 80, 100, 436–7, 439–40, 442, 444, 446, 460
balloon, 7, 103–4, 446
Baldwin-Lima-Hamilton Corp., 176
baryon, 19, 465, 632, 634–6, 704–5, 708–9